McGRAW-HILL
YEARBOOK OF
SCIENCE &
TECHNOLOGY

2009

McGRAW-HILL
YEARBOOK OF
SCIENCE &
TECHNOLOGY

2009

Comprehensive coverage of recent events and research as compiled by
the staff of the McGraw-Hill Encyclopedia of Science & Technology

New York   Chicago   San Francisco   Lisbon   London   Madrid   Mexico City

Milan   New Delhi   San Juan   Seoul   Singapore   Sydney   Toronto

On the front cover
The location of the strongest hydrogen-binding sites (red-green areas) in a metal-organic cage structure (ball-and-stick model superimposed with color-coded atoms: white = hydrogen; gray = carbon; red = oxygen; purple = zinc). The neutron diffraction data on which the image is based were taken at the NIST Center for Neutron Research. (*Taner Yildirim*)

ISBN   978-007-160562-5
MHID   0-07-160562-2
ISSN   0076-2016

1 2 3 4 5 6 7 8 9 0   DOW/DOW   0 1 3 2 1 0 9 8

This book was printed on acid-free paper.

*It was set in Garamond Book and Neue Helvetica Black Condensed by
Aptara, Falls Church, Virginia. The art was prepared by Aptara.
The book was printed and bound by RR Donnelley.*

Contents

Editorial Staff

Editing, Design, and Production Staff

Consulting Editors

Dr. Donald Platt. *Micro Aerospace Solutions, Inc., Melbourne, Florida.* SPACE TECHNOLOGY.

Dr. Kenneth P. H. Pritzker. *Pathologist-in-Chief and Director, Head, Connective Tissue Research Group, and Professor, Laboratory Medicine and Pathobiology, University of Toronto, Mount Sinai Hospital, Toronto, Ontario, Canada.* MEDICINE AND PATHOLOGY.

Prof. Justin Revenaugh. *Department of Geology and Geophysics, University of Minnesota, Minneapolis.* GEOPHYSICS.

Dr. Roger M. Rowell. *USDA Forest Service, Forest Products Laboratory, Madison, Wisconsin.* FORESTRY.

Dr. Thomas C. Royer. *Department of Ocean, Earth, and Atmospheric Sciences, Old Dominion University, Norfolk, Virginia.* OCEANOGRAPHY.

Prof. Ali M. Sadegh. *Director, Center for Advanced Engineering Design and Development, Department of Mechanical Engineering, The City College of the City University of New York.* MECHANICAL ENGINEERING.

Prof. Joseph A. Schetz, *Department of Aerospace and Ocean Engineering, Virginia Polytechnic Institute & State University, Blacksburg.* FLUID MECHANICS.

Dr. Alfred S. Schlachter. *Advanced Light Source, Lawrence Berkeley National Laboratory, Berkeley, California.* ATOMIC AND MOLECULAR PHYSICS.

Prof. Ivan K. Schuller. *Department of Physics, University of California-San Diego, La Jolla, California.* CONDENSED-MATTER PHYSICS.

Jonathan Slutsky. *David Taylor Model Basin, Naval Surface Warfare Center, Carderock Division, West Bethesda, Maryland.* NAVAL ARCHITECTURE AND MARINE ENGINEERING.

Dr. Arthur A. Spector. *Department of Biochemistry, University of Iowa, Iowa City.* BIOCHEMISTRY.

Dr. Anthony P. Stanton. *Tepper School of Business, Carnegie Mellon University, Pittsburgh, Pennsylvania.* GRAPHIC ARTS AND PHOTOGRAPHY.

Dr. Michael R. Stark. *Department of Physiology, Brigham Young University, Provo, Utah.* DEVELOPMENTAL BIOLOGY.

Dr. Daniel A. Vallero. *Adjunct Professor of Engineering Ethics, Pratt School of Engineering, Duke University, Durham, North Carolina.* ENVIRONMENTAL ENGINEERING.

Dr. Sally E. Walker. *Associate Professor of Geology and Marine Science, University of Georgia, Athens.* INVERTEBRATE PALEONTOLOGY.

Prof. Pao K. Wang. *Department of Atmospheric and Oceanic Sciences, University of Wisconsin, Madison.* METEOROLOGY AND CLIMATOLOGY.

Dr. Nicole Y. Weekes. *Department of Psychology, Pomona College, Claremont, California.* NEUROPSYCHOLOGY.

Prof. Mary Anne White. *Department of Chemistry, Dalhousie University, Halifax, Nova Scotia, Canada.* MATERIALS SCIENCE AND METALLURGICAL ENGINEERING.

Dr. Thomas A. Wikle. *Department of Geography, Oklahoma State University, Stillwater.* PHYSICAL GEOGRAPHY.

Prof. Jonathan Wilker. *Department of Chemistry, Purdue University, West Lafayette, Indiana.* INORGANIC CHEMISTRY.

Article Titles and Authors

Preface

The 2009 *McGraw-Hill Yearbook of Science & Technology* appears at a moment when authoritative scientific and technical information is required more than ever in order to understand and act upon many of the burning issues of our time. One need only consider, for example, the debates taking place in the public arena over dealing with global climate change and the need to secure new energy supplies to see the dependence of truly crucial societal decisions on scientific knowledge. The ways we communicate, learn, do business, and even entertain ourselves are changing dramatically with the rapid advances in wireless technology, telecommunication systems, and computing among other technologies. Even words such as nanotechnology, broadband, and DNA have entered everyday language. This edition of the *Yearbook* strives to provide a broad overview of important recent developments in science, technology, and engineering as selected by our distinguished board of consulting editors. It satisfies the nonspecialist reader's need to stay informed about important trends in research and development that will advance our knowledge in fields ranging from astrophysics to zoology and lead to important new practical applications.

In the 2009 edition, we report on the rapid advances in cell biology and genetics with articles on topics such as gene targeting, cloning research, microRNA, and whole-genome association studies. Reviews in topical areas of biomedicine such as androgenic-anabolic steroids in athletes; diabetes and its causes; the epidemic of obesity; epilepsy; human papillomavirus and the impact of cervical cancer vaccine; phytochemicals as human disease prevention agents; and the "hygiene hypothesis" are presented, as well as articles on consciousness and conscious awareness; hallucinations; neuroeconomics; positive psychology and human happiness; and the role of the frontal lobe in violence in the fields of psychiatry and psychology. Major developments in the plant sciences are covered, for example, in articles on plant defenses against pathogens and herbivorous insects, the regulation of fruit ripening, as well as engineering minichromosomes in plants. Advances in computing and communication are documented in articles on audio compression; assisted GPS and location-based services; human-computer interface design; nonvolatile memory devices; scientific workflows; service-oriented computing; and WiMAX broadband wireless communication. In the developing field of nanotechnology, we review the synthesis of new nanocarbon materials; polymer nanocomposites; graphene; and the assessment of risks from nanomaterials. In energy and research on alternative energy sources we review advanced batteries; artificial photosynthesis; bioethanol production; in-stream tidal power generation; the International Tokamak Experimental Reactor (ITER); space-beamed solar power; syngas from biomass; underground processing of oil shale; and wind turbines. Noteworthy developments in engineering and technology are reported in articles such as absorbable orthopedic implants; auxetic materials; the economics of low-tech innovation; progressive collapse of building structures; heat-treated wood; imaging mechanical defects; MEMS sensors; midrange wireless power transfer; and the virtual testing of structures. In the physical sciences and astronomy we report on alpha and proton radioactivity; the cyclic universe theory; femtosecond phenomena; the Gravity Probe B mission; the Large Hadron Collider; the MESSENGER mission; space flight; and table-top synchrotron systems. And reviews on air pollution monitoring; Argo, the ocean observing network; assisted migration for species preservation; childhood lead exposure and lead toxicity; corals and ocean acidification; extratropical cyclone occlusion; and the hydrological consequences of global warming are some of the topics covered in the earth and environmental sciences.

Each contribution to the *Yearbook* is a concise yet authoritative article authored by one or more authorities in the field. The topics are selected by our consulting editors, in conjunction with our editorial staff, based on present significance and potential applications. McGraw-Hill strives to make each article as readily understandable as possible for the nonspecialist reader through careful editing and the extensive use of specially prepared graphics.

Librarians, students, teachers, the scientific community, journalists and writers, and the general reader continue to find in the *McGraw-Hill Yearbook of Science & Technology* the information they need in order to follow the rapid pace of advances in science and technology and to understand the developments in these fields that will shape the world of the twenty-first century.

Mark Licker
Publisher

McGRAW-HILL
YEARBOOK OF
SCIENCE &
TECHNOLOGY

2009

Absorbable orthopedic implants

There are approximately 8 million bone fractures a year in North America. Approximately 90% of these are simple fractures that are treated with a cast or brace. The remaining 10% are more severe and require some kind of intervention to facilitate healing. Today, surgeons have a wide range of implant options, including metal plates, screws, nails, and external fixation frames. These innovations have greatly improved the clinical outcome and lives of countless patients. In the ideal case, an implant would be removed upon successful fracture healing. In reality, this is rarely done because it subjects the patient to an additional operation and associated risks.

A metal device is designed to carry the load of the broken bone. It will continue to carry some load after the fracture has healed, causing what is known as stress shielding of the bone. As a result, the strength of bone restored at the fracture site will not be as high as normal bone. This is because bone remodels according to Wolff's law—by adapting to the mechanical environment it experiences. If a metallic device is removed, there is an increased chance of refracture while the bone remodels to the daily loads it encounters. Bioresorbable materials have the potential to overcome these issues because as the device is slowly broken down and absorbed by the body, its mechanical properties will gradually decrease. This would result in the gradual transfer of the load from the device to the new bone, which could result in improved quality of the new bone.

Degradable biomaterials. The concept and use of degradable materials is not new, as degradable sutures were first used in the late 1960s. In the 1980s, degradable materials started to be formed into medical devices, such as suture anchors and screws for ligament reattachment. The use of bioresorbable materials for repairing bone fractures has been a long-term goal of many academic and industrial research groups.

The most commonly used biomaterials are homopolymers or copolymers of lactic and glycolic acid. Some of the early devices were prepared from glycolic acid–based polymers. Upon degradation of these polymers, it was found that in some cases there would be a local reaction to the breakdown products. This would appear as a localized redness of the skin above the implant. In very severe cases, a draining sinus would occur. The visual appearance of this phenomenon is similar to that of a local tissue infection. Microbiological testing of the area would show no infection, which led to the term sterile sinus being used to describe this effect. For poly(glycolic acid) or copolymers composed of predominantly glycolic acid, sterile sinus occurrence is in the range of 4–8%.

For polymers based on lactic acid, the reported incidence of sterile sinus is well below 1%. This has led to current products and research interests based on lactic-acid polymers. A wide range of products made from these polymers are commercially available. Typically, they are injection molded into the required shapes, such as screws or plates. These injection molded products are typically used in parts of the body that experience low levels of mechanical load. The most common applications are the use of screws to fix fractures in the foot and ankle, and screws and plates for craniomaxillofacial use.

Higher-strength materials. There are many published studies that show good clinical outcomes for biodegradable devices. Even with these results, the use of bioresorbable materials still remains at a low level. One reason for this is the difference in the mechanical properties between metals and plastics. While bioresorbable materials are good enough for low-load-bearing applications, their mechanical properties prevent their use in high-load-bearing applications, which are the type of fracture that are more likely to require surgical intervention to heal. This deficit in mechanical properties has led many research groups to work on improving the properties of biodegradable materials through processing technologies or the development of new materials.

To manufacture high-strength materials, researchers such as Y. Shikinami have successfully exploited the orientation of the degradable polymers (for example, by drawing or stretching) to form devices with superior mechanical properties. These materials have been shown to be usable in higher-load-bearing applications, such as screws for hip fractures. However, one result of orientating the polymers is to slow their degradation rate. While these materials are stronger than their injection-molded counterparts, they take many years to degrade.

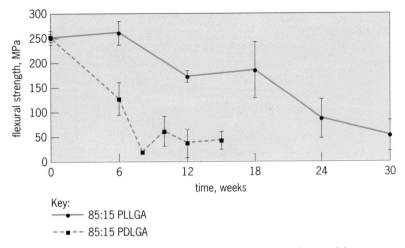

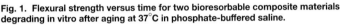

Key:
——•—— 85:15 PLLGA
- - ■- - - 85:15 PDLGA

Fig. 1. Flexural strength versus time for two bioresorbable composite materials degrading in vitro after aging at 37°C in phosphate-buffered saline.

Outlook. Using process technologies developed for the aerospace industry, it is possible to fabricate devices, such as screws, using degradable composite materials (**Fig. 3**). Recent materials developments are advancing the mechanical properties and the biological interaction of biodegradable polymers. With these new polymers, the applications and clinical acceptance of bioresorbable materials will continue to increase. Looking to the future, researchers are now starting to think about incorporating active compounds into biodegradable devices. As the material degrades, it will release the active compound into the body. This will result in devices that not only stabilize the bones to enable fracture repair, but also

Degradable composites. To achieve faster degradation without compromising the mechanical gains achieved through orientation, J. Rose and coworkers have recently developed degradable composite materials. In these composites, the reinforcing component is a poly(lactic acid) [PLLA] fiber that has been highly drawn to achieve a tensile strength in excess of 1 GPa. These high-strength fibers are then held together in the shape of the device in a matrix component consisting of another degradable material. The initial strength of these composites is mostly influenced by the strength of the reinforcing fiber, but the time taken for the composites to lose their strength is determined by the selection of the matrix material. Once the matrix material has lost its strength, the composite will also lose its strength. This is illustrated in **Fig. 1**, which shows the flexural strength for two composite materials, each with a different matrix polymer. Composite samples were aged at 37°C in a phosphate-buffered saline solution to simulate the effects of implantation into the body. One composite was made with a fast-degrading matrix, based on a copolymer of the cyclic esters DL-lactide and glycolide [poly(D,L-lactide-co-glycolide)] in the ratio of 85:15 (PDLGA 85:15). A second composite was made with a slower degrading matrix, which is a copolymer of L-lactide and glycolide [poly(L-lactide-co-glycolide); PLLAGA 85:15]. Figure 1 shows, as composite theory would predict, that the initial strengths of the composites are identical but the time at which the materials lose their strength is different. Using composites in which the matrix degrades at a different rate from the reinforcing fiber may provide an additional benefit, in that as the matrix is degrades, the remaining fibers will act as a scaffold to facilitate the replacement of bone at the implant site. This concept is shown in **Fig. 2**. In addition to allowing the implant to degrade to a scaffold, the concept also allows the mass loss of the device to occur in two steps. This has the potential to reduce the chance of sterile sinus formation, as the breakdown products are released over a longer timeframe.

Time	Comment	
Immediately after implantation	Composite has full mechanical properties. Fibers (black) and matrix (grey) are intact.	
After 12–36 weeks in vivo	Mechanical strength of the composite is lost when the matrix component looses its strength.	
After 12–18 months in vivo	Mass loss of the matrix starts to occur. New bone (chequered region) starts to grow around the remaining fibers. Fibers start to lose their strength.	
After 24 months in vivo	Complete mass loss of matrix has occurred. Bone has grown around the fibers. Fibers are starting to lose their mass.	
After 48 months	Mass loss of the fibers is complete. There is no trace of the degradable composite left and the implantation site has been replaced by bone.	

Fig. 2. Sequence showing composite degradation and replacement by bone.

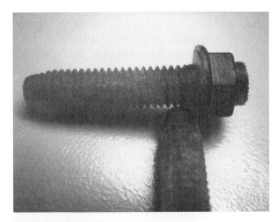

Fig. 3. Biodegradable composite screws.

accelerate the healing process. This combination of technologies has the potential to greatly enhance the treatment of bone fractures, enabling patients to return to a normal lifestyle in a shorter time.

For background information *see* BIOMEDICAL ENGINEERING; BONE; COMPOSITE MATERIAL; POLYESTER RESINS; POLYMER; POLYMER COMPOSITES; PROSTHESIS in the McGraw-Hill Encyclopedia of Science & Technology.   John Rose; Abraham Salehi

Bibliography. M. Bettenga and J Rose, Dependence of matrix selection on the flexural strength retention of biodegradable composite pins during in vitro degradation, *Proceedings of 54th Annual Meeting of the Orthopaedic Research Society,* San Francisco, 2008; G. Scott and D. Gilead (eds.), *Degradable Polymers: Principles and Applications,* Chapman & Hall, London, 1995; Y. Shikinami and M. Okuno, Bioresorbable devices made of forged composites of hydroxyapatite (HA) particles and poly-Image L-lactide (PLLA): Part I. Basic characteristics, *Biomaterials,* 20(9):859–877, 1999; E. Waris et al., Bioabsorbable fixation devices in trauma and bone surgery: Current clinical standing, *Expert Rev. Med. Devices,* 1(2):229–240, 2004.

Advanced batteries

Numerous rechargeable batteries have been considered for use in vehicular applications over the years, including lead-acid, nickel-metal hydride (Ni-MH), and lithium-ion (Li-ion) batteries. Each of these differs from the other in the choice of the anode, cathode, and electrolyte, with these choices leading to differences in the metrics that are used to gauge batteries. For batteries for vehicles, the important metrics include energy (representing how much useful work can be performed), power (representing how quickly this energy can be released), cycle life (representing how many times the battery can be charged and discharged before its capacity fades), calendar life (how many years the battery can last before its capacity fades), self-discharge (how many days the battery holds its charge), cost, and safety. Capacity represents how much electricity (electric charge, or electric current integrated over time) can be obtained from the battery; energy is the capacity times the voltage. Depending on the application, these metrics change. For example, while batteries for hybrid electric vehicles (HEVs) require large power capability with energy being less critical, batteries for electric vehicles (EVs) require large energy to maximize the amount of driving time before recharging.

Performance. Energy and power are often used as means of distinguishing batteries from each other. On this basis, the comparison of various energy storage and conversion devices is shown in **Fig. 1,** which is called a Ragone plot. Here the abscissa is the specific power (in W/kg). One can interpret it as the acceleration that can be obtained in an EV. The ordinate represents the specific energy (in Wh/kg) and represents the miles that can be driven before recharge. The graph plots these quantities for three

battery types, electrochemical capacitors, fuel cells, and the internal combustion (IC) engine. Comparing the performance of batteries to the IC engine suggests that while advanced batteries have similar power to gasoline cars, the energy is significantly lower. This translates to low range for EVs, which is one of the problems that have prevented commercialization of this concept. The cross lines represent the time for charge or discharge. For a battery that is used with a solar cell, one would look for systems that can charge and discharge over many hours. For HEVs, where the battery is used in bursts that last ~10 s, systems that operate to the right of the diagram are more useful. Figure 1 shows that lithium batteries show the most promise in all applications, when compared to lead-acid and Ni-MH batteries. It is for this reason that considerable attention has been focused on Li-ion batteries, especially for vehicular applications.

Li-ion batteries. While Fig. 1 focuses on two of the metrics needed for commercialization, closer analysis of the other metrics suggests that while Li-ion batteries are ideally suited for use in HEVs, a significant increase in energy is needed for long-range EVs. Further, EVs suffer from very large charge times (4–8 h). Plug-in hybrid-electric vehicles (PHEVs), where a large battery (which is externally recharged) is used with a gasoline engine, strike an ideal compromise between the minimal fuel savings of a HEV and the range-time and charge-time issues of an EV. Present-day Li-ion batteries appear to have the energy needed to achieve 20–30 mi (30–50 km) of equivalent electric range. Therefore, research is focused on improving the energy of the battery to enable PHEVs (with higher electric range) and EVs. Note that increasing the energy of the battery is a means of decreasing the cost via a decrease in the number of cells used (that is, higher energy results in the use of fewer cells in a battery pack for a given energy requirement). A typical rule of thumb is a sedan requires 300 Wh of battery for every mile of electric range. So a 200-mi (320-km) EV requires 60 kWh. Higher-energy batteries could help decrease the cost per

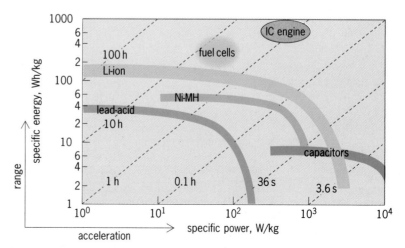

Fig. 1. Ragone plot of specific power density in W/kg versus specific energy density in Wh/kg of various electrochemical energy storage and conversion devices.

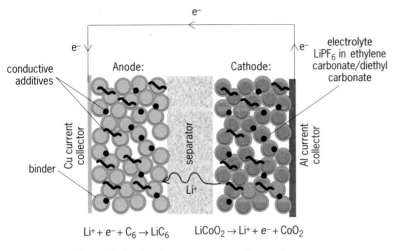

Fig. 2. Typical Li-ion cell indicating the processes occurring on charge.

kilowatt hour, thereby decreasing the system cost. In addition, the cycle/calendar life of batteries and their safety characteristics remain a matter of concern.

A typical Li-ion cell is shown in **Fig. 2**, consisting of a graphite anode and a lithium cobalt oxide (LiCoO$_2$) cathode. The electrodes consist of active materials bound together with an electronically insulating binder and conductive additives. Each electrode is pasted onto the current collectors. During charge, Li is removed from the cathode, transferred through the separator via the electrolyte, and inserted into the anode (this process is called intercalation). The reverse occurs on discharge. **Figure 3** shows the typical steady-state charge of the anode and cathode of a Li-ion cell with a graphite anode and a LiCoO$_2$ cathode in an organic electrolyte consisting of a Li salt (lithium hexafluorophosphate, LiPF$_6$) in a solvent (for example, ethylene carbonate and diethyl carbonate). This is the battery used in laptops and cell phones. The voltage

of each electrode is represented with respect to a Li-metal reference electrode. As the Li is removed from the cathode, its potential increases, while the potential of the anode decreases with insertion of Li. The voltage of the battery is the difference in the voltage of the cathode and the anode, which increases as charge proceeds. The abscissa represents how much Li is stored in the cell, while the ordinate shows at what voltage the Li is inserted or removed from the materials. Three avenues can be pursued to increase the energy of the battery, namely (1) increase the voltage of the cathode, (2) decrease the voltage of the anode, and (3) increase the capacity of the cell. However, the thermodynamics of electrochemical reactions other than the intercalation of Li (referred to as side reactions) limit these quantities. The three side reactions worth mentioning in this plot are the oxidation of the solvent that occurs above ~4.2 V versus Li, Li-metal deposition that occurs below 0.0 V versus Li, and solvent reduction that occurs below 1 V versus Li. These reactions not only limit the energy of the cell, but also are implicated in the life and safety problems associated with Li-ion batteries. Staying within the voltage window allows these problems to be minimized, at the loss of energy. For example, while the capacity of LiCoO$_2$ is limited to ~140 mAh/g when restricting the voltage, its theoretical capacity is ~275 mAh/g. In other words, practical LiCoO$_2$ cells only use about 50% of the theoretically available capacity of the cathode, which translates to using only a fraction of the theoretically available energy of the cell. Increasing the voltage limit to access this energy would severely restrict the life of the battery and lead to safety issues. The tradeoffs that are needed to balance these various parameters are captured by Fig. 3.

Innovation in Li-ion batteries can occur via (1) engineering advances that reduce, for example, the thickness of the separator and (2) innovation in the materials used as the active material. For example, the LiCoO$_2$/graphite battery has a theoretical energy density of ~360 Wh/kg. (This is calculated by restricting the capacity of the cathode to 140 mAh/g, and accounts for only the weight of the active material and not the weights of the other components in the cell such as the current collectors, electrolyte, binders, and cell packaging.) The practical energy density of a packaged cell is ~190 Wh/kg. In the early days of Li-ion batteries, this value was ~90 Wh/kg, using the same material sets. In other words, engineering advances have resulted in a doubling of the energy density of Li-ion batteries over the last 15 years. In the future, it is expected that improvements in performance will occur by moving to new, higher-energy materials.

It has been observed that Li can intercalate into many different anode and cathode materials. At present, three classes of cathodes, four classes of anodes, and four classes of electrolytes are being considered for use in Li-ion cells. Depending on the combination of the anode, cathode, and electrolyte, one can have a completely new battery with changes

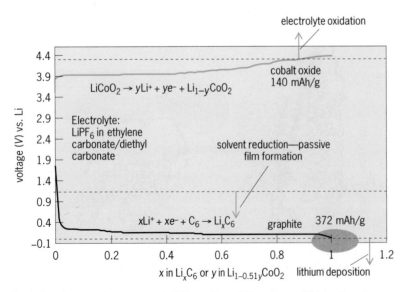

Fig. 3. Steady-state charge curve of a Li-ion cell consisting of a graphite anode and a LiCoO$_2$ cathode. The graph shows the half-cell potentials and the thermodynamic potentials for various side reactions.

to the energy, power, life, safety characteristics, and low-temperature performance. With each choice, it has been observed that some characteristics are improved and others prove lacking. To date, no ideal Li-ion chemistry has been found.

Research. Numerous techniques are being pursued to improve the energy of Li-ion batteries, with long life and enhanced safety. Ideas include finding new cathode and anode materials that promise significantly higher energy by enhancing the capacity and increasing the voltage. In addition, manipulation of materials at the nanoscale has proven to be an effective strategy in enabling the use of materials that were traditionally thought to be ineffective for energy storage. Examples include the lithium iron phosphate cathode and silicon anodes. Synthesis of new electrolytes, which have a higher stability window (allowing cathodes to operate at voltages greater than 4.2 V) and possess low flammability, could allow existing electrode materials to be improved for use in vehicles. Techniques to protect the interface between the electrode and the electrolyte (where the reaction occurs) via surface modifications can prove very useful in enhancing the reaction of interest and retarding the side reactions, thereby increasing the energy and maintaining life and safety.

Research in these areas is expected to lead to the continued improvement of lithium batteries, allowing for their use in vehicular applications. The specific energy of practical systems being studied today is far from the limits for battery storage. Systems exist that promise theoretical specific energies in the 2000–5000 Wh/kg range, such as lithium-sulfur and lithium-air. However, numerous problems plague these concepts, and many years of fundamental research are needed to evaluate their usefulness in real systems. While other chemistries exist that do not use lithium as the ion source (such as magnesium or sodium), these systems appear to be even farther from commercialization.

Outlook. With advances, it is expected that the Li-ion battery will enter the HEV market in the very near future, while use in PHEVs will be more gradual and dependent on solving life and cost challenges. Batteries for EVs remain problematic because of range and charging-time issues. Safety remains a challenge for Li-ion batteries, and their use in vehicular applications is critically dependent on controlling safety incidents via the use of safer materials and engineering solutions, such as overcharge protection mechanisms.

For background information *see* BATTERY; ELECTRIC VEHICLE; ELECTROCHEMISTRY; ELECTROLYTE; ELECTROMOTIVE FORCE (CELLS); ENERGY STORAGE; FUEL CELL; INTERCALATION COMPOUNDS; SOLID-STATE BATTERY in the McGraw-Hill Encyclopedia of Science & Technology.   Venkat Srinivasan

Bibliography. R. J. Brodd et al., Batteries, 1977 to 2002, *J. Electrochem Soc.*, 151(3):K1–K11, 2004; D. Linden and T. B. Reddy (eds.), *Handbook of Batteries*, 3d ed., McGraw-Hill, 2002; D. A. Scherson and A. Palsencsar, Batteries and electrochemical capacitors, *Electrochem. Soc. Interf.*, 15(1):17–22, 2006; W. Van Schalkwijk and B. Scrosati (eds.), *Advances in Lithium-Ion Batteries*, Kluwer Academic/Plenum Publishers, 2002; M. Winter and R. J. Brodd, What are batteries, fuel cells, and supercapacitors?, *Chem. Rev.*, 104(10):4245–4270, 2004.

Advances in electrophotography

Electrophotography, also known as xerography, is a printing technique that uses static electricity to create images and transfer them to substrates ranging from office papers to foils and films. In 1938, Chester Carlson produced the first image by xerography (**Fig. 1**). Since then, xerography has grown to a $100 billion business, annually. Some have compared the development of xerography and its evolution as a technology to the development of the printing press because it has democratized the reproduction of all types of documents by giving that power to any person.

Process fundamentals. Key to this process is the use of electrostatics. Throughout the process, electrostatic forces act as a temporary "glue" to attract and move colored ink particles from one surface to another. The other core element in the xerographic process is a photoconductive material that is an electrical insulator in the dark and conductive when illuminated. While Carlson did not discover or invent either photoconductivity or electrostatics, his breakthrough was to combine them and create a method for copying any original onto plain paper.

Although the technology has evolved significantly, the fundamental process steps that Carlson developed still stand. And while many variants of the electrophotographic process exist, each uses the following fundamental sequence of steps (**Fig. 2**). A photoconductive material coated on a conductive drum (or belt) is (1) charged uniformly in the dark and then selectively discharged by (2) exposure to light in the pattern of the desired image. The latent image, described as a pattern of electrical charge, is made visible by (3) development with a colored powder that has been given the opposite charge. The resultant image of colored particles is (4) transferred to paper, or another substrate, by charging the back of that medium to such a degree that the attraction

Fig. 1. The first xerographic image produced by Chester Carlson and his assistant, the physicist Otto Kornei, at their Astoria, Queens, laboratory in New York City.

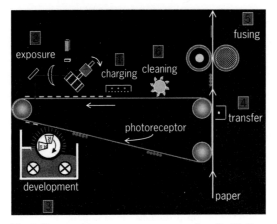

Fig. 2. Xerographic process sequence, as described in the text.

of the colored particles to the photoconductive material is overwhelmed by their attraction to the receiver. Heat and/or pressure are applied, (5) fusing the transferred image to permanently fix it to the substrate. Finally, the photoreceptor is (6) cleaned of any remaining material with a rotating brush, returning the system to its original state so that the process can be repeated. The ability to repeat the process from any original and the ability to print on standard media are the keys to xerographic printing.

Evolution of xerography. Over the past 50 years, there have been three fundamental technology waves in xerographic document production.

Analog plain-paper copying. The first was the development by the Haloid Corporation (later named Xerox) of the office plain-paper copier, which was a breakthrough that provided a dry, low-cost method for using reflected light to copy any original onto standard papers. This revolutionized office work practices.

Laser printing. The second wave was the development of laser printing, which extended the technique from copying to creating hard copy from a digital file.

Document publishing. The third and current wave is the creation of systems that enable mass customization of publishing and printing. Systems developed

in the current wave enable print-on-demand publishing of single, highly personalized versions of books, marketing materials, catalogs, and many other documents, such as photo books and calendars.

Xerographic technology continues to evolve significantly, and some key areas of its current evolution include color, productivity, cost, and energy and the environment.

Color. Affordable, high-quality color xerographic systems became available in the 1990s. As a printing process, xerography uses a subtractive color model based on mixing different amounts of cyan, yellow, magenta, and black toners to achieve a large color gamut that is compatible with industry standards for color reproduction. Current high-end production digital color presses achieve natural, high-quality color that is equivalent to that achieved with lithographic printing. Key technologies are color mixing, enabled by nanoscale color pigments embedded in the toner; accurate color rendering through image processing; and incorporation of modern controls and sensors into the overall print architecture of the system.

There are a number of ways of designing color devices to achieve the buildup of the color image from the four separations. In image-on-image architectures (**Fig. 3***a*), the color separations are assembled on an intermediate receiver and then transferred to paper in one step. In tandem architectures (Fig. 3*b*), each of the four color images is produced independently and then transferred sequentially, separation by separation, directly to the paper. To avoid color errors in the print, registration of the four individual images to each other must be maintained at 40 μm or less. The most effective architecture for minimizing these errors is image-on-image (Fig. 3*a*), where the four images are exposed and developed on top of each other before transfer occurs. In either case, achieving the highest quality requires closed-loop control feedback and precision position control systems for repeatability at the several micrometers level.

Productivity. A long-term trend in xerographic systems is increased speed or productivity. In commercial document production, printing is best viewed

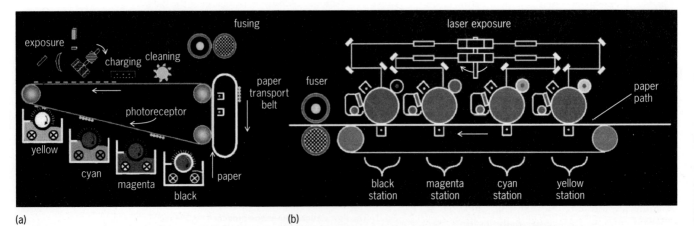

(a)   (b)

Fig. 3. Architectures for color xerographic printing devices: (*a*) image-on-image and (*b*) tandem.

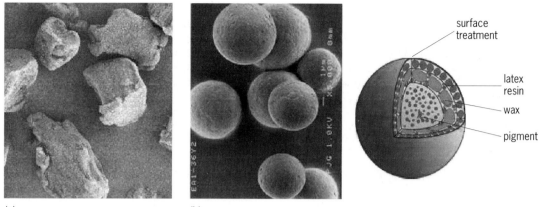

(a)   (b)

Fig. 4. Toners: (*a*) conventional and (*b*) chemically prepared, showing cross section.

as a manufacturing process, where productivity is a function of both raw speed and reliability, or uptime. Cut-sheet systems have been capable of up to 180 prints per minute, with failure rates as low as 3 per million prints. This performance has been enabled by high-speed laser scanning systems, low-wear and larger photoreceptors, and closed-loop paper-handling systems. A new trend in xerographic system design is the development of integrated multiengine systems to achieve greater speed and uptime.

Cost. Desktop laser printers have undergone radical price declines, with color laser printers now approaching the price of inkjet printers. Key technological drivers of low cost include miniaturization and the reuse of technologies developed in the high-volume electronics industry. For example, the scanning laser system reuses high-speed, low-cost drives, air bearings, and lasers from the disk-drive industry, as well as low-cost image-processing technologies developed for digital cameras. Miniaturization and the use of low-cost, powerful computation will continue to drive down costs.

Energy and the environment. Since xerography is a dry-powder printing process that works on plain papers, it has some inherent environmental advantages compared to standard printing processes. However, because of its low cost and broad use, the impact of paper use is critical. The printing and paper industries have developed and deployed sustainable forestry practices, such as eco-friendly pulping processes and recycling, to improve the sustainability and the "green" footprint of their business. The drive to reduce the environmental impact of xerographic printing has meant printing on sheets with high-recycled-fiber content, which can have a negative effect on the print quality. Managing this challenge has led to many technological improvements, including the development of advanced transfer systems that use acoustic energy to help move the toner from the photoreceptor to the paper. Additional environmental attention recently has been drawn to the toner fusing process, as it is the most energy-intensive step in xerographic printing. In the past, fusers have consumed power both during printing

and during standby. Minimizing the fusing energy requires a systems approach, including rapid heating technologies and toner materials that melt at lower temperatures. Technological advances include "instant-on" fusers that draw little or no power except when printing. These are based upon inductive heating technologies and toner materials that rely on nanotechnology to reduce the energy needed. Conventional toners are made by grinding colored plastics mechanically and then treating them with materials to control their charge and flow properties. Recently, toners called "chemically prepared toners" have emerged that are grown from nanoscale precursors that have precisely defined and nonhomogeneous structures and materials (**Fig. 4**). The ability to precisely control the toner properties has resulted in a new generation of materials that are more energy efficient to both manufacture and fuse to the paper. Additionally, chemically prepared toners have enabled many other improvements in the printing process, including better image quality driven by smaller, more consistent particle sizes (**Fig. 5**) and better charging performance, which enables printing in more extreme environments of temperature and humidity.

The environmental focus is being extended from device energy consumption, driven by fusers, toners, and low-environmental-impact papers, to lowering the overall cradle-to-grave carbon footprint. Key trends are significant improvement in the lifetime of parts that wear out (such as the printer cartridge) and miniaturization.

(a)   (b)

Fig. 5. Image quality with (*a*) conventional and (*b*) chemically prepared toners.

For background information *see* ELECTROSTAT-ICS; PAPER; PHOTOCONDUCTIVITY; PHOTOCOPYING PROCESSES; PRINTING in the McGraw-Hill Encyclopedia of Science & Technology.   Steve Hoover; Rick Lux; George Gibson

Bibliography. C. Anderson, *The Long Tail: Why the Future of Business Is Selling Less of More*, Hyperion, New York, 2006; D. Owen, *Copies in Seconds*, Simon & Schuster, New York, 2005; D. M. Pai and B. E. Springett, Physics of electrophotography, *Rev. Mod. Phys.*, 65:163–211, 1993; L. B. Schein, *Electrophotography and Development Physics* (Springer Series in Electrophysics), vol. 14, Springer-Verlag, Berlin, 1988.

Air pollution monitoring site selection

Criteria air pollutants, including particulate matter, sulfur dioxide, nitrogen oxides, volatile organic compounds (precursors of ground-level ozone), carbon monoxide, and lead, and toxic air pollutants are a global concern. Historically, the Clean Air Act was concerned only with criteria air pollutants, the six common air pollutants for which the U.S. Environmental Protection Agency (EPA) had set standards. In 1990, the Clean Air Act was amended to address 188 chemical classes of air toxics. A particular scenario that is receiving increased attention in the research and regulatory community is exposure to these compounds in near-road settings. Mobile source air toxics (MSATs) are emitted by vehicles either directly from exhaust systems or indirectly, such as from reentrainment of particle matter from roads. Addressing MSATs requires the combined expertise of engineering (for example, civil and mechanical engineering as applied to highway design and vehicle performance, respectively), the physical sciences (for example, particulate and gas phase partitioning of chemical compounds), and the social sciences (for example, decision theory as applied to selecting sites for representative samples from which to infer possible exposures).

Multiple-criteria decision analysis. Multiple-criteria decision analysis (MCDA) considers both quantitative and qualitative criteria when comparing alternative solutions to a problem. Decisions can be made by comparing various options and removing those that do not meet certain criteria. Of those that remain, possible interdependencies among alternatives are identified, sorted, and prioritized based on factors unique to a given decision. Ultimately, the MCDA process leads to selection of the best option meeting the decision criteria.

To collect air pollutant data in a near-road setting, the proper site must be chosen based on a set of relevant criteria. The purpose of any site selection process is to gather and analyze data that would lead one to draw informed conclusions regarding the most appropriate location of monitoring instruments.

As shown in **Table 1** and **Fig. 1**, the site selection process is a series of steps, with each step having varying degrees of complexity as a result of real-world issues. The first step is to develop a set of site selection criteria. For example, if the project is attempting to collect air pollutant emissions from highway vehicles, then a criterion might be to locate a section of highway that has significant traffic volume. Ideally, one would develop a set of site selection criteria that would maximize positive and minimize negative impacts on the goals of the data collection activity. Maximizing a positive impact might be the inclusion of meteorology (that is, wind speed and wind direction) as a criterion. One would not want to select a location in which the air pollutant emissions would not be "blowing" toward the monitoring instruments. Minimizing a negative impact might be to ensure that the site is located away from other nearby sources of emissions, such as an industrial source or power plant.

Additional steps in the site selection process include (1) developing a list of candidate sites and supporting information, (2) applying site selection filters (coarse and fine), (3) visiting the site(s), (4) selecting candidate site(s) via team discussion, (5) obtaining site access permission(s), and (6) implementing site logistics.

Candidate sites. A list of candidate sites is developed using the site selection criteria established in Step 1. Geographic information system (GIS) data, tools, and techniques, as well as on-site visits by

TABLE 1. Site selection process steps

Site Selection Steps	Method	Comment
1. Develop site selection criteria	Team discussions, management input	Project-specific: may change from project to project or site to site
2. Develop list of candidate sites	GIS data; on-site visit(s)	Additional sites added as information is developed
3. Apply coarse site selection filter	Team discussions, management input	Eliminate sites below acceptable minimums
4. Site visit	Field trip	
5. Select candidate site(s)	Team discussions, management input	Application of fine site selection filter
6. Obtain site access permissions	Contact property owners	If property owners do not grant permission, then the site is dropped from further consideration
7. Implement site logistics (i.e., physical access, utilities: electric and communications)	Site visit(s), contact utility companies	

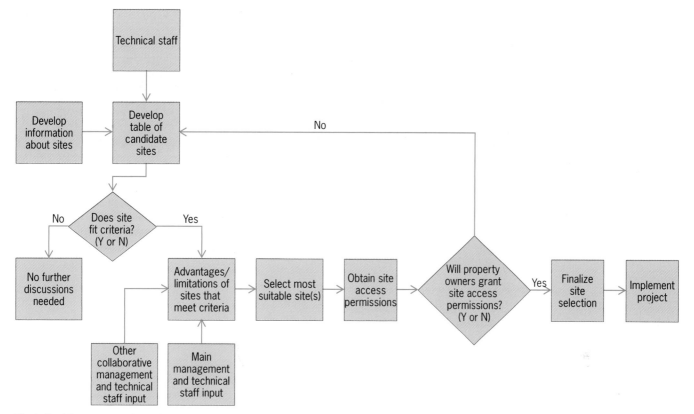

Fig. 1. Decision process schematic.

project team members, are the means of developing supporting information regarding potential sites. Other types of spatial data (for example, a street network) may be located or downloaded from GIS and other relevant Web sites. Nonspatial data (for example, meteorological data) may be downloaded from the National Climatic Data Center or other relevant Web sites. Site visits will provide information that is not readily available or provided or that is not easily gained from site maps.

Coarse filter. Team meetings are necessary to choose the most promising sites. This step involves a review of all the candidate sites, eliminating those that obviously do not meet the minimum criteria requirements. After applying the selection criteria as a set of filters, it may be possible to eliminate most of the candidate sites. For example, an obvious filter for a highway site would be to eliminate those sites with low traffic volume. Other filters should be applied to eliminate even more sites. Ideally, through the use of these filters, unsuitable sites are eliminated and the most suitable ones remain.

Ground truthing. An important component of ground truthing, or site visit, is to obtain information from local sources. Depending on the nature of the air-monitoring project, team members may need to meet with state/local air-quality staff, traffic engineers, or other relevant parties. All of these groups can provide valuable information regarding local conditions that would be difficult, if not impossible, to obtain otherwise. Too often, local resources are overlooked, and this can lead to poor decision making.

Site access. Another important consideration is site access. While site access may not be explicitly included in the selection criteria, it is critical. Even though a project may not directly affect property owners, they may be very reluctant to allow access to their property. While we would like to believe that property owners would permit access out of a sense of corporate or civic responsibility, they sometimes have a different perspective. Property owners may be reluctant to grant access for a variety of reasons, including liability, financial issues, suspicion of government activities, and so on. Therefore, access to any given property is not guaranteed, and researchers should be prepared for a long, involved process.

Logistics. Site logistics includes, but is not limited to, gaining access to electric power and communications connectivity, and arranging for security fencing. Any location where site logistics is restricted or prohibited because of administrative or physical issues is highly problematic and should be eliminated from further development.

Spatial tools. The use of spatial tools in decision processes is increasing and will continue to increase. Historically, the use of spatial tools (GIS) in decision processes has been somewhat problematic because of the magnitude of the data required by a GIS, the perception and reality of operating GIS software, and the level of knowledge required by end users to manipulate data in a GIS. In the last 15 years, GIS data have become more readily available in both quantity and quality, while the availability of a GIS in the Microsoft Windows® operating system environment,

TABLE 2. Spatial and nonspatial data inputs		
Data layer/input	Type (example)	Source
Spatial data		
Point data	Major point sources of emssions	Federal, state, local Web sites
Line data	Road/street network	Federal, state, local DOTs
Polygon data	Cadastral data (tax parcel) Administrative boundaries	Local (county) GIS Web sites
Raster data	Elevation data	USGS, local (county) GIS Web sites
	Aerial photos	Commercial Web sites, Google Earth
Nonspatial data		
Meteorology	Wind speed, wind direction	National Climatic Data Center

the availability of low-cost computer hardware, and the development and implementation of easy-to-use GIS tools (in a Windows environment) has made implementation of GIS-based decision-support tools more practical.

Typically, quantitative weighting criteria are associated with siting criteria, as well as elements of the GIS data layers (for example, certain types of highways would be more suitable than others and thus would have applicable quantitative values). It may

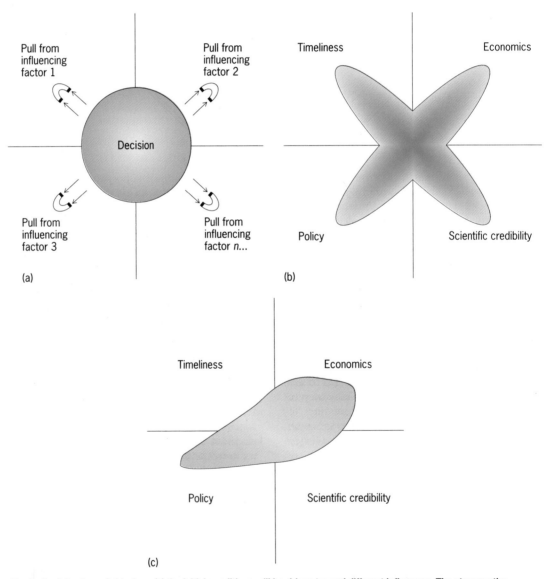

Fig. 2. Decision force field where (*a*) the initial conditions will be driven toward different influences. The stronger the influence of a factor (for example, high traffic volume), the greater the decision will be drawn to that perspective. Decision force field where (*b*) a number of factors have nearly equal weighting in a design decision. For example, if the site selection criteria are somewhat ambiguous, a number of alternatives are available, costs are flexible, and scientific credibility is minimally affected, the design has a relatively large degree of latitude and elasticity. Force field for (*c*) a decision that is most strongly influenced by policy/regulatory constraints and drivers. Note that all factors drive the decision, but the site selection criteria and other policy and regulatory issues have the greatest pull.

not always be possible to explicitly assign quantitative values to selection criteria. However, as the site selection process proceeds, it may become apparent that quantitative values are being implicitly assigned by team members. For example, sites with high traffic volume are more highly valued than sites with lower traffic volume. Thus, while it may appear that the decision process is based on a high degree of subjectivity, given that one might not explicitly assign quantitative values to selection criteria, the process of intrateam communication and application of selection criteria does lead to the selection of an appropriate site for the project.

GIS software may be used to create the maps for the site selection process. Spatial data may be located and downloaded from GIS or other relevant Web sites. **Table 2** shows some typical sources of data that may be used for the site selection process. The use of maps for the site selection process is only a tool.

Selection. Following the application of the selection criteria, candidate sites may be further prioritized during team discussions, with the pros and cons of each site being debated. Typically, a primary site is chosen along with a backup site. This is to ensure that if something unforeseen occurs regarding the primary site during the early stages of project implementation, then the backup site is available for deploying the air-monitoring instruments.

In an air pollutant monitoring project, numerous organizations may be involved in the process. Some of these may be policy/regulatory groups that have a different mission and perspective pertaining to overall project implementation and more specifically site selection. Under certain conditions, policy requirements may supplant the researchers' inherent quest for knowledge as a reason to investigate certain phenomena. However, policy must never be divorced from sound science.

There is no perfect air monitoring site, as compromises will have to be made for any environmental study conducted within any area. It is a question of balancing benefits with risks and costs. The selection may be further complicated by external constraints. A principal constraint may be the legal mandate of a regulatory policy. Few, if any, design decisions can be made exclusively from a single perspective. These decisions can be visualized as attractions within a force field, where the center of the diagram represents the initial condition with a magnet placed in each sector at points equidistant from the center of the diagram as shown in **Fig. 2a**. If the factors are evenly distributed and weighted, the diagram might appear as in Fig. 2b. But as the differential in magnetic force increases, the stronger factor(s) will progressively drive the decision. In a particular case study, the decision might be most directly influenced by legal requirements, but it also needs to be scientifically credible and economically feasible (Fig. 2c).

[Disclaimer: The U.S. Environmental Protection Agency through the Office of Research and Development funded and managed the research described here. The present article has been subjected to the agency's administrative review and has been approved for publication.]

For background information *see* AIR POLLUTION; DECISION ANALYSIS; ENVIRONMENTAL ENGINEERING; GEOGRAPHIC INFORMATION SYSTEMS in the McGraw-Hill Encyclopedia of Science & Technology.
Sue Kimbrough; Dan Vallero

Bibliography. Editors of ESRI Press, *Understanding GIS: The ARC/INFO Method*, ESRI Press, Redlands, CA, 1995; S. French and J. Geldermann, The varied contexts of environmental decision problems and their implications for decision support, *Environ. Sci. Policy*, 8(4):378–391, 2005; C. D. Gamper and C. Turcanu, On the governmental use of multicriteria analysis, *Ecol. Econ.*, 62(2):298–307, 2007; S. Kimbrough et al., Multi-criteria decision analysis for the selection of a near road ambient air monitoring site for the measurement of mobile source air toxics, *J. Environ. Eng.*, in press; I. Linkov and A. B. Ramadan, *Comparative Risk Assessment and Environmental Decision Making*, NATO Advanced Research Workshop, Rome (Anzio), Italy, Springer, 2004; J. Malczewski, *GIS and Multicriteria Decision Analysis*, John Wiley & Sons, New York, 1999; V. R. Sumathi, U. Natesana, and C. Sarkar, GIS-based approach for optimized siting of municipal solid waste landfill, *Waste Manage.*, corrected proof, in press; S. Xenarios and I. Tziritis, Improving pluralism in Multi Criteria Decision Aid approach through Focus Group technique and Content Analysis. *Ecol. Econ.*, 62(3–4):692–703, 2007.

Alpha and proton radioactivity above tin-100

Only 263 nonradioactive nuclei occur naturally. These isotopes of various elements are the combinations of Z protons and N neutrons that are the most stable. In all other nuclei, including about 3000 known today, the ratio of proton-to-neutron numbers is nonoptimal, and the isotopes decay to more stable ones, emitting several kinds of radiation. The terms proton radioactivity and alpha radioactivity refer to the spontaneous emission of a charged particle, either a single proton or an alpha (α) particle, from an atomic nucleus. A clustering of two protons and two neutrons is needed to create an alpha particle inside an atomic nucleus, before its emission. This so-called preformation probability reduces the alpha emission rate. Even today, 100 years after understanding of the nature of alpha particles was achieved, the dynamics of the preformation of the alpha particle inside the atomic nucleus is not well understood.

Prediction of superallowed alpha decay. It was postulated in 1965 by R. D. Mcfarlane and A. Siivola that the decay of tellurium-104 (^{104}Te), with $Z = N = 52$, into tin-100 (^{100}Sn), with $Z = N = 50$, would proceed via a "superallowed" alpha transition. The ^{100}Sn nucleus is expected to have "doubly magic" properties in the spherical shell model, that is, to be

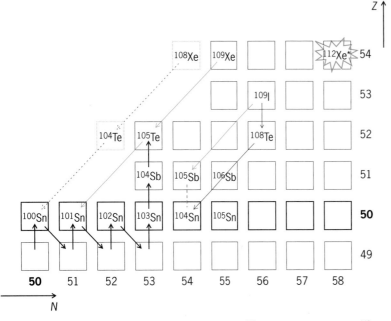

Fig. 1. Chart of the nuclei near the doubly magic nucleus ^{100}Sn. The excited nucleus ^{112}Xe, produced in a fusion of ^{58}Ni projectiles and the ^{54}Fe target nuclei, is marked with an exploding star. The fast double-alpha decay chain of ^{109}Xe → ^{105}Te → ^{101}Sn and a weak alpha transition in the decay of ^{109}I (marked in color) were discovered at Oak Ridge National Laboratory. The potential *rp*-process termination cycle, with a crossing of the $Z = 50$ magic gap through the proton capture on ^{103}Sn, is indicated by heavy black arrows. (*Courtesy of Chiara Mazzocchi*)

particularly strongly bound and of spherical shape, with 50 protons and 50 neutrons closing respective nuclear shells. The ^{104}Te nucleus can be described as a nuclear molecule with a ^{100}Sn doubly magic core plus one valence alpha particle (**Fig. 1**). The two protons and two neutrons forming this alpha particle have similar properties, because both protons and neutrons occupy identical quantum states, the first levels above the $N = 50$ and $Z = 50$ shell closures. Therefore, the alpha preformation probability can be expected to have the largest value in ^{104}Te. The ^{104}Te → ^{100}Sn alpha-decay strength, when measured, should replace our current reference strength derived from ^{212}Po → ^{208}Pb decay properties. The nucleus lead-208 (^{208}Pb) is also doubly magic, with $Z = 82$ and $N = 126$ closing proton and neutron shells. However, because these magic numbers are different, the proton and neutron states above the closed shells in ^{208}Pb are also different. In contrast to the ^{100}Sn region, the alpha particle above the ^{208}Pb core is made out of protons and neutrons of opposite parity. One can picture a pair of protons moving in the opposite direction to the pair of neutrons in polonium-212 (^{212}Po). In ^{104}Te, protons and neutrons are in states of identical parity, with a better chance to overlap and preform an alpha particle. However, ^{104}Te is extremely difficult to produce and has never been observed so far.

Discovery of tellurium-105. Scientists at Oak Ridge National Laboratory (ORNL) and at Argonne National Laboratory have recently and independently discovered tellurium-105 (^{105}Te), a nucleus next to ^{104}Te in the chart of the nuclei, and the lightest-mass alpha radioactivity known to date. The ^{105}Te nucleus

has a half-life as short as 0.6 microseconds. Its alpha decay populates tin-101 (^{101}Sn), the closest neighbor to ^{100}Sn. The experiment at ORNL used the fusion of two heavy ions, an energetic nickel-58 (^{58}Ni; $Z = 28$, $N = 30$) projectile accelerated to 220 megaelectronvolts kinetic energy, and an iron-54 (^{54}Fe; $Z = 26$, $N = 28$) target, to produce nuclei of xenon-112 (^{112}Xe). These ^{112}Xe compound nuclei, with $Z = 54$ and $N = 58$, had a kinetic energy of about 100 MeV, allowing them to fly out of the target foil within a few tens of femtoseconds (1 fs = 10^{-15} s). In addition to the kinetic energy, the ^{112}Xe nuclei were created with an excitation energy of about 50 MeV, which is above the energy needed to separate a proton ($\sim$2 MeV) or a neutron ($\sim$14 MeV) from ^{112}Xe. Therefore, the excited nucleus ^{112}Xe quickly cooled down through the evaporation of several nucleons, most likely four to five protons. The nucleon evaporation occurred immediately, that is, at a time scale of 10^{-17} s, after the production of ^{112}Xe in the target foil. In a very rare case, ^{112}Xe released its excitation energy with an emission of exactly three neutrons and several energetic photons, creating ^{109}Xe in its lowest energy state (Fig. 1). The emission of three neutrons from ^{112}Xe did not effectively change the kinetic energy. The ^{109}Xe velocity was therefore about 4% of the speed of light. The fast-moving ^{109}Xe recoiled from the ^{54}Fe target material as an ion, which was usually stripped of about half of the 54 electrons that are present in a neutral xenon atom. The proper combination of magnetic and electric fields in the ORNL device, called a recoil mass spectrometer (RMS), allowed physicists to select and guide desired ions to the detector setup. The ^{109}Xe ion was implanted into a silicon detector after only 2 microseconds time of flight through the 30-m-long (100-ft) RMS.

The ^{109}Xe nucleus can be pictured as a molecule, again with a ^{100}Sn core, and two alpha particles and one neutron outside this core. The alpha particle is unbound in ^{109}Xe, triggering its fast decay to ^{105}Te, which is even less bound against alpha emission (Fig. 1). Hence, the decay chain ^{109}Xe → ^{105}Te → ^{101}Sn produces two alpha particles, whose emissions are separated in time by only the very short half-life of the intermediate nucleus ^{105}Te. The development of a new detection technique, based on digital signal processing, was needed to observe simultaneously the signals induced by both emitted alphas after the implantation of ^{109}Xe into a silicon detector. Over 100 digital images of the pulses corresponding to the two overlapping alpha signals representing the decay chain ^{109}Xe → ^{105}Te → ^{101}Sn were collected in this pioneering study (**Fig. 2**). The short time difference between the pulses directly reflects a very short half-life of about 0.6 μs for ^{105}Te. This fast alpha emission was compared to the ^{213}Po → ^{209}Pb alpha decay. The structure of ^{213}Po resembles that of ^{105}Te but with a doubly magic nucleus ^{208}Pb acting as a core, and also with one alpha particle and one neutron coupled to the core. The comparison has demonstrated that indeed the alpha transition occurring in the ^{105}Te → ^{101}Sn decay is about three times faster

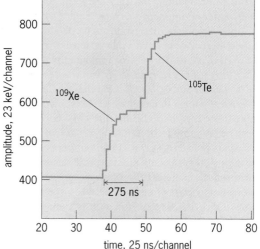

Fig. 2. Digital image of a detector signal showing two overlapping alpha pulses, representing the alpha decay of ^{109}Xe ($Z = 54$) followed by alpha emission from ^{105}Te ($Z = 52$) about 275 ns later. (*Courtesy of Sean Liddick*)

than the ^{213}Po $\rightarrow$ ^{209}Pb alpha decay. It was the first experimental evidence for the existence of superallowed alpha decay near ^{100}Sn. An investigation aiming at the identification of the "reference" superallowed alpha transition in the ^{108}Xe $\rightarrow$ ^{104}Te $\rightarrow$ ^{100}Sn decay chain will be attempted in the near future at ORNL. The half-life of ^{104}Te, estimated to be about 10 nanoseconds, poises a real challenge to the successful performance of this experiment.

Determination of the rp-process. Proton emission represents an inverse process to the fusion of a proton projectile with a heavy target nucleus. The latter reaction, leading effectively to proton capture, plays an important role during the nucleosynthesis process. In particular, the so-called rapid proton capture process (*rp*-process) creates heavy nuclei along a path located near the proton drip line, up to the doubly magic ($N = Z = 50$) nucleus ^{100}Sn (Fig. 1). Further proton captures, advancing atomic numbers of created nuclei above $Z = 50$, seem to be impossible, because the $Z = 51$ antimony isotopes ^{101}Sb, ^{102}Sb, ^{103}Sb, and ^{104}Sb are predicted to be proton-unbound, and ^{105}Sb was even reported as an identified proton radioactivity, emitting protons of 0.5 MeV kinetic energy. The proton capture rate on a proton-unbound isotope is very small, and a subsequent proton emission reverses the capture effect quickly.

The determination of the actual path, time scale, and the termination region of the *rp*-process is still among the most important scientific questions to be addressed in nuclear physics and astrophysics studies. Evidently, the creation of heavier elements in the *rp*-process slows down after reaching $Z = 50$. The beta (β) decays of very proton-rich tin isotopes populate $Z = 49$ indium isotopes of the same mass number, for example, ^{100}Sn $\rightarrow$ ^{100}In. These daughter indium isotopes subsequently capture a proton and create the tin isotope with a mass number larger by one, for example, ^{100}In $+$ p $\rightarrow$ ^{101}Sn. Such a beta decay–proton capture sequence was earlier ex-

pected to follow up to ^{105}Sn. The proton capture by ^{105}Sn would create ^{106}Sb, with $Z = 51$, which is known to be bound against proton emission. This scenario requires the *rp*-process with a large density of energetic protons to last a relatively long time, comparable to the beta-decay lifetimes of the involved tin isotopes, from ^{100}Sn to ^{104}Sn. (The half-life of ^{104}Sn is 21 s.)

To verify (or modify) this scenario and to estimate the time scale needed to cross the $Z = 50$ magic proton number, data on the proton separation energies S_p in lighter antimony isotopes are needed. Researchers at ORNL have recently determined that ^{105}Sb is actually bound with respect to proton emission by over 130 keV more than was earlier reported. This determination was performed by measuring the ultraweak alpha decay of iodine-109 (^{109}I) to ^{105}Sb and checking the loop of nuclear masses involved. Starting from ^{109}I, one can reach ^{104}Sn through a proton emission from ^{109}I to ^{108}Te (known) and the following alpha transition from ^{108}Te to ^{104}Sn (also known). Alternatively, the alpha decay of ^{109}I populates ^{105}Sb (just discovered), and the mass loop is closed by determining the proton separation energy $S_p = -356$ keV for ^{105}Sb (Fig. 1).

The new information on the S_p value of ^{105}Sb opened up another possibility for the *rp*-process path. The S_p values measured for neighboring pairs of heavier proton and alpha-emitting nuclei in the ^{100}Sn region (such as ^{108}I–^{109}I and ^{112}Cs–^{113}Cs) suggest that ^{104}Sb might be more strongly bound than ^{105}Sb, and might even be stable against proton emission. In that case, the *rp*-process would have a chance to cross the $Z = 50$ gap via proton capture on ^{103}Sn to ^{104}Sb. The longest beta half-life among tin isotopes on this path is less than 4 s (^{102}Sn). Proton capture to create nuclei above the $Z = 50$ magic number may occur earlier to terminate nucleosynthesis within the *rp*-process in a much faster cycle than previously was thought to operate. The two subsequent proton captures on ^{103}Sn will produce the very short-lived nuclide ^{105}Te, which will decay back to ^{101}Sn to close the *rp*-process termination cycle. Further proton- and alpha-radioactivity studies in the ^{100}Sn region planned at ORNL are aimed at determining the actual nucleosynthesis path within the *rp* process in hot stars. As might be expected, these studies include the measurement of the proton separation energy value in ^{104}Sb.

For background information *see* ALPHA PARTICLES; MAGIC NUMBERS; NUCLEAR STRUCTURE; NUCLEOSYNTHESIS; PARITY (QUANTUM MECHANICS); RADIOACTIVITY in the McGraw-Hill Encyclopedia of Science & Technology.   Krzysztof P. Rykaczewski;
Robert K. Grzywacz

Bibliography. S. N. Liddick et al., Discovery of ^{109}Xe and ^{105}Te: Superallowed α decay near doubly magic ^{100}Sn, *Phys. Rev. Lett.*, 97:082501, 2006; C. Mazzocchi et al., α decay of ^{109}I and its implications for the proton decay of ^{105}Sb and the astrophysical rapid proton-capture process, *Phys. Rev. Lett.*, 98:212501, 2007; R. D. Mcfarlane and A. Siivola, New region of alpha radioactivity, *Phys. Rev. Lett.*, 14:114–115, 1965;

E. Rutherford, The nature and charge of the α particles from radio-active substances, *Nature*, 79:12–15, 1908; H. Schatz et al., End point of the *rp* process on accreting neutron stars, *Phys. Rev. Lett.*, 86:3471–3474, 2001; D. Seweryniak et al., α decay of ^{105}Te, *Phys. Rev. C*, 73:061301(R), 2006.

Ambient noise seismic imaging

Traditional seismic imaging of large-scale structures within the Earth's interior is based on observations of surface displacements following earthquakes or human-caused explosions. Explosive sources are used less often because of their expense and environmental impact. These methods measure body and surface wave travel times as well as whole waveforms typically following earthquakes. Such measurements are unraveled (inverted) to reveal the isotropic and anisotropic variation of compressional (Vp) and shear (Vs) wave speeds in the Earth's crust, mantle, and core, which are then interpreted in terms of temperature, composition, and fluid content. The ability of earthquake-based methods to resolve structural features within the Earth degrades during the propagation of the wave over long (teleseismic) distances. For seismic surface waves, such as Rayleigh waves (vertically polarized waves) and Love waves (horizontally polarized waves transverse to the direction of motion), teleseismic transmission results in the loss of the high frequencies needed to infer information about the Earth's crust and uppermost mantle. A recent innovation in seismic imaging based on using long time sequences of ambient seismic noise moves beyond some of the limitations imposed on earthquake-based methods to reveal higher-resolution information about the crust and uppermost mantle. This method is called ambient noise tomography (ANT), and it has been applied predominantly to seismic surface waves. With the application of ANT to data from ambitious new deployments of seismic arrays, such as the EarthScope USArray in the United States, improved seismic models of the Earth's crust and uppermost mantle are rapidly emerging at unprecedented resolution.

Ambient noise tomography. Between earthquakes, seismometers continuously record surface displacements with a wide range of causes, such as wind, atmospheric pressure variations, fluid flows beneath and on the surface, human and animal motions, and ocean waves. Seismic waves produced by ocean waves, called microseisms, are particularly well studied and are observed to propagate deep into continental interiors. Microseismic amplitudes peak near periods of 8 and 16 s, but extend to longer periods, merging into the somewhat more enigmatic, but increasingly well studied, "earth hum" at periods above 20 s. Debate continues as to whether earth hum is generated predominantly in shallow waters, like microseisms, or in deep waters. Recent evidence presented by B. Romanowicz and collaborators as well as others indicates that it is predominantly a shallow-

water phenomenon, but this does not preclude a deep-water component.

Any mechanism that produces waves that propagate coherently between a pair of seismometers can be used as a basis for seismic tomography. This idea has a long history in seismology, but was resurrected by R. Weaver and O. Lobkis and other researchers in a series of papers beginning in 2001, showing in the laboratory and theoretically that cross-correlations between recordings of diffuse waves at two receiver locations yield the "Green's function" between these positions. The Green's function is a seismic waveform that contains all the information about wave propagation in the medium between the two stations. Once the Green's function is estimated, traditional seismic methods of tomography then can be applied to it to recover information about the medium of transport.

The relevance of these results to large-scale earth imaging was not immediately clear, because the Earth's ambient noise field containing energetic microseismic energy is not diffuse, it probably is not homogeneously distributed in the direction of propagation (azimuth), and its frequency content was poorly understood. In 2004, N. Shapiro and M. Campillo showed that coherent Rayleigh surface waves can be extracted from the Earth's ambient noise field and that the primary frequency content of the waves lies in the microseismic and earth-hum bands from periods of about 6–100 s, with the highest amplitudes in the microseismic band. Subsequent studies have confirmed that the full Green's function does not emerge from cross-correlating seismic data because the cross-correlations are three dominantly surface waves, with Love waves also being observable. Nevertheless, the ability to constrain surface-wave speeds at periods from 6–20 s, which are sensitive to crustal depths but difficult to measure from teleseismic earthquakes, provided much of the early interest in the method.

Observations of broadband surface waves. Seismic surface waves, in contrast with body waves, are waves that propagate in a layer near the Earth's surface. The depth extent of the layer depends on wavelength, with a fair approximation being about one-third of a wavelength. Thus, surface waves with periods below about 20 s are sensitive to the crust, and waves with periods between 20 and 100 s are sensitive predominantly to the uppermost mantle to a depth of about 150 km (90 mi). Both Rayleigh and Love waves are dispersive. Their speeds depend on frequency, with lower frequencies typically traveling faster than higher frequencies.

Surface waves appear strongly on cross-correlations of ambient noise, and their dispersion characteristics are readily identifiable (**Fig. 1**). G. Bensen and collaborators presented a primer on ambient noise data processing in 2007. They provide methods for removing earthquakes and instrumental irregularities from seismograms prior to cross-correlation and show that longer time series (a year or more) homogenize the azimuthal content of ambient noise, that reliable measurements require

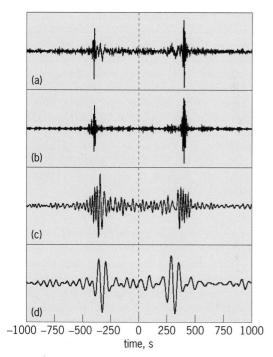

than 400 broadband seismometers deployed concurrently with a station separation of about 70 km (40 mi) and is presently sweeping across the United States. F. Lin and collaborators showed that the resolution of ANT applied to EarthScope TA data is better than the interstation spacing, which is unprecedented over an area the size of the western United States. The construction of similar large-scale deployments of seismometers is occurring or planned in China and Europe.

There are remaining pockets of concern among seismologists that ambient noise in the Earth does not meet the theoretical conditions on which ANT rests. In particular, they worry that the azimuthal inhomogeneity of ambient noise may, at worst, vitiate

Fig. 1. Cross-correlation between two years of ambient noise recorded on the vertical components of two seismic stations in the western United States. The stations are EarthScope/USArray transportable array stations M03 (McCloud, California) and Y14A (Wickenburg, Arizona), separated by a distance of 1144 km (711 mi). Arrivals at positive and negative times are for waves traveling in opposite directions between the stations. Rayleigh waves arrive at times between 200 and 500 s. (*a*) The broadband cross-correlation is shown. (*b–d*) Bandpass filters are applied to the broadband cross-correlation centered on periods of 10, 20, and 50 s, respectively. The longer periods are seen to travel faster, indicative of the dispersive nature of the Rayleigh wave. (*Courtesy of Morgan Moschetti*)

a station separation of at least two wavelengths, and that uncertainties can be estimated from the temporal repeatability of the measurements.

The production of maps of the speed of Rayleigh or Love waves as a function of frequency is called surface-wave tomography. What has come to be known as ambient-noise tomography (ANT) is the generation of such maps from interstation ambient noise cross-correlations. The first ambient noise tomographic images of Rayleigh-wave group speeds in the micrsoseismic band were presented simultaneously by N. Shapiro and collaborators and K. Sabra and collaborators in 2005, based on one to several months of data from stations in southern California. These studies were followed by a multiplicity of applications around the world, including studies in Europe, New Zealand, South Africa, Korea, Japan, Iceland, Canada, Australia, and China, in addition to the United States. Both Rayleigh- and Love-wave dispersion maps are now commonly obtained at periods from 6 to 100 s, with the spatial extent of the study ranging up to the continental scale and time series lengths of more than four years being used in some cases. ANT is most powerfully applied to large deployments of seismometers, such as the transportable array (TA) component of EarthScope/USArray (**Fig. 2**), which includes more

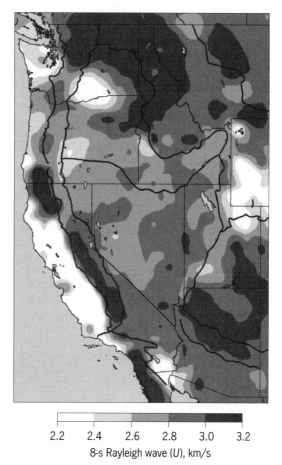

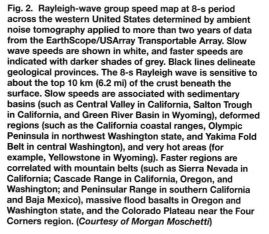

Fig. 2. Rayleigh-wave group speed map at 8-s period across the western United States determined by ambient noise tomography applied to more than two years of data from the EarthScope/USArray Transportable Array. Slow wave speeds are shown in white, and faster speeds are indicated with darker shades of grey. Black lines delineate geological provinces. The 8-s Rayleigh wave is sensitive to about the top 10 km (6.2 mi) of the crust beneath the surface. Slow speeds are associated with sedimentary basins (such as Central Valley in California, Salton Trough in California, and Green River Basin in Wyoming), deformed regions (such as the California coastal ranges, Olympic Peninsula in northwest Washington state, and Yakima Fold Belt in central Washington), and very hot areas (for example, Yellowstone in Wyoming). Faster regions are correlated with mountain belts (such as Sierra Nevada in California; Cascade Range in California, Oregon, and Washington; and Peninsular Range in southern California and Baja Mexico), massive flood basalts in Oregon and Washington state, and the Colorado Plateau near the Four Corners region. (*Courtesy of Morgan Moschetti*)

the method and, at best, generate biased measurements. Studies of the directionality of ambient noise published in 2006 and 2008 by L. Stehly and Y. Yang and their respective collaborators demonstrate that with the use of time series of a year or more in length, ambient noise propagates across a wide range of azimuths, although there may be some preferred directions. Simulations of the observed azimuthal content of ambient noise establish that measurement bias is small relative to other sources of measurement error.

3D images of Earth's interior. The purpose of ANT is not just to reveal the speed of surface waves at different periods (Fig. 2), but to use this information to unveil the three-dimensional (3D) variation of seismic waves in Earth's interior in order to advance knowledge of temperature, composition, and fluid content, which hold the key to the understanding of Earth processes. Recent studies, such as that by Y. Yang and collaborators in 2008 for the western United States, which inverted ambient noise- and earthquake-derived information simultaneously, are now providing 3D images of the crust and uppermost mantle over large areas in unprecedented detail (**Fig. 3**). Not only does ANT provide better lateral resolution than traditional surface-wave methods in regions with good station coverage, but also its broad frequency content, which extends to periods below 10 s, gives the vertical resolution needed to resolve crustal from mantle structures clearly.

Applications other than earth imaging. Ambient noise can be exploited constructively in other contexts than Earth imaging. Other bodies in the solar system, for example, have been targets for the

method. T. Duvall and collaborators in 1993 established time-distance seismology on the Sun (helioseismology) based on cross-correlating intensity fluctuations observed on the solar surface. In 2005, E. Larose and collaborators correlated seismic noise on the Moon's surface taken from the *Apollo 17* Lunar Seismic Profiling Experiment, estimated Rayleigh-wave group speeds between frequencies of 4 and 11 Hz, and inverted them to provide new information about the lunar regolith. They also established that the Sun actively generates the lunar seismic noise because of strong thermal gradients induced during the lunar day. These results suggest the extension of ambient noise tomography to planetary exploration, where the origin of the noise may be quite different from that on Earth. On Earth, variations in cross-correlations between stations can provide information about the changing state of the shallow crust that may, for example, precede volcanic activity or possibly earthquakes. In 2008, F. Brenguier and collaborators showed how seismic wave speeds determined from ambient noise decreased before eruptions of the Piton de Fournaise volcano on La Réunion Island in the Indian Ocean, presumably attributable to the preeruptive inflation caused by increased magma pressure.

For background information *see* COMPUTERIZED TOMOGRAPHY; EARTH INTERIOR; EARTHQUAKE; HELIOSEISMOLOGY; SEISMOGRAPHIC INSTRUMENTATION; SEISMOLOGY in the McGraw-Hill Encyclopedia of Science & Technology.   Michael H. Ritzwoller

Bibliography. G. D. Bensen et al., Processing seismic ambient noise data to obtain reliable broad-band surface wave dispersion measurements, *Geophys. J. Int.*, 169:1239–1260, doi: 10.1111/j.1365-246X.2007.03374.x, 2007; E. Larose et al., Lunar subsurface investigated from correlation of seismic noise, *Geophys. Res. Letts.*, 32:L16201, doi:1029/2005GL023518, 2005; O. I. Lobkis and R. L. Weaver, On the emergence of the Green's function in the correlations of a diffuse field, *J. Acoust. Soc. Am.*, 110(6):3011–3017, 2001; N. M. Shapiro et al., High resolution surface wave tomography from ambient seismic noise, *Science*, 307(5715):1615–1618, 2005; N. M. Shapiro and M. Campillo, Emergence of broadband Rayleigh waves from correlations of the ambient seismic noise, *Geophys. Res. Lett.*, 31:L07614, doi:10.1029/2004GL019491, 2004; Y. Yang et al., The structure of the crust and uppermost mantle beneath the western U.S. revealed by ambient noise and earthquake tomography, submitted to *J. Geophys. Res.*, 2008.

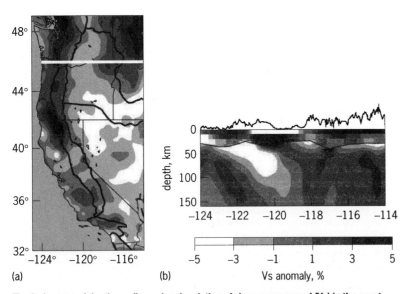

Fig. 3. Images of the three-dimensional variation of shear-wave speed (Vs) in the crust and uppermost mantle determined from ambient noise and earthquake information. (*a*) Horizontal slice at 100 km (620 mi) depth. (*b*) Vertical profile underlying the white line in *a*. Vertically exaggerated surface topography is presented at top, and the black line indicates the Mohorovicic discontinuity, separating the crust from the mantle. In both panels, Vs is presented as the perturbation in percent from the average at each depth across the model. Numerous features are imaged in the crust and mantle. For example, in the mantle, the subducting Juan de Fuca and Gorda plates are seen as high Vs beneath Northern California, Oregon, and Washington in both *a* and *b*. In *b*, the high Vs subducting plate is overlain by low Vs speeds (high temperature and volatile content) beneath active volcanoes in the Cascade Range. (*Courtesy of Yingjie Yang*)

Androgenic-anabolic steroids and athletes

Androgenic-anabolic steroids (AAS) are synthetic manufactured drugs derived from the male sex hormone testosterone. All AAS exert both androgenic and anabolic effects. "Androgenic" refers to the masculinizing effects, for example, growth of genital size, muscle growth, and male body hair distribution. The "anabolic" effects are characterized by the

TABLE 1. Commonly used androgenic-anabolic steroids

Oral agents	Intramuscular agents	Transdermal agent
Chlorodehydromethyltestosterone	Boldenone undecylenate	Testosterone
Danazol	Drostanolone	
Drostanolone	Methenolone enanthate	
Ethylestrenol	Nandrolone decanoate	
Fluoxymesterone	Nandrolone phenpropionate	
Mesterolone	Nandrolone undecanoate	
Methandienone	Stanozolol	
Methandrostenolone	Testosterone aqueous suspension	
Methenolone	Testosterone cypionate	
Methyltestosterone	Testosterone enanthate	
Mibolerone	Testosterone ester blends	
Oxandrolone	Testosterone propionate	
Oxymetholone	Testosterone undecanoate	
Stanozolol	Trenbolone acetate	
Striant (buccal delivery of testosterone)	Trenbolone hexahydrobencylcarbonate	

TABLE 2. Example of AAS abuse scheme of an athlete

AAS	Week							
	1	2	3	4	5	6	7	8
Nandrolone decanoate (im)	50	100	200	400	400	200		
Testosterone enanthate (im)		50	100	200	200	100	100	100
Methenolone (po)					100	100	100	100
Stanozolol (im)	50	100	50	100	50	100	50	100
Stanozolol (po)			25	50	75	75	75	25

Numbers represent weekly dose (mg) of each substance; im, intramuscularly; po, per os (orally).

building properties on several organ systems, for example, muscle and bone growth and enlargement of organs such as the heart, kidney, and prostate. Manipulation of the chemical composition of these substances to dissipate both types of effects has failed, but has resulted in a number of compounds with different androgenic and anabolic properties (see **Table 1**).

Abuse of AAS by athletes is widespread because of their putative muscle-building and performance-enhancing effects. The abuse is not restricted to elite or professional athletes, but is also common among amateur and recreational sports enthusiasts. In some sports, such as weight lifting, bodybuilding, and American football, up to 68% of athletes admitted to having used AAS at least once to improve performance. Even at the high school and college level in the United States, 12–15% of athletes were found to have used these substances.

Athletes purchase these drugs mainly illegally, which implies that quality is not guaranteed. The use of AAS by these athletes is characterized by the administration of several substances simultaneously (both oral and parenteral) in doses greatly exceeding the maximal therapeutic dose (see **Table 2**). From a medical point of view, the adverse effects on health of such unrestrained AAS abuse are of great concern.

The effects of AAS on body composition and performance (see **Table 3**), and the deleterious health effects (see **Tables 4** and **5**), are reviewed here. Moreover, the therapeutic use and mechanisms of action are discussed.

Therapeutic use of AAS. The medical use of AAS is limited. Medical treatments that employ AAS include disease conditions such as androgen deficiency states, delayed male puberty, lichen sclerosis (a painful skin condition that occurs most often on the genital area of females), dystrophy of the vulva, postmenopausal osteoporosis, aplastic anemia, and anemia in chronic renal insufficiency. New areas of application, particularly chronic wasting conditions such as chronic obstructive pulmonary disease (COPD) and human immunodeficiency virus/acquired immune deficiency syndrome (HIV/AIDS), are under active research.

Mechanisms of action. AAS exert their actions on the different organ systems via several pathways. These substances pass through the cell membrane and bind to androgen receptors. The steroid–receptor complex that is formed acts in the cell nucleus directly after binding to deoxyribonucleic acid (DNA). The binding to androgen receptors is mediated by the intracellular enzymes, 5-α-reductase and aromatase. The former converts AAS

TABLE 3. Physiologic and ergogenic (performance-enhancing) effects of AAS use in adult athletes

Body mass	Increase
Lean body mass	Increase
Fat mass	No change
Muscle mass	Increase
Muscle fiber dimensions	Increase
Muscle strength	Increase
Aerobic performance	No change
Combined aerobic/strength performance	Increase
Recovery from physical exercise	No change (or slight increase?)

TABLE 4. Relatively common possible side effects of AAS use in adult athletes

Reproductive system
  Suppression of hypothalamic–pituitary–gonadal axis
  Reduced fertility
  Infertility
    Male-specific
    Testicular atrophy
    Reduction of sperm production
    Reduced sperm quality

    Female-specific
    Menstrual irregularities
    Breast atrophy
    Clitoral enlargement
    Voice deepening
    Male hair pattern
Gastrointestinal system
  Elevation of liver enzymes ASAT and ALAT*
    (particularly when using oral AAS)
Cardiovascular system
  Unfavorable alteration of serum lipid and
    lipoprotein profile
Hematology
  Increased red blood cell count
  Increased hematocrit
  Hemostasis alterations
Psyche and behavior
  Increased aggression
  Increased hostility
  Mood imbalance
  Libido alterations

*ALAT, alanine aminotransferase; ASAT, aspartate aminotransferase.

TABLE 5. Less common and rare side effects associated with AAS use in adult athletes

Reproductive system
  Gynecomastia
  Priapism
  Hirsutism
Other endocrine systems
  Decreased glucose tolerance
  Disturbance of thyroid function
Gastrointestinal system
  Hepatitis
  Cholestasis
  Peliosis hepatis
  Hepatocellular adenoma
  Hepatocarcinoma
Urinary system
  Renal adenocarcinoma
  Prostatic adenocarcinoma
  Wilms' tumor
Cardiovascular and hematological system
  Blood pressure elevation
  Decreased cardiac diastolic function
  Repolarization alterations on electrocardiogram
  Heart rhythm alterations
  Cardiac hypertrophy
  Cerebrovascular thrombosis
  Cardiomyopathy
  Bleeding esophageal varices
  Myocardial infarction
  Sudden cardiac arrest
  Hemorrhagic stroke
  Myocardial fibrosis
  Pulmonary embolism
Musculoskeletal system
  Bone fracture
  Tendon injury
  Muscle injury
Psyche and behavior
  Affective disorder
  Manic-depressive disorder
  Depression
  Psychosis
  Hypomania
  Schizophrenia
  Suicide
  (Near-)Homicide
  AAS dependence
  Withdrawal disorder
Dermatological alterations
  Acne vulgaris
  Acne fulminans
  Coproporphyria
  Linear keloid formation
Other
  Chickenpox pneumonitis
  Rhabdomyolysis
  Exacerbation of psoriasis
  Sleep apnea syndrome

to dihydrotestosterone and plays an important role in expression of the male sex hormones, whereas the latter converts AAS into female sex hormones and is active mainly in the presence of larger concentrations of AAS. When the androgen receptors are saturated, AAS may bind to glucocorticoid receptors. This counteracts the glucocorticoid-induced breakdown of proteins and inhibits catabolism (metabolic breakdown of complex molecules into simpler ones, often with the liberation of energy).

AAS induce muscle growth by enhancement of protein synthesis and via the formation of new muscle fibers. In this process, satellite cells, which are progenitor cells of muscle, and androgen receptors play an important role. The number of satellite cells increases and their incorporation in preexisting muscle fibers is accelerated by AAS. Because androgen receptors are located predominantly in the neck and shoulder girdle, the largest muscle growth can be observed at these sites.

The effects on the cardiovascular system are mediated by at least four pathways. Stimulation of the enzyme hepatic triglyceride lipase results in decreased serum HDL (high-density lipoprotein)–cholesterol levels and elevation of serum LDL (low-density lipoprotein)–cholesterol levels, which promotes atherogenic plaque formation in vessel walls. AAS affect the hemostatic system unfavorably, especially causing platelet aggregation, which can lead to an increased risk for thromboembolic diseases. In smooth muscles and arteries, nitric oxide, an endothelium-derived relaxing factor, may be negatively affected by AAS, leading to reduced relaxation of vessel wall muscles or induction of vasospasm. Finally, AAS can injure myocardial cells and even may cause myocardial cell death.

Effects on body composition. In athletes participating in a strength training program, the average body-mass gain may be considered to be 2–5 kg (4.4–11 lb). Novice athletes and nontraining healthy subjects may also undergo similar body-mass alterations. The increase of body mass can be attributed to an increase of fat-free mass, whereas fat mass remains unaltered. Fat-free mass increments show a strong dose–response relationship and reflect mainly real muscle mass growth (see Table 3). Water retention may play only a minor role.

Skeletal muscle mass. AAS possess strong muscle-building properties both in athletes and in healthy subjects who are not under physical training. In the former population, the effects of AAS are greater. In nontraining males, AAS increased thigh muscle volume by 15% during testosterone enanthate (TE) administration (600 mg per week, intramuscularly) for 20 weeks. However, triceps brachii muscle mass of experienced strength athletes increased by 31.4% after administration of even a lower dose of TE given for only 12 weeks.

Muscle fiber. Muscle mass increment can be attributed to enlargement of muscle fibers and an increase of the number of muscle fibers. AAS influence muscle fiber in a dose-dependent way, with androgenic substances being more effective than anabolic agents. Nevertheless, polypharmacy (multidrug use) has the most profound effects. Therapeutic doses of nandrolone decanoate, a more anabolic agent, did not affect muscle fiber dimensions in strength athletes. However, the administration of both androgenic and anabolic substances in supratherapeutic doses increased muscle fiber dimensions to the greatest degree. Short-term administration (8–10 weeks) affected muscle fiber dimensions less compared to long-term administration (24 weeks), with the longer administration period leading to an additional increase of muscle mass by approximately 50%. In healthy, nontraining young males, 20 weeks of TE administration at a dose of 50 mg per week promoted approximately 20% muscle fiber growth, whereas a dose of 600 mg per week increased muscle fiber dimensions by 50%.

Effects on performance. The effects of AAS on performance can be measured in many ways and in relation to many parameters (see Table 3).

Strength. Athletes began to self-administer AAS because of their putative strength-building properties. For a long period of time, it was assumed that AAS may promote muscle strength only in combination with a strength training program in experienced strength trainers. However, recent research elucidated that these substances are able to improve muscle strength in novice athletes and even in nontraining healthy males, although higher doses of AAS are required in these populations.

Endurance. Because AAS affect the hematological system, they were considered to affect endurance performance beneficially. This theory could not be proven-AAS failed to improve endurance performance in all studies of endurance athletes.

Mixed strength–endurance. Only one study has investigated the effects of AAS on a mixed strength-endurance type of sport, in this case, canoeing. Both strength and endurance performance improved during dehydrochlormethyltestosterone administration for 6 weeks, by 6% and 9%, respectively.

Recovery. Athletes may use AAS to enhance recovery because of their anabolic and anticatabolic properties. Most studies have investigated indirect parameters for recovery, such as the response of enzymes (for example, creatine phosphokinase and aspartate aminotransferase), heart rate response, steroid hormone responses, and lactate response, but the results are inconsistent and do not allow drawing firm conclusions.

Health effects. Because AAS are strong hormonal agents that may affect several organ systems, their abuse may result in unwanted effects (see Tables 4 and 5). In particular, the side effects on the reproductive system, cardiovascular system, brain, and liver are of concern.

Reproductive system. The hypothalamic–pituitary-gonadal axis (HPG axis) may be suppressed by AAS administration, thereby lowering serum levels of testosterone, follicle-stimulating hormone (FSH), and luteinizing hormone (LH). Male athletes may experience testicular atrophy, alterations of libido, subfertility (diminished ability to reproduce), gynecomastia (abnormal enlargement of the mammary glands in the male), and prostate and erectile dysfunction. Females may suffer mainly from masculinization, menstrual irregularities, breast atrophy, clitoris hypertrophy, and voice deepening. Complete suppression of the HPG axis leads to hypogonadotropic hypogonadism (reduced function of the gonads), and full recovery may take many months after AAS withdrawal.

Cardiovascular system. AAS change the cardiovascular risk profile unfavorably and may affect heart muscle. Serum lipid and lipoprotein levels change, partially dependent on the drugs used. These agents may suppress serum levels of HDL-cholesterol and its subfractions (HDL2- and HDL3-cholesterol) and apolipoprotein-B1, but elevate LDL-cholesterol and apolipoprotein-A. On the other hand, serum lipoprotein(a) levels may be affected beneficially. The effects on serum total cholesterol seem less predictable, and serum triglyceride levels do not seem to change.

Blood pressure alterations due to AAS may be limited and remain mostly within medical reference values. Some individuals seem to be more susceptible to blood pressure elevations under AAS than others.

The effects of AAS on heart function and structure in humans are so far inconclusive. In male AAS-using athletes, cross-sectional echocardiographic examinations revealed larger left ventricular wall thickness, posterior wall thickness, and interventricular septum thickness compared to nonusing counterparts; in addition, diastolic function may be impaired. However, in short-term prospective studies, no alterations of heart structure and function in humans could be observed at all. AAS, though, are likely to affect heart muscle, because they induce heart cell damage within days or weeks in animals.

Liver. In athletic populations, AAS did not induce liver damage as measured by serum liver function enzymes. Only slight transient elevations of serum liver enzymes alanine aminotransferase (ALAT) and aspartate aminotransferase (ASAT) were observed, mainly after administration of oral agents, but these effects may reflect, at least in part, muscle damage. Nevertheless, AAS may induce deleterious effects on the hepatic system, including impaired excretion function, cholestasis, peliosis hepatis, hepatocellular hyperplasia, and carcinomata. In particular, the so-called 17-alpha-alkylated steroids (oxymetholone and methyltestosterone) are notorious for their liver toxicity.

Psyche and behavior. AAS may affect psychological state and behavior, although only a small number of AAS users may have serious problems. Psychiatric symptoms such as increased aggression and hostility, as well as mood disturbances (for example, mania, psychosis, and depression), may be particularly experienced after abuse of higher doses of AAS. Others may develop criminal behavior and may resort to assault and homicide. Interestingly, many AAS users expose a narcissistic personality and are dissatisfied with their body, with both factors predisposing individuals to develop AAS dependence.

Other health effects. Many other side effects have been associated with AAS use. Reported effects include disturbance of glucose metabolism and thyroid function, occurrence of acne vulgaris and fulminans, sebaceous gland alterations, reduction of immune function, myocardial infarction, suicide attempts, musculoskeletal injuries, bladder and prostate dysfunction, and the formation of several types of tumors (e.g., renal cell carcinoma and Wilms' tumor).

For background information *see* ANDROGENS; BIOMECHANICS; HORMONE; MUSCLE; MUSCULAR SYSTEM; PHARMACY; REPRODUCTIVE SYSTEM; SPORTS MEDICINE; STEROID; STEROL in the McGraw-Hill Encyclopedia of Science & Technology.   Fred Hartgens

Bibliography. N. A. Evans, Current concepts in anabolic-androgenic steroids, *Am. J. Sports Med.*, 32(2):534–542, 2004; K. E. Friedl, Effect of anabolic steroid use on body composition and physical performance, pp. 139–174, in C. Yesalis (ed.), *Anabolic Steroids in Sport and Exercise*, 2d ed., Human Kinetics, Champaign, IL, 2000; K. E. Friedl, Effects of anabolic steroids on physical health, pp. 175–224, in C. Yesalis (ed.), *Anabolic Steroids in Sport and Exercise*, 2d ed., Human Kinetics, Champaign, IL, 2000; R. C. Hall and R. C. Hall, Abuse of supraphysiologic doses of anabolic steroids, *South. Med. J.*, 98(5):550–555, 2005; F. Hartgens and H. Kuipers, Effects of androgenic-anabolic steroids in athletes, *Sports Med.*, 34(8):513–554, 2004.

Angiosperm phylogenetics

Reconstructing the actual relationships within the angiosperm Tree of Life (the tree of evolutionary relationships of all flowering plants) provides a necessary framework for addressing questions of evolutionary importance in flowering plants, such as the origin and evolution of the flower and the age of angiosperms. Over the past two decades, our understanding of angiosperm phylogeny has been revolutionized by a shift from traditional morphology-based assessment of relationships (in other words, based on anatomical structures) to the phylogenetic analysis of deoxyribonucleic acid (DNA) sequence data. This molecular phylogenetic revolution has enabled plant systematists to use DNA sequences from several genes to confidently reconstruct much of the angiosperm Tree of Life. Despite this progress, relationships among many of the most important ancient lineages of angiosperms have proved difficult to resolve with just a few genes.

To reconstruct these remaining recalcitrant relationships, plant systematists have turned to sequencing and analyzing complete plastid genome sequences. The typical angiosperm plastid genome is approximately 160,000 nucleotides in length (**Fig. 1**), allowing much larger and therefore more powerful phylogenetic data sets to be constructed compared to previous analyses. The recent arrival of a new generation of genome-scale DNA sequencing technologies has facilitated construction of these large data sets by allowing systematists to rapidly and inexpensively sequence complete plastid genomes. This revolution in plastid genomics has enabled the development of DNA sequence alignments for more than 80 plastid genes, or roughly 10 times more genes than were present in earlier angiosperm phylogenetic data sets. These genomic data sets have helped to clarify the relationships among several important lineages of angiosperms with a much higher degree of confidence than in previous analyses of only a few genes. Several of these newly resolved relationships are described below, along with their broader evolutionary implications.

Earliest angiosperm diversification. A decade ago, plant systematists using DNA sequence data from a handful of genes identified the genus *Amborella*, which consists of a single species of large shrub endemic to the island of New Caledonia, and the order Nymphaeales (water lilies) as the earliest diverging extant angiosperm lineages. These results were surprising, considering that earlier taxonomists had proposed that other groups, such as Magnoliales, Chloranthaceae, and *Ceratophyllum*, occupied such a position. Although early phylogenetic analyses strongly supported *Amborella* and Nymphaeales as isolated lineages of angiosperms, they did not confidently resolve relationships among these lineages and all remaining angiosperms. Most of these early analyses favored one of two scenarios: (1) *Amborella* alone as sister to all other angiosperms or (2) a clade of *Amborella* + Nymphaeales as sister to all other angiosperms.

To resolve these relationships, plant systematists recently sequenced complete plastid genomes for representatives of all the major basal lineages of angiosperms. Phylogenetic analyses of these genome-scale data provide strong support for the sister relationship of *Amborella* alone to all other angiosperms (**Fig. 2**), suggesting that the earliest branches of the angiosperm phylogenetic tree are now resolved.

A few unusual morphological traits of *Amborella* seem to support its position as sister to all other angiosperms. *Amborella* is one of a few angiosperms to lack vessels in its xylem. Vessels are relatively large-diameter tracheary elements that allow for efficient

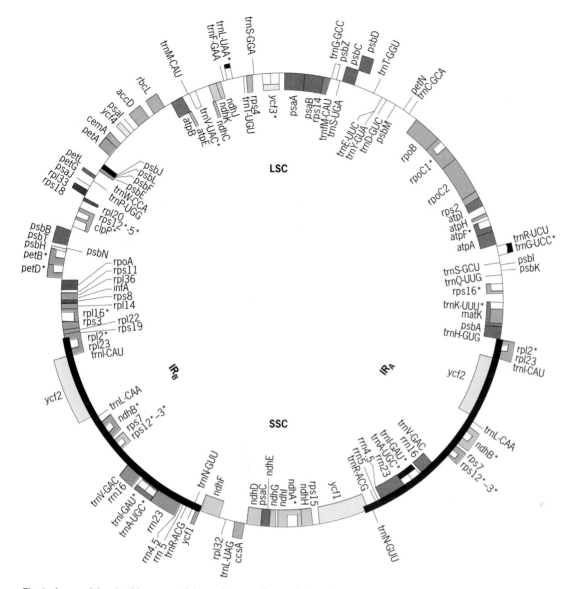

Fig. 1. A map of the plastid genome of the angiosperm *Ceratophyllum demersum*, showing the typical gene content and arrangement present in nearly all angiosperms. The typical plastid genome is divided into four regions: the large single-copy (LSC) region, the small single-copy (SSC) region, and two copies of a large inverted repeat (IR$_A$ and IR$_B$). Boxes with associated gene names show the position and extent of individual genes. Intron-containing genes are indicated by asterisks; the extent of introns is indicated by white boxes within genes.

water transport through the xylem. It is clear from the phylogenetic pattern of vessel presence and absence in angiosperms that all angiosperm lineages lacking vessels except *Amborella* originally possessed vessels and then lost them during their evolution. In contrast, vessel absence in *Amborella* may represent the ancestral condition in angiosperms, especially given that almost all gymnosperms, which are the closest living relatives to angiosperms, also lack vessels.

Features of *Amborella* reproductive biology are also congruent with its isolated position among angiosperms. *Amborella* is unique among angiosperms in having an 8-celled, 9-nucleate embryo sac, in contrast to almost all other angiosperms, which possess a 7-celled, 8-nucleate embryo sac. The next two angiosperm lineages to diverge, the Nymphaeales and Austrobaileyales, also differ from the typical an-

giosperm pattern by having a 4-celled embryo sac. *Amborella* pollen grains are also unique among angiosperms in lacking a reticulate tectum (a rooflike structure resembling a network of fibers or lines) on their outermost covering. The differences between *Amborella* and other angiosperms in such fundamental reproductive traits strongly reinforce the molecular phylogenetic data and demonstrate the power of a molecular phylogenetic framework in interpreting morphological evolution.

Another surprising recent phylogenetic discovery in the earliest angiosperms was the identification of the small, enigmatic, aquatic flowering plant family Hydatellaceae as one of these isolated angiosperms. Formerly classified within monocots, Hydatellaceae are now firmly placed by newly obtained DNA sequence data within Nymphaeales, as sister to all other water lily families (Fig. 2). At

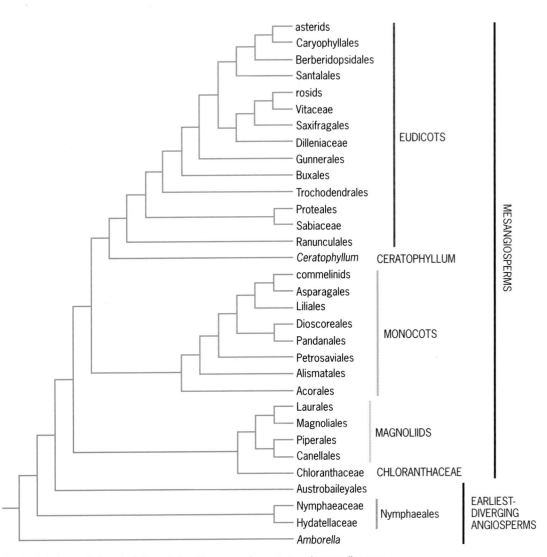

Fig. 2. **A phylogenetic tree depicting relationships among important angiosperm lineages.**

first glance, members of Hydatellaceae appear to share little in common morphologically with other Nymphaeales. However, when reevaluated in light of the new molecular phylogenetic evidence, Hydatellaceae flowers have been found to share several important diagnostic features with other Nymphaeales, most notably the presence of a 4-celled embryo sac.

Mesangiosperm diversification. Almost all flowering plants (>99.9%) fall into one of five major lineages that diverged subsequent to the earliest-diverging lineages of angiosperms (*Amborella*, Nymphaeales, and Austrobaileyales). These five lineages—Chloranthaceae, magnoliids, monocots, *Ceratophyllum*, and eudicots—are collectively referred to as mesangiosperms.

Relationships among the five mesangiosperm groups had been impossible to resolve phylogenetically using DNA sequence data from only a few genes. However, recent phylogenetic analyses based on complete plastid genome sequences of all major mesangiosperm lineages have provided the first consistent support for a fully resolved mesangiosperm tree (Fig. 2). These analyses support a sister rela-

tionship between Chloranthaceae, a small, mainly tropical group of trees and herbs with unusual flowers that lack petals and sepals, and magnoliids, a much larger and more diverse group that contains several important families including the magnolia, laurel, and black pepper families. Together, Chloranthaceae and magnoliids are sister to the remaining mesangiosperms.

Plastid genome analyses also support an unexpectedly close relationship between the two largest groups of mesangiosperms—the monocots (~22% of all angiosperms) and eudicots (~75% of all angiosperms). Both these groups contain many familiar and economically important plants. For example, monocots include grasses (and thus all cereal crops), pineapples, asparagus, bananas, and several horticulturally important groups such as orchids, lilies, irises, and palms. Eudicots include important vegetable, fruit, and nut crops such as potatoes, tomatoes, sunflowers, carrots, mustards, legumes, apples, peaches, blueberries, and pecans, as well as cotton and such horticulturally important groups as the composite, mint, snapdragon, carnation, and cactus families.

Also associated with monocots and eudicots in plastid genome analyses is *Ceratophyllum*, a phylogenetically isolated genus composed of a few fully aquatic freshwater species with reduced floral morphology. *Ceratophyllum* has caused a disproportionate share of problems in angiosperm phylogenetic analyses because of its relatively high rate of evolution, which apparently has partially scrambled the phylogenetic signal in its DNA sequence. However, plastid genome analyses using mathematical models of DNA sequence evolution that are designed to circumvent this problem consistently support a sister relationship of *Ceratophyllum* with eudicots, with these two groups together being sister to monocots (Fig. 2).

In addition to clarifying phylogenetic relationships among mesangiosperms, plastid genome-scale analyses suggest that the five major lineages of mesangiosperms diversified relatively rapidly during the Early Cretaceous Epoch, perhaps as early as 145 million years ago. Sophisticated analyses that convert the amount of evolutionary change inferred to have occurred among mesangiosperms into estimates of time imply that all five mesangiosperm lineages diverged over approximately 3 million years. Such rapid diversification often causes problems in phylogenetic analyses and has contributed significantly to the difficulty in reconstructing mesangiosperm relationships.

Identifying unique morphological features that are clearly shared among these newly determined mesangiosperm groups, and that would therefore serve to corroborate the plastid genome phylogenetic results, has been difficult. It is possible that a two-whorled perianth (in other words, the presence of two distinct groups of nonreproductive floral parts, such as sepals and petals) represents a diagnostic floral feature of the group including monocots and eudicots. However, the lack of a perianth in *Ceratophyllum*, combined with some ambiguity in the number of perianth whorls in several groups of basal eudicots, prevents drawing a solid conclusion at present. The rapid diversification of mesangiosperms provides a plausible explanation for the lack of morphological support among major mesangiosperm lineages: it appears that these lineages diversified so quickly that novel, diagnostic morphological features may not have had enough time to evolve.

Phylogenomic era. Plastid genome-scale data sets have clearly demonstrated the power of a phylogenomic approach in helping to resolve angiosperm relationships. Plant systematists are currently extending this plastid phylogenomic approach to other areas of the angiosperm Tree of Life in an effort to settle relationships among such difficult groups as the earliest-diverging lineages of eudicots and several groups of core eudicots. Although the plastid genome will always remain important phylogenetically, future phylogenomic studies will expand to include the underutilized nuclear and mitochondrial genomes of angiosperms as DNA sequencing costs continue to decline. The phylogenetic power of combining information from all three plant genomes promises to allow even greater insight into angiosperm evolution.

For background information *see* CELL PLASTIDS; DEOXYRIBONUCLEIC ACID (DNA); FLOWER; GENOMICS; MAGNOLIOPHYTA; NYMPHAEALES; PLANT EVOLUTION; PLANT KINGDOM; PLANT PHYLOGENY; PLANT TAXONOMY; SYSTEMATICS; XYLEM in the McGraw-Hill Encyclopedia of Science & Technology.   Michael J. Moore

Bibliography. M. W. Frohlich and M. W. Chase, After a dozen years of progress the origin of angiosperms is still a great mystery, *Nature*, 450:1184–1189, 2007; D. E. Soltis et al., *Phylogeny and Evolution of Angiosperms*, Sinauer Associates, Sunderland, MA, 2005; P. F. Stevens, Angiosperm Phylogeny Website, Version 8, 2007 (updated continuously).

Apical development in plants

All aboveground organs of plants, including stems, leaves, branches, and flowers, are ultimately derived from a small pool of cells called the shoot apical meristem (SAM). Located at the tip of the growing stem, the SAM contains a pool of pluripotent stem cells that differ from animal stem cells owing to their intrinsic ability to sustain indeterminate growth throughout the life cycle of the plant. This unique property of the SAM to maintain embryolike growth during adult stages accounts for a major difference in developmental strategies between animals and plants. Whereas most animals cease organogenesis (organ formation) quite early in development, plants continue to grow via the addition of newly developed stems and leaves on top of stems and leaves formed earlier in development. As a result, the SAM must maintain a precise equilibrium whereby cells lost during organogenesis are replenished by stem cells that divide to maintain the SAM and furnish new cells toward the formation of additional organs. Thus, the SAM performs two essential functions, namely (1) organogenesis and (2) self-maintenance. Research on SAM biology is focused on understanding the genetic and biochemical parameters controlling these two fundamental components of SAM function.

Conserved SAM structure correlates with function. A survey of plant taxa reveals widespread variation in SAM size and shape, ranging from the small conical SAM found in *Arabidopsis* and *Ginkgo*, to the elongated SAM of maize and other grasses, to the flattened, nearly concave SAM seen in sunflower (**Fig. 1**). Seedless plants such as ferns and *Equisetum* feature an enlarged apical-initial cell at the meristem summit that is purported to ultimately give rise to all the cells in the shoot (Fig. 1). Initially described in 1759 by the German anatomist Caspar Wolff, the first to recognize the SAM as the morphogenetic center of the plant shoot, the apical-initial cell was once believed to be an essential governing feature of SAM function. However, this notion was dispelled when subsequent analyses of SAM anatomy revealed that gymnosperms and angiosperms (that

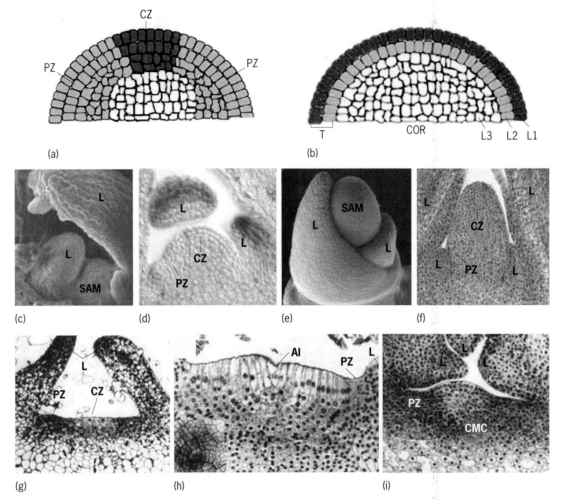

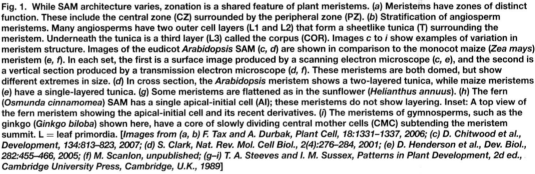

Fig. 1. While SAM architecture varies, zonation is a shared feature of plant meristems. (*a*) Meristems have zones of distinct function. These include the central zone (CZ) surrounded by the peripheral zone (PZ). (*b*) Stratification of angiosperm meristems. Many angiosperms have two outer cell layers (L1 and L2) that form a sheetlike tunica (T) surrounding the meristem. Underneath the tunica is a third layer (L3) called the corpus (COR). Images *c* to *i* show examples of variation in meristem structure. Images of the eudicot *Arabidopsis* SAM (*c, d*) are shown in comparison to the monocot maize (*Zea mays*) meristem (*e, f*). In each set, the first is a surface image produced by a scanning electron microscope (*c, e*), and the second is a vertical section produced by a transmission electron microscope (*d, f*). These meristems are both domed, but show different extremes in size. (*d*) In cross section, the *Arabidopsis* meristem shows a two-layered tunica, while maize meristems (*e*) have a single-layered tunica. (*g*) Some meristems are flattened as in the sunflower (*Helianthus annuus*). (*h*) The fern (*Osmunda cinnamomea*) SAM has a single apical-initial cell (AI); these meristems do not show layering. Inset: A top view of the fern meristem showing the apical-initial cell and its recent derivatives. (*i*) The meristems of gymnosperms, such as the ginkgo (*Ginkgo biloba*) shown here, have a core of slowly dividing central mother cells (CMC) subtending the meristem summit. L = leaf primordia. [*Images from (a, b) F. Tax and A. Durbak, Plant Cell, 18:1331–1337, 2006; (c) D. Chitwood et al., Development, 134:813–823, 2007; (d) S. Clark, Nat. Rev. Mol. Cell Biol., 2(4):276–284, 2001; (e) D. Henderson et al., Dev. Biol., 282:455–466, 2005; (f) M. Scanlon, unpublished; (g–i) T. A. Steeves and I. M. Sussex, Patterns in Plant Development, 2d ed., Cambridge University Press, Cambridge, U.K., 1989*]

is, naked seed plants and flowering plants) lacked apical-initial cells. In contrast, gymnosperms such as ginkgo and pine form a cluster of large, highly vacuolated cells near the SAM subsurface called central mother cells. An anatomical innovation that is specific to angiosperms and a limited number of gymnosperms is the stratification of the SAM into distinct tissue layers, the so-called tunica-corpus structure. In these meristems, the tunica forms one or more peripheral cell layers that divide only within their own layer to create a sheet of cells of identical ancestry. The sheathing tunica surrounds the corpus, the multilayered body of the SAM in which cell divisions occur in all planes. Cell divisions within the tunica extend the surface area of the stratified SAM, whereas growth in the corpus contributes to meri-

stem volume. Likewise, during organogenesis, the internal tissues of plant organs are ultimately derived from cells in the SAM corpus, and the tunica generates the epidermis.

Despite these disparities in meristem morphology, plant shoot meristems do share a common pattern of histological zonation that is intrinsically linked to SAM function (Fig. 1). The central zone (CZ) contains relatively slowly dividing cells located near the meristem summit, comprising a population of stem cells that function to maintain the SAM. Surrounding the CZ is a region of smaller, more rapidly dividing cells, comprising the peripheral zone (PZ). Cells occupying the PZ are recruited to form lateral organs such as leaves. The study of the mechanisms whereby the dual SAM functions of stem cell maintenance and lateral organ initiation are regulated and relegated to distinct meristematic zones is an especially active area of plant research, as outlined later.

Tight regulation of meristem size. In order to maintain SAM function throughout the plant life cycle, the loss of cells from the PZ during organogenesis must be compensated by the addition of new stem cells to the CZ. Failure to regenerate the stem cell pool will result in consumption of the SAM and the cessation of plant development. On the other hand, overproduction of stem cells will lead to SAM enlargement, which can result in meristem bifurcation, altered phyllotaxy (arrangement of leaves on a stem), and increased organ number. Accordingly, an intricate balance between SAM proliferation and consumption is maintained via the precise regulation of an interactive complex of genetic networks and hormone signaling (**Fig. 2**).

The molecular mechanisms that control SAM size are best characterized in *Arabidopsis*, a small plant within the mustard family. With its sequenced genome, short generation time, and ease of genetic manipulation, *Arabidopsis* is an ideal organism for developmental genetic analyses. Studies of a growing number of *Arabidopsis* mutants with genetic defects in SAM function have demonstrated that meristem size is regulated by an overlapping array of negative feedback loops—self-correcting systems in which the product of a genetic pathway functions to ultimately inhibit its own production. In particular, a negative feedback loop involving at least three genes termed *CLAVATA1* (*CLV1*), *CLAVATA2* (*CLV2*), and *CLAVATA3* (*CLV3*), as well as a fourth gene named *WUSCHEL* (*WUS*), plays a central role during regulation of the stem cell pool in the *Arabidopsis* SAM (Fig. 2). The *WUS* gene is expressed within a small domain just under the CZ known as the SAM organizing center, where it generates a mobile signal promoting stem cell formation in the CZ above. WUS function is also self-correcting in that WUS promotes expression of *CLV3*, thereby activating a second signaling pathway involving *CLV1* and *CLV2* that represses the domain of *WUS* expression. Mutant SAMs that lose WUS function are unable to maintain the CZ and are eventually consumed. Conversely, loss of CLV function results in an enlarged meristem because of the uncontrolled expansion of the *WUS* expression domain and the overproliferation of stem cells.

Another class of genes of central importance during meristem maintenance includes the *KNOTTED1-LIKE HOMEOBOX* (*KNOX*) genes (Fig. 2). The founding member of this class of genes, *KNOTTED1* (*KN1*), was identified in maize. A gene with the same function, named *SHOOTMERISTEMLESS* (*STM*), was later discovered in *Arabidopsis*. Both *KN1* and *STM* are expressed within the SAM, but are excluded from initiating leaves. *KNOX* genes function to prevent differentiation of SAM cells, thereby maintaining the indeterminate, stem cell nature of the meristem. Mutations that induce abnormal expression of *KN1* in maize leaves cause conspicuous outgrowths, or knots, of undifferentiated tissue along the leaf veins.

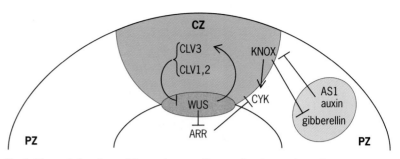

Fig. 2. The main functions of the meristem, self-renewal and organ formation, are tightly controlled by the interaction of genetic pathways and hormonal signaling. Leaves are initiated from cells in the peripheral zone (PZ). This process requires position-dependent signaling. Terms: CLV, *CLAVATA*; WUS, *WUSCHEL*; AS1, *ASYMMETRIC LEAVES1*; ARR, *Arabidopsis Response Regulator of cytokinin*; KNOX, *Knotted-like homeobox* genes; CYK, cytokinin; CZ, central zone. Arrows represent positive regulation, while T-bars indicate inhibitory effects.

Furthermore, loss of KN1/STM function results in aborted shoot growth and failure to maintain the SAM, phenotypes that clearly illustrate the role of *KNOX* genes in promoting meristem indeterminacy.

SAM function is also regulated by plant hormonal signaling; the roles of cytokinins, auxins, and gibberellins are particularly well studied. Cytokinins (CYK) promote cell division and are required for SAM maintenance. Mutations in cytokinin biosynthetic genes reduce SAM size and arrest plant development. In contrast, auxins and gibberellins promote differentiation and are necessary for organ formation. Thus, the SAM CZ contains high levels of cytokinins and lower levels of auxins, and shoot development can be induced in tissue culture by exposing plant callus (undifferentiated tissue that forms over a damaged plant surface) to high ratios of cytokinin/auxin concentration. Logically, genetic pathways involved in meristem maintenance are integrated at the level of hormonal regulation (Fig. 2). Thus, downstream targets of WUS repression include the *ARABIDOPSIS RESPONSE REGULATOR*s (*ARR*s), a family of cytokinin response genes that in turn repress the further synthesis of cytokinins. By repressing the expression of genes that suppress cytokinin synthesis, WUS functions in yet another negative feedback network to ultimately increase cytokinin levels in the SAM. Similarly, KNOX proteins promote the expression of cytokinin biosynthetic genes, while simultaneously repressing gibberellin production and promoting gibberellin degradation. In this way, KNOX function mediates cytokinin accumulation in the SAM, while repressing gibberellin-induced patterns of determinate growth.

Leaf initiation: indeterminate to determinate growth. Organogenesis, such as in leaf initiation, is accompanied by rapid changes in both the rate and the orientation of cell divisions at the meristem flank. Prior to any visible changes in SAM morphology, however, cells occupying the PZ of the SAM are recruited to undergo a developmental switch from indeterminate meristematic cells to determinate leaf founder cells, a process that involves position-dependent intercellular signaling. Founder cell number varies greatly among plant species; approximately 250 cells are recruited for maize leaf development, whereas leaves

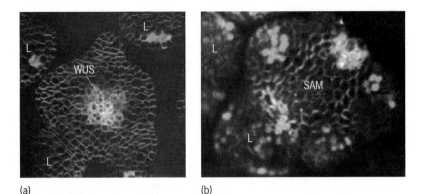

(a)   (b)

Fig. 3. The dynamics of meristem biology can be captured using live imaging. (*a, b*) The interior of the *Arabidopsis* SAM is digitally recreated from images taken above the live meristem. (*a*) The WUS domain (made green by fusing the *WUS* gene with a green fluorescent protein) is confined to a small area in the center of the meristem. (*b*) As shown by the auxin-responsive reporter DR5 (bright areas), auxin accumulates in the periphery of the meristem at sites of leaf initiation and within newly formed leaves (L). [*Images from (a) H. Jönsson et al., Bioinformatics, 21(suppl. 1):i232–i240, 2005, and (b) M. Heisler et al., Curr. Biol., 15(21):1899–1911, 2005*]

of the basal vascular plant *Selaginella kraussiana* are derived from just two initial cells. As is the case for the regulation of SAM maintenance described previously, leaf initiation involves a complex network of genetic pathways and hormonal signaling.

The earliest known signal for leaf initiation is the transport of the phytohormone (plant hormone) auxin to the PZ. Auxin represses *KNOX* gene expression, so that the localized accumulation of auxin at the site of leaf initiation results in the concomitant accumulation of gibberellins, which promote cell differentiation. When auxin transport is blocked by treatment with chemical inhibitors, new leaf initiation is arrested. Application of auxin to the PZ of the chemically arrested SAM is sufficient to induce new leaf formation, thereby demonstrating the vital role of localized auxin accumulation during leaf initiation. Following auxin transport to the PZ, the leaf developmental gene *AS1* is expressed in a pattern that is directly opposite that of KNOX accumulation. Mutations in *AS1* cause abnormal KNOX accumulation in leaves and resultant defects in leaf differentiation, suggesting that AS1 may promote leaf development by preventing the expression of *KNOX* genes in leaves.

Following leaf initiation, growth and differentiation proceed along three developmental axes, corresponding to the proximal-distal (tip to base), dorsiventral (top to bottom), and mediolateral (side to side) planes of the mature leaf. Intriguingly, the expression of genes controlling differentiation along these three leaf axes is seen in the PZ, prior to the outgrowth of the leaf primordium (organ or tissue in its earliest identifiable stage of development). These data suggest that patterning of leaves begins at their inception, as founder cells within the SAM.

Live imaging. Many studies of SAM function have relied upon examinations of genetic mutants or static, two-dimensional images obtained from fixed, dead plants. Although these studies are extremely valuable, a better understanding of meristem biology may be achieved by examining the dynamic

and interactive patterns of gene expression within living meristems in real time. Recent technological advances have enabled the use of fluorescently labeled proteins and advanced imaging technology to monitor stochastic changes in gene expression and protein accumulation during cell division and organogenesis within live meristems. For example, researchers were able to fluorescently label and quantify WUS accumulation in the stem cell organizing center of the SAM (**Fig. 3***a*). A second experiment simultaneously visualized the dynamic accumulation pattern of the auxin-responsive reporter DR5 in live *Arabidopsis* meristems. These analyses illustrate that auxin accumulation and response coincide maximally at the sites of new organ initiation along the meristem periphery (Fig. 3*b*). Studies such as these will prove invaluable in obtaining an accurate understanding of how meristems balance the activities of self-renewal and organogenesis throughout development.

For background information *see* APICAL DOMINANCE; APICAL MERISTEM; AUXIN; CYTOKININS; GENE; GIBBERELLIN; LATERAL MERISTEM; LEAF; PLANT DEVELOPMENT; PLANT GROWTH; PLANT HORMONES; PLANT ORGANS; PLANT TISSUE SYSTEMS; STEM; STEM CELLS in the McGraw-Hill Encyclopedia of Science & Technology.   Michael J. Scanlon; John B. Woodward

Bibliography. R. F. Evert, *Esau's Plant Anatomy: Meristems, Cells, and Tissues of the Plant Body— Their Structure, Function, and Development*, 3d ed., Wiley Interscience, Hoboken, NJ, 2006; D. Kwiatkowska, Flowering and apical meristem growth dynamics, *J. Exp. Bot.*, 59(2):553–557, 2008; O. Leyser and S. Day, *Mechanisms in Plant Development*, Blackwell Science, Oxford, 2003.

Argo (ocean observing network)

Argo is a new ocean monitoring program that is entirely different from anything previously undertaken in the oceans of the world. To understand why it is so different, consider a previous attempt to survey the climate of the world's oceans.

The World Ocean Circulation Experiment (WOCE) was designed to supply a snapshot of the state of the oceans. Observations took place in the early 1990s. Ships from 25 nations traveled the world's oceans and took measurements over a 6-year period. The ship time alone has been estimated at 25 ship-years, spread across an international research fleet at a cost of about $200 million. During this experiment, there was an El Niño and a La Niña, major perturbations in the state of the ocean originating in the equatorial Pacific, which made this "snapshot" rather fuzzy.

Most of the available heat energy in the seasonal climate system is in the oceans. If we want to improve seasonal climate forecasts, we need seasonal observations of the climatic state of the oceans. But if surveys using ships require 6 years, it is clear that seasonal climate forecasting will remain a major challenge. A new approach was needed that would allow the state of the ocean climate to be surveyed much faster, and this need generated project Argo.

In 1999, the Argo concept sprang from discussions among a few scientists meeting in Easton, Maryland, who drafted a prospectus that envisioned launching an armada of small robotic devices (1.5 m long and weighing about 22 kg). They could be launched from research vessels, merchant vessels, and aircraft; even an ocean-going rowboat would be able to launch one. Each float would be designed to adjust its buoyancy so that it could sink to a predetermined depth of, typically, 1000 m, as shown in **Fig. 1**. Every 10 days, the float would change its buoyancy, dive to a depth of 2000 m, and then rise to the surface over a period of about 6 h, measuring properties of the ocean descriptive of its climatic status (originally, only temperature and salinity versus pressure were imagined) on the way up. Once at the sea surface, the robot would transmit its accumulated data to land stations via satellite. The data would be put through some automated checks and then made available to users within 24 h. The float would then dive to 1000 m to start a new cycle. In this way, the complete array would supply a global view of the climatic state of the oceans of the world every 10 days. These basic climate properties would then allow computation of ocean-current speeds.

Float design. One of the problems that had to be resolved to make the Argo concept work was to build a device that is capable of adjusting its buoyancy on command. Elementary physics tells us that something floats in water if its density is less than or the same as the density of the water surrounding it. Water is very slightly compressible, so its density increases with increasing pressure, or depth, in the ocean. Therefore, one can build a device with a density higher than surface waters that will sink until it reaches a depth or pressure where its density is the same as that of the water surrounding it. It will then float at that depth. The density of any object is given by the equation: density = mass/volume.

The design of an Argo float allows us to keep the mass constant and vary the volume. There are a variety of designs for Argo floats, but most contain a piston. As the piston is pushed outward by battery power, it pushes mineral oil out of the float. The oil is isolated from contact with seawater by a rubber bladder, and as the oil is expelled, the bladder expands, increasing the volume of the float. When the piston is withdrawn, oil is allowed to move back into the body of the float, the bladder collapses, and the volume of the float decreases. Thus, pushing the piston outward increases the volume and decreases the density of the float, allowing it to rise. Withdrawing the piston decreases the float volume, increases its density, and allows the float to sink. As shown in **Fig. 2**, on an actual float, the rubber bladder is protected with a plastic case to prevent damage if the float hits the bottom. Partway up the outside of the float, a damper plate helps keep the antenna above water if the float surfaces in large waves. The sensors are at the top of the float so that they will be unaf-

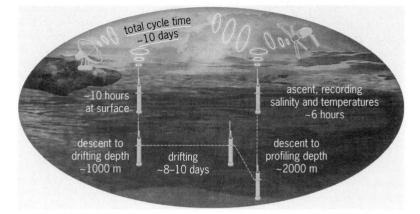

Fig. 1. A typical duty cycle for an Argo float.

fected by turbulence in the wake of the device as it ascends through the water column.

Program design. The initial prospectus for Argo called for 3000 floats to be deployed. This was based on the need for seasonal climate forecasts. From the beginning, it was recognized that the deployment of such a large array with even coverage over the face

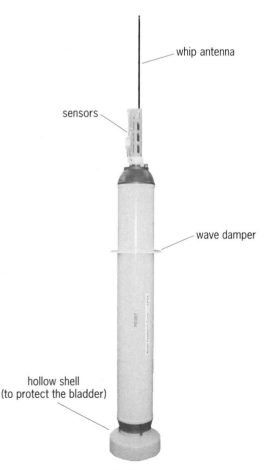

Fig. 2. An APEX float, one of the three major designs used in the Argo armada. This float is manufactured by the Webb Research Corporation. Two other designs operate on essentially the same principles. One is the SOLO, which is made at two float factories at the Woods Hole Oceanographic Institution and Scripps Institute of Oceanography. The other is the PROVOR float, manufactured by the Martec Corporation in France.

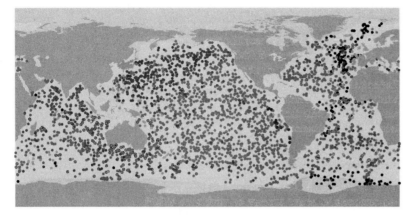

Fig. 3. The distribution of 3129 active Argo floats in March 2008. The dots represent different countries of origin.

of the Earth was well beyond the capability of any one country, so collaboration began in 1999 with a core group of eight nations. Deployments began in earnest in 2001, and gradually more countries have acquired floats to be deployed in support of the global array. By the end of 2007, the original goal had actually been achieved. And by March 2008, there were 3129 floats responding that had been deployed by 24 nations, as shown in **Fig. 3.**

Some assumptions went into the notion of a 3000-float array, notably that mapping of the state of the ocean could be done with a nearest-neighbor separation of about 300 km and an even distribution over the deep ice-free areas of the globe. It was computed that the 3000 floats would supply an average nearest-neighbor distance of 300 km, which meteorologists believed they needed for seasonal climate forecasting. However, since the Argo program started, floats have been deployed in marginal seas, and new technical developments have allowed the deployment of floats in ice-infested waters. These developments now require a rather larger array of perhaps 3400 floats to carry out the climate mapping function globally.

Data access. One of the decisions that took Argo away from the historical track established by previous large-scale surveys was to have an open-data policy. All 24 of the nations presently deploying floats in support of Argo have agreed to process all data with the same quality control, following the same principles and protocols. Further, all nations agreed to deliver their data in the same format to two "global data servers" hosted by IFREMER (Institut Français de Recherche pour l'Exploitation de la Mer) in Brest, France, and at the U.S. Fleet Numerical Meteorology and Oceanography Center in Monterey, California.

The result of this decision is that users from around the world, whether their nations deploy floats in support of Argo, can access either one of the two Web sites and download Argo data for their areas of interest. Free graphical software exists, called Ocean DataView, that allows an easy route to visualize Argo data for any particular region and offers options for mapping Argo data in two-dimensional horizontal or vertical slices.

The creation of the Argo array has involved a degree of collaboration that is unprecedented in oceanography. Two-thirds of the oceans are south of the equator, whereas the overwhelming majority of oceanographic laboratories are in the northern hemisphere. Thus, the goal of achieving global coverage has required that national Argo programs set their regional priorities second to the global objective.

Use of Argo. Argo data are now being widely used for a wide range of activities. The original objective was to support seasonal climate forecasting. In 2008, there were at least four operational forecast systems using advanced data assimilation techniques to compute the coupled state of the ocean and the atmosphere, initializing a computer model that will make the optimal climate forecast over periods of 3 to 6 months. However, the Argo armada represents a global archive of proportions that could not have been imagined 15 years ago. It has been estimated that in 2007, more ocean profiles were gathered in the Southern Ocean (Antarctic Ocean) by Argo than have been gathered by the sum of all previous research activities there. For the first time, we know what is happening in the Southern Ocean, or any other ocean, and we can monitor the changing patterns of circulation every 10 days.

Today, a Ghanaian or Nigerian scientist can monitor temperature and salinity in the Gulf of Guinea, and a Sri Lankan ocean scientist can track changing conditions in the Bay of Bengal. The Argo array is an asset designed to be exploited by scientists from all nations.

For background information *see* ANTARCTIC OCEAN; CLIMATE MODELING; CLIMATIC PREDICTION; DENSITY; EL NIÑO; INSTRUMENTED BUOYS; OCEAN; OCEAN CIRCULATION; OCEANOGRAPHY; SEAWATER in the McGraw-Hill Encyclopedia of Science & Technology.   Howard Freeland

Bibliography. H. J. Freeland and P. Cummins, Argo: A new tool for environmental assessment and monitoring of the world's oceans, *Prog. Oceanogr.,* 64(1):31–44, 2005; A. Matthews, P. Singhruck, and K. Heywood, Deep ocean impact of a Madden-Julian Oscillation observed by Argo floats, *Science,* 318:1765–1769, 2007; D. Roemmich et al., Decadal spinup of the South Pacific subtropical gyre, *J. Phys. Oceanogr.,* 37(2):162–173, 2007.

Artificial photosynthesis

The harnessing of sunlight in the form of a fuel through the direct conversion of visible photons to chemical energy by a synthetic device made of inorganic, organic, or hybrid materials is an attractive goal. Such artificial photosynthetic devices, in contrast to the synthesis of biofuels via plants, would not be restricted by the availability of arable land. Moreover, many efficiency limitations of biological photosynthesis, such as the tenfold attenuation of the photon conversion yield at high solar light intensity (at noon) by light regulation mechanisms acting to prevent photooxidative damage of plant cell material, would be absent in technological systems. Proof of concept for the efficient conversion of sunlight to hydrogen and oxygen by water splitting was demonstrated in a single integrated device by John Turner of the National Renewable Energy Laboratory in the late 1990s. The device is a monolithic, multilayer semiconductor material capable of splitting water with a conversion efficiency in excess of 10% for sunlight to chemical energy of hydrogen (H_2). However, the material degrades under use within days, is made of elements that are not sufficiently abundant, and employs fabrication processes akin to making a computer microchip, all of which make the device unsuitable for manufacturing on the scale needed for global fuel generation. Therefore, the most pressing need for making practical artificial solar fuel generators a reality is to develop robust systems that are made from abundant elemental components using scalable processes.

The essential components of an artificial photosynthetic system would carry out light absorption, charge separation, and catalytic conversion of water or carbon dioxide (CO_2) to fuel molecules (**Fig. 1**). Major scientific and technical breakthroughs are needed to make each of these components robust, efficient, and composed of abundant materials. The most daunting scientific gaps persist for the fuel-forming catalysts, namely for proton reduction to diatomic molecular hydrogen ($H^+ \rightarrow H_2$) and for the reduction of CO_2 to liquid fuels such as methanol

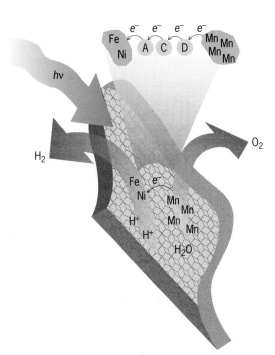

Fig. 1. An idealized diagram showing important aspects of artificial photosynthesis. The light absorber (C) and the charge separator, consisting of acceptor (A) and donor (D), are embedded across a membrane and electronically connected to multielectron reductive and oxidative catalysts (represented by FeNi and a Mn cluster, respectively). For clarity, the unit is displaced from the membrane and shown on an enlarged scale. When light is absorbed, charge is separated and utilized at the catalysts to form H_2 and O_2. The membrane further serves to instantaneously separate H_2 and O_2 as they are made.

or a higher-order alcohols ($CO_2 \rightarrow CH_3OH \rightarrow C_nH_{2n+1}OH$). Moreover, the reducing electrons needed for making hydrogen or a liquid fuel from CO_2 have to be taken from water. Using water as an electron source ensures that artificial photosynthesis and the subsequent use of the fuel constitute a full cycle, with no by-products accumulating in the process. Yet, reaching the goal of efficient and robust catalysts for the complex, multistep chemistries of water oxidation and proton/CO_2 reduction awaits major breakthroughs in the field of catalysis. Furthermore, the coupling of catalysts to solar photons, and integrating components into artificial photosynthetic systems that generate fuels under minimal loss of energy or charge, require major progress in chemistry and the science of nanomaterials. The highlights discussed here show that substantial advances in these areas have been achieved recently.

Development of robust oxidation catalysts. Well-defined molecular catalysts capable of evolving oxygen from water feature one or several transition-metal atoms (elements broadly associated with groups 3–11 of the periodic table plus the lanthanides) surrounded by organic ligands (carbon-containing molecules binding to the metals) that stabilize the catalytic core. While a definitive mechanistic role has been attributed to the typically heteroaromatic ligands (organic ligands having a closed ring structure containing at least one noncarbon atom in the ring), in some cases, it is desirable to replace the organic ligands with an all-inorganic environment. Since carbon-free ligands are substantially less oxidizable, the resulting complexes are expected to be much more stable to the high oxidation potentials needed to form oxygen from water.

Two teams have independently reported nearly identical molecules for water oxidation based on the all-inorganic ligand $[SiW_{10}O_{36}]^{8-}$. Two of these polyoxotungstenate ligands were used to stabilize a tetra-ruthenium core (Ru_4O_6, with adamantane-like structure) in which the four ruthenium centers span a distorted tetrahedron (**Fig. 2**). When exposed to oxidizing equivalents, either $Ce(NO_3)_4$ or $Ru(bpy)_3^{3+}$ [bpy = bipyridyl ligand], the complexes catalytically oxidize water to form oxygen. These catalysts appear to be stable during catalysis, maintaining their spectroscopic features after oxygen evolution over extended periods of time.

The rate-limiting step in these systems is thought to be the charge transfer to the oxidant rather than an intrinsic limit of water oxidation catalysis itself. This implies that rates can be improved by coupling the catalyst more efficiently to an electron or photon source, which is very encouraging. While these systems need further study to optimize charge transfer and unravel the underlying catalytic mechanism, the stability of these compounds, engendered by the all-inorganic nature, makes them an important step toward viable water oxidation catalysts for artificial photosynthesis.

Direct reduction of CO_2 to liquid fuel. A liquid solar fuel is attractive in terms of ease of distribution and transportation. Yet, less than a handful of

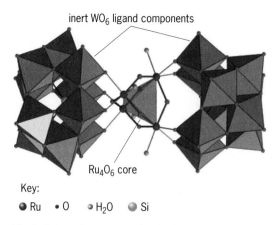

inert WO$_6$ ligand components

Ru$_4$O$_6$ core

Key:

● Ru   • O   ● H$_2$O   ● Si

Fig. 2. Schematic representation for the x-ray structure of [SiW$_{10}$O$_{36}$][Ru$_4$(O)$_4$(OH)$_2$(H$_2$O)$_4$][SiW$_{10}$O$_{36}$]. The tetrahedron in the center represents the catalytically active Ru$_4$O$_6$ core, with the octahedra on either side showing the inert WO$_6$ components of the stabilizing polytungstenate (SiW$_{10}$O$_{36}$) ligands. This all-inorganic molecule catalytically oxidizes water to oxygen when treated with strong oxidants such as Ce(IV) or Ru(III). Ru, O, H$_2$O, and Si are identified in the key; hydrogen atoms are omitted for clarity. (*Figure adapted from Y. V. Giletti et al., Angew. Chem. Ind. Ed., 47:3896–3899, 2008*)

catalytic processes are known to directly convert carbon dioxide and water to methanol or higher alcohol molecules. However, a method for the direct reduction of CO_2 to methanol with essentially no by-products has been developed. The reaction runs with a mild underpotential (220 mV) and high current efficiency at a semiconductor electrode in aqueous solution using the pyridinium ion as a mediator.

While the step-by-step process by which pyridinium accomplishes the six-electron reduction of carbon dioxide to methanol remains to be established, the work points to as-yet undiscovered paths that lead directly to a liquid fuel. In particular, insertion of carbon-carbon bond-forming steps into the mechanism would open up the possibility of developing practical methods for producing fuels approaching the high energy density of petroleum, while bypassing the need for other higher-energy intermediates as utilized by nature (NADPH), or hy-

drogen. Moreover, the reaction can use visible light as the sole energy source. The semiconductor electrode used is made of gallium phosphide (GaP), which is an indirect-bandgap semiconductor material (2.24 eV). While the researchers achieved the highest quantum efficiency for methanol production ever reported, exploration of more efficient direct-bandgap semiconductor electrodes is likely to afford substantial improvement over the current performance of the system.

Integrated photocatalytic unit for H$_2$ generation. Coupling of the light-harvesting components with fuel-generating catalysts with minimal energy or charge loss, while transferring electrons at rates that keep up with the solar light flux, is a formidable task. Meeting the challenge requires tight control of the spatial arrangement and electronic interaction of the chromophore (light absorber) and catalytic site. A unit has been prepared that features an organometallic ruthenium complex covalently linked via an electron-rich bridge to a cobalt complex that acts as an efficient catalyst for the reduction of water to hydrogen (**Fig. 3**).

While turnover number (a measure of stability) and frequency (a measure of reaction rate) are still modest, they are much higher than in systems where the ruthenium sensitizer and cobalt catalyst are not positioned relative to one another through a covalent linkage. In light of the tremendous progress in recent years in the understanding of photo-induced electron transfer across organic linkers, knowledge is now available for structurally modifying the bridge between the light harvesting and catalytic components for optimal coupling and efficient use of sunlight for hydrogen generation. This photocatalytic system features a first-row transition metal, cobalt, for the catalytic site, another important step toward engaging inexpensive abundant elements for artificial photosynthesis.

Chromophore-catalyst coupling in a nanoporous scaffold. A complete artificial photosynthetic system requires the arrangement and coupling of light absorption, charge separation, and catalytic units in nanostructured membranes. The high surface area of such supports is necessary to reach a sufficient density of photocatalytic sites so that the fuel-forming chemistry can keep up with the photon flux at high solar intensities. Membranes also offer nanostructural physical barriers that separate the reduced fuel products from evolving oxygen, minimizing inefficient back reactions.

With the goal of combining the advantages of molecular components with the robustness of inorganic materials, researchers developed selective synthetic methods for the assembly of inorganic polynuclear photocatalytic units in inert nanoporous silica scaffolds. The chromophore is an all-inorganic oxo-bridged (an oxo unit has the structure metal=O) binuclear unit, Ti^{IV}OCr^{III}, covalently anchored on the nanopore silica surface. It can then act as a visible light-powered electron pump. When it is coupled to oxygen-evolving multielectron catalysts, such as iridium oxide nanoclusters inside the nanopores,

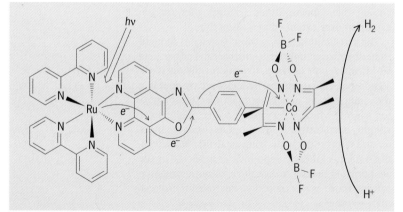

Fig. 3. An efficient catalyst for the reduction of water to hydrogen. The organometallic ruthenium complex (left-hand side) is linked to a cobalt complex (right-hand side) via an electron-rich bridge.

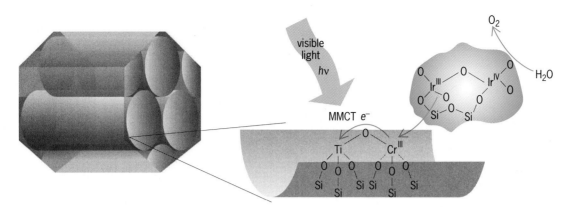

Fig. 4. Diagram of a light-driven oxidative half-reaction on the surface of the nanoporous silicate MCM-41. The metal-to-metal charge transfer (MMCT) chromophore (Ti^{IV}-O-Cr^{III}) serves to drive an iridium oxide multielectron oxidation catalyst, forming oxygen from water.

oxidation of water is observed when the material is suspended in aqueous solution and illuminated with visible light (**Fig. 4**). The molecular nature of the chromophore allows for precise tuning of the redox properties of the donor and acceptor by selecting appropriate metals and oxidation states. Such control is key for the optimization of light absorption, thermodynamic efficiency, and directional charge flow in photocatalytic devices. *See* POROUS ALUMINA.

To investigate electron transfer inside silica nanopores, researchers used a structural model of molecular water oxidation catalysts, namely the di-μ-oxo bridged dinuclear manganese complex $(bpy)_2Mn^{III}(\mu\text{-}O)_2Mn^{IV}(bpy)_2$. This system takes advantage of the spectroscopic precision of differing chemical and electronic states in discrete molecules at levels not readily attainable with heterogeneous metal oxide nanocluster catalysts. When loaded inside silica nanopores, the Mn_2O_2 unit is electronically coupled to a chromium-based visible-light-excited chromophore. The initial and final redox states of the Mn_2O_2 core and chromium chromophore could be definitively observed upon light-induced electron transfer, laying the foundation for detailed evaluation of the kinetics and energetics of light-induced electron flow between catalysts and molecular charge-transfer chromophores inside high-surface-area nanostructured supports. Such advances will allow for the design and control of artificial photosynthetic systems.

For background information *see* CATALYSIS; LIGAND; ORGANOMETALLIC COMPOUND; PERIODIC TABLE; PHOTOCHEMISTRY; PHOTOSYNTHESIS; TRANSITION ELEMENTS in the McGraw-Hill Encyclopedia of Science & Technology. Walter W. Weare; Heinz Frei

Bibliography. E. E. Barton, D. M. Rampula, and A. B. Bocarsly, Selective solar-driven reduction of CO_2 to methanol using a catalyzed p-GaP based photoelectrochemical cell, *J. Am. Chem. Soc.*, 130:6342–6344, 2008; A. Fihri et al., Cobaloxime-based photocatalytic devices for hydrogen production, *Angew. Chem. Int. Ed.*, 47:564–567, 2008; Y. V. Geletii et al., An all-inorganic, stable, and highly active tetraruthenium homogeneous catalyst for water oxidation, *Angew. Chem. Int. Ed.*, 47:3896–3899, 2008; H. Han and H. Frei, In-situ spectroscopy of water oxidation at Ir oxide nanocluster driven by visible TiOCr charge-transfer chromophore in mesoporous silica, *J. Phys. Chem.*, in press; A. Sartorel et al., Polyoxometalate embedding of a tetraruthenium(IV)-oxo core by template-directed metalation of $[\gamma\text{-}SiW_{10}O_{36}]^{8-}$: A totally inorganic oxygen-evolving catalyst, *J. Am. Chem. Soc.*, 130:5006–5007, 2008; W. W. Weare et al., Visible light-induced electron transfer from di-μ-oxo bridged dinuclear Mn complexes to Cr centers in silica nanopores, *J. Am. Chem. Soc.*, 130:11355–11363, 2008.

Assessing risks from nanomaterials

Nanotechnology is being used in exciting and potentially revolutionary applications, ranging from sunscreens and drug delivery to more efficient solar cells and the production of stronger and lighter tires. As nanomaterials are produced and incorporated into products and processes, they will inevitably enter the environment and come in contact with living organisms. The novel properties of materials with at least one dimension smaller than 100 nanometers can cause them to behave differently from their bulk counterparts. These special characteristics and interactions are the reason nanomaterials are advantageous for new or improved applications, as well as why they may interact with organisms and the environment in unexpected ways. Some of these unpredictable impacts will be positive, and some will be potentially harmful. Regulators, companies, researchers, and the public agree that methods for assessing the relative safety of nanomaterials are needed immediately. The potential for negative impacts must be addressed if we are to realize the positive gains. To this end, frameworks are being developed for assessing risk prior to a complete understanding of nanoparticle behavior.

Why nano needs a new approach. The risk associated with a material is a function of hazard and exposure, where hazard refers to an unwanted end result such as toxicity, and exposure refers to a material's coming in contact with a target and rendering that toxicity possible. Assessing the risks of materials in their bulk form is not a new concept. These processes require a knowledge of which measures can be used to predict the behavior of a substance, including quantitative or qualitative parameters such as size, partitioning coefficients, lethal concentration (LC50) toxicity values, and physical quantities. These factors are incorporated into mathematical models that predict the behavior of the substance in terms of hazards to health and the environment (toxic effects) and expected exposures (fate and transport). In their nanoscale forms, new properties emerge for some materials, such as large comparative surface areas, differing optical effects, or a propensity to produce reactive oxygen species. This can change how a material interacts with the environment, and casts doubt on whether the same parameters are adequate to predict its risk. For example, if the reactivity of a chemical is based on its available bond sites, and if that is based on its relative surface area, then the mass concentration of the chemical would no longer adequately predict its interactions because the same mass of the nanoscale material would have more reactivity than it would on the bulk scale. We need to know what characteristics determine the mechanisms of nanomaterials' toxic effects, their transport behavior, and their fate in the environment. And since the term "engineered nanomaterials" encompasses a vast range of different chemicals whose primary common trait is their small size, these studies by necessity begin on a case-by-case basis but must move toward identifying parameters that could more generally predict their behaviors.

Prioritization. By 2005, several major reports had identified information gaps that needed to be filled to adapt risk frameworks for nanomaterials. Since then, the number of groups working on such issues has grown, and some approaches to risk assessment have been proposed. This developing body of literature has shown a general agreement around two overarching concepts. First, there is a need to identify and prioritize the data requirements for predicting nanomaterial behavior, and to commit resources to filling those data gaps. Standardization and categorization agreements will be required to

organize these research goals and enable decision making. To consider environmentally relevant nanomaterial effects beyond the laboratory, differences must be noted between engineered and unintentionally produced nanomaterials; free nanomaterials and those embedded in a matrix; coated, functionalized, and unmodified nanomaterials; and short-lived engineered nanoparticles and more durable engineered nanoparticles. Second, some existing methodologies for approaching risk assessment can be applied now to understand nanomaterials' behavior, before filling all the data gaps. These two conceptual thrusts of nanomaterials risk research must be maintained in parallel. A few such proposed methods will be reviewed here.

There are multiple ways of translating the concept of assessing risk into adoption of policies, given the current state of ambiguity as to the ultimate effects of these materials on health and the environment. The precautionary principle would advocate careful vetting of new technologies prior to their widespread adoption, erring on the side of caution with respect to possible negative environmental health effects. In Europe, this approach is the default, as one can see manifested in the multiple calls for regulation and research through government bodies and insurance companies. Adherence to this way of thinking is intertwined with public perception and cultural acceptance of risk. One example of the precautionary principle applied to a new technology is the ban on genetically modified organisms or foods in Europe. In the United States, a similar funding push is being made on behalf of the need to understand nanomaterial effects. However, the precautionary principle is less likely to be applied because the cultural risk default in the United States is to err on the side of caution with regard to possible negative economic effects. In this approach, it is still important to understand potential liabilities, but it's more socially plausible to let a revolutionary technology go to market until proven dangerous, instead of benching it until proven safe.

Approaches. A logical first step in assessing the risks of nanomaterials is to look at the established methods of characterizing the risks posed by materials for which we have more information. To date, most approaches propose assessing risk from a product-oriented environmental management viewpoint, applying life-cycle assessment (LCA) tools. This type of cradle-to-grave analysis takes into account the whole system of inputs, products, and waste streams involved with the many stages of a material's life. This is a viable starting point, because even as questions persist about the toxicity and exposure potential of nanomaterials, products incorporating such materials are being produced and used. Already the systems can be defined in terms of boundaries, expected flows of inputs and wastes, and product type; the manufacturing processes that are likely to be scaled up industrially can be studied; and probable exposure pathways can be identified for further scrutiny. The EPA has combined product LCA with a risk assessment framework to create

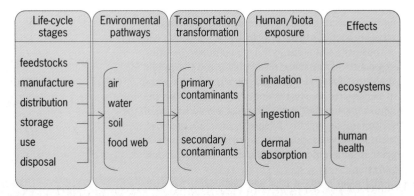

Comprehensive environmental assessment (CEA) structure. (*Adapted from J. M. Davis and V. M. Thomas, 2006*)

a comprehensive environmental assessment (CEA) tool. Developed partially in response to the problems brought about by methyl tertiary-butyl ether (MTBE) and asbestos and in consideration of the reality that complex technological decisions may require taking action in the face of unknowns, this tool lays out the considerations necessary to assess the ecological and human health impacts of nanomaterials. Its stated goal is to help identify and prioritize research efforts for nanoscale products. The **illustration** represents the basic structure of their proposed risk framework.

Other efforts have yielded similar conclusions about the use of existing tools for assessing nanomaterials by identifying and characterizing hazards, determining key exposure pathways, and deciding how to characterize the risks. Another approach to prioritizing data needs used expert elicitation to assemble and organize a network of factors expected to govern nanomaterial impacts. The variables collected from a wide range of experts were assembled into an influence diagram, linking the components to create a more qualitative understanding of how these experts viewed the relationships between factors affecting nanomaterial impacts.

Other existing risk assessment tools can be applied as well, such as those used by insurance companies to rank the relative risk of production processes. One such study assessed the manufacturing risks of several nanomaterials using an insurance industry protocol. Considering the inputs and by-products without the final nanomaterial, the production processes themselves were shown to pose equivalent or less risk than a host of already established industrial processes. Further adaptation of insurance assessment tools could prove useful in creating a baseline of quantitative risk information for the life-cycle impacts of engineered nanomaterials, as well as giving nanomaterial producers a glimpse of the type of assessments that are likely to be imposed as insurers deal with how best to qualify the risks of these new processes.

Regulations are the outgrowth of such assessments, and the beginnings of some regulatory actions are starting to be seen. For example, a local voluntary reporting process for nanomaterial production was adopted in Berkeley, California, in 2007. There have also been calls for registering each new nanomaterial as a separate chemical in the Chemical Abstracts Service (CAS) registry to denote that it is fundamentally different in its nanoscale form. Some promising instances of incorporating risk assessment along with self-regulation are emerging, as collaborative studies appear from corporations and governments. Such is the case with the joint risk assessment methodology created by DuPont and Environmental Defense, a framework built on previous risk assessment methodologies that adds the steps of "document, decide, and act" and "review and adapt." In this way, the decisions about producing and handling nanomaterials will be made with impacts in mind, and will be documented, justified, and reassessed as new data become available.

Outlook. All these risk assessment approaches help in understanding and beginning to control the impacts of nanomaterials. As a body of literature, these frameworks also demonstrate the growing international consensus among a wide variety of stakeholders about the priorities and guidelines that ought to be followed in the responsible development of nanotechnology.

For background information *see* ENVIRONMENTAL ENGINEERING; ENVIRONMENTAL TOXICOLOGY; MESOSCOPIC PHYSICS; NANOPARTICLES; NANOTECHNOLOGY; RISK ASSESSMENT AND MANAGEMENT; TOXICOLOGY in the McGraw-Hill Encyclopedia of Science & Technology.   Christine Ogilvie Robichaud

Bibliography. J. M. Davis, How to assess the risks of nanotechnology: learning from past experience, *J. Nanosci. Nanotechnol.*, 7(2):402–409, 2007; J. M. Davis and V. M. Thomas, Systematic approach to evaluating trade-offs among fuel options: the lessons of MTBE, *Ann. N.Y. Acad. Sci.*, 1076:498–515, 2006; *Nanotechnologies: A Preliminary Risk Analysis on the Basis of a Workshop*, Community Health and Consumer Protection Directorate General of the European Commission, Brussels, Belgium, pp. 11–29, 2004; T. Medley and S. Walsh, *Nano Risk Framework: Environmental Defense—DuPont Nano Partnership*, 2007; K. Morgan, Development of a preliminary framework for informing the risk analysis and risk management of nanoparticles, *Risk Anal.*, 25(6):1621–1635, 2005; *Nanotechnology White Paper*, United States Environmental Protection Agency, Washington D.C., 2005; *Opinion on the Appropriateness of Existing Methodologies to Assess the Potential Risks Associated with Engineered and Adventitious Products of Nanotechnologies*, European Commission, Scientific Committee on Emerging and Newly Identified Health Risks, pp. 41–58, 2005; C. O. Robichaud, D. Tanzil, and M. Wiesner, Assessing life-cycle risks of nanomaterials, in M. Wiesner and J. Bottero (eds.), *Environmental Nanotechnology: Applications and Impacts of Nanomaterials*, pp. 481–522, McGraw-Hill, New York, 2007; J. S. Tsuji et al., Research strategies for safety evaluation of nanomaterials, Part IV: Risk assessment of nanoparticles, *Toxicol. Sci.*, 89(1):42–50, 2006.

Assisted GPS and location-based services

As recently as 2004, the Global Positioning System (GPS) receiver most widely used by the U.S. military was the PLGR (Precision Lightweight GPS Receiver). It is a five-channel receiver that operates on only one of the three GPS frequencies (L1), has a typical startup time of over a minute, and costs about $2000. The PLGR receives encrypted military signals, is waterproof, and weighs about 1.4 kg (3 lb), making it far more rugged than any modern mobile telephone. But many of those mobile phones today have 12-channel A-GPS (Assisted GPS), which can compute a position within a second and acquire satellite signals that are more than 100 times weaker than can be acquired by the PLGR, and which adds less than $5 to the cost of the phone.

This article provides an overview of A-GPS and why it has led to the adoption of a GPS architecture known as host-based GPS. Host-based GPS has significantly reduced the size, cost, and power consumption of GPS receivers, thereby enabling both the mass-market adoption of the technology in mobile phones and the emergence of location-based services.

A-GPS overview. A-GPS is the technique for improving standard GPS performance by providing the GPS receiver with information that it ordinarily would have to receive from the satellites themselves. A-GPS does not excuse the receiver from receiving and processing signals from the satellites; it simply makes this task easier and minimizes the amount of time and information required from the satellites.

GPS was originally designed to guide bombs, aircraft, soldiers, and sailors. In all cases, the GPS receiver was expected to be outdoors with a relatively clear view of the sky. The system was designed with a start-up time of something like a minute in mind, after which it would operate continuously. Today GPS is used for many more civilian than military purposes. (It is estimated that more GPS receivers are now sold in mobile phones each year than the cumulative number ever used in military service.) Paradoxically, the system demands of these civilian applications far exceed those seen previously. GPS is now expected to work almost anywhere (even, sometimes, indoors), "push-to-fix" applications have emerged where a position is expected almost instantly, and all of this must be delivered in a way that adds little or no cost, size, or power consumption to the host device. These demands are what drove the development of A-GPS.

To calculate a position (or "fix"), a GPS receiver must first find and acquire the signal from four or more satellites, and then decode data from each satellite. The sequence can be understood by an analogy with FM radio. Finding the signal in the first place is rather like finding a new FM radio station as you drive on a long journey. Each GPS satellite appears on a different frequency, thanks to the Doppler shift induced by the high speeds at which the satellites move (5 km/s, or 3 mi/s). The observed Doppler shift is a function of your location. Until your receiver knows where it is, it cannot calculate the correct Doppler shift; hence standard GPS receivers exhaustively search all possible frequencies in much the same way that you might scan the dial of your FM radio. Having found a signal, standard receivers must decode the data modulated on the satellite signal to find the position of the satellite. This is analogous to waiting for the FM station identification to know what you have found once you find it. Only after these satellite position data are decoded can a standard GPS receiver compute your position.

A-GPS uses cell phone messages to provide the information that allows the GPS receiver to know what frequencies to expect before it even tries, as well as the data bits from which the satellite location is determined. Having acquired the satellite signals, all that is left for the receiver to do is to take range measurements from the code phase of the satellite signal (this takes milliseconds, not minutes), and then the A-GPS receiver can compute your position. The time to first fix is reduced from the order of 1 minute to the order of 1 second.

The GPS signal is composed of data bits that are superimposed on a wideband pseudorandom bit stream. In **Fig. 1**, the satellite data are represented by a square wave, and the pseudorandom noise (PRN) code is represented by a sinusoid. The PRN code is a binary code that modulates the carrier and enables satellite range measurements. Figure 1 illustrates the fact that in A-GPS, the satellite data can be provided from a separate source (the cell tower), and that the PRN code phase can still be measured in circumstances where the data from the satellite cannot be decoded.

Because the A-GPS receiver is designed to know in advance what frequencies to search, it can allocate its signal processing to allow longer dwell times than a conventional receiver. The longer dwell times increase the amount of energy received at each particular frequency, thereby increasing the sensitivity of the A-GPS receiver. This increased sensitivity allows the receiver to acquire signals at much lower signal strengths than a conventional military receiver.

Standards have evolved for specifying how the GPS assistance data are provided from the telephone network to the handsets. The GPS chip vendor usually supplies the network communications protocol software that implements these standards. Because this protocol software is a necessary part of an A-GPS implementation, a relatively small change in the software allows the host (the mobile phone) to take over

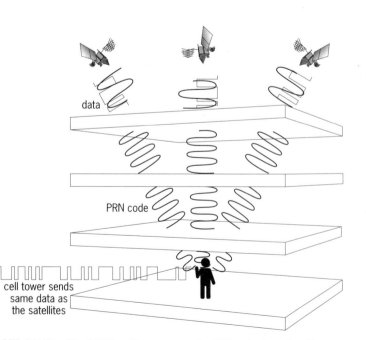

Fig. 1. A-GPS data flow. The A-GPS receiver measures the PRN code phase from the satellites and gets data (for example, the satellite position) from the cell tower.

much of the functionality traditionally performed within the GPS module itself. This gives rise to the host-based GPS architecture.

Host-based GPS. The architecture of a host-based system will be explained by comparing it to the traditional system-on-chip (SOC) approach used in the design of conventional GPS equipment. In the system-on-chip architecture, the entire GPS system is integrated within a single device. The system-on-chip contains three major building blocks: a radio-frequency tuner block, a baseband processing block, and a central processing unit (CPU) subsystem that runs a complete GPS software application. The output of the system-on-chip is position, velocity, and time (PVT) data that are sent to the host device.

One of the key drawbacks of system-on-chip architecture is the complexity of its hardware, which largely determines the size and cost of the chip. The host-based alternative offers reduced hardware complexity, which translates into a smaller, less expensive chip. In the host-based approach, the on-chip CPU subsystem is eliminated, leaving only the radio-frequency tuner and the GPS baseband processor. The output of the GPS baseband processor is raw measurement data, which are streamed to the host.

The host software application includes a module that computes GPS navigation data. This module is provided as part of the host-based GPS solution. The input to the navigation processing module is raw measurement data from the GPS chip; the output is PVT data identical to those produced by the system-on-chip.

Figure 2 compares the hardware contents of host-based and system-on-chip GPS chips. The three die shown in the system-on-chip are typical of today's GPS chip offerings, versus a single die for the host-based chip. The system-on-chip requires a separate die for flash memory to hold the programming code for the chip. Since flash cannot be integrated easily with CMOS logic, it is likely that this will remain a separate die for some time to come. The system-on-chip also uses a separate die for the radio-frequency section; again, this is typical of most system-on-chip solutions available today. The host-based GPS chip has no CPU or flash memory, so the task of integrating the radio-frequency and baseband sections in a single die is simplified. *See* NONVOLATILE MEMORY DEVICES.

The benefit of having less hardware is obvious from examining the footprint sizes of commercial GPS chips. One of the smallest and most widely used system-on-chip products measures 7×10 mm (0.28×0.40 in.), compared to less than 4×4 mm (0.16×0.16 in.) for some host-based devices. Both chips include blocks for the GPS radio-frequency front end and GPS baseband processing.

Software GPS. Host-based GPS is sometimes confused with software GPS. In fact, the two architectures are quite different. In a software-GPS architecture, all of the GPS signal processing, in addition to the navigation functions, is performed in the host. Software GPS is useful from the research and development perspective because the signal processing algorithms are executed in software and are therefore

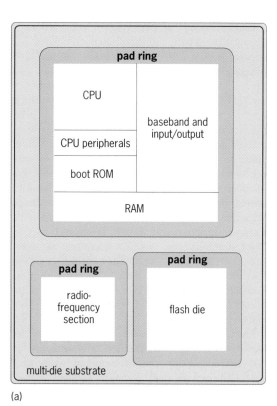

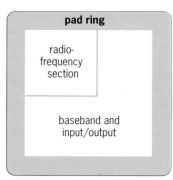

(a)

(b)

Fig. 2. Hardware comparison of (a) system-on-chip (SOC) GPS and (b) host-based GPS.

more accessible for development and testing. Some software GPS products are targeted exclusively for research and development, but the use of software GPS in commercial applications faces significant obstacles and has not been broadly adopted.

One of the obstacles hindering commercial adoption of software GPS is the immensity of GPS signal processing requirements: multi-megabits of bandwidth and hundreds of millions of instructions per second (MIPS) are needed. Intuitively one might assume that a software-GPS solution is less expensive, but the reality of current semiconductor technology is that cost and size are often driven as much by interfacing requirements (for example, pin count) as by the core processing area. With its simple interfaces, a host-based solution can actually result in a smaller chip than a software-based solution, and this will be increasingly true as the industry migrates to 65-nm CMOS technology.

Location-based services (LBS). The advances brought by A-GPS and host-based GPS have led to size, power, and cost reductions that are now bringing GPS into the mass market, including cell phones, and in turn, this development enables a myriad of location-based services (LBS). So far there are two "killer applications": E911 (which has been mandated) and in-car navigation (which has been market driven). The E911 mandate in the United States requires that all 911 emergency calls include the location of the device. Just a few years ago, it was assumed that network-based location techniques would dominate major mobile communications standards such as GSM (Global System for Mobile Communications) and UMTS (Universal Mobile Telecommunications System) telephone services, but A-GPS has taken over almost entirely. A-GPS is also proving synergistic with in-car navigation; as real-time traffic and map corrections become more common, the information bearer carries the GPS assistance data as well.

LBS applications that have been discussed for years are now starting to get traction, most notably social networking applications including buddy finders, child trackers, teenage-driver monitors, and elderly parent trackers. Major network operators are offering such services based on A-GPS.

Other niche applications that are moving to the mainstream include geocoding of digital photographs. More esoteric applications are becoming visible in small volumes: A-GPS in dog collars, A-GPS in sneakers, and running/walking/cycling/skiing applications in A-GPS–enabled mobile phones.

A-GPS–enabled location-based services have developed rather as Bluetooth did: first hyperbole, then a long wait, then the reality of products, and finally ubiquitous deployment.

Adoption. The move to A-GPS was driven by consumer demands for increased GPS performance: instant time to fix and high sensitivity (including, sometimes, indoor operation). To support A-GPS data transfer, chip vendors supply network communications protocol software, and this in turn has encouraged the widespread adoption of host-based GPS architectures. Host-based GPS has dramatically lowered the size, cost, and power consumption of GPS implementations, so that we are now seeing the mass adoption of GPS in mobile handsets and the emergence of location-based services.

Perhaps the best indicators of these trends are provided by the actions of the largest companies in the communications industry: the world's largest manufacturers of both mobile handsets and personal navigation devices (PNDs) have adopted host-based GPS solutions, and the most popular Internet corporation has entered the location-based services arena, offering free maps for mobile phones.

For background information *see* DOPPLER EFFECT; FREQUENCY-MODULATION RADIO; INTEGRATED CIRCUITS; MOBILE COMMUNICATIONS; SATELLITE NAVIGATION SYSTEMS; SEMICONDUCTOR MEMORIES in the McGraw-Hill Encyclopedia of Science & Technology.

Frank van Diggelen

Bibliography. C. Abraham, GNSS solutions: host-based processing, *InsideGNSS*, pp. 30–36, May/June 2007; C. Abraham and F. van Diggelen, Indoor GPS: The no-chip challenge, *GPS World*, September 2001; P. Misra and P. Enge, *Global Positioning System: Signals, Measurements, and Performance*, 2d ed., Ganga-Jumana Press, 2006.

Assisted migration for species preservation

Emissions of greenhouse gases from the burning of fossil fuels are altering Earth's climate. This human-driven (anthropogenic) climate change (or "global warming") presents yet another threat to species' survival. The current episode of global climate change has caused measurable geographic shifts in climate zones, principally poleward and upward. Many wild species are already showing changes in their distributions that are, to varying degrees, tracking the shifting climate zones. One of the main conservation concerns is that, as climate zones shift across the landscape, our current preserve network will no longer contain appropriate climates for the species for which those preserves were designed. Furthermore, human domination of the landscape creates barriers to natural movements of species toward new geographic areas that have only recently become climatically suitable. If species within preserves (or other undisturbed habitats) experience degradation of their local climate, their natural dispersal abilities may be insufficient to allow them to cross agricultural lands and urban areas to successfully colonize newly formed habitats outside their current range. In these situations, it has been suggested that human-assisted translocation of individuals, often termed "assisted migration" or "assisted colonization," may be necessary to ensure colonization of new geographic regions as parts, or all, of the species' historic range becomes climatically unsuitable.

Need for translocation of species. Species translocations have been a frequent tool employed by conservation biologists to increase the numbers of populations of rare and endangered species. These translocations typically take the form of reintroductions of individuals into an area where the species had been extirpated, primarily as a result of human activities. By far, habitat destruction and introduction of invasive species have had the greatest negative impacts on global biodiversity over the past 200 years. A driving philosophy of modern conservation is that the protection of one species should not come at the detriment of another species. Specifically, management for one species should not result in the introduction of an exotic species into a new community in which it may become invasive and cause the demise of native species. Therefore, a golden rule in the conservation community has been to never introduce a species into an area where it did not exist in recent history (that is, in the past few hundred years). With this foundation, it is not surprising that the concept of assisted movement of species into novel geographic areas as a climate change adaptation strategy is anathema to many conservation biologists.

A counterpoint to the view that traditional conservation philosophies should be adhered to is that climate change is distinctly different from traditional anthropogenic drivers of biodiversity loss, and thus requires novel approaches. Given sufficient budget and personnel, traditional threats can often be mitigated, or even neutralized, at the local level. Invasive species can be exterminated, and degraded lands can be restored. One of the ways in which climate change is unique is that it is truly global in nature. Although a few threats approach the global scale (for example, acid rain from sulfur oxide pollution and weakening of the protective ozone layer from industrial chemicals), they have been relatively easily reduced in severity through international efforts focused on reducing the output of the chemical pollutants causing the problem. Local actions to reduce greenhouse pollutants will have little local mitigating effect, however. Successful reduction of the magnitude of the climate change threat can come about only through well-coordinated and universal actions across all nations. However, the greenhouse gas pollutants causing climate change are much harder to reduce than other global pollutants, both because they have many more, and more diverse, point sources and because there are often no substitutes.

Another way in which climate change differs from traditional threats is that there is a long lag time between actions and results. One of the main greenhouse gases, carbon dioxide, is extremely stable in the atmosphere and continues to affect the global climate for hundreds of years after it has been released. In addition, the climate system itself has a lagged response: it will be another hundred years before the full effects of the current increased levels of greenhouse gases in the atmosphere are fully reflected in atmospheric temperatures, precipitation, and other climate patterns, and sea surface temperature could take several hundred years to stabilize. These unique traits of anthropogenic climate change have led many in the conservation community to consider that this is a threat that will become increasingly severe over time, and for which standard management actions currently used by reserve managers will produce little benefit.

Debate on assisted migration. The unique nature of the climate change threat has spurred the conservation community to suggest creative new approaches and tools for the preservation of local and global biodiversity. There is a consensus that more emphasis should be placed on some of the traditional conservation actions that could aid species in tracking the shifting climate, such as improved connectivity of reserve networks through the creation and preservation of dispersal corridors and "stepping-stone" habitats (small reserves placed in between large preserves). Deference to climate change would give priority to corridors that encourage elevational and latitudinal movements, and to reserve designs that emphasize elevational and latitudinal gradients, as well as topographic diversity in general. In contrast, there is an active debate about the use of unconventional approaches, and assisted movement is at the core of this debate.

The debate on assisted movement is not just about whether this should be undertaken, but on the details of when, how, and for which species this action should be considered. The most mild form would be assisted movement of individuals to novel habitats that are still within the broad historic range of the species. For example, the American pika (*Ochotona princeps*) is a mountain-restricted species that already appears to be contracting its range upward. Its habitat is relatively rare and patchy [talus, or scree, slopes above about 7500 ft (2300 m)]. A mild form of assisted migration might be to transplant individuals from the lowest-elevation mountains to nearby, higher-elevation mountains that the pika had not naturally colonized, but that would allow greater "natural" elevational shift in the future. In this hypothetical example, the lack of a population on the higher-elevation mountains is likely to be due to chance (historical flukes of past colonization events), and natural colonization could have occurred at any time—hence, human-assisted movement is in concert with natural processes.

A more controversial category of assisted movement is the transplantation of individuals outside of the geographic region in which they have historically been documented. (Note that "historically" generally refers to the past several thousand years of the current interglacial period.) There is an active movement to do just this for the endangered Florida torreya tree (*Torreya taxifolia*), spearheaded by a group known as the "Torreya Guardians." The Florida torreya has been thriving in a Georgia garden for some 70 years, far outside its native range. The debate is over whether to actively transplant individuals into surrounding native communities, where the local native plant species have no history of interacting with torreya. There is always the risk that such a novel species could have unforeseen negative impacts on the local community, as witnessed by the devastating effects of many introduced exotic species, such as the kudzu vine in the southeastern United States, and rabbits and the cane toad in Australia.

An alternative between the most mild and the most drastic forms of assisted movement (exemplified above) is to actively create suitable conditions for reproduction outside the current species' range, but then allow natural colonization of the newly created sites. Excellent candidates for this approach are coral reef systems. It has been demonstrated that artificial structures can promote the establishment of coral reefs and their associated plant and animal communities. Therefore, it could be possible to promote poleward expansion of coral reefs by placing artificial structures in a stepping-stone pattern outside their current range. Such expansion may be ultimately limited by other constraints (such as light availability), but this action could promote reef persistence in the near term, while having no foreseeable negative impacts on nontarget biodiversity.

Future outlook. In summary, in considering whether assisted movement is a wise conservation measure, the most suitable scenario is one in which the risk of extinction of the target species in its historic range is high, but the risk to the community into which the species will be imported is low; and in which the likelihood of successful colonization is high, but the time and cost to perform the transplantation is low. Thus, passively assisting coral reef migration may be acceptable, but transplanting polar bears to Antarctica, where they would be likely to drive native penguins to extinction, would not be acceptable. Ultimately, the decision as to whether to actively assist the movement of a species into new territories will rest on ethical and esthetic grounds as much as on hard science. Decisions to move or not to move are bound to be heavily influenced by the social and cultural value placed on the target species as compared to the value placed on the recipient system. Decisions to move charismatic species (that is, species having great popular appeal, such as gorillas or giant panda) into degraded lands that have been restored explicitly for that target species will be far more likely to be approved than efforts to move small species that lack popular appeal into highly prized, biologically rich areas. The level of intervention taken will effectively decide which species will be allowed to go extinct in situ (that is, with no attempt to assist in a range shift) because of overriding costs, potential risk to other species, or other unknown ecosystem consequences. Conservation has never been an exact science, but species preservation in the face of climate change is likely to require a fundamental rethinking of what it means to preserve biodiversity.

For background information *see* BIODIVERSITY; CLIMATE MODIFICATION; ECOLOGICAL COMMUNITIES; ECOLOGY; ENDANGERED SPECIES; EXTINCTION (BIOLOGY); GLOBAL CLIMATE CHANGE; POPULATION DISPERSAL; POPULATION DISPERSION; POPULATION ECOLOGY in the McGraw-Hill Encyclopedia of Science & Technology.   Camille Parmesan

Bibliography. O. Hoegh-Guldberg et al., Assisted colonization and rapid climate change, *Science*, 321:345–346, 2008; M. L. Hunter, Climate change and moving species: Furthering the debate on assisted colonization, *Conserv. Biol.*, 21:1356–1358, 2007; IPCC, *Climate Change 2007: Impacts, Adaptations and Vulnerability*, Contribution of Working Group II to the Fourth Assessment Report of the Intergovernmental Panel on Climate Change, Cambridge University Press, Cambridge/New York, 2007; IPCC, *Climate Change 2007: The Physical Science Basis*, Contribution of Working Group I to the Fourth Assessment Report of the Intergovernmental Panel on Climate Change, Cambridge University Press, Cambridge/New York, 2007; J. S. McLachlan, J. J. Hellmann, and M. W. Schwartz, A framework for a debate of assisted migration in an era of climate change, *Conserv. Biol.*, 21:297–302, 2007; C. Parmesan, Ecological and evolutionary responses to recent climate change, *Annu. Rev. Ecol. Systematics*, 37:637–669, 2006; P. J. Seddon, D. P. Armstrong, and R. F. Maloney, Developing the science of reintroduction biology, *Conserv. Biol.*, 21:303–312, 2007.

Audio compression

The role of audio compression or audio coding has become ubiquitous, as most people are exposed to the benefit of audio compression almost every day without even realizing it. Audio compression is used for MP3 players, satellite radio, Internet sites with streaming music, ringtones on mobile devices, digital video disk (DVD) movies, and high-definition television (HDTV).

Audio compression originated in the 1980s with the motivation of reducing the bit-rate requirement for compact disk (CD) audio without sacrificing its quality. The CD format with the bit-rate requirement of 1.44 Mbits/s [two-channel, 16-bit, PCM (pulse-code modulation) audio samples, stored at a sampling rate of 44.1 kHz] had previously emerged as the standard for high-fidelity consumer music. During the early days of audio compression, the goal was to achieve 4-fold to 12-fold compression (bit rates of 128–760 kbps) with what was termed transparent audio quality. The term "transparent audio quality" roughly implies that a listener will perceive either no difference between the original and the compressed audio or only a very slight nonannoying difference, so that the overall listening experience will be identical to that of listening to the original.

Even in those early days, it was clear that audio compression had some fundamental differences in comparison to the already established area of speech compression. It was realized that, unlike its speech counterpart, audio compression can rely less on signal production models because of the diversity of audio sounds, such as musical instruments, vocal singing or other human sounds, and the chirping of birds. As a result, more elaborate models for the perception of the audio signal were used to take advantage of the so-called masking phenomenon. It was noticed that because of the limited resolution of hearing, the signal contains perceptually irrelevant information, and this circumstance can be exploited to shape the quantization noise so that it is masked by the signal. Hence, the field of perceptual audio coding was born, and masking models came into use. Over time, many new applications have emerged, with increased demand for higher compression efficiency. In fact, there appears to be a proliferation of applications demanding CD-quality stereo at bit rates of 48 kbps and lower and high-quality frequency modulation (FM) grade mono/stereo at bit rates of about 20–24 kbps. These, in turn, continue to spur the demand for newer tools for audio coding. This article will take a closer look at some of the key tools used in audio compression algorithms, popularly known as audio codecs. It will also take a brief look at the available audio codecs and some of the standardization activities in recent times,

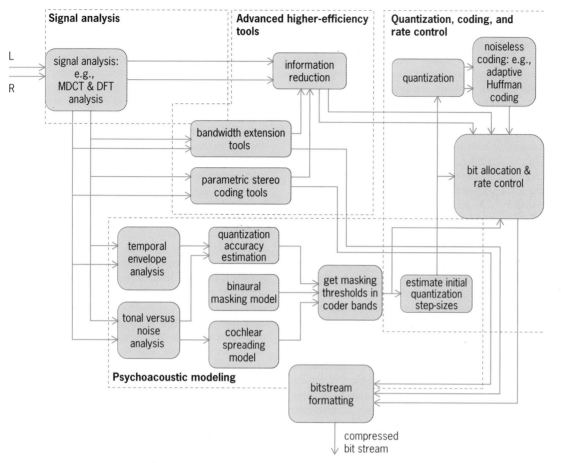

Fig. 1. Overview of a stereo perceptual audio coder.

such as those of the Moving Picture Experts Group (MPEG).

Overview of a perceptual audio coder. Figure 1 shows the high-level structure of a typical audio compression algorithm. Usually, the encoder is substantially more complex than the decoder. There are four key functional areas in the encoder: (1) signal analysis tools, (2) psychoacoustic modeling, (3) enhanced coding efficiency tools, and (4) quantization, coding, and rate-control mechanisms.

Signal analysis tools. The role of signal analysis tools is twofold. First, they are used to provide a compact representation of the signal. Second, they are used to transform the signal into a domain in which the principles of psychoacoustics can be easily applied. Although analysis tools for signal representation and psychoacoustic modeling may differ, typically it is found advantageous to use identical tools for both purposes. Frequency-domain analysis tools known as filterbanks (or overlapped transforms) have emerged as the primary choice for representation in audio compression. These typically yield a compact representation for audio signals, which tend to be rich in tonal components or harmonics. Furthermore, since the human ear acts as a frequency analysis tool, the concepts of perceptual masking are best used in the frequency domain. Specifically, the modified discrete cosine transform (MDCT) has emerged as the

filterbank of choice. It is similar to the better known discrete frequency transform (DFT), but it has important differences, such as 50% overlap between consecutive analysis windows and critical sampling, which implies that analysis over a $2N$-sample block yields N transform samples.

Psychoacoustic modeling. The centerpiece of psychoacoustic or perceptual modeling is the concept of auditory masking. The goal is to quantize the audio signal in such a way that the quantization noise is either fully masked or rendered less annoying through masking by the audio signal. Building of a perception model in an audio codec typically involves the use of (1) simultaneous masking, (2) temporal masking, (3) frequency spread of masking, and (4) asymmetry of tone versus noise masker. While simultaneous masking is a phenomenon whereby a masker masks the perception of a "maskee" at the same time, temporal masking refers to a different mechanism in which a masker masks a maskee prior to or after its occurrence. Frequency spread of masking implies that a masker at a certain frequency has a masking potential at neighboring frequencies as well. The masking potential of a masker is strongly dependent on its tone versus noiselike nature, with a noiselike masker providing a substantially higher level of masking. These factors are used to estimate desired quantization accuracy, or signal-to-mask ratio (SMR), within carefully chosen bands of frequencies.

Advanced coding efficiency tools. To reduce the bit rate requirement, several parametric coding approaches have been proposed. Two approaches that have proven to be particularly effective are audio bandwidth extension techniques and parametric stereo coding techniques.

1. *Audio bandwidth extension.* The key idea behind audio bandwidth extension is that only a low-pass filtered version of the signal is directly coded using the conventional perceptual coding paradigm. The high-frequency portion of the spectrum is recreated at the decoder by a mapping generated from the low-frequency spectrum. Typically, an attempt is made to match the reconstructed spectrum to the original high-frequency spectrum. In practice, significant mismatch may remain between the two. However, the philosophy is that the increased naturalness of the higher-bandwidth signal compensates for any other perceived distortion in the (reconstructed) higher frequencies. One such audio bandwidth extension technique is the spectral band replication (SBR) approach, which is popular in a number of MPEG and other standards. In SBR, high frequencies are regenerated by making replicas of the baseband to achieve the desired extended bandwidth. The temporal envelope of the reconstructed higher frequencies is then shaped using independently transmitted side information. Another bandwidth extension technique is embodied in the audio bandwidth extension toolkit (ABET), which consists of tools for regenerating higher frequencies using a self-similarity model. This model involves techniques to synthesize tonal components with high accuracy and techniques to shape the time-frequency envelope of the signal. ABET also includes a blind-bandwidth extension mode that entails no side information.

2. *Parametric stereo coding.* These techniques, which create stereo (two-channel or left/right) separation from a mono (one-channel) signal using a parametric model that requires only a small number of bits, have become popular at lower bit rates. Parametric stereo coding can be used for all or part of the spectrum. In general, parametric coding techniques attempt to code interchannel (stereo image, that is, the spread of musical sounds over the front-lateral plane when using a two-speaker stereo system) localization cues using as few bits as possible. These cues include interchannel intensity or level difference (IID), interchannel time difference (ITD), and interchannel coherence or correlation (ICC) cues.

Intensity stereo, perhaps the most commonly found form of parametric stereo coding, is based solely on IID cues. Another popular format called binaural cue coding (BCC) purports to apply all these cues. Applying the ITD cues of significance below 1500 Hz is a challenging proposition and can create image instability (that is, randomly shifting sounds). Therefore, it is rarely used. Generally, a combination of IID and some method to create incoherence or diffuseness between left and right channels (so that stereo sounds more open) is sufficient.

Quantization, coding, bit allocation, and rate control. Another important functional block in an audio codec consists of bit allocation, quantization, coding, and rate-control modules. The target quantization accuracy information derived from the psychoacoustic model is used to quantize the frequency representation of the signal. Quantization is typically followed by a round of noiseless coding, such as adaptive Huffman coding, for further compression. The use of noiseless coding implies that perceptual audio coders are inherently variable bit rate in nature. For constant-bit-rate applications, a combination of bit allocation and rate-control mechanisms is used. The role of bit allocation is to decide the bit budget for an individual coding unit, that is, a frame of audio. The rate control then ensures that the final bit consumption fits within the allocated budget. The rate-control mechanism may necessitate that quantization accuracy be worse than the initial target, a task accomplished using psychoacoustic principles.

Statistical multiplexing for multiprogram encoding. In many applications, multiple audio programs are encoded and transmitted to a receiver via a fixed-bandwidth transmission system. In satellite radio, for example, a joint encoding technique called statistical multiplexing may be used. Statistical multiplexing operates by encoding multiple channels together, so that channel capacity can be adaptively steered from one channel to the other. This improves channel use, as bandwidth is divided based on need. **Figure 2** shows that statistical multiplexing is a simple, yet powerful concept. The individual curves in Fig. 2 represent the cumulative density/distribution function (CDF) for the distribution of per-frame bit demand for a particular audio channel. The randomness of the individual curves in Fig. 2 illustrates the

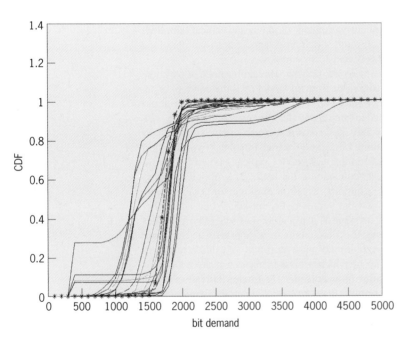

Fig. 2. Cumulative density functions (CDFs) for joint encoding and individual encoding for randomly chosen audio channels, illustrating the principle of statistical multiplexing. As expected, there are wide variations between the CDF curves for bit demands of individual channels. On the other hand, the smoothness of the CDF curve for average bit demand (marked with *) illustrates the statistical multiplexing gain, which was computed to be about 13 dB for this combination of audio channels.

difficult tasks encountered in deciding what capacity or bit rate should be allocated to an audio channel to ensure consistent quality. It is obvious that a suboptimal level of capacity allocation is almost guaranteed, as one is forced to take a somewhat conservative approach to ensure quality. The smooth curve (* symbols) is the CDF of the average of the individual bit demands. The smoothness of this curve illustrates how a fixed capacity can be more easily allocated to a cluster of jointly encoded audio channels while maintaining consistent quality. Statistical multiplexing is emerging as a powerful tool for use in multiprogram broadcast systems.

Multichannel and parametric multichannel coding. With the proliferation of home theater systems, there is an increased interest in multichannel audio coding. The straightforward approach to multichannel audio coding is to code the channels as multiple stereo pairs (for example, a front pair, a surround pair, and a center channel), or to use enhanced interchannel prediction. High-quality compression schemes using this approach operate in the range of 192–320 kbps. Recently, there has been much interest in parametric multichannel audio coding. These techniques, which may be viewed as an extension of parametric stereo coding techniques to a five-channel scenario, create five-channel audio from a stereo downmix using only a small number of bits (0–16 kbps). Schemes based on BCC are used in MP3 surround and MPEG spatial audio coding standards. A newer technique, called the immersive sound-field rendition (ISR) system, is based on the idea that the spatial localization audio cues are accurately reproduced with the aid of a multiband temporal envelope generated on a detailed time-frequency grid. ISR places emphasis on the creation of acoustic diversity between the front and the surround channels.

Audio compression formats and standardization. A number of mature audio coding technologies now exist. These include proprietary and standard-based schemes. MPEG has been active in the area of standardizing audio compression schemes since the 1980s, and MPEG-1, MPEG-2, and MPEG-4 standards exist. Audio compression schemes have also been standardized as part of advanced mobile standards, such as 3GPP2, and digital broadcast standards. The ubiquitous MP3 format is actually the MPEG-1 Layer III audio compression standard and is widespread in Internet music and portable players. The advanced audio coder (AAC) is increasingly used as an MPEG-4 standard, as are its slight variants, as used in iPods. MPEG-4 specifies the so-called high-efficiency profile of AAC, which includes SBR in combination with AAC and parametric stereo coding. The Dolby AC-3 compression standard is used for DVD audio and HDTV broadcast in the United States. The perceptual audio coder (PAC) is used in the Sirius Satellite Radio Broadcast System.

For background information *see* ACOUSTIC NOISE; ACOUSTIC SIGNAL PROCESSING; ACOUSTICS; BANDWIDTH REQUIREMENTS (COMMUNICATIONS); DATA COMPRESSION; ELECTRIC FILTER; INTEGRAL TRANSFORM; MASKING OF SOUND; MULTIPLEXING AND MULTIPLE ACCESS; PSYCHOACOUSTICS; SIGNAL PROCESSING; SOUND; SOUND-REPRODUCING SYSTEMS in the McGraw-Hill Encyclopedia of Science & Technology.

Deepen Sinha

Bibliography. J. Blauert, *Spatial Hearing*, rev. ed., MIT Press, Boston, 1996; International Organization for Standardization (ISO)/International Electrotechnical Commission (IEC), *Information Technology: Coding of Audio-Visual Objects, Part 3: Audio*, ISO/IEC 14496-3:2005 (MPEG-4 Audio), 2005; B. C. J. Moore, *An Introduction to the Psychology of Hearing,* 5th ed., Academic Press, New York, 2003; K. C. Pohlmann, *Principles of Digital Audio*, 5th ed., McGraw-Hill, New York, 2005; J. Princen and A. Bradley, Analysis/synthesis filter bank design based on time domain aliasing cancellation, *IEEE Trans. Acoust. Speech Signal. Proc.*, ASSP-34(5):1153–1161, 1986.

Auxetic materials

The word auxetic was coined by Ken Evans of the University of Exeter (UK) and derives from the Greek word *auxetos*, referring to something "that may be increased in size." It is used in science and engineering to describe materials that expand laterally when stretched, thus defying everyday experience, which suggests that the contrary should happen. These materials, commonly known as auxetics, are characterized by their negative Poisson's ratio (a property that is usually positive and describes the extent by which a material contracts when uniaxially stretched, **Fig. 1**).

Negative Poisson's ratios were first reported in the first half of the twentieth century in single crystals of iron pyrites by Woldemar Voigt. The phenomenon was attributed to twinning defects and given very little importance. For decades there was little advancement in the science of auxetics. In fact, one finds very few isolated references to this property in the scientific literature. All this changed in the 1980s, particularly following the pioneering work of researchers such as Ken Evans and his group in Exeter with their work on auxetic polymers, Rod Lakes from the University of Wisconsin, who manufactured the first purposely made auxetic material (a foam), and Krzysztof Wojciechowski of the Polish Academy of Sciences, who developed some of the earlier models on auxetics. These major breakthroughs paved the way for other studies on similar materials. This led to a clearer understanding of the requirements for materials to be defined as auxetic, as well as of the several useful properties that characterize such materials.

The available knowledge about auxetics has grown considerably over the last 20 years, and negative Poisson's ratios have been predicted, discovered, or deliberately introduced in several classes of naturally occurring and synthetic materials, including foams, nanostructured and liquid-crystal polymers, composites, metals, alloys, silicates, and zeolites. It was found that in all of these cases, this unusual property can be described in terms of models based

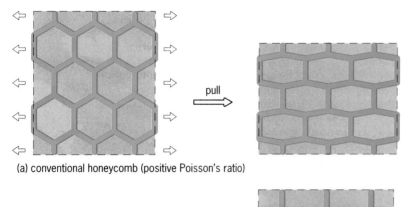

(a) conventional honeycomb (positive Poisson's ratio)

(b) auxetic honeycomb (negative Poisson's ratio)

Fig. 1. **Conventional versus auxetic materials. When a honeycomb structure deforming through hinging is stretched, it exhibits (a) conventional behavior if constructed from regular hexagon cells and (b) auxetic behavior if constructed from re-entrant cells. (D. Attard, University of Malta)**

on the geometric features present in the material's internal structure and the way these deform in response to applied loads. Such deformations can take place at any scale, ranging from the molecular to the macrolevel (in the latter case, the auxetic system is actually a structure rather than a material). The study of auxetics has developed in a manner that focuses on mechanistic models. Among the first to be studied were the hexagonal honeycomb structures, deform-

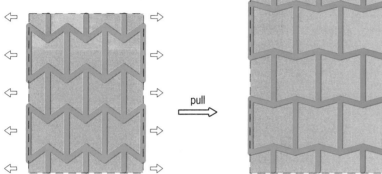

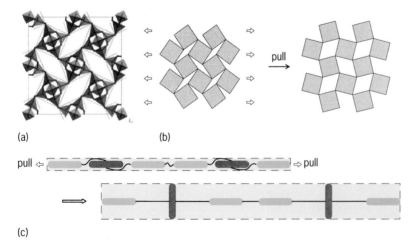

Fig. 2. **Examples of auxetics. (a) The crystal structure of natrolite showing the tetrahedral three-dimensional arrangement of the atoms within the unit cell and the "rotating squares" configuration when viewed in the xy plane. (b) An idealized representation of the "rotating rigid squares" model. (c) A schematic illustration of the deformation mechanism that gives rise to auxeticity in liquid-crystal polymers. (D. Attard, University of Malta)**

ing through flexure or hinging, which can be made auxetic if the shape of the honeycomb is changed in such a way that the Y-shaped joints normally found in honeycombs are transformed into arrow-shaped joints (that is, the honeycomb becomes re-entrant, Fig. 1). This model, together with other, more complex models that are based on the same principles, can be used to explain the behavior of auxetic open-cell foams. These are auxetic materials that are manufactured from conventional foams through a process involving compression of the original materials to around two-thirds of their original volume, heating them in the compressed form to just above their softening temperature, and then leaving them to cool in the compressed state.

Another mechanism that is commonly associated with auxetics involves rotation of rigid or semirigid units. For example, the experimentally measured auxetic behavior in the zeolite, natrolite, has been described through a "rotating squares" model developed by Joseph Grima of the University of Malta and coworkers, where the squares represent the two-dimensional (2D) projection of the natrolite crystal structure in the auxetic (001) plane (**Fig. 2a**). Similarly, the experimentally measured auxetic behavior in the silicate α-cristobalite has been described through models involving rotating silica tetrahedra and in terms of "rotating rectangles," which once again represent 2D projections of the cristobalite crystal.

Considerable effort is also being directed toward synthesizing auxetic materials, in which the auxetic effect results from molecular-level deformations. Some very interesting results have been achieved with liquid-crystal polymers, some of which have been reported to be auxetic. These auxetic polymers are composed of rigid rodlike units connected to each other by a flexible chain. Most of the rigid units are terminally connected, but around one in five is laterally connected. In the undeformed state, the laterally attached units align themselves parallel to the main polymer chain. The units form a highly ordered and dense polymer. However, when such polymers are subjected to a tensile stress (stretching), the flexible units extend, and in so doing, force the laterally attached rods to rotate and reorient themselves in a direction perpendicular to that of the main chain. This causes neighboring chains to be pushed apart so that the polymer expands laterally, with the result that auxetic behavior is observed (Fig. 2b).

Another achievement in the manufacture of auxetics was recently made by the Alderson group at the University of Bolton (UK). They developed a novel method in the production of auxetic fibers, a process they adapted from conventional melt extrusion techniques. These fibers can be used in the manufacture of fiber-reinforced composite materials, in which the fibers are more difficult to pull out than conventional ones, thus resulting in high-performance composite materials, which are significantly superior to existing composites.

Properties and applications. The ability of auxetic materials to expand laterally when uniaxially stretched and to contract under compression gives rise to several interesting applications in which significant use is made of their ability to surpass the properties of conventional materials. For example, auxetic materials are ideal to use in press-fit fasteners, because the compressive force used to insert the fastener into a hole causes a lateral contraction, thereby facilitating insertion. Trying to pull a fastener back out will cause the material to expand within the hole, making its emergence more difficult if some heavy object is attached to it. Similarly, auxetic materials can be used in gaskets where, as opposed to conventional materials that are squeezed out when pressed between the flanges of two pipes, they will contract to yield an even tighter fit. They also can be used in seat belts to reduce the pressure exerted upon impact by becoming wider when the passenger is thrown forward during a collision, thereby enhancing the restraining force.

Auxetics have been shown to be useful in the manufacture of composites and systems that have an overall Poisson's ratio of zero. For example, Mitsubishi has patented a composite bullet having an auxetic component, such that the overall Poisson's ratio is zero. Consequently, the pressure exerted by the trigger does not cause radial expansion, ensuring that friction at the walls of the barrel is kept at a minimum.

Auxetics are useful in applications other than those that require the property of becoming thicker when stretched. This is because a negative Poisson's ratio imparts several additional enhanced characteristics in materials, which arise as an indirect consequence of auxeticity. For example, auxetic materials benefit from an increased shear modulus, high-plane fracture toughness, and increased indentation resistance, and have a natural ability to form dome-shaped surfaces (synclastic behavior) as opposed to conventional materials, which tend to take on a saddle-shaped configuration (anticlastic behavior) when bent. These latter two properties are particularly desirable in foams used in the manufacture of seats, packing, and safety equipment. It has been suggested that this observed resistance to indentation arises from the geometric features of the foam's microstructure, which can be trivially described in terms of honeycomb models. As illustrated in **Fig. 3**, during events of large compressive stresses, conventional honeycomb structures buckle and eventually fail to provide a cushioning effect. By contrast, reentrant honeycombs densify at locations under the indenter as the cells fold up and undergo elastic collapse with no rib failure, so that upon removing the load, the honeycomb returns to its original shape. The extent of densification is strain-dependent, allowing for the impact of a sudden blow to be reduced in a gradual manner. Furthermore, the ability to form dome-shaped surfaces makes auxetic foams better for the manufacture of cushions and mattresses to accommodate body curvatures in a more versatile way, allowing for enhanced comfort. Auxetic composites also benefit significantly from these properties and are ideal for the manufacture of curved body parts

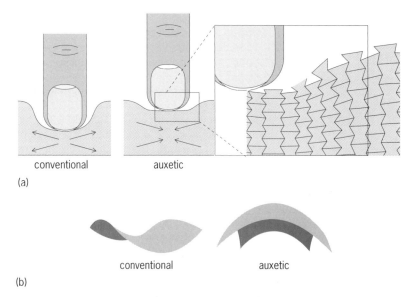

(a)

(b)

Fig. 3. Enhanced characteristics of auxetics. (*a*) Indentation in conventional materials causes the material to flow from beneath the load, where as in auxetic materials the material flows to the site of impact through buckling of reentrant cells. (*b*) Auxetic materials can dome in a synclastic manner, a property that is difficult to achieve with conventional materials, which bend in an anticlastic manner. (*D. Attard, University of Malta*)

used in the construction of aircraft, naval vessels, and automobiles.

It has also been shown that auxetic materials are better at absorbing sound waves and vibrations, and therefore may be used for acoustical damping in soundproofing systems and as shock absorbers; for example, in car bumpers to dampen impact energy or in buildings in areas susceptible to earthquakes.

Another interesting application of auxetics is in the field of smart, adjustable filters, where the pore size can be set at a particular dimension by applying a strain across the material. Such filters can be easily cleaned simply by applying a stress to open up the pores, allowing any residue to pass through. This property can also lead to some other very interesting applications, such as the manufacture of smart medical dressings made from an auxetic porous material impregnated with medication (**Fig. 4**). When a

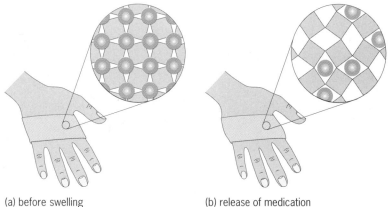

(a) before swelling

(b) release of medication upon swelling

Fig. 4. Applications of auxetics. Auxetic materials can be used in smart bandages. The bandage, impregnated with medication, is (*a*) applied to the wound; and (*b*) as the wound swells, the bandage stretches and the pores within open up to release the medication. (*D. Attard, University of Malta*)

patient experiences swelling, the dressing will stretch to release the medication in a proportion related to the extent of swelling. And as the swelling subsides, the dressing no longer remains in the stretched state and the medication is no longer released, thus minimizing the exposure of the patient to unnecessary medication.

Outlook. It is clear that auxetic materials offer numerous benefits in the manufacture of a range of products and materials. In the context of the contribution of auxetic materials to improved lifestyle, well-being, and production techniques, the expectation is that scientists and engineers working in the field of auxetics will be able to maintain and increase the momentum of research and development. With the growing demand for improved structural and functional materials, auxetics will likely come to shape everyday products and contemporary living.

For background information *see* COMPOSITE MATERIALS; CRYSTAL STRUCTURE; FOAM; METAL, MECHANICAL PROPERTIESN OF; NATROLITE; STRESS AND STRAIN; ZEOLITE in the McGraw-Hill Encyclopedia of Science & Technology.   Joseph N. Grima

Bibliography. A. Alderson, A triumph of lateral thought, *Chem. Ind.*, May, pp. 384–391, 1999; R. H. Baughman, Avoiding the shrink, *Nature*, 425:667–667, 2003; K. E. Evans and A. Alderson, Auxetic materials: Functional materials and structures from lateral thinking, *Adv. Mater.*, 12:617–628, 2000; R. Lakes, Deformation mechanisms in negative Poisson ratio materials—Structural aspects, *J. of Mater. Sci.*, 26, 2287–2292, 1991; W. Yang et al., On auxetic materials, *J. of Mater. Sci.*, 39:3269–3279, 2004.

Barge technology

The U.S. tug and barge fleet has grown to nearly 4000 modern tugboats and towboats and more than 27,000 barges. The industry moves over 800 million tons (725 million metric tons) of raw materials and finished goods each year. A recent study by the Texas Transportation Institute and Texas A&M University compared inland waterway capacity to other transport modes (truck and rail) and noted that the inland waterway alone accounts for "about 624 million tons of waterborne cargo transit ... a volume equal to about 14% of all intercity freight. This commerce has an overall value of about $70 billion, substantially contributing to America's economic strength." The study concluded that inland waterways "have lots of unused capacity—capacity which could be used to relieve growing transportation congestion, with lesser impacts on air quality, public safety and the environment than the other modes."

Regulatory, physical, and economic constraints influence the development of barges and barge transportation technology. Inland and offshore barges have differing constraints, and these have resulted in differing technological developments.

Barges carry a wide variety of liquid and dry cargoes, and are often specially designed and fitted with cargo-handling systems for a particular product. Tank barges carry liquids in internal compartments and have installed pumping systems for loading and unloading petroleum or chemical liquid cargos. Hopper barges carry dry bulk cargoes, especially on inland waters. Hoppers with weather-tight hatch covers carry perishable cargoes such as grains or fertilizers. Open hoppers carry bulk commodities, for example, coal, aggregates, or wood chips. Flat-deck cargo barges carry packaged cargos, such as freight containers or bundled building materials, and are loaded by crane or forklift truck, or cargo driven aboard (**Fig. 1**).

Physical constraints. Inland systems are, of course, subject to more physical constraints than coastwise systems. These constraints include the water depth, and the width and length of navigation locks. The largest inland system is on the Mississippi and Ohio Rivers, and has developed a voluntary industry-standard barge size that allows barges from many different companies to be rafted together. Standard length is 195 ft (59.4 m), width is 35 ft (10.7 m), and depth from the bottom to the deck is usually 12 ft (3.7 m). Operating draft in some seasons is limited to 8 ft (2.5 m). The more modern navigation locks have inside dimensions of 1200 by 110 ft (366 by 33.5 m). This allows a raft or "tow" of 15 barges, 3 wide by 5 long, plus a towboat to lock through together (**Fig. 2**). Some of the older locks on the Ohio River are 600 by 110 ft (183 by 33.5 m), so that larger tows need to be broken into two units to pass through. On the uppermost part of the Mississippi, older locks are only 56 ft (17 m) wide.

Other river systems have developed different standards. For instance, the first major lock and dam on the Columbia River was Bonneville, completed in 1937. A second, larger lock was completed there in 1993, so that all of the Columbia and Snake River locks are now 675 by 86 ft (206 by 26 m) with a controlling water depth of 12 ft (3.7 m). A standard

Fig. 1. Liner barge carrying highway trailers to Puerto Rico. (*Courtesy of Crowley Maritime Corporation*)

Fig. 2. Large river barge tow exiting a navigation lock.

barge size on this river system is 252 by 42 by 14 ft (77 by 12.8 by 4.3 m).

The Panama Canal locks limit larger ocean barges to a width of 105 ft (32 m). A widening project has begun that will have new locks 150 ft (46 m) wide, but they will not be open to traffic until at least 2015.

Resistance and propulsion. Resistance of inland barge tows to movement through the water would be difficult to improve on because of the limited water depth and the need to move maximum tonnage within lock size dimensions. These limitations lead to nearly boxlike shapes. There have been advances in river towboat propulsive efficiency in the last half-century, such as propeller tunnels recessed into the hull that permit larger-diameter, slower turning propellers for greater efficiency. Propeller nozzles, rings that surround the propeller tips, also increase efficiency. However, it seems unlikely that there are opportunities for any great improvements in inland barge propulsion or fuel consumption rates. As it is, fuel consumption per ton of cargo per mile is lower for barges than for competing rail or truck transportation modes (see **table**; similar figures are not available for ocean towing.)

Ocean barges have undergone improvements as well. The traditional propulsion method has been that of towing by a long wire towline, which isolates the different motions of the tug and barge and prevents shock loads in the wire. Often a section of large

Summary of fuel efficiency		
Mode	Ton-miles/ gallon	Metric ton-kilometers/ liter
Inland towing	576	222
Railroads	413	159
Truck	155	60

SOURCE: From Center for Ports and Waterways, Texas Transportation Institute, *A Modal Comparison of Domestic Freight Transportation Effects on the General Public*, 2007.

chain, called a surge chain, is added at the connection to the barge to further reduce shock. All of this gear in the water adds to the overall resistance.

Towed barges have a tendency to be directionally unstable, especially when deeply loaded; that is, they will take large off-track excursions to one side and then to the other. This instability not only increases resistance, it poses navigation safety risks. Various types of skegs or foil-type devices are fitted at the stern of the barge to provide directional stability. Most types of skegs also increase resistance, often by as much as 20%. Considerable research has been done on the resistance and directional stability of towed barges.

The resistance of the tugboat itself is significant as well, because it is moving through the water relatively quickly for its waterline length, which increases wave-making resistance. All of these factors combine to make towed barges rather inefficient to move through the water compared to self-propelled ships. *See* TUGBOATS.

Tug–barge connections. Another design development in the last few decades has been alternative connection devices that allow a tug to push a barge in rough seas. Although these devices are proprietary, they can be grouped into two categories: articulated tug–barge units (ATBs) and integrated tug–barge units (ITBs).

Articulated units involve a connection, such as pins on the side of the tug that engage holes at the stern of the barge, so that at least one degree of freedom is unrestrained, usually pitch. As a minimum, fore-and-aft relative motion (surge) should be restrained to transmit propelling force, as should lateral rotation (yaw) to provide steering control. Usually the tug is engaged in a deep notch at the stern of the barge (**Fig. 3**). The flow of water off the

Fig. 3. Coastwise articulated tug–barge unit carrying petroleum. (*Courtesy of Crowley Maritime Corporation*)

stern of the barge and around the tug is made as clean a possible, so that the propeller does not operate in turbulent water.

Integrated tug–barge units have all six degrees of freedom restrained. They generally engage sloping or wedged surfaces, and the tug is firmly latched into place.

In normal operation these units function more like a ship than a tug and barge. Nevertheless, the U.S. Coast Guard regulators will treat such units as a tug and barge rather than a ship, with regard to crew levels and safety equipment, provided that they meet two tests: they must be able to disengage quickly in an emergency, and the tug must be stable and seaworthy on its own.

Most of the ATB and ITB units in service are liquid-petroleum carriers, because the cargo is carried below deck and is quite dense, making it is easier to achieve good visibility from the tug's wheelhouse. Lower-density cargos carried on deck, such as freight containers, are still carried mostly on towed barges.

Regulatory environment. Ocean-going ships and inland steamboats have a long history of government regulation in response to catastrophic accidents that occurred early in their development. Regulations govern crew size, crew qualifications, and life-saving equipment, as well as many other technical aspects of design, construction, and operation. Though well intended, extensive regulation is slow to respond to technical advances and may stifle innovation.

The industry of modern, diesel-propelled, coastwise tugboats and transport barges developed rather quickly in the mid-twentieth century; consequently, the industry flourished in an environment of minimal regulation. This new industry established a good safety record and escaped the scrutiny of regulators. In addition, regulations that have developed more recently recognize technical advances such as engine room automation and advanced life-saving equipment. This regulatory environment did not focus on tug crew levels, which led to much smaller crew sizes than on ships. Similarly, on inland waterways, following a number of early boiler casualties, regulations focused on steam plant safety. In contrast, diesel-driven vessels remain unregulated.

Structural design codes. The American Bureau of Shipping (ABS) publishes the most commonly used design code for barges in the United States. The ABS is an independent nonprofit institution, originally formed to protect the interests of marine insurance underwriters. It reviews plans for conformance to ABS rules and inspects vessels during construction. The organization also performs periodic safety inspections of vessels during their life and has been delegated certain regulatory authority by the U.S. government—for example, enforcing certain international treaty requirements covering the design and construction of vessels. In the case of larger barges or unusual structural features, the ABS may also require first-principles engineering such as finite-element analysis (FEA) before approving a de-

sign. Many other maritime nations have similar organizations, called classification societies, for example: Lloyds Register in Great Britain, Germanischer Lloyd in Germany, Det Norske Veritas in Norway, Bureau Veritas in France, the Russian Register of Shipping, and Nippon Kaiji Kyokai in Japan.

Economic influences. Coastwise transportation systems hinge on three principal cost categories: capital costs, consumables, and crew costs. Fuel is the principal consumable cost, followed by lubricating oil and parts. Crew costs, in addition to wages, include benefits, wage-related insurance, food, and transportation.

Following World War II, fuel costs were low and crew costs were rising. Surplus hulls such as LST (landing ship tank) landing craft were readily available, as were army and navy surplus tugboats. Consequently, the capital costs of entering a tug and barge business were relatively low. With the much smaller crew sizes allowed on tugboats, coastwise tugs and barges overtook coastwise shipping, especially on the west coast of the United States.

Trends. As previously stated, inland water transportation is quite efficient and standardized, and no dramatic new developments are foreseen. Revisiting the cost categories discussed above (capital, crew, and fuel), it is clear that today's rising fuel prices work against the economy of towed coastwise barges. ATB and ITB solutions offset some of the disadvantages of towed barges for some higher-density cargos. However, many now think that with such high fuel costs, and with only modest rationalization of the crew-level regulations, self-propelled ships may return to U.S. coastal trade routes.

For background information, *see* BOAT PROPULSION; INLAND WATERWAYS TRANSPORTATION; MARINE ENGINE; PROPELLER (MARINE CRAFT); SHIP POWERING, MANEUVERING, AND SEAKEEPING in the McGraw-Hill Encyclopedia of Science & Technology.

Thomas Bringloe

Bibliography. American Bureau of Shipping, *Rules for Building and Classing Steel Barges*, 2003; Center for Ports and Waterways, Texas Transportation Institute, *A Modal Comparison of Domestic Freight Transportation Effects on the General Public*, 2007.

Bioethanol production

Ethanol is commonly referred to as bioethanol when it is manufactured from agricultural sources, for example, corn or wood. The production of bioethanol from biomass is a proven industrial process for producing fuel from a renewable source. It can be directly mixed with gasoline (petrol) and used in today's automotive vehicles, or used as a fuel for the generation of electricity. Currently, there are two types of blends of ethanol and gasoline on the market: E10, which is 10% ethanol and 90% gasoline, and E85, which is 85% ethanol and 15% gasoline (E85). In the United States, many states currently mandate E10. It is generally accepted that bioethanol gives a 70% carbon dioxide reduction (compared to unblended gasoline), which means 7% in an E10 blend or 50% in an E85 blend. Increased bioethanol usage could reduce U.S. greenhouse gas emissions to 1.7 billion tons/year (22% of the 2002 emissions). Recent investigations have established that polysaccharides in biomass can be hydrolyzed enzymatically into glucose sugar that can be fermented to bioethanol. Fungi, including yeast, are a key source for some of the industrially important enzymes used in this process.

Bioethanol. Theoretically, bioethanol production should be able to yield 0.5 g of ethanol per gram of raw biomass, which translates into an energy recovery of approximately 90%. Corn is a common substrate for bioethanol manufacture because the process is relatively free of technical obstacles. Microorganisms are involved in the transformation of corn to bioethanol in two ways: they catalyze the hydrolysis of starches using amylases and amyloglucosidases, and they ferment the resulting sugars to bioethanol. The fermentation step is generally carried out by yeast, but certain strains of bacteria, including *Zymomonas mobilis* and recombinant strains of *Escherichia coli* and *Klebsiella oxytoca*, are also capable of producing high yields of bioethanol.

Bioethanol can be derived from renewable sources besides corn, typically plant feedstocks such as wheat, sugar beets, straw, and wood. It is possible that even household wastes may be economically converted to bioethanol. **Table 1** shows ethanol production in different continents. **Table 2** shows the top 10 bioethanol producers.

Because bioethanol is a liquid, it fits into the current fuel infrastructure, although its transport requires special handling to prevent water accumulation. Furthermore, bioethanol as a fuel additive or substitute can be cost-competitive with petroleum. However, using solid substrates in converting lignocellulose (any of a group of substances in woody plant cells consisting of cellulose and lignin) to sugars poses a dilemma in bioethanol production. Cellulose and lignocellulose are in much greater supply than starch and sugars and therefore are preferred substrates for ethanol production, yet producing ethanol from cellulose and lignocellulose is comparatively difficult and expensive.

Lignocellulose is obtained from such diverse sources as switchgrass, cornstalks, and wood chips. Fungal enzymes and fermentative yeasts are then used to transform lignocellulose first to sugars and then to bioethanol. The low lignocellulose reactivity limits production. The expense of enzyme production is the biggest economic barrier to lignocellulose conversion to bioethanol.

Use of enzymes. Biological pretreatments use fungi to solubilize the lignin. Biodelignification is the biological degradation of lignin by microorganisms; it was mentioned in the 1980s as being possibly useful in the future, although at that time production was insufficient and expensive, it required a long process time, and the microorganisms were poisoned by lignin derivatives. Newer technologies employing biodelignification could greatly simplify pretreatment, but so far the rates are slow, yields are low, and little experience with such approaches has been developed.

Recent research has focused on enzyme catalysts called cellulases that can attack the cellulose chains more efficiently, leading to very high yields of fermentable sugars. Fungal cellulases and β-glucosidases produced in separate aerobic reactors can be extracted in very high yields; however, because these enzymes have low specific activities, they must be used in large quantities to achieve lignocellulose conversion.

Fungi produce a plethora of enzymes that are used to degrade complex polysaccharides and proteins into simpler sugars and amino acids. They have long been established as a key source of a wide variety of industrially important enzymes. Some of these enzymes have already been harnessed in releasing fermentable sugars from a variety of biomass feedstocks, including wastepaper, foodstuffs, cereals, sugar crops, grains, and woods.

Both bacteria and fungi can produce cellulases for the hydrolysis of lignocellulosic materials. These microorganisms can be aerobic or anaerobic, mesophilic [growing best at moderate temperature, neither too hot nor too cold, typically 25–40°C (77–104°F)] or thermophilic [growing best at high temperatures, typically 40.5–50°C (105–122°F)]. Bacteria belonging to *Clostridium, Cellulomonas, Bacillus, Thermomonospora, Ruminococcus, Bacteroides, Erwinia, Acetivibrio, Microbispora,* and *Streptomyces* can produce cellulases.

The widely accepted mechanism for enzymatic cellulose hydrolysis involves synergistic actions by endoglucanases or endo-1,4-β-glucanases (EGs), exoglucanases or cellobiohydrolases (CBHs), and β-glucosidases (BGLs). EGs play an important role in cellulose hydrolysis by cleaving cellulose chains randomly and thus encouraging strong

TABLE 1. Ethanol production in different continents (billion liters/year)

North and South America	Asia	Europe	Africa	Oceania
22.3	5.7	4.6	0.5	0.2

TABLE 2. Top 10 bioethanol producers (billion liters/year)

Country	2004	2005	2006
United States	13.40	16.13	18.36
Brazil	15.10	16.01	17.00
China	3.63	3.79	3.86
India	1.74	1.70	1.89
France	0.83	0.90	0.95
Germany	0.26	0.42	0.76
Russia	0.76	0.76	0.64
Canada	0.23	0.23	0.57
South Africa	0.42	0.38	0.38
Thailand	0.26	0.30	0.34

degradation. They hydrolyze accessible intramolecular 1,4-β-glucosidic bonds of cellulose chains randomly to produce new chain ends. Exoglucanases cleave cellulose chains at the ends to release soluble cellobiose or glucose in a processive manner (that is, after a cellobiose/glucose residue is cleaved from the cellulose chain, the enzyme will not dissociate from the substrate but continues to hydrolyze the same chain through many cycles), while BGLs hydrolyze cellobiose to glucose in order to eliminate cellobiose inhibition. BGLs complete the hydrolysis process by catalyzing the hydrolysis of cellobiose to glucose.

Filamentous fungi are the major source of cellulases and hemicellulases. Mutant strains of *Trichoderma* sp. (*T. viride, T. reesei, T. longibrachiatum*) have long been considered to be the most productive and powerful destroyers of crystalline cellulose. CBH I and CBH II are the major *T. reesei* enzymes; the content of CBH I comprises up to 60% of the total cellulolytic protein, whereas the content of CBH II is about 20%. Similarly, EG I and EG II are the dominant EGs in *T. reesei*, and presumably act as important partners to CBH I in nature. Such protein yields are comparable to or exceed the respective parameters for the best *Trichoderma* sp. strains (35–40 g/L). Yeast and fungi also tolerate a pH range of 3.5–5.0. The ability to lower pH below 4.0 (that is, to levels where bacteria rarely grow) offers a method for present operators using yeast in less than aseptic equipment to minimize loss due to bacterial contaminants.

Thus, fungal lignocellulolytic enzymes have been employed to convert lignocellulosic biomass to fermentable sugars for the production of bioethanol. To aid this process, wheat straw has been pretreated with dilute sulfuric acid followed by steam explosion (a pretreatment process in which the biomass is exposed to high-pressure steam, and the resulting product is then explosively discharged to an atmospheric pressure, making it more readily digestible by enzymes). In addition, several enzymatic treatments implementing hydrolases [cellulases and xylanases from *T. reesei*, recombinant feruloyl esterase (FAE) from *Aspergillus niger*, and oxidoreductases (laccases from *Pycnoporus cinnabarinus*)] have been investigated with regard to the saccharification of exploded wheat straw. A synergistic effect between cellulases, FAE, and xylanase was proven under a critical enzymatic concentration (10 U/g of cellulases, 10 U/g of FAE, and 3 U/g of xylanase). The yield of enzymatic hydrolysis was enhanced by increasing the temperature 37–50°C (98.6–122°F) and adding a nonionic surfactant (Tween 20).

Outlook. Fungal enzymes are naturally occurring plant proteins that cause certain chemical reactions to occur. As such, they can be utilized to produce good yields of bioethanol from saccharides (glucose, fructose, cellobiose, arabinose, galactose, mannose, and ribose), as well as from the simultaneous saccharification and fermentation of cellulose with hemicelluloses of plant biomass. Hence, fungal enzymatic hydrolysis and fermentation of biomass is a promising method in future applications of bioethanol production.

For background information *see* ALCOHOL FUEL; ALTERNATIVE FUELS FOR VEHICLES; BIOMASS; CELLULOSE; CORN; ENZYME; ETHYL ALCOHOL; FERMENTATION; FUNGAL BIOTECHNOLOGY; FUNGI; GASOLINE; LIGNIN; RENEWABLE RESOURCES; YEAST in the McGraw-Hill Encyclopedia of Science & Technology.

Ayhan Demirbas

Bibliography. A. V. Gusakov et al., Design of highly efficient cellulase mixtures for enzymatic hydrolysis of cellulose, *Biotechnol. Bioeng.*, 97:1028–1038, 2007; A. V. Gusakov et al., Purification, cloning and characterisation of two forms of thermostable and highly active cellobiohydrolase I (Cel7A) produced by the industrial strain of *Chrysosporium lucknowense, Enzyme Microb. Tech.*, 36:57–69, 2005; Y. Sun, Enzymatic hydrolysis of rye straw and bermudagrass for ethanol production, Ph.D. thesis, Department of Biological and Agricultural Engineering, North Carolina State University, 2002; M. G. Tabka et al., Enzymatic saccharification of wheat straw for bioethanol production by a combined cellulase xylanase and feruloyl esterase treatment, *Enzyme Microb. Tech.*, 39:897–902, 2006; P. Väljamäe, G. Pettersson, and G. Johansson, Mechanism of substrate inhibition in cellulose synergistic degradation, *Eur. J. Biochem.*, 268:4520–4526, 2001; Y. H. P. Zhang, M. E. Himmel, and J. R. Mielenz, Outlook for cellulase improvement: Screening and selection strategies, *Biotechnol. Adv.*, 24:452–481, 2006.

Buddenbrockia

Buddenbrockia plumatellae was described as a species in 1910 by O. Schröder, when he observed strange wormlike parasites in the body cavity of freshwater colonial invertebrates called bryozoans. These worms actively wriggle with sinusoidal movements inside their host and grow to 1–3 mm (0.04–0.12 in.) in length. Their body plan is extremely simple, lacking a gut, nervous system, or even external sensory structures. In fact, the worms appear to be morphologically identical from both ends. Their bryozoan hosts grow by budding new zooids, the individuals that collectively form the colony, and *Buddenbrockia* proliferates along with the growth of its host to form dense infections of worms (**Fig. 1a**).

Although *Buddenbrockia* has been described for nearly 100 years and has been observed in countries in Europe, Asia, North America, and South America, its biology and relationships to other animals have, until recently, been poorly understood due to its extreme morphological simplification. Indeed, in 2001, *Buddenbrockia* was still referred to as one of the last five enigmatic animal taxa.

General morphology. Probably as a result of parasitism, *Buddenbrockia* worms have lost many traits. Their lack of a gut and a nervous system means that there is no recognizable anterior and posterior nor dorsal and ventral. They also lack cilia and gametes (eggs and sperm). They do, however, possess four sets of longitudinal muscles that run the length of the worm and generate their characteristic sinusoidal (S-shaped) movements. One could say that *Buddenbrockia* is basically a tube lined with muscle. As the worm matures, the internal space becomes filled with cellular constituents that eventually develop into multicellular entities, which Schröder believed to be embryos. Recent studies, though, have revealed that these stages are actually infective spores and their presence is an important clue to the origin of *Buddenbrockia*.

Early speculations about phylogenetic placement. Schröder first proposed that *Buddenbrockia* was a mesozoan, a group of tiny parasites of marine invertebrates. Mesozoans possess a layer of outer cells around one or more inner reproductive cells, but they possess cilia, lack muscles, and do not develop into elongate worms. Schröder later thought that *Buddenbrockia* was related to the nematodes, a group of free-living and parasitic worms with a similar arrangement of four longitudinal muscles. However, *Buddenbrockia* lacks a number of characteristics of parasitic nematodes, such as a cuticle, a sucking pharynx, and a gut. It has also been suggested that *Buddenbrockia* might represent the sporocyst larval stages of trematodes, a group of parasitic flatworms; like mesozoans, though, sporocysts lack muscles and are not elongate. They also typically develop in mollusks.

Resolving the affinities of Buddenbrockia. Studies based on gene sequencing and electron microscopy have now solved the long-standing enigma of *Buddenbrockia*. This progress has been achieved in two steps.

The first breakthrough revealed that *Buddenbrockia* belongs to the Myxozoa, a group of extremely morphologically simplified microparasites of vertebrates and invertebrates. The Myxozoa comprises well over 1300 species and is divided into two subgroups: the Myxosporea (including the majority of species) and the Malacosporea (with only a handful of species). In 2002, analyses of ribosomal ribonucleic acid (rRNA) clearly showed that *Buddenbrockia* is closely related to sac-forming parasites of freshwater bryozoans belonging to the Malacosporea. The genes that code for rRNA are frequently used for understanding the evolutionary relationships amongst organisms. Concurrently, electron microscopic investigations revealed that *Buddenbrockia* possessed a key myxozoan trait—namely, the presence of tiny intracellular organelles called polar capsules. By focusing a beam of electrons on objects, electron microscopy is able to visualize much smaller features than can be observed by ordinary light microscopy, and it clearly revealed polar capsules in *Buddenbrockia*.

Polar capsules comprise a capsule that is formed within a cell and that contains a coiled, eversible filament. They are present in infective spores of myxozoans. Upon contact with a new host, the polar capsule filament everts and attaches the spore to the host. Amoeboid cells within the spore then invade the host to cause infection. Electron microscopy

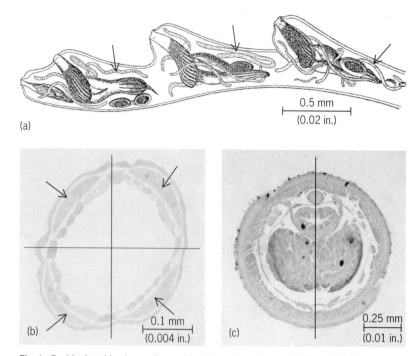

(a)

0.5 mm
(0.02 in.)

(b)   0.1 mm
             (0.004 in.)

(c)   0.25 mm
             (0.01 in.)

Fig. 1. *Buddenbrockia plumatellae* and its internal morphology. (*a*) Dense infection of *Buddenbrockia* worms (arrows) in a bryozoan colony (*original figure from* O. Schröder, Buddenbrockia plumatellae, *eine neue Mesozoenart aus* Plumatella repens L. und Pl. fungosa Pall, Z. Wiss. Zool., 96:525–537, 1910). (*b*) Cross section of *Buddenbrockia* showing four longitudinal muscle blocks (arrows) and lack of gut. The two crossed lines indicate two planes that bisect *Buddenbrockia* to make two mirror images, indicating its tetraradial symmetry. (*c*) Cross section of a relative of the earthworm, the oligochaete *Ocnerodrilus occidentalis*, showing one plane that bisects this bilaterally symmetrical worm.

also demonstrated that the multicellular bodies that Schröder believed to be embryos were in fact typical malacosporean spores, with several valve cells enclosing four cells each with intracellular polar capsules and two infective amoeboid cells.

The demonstration that *Buddenbrockia* is a myxozoan was astonishing since myxozoans were known to occur as relatively amorphous plasmodia or pseudoplasmodia with no real tissue layers or as inert sacs. Myxozoans also display the bizarre process of endogeny, whereby proliferation can occur via cells developing within cells, a trait characteristic of certain protistan parasites. Nevertheless, while this placement solved the immediate affinities of *Buddenbrockia*, it did not help to understand its status within the animal kingdom since the Myxozoa themselves were enigmatic. Their phylogenetic placement has been highly debated due to their morphological simplicity and propensity for endogeny, characteristics that had contributed to their classification as protists for many years. However, the possibility that myxozoans are multicellular animals was proposed in 1938 by R. Weill, who noted the great similarity between polar capsules in myxozoans and cnidarian nematocysts or stinging cells. The latter function by capturing prey through the similar eversion of a filament from an intracellular capsule.

The next breakthrough came via a large-scale phylogenetic analysis of 50 protein-coding genes of *Buddenbrockia*. These genes were compared

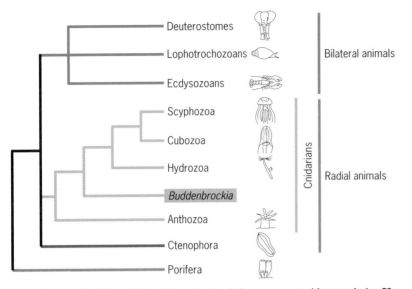

Fig. 2. Schematic phylogenetic tree based on the phylogeny recovered from analyzing 50 protein-coding gene sequences, showing the position of *Buddenbrockia* within the Cnidaria.

to a database containing 129 protein-coding gene sequences for 46 animals from across the animal kingdom along with gene sequences for 13 nonanimal taxa. This large-scale study was finally able to resolve the origins of *Buddenbrockia* and the relationship of the Myxozoa to other taxa by showing that *Buddenbrockia* is a metazoan (a multicellular animal) and confirming Weill's hypothesis that it is a member of the phylum Cnidaria. Moreover, the study showed that *Buddenbrockia* is more closely related to the Medusozoa, cnidarians that generally have a medusa (or jellyfish-like) stage in the life cycle, than to the more primitive Anthozoa, which occur only as attached polyp stages (**Fig. 2**). The medusozoans include the Scyphozoa (the jellyfish), the Hydrozoa (a diverse group containing colonial and solitary forms such as the freshwater *Hydra*), and the Cubozoa (the box jellyfish), whereas the anthozoans include the sea anemones and corals. This finding that *Buddenbrockia* is a cnidarian has demonstrated that myxozoans represent a hitherto unappreciated and spectacular radiation of cnidarians that have evolved to exploit marine, freshwater, and terrestrial hosts.

Body plans and the evolution of worms. The confirmation that *Buddenbrockia* is a cnidarian has also established a major novelty in animal body plan, and this has significant evolutionary implications. Up until this discovery, no cnidarian was known to possess a true wormlike (vermiform) shape. Such vermiform body plans had previously been regarded as exclusive to the higher animals known as the bilaterians, which possess axes specified along anterior-posterior and dorsal-ventral planes. As the name implies, bilaterians are bilaterally symmetrical and thus can only be bisected in one plane to produce two mirror images (as in, for example, vertebrates). In contrast, cnidarians and ctenophores (the comb jellies) are nonbilaterian animals characterized by radial symmetry and thus have a similar distribu-

tion of parts about a central axis. Animals with a radial symmetry can be bisected at more than one plane to produce two mirror images (as in, for example, flowerlike sea anemones). Figure 1*b* and *c* illustrates the tetraradial symmetry of *Buddenbrockia* and the bilateral symmetry of an oligochaete worm.

Traditional views about animal evolution have linked certain designs or body plans with particular lifestyles. Thus, a radial symmetry is viewed to be adaptive for animals that are attached (for example, sea anemones) or that float in the water column (for example, jellyfish). In these situations, the direction of encountering food is unpredictable and a radial deployment of food-capturing surfaces will maximize food capture. In contrast, bilateral symmetry is traditionally regarded as related to directed movement through the environment as bilateral animals are more streamlined and have a centralized nervous system at the anterior end that allows them to perceive the environment as they move through it.

A recent hypothesis, however, is that bilateral symmetry may be an adaptation to improve the efficiency of internal circulation and could have evolved prior to the origin of directed locomotion. The subtle bilateral symmetry in some attached cnidarians supports this scenario. By bucking the trend and being an active wormlike animal with radial symmetry, *Buddenbrockia* also supports this scenario. Furthermore, since *Buddenbrockia* is the only known cnidarian to possess a wormlike morphology, this appears to be a secondarily acquired rather than a primitive or basic cnidarian shape. Nevertheless, *Buddenbrockia* unequivocally demonstrates that locomotion and bilateral symmetry are not necessarily linked. Since it is now clear that animals with both bilateral and radial symmetry can evolve to become active worms, one of the most interesting questions is whether the same genes are used to produce worms in animals with very different modes of development.

For background information *see* ANIMAL EVOLUTION; ANIMAL KINGDOM; ANIMAL SYMMETRY; BILATERIA; BRYOZOA; CNIDARIA; METAZOA; PARASITOLOGY; PHYLOGENY; SYSTEMATICS in the McGraw-Hill Encyclopedia of Science & Technology.   Beth Okamura; Eva Jiménez-Guri

Bibliography. J. Finnerty, Did internal transport, rather than directed locomotion, favor the evolution of bilateral symmetry in animals?, *Bioessays*, 27:1174–1180, 2005; E. Jiménez-Guri et al., *Buddenbrockia* is a cnidarian worm, *Science*, 317:116–118, 2007; A. S. Monteiro, B. Okamura, and P. W. H. Holland, Orphan worm finds a home: *Buddenbrockia* is a myxozoan, *Mol. Biol. Evol.*, 19:968–971, 2002; C. Nielsen, *Animal Evolution: Interrelationships of the Living Phyla*, Oxford University Press, 2001; B. Okamura et al., Ultrastructure of *Buddenbrockia* identifies it as a myxozoan and verifies the bilaterian origin of the Myxozoa, *Parasitology*, 124:215–223, 2002; R. Weill, L'interpretation des Cnidosporidies et la valeur taxonomique de leur cnidome: Leur cycle comparé à la phase larvaire des Narcomeduses cuninides, *Trav. Station Zool. Wimereaux*, 13:724–744, 1938.

Childhood lead exposure and lead toxicity

Elevated blood lead levels (EBLLs) are one of the major environmental problems presently facing children in the United States and elsewhere. In 1991, the Centers for Disease Control and Prevention (CDC) identified 10 μg/dL as an EBLL that should prompt public health actions to identify and manage exposure to lead. [A commonly referenced unit of measure is the blood lead value in μg/dL (micrograms per deciliter); 10 μg/dL is equivalent to 0.48 μmol/L (micromoles per liter); the conversion factor of 0.04826 is multiplied by the lead value in μg/dL to give the lead value in μmol/L.] The level had been reduced from 40 in 1971 to 10 in 1991. It has also been termed a "level of concern," which is an unfortunate wording because there has been concern about harmful effects below this level since before 1991. In addition, this level was never seen as a threshold level (a level below which adverse effects would not occur). Older terms, lead toxicity and lead poisoning, were less specific, and generally referred to adverse or toxic effects from lead exposure that were not specific to a particular blood lead level (BLL). The prevalence of EBLLs has decreased dramatically for the entire population and for children aged 1–5 years in the United States. Data from the National Health and Nutrition Examination Surveys (NHANES) show a decline in the geometric mean (average) BLLs from 15.2 (1976–1980) to 1.9 (1999–2002), with a similar decline in the prevalence (or percentage) of children with BLLs of 10 μg/dL or higher from 88.2% to 1.6% over the same time periods. Although the numbers have decreased significantly, this still represents an estimated 310,000 children with BLLs in this category.

Children's exposure to lead. For U.S. children, the most common lead exposure source is lead-based paint (LBP) in older housing, through ingestion of dust and soil contaminated by lead paint, or occasionally through ingestion of paint chips. Any housing constructed before 1978 may have LBP; however, the older the housing, the more likely is the presence of LBP and at a higher concentration. Housing in low-income communities tends to be more deteriorated because of deferred maintenance and repairs. The prototypical child who may be exposed to lead is a young, mobile child with significant hand-to-mouth activity living in older, high-risk housing, as described above. Other sources of lead include many products used by ethnic populations (cosmetics, folk remedies, foods, candies, and ceramic cookware); consumer products including toys, jewelry, lunch bags, and other products used by children; and emissions from industries with point sources of lead, such as mines and smelters. There were 29 distinct product recalls for children's toys because of lead contamination from the Consumer Product Safety Commission in 2007, and a number of recalls in 2008. Children's exposure to lead from use of consumer products is becoming a growing concern for U.S. children. Lastly, exposure to lead through drinking water can occur as a result of lead in older water pipes, including service lines from municipal water systems; in solder used in plumbing; from fixtures containing lead (such as brass fixtures); from wells contaminated by lead; and from items used to heat water such as kettles or dishes made from pewter or crystal or with lead glazes.

Children are exposed to lead through several pathways. With ingestion, lead is swallowed after putting contaminated dust or soil, paint chips, toys, or other lead-containing objects into the mouth. [With regard to contaminated dust or soil, children may touch these items with their hands and then put their hands in their mouths.] Inhalation is less common but can occur by breathing lead fumes or dust when lead is heated or sanded. This may occur with renovation, rehabilitation, or lead hazard control work in a property when a child is present in the work area. Transplacental transfer occurs when lead passes freely from mother to fetus during pregnancy, with generally similar BLLs.

Clinical considerations. As most patients with EBLLs do not have symptoms and are identified from routine blood lead screening activities, screening is very important. At one time there were national recommendations for lead screening. In 1997, the CDC issued guidance for state and local health departments to create their own unique lead screening recommendations, based on local conditions and lead hazards. The CDC also had a default recommendation for health care providers to do blood lead screening at 12 and 24 months of age, and during 3–6 years of age when no record of prior lead screening is available. Thus, the recommendation for any one child depends on the local or state recommendations.

Clinical management of children with EBLLs includes frequent monitoring of BLLs (with a schedule for this set by the CDC based on lead levels), communication with the local or state health department to ensure evaluation of all environments of children with elevated levels, and consideration of chelation therapy (giving a medicine that binds metals, including lead) for those with very elevated BLLs. Chelation therapy is recommended at a venous lead level of 45 μg/dL or higher, or if a patient is symptomatic [with a complex of many nonspecific symptoms such as abdominal pain, vomiting, constipation, lethargy, irritability, and developmental regression (loss of developmental milestones)] at lower levels. A patient with signs and symptoms of lead encephalopathy (acute poisoning of the brain leading to convulsions, unconsciousness, coma, brain swelling, and sometimes brain death) should be treated in an intensive care setting by experts in lead poisoning management and intensive care therapy. Lead encephalopathy is very rare; there have been two cases reported by the CDC since the year 2000 (one case reported in 2000 and one in 2006).

Harmful effects of lead exposure. Acute exposure to lead can affect many body organ systems, particularly the central nervous system (CNS), but also the kidneys, and the cardiovascular, hematological (blood-forming), digestive, and reproductive systems. Effects on the CNS are most important in children, because their nervous systems continue to develop for years after birth. Chronic, low-level exposure can lead to subtle neuropsychological effects (effects on the nervous system and behavior) that can have an impact on children's future education and employment status. Sometimes lead-related effects are not recognized until older grades, when schoolwork requiring higher-level cognitive functions is introduced. Newer studies indicate that there is no apparent threshold for these effects.

Neuropsychological effects of lead exposure. Earlier studies looked at numerous adverse effects in children with BLLs ≥10 μg/dL. Many hundreds, if not thousands, of studies have been done in a number of countries and with different populations. There have been cross-sectional studies [which look at lead exposure and different outcomes such as intelligence quotient (IQ) at one point in time] and cohort or prospective studies [which follow a group for a period of time, looking at both lead exposure and outcomes as they present over time]. Some of the newer studies have looked at groups of children whose BLLs never went above 10 μg/dL. In general, there has been an inverse relationship between indices of lead exposure and IQ (cognition or the ability to think). More recent studies have found a 4–8 IQ-point decrease as BLL increases to 10 μg/dL, followed by an additional decrement of 1–5 points as BLLs reach 20 μg/dL, depending on the specific study. Studies have found a number of other neuropsychological effects from lead exposure, including a decrease in proficiency in basic academic skills (math, reading); decreased school achievement; poor organizational skills; association with learning, behavioral (such as distractibility and hyperactivity), and attention problems; and increased risk for adolescent antisocial and/or delinquent behavior.

The studies of lead effects included animal and human studies, using different methodologies and conducted over various time periods. All utilized some index of lead exposure, but there were differences in measures of blood or bone lead levels. Studies utilized different outcome measures. Those looking at IQ and cognition used various psychometric tests, many of which are age-specific for use with children. These studies included and adjusted for (to isolate the effect of lead) some factors related to the home environment that could influence the IQ score, such as maternal intelligence, family income, parental intellectual stimulation and care, parental education, marital status, maternal drug use (tobacco, alcohol, and drugs), and maternal mental illness. These factors can confound the study of lead exposure and IQ. It is important to keep in mind that current lead exposure accounts for a small amount of variance in cognitive ability (about 1–4%), whereas social and parenting factors account for about 40%. However, lead exposure is important in that it represents an adverse influence on intelligence that can be prevented.

Recent studies from 2000 and later suggest that the harmful effects of lead can occur to a stronger degree at relatively lower BLLs (in general, those below 10 μg/dL) when compared to the harmful effects seen at BLLs of 10 and higher. The CDC created a work group to study this question, and the results were published in 2005. The group reviewed a number of current studies looking for an association between BLLs < 10 μg/dL and adverse health outcomes, both for cognition and for a variety of other effects, as well as for possible causality. For cognitive effects, the review found an inverse association of BLLs < 10 μg/dL and cognitive function of children, with no indication of a threshold for effects. For some studies, the decrement in IQ score was more pronounced at lower BLLs within the range up to 10 μg/dL. Effects were seen with varied study types and in varied study populations. The association was judged to be causal (that is, lead exposure caused the cognitive impairment). However, the strength and the shape of this relationship (how much of an effect there is) are uncertain because of some of the limitations of the studies. For other health effects (stature, sexual maturation, dental caries, etc.), there was an association between BLLs < 10 μg/dL and poorer health indicators, but not enough information to determine whether this was cause and effect. A study that pooled data from seven international study groups and from 1333 children also found an inverse relationship between BLLs and IQ scores. Of the pooled group, 18% had maximal BLLs < 10 μg/dL and 8% had maximal BLLs < 7.5 μg/dL. The study found a decrease in IQ of 3.9 points going from a BLL of 2.4 to 10 μg/dL, a 1.9-point decrease from 10 to 20 μg/dL, and a 1.1-point decrease from 20 to 30 μg/dL. The study found deficits in IQ at a maximal BLL of <7.5 μg/dL. Several further studies have been published since 2005 that suggest a substantial decline in IQ points or achievement test scores at BLLs < 10 μg/dL, with a steeper response at the lower end of the range from 0 to 10 μg/dL.

Long-term impact of lead toxicity. Some studies have looked at medical and special education costs, and loss of future earning potential costs of EBLLs. These studies give data to support the concern about the long-term impact of lead on a child's learning and educational achievement level, which can then presumably have negative effects on long-term education and occupational goals.

Prevention of lead exposure. With studies not showing any cutoff BLL below which adverse health effects do not occur, it becomes increasingly more important to prevent children being exposed to lead from the start. Essentially, no level of lead is good for a child. Primary prevention of lead exposure (keeping children from being exposed to lead) is recommended and promoted in the United States by the CDC, the Environmental Protection Agency (EPA), and the Department of Housing and Urban Development (HUD). Activities include screening housing for possible lead hazards and correcting any that are found; doing renovation and maintenance work safely so as not to disturb painted areas with LBP; and keeping any sources of lead (including consumer or ethnic products) away from children. There will still be a need to do blood lead screening in certain groups of children, as defined by state and local health departments. Parents can work to keep their children from being exposed to housing-based lead, consumer product-based lead, and other sources of lead. Teachers, doctors, nurses, children's advocates, public health professionals, elected officials, informed citizens, and others can advocate for primary prevention to keep our children safe from exposure to lead, which can lead to unnecessary harmful effects on their development and potential for future achievement.

For background information *see* BEHAVIORAL TOXICOLOGY; COGNITION; ENVIRONMENTAL TOXICOLOGY; LEAD; PAINT AND COATINGS; POISON; PUBLIC HEALTH; TOXICOLOGY; TOXIN in the McGraw-Hill Encyclopedia of Science & Technology.   Carla Campbell

Bibliography.   H. J. Binns et al., Interpreting and managing blood lead levels of less than 10 μg/dL in children and reducing childhood exposure to lead: Recommendations of the Centers for Disease Control and Prevention Advisory Committee on Childhood Lead Poisoning Prevention, *Pediatrics*, 120:e1285–e1298, 2007; Centers for Disease Control and Prevention, *Managing Elevated Blood Lead Levels among Young Children*, CDC, Atlanta, 2002; Centers for Disease Control and Prevention, *Preventing Lead Exposure in Young Children: A Housing-Based Approach to Primary Prevention of Lead Poisoning*, CDC, Atlanta, 2004; Centers for Disease Control and Prevention, *Preventing Lead Poisoning in Young Children*, CDC, Atlanta, 2005; Committee on Environmental Health—American Academy of Pediatrics, Lead exposure in young children, *Pediatrics*, 116:1036–1046, 2005.

Cholesterol and the SREBP pathway

Cholesterol is an essential component of mammalian cell membranes. (In general, cholesterol and structurally related compounds made by plants and fungi are collectively termed "sterols." Virtually all eukaryotic organisms require some sort of sterol in their membranes, and the presence of sterols is evidence for the existence of eukaryotes in the fossil record.) Without sufficient cholesterol, membranes cannot work properly and cells die. Conversely, too much cholesterol is also lethal to cells and organisms, and high levels of cholesterol in blood are a major cause of atherosclerosis in humans. Therefore, cells and organisms must maintain an exquisite balance between the cholesterol supply and cellular demand. This balancing act is regulated at just about every conceivable level, from gene transcription to enzyme activity. The global coordination of cellular cholesterol metabolism, however, is largely effected at the level of transcription of the genes of cholesterol synthesis and uptake [for example, the low-density lipoprotein (LDL) receptor, 3-hydroxy-3-methylglutaryl coenzyme A (HMG CoA) reductase, and HMG CoA synthase] through the action of the sterol regulatory element

binding proteins (SREBPs). In mammals, there are three different SREBPs encoded by two different genes. SREBP-1a and -1c result from the utilization of alternative transcription initiation sites of the *SREBF-1* (sterol regulatory element binding transcription factor-1) gene; SREBP-2 is encoded by the *SREBF-2* gene. In adult animals, SREBP-1c and -2 are the predominant species, with the former preferentially targeting genes of fatty acid metabolism and the latter preferentially targeting genes of cholesterol metabolism.

The SREBP pathway. The SREBPs are transcription factors that bind to deoxyribonucleic acid (DNA) sequences [the sterol regulatory elements (SREs)] in the upstream region of target genes to promote initiation of transcription of those genes. Unusually for proteins that function in the nucleus, the SREBPs are made as integral membrane proteins of the endoplasmic reticulum (ER; a vacuolar system of the cytoplasm in differentiated cells that functions in protein synthesis and sequestration). The transcriptionally active portion of each protein lies amino-terminally to the two membrane-spanning helices that anchor the precursor in the membrane. The carboxy-terminal domain plays a regulatory role. The transcription factor and regulatory domains reside in the cytoplasm, whereas the short, water-soluble sequence that separates the two helices is located within the lumen of the ER. The topology of the precursor is thus like a hairpin sticking through the membrane.

When cells need to make more cholesterol (and take it up from the bloodstream via LDL), the active transcription factor portion of the SREBP must get to the nucleus and mediate the increased transcription of target genes. Once in the nucleus, active SREBP is rapidly degraded; continued transcription of target genes requires continued production of active SREBP.

Active SREBP is produced (see **illustration**) when the amino-terminal portion of the precursor is cleaved from the membrane-spanning helices, releasing it to the cytoplasm and freeing it to travel to the nucleus. Release requires separate cleavages by two distinct proteases that reside in the Golgi apparatus. The first cleavage, at site 1, occurs in the short luminal sequence, cutting the precursor in two. The protease responsible, termed site-1 protease (S1P), is a serine protease that is attached to the membrane by a single membrane-spanning helix. Its active site lies within its large luminal domain, as expected for a protease that cleaves a luminal substrate. Both halves of SREBP resulting from cleavage by S1P remain bound to the membrane, since each retains a membrane-spanning helix.

Following cleavage at site 1, the amino-terminal half is cleaved at site 2, which lies within the first membrane-spanning helix. It is this second cleavage that releases SREBP from the membrane. The site-2 protease (S2P) is a highly hydrophobic metalloprotease with numerous membrane-embedded helices. Structural studies of a related bacterial protease indicate that the active site lies within the plane of

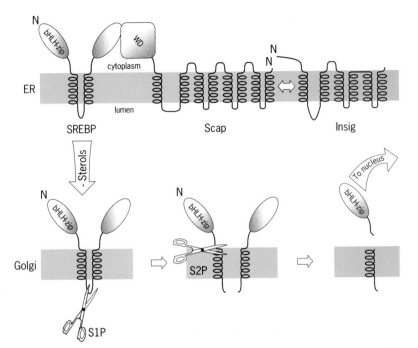

Activation of SREBP. SREBPs are synthesized as membrane-bound precursors that reside in the endoplasmic reticulum (ER). There, they interact with Scap via their respective carboxy-terminal domains. This interaction occurs in both the absence and the presence of sterols. When sterols are present, Scap, in turn, interacts with Insig, a resident ER membrane protein. The Scap:Insig interaction serves to retain the SREBP:Scap complex in the ER. Thus, SREBPs are not cleaved in the presence of sterols. In the absence of sterols, Scap no longer binds to Insig, and the SREBP:Scap complex is free to traffic to the Golgi apparatus. There, SREBP undergoes two sequential cleavages. The first cleavage, in the luminal loop, separates the two halves of SREBP. It is the work of a membrane-bound serine protease, S1P. This cleavage produces the substrate for S2P, which cleaves the amino-terminal portion of SREBP within the first membrane-spanning helix. This liberates the nuclear form, freeing it to travel to the nucleus, where it mediates the increased transcription of target genes. "N" indicates the amino-terminus of the protein. The lipid bilayer is indicated by the color rectangle. The cartoons reflect the topology of SREBP, Scap, and Insig. Scissors indicate the sites of proteolysis.

the lipid bilayer, as does its substrate. Both S1P and S2P reside in the Golgi apparatus, so that, in order to be cleaved, the SREBP precursor must change compartments, moving from the ER to the Golgi apparatus. This movement requires SREBP cleavage activating protein (Scap), a large integral membrane protein that binds to the regulatory domain of SREBP and essentially carries the SREBP to the Golgi apparatus. Analysis of the Scap sequence revealed seven predicted membrane-spanning helices in the amino-terminal half of the protein, whereas multiple WD-40 protein-protein interaction motifs (characteristic protein-binding motifs with residue lengths of ~40 amino acids) made up the remainder. Helices 2–6 bore striking similarity to the membrane-associated region of HMG CoA reductase (the rate-limiting enzyme for cholesterol biosynthesis, whose product is mevalonate), which was known to be required for cholesterol-regulated stability of that enzyme. Accordingly, this region was termed the sterol-sensing domain (SSD). Recent studies demonstrate that cholesterol binds to the SSD and promotes interaction between Scap and Insig (insulin-induced gene) proteins. (Insig is an ER-resident protein found in two isoforms, Insig-1 and Insig-2, encoded by separate genes with distinct expression characteristics;

however, the two proteins are functionally indistinguishable.)

When cellular sterol levels are adequate, the SREBP:Scap complex remains in the ER because interaction between Scap and Insig renders a hexapeptide sorting signal in Scap (MELADL) inaccessible to coatamer protein II complexes (COPII) that mediate ER-to-Golgi vesicular transport. When sterol levels fall, the interaction between Scap and Insig is disrupted, the MELADL signal is freed to interact with COPII, and the SREBP:Scap complex exits the ER via COPII-coated vesicles. Once the SREBP:Scap complex enters the Golgi apparatus, sequential cleavage by S1P and S2P releases active SREBP. Thus, when cellular demand for sterols rises, more of the active SREBP is produced, and the transcription of the genes of cholesterol synthesis and uptake increases.

As sterol levels rise as a result of increased synthesis and uptake, interaction between Insig and Scap is restored, and the sorting signal becomes inaccessible. The SREBP:Scap complex remains in the ER, preventing cleavage by the Golgi proteases. The rapid degradation of active, nuclear SREBP means that transcription of target genes falls almost as soon as cleavage stops. Thus supply is finely tuned to demand.

Identifying the players. The SREBPs were identified biochemically as proteins purified from the nuclei of human HeLa cells (an immortalized, continuously cultured cell line of human cancer cells) that bound to wild-type SREs, but not to mutant versions that were defective in promoting cholesterol-sensitive transcription of linked genes. Cloning of complementary DNAs (cDNAs, which are DNA molecules synthesized by reverse transcriptase from a ribonucleic acid template) for these proteins revealed two different encoded nuclear proteins that were fragments of much larger precursors, termed SREBP-1 and -2, each harboring two membrane-spanning helices. Subsequent studies showed that proteolytic release of active SREBP was regulated by cellular sterol levels.

The machinery required for proteolytic activation was identified using mutant cells that had various defects in their ability to regulate cholesterol metabolism. Scap was cloned from mutant hamster cells that could not shut down transcription in the presence of adequate cholesterol and thus accumulated massive amounts of fat and cholesterol. When cDNAs made from the mutant cells were expressed in wild-type cells, the wild-type cells that received a copy of the mutant gene also failed to shut down transcription. Subsequent rounds of enrichment enabled the Scap cDNA to be identified and sequenced.

Next, S2P was cloned by using human genomic DNA to complement mutant hamster cells that could not upregulate transcription in the absence of cholesterol. In consequence of their inability to synthesize cholesterol or the LDL receptor, these cells required free cholesterol added to their growth medium; without it, they died. When wild-type human DNA was transfected onto these mutant cells, the only mutant cells that could survive without free cholesterol were those that received a working human copy of the defective hamster gene.

S1P was identified from mutant hamster cells that had the same phenotype as the S2P mutants, but owing to mutation of a different gene. A sensitive assay for S1P activity was developed and used to screen mutant cells that had been transfected with wild-type cDNAs. Again, only mutant cells that received a working copy of S1P would register in this assay, permitting the S1P cDNA to be identified.

Biochemical methods revealed the role of Insig in the SREBP pathway. *Insig-1* was first identified as a gene upregulated by insulin treatment of regenerating liver. It was found that Insig proteins (Insigs) bound to the SSD of Scap. Further studies revealed that multiple independent mutations within the SSD rendered Scap insensitive to normal sterol-responsive regulation acted by blocking interaction between the SSD and Insigs.

Isolation of hamster cells lacking Insigs demonstrated their essential role in retaining the SREBP:Scap complexes in the ER. In the absence of Insigs, Scap continually moved from the ER to the Golgi, dragging SREBP with it. Once in the Golgi, S1P and S2P acted sequentially to release active SREBP. Just as in cells harboring mutant versions of Scap with defective SSDs, these cells could not shut down transcription even in the presence of very high levels of cholesterol.

Oxysterols. Oxysterols are oxygenated derivatives of cholesterol, such as 25-hydroxycholesterol, that contain an additional hydroxyl (or keto) group, typically on the isooctyl side chain. Oxysterols are potent inhibitors of cellular cholesterol synthesis. A major portion of this effect is due to suppression of SREBP cleavage and thus of transcription of target genes. Additionally, oxysterols accelerate the degradation of the key enzyme of cholesterol biosynthesis, HMG CoA reductase. Both actions of oxysterols may be attributed to Insigs, which bind oxysterols. This binding promotes interaction between the SSD and Insig, much as binding of cholesterol to the SSD itself does.

In the case of SREBP:Scap complexes, interaction promoted by oxysterols blocks proteolysis by keeping the complex in the ER. In the case of HMG CoA reductase, interaction with Insig promotes ubiquitination (addition of ubiquitin, a necessary step for protein degradation) by a membrane-anchored ubiquitin ligase, leading to proteasome-mediated degradation of the enzyme.

For background information *see* CELL MEMBRANES; CHOLESTEROL; ENDOPLASMIC RETICULUM; GENE; GOLGI APPARATUS; LIPID; LIPID METABOLISM; LIPOPROTEIN; METABOLIC DISORDERS; STEROL; TRANSCRIPTION in the McGraw-Hill Encyclopedia of Science & Technology.   Robert B. Rawson

Bibliography. M. T. Bengoechea-Alonso and J. Ericsson, SREBP in signal transduction: Cholesterol metabolism and beyond, *Curr. Opin. Cell Biol.*, 19(2):215–222, 2007; M. S. Brown and J. L. Goldstein, A proteolytic pathway that controls the cholesterol content of membranes, cells, and blood, *Proc. Natl. Acad. Sci. USA*, 96(20):11041–11048, 1999; J. L. Goldstein, R. B. Rawson, and M. S. Brown, Mutant mammalian cells as tools to delineate the sterol regulatory element-binding protein pathway for feedback regulation of lipid synthesis, *Arch. Biochem. Biophys.*, 397(2):139–148, 2002; R. B. Rawson, The SREBP pathway—insights from Insigs and insects, *Nat. Rev. Mol. Cell Biol.*, 4(8):631–640, 2003; N. G. Seidah, A. M. Khatib, and A. Prat, The proprotein convertases and their implication in sterol and/or lipid metabolism, *Biol. Chem.*, 387(7):871–877, 2006.

Cloning research

The procedure of cloning or nuclear transfer (NT) is defined as the transfer of a nucleus from one cell into another cell from which the nucleus has been removed. NT was first devised by the German Nobel laureate Hans Spemann in 1938, when he proposed an experiment involving the insertion of a nucleus into an enucleated oocyte, that is, an egg that has had its nucleus removed. The idea was not pursued, however, because Spemann did not have the equipment required to perform such an experiment. Robert Briggs and Thomas J. King were the first to utilize NT successfully in the production of live offspring from metazoan cells. They reported the successful production of Northern leopard frog (*Rana pipiens*) tadpoles via NT in 1952. Continued research by the same group later concluded that the developmental potential of NT embryos declined with increased embryonic cell age. In 1966, frog larval nuclei were used successfully to produce fertile *Xenopus* frogs and demonstrated proof in principle that somatic cells (any cells of an organism except the germ cells) from a variety of sources, including skin, lymphocytes, erythrocytes, leukocytes, and erythroblasts, were able to regress from a specialized or differentiated cell to a simpler, unspecialized cell and yield morphologically normal tadpoles; however, none of these tadpoles survived to adulthood. These results showed the potential of differentiated cells to derive numerous different cell types in a complex organism; the question remained, though, whether adult cells could be dedifferentiated or reprogrammed to a totipotent state (capable of differentiating into every type of cell found in an organism and of forming the entire organism) and consequently regain the ability to give rise to any cell type.

Success with NT in mammals was not reported until the 1980s. Initially, experiments involving the transfer of pronuclei, the precursor of a nucleus, from one mouse embryo to another proved successful in producing live births. However, the next step was to produce viable embryos with the capacity to develop past the blastocyst stage (the developmental stage when differentiation occurs) using blastomeres—individual embryonic cells isolated from the developing embryo—as the nuclear donors. Finally, in 1986, Steen Willadsen reported the production of completely viable sheep embryos derived from the transfer of 8- and 16-cell blastomeres to enucleated oocytes. In 1987, cattle were produced by the same procedure.

Based on early work with somatic cells, it was believed that it was not possible to produce viable offspring from adult cells. However, in 1996, Ian Wilmut, Keith Campbell, and colleagues at the Roslin Institute in Scotland announced the production of five cloned sheep from in vitro cultured, differentiated cells derived from a blastocyst inner cell mass (ICM; the population of cells within the embryo from which the fetus is derived). The success with cultured cells was closely followed by the announcement of the birth of the sheep Dolly, the first cloned animal derived from an adult cell. The announcement of Dolly was significant in that it demonstrated that a differentiated (somatic) mammary cell derived from an adult animal was able to be reprogrammed to an embryonic state and give rise to a complete and healthy animal. Since the first successful somatic cell nuclear transfer (SCNT) experiments in sheep, the technology has been applied to the production of a number of other species, including mice, cattle, goats, pigs, mouflon sheep, rabbits, mules, cats, rats, horses, ferrets, a deer, and a dog, with the list continuing to grow (**Fig. 1**).

Methods of manipulation and activation. A great deal of research has evaluated numerous manipulation and activation protocols in an effort to develop methods resulting in improved SCNT efficiencies. SCNT typically involves the removal of deoxyribonucleic acid (DNA) from a mature oocyte, followed by the transfer of a donor cell or nucleus either to the perivitelline space (the space between the oocyte plasma membrane and the zona pellucida, the shell that surrounds the egg) or directly into the oocyte cytoplasm. Injection of the cell into the perivitelline space requires a subsequent fusion step in order to fuse the donor cell and oocyte membranes and introduce the donor nucleus into the oocyte cytoplasm. Usually following NT, but sometimes preceding it, activation of the embryo is required in order to signal the oocyte to initiate cell division. In the case of natural fertilization, the interaction between the sperm and the oocyte triggers this activation event. However, with SCNT, a synthetic activation is required. A variety of methods have been employed for manipulation and activation of SCNT embryos, with varying degrees of success (**Fig. 2**).

The most common method for enucleation employs the use of a small, polished-glass holding pipette to keep the oocyte stationary and an enucleation pipette that is used to pierce the zona pellucida and aspirate the DNA from the egg. Other methods include chemically induced enucleation using various inhibiting agents (etoposide, etoposide in conjunction with cycloheximide, and ethanol with demecolcine). Although these methods greatly facilitate the enucleation process, development of NT embryos following chemically induced enucleation remains less successful than development of mechanically enucleated oocytes. Zona-free cloning methods have also been employed successfully in bovine (cow) and porcine (pig) SCNT protocols. These latter methods do not require micromanipulation and have the potential to be automated, but the culture requirements for zona-free embryos are more problematic than for manually manipulated embryos.

Researchers have evaluated the effect of manipulation and activation of oocytes in various stages of meiosis on SCNT efficiency. Two predominant methods have been successfully utilized for NT. The first is a protocol in which the donor nucleus is transferred into a preactivated, enucleated cytoplast. The other protocol involves the transfer of a donor nucleus into an M2 phase-arrested cytoplast followed by subsequent activation. The latter protocol results in more efficient development to the blastocyst stage in bovine NT and is therefore most frequently used.

In addition to the effects of manipulation methods and timing on SCNT efficiency, the timing and the method of activation have been shown to affect efficiency. Several groups have shown that the duration of exposure of the donor nucleus to oocyte cytoplasm affects in vitro development. The mechanisms underlying the differences observed in development rates based on the duration of cytoplasmic exposure prior to activation remain obscure.

Following normal fertilization, activation by the sperm elicits regular, repetitive intracellular calcium transients. Activation results in resumption of meiosis, cortical granule release, decondensation of the sperm nucleus, and formation of male and female pronuclei. Because the donor cell does not have

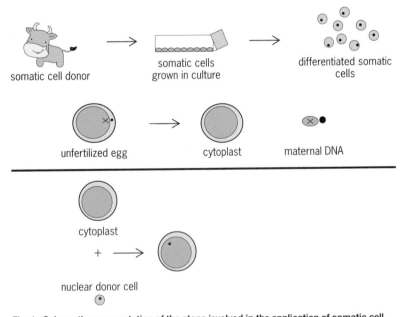

Fig. 1. Schematic representation of the steps involved in the application of somatic cell nuclear transfer to produce multiple copies of genetically identical animals. Tissue is collected from an animal with superior or unique trait(s) and placed in culture to establish a primary somatic cell culture; these will be the nuclear donor cells. Immediately prior to use for nuclear transfer, individual cells are harvested from cell culture flasks and prepared for use. Microsurgery is conducted with mature unfertilized eggs, which have a first polar body (PB; dot) and nuclear chromatin material ("x") adjacent to and inside the egg cytoplasm, respectively, to remove all of the genetic material (PB and chromatin); the egg without genetic material is now referred to as a "cytoplast." The nuclear transfer is carried out and the nucleus from the donor cell is incorporated into the cytoplast, producing the resultant nuclear transfer zygote.

somatic cell donor

somatic cells grown in culture

differentiated somatic cells

unfertilized egg

cytoplast

maternal DNA

cytoplast

+

nuclear donor cell

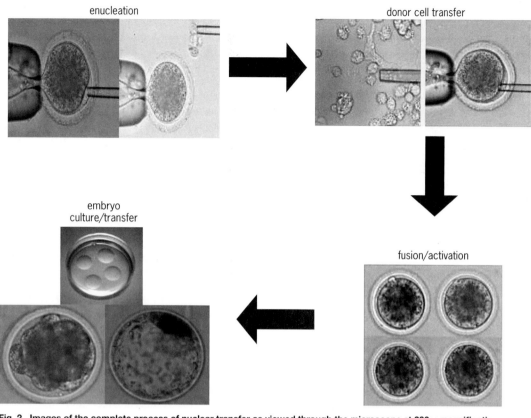

enucleation

donor cell transfer

embryo
culture/transfer

fusion/activation

Fig. 2. Images of the complete process of nuclear transfer as viewed through the microscope at 200× magnification. Enucleation is the term used to describe the removal of the first polar body and adjacent chromatin from the egg, rendering the egg devoid of any nuclear genetic material (cytoplast). The holding pipette to the left holds the egg in place, while the nuclear pipette removes the first polar body and egg chromatin. These two structures can be seen as they are expirated from the nuclear pipette following the completion of the process. Donor cell transfer refers to the selection of a donor cell (left) to be aspirated into the nuclear pipette and transferred into the space between the plasma membrane of the cytoplast and the shell that surrounds the cytoplast. Fusion/activation illustrates the process of fusion. Once the donor cell transfer is complete, the resulting nuclear transfer couplet (donor cell + cytoplast) is placed between two electrodes and an electrical pulse is delivered to the couplet. This induces membrane fusion between the membrane of the donor cell and the membrane of the cytoplast, resulting in the nucleus of the donor cell entering the cytoplasm of the cytoplast. The top left plate in this image shows the donor cell at "11:00" just under the shell immediately after the electrical pulse. The next plate to the right (5–6 min after pulse) shows that the donor cell is less distinct, indicating the fusion process has initiated. The plate on the bottom right (10–12 minutes after pulse) indicates almost complete incorporation into the cytoplast. The final plate at the bottom left shows the fusion completed. After fusion is completed, the nuclear transfer embryos "rest" for 2–4 hours and then are "activated" to initiate growth of the embryo. Embryo culture/transfer shows the culture drops that support embryonic development and the nuclear transfer embryos after several days in culture. The culture period for cattle embryos is 7 days, at which time those embryos that are viable have grown to the morula (bottom left; 40–60 total cells) or blastocyst (bottom right; 80–120 cells) stage and are ready to be transferred into a surrogate mother.

the capacity to activate the oocyte, artificial means of activation are required. Early on, it was discovered that mature oocytes could undergo parthenogenetic activation in the absence of the male gamete using a number of physical and chemical methods (parthenogenesis is a special type of reproduction in which an egg develops without entrance of a sperm). A number of parthenogenetic activation protocols have been applied successfully to SCNT. A short, high-voltage electrical pulse can be used to create transient pores in cellular membranes, allowing the influx of calcium from extracellular pools. Treatment with Ca^{2+} ionophores such as ionomycin results in the influx of Ca^{2+} as well as the release of Ca^{2+} from intracellular stores. Inhibition of protein synthesis using agents such as cycloheximide or puromycin induces activation in mouse and human oocytes. More efficient activation occurs with a combination of calcium stimulus in conjunction with protein synthesis inhibition. There is certainly a need for contin-

ued research pertaining to oocyte activation in SCNT protocols.

Status of SCNT technology. An incredible amount of research has focused on SCNT, and progress continues, but the molecular events underlying the successful conversion of a differentiated somatic cell nucleus to a totipotent embryonic nucleus with the capacity to derive a healthy and normal animal remain poorly understood. Furthermore, the efficiency with which this process occurs successfully remains very low. Although it is difficult to ascertain the overall efficiencies because of differences in protocols, embryo transfer criteria, and data presentation, the overall efficiency of SCNT across species based on the number of embryos produced is believed to be less than 5%. In cattle, approximately 10–15% of SCNT embryos transferred develop to term.

In addition to the problems associated with poor efficiency following SCNT, including lower rates of development to blastocyst in vitro, lower rates of pregnancy establishment, and higher rates of pregnancy loss, a number of other differences between SCNT and control embryos and fetuses have been reported. Given the complexity of the SCNT process, it is not surprising that a variety of different factors can affect efficiency.

Factors affecting SCNT efficiency. The inefficiencies associated with SCNT likely stem largely from deficiencies in the reprogramming process following NT. Following the transfer of a differentiated cell or nucleus into an enucleated oocyte, the DNA must be reprogrammed from a cell-type–specific gene expression pattern to a totipotent embryonic-cell state (in a process called epigenetic reprogramming). Modifications to the epigenetic state (the developmental, gradual stages of differentiation) of the DNA (involving DNA methylation and histone protein modifications) are required in order for this to occur.

Numerous factors can affect the efficiency of nuclear reprogramming following NT. These include the state and source of the donor cell, cytoplast source and quality, timing and methods of manipulation and activation, and embryo culture conditions. Evaluation of the literature associated with SCNT suggests that most if not all deficiencies associated with the low efficiency in cloning stem from failures or deficiencies in epigenetic reprogramming.

Future outlook. Somatic cell nuclear transfer offers promise for many different applications, including rescue of endangered species, production of animals with genetically superior traits, biopharmaceutical production, xenotransplantation applications, and stem cell production. As SCNT efficiency increases, the utility of the process will lead to further advances in these applications. In addition, an understanding of the factors that affect SCNT efficiency will offer insights into the complex and poorly understood field of epigenetic reprogramming. The complex processes by which epigenetic modifications are initiated and propagated and the mechanisms by which these modifications effect gene expression are beginning to be characterized. There remains, however, much to be learned in this area, including understanding how specific environmental cues function to bring about epigenetic changes and how specific genes are targeted for silencing or activation by epigenetic controls. Continued research in the field of epigenetics will undoubtedly open doors to increased understanding in many related fields.

For background information *see* BIOTECHNOLOGY; CELL CYCLE; CELL DIFFERENTIATION; CELL DIVISION; CELL NUCLEUS; CLONING; GENE; GENETIC ENGINEERING; SOMATIC CELL GENETICS; STEM CELLS in the McGraw-Hill Encyclopedia of Science & Technology.

Kenneth L. White; Kenneth I. Aston; Benjamin R. Sessions

Bibliography. R. Briggs and T. J. King, Transplantation of living nuclei from blastula cells into enucleated frogs' eggs, *Proc. Natl. Acad. Sci. USA*, 38(5):455–463, 1952; K. H. Campbell et al., Sheep cloned by nuclear transfer from a cultured cell line, *Nature*, 380(6569):64–66, 1996; J. McGrath and D. Solter, Inability of mouse blastomere nuclei transferred to enucleated zygotes to support development in vitro, *Science*, 226(4680):1317–1319, 1984; R. S. Prather et al., Nuclear transplantation in the bovine embryo: Assessment of donor nuclei and recipient oocyte, *Biol. Reprod.*, 37(4):859–866, 1987; H. Spemann, *Embryonic Development and Induction*, Yale University Press, New Haven, CT, 1938; S. M. Willadsen, Nuclear transplantation in sheep embryos, *Nature*, 320(6057):63–65, 1986; I. Wilmut et al., Viable offspring derived from fetal and adult mammalian cells, *Nature*, 385(6619):810–813, 1997.

Composite material systems and structures

Composites combine the properties of two or more materials (constituents). Any two materials, such as metals, ceramics, polymers, elastomers, and glasses, can be combined to make a composite. They may be mixed in many geometries (particulate, chopped-fiber, woven, unidirectional fibrous, and laminate composites) to create a system with a property profile not offered by any monolithic material. In mechanical design, this is often done to improve the stiffness-to-weight ratio, strength-to-weight ratio, or toughness, while in thermomechanical design, it is to reduce thermal expansion, maximize heat transfer, or minimize thermal distortion. Composites have gained popularity in high-performance products that need to be lightweight yet strong enough to take high loads, such as aerospace structures, space launchers, satellites, and racing cars. Their growing use has arisen from their high specific strength and stiffness when compared to metals, and the ability to shape and tailor their structure to produce more aerodynamically efficient configurations.

Fiber-reinforced polymers, especially carbon-fiber-reinforced plastics (CFRP), can and will in the near future contribute more than 50% of the structural mass of an aircraft. The main advantages provided by CFRP include mass and part reduction, complex shape manufacture, reduced scrap, improved fatigue life, design optimization, and generally improved corrosion resistance. The main challenges restricting their use are material and processing costs, damage tolerance, repair and inspection, dimensional tolerance, and conservatism associated with uncertainties about relatively new and sometimes variable materials. For secondary structures, weight savings approaching 40% are feasible by using composites, while for primary structures, such as wings and fuselages, 20% is more realistic.

Fiber-reinforced plastics. The adoption of composite materials as a major contributor to aircraft structures followed the discovery of carbon fiber in 1964. In the late 1960s, these materials began to be applied, on a demonstration basis, to military aircraft in such components as trim tabs, spoilers, rudders, and doors. With increasing application and experience with their use came improved fibers and resins resulting in composites with improved mechanical properties, allowing them to displace aluminum and titanium alloys for primary structures.

Modern carbon fibers. High-strength, high-modulus carbon fibers are about 6 micrometers in diameter and consist of small crystallites of "turbostratic" graphite, one of the allotropic forms of carbon. Refinements in fiber process technology have led to considerable improvements in tensile strength ($\sim$4.5 gigapascals) and in strain to fracture (more than 2%) for PAN-based (polyacrylonitrile) fibers. These come in three basic forms: high modulus (HM, $\sim$380 GPa); intermediate modulus (IM, $\sim$290 GPa); and high strength (HS), with a modulus of around 230 GPa, tensile strength of 4.5 GPa, and strain of 2% before fracture. The selection of the appropriate fiber depends very much on the application. For military aircraft, both high modulus and high strength are desirable. Satellite applications, in contrast, benefit from the use of high fiber modulus, improving stability and stiffness for reflector dishes, antennas, and their supporting structures.

Rovings are the basic forms in which fibers are supplied, a roving being a number of strands or bundles of filaments wound into a package or reel. Rovings or tows can be woven into fabrics, and a range of fabric constructions are available commercially, such as plain weave, twills, and various satin weave styles, woven with a choice of roving or tow size depending on the weight or areal density of fabric required. Fabrics can be woven with different kinds of fiber, for example, carbon in the weft (fiber running perpendicular to the lengthwise fibers) and glass or Kevlar® in the warp (lengthwise) direction, and these options increase the range of available properties. One advantage of fabrics for reinforcing purposes is their ability to drape or conform to curved surfaces without wrinkling. It is possible, with certain types of knitting machines, to produce fiber preforms tailored to the shape of the eventual component.

Fiber-matrix interface. The fibers are surface treated during manufacture to prepare adhesion with the polymer matrix, whether thermosetting or thermoplastic. The fiber surface is roughened by chemical etching and then coated with an appropriate size to aid bonding to the specified resin. Whereas composite strength is primarily a function of fiber properties, the ability of the matrix to both support the fibers and provide out-of-plane strength is, in many situations, equally important. The aim of the material supplier is to provide a system with a balanced set of properties. While improvements in fiber and matrix properties can lead to improved lamina (single ply or layer) or laminate (layered) properties, the fiber-matrix interface is all-important. The load acting on the matrix has to be transferred to the reinforcement via the interface. Thus, fibers must be strongly bonded to the matrix if their high strength and stiffness are to be imparted to the composite. The fracture behavior is also dependent on the strength of the interface. A weak interface results in low stiffness and strength but high resistance to fracture, whereas a strong interface produces high stiffness and strength but often a brittle fracture. Conflict

therefore exists, and the designer must select the most suitable material. Resistance to creep, fatigue life, and environmental degradation are also affected by the characteristics of the interface.

Matrix materials. The matrix is the weak point of the composite system and limits the fiber from exhibiting its full potential in terms of laminate properties. The matrix performs several functions, among which are stabilizing the fiber in compression, translating the fiber properties into the laminate, minimizing damage due to impact, and providing out-of-plane properties to the laminate.

Conventional epoxy aerospace resins (thermosets) are designed to cure at 120–135°C (248–275°F) or 180°C (356°F), usually in an autoclave at pressures up to 8 bar (0.8 megapascal), occasionally with a postcure at higher temperature. The resins must have a room-temperature life beyond the time it takes to lay-up (placing individual layers, or plies, on top of each other to create the laminate, or layered composite plate) a part and have time, temperature, and viscosity properties suitable for handling. The resultant resin characteristics are normally a compromise. For example, improved damage tolerance performance usually causes a reduction in hot-wet compression properties, and if this is attained by an increased thermoplastic content, then the resin viscosity can increase significantly. Increased viscosity is especially undesirable for a resin transfer molding (RTM) process.

The first generation of composites introduced to aircraft construction in the 1960s and 1970s employed brittle epoxy resins, leading to structures with a poor tolerance to low-energy impact caused by runway debris thrown up by aircraft wheels or the impacts occurring during manufacture and subsequent servicing operation. Although the newer toughened epoxies provide improvements in this respect, they are still not as damage tolerant as thermoplastic materials. Polyetheretherketone (PEEK) is a relatively costly thermoplastic with good mechanical properties. Carbon fiber/PEEK is a competitor with carbon fiber/epoxies and aluminum alloys in the aircraft industry. On impact at relatively low energies (5–10 joules), carbon fiber-PEEK laminates show only an indentation on the impact site, while in carbon/epoxy systems, ultrasonic C scans (a nondestructive inspection technique utilizing short pulses of ultrasonic energy) show that delamination (layer separation) extends a considerable distance, affecting more dramatically the strength and stiffness properties of the composite. In the effort to improve the through-the-thickness strength properties, the composites industry has moved away from brittle resins and progressed to thermoplastic resins, toughened epoxies, Z-fiber (carbon, steel, or titanium pins driven through the *z*-direction to improve the through-thickness properties), stitched fabrics, and stitched preforms. The focus is now on affordability, particularly affordable processing methods such as RTM processing, nonthermal electron beam curing by radiation, and cost-effective fabrication.

Design and analysis. Aircraft design since the 1940s has been based primarily on the use of aluminum alloys. With the introduction of laminated composites that exhibit anisotropic properties (properties varying with the direction of applied load) the methodology of design had to be reviewed and in many cases replaced. It is accepted that designs in composites should not merely replace the metallic alloy but should take advantage of exceptional composite properties if the most efficient designs are to evolve. Of course, the design should account for through-thickness effects that are not encountered in the analysis of isotropic materials. For instance, in a laminated structure, since the layers (laminae) are elastically connected through their faces, shear stresses are developed on the faces of each lamina. These stresses can be large near a free boundary (free edge, cut-out, or open hole) and may influence the failure of the laminate.

The lay-up geometry and stacking sequence of a composite strongly affects not only crack initiation but also crack propagation, with the result that some laminates appear highly notch sensitive, whereas others are totally insensitive to the presence of open holes. The selection of fibers and resins, the manner in which they are combined in the lay-up, and the quality of the manufactured composite must all be carefully controlled if optimum toughness is to be achieved. Compared with fracture in metals, research into the fracture behavior of composites is in its infancy. Much of the necessary theoretical framework is not yet fully developed, and there is no simple recipe for predicting with certainty the toughness of all composites.

Fracture in composite materials seldom occurs catastrophically without warning, but tends to be progressive, with substantial damage widely dispersed through the material. Tensile loading can produce matrix cracking, fiber bridging, fiber pull-out, fiber/matrix debonding, and fiber rupture, which provide extra toughness (energy sinks) and delay failure. Compression failures can occur either at the macroscale or at each individual reinforcing fiber, as occurs in compression buckling (fiber microbuckling). The fracture behavior of the composite can be reasonably well explained in terms of some summation of the contributions from these mechanisms, but as noted above, it is not yet possible to design a laminated composite to have a given toughness.

Other important modeling issues relate to shock, impact, or repeated cyclic stresses (fatigue) that cause the laminate to separate (delaminate) at the interface between two layers. In contrast to metals, in which fatigue failure generally occurs by the initiation and propagation of a single crack, the fatigue process in composites is complex and involves several damage modes, including fiber/matrix debonding, matrix cracking, delamination, and fiber fracture. By a combination of these processes, widespread damage develops throughout the bulk of the composite and leads to a permanent degradation in laminate stiffness and strength.

Fig. 1. V22-Osprey tilt-rotor plane. Fiber-placement technologies are being used to build this aircraft.

Although these complexities lengthen the design process, they are more than compensated for by the mass savings and improvements in aerodynamic efficiency that result. Also, the introduction of virtual manufacturing will play an enormous role in further reducing overall cost. The use of virtual reality in design prior to manufacture to identify potential problems is relatively new but has already demonstrated great potential. Virtual manufacturing validates the product definition and optimizes the product cost; it reduces rework and improves learning.

Manufacturing techniques. A number of techniques have been developed for the accurate placement of the material, ranging from labor-intensive hand lay-up techniques to those requiring high capital investments in automatic tape-laying and fiber-placement machines (**Fig. 1**). Once the component is laid-up on, the mold is enclosed in a flexible bag tailored to the desired shape, and the assembly is enclosed, usually in an autoclave and fitted with a means of raising the internal temperature to that required to cure the resin. The flexible bag is first evacuated, thereby removing trapped air and organic vapors from the composite, after which the chamber is pressurized (to 15 bar or 1.5 MPa) to provide additional consolidation during cure. The process produces structures of low porosity (less than 1%) and high mechanical integrity.

Alternatively, inexpensive nonautoclave processing methods can be used, such as vacuum molding, resin transfer molding (RTM), vacuum-assisted RTM, and resin film infusion (RFI). Vacuum molding makes use of atmospheric pressure to consolidate the material while curing (at 60–120°C or 140–248°F), thereby obviating the need for an autoclave or a hydraulic press. Vacuum-assisted RTM, a liquid-resin infusion process, is considered by the aircraft industry to be the favored low-cost manufacturing process for the future. Further cost reduction can be achieved by reducing the assembly cost, by moving away from fastening (drilling of thousands of holes followed by fastener insertion and sealing) toward bonding. Of course, certification challenges must be addressed with an adhesively bonded joint for a primary aircraft structure application.

Applications. Composites have gained popularity, especially in high-performance products that need to be lightweight but strong enough to sustain harsh loading conditions, such as aircraft components (tails, wings, fuselages), ship hulls, bicycle frames, and racing car bodies. Other uses include fishing rods and sports equipment such as tennis rackets and golf clubs. Carbon composite is a key material in launch vehicles and spacecraft. It is widely used in satellite applications for reflector dishes, antennas, and their supporting structures.

Current civil aircraft applications have concentrated on replacing the secondary structure with fiber-reinforced epoxies whose reinforcement media have been carbon, glass, Kevlar, or hybrids of these materials. Typical examples of the extensive application of composites in this manner are the Boeing 757, 767, and 777 and the Airbus A310, A320, A330, and A340 airliners. The A310 carries a vertical stabilizer 8.3 m (27 ft) high by 7.8 m (26 ft) wide at the base, fabricated in its entirety from carbon composite with a total weight saving of almost 400 kg (880 lb) when compared with the aluminum unit previously used. The A320 has extended the use of composites to the horizontal stabilizer in addition to secondary control surfaces leading to a weight saving of 800 kg (1760 lb) over aluminum alloy skin construction. It has been estimated that 1 kg (2.2 lb) weight reduction saves over 2900 liters (766 gal) of fuel per year. Larger amounts of composites are used in the larger A330 and A340 models, and in the A380 super jumbo jet (**Fig. 2**).

The Boeing 787 Dreamliner structure is made of over 50% composites (80% by volume). The all-composite fuselage makes it the first composite airliner in production. Each fuselage barrel is manufactured in one piece about 13.5 m (45 ft) long, eliminating the need of more than 50,000 fasteners used in conventional aircraft building. However, there are major assembly issues with the composite fuselage sections and there are problems with what is coming out of the autoclave. There is also concern about electromagnetic hazards such as lightning strikes, since the material does not conduct away electric energy.

Composites are also extensively applied in the A400M military cargo plane and the tail of the C17 strategic airlifter. Agile fighter aircraft currently being designed or built in the United States and Europe contain roughly 40% of composites in their structural mass, covering 70% of the surface area of the aircraft. The essential agility of the aircraft would be lost if this amount of composite material were not used because of the consequential mass increase.

For background information *see* AIRCRAFT DESIGN; COMPOSITE LAMINATES; COMPOSITE MATERIAL; POLYMER COMPOSITES in the McGraw-Hill Encyclopedia of Science & Technology.   Constantinos Soutis

Bibliography. *Aerospace Composite Structures in the USA*, Report for the International Technology Service (Overseas Missions Unit) of the DTI, United Kingdom, 1999; A. Kelly (ed.), *Concise Encyclopaedia of Composite Materials*, Pergamon, 1994; F. L. Matthews et al., *Finite Element Modelling of Composite Materials and Structures*, Woodhead Publishing Ltd., 2000; C. Soutis, Failure of notched CFRP laminates due to fiber microbuckling: A topical review, *J. Mech. Behav. Mater.*, 6(4):309–330, 1996; C. Soutis and M. Kashtalyan, Delamination growth and residual properties of cracked orthotropic laminates under tensile loading. *J. Thermoplast. Compos. Mater.*, 15(1):13–22, 2002.

Consciousness and conscious awareness

Consciousness is central to human existence, yet it has proved difficult to define and characterize. A useful criterion in humans is as follows: if one can communicate with others about an event or a perception, it is conscious; otherwise, it is not.

There are several meanings of consciousness in normal discourse. It can mean awareness of an event; if the event is an internal process rather than an occurrence in the world, it becomes self-consciousness. In modern psychology, external and internal events are similar because both change ongoing brain processes and are dealt with as such. Thus, events such as "I see a red book" and "I feel anger" both refer to changes in parts of the brain. The difference between consciousness of external events and self-consciousness becomes one of content, not kind. This conception is also extended to our ongoing awareness of ourselves as individuals, a package of coherent memories of life experiences.

Psychology and consciousness. The study of consciousness was the original goal of psychology. Many nineteenth-century psychologists assumed that the workings of the brain were accessible to consciousness, making psychology a matter of thinking carefully about one's own mental processes. The goal of this introspective method was to describe brain function by analyzing conscious experience. It turned out, however, that introspection could not form a basis for psychology—what seemed right to one psychologist was nonsense to another. Further, it was discovered that most of the processes in the brain were inaccessible to consciousness. More fundamentally, introspection concerns private opinions, not public replicable data. It could not form the basis of a

Fig. 2. Tail of the Airbus A380, entirely built out of carbon-fiber-reinforced plastic.

science. While introspection remains the method of choice for philosophers, it was banished from American psychology by behaviorism, which admitted only observations of behavior. Until the 1960s, behaviorism excluded the study of consciousness from psychology. Contemporary psychology and neuroscience have brought it back with new methods, recognizing that consciousness consists of processes occurring in the brain, processes that cease with death. There is nothing magical or nonphysical about it.

Function. Does consciousness have a function, or is it just an inevitable by-product of a large, complex brain? Evolutionary theory indicates that it does have a function. Since consciousness is culturally universal, we can infer that it is supported by genetically influenced brain mechanisms. The corresponding genes must have been selected for during human evolution and are maintained by evolutionary selection at present. Since brain tissue is energetically expensive, these genes and their resulting brain tissue would have been selected against if consciousness bestowed no benefit. An analogous process is more visible in organisms such as fish that become trapped in caves, living generation after generation in darkness. Released from the natural selection that normally maintains the structure of their eyes, these fish gradually evolve to become blind. The fact that we retain the capability for consciousness indicates that natural selection is maintaining it in us; it has a function. The argument from evolution, though, does not tell us what that function is.

If consciousness has an independent function, there must be human capabilities that can be realized only through conscious intervention. In an experiment demonstrating one such capability, subjects were briefly exposed to a word on a screen; then they were given the first three letters of the word and were asked to complete the three-letter stem with any word other than the one that had just been displayed. They could be shown the word "primate," for example, and immediately see the stem "pri _ _ _ _ ." A correct completion would be "private," "printer," or some other seven-letter word beginning with the letters "pri." If the subjects were shown the target word for a relatively long time, such as a quarter of a second, they usually saw it and were able to complete the task successfully. After a shorter exposure, however, such as a twentieth of a second, they were more likely to produce the target word than to choose another one. In fact, they picked the target word more often than would be expected at random. Picking the target word indicates that information about it had entered the nervous system; however, without a consciousness of that event, the subjects were unable to prevent it from affecting their behavior as instructed. A function of consciousness, then, is to prevent behavior from bending to the momentary availability of information in the environment. It allows behavior to be driven by one's own plans, escaping the tyranny of the environment.

Memory and anesthesia. Basing behavior on plans rather than stimulus-response contingencies re-

quires memory, both of the plans themselves and of relevant events in the environment. The central role of memory in consciousness is revealed in studies of anesthesia, a prime example of induced lack of consciousness. An anesthetic need not deaden pain—it must only suppress memory, and also suppress reflexes so that the patient doesn't jump off the operating table at the first touch of the scalpel. Under these conditions, surgical pain might be severe, but afterward neither the surgeon nor the patient would have any way of knowing that it had occurred. The surgeon would encounter no avoidance responses. When asked later about the anesthesia, the patient will say that everything was fine; he/she didn't feel a thing. In reality, though, the patient didn't *remember* a thing.

Some anesthetics are even more limited; the drug scopolamine, an anticholinergic drug that was once popular during childbirth, left the mother awake and able to cooperate during a delivery, but prevented her from remembering the episode. Afterward, it is as though the pain, and the joy, had never existed. Recent research has shown that even general anesthesia does not always succeed in suppressing consciousness. These studies use the "isolated forearm technique": a tourniquet is tied around the forearm to stop blood flow, preventing a paralyzing agent from reaching the arm during surgery. The patients can then respond with prearranged signals during the operation. Often, they demonstrate high levels of conscious awareness without conscious recall afterwards. Sometimes, though, patients develop postoperative psychological disturbances that indicate the influence of unconscious memory on moods and attitudes.

Neural view. Another technique investigates the neural mechanisms of consciousness by contrasting the brain's responses to stimuli of which a subject is aware with its responses to identical stimuli that escape awareness. Brain researchers have hypothesized that brain waves oscillating about forty times per second are necessary for binding together disparate parts of the brain to coordinate conscious activity. Brain waves have been studied in patients implanted with electrodes to diagnose epilepsy. While the electrodes were in place, but when patients were not experiencing seizures, they could be given a small, painless stimulus to a finger. If the patient noticed the stimulus, an evoked potential (an electrical potential recorded from the central nervous system in response to a stimulus) accompanied by high-frequency gamma waves (electrical brain impulses associated with perception and consciousness) would appear in the hand region of the sensory cortex. An identical but unnoticed stimulus would evoke similar potentials, but without the gamma waves.

The neural correlates of consciousness at a finer scale of brain function have proved elusive. Tracking a stimulus through the visual system is a good way to illustrate the role of consciousness; it is less than one would think. First, you cannot see what is on your retina, and it is fortunate that you cannot. The retinal images are a mess, very different from the accuracy of visual perception. The retina is a curved, spherical surface, and it delivers two independent images, while we perceive only one visual world. The blind spot is not perceived (it was discovered by anatomical methods), and blood vessels course in front of the retinal surface. Color coding and acuity are different in various parts of the retina. None of these distortions reach our conscious perception. The next level of processing, in the thalamus, is no better; the two retinal images are still coded separately.

The visual cortex is where we would expect to begin to see neural correlates of consciousness. There is not a single projection of the retinas on the cortex, though, but dozens. The first and largest of these, the primary visual cortex, occupies a special place in visual processing because it forms the first binocular representation. However, its receptive fields, the areas and patterns to which its neurons are sensitive, do not match our conscious perceptions. The neurons report color balance only in their very restricted regions of the visual field, for instance, while we consciously experience color constancy; colors and their relationships appear the same despite large differences in illumination.

Beyond the primary visual cortex, a welter of specialized areas handles color, motion, and other aspects of vision. None of them support the perception of a high-acuity, full-color visual world. The world that we experience is not localized in a particular area or a particular set of neurons; rather, it reflects the finished product of participation by a large and distributed neural network. It is likely that other aspects of consciousness follow a similar pattern, echoing the results of extensive preconscious processing in the brain. When the final answer comes in about how much of the brain's work is accessible to consciousness, the proportion will be frighteningly small.

For background information see ANESTHESIA; BRAIN; COGNITION; CONSCIOUSNESS; INFORMATION PROCESSING (PSYCHOLOGY); MEMORY; NEUROBIOLOGY; PERCEPTION; PSYCHOLOGY; VISION in the McGraw-Hill Encyclopedia of Science & Technology.
Bruce Bridgeman

Bibliography. B. J. Baars, *In the Theater of Consciousness*, Oxford University Press, New York, 1997; F. C. Crick, *The Astonishing Hypothesis*, Charles Scribner's Sons, New York, 1994; A. R. Damasio, *The Feeling of What Happens: Body and Emotion in the Making of Consciousness*, Harcourt Brace, New York, 1999; C. Koch, *The Quest for Consciousness: A Neurobiological Approach*, Roberts & Company, Englewood, CO, 2004.

Coral reef complexity

Coral reefs occupy some 600,000 km² (232,000 mi²) of tropical seafloor. Like rain forests, they maintain high diversity by recycling limited resources within a tight loop. This allows high productivity at a low cost, but potentially creates a susceptibility to perturbation by even small disturbances. Nevertheless, reefs have shown remarkable stability when considered on larger time scales over the past few million years. Historically, they have likewise shown surprising resilience to increasing human exploitation. Recently, however, reefs have changed dramatically, from coral dominance to overgrowth by macroalgae (seaweeds and kelps; **Fig. 1**). A typical example of this problem has been observed in a study of the decline in Jamaican reefs between 1975 and 1995 (**Fig. 2**), a pattern that has been repeated in every ocean where coral reefs occur. Suggested causes include overfishing, increased sedimentation, pollution, nutrient loading (quantity of nutrients entering an ecosystem in a given period of time), and climate change. The persistence of high coral cover until the 1970s, despite increasing stress, highlights an apparent nonlinearity in the response by the world's reefs, a factor that further complicates our ability to predict future scenarios.

Evolution of the problem. When ecological studies began in the 1950s, coral reefs could be largely explained by natural factors. Coral zonation followed orderly rules that integrated the effects of light, wave energy, and sedimentation along a depth gradient. Regional patterns were similarly well behaved, and reefs were typically characterized as "pristine." While recent historical investigations have shown that reefs were already affected by stresses dating to at least the sixteenth century, coral cover remained high. Even in the early 1990s, reports described reefs as largely in "good condition" because they were either "remote from population centers" or "under good management."

In 1981, the International Coral Reef Symposium started a trend that directed our focus toward management and regulation at the local level. Stressors fell into two principal categories. "Top-down" factors, like overfishing, created downward-cascading impacts that disrupted the coral community structure, and particularly the ability of grazers to keep reef macroalgae in check. In contrast, "bottom-up" problems were those that affected the ability of the reef habitat to support the organisms that lived in, on, and above it. For example, increased nutrients from coastal development, agriculture, and forest clearing encouraged the proliferation of macroalgae, which shaded corals and inhibited the settlement of new coral larvae. Also, because macroalgae were less palatable to grazers, this created a weak link in the food chain.

Starting in the late 1970s, a combination of converging events led to a sudden and precipitous decline in coral reefs that has charged a growing debate over both the causes of this trend and the best ways to mitigate or reverse it. The common link was an increase in temperature tied to fossil-fuel emissions. Corals live close to their upper thermal threshold. When this level is exceeded in late summer, they expel the zooxanthellae (photosynthetic dinoflagellates) harbored within their epithelial tissues. The loss of pigment carried by the expelled endosymbionts exposes the white skeleton beneath—hence

Fig. 1. Underwater photographs of eastern Cane Bay (St. Croix, U.S. Virgin Islands) in 1980 and 2007. Note the common overgrowth of fleshy macroalgae between the sparsely distributed corals in 2007 compared to the more open substrate in 1980 between healthy stands of coral dominated by *Montastraea* spp. (*Photograph from 2007 by Karl Wirth*)

coral bleaching. If temperatures remain high for 2–3 weeks, the metabolically compromised corals die. Originally described in the first quarter of the twentieth century by Alfred Mayer in the Dry Tortugas, Florida, coral bleaching reappeared in the 1980s (Fig. 2). The number of reported cases rose from 3 in 1997 to nearly 30 in 1998 as a result of El Niño warm-

ing events. The frequency of bleaching has progressively increased to the point where a year without significant bleaching is rare.

This was accompanied by the emergence of coral diseases (Fig. 2). In the 1970s, such coral-debilitating diseases as "Black Band" and "White Band" were described in Florida and the Caribbean. By the mid-1980s, new diseases were reported in the Indo-Pacific, followed by the Great Barrier Reef in the mid-1990s. While the proliferation of diseases has tracked recent warming, the link remains elusive. Has rising temperature encouraged new diseases? Have warmer waters strengthened existing ones? Have the myriad stresses related to increased temperature and pollution simply left corals susceptible? To understand these factors, much research is necessary. Most recently, many corals that survive bleaching subsequently succumb to disease.

Agreeing on causes. The recent debate has centered on which impacts are most important to reef decline and, therefore, central to mitigation. One extreme argues that macroalgae have simply moved into space vacated by bleached and diseased corals; thus, local solutions will be ineffective until global warming is reversed. Conversely, removing local stressors may make corals more resistant to warming effects.

Figure 2 provides a broad overview of the factors that have been linked to reef decline. The recent pattern for Jamaica reflects monitoring measurements. Other information is drawn from various publications and ideas raised in the ongoing debate concerning reefs and global warming. Humans probably had measurable effects on the global carbon cycle long before industrialization and the rise of fossil fuels. CO_2 emissions and methane production increased as humans shifted from a hunter-gatherer society to one that cleared forests for agriculture, introduced irrigation, and domesticated animals. However, while humans have lived near reefs for at least the last millennium, their impacts were limited.

The problems with choosing a dominant stressor are related to timing and scale. Both larger predators, such as sharks, and smaller grazing fish were

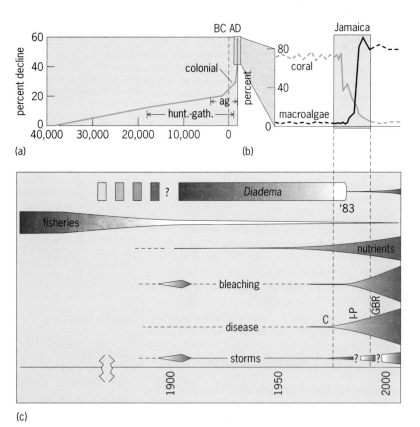

Fig. 2. Changes in the factors related to recent coral reef decline. (*a*) The percent decline over the past 44,000 years is based on Pandolfi et al. (2003). Significant decline increased within the colonial period as human populations spread out from Europe and Africa. (*b*) The most recent, and most severe, decline (that is, the data from the north coast of Jamaica between 1975 and 1995) is typified by the rapid phase shift from coral-dominated reefs to those covered by fleshy macroalgae. Data based on Hughes (1994). (*c*) Historical changes in the main factors thought to be related to reef decline since the colonial period. The emergence of temperature-related responses (bleaching and disease) is coincident with both the loss of urchins and the increase of nutrient loading.

(a)   (b)   (c)

Fig. 3. Three dominant shallow-water communities on Easter Island. Since the 1950s, fleshy algae such as the (a) *Sargassum* **were the dominant cover, and coral was rare in shallow water. In the 1980s, warming drove a rapid shift to a low-diversity coral community dominated by (b)** *Pocillopora verrucosa* **(***photo from 1999 by Henry Tonnemacher/7-Seas-Ltd.***). (c) The newly emerged coral community was decimated by bleaching in 2000 (***photo by Michel Garcia***).**

significantly depleted prior to 1900 (Fig. 2). Nevertheless, the abundance of reef-building corals remained high for at least another 70 years. It has been proposed that the loss of Caribbean grazing fish was offset by rapidly expanding urchin populations (*Diadema antillarum*), which kept macroalgae in check until their sudden regional demise in 1983. However, accounts of high urchin abundance prior to 1900 argue against this. For example, maps from the eighteenth century reflect a huge abundance of urchins in the shallow waters surrounding Barbados—one reef was named "Cobblers' Reef," after the local home for urchins. While nutrification (excessive addition of nutrients to a water body) seems like a good temporal fit, it is insufficient to explain the synchronicity of a global decline that includes reefs removed from obvious nutrient sources. A warming threshold is the easiest to envision because corals live so close to their upper thermal limits. However, the loss of corals will not result in a takeover by macroalgae in the absence of the elevated nutrient levels required to support them. Any solution will have to recognize the complexity of the relationships shown in Fig. 2.

Agreeing on solutions. Until the recent emergence of disease and bleaching, solutions focused on "No Take Areas" to encourage the recovery of grazers and top-end predators. These were often surrounded by "Marine Protected Areas (MPAs)," where activities perceived to be detrimental to recovery within the exclusionary core were restricted or regulated. The focus was often on numbers of species, while recognizing that surrounding areas contributing to these "biodiversity hot spots" were equally important. Success rested on determining the minimum size needed for adequate buffering and the distance between MPAs required to maintain interconnectivity. All of this is complicated further by a recent recognition that the effects of encouraging predators will differ depending on the extent to which algal overgrowth is the main problem on a specific reef.

We have limited economic, political, and intellectual capital to expend on this problem. Thus, our strategy will involve triage. Which of the areas that can be saved will actually make a difference? Do we focus on protecting reefs that remain more "pristine"

or do we try to reverse the decline of those that are on the brink of irrecoverable decline?

Clearly, global warming has changed the landscape of both coral reefs and the politics that will determine their future. However, the most aggressive proposals to reduce carbon emissions will still result in CO_2 concentrations leveling off in the vicinity of 500 parts per million (ppm). Curbing emissions is clearly the most pressing imperative of this century, but remediation strategies that involve lower absolute temperatures in the near future are not realistic. The remaining options may be limited to locally reversing trends on individual reefs while climate change is addressed on a larger stage. One immediate challenge is to identify those species that are critical to maintaining the most important reef functions. *Acropora palmata* appears to have been largely absent from Caribbean reefs for centuries at a time over the past 10,000 years. What does this mean, and how does it relate to decisions to focus on particular species as "keystones," without which reefs will cease to function efficiently? On Easter Island, macroalgae dominated in shallow water until they were killed in the early 1980s and replaced by dense stands of the coral *Pocillopora verrucosa*— which in turn were decimated by bleaching in the year 2000 (**Fig. 3**). Which, if any, of these was the "normal" or "healthy" population, and how does this turbulent scenario compare to the 1500 years of previous occupation? Until we understand the implications of these and similar changes, we will remain ill-equipped to choose what to protect or not. Ultimately, we are left with the challenge of designing strategies to protect reefs that will have to adapt to life in a CO_2 world, the likes of which have never been seen by *Homo sapiens*.

For background information *see* ALGAE; CLIMATE MODIFICATION; ECOLOGICAL SUCCESSION; ECOSYSTEM; ENVIRONMENTAL TOXICOLOGY; GLOBAL CLIMATE CHANGE; MARINE CONSERVATION; MARINE ECOLOGY; PALEOECOLOGY; REEF; RESTORATION ECOLOGY in the McGraw-Hill Encyclopedia of Science & Technology.   Dennis Hubbard

Bibliography. R. B. Aronson et al., Causes of coral reef degradation, *Science*, 302:1502–1504, 2003; T. P. Hughes, Catastrophes, phase shifts and large-scale degradation of a Caribbean coral reef, *Science*, 265:1547–1551, 1994; T. P. Hughes et al., Climate change, human impacts, and the resilience of coral reefs, *Science*, 301:929–933, 2003; J. B. C. Jackson, Reefs since Columbus, *Coral Reefs*, 16(suppl.):S23–S32, 1997; P. J. Mumby et al., Fishing, trophic cascades, and the process of grazing on coral reefs, *Science*, 311:98–101, 2006; J. M. Pandolfi et al., Global trajectories of the long-term decline of coral reef ecosystems, *Science*, 301:955–958, 2003; W. F. Ruddiman, How did humans first alter global climate?, *Sci. Am.*, 292:46–53, 2005.

Corals and ocean acidification

Living corals classified as order Scleractinia are tiny solitary and colonial marine animals with delicate soft polyps. Most shallow reef corals harbor symbiotic algae called zooxanthellae that are not only essential for survival and metabolism but also increase the coral's skeletal growth. By secreting massive skeletons of calcium carbonate ($CaCO_3$), scleractinians build fantastic undersea structures called coral reefs. Greenhouse warming from atmospheric carbon dioxide (CO_2) is causing massive coral bleaching, signaling the loss of zooxanthellae. In addition to a rise in coral disease and direct human impacts, the mixing of anthropogenic CO_2 in seawater is negatively affecting the calcification of coral skeletons and thereby undermining the ability of coral to build reefs. Biologists are testing the effects of ocean acidification on corals, while paleontologists are exploring the geological record of mass extinction to uncover clues of ancient episodes of acidification. The "naked coral" hypothesis (see below and **Fig. 1**) helps explain enigmas of coral evolution and the response of corals and reefs to ocean acidification.

Fig. 1. Specimens of living *Oculina patagonica* following the aquaria dissolution experiments. This illustrates living naked polyps that remained after the coral skeleton was completely dissolved in low-pH conditions. (*Image courtesy of Maoz Fine, Bar-Ilan University, Israel*)

Sea chemistry changes and ocean acidification. Since the Industrial Revolution, the burning of fossil fuels has markedly increased the amounts of CO_2 in the atmosphere, to a point where 22 million tons are released each day. The oceans are a natural sink for this greenhouse gas, which is increasingly being dissolved in seawater. Increasing the partial pressure of CO_2 in the atmosphere (pCO_2) affects carbonate solubility and saturation, thereby decreasing $CaCO_3$ precipitation or secretion by calcifying animals. Corals in the future high-CO_2 world are likely to develop thin, fragile skeletons and experience declines in growth, hampering their ability to build reefs. Rising CO_2 is changing the ocean's pH, and acidification is the result. It can be illustrated by the following equation for calcium carbonate:

$$Ca^{2+} + 2HCO_3^- \leftrightarrow CaCO_3\downarrow + H_2O + CO_2\uparrow$$

The balance between the two sides of the equation relates to pH, either becoming more alkaline (increase of pH; equation driven to right) or becoming more acidic (decrease of pH; equation driven to left). On a scale of 14 to 0, a pH of 7 is neutral. When pH decreases in the ocean, scientists call it acidification. As the world's oceans continue to absorb CO_2, they are shifting toward the acid range. Since the Industrial Revolution, pH has decreased 0.11 of a pH unit. Although this may seem a trivial amount, the pH scale is logarithmic, so a value of 7.0 is 10 times more acidic than 8.0. Ocean acidification is dissolving minute calcified phytoplankton called coccolithophores and other floating calcified life that are part of the essential food chain, and now scleractinian corals and other calcifying creatures are feeling the effects of ocean acidification. If this trend continues, nutrification reefs, already under attack by bleaching, nitrification, and disease, are predicted to collapse globally as early as the end of the twenty-first century.

Problems of the origin of modern corals. Living and fossil Scleractinia secrete $CaCO_3$ as the mineral aragonite. Their extensive geological record goes back to the Middle Triassic period [240 million years ago (Mya)], and soon afterwards they became reef builders. Extinct corals of the Paleozoic, including Rugosa, Heterocorallia, and Tabulata (**Fig. 2**), were reef builders as well, especially during the Silurian and Devonian periods. However, by 251 Mya, the time of the great end-Permian mass extinction, these corals and reefs had all become extinct. Paleozoic corals secreted calcite and were morphologically different from order Scleractinia, which did not appear until some 11–14 million years after the mass extinction, long after the demise of Paleozoic corals. The enigma of scleractinians is not only this large gap in geologic time without skeletonizng corals, but the failure to find any likely ancestors in the fossil record. Also curious was the discovery that the earliest scleractinians morphologically were not simple but complex and were characterized by diversity levels greater than expected for a newly evolved group. Biologists puzzling over these problems had suspected a kinship, based on anatomical similarities, between scleractinians and some soft-bodied anemone-like groups—groups traditionally classified into orders separate from Scleractinia. Some genetic molecular data indicated that all members of Scleractinia shared a common ancestor, but results were ambiguous and failed to resolve this ancestor. In an attempt to rectify these problems, the "naked coral" hypothesis was offered. It postulates that scleractinian skeletons are capable of being lost or acquired at different times, depending on the seawater chemistry. This hypothesis challenges the traditional notion of a single common ancestor by predicting multiple, independent origins for all calcified corals called "Scleractinia." The "naked coral" hypothesis also implies that the traditional classification based on the presence or absence of a skeleton is artificial, and that some anemone-like animals are best grouped with calcified corals.

Recent molecular data strongly indicate that there was more than one common ancestor for living scleractinians, suggesting as many as four separate origination events. Furthermore, the data derived from the DNA clock method [a dating technique derived from analyses on extant species and based on the assumption that mutations in the mitochondrial genome (mtDNA) accumulate over time at rates that can be calibrated] do not agree with the fossil record data; that is, the molecular data indicate that scleractinians must have evolved in the Paleozoic, whereas the fossil record indicates their evolution in the Middle Triassic period. This pushes scleractinian origins back to the Paleozoic, when now-extinct corals with calcite skeletons dominated. Rare discoveries in Permian and Ordovician rocks of very strange scleractinian-like fossils with aragonite skeletons, dubbed "scleractiniamorphs" (Fig. 2), were regarded as either scleractinian progenitors or aberrant, dead-end Paleozoic coral offshoots. It seems more reasonable in light of the above hypothesis that scleractiniamorphs were brief experiments by "naked corals" at calcification during a time when aragonite was not the favored skeletal mineral.

Experimental results and naked corals. How will living corals respond to rising levels of CO_2? Experiments that artificially increased pCO_2 on corals in aquaria confirmed the predictions of decreased calcification rates. For example, increasing atmospheric CO_2 from only 330 to 560 parts per million (ppm) would decrease world coral growth some 40%. In addition, corals may respond dramatically to pH effects. Both expected and unexpected results have been found in experiments on corals in which seawater pH was altered to resemble that expected of a higher-CO_2 world. In this test, the survival of five species of colonial scleractinians based on increased pH was analyzed. In marine aquaria, healthy colonial zooxanthellate corals were subjected to prolonged exposure to pH levels of 7.3 to 7.6 and 8.0 to 8.3 (ambient) for 12 months (pH 8.2 was normal for these corals). Soon coral skeletons started dissolving. After 1 month, corals in the lower pH aquaria experienced complete skeletal dissolution, but surprisingly they

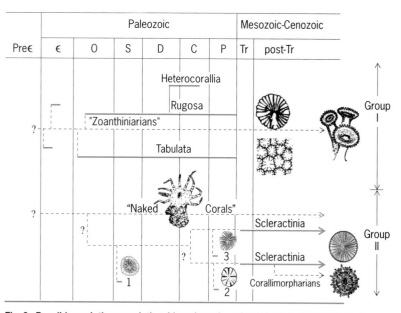

Fig. 2. Possible evolutionary relationships of corals and soft-bodied groups (dashed lines = "naked," noncalcified groups; solid lines = calcified corals). Geologic periods are abbreviated (full names are given in Fig. 3). Group I: extinct calcitic coral groups and living zoanthiniarian anemones, some of which may have given rise to Paleozoic corals. Group II: calcified corals and anemone-like groups discussed in the text. "Naked corals" represent a generalized group or groups of noncalcified organisms that left no fossil record. Numbers 1, 2, and 3 denote scleractiniamorphs discussed in the text. The diagram illustrates multiple origins of "Scleractinia" and their sudden appearances from "naked" groups. (*Modified from W. A. Oliver, Jr., Origins and relationships of Paleozoic coral groups and the origin of the Scleractinia, Paleontol. Soc. Papers, 1:107–134, 1996*)

did not die. Rather, they continued living as tiny soft polyps (Fig. 1). Even after 12 months in these low-pH conditions, the soft polyps continued to metabolize and reproduce without benefit of protective skeletons. Some colonial polyps even retained their symbiotic zooxanthellae. The test corals survived at low pH for over a year (minus the skeleton). After a year, the pH was changed back to normal levels. Astonishingly, the coral polyps responded by reacquiring skeletons. By experimentally demonstrating the effects of acidification on skeletal dissolution and nonskeletal survival, support grew for "naked" or ephemeral corals.

The "naked coral" hypothesis also received validation from molecular techniques. Mitochondrial DNA obtained from both extant corals and anemone-like species pinpointed for the first time a particular scleractinian group with an unambiguous connection to the order Corallimorpharia, a group of living "naked corals" without any fossil record. An ancient scleractinian progenitor was identified, revealing more molecular correspondence to Corallimorpharia than to other living scleractinians. Molecular methods dated the progenitor to the Cretaceous "greenhouse" world, about 110–132 Mya. In these high-CO_2 oceans, the ancestor presumably lost its skeleton and survived "naked" for more than 100 million years.

Mass extinctions and ancient CO_2 levels. The geological record indicates the global collapse of reefs during mass extinctions ranging from the Ordovician through the Cenozoic. In many cases, a high correspondence exists between high CO_2 and mass

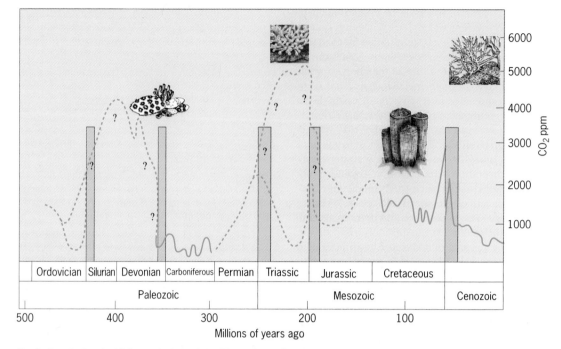

Fig. 3. Rough sketch of CO_2 trends through the Phanerozoic. Question marks and dashed lines emphasize many uncertainties about the levels of CO_2. The dark bars mark the often-lengthy postextinction eclipse of reefs. Artistic sketches above the CO_2 curves show reconstructions of coral reef organisms from geological time intervals discussed in the text. Upright rudist bivalves are depicted during the Cretaceous Period. (*Adapted from J. E. N. Veron, A Reef in Time, Belnap Press, Cambridge, MA, 2008*)

extinctions (**Fig. 3**). The CO_2 curves generated for Phanerozoic time appear extraordinary, indicating high levels of CO_2, up to 20 times that of the present day. Results from scant paleo-CO_2 isotope data are fraught with difficulties, including inconsistent quality and a small number of data points distributed irregularly over 500 million years of an incomplete geological record. Inability to resolve geological time with the precision needed to assess climate cycles and CO_2 levels is frustrating, but the CO_2 trends thus far revealed have received support by independent paleoclimate indicators. Some calcifying marine organisms building reefs, such as during the Silurian-Devonian, may have adapted to long intervals of high CO_2, but the geologically sudden release or fall of greenhouse gases such as well documented at the Paleocene-Eocene may have resulted in a much greater extinction response.

The largest mass extinction of all time at the end of the Permian period caused the loss of 92% of species and wreaked great havoc in the oceans. It ended with total extinction of all corals and reefs, tremendous losses among the shelly calcifying biota, and cessation of carbonate deposition. Volcanic eruptions produced high CO_2 levels and global warming, possibly with injection of large amounts of methane, a greenhouse gas 20 times worse than CO_2. It is known that global warming can affect ocean circulation. Stagnant ocean circulation persisted for more than 11 million years of the Early Triassic aftermath. With imbalances in carbon cycling leading to anoxia (oxygen depletion) and ocean acidity, conditions were certainly inimical for corals with skeletons. All Paleozoic calcified corals died out, but the "naked coral"

hypothesis posits that nonskeletonized, anemone-like forms survived. Clinging to isolated refugia, they left no fossil record. However, much later in the Middle Triassic, some acquired calcified skeletons. This is why the aquaria experiments are of such interest. The conundrum of the sudden and unexplained appearance of scleractinians is easy to explain with the "naked coral" hypothesis. The sudden appearance of Middle Triassic corals was likely an adaptive response to ameliorating ocean chemistry.

The succeeding end-Triassic mass extinctions witnessed the collapse of reefs and the disappearance of most heavily calcified corals. It was followed by an Early Jurassic gap during which reefs are scarce. Causes included volcanism, high CO_2, and ocean acidification. During the Cretaceous "greenhouse" conditions, CO_2 was 5–10 times that of today. Reefs of scleractinians were replaced during Cretaceous time by strange clams called rudists (Fig. 3). The end-Cretaceous "K/T" mass extinction that wiped out dinosaurs also destroyed more than 50% of marine life, including all rudists and 70% of coral genera. Associated with a meteorite impact, the extinction also coincided with vast volcanic eruptions and injection into the atmosphere of much CO_2. Extinctions among corals and other calcified marine life likely coincided with low carbonate saturation and acidification. Warm "greenhouse" conditions continued into the Cenozoic during the lengthy reef gap (Fig. 3). Scleractinians recovered later in the Paleogene period, only to experience an episode of ocean warming followed by cooling near the end of that period. The following Neogene period ushered in the "icehouse" world of today and initiated a renewed phase of reef building, producing many new coral lineages. However, gaps between some lineages and failure to find ancestor-descendants are explained by the "naked coral" idea.

Outlook. CO_2 and pH experiments showing dissolution of calcareous skeletons and reefs provide insight and direction into where the present climate is moving. Coral survival and amazing recovery in the high-pH aquaria experiments cannot justify relaxing concerns over CO_2 and ocean acidification. The fossil record of past mass extinctions of corals and reefs indicates a possible dark future for the reefs of today. The predicted world of the future, without tropical reefs, fishes, or complex interactions to maintain Earth's most vital ecosystems, could hardly hold a place for humans.

For background information *see* ARAGONITE; CARBON DIOXIDE; CLIMATE HISTORY; CLIMATE MODIFICATION; CLIMATE PREDICTION; CORALLIMORPHARIA; EXTINCTION (BIOLOGY); GEOLOGIC TIME SCALE; MARINE ECOLOGY; OCEAN; PALEOECOLOGY; REEF; SCLERACTINIA; SEAWATER in the McGraw-Hill Encyclopedia of Science & Technology.   George D. Stanley, Jr.

Bibliography. M. Fine and D. Tchernov, Scleractinian coral species survive and recover from decalcification, *Science*, 315:1811, 2007; O. Hoegh-Guldberg et al., Coral reefs under rapid climate change and ocean acidification, *Science*, 318:1737–1742, 2007; M. Medina et al., Naked corals: Skeleton loss in Scleractinia, *Proc. Nat. Acad. Sci. USA*, 103:9096–9100, 2006; W. A. Oliver, Jr., Origins and relationships of Paleozoic coral groups and the origin of the Scleractinia, *Paleontol. Soc. Pap.*, 1:107–134, 1996; G. D. Stanley, Jr., The evolution of corals and their early history, *Earth-Sci. Rev.*, 60:195–225, 2003.

Cyclic universe theory

The cyclic universe theory is a model of cosmic evolution according to which the universe undergoes endless cycles of expansion and cooling, each beginning with a "big bang" and ending in a "big crunch". The theory is based on three underlying notions: First, the big bang is not the beginning of space or time, but rather a moment when gravitational energy and other forms of energy are transformed into new matter and radiation and a new period of expansion and cooling begins. Second, the bangs have occurred periodically in the past and will continue periodically in the future, repeating perhaps once every 10^{12} years. Third, the sequence of events that set the large-scale structure of the universe that we observe today took place during a long period of slow contraction before the bang; and the events that will occur over the next 10^{12} years will set the large-scale structure for the cycle to come. Although the cyclic model differs radically from the conventional big bang–inflationary picture in terms of the physical processes that shape the universe and the whole outlook on cosmic history, both theories match all current observations with the same degree of precision. However, the two

pictures differ in their predictions of primordial gravitational waves and the fine-scale statistical distribution of matter; experiments over the next decade will test these predictions and determine which picture survives.

Cosmic evolution in the cyclic model. In the cyclic picture, the universe immediately after the big bang is filled with hot, dense matter and radiation with a temperature that is about 10^{20} times the temperature in the core of the Sun. This temperature is hot enough that the matter is broken down into its elemental constituents (quarks, electrons, photons, and so forth), although the temperature is modest compared to the divergent temperatures expected in the usual big bang.

The expansion and cooling over the next 9×10^9 years lead to the successive clumping of the elemental constituents into protons and neutrons, then into atoms, molecules, dust, planets, stars, galaxies, and larger-scale structures, just as in the conventional big bang picture. At this point, the matter density has become so dilute that a new form of energy, known as dark energy, overtakes the universe. Unlike matter and more familiar forms of energy, which are gravitationally self-attractive and resist the expansion of the universe, dark energy is gravitationally self-repulsive and causes the expansion of the universe to speed up. This is consistent with observations of distant supernovae and the cosmic background radiation showing that the expansion of the universe has been accelerating for the past 5×10^9 years. Over the next 10^{12} years, the accelerated expansion continues, rapidly diluting the matter and lumpy structures formed in the universe since the last big bang. In this way, dark energy naturally and efficiently brings space back to a simple, uniform, pristine state leading up to each big bang.

According to the cyclic model, the dark energy is unstable, decaying near the end of the cycle into a form of very high-pressure energy (pressure exceeding the energy density) that causes the universe to contract ultra-slowly. A slowly contracting universe filled with this high-pressure form of energy produces two important effects. First, space becomes increasingly smooth and flat as the contraction proceeds, so that these conditions are set leading up to and immediately after the big bang. Second, random fluctuations of the high-pressure form of energy due to quantum physics produce tiny variations in the rate of contraction and the time when different regions of space are filled with new matter and radiation. This combination of effects results in a nearly perfectly smooth universe after the bang except for a pattern of tiny temperature variations just like those observed in the cosmic microwave background radiation and tiny variations in the concentration of matter just like what is needed to explain the distribution of galaxies.

Braneworlds and the cyclic model. The cyclic model was inspired by developments in string theory, especially the ideas of braneworlds and extra dimensions. Although the cyclic model does not require these ideas, or string theory, for that matter, they suggest an intuitive and appealing geometrical interpretation of the cyclic model. According to a version of string theory known as M theory, our three-dimensional world is actually a braneworld embedded in a space with an extra spatial dimension separated by a microscopic distance from a second three-dimensional braneworld. (In the full version of string theory, there are six additional dimensions, but they play no role in the cyclic cosmology and can be ignored.) The braneworlds are elastic—they can expand, wiggle, warp, and move. (The word braneworld, sometimes called simply brane, is short for "membrane-like world.") Matter on our brane is constrained, by the laws of string physics, to move only in our three dimensions. Consequently, we cannot reach into the fourth spatial dimension, nor can we touch, feel, or see the other braneworld. However, the braneworlds are attracted to one another by gravity, and there can be additional fields that generate springlike forces between braneworlds.

In the braneworld depiction, the cyclic model is described as the regularly repeating collisions between the two braneworlds caused by gravity and the springlike forces drawing them together. The big bang is the collision itself. This picture makes it immediately clear that the big bang is not the "beginning," since a collision necessarily implies that the braneworlds—that is, space and time—exist before and after that collision. The big bang is merely the moment when some of the gravitational energy and braneworld kinetic energy are converted into hot matter and radiation, returning the universe to high temperature and matter density. After the bang, the branes bounce apart and begin to stretch along the three large dimensions, causing the matter and radiation to expand, cool, and clump into the structure observed today. The dark energy is the potential energy stored in the springlike force drawing the branes together. Once the branes have stretched enough that the matter density is less than the stored energy density, the stretching of the branes speeds up, thus explaining the accelerated expansion observed today. However, this phase does not last forever. Eventually, after perhaps 10^{12} years, the springlike force draws the branes together and converts the stored energy into braneworld kinetic energy. At the next collision, the kinetic energy plus gravitational energy is converted partially into new matter and radiation at the collision. Because of the regular contribution of gravitational energy, which is an unlimited source of energy, the spring never winds down and the collisions can continue forever (see **illustration**).

In the colliding braneworld picture, the cycling is associated with expansion and contraction of the extra dimensions. The two braneworlds themselves are expanding throughout. In this sense, the cyclic model is really a hybrid of periodic oscillations along the extra dimension and steady growth along the usual three dimensions. The fact that the braneworlds are always expanding means that, unlike earlier oscillatory models of the 1920s and 1930s, the entropy created in previous cycles is

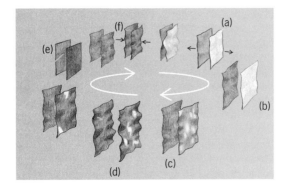

The cyclic universe can be described as two branes engaging in an endless cycle of collision, rebound, stretching, and collision once again. (*a*) Wrinkled branes collide (the big bang), create slightly nonuniform hot plasma, and rebound. (*b*) A microsecond after the bang, branes reach maximum separation but continue to stretch rapidly, filled with radiation. (*c*) Radiation dilutes away; matter dominates and clusters around nonuniformities to form galaxies and stars. (*d*) In the present epoch, dark energy has taken over, driving accelerated expansion that begins to spread out galaxies and matter. (*e*) At 10^{12} years after the bang, branes are empty, flat, and parallel. (*f*) The interbrane force draws branes together, amplifying quantum wrinkles.

not reconcentrated at the end of each cycle. It is diluted during the 10^{12} years of accelerated expansion and remains dilute at the bang. If it were reconcentrated, as in the early oscillatory models, the increasing entropy density from cycle to cycle would mean that the cycles could not have gone on indefinitely into the past—this was the "entropy problem" with the earlier models that led to their undoing, as first pointed out by Richard Tolman in the 1930s. Instead, the entropy produced in each cycle is diluted by the stretching of the braneworlds and the creation of new space, so that, by the time the next bang occurs, the earlier entropy is too dilute to have any cosmological effect.

Distinguishing models. The past history and future evolution of the universe are remarkably different in the cyclic model and the big bang inflationary model. In the big bang inflationary picture, the big bang is the beginning of time, so the universe is only 14×10^9 years old. The large-scale structure of the universe was set by a period of inflation after the big bang. Dark energy has to be introduced ad hoc to explain the observed accelerated expansion of the universe, but otherwise it serves no needed role. Once introduced, though, dark energy dominates the future of the universe, turning the space we observe into an eternal, dark, vacuous wasteland. By contrast, according to the cyclic picture, 14×10^9 years represents only the time since the last big bang and the creation of the matter and radiation that we see. However, the universe has had many cycles, perhaps infinitely many, prior to the present one, so that the actual age of the universe is much greater than 14×10^9 years. There is no need for inflation because large-scale structure is set by events leading up to the bang, as described previously. Dark energy is an essential element that is integrated into the cosmology. The future is more hopeful: Space will become dark and vacuous for a period, but then a new cycle will begin, and the space we observe will be infused with new matter, galaxies, stars, planets, and life.

Although the two theories are equally capable of explaining all astronomical and cosmological observations collected thus far, there are two key tests that can distinguish them. First, inflation produces a spectrum of gravitational waves, wrinkles in space that propagate through the universe that should produce a detectable polarization pattern in the cosmic background radiation. The cyclic model has no period of inflation and produces gravitational waves that are far too weak to be detected. Second, the inflationary picture predicts that the statistical distribution of temperature variations in the cosmic background should be nearly perfectly gaussian (that is, the distribution about the mean temperature should follow a bell curve). The cyclic model predicts a measurable deviation from gaussianity.

A number of experiments capable of detecting the polarization of the cosmic background radiation and the deviations from gaussianity, if either are present, will be conducted over the next decade. The outcome will determine which cosmic history is correct and whether the future will be wasteland or periodic rebirth.

For background information *see* ACCELERATING UNIVERSE; BIG BANG THEORY; COSMIC BACKGROUND RADIATION; COSMOLOGY; DARK ENERGY; DISTRIBUTION (PROBABILITY); ENTROPY; GRAVITATIONAL RADIATION; INFLATIONARY UNIVERSE COSMOLOGY; SUPERSTRING THEORY; UNIVERSE in the McGraw-Hill Encyclopedia of Science & Technology.   Paul J. Steinhardt

Bibliography. A. H. Guth, *Inflationary Universe: The Quest for a New Theory of Cosmic Origins*, Perseus, New York, 1997; J. P. Ostriker and P. J. Steinhardt, The quintessential universe, *Sci. Amer.*, 284(1):46–53, January 2001; P. J. Steinhardt and N. Turok,, *Endless Universe: Beyond the Big Bang*, Doubleday, New York, 2007.

DARPP-32

Dopamine- and cAMP-regulated phosphoprotein of molecular weight 32 kDa (usually known by its acronym, DARPP-32) is a master molecular regulator in neurons that receive the neurotransmitter dopamine. For over two decades, an intense research effort has demonstrated that DARPP-32 occupies a unique position as a central molecular control point, integrating multiple information streams that converge through a variety of neurotransmitters, neuromodulators, neuropeptides, and steroid hormones onto dopaminoceptive neurons. Since dopaminergic neurotransmission is critical for motivated behavior, working memory, and reward-related learning and is implicated in, among other conditions, schizophrenia, alcoholism, Parkinson's disease, and pathological gambling, DARPP-32 has received considerable attention not only in basic neuroscience, but also in studies of the pathogenesis of these disorders and as

a promising target for innovative drug treatment in psychiatry. In addition, DARPP-32 is found in nonneuronal tissues and may have relevance for certain types of cancer.

Molecular action. DARPP-32 was initially identified as a major target for an enzyme, dopamine-activated adenylyl cyclase, in the striatum (the part of the basal ganglia of the brain that includes the caudate nucleus and putamen). Subsequent work has shown that DARPP-32 is essentially a molecular switch that is regulated by whether or not certain of its amino acid residues are phosphorylated (see **illustration**). Two of these sites (Thr^{34} and Thr^{75}) are in the N-terminus of the protein. Through signals that increase cGMP or cAMP [two major forms of cyclic nucleotides, 3′,5′-cyclic guanosine monophosphate (cyclic GMP or cGMP) and 3′,5′-cyclic adenosine monophosphate (cyclic AMP or cAMP)] in the cell, activation of protein kinase A (PKA) leads to phosphorylation of DARPP-32 at Thr^{34}. This turns DARPP-32 into a potent inhibitor of protein phosphatase 1 (PP-1). Because PP-1 dephosphorylates and thereby deactivates many molecules in the cell, the inhibition of PP-1 through DARPP-32 amplifies the action of many transcription factors, receptors, ion channels, and protein kinases. At baseline, DARPP-32 is also usually phosphorylated at Thr^{75} by the enzyme Cdk5, which makes DARPP-32 into an inhibitor of PKA. Two other regulatory sites lie at the C-terminus of the molecule: DARPP-32 is highly phosphorylated at Ser^{102} (by the enzyme CK2) and Ser^{137} (by the enzyme CK1) under basal conditions. The overall consequence of this is to increase the phosphorylation status of Thr^{34} because phosphorylation of Ser^{102} of DARPP-32 increases the efficiency of phosphorylation of Thr^{34} by PKA, and phosphorylation of Ser^{137} decreases the rate of dephosphorylation of Thr^{34} by another phosphatase, PP-2B. DARPP-32 thus is a point of convergence for at least four kinases and three phosphatases that integrate events from

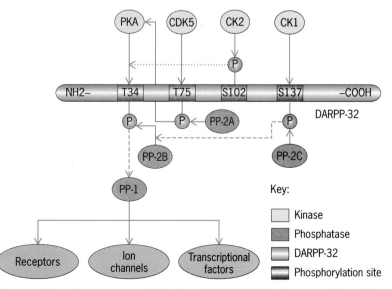

Phosphorylation sites, enzymes, and effectors of DARPP-32. Dotted arrow: positive effect; dashed arrow: negative effect; solid arrow: biochemical pathway. (*From H. J. Reis et al., Is DARPP-32 a potential therapeutic target?, Expert Opin. Ther. Tar., 11(12):1649–1661, 2007, with permission*)

multiple messenger systems; it has been called "a network within one protein."

DARPP-32 is regulated by dopamine in a bidirectional manner: stimulation of D1 (dopamine subtype 1) receptors, through Gs-protein, increases the activity of adenylyl cyclase and consequently PKA activity, which increases phosphorylation of DARPP-32 at Thr^{34}. Conversely, D2 receptor stimulation inhibits adenylyl cyclase and PKA and attenuates DARPP-32 inhibition of PP-1. These pathways are critical for mediating not only the cellular actions of dopamine, but also the interactions of this neurotransmitter system with glutamatergic, serotonin, adenosine, and GABAergic (γ-aminobutyric acid–transmitting or -secreting) systems, among others.

Brain systems involved in DARPP-32 effects. DARPP-32 is expressed in regions receiving dopaminergic innervation. By far the highest levels are found in the neostriatum (caudate and putamen), where DARPP-32 is expressed in GABAergic medium-sized spiny neurons. However, DARPP-32 is found in other regions of the brain as well, such as the cerebral cortex and hippocampus. In a topographically well-organized manner, the neostriatum receives excitatory glutamatergic projections from the cortex and thalamus, integrates them with monoaminergic inputs, and sends them via the globus pallidus and substantia nigra pars reticulata to the thalamus, which projects back to the cortex. These parallel processing loops are critical for the ongoing processing of sensorimotor, cognitive, and emotional information. Of particular interest for neuropsychiatry is a circuit linking the dorsolateral prefrontal cortex (DLPFC) with the rostral striatum. One function attributed to these prefrontal-striatal interactions is that of acting as a "filter" of information competing for prefrontal cortical processing.

DARPP-32 and human cognition. Recently, using a translational genetic approach, a genetic variation was identified in the gene encoding DARPP-32, *PPP1R1B*, through resequencing. The common variants, none of which were coding, were then genotyped in a large Caucasian family-based data set, identifying genetic variants [single nucleotide polymorphisms (SNPs, single base-pair differences between copies of a DNA sequence from two individuals) and haplotypes (a set of alleles of closely linked loci on a chromosome that tend to be inherited together)] that were associated with performance on a range of cognitive tests dependent on frontostriatal function, such as attention, psychomotor speed, and working memory. Since this suggested an impact on neuronal function, an investigation using an independent postmortem data set was undertaken to study the influence of genetic variation in *PPP1R1B* on *PPP1R1B* mRNA levels, as transcriptional regulation is a frequently discussed mechanism for the impact of nonexonic (nonprotein-coding) genetic variation on neuronal function in humans. It was found that the same variants that affected cognitive function predicted mRNA expression of *PPP1R1B* isoforms in postmortem human brain. Together with the cognitive data, this indicated a molecular mechanism for

the functional influence of *PPP1R1B* genetic variation on striatal neurons and their likely interactions with prefrontal cortex. In an independent and large sample of healthy Caucasian subjects, this possibility was tested on the neural systems level using multimodal neuroimaging. It was found that these same genetic variants in *PPP1R1B* affected neostriatal structure and activation, as well as structural and functional connectivity of striatum with prefrontal cortex. A compatible finding showing that the same genetic variation contributes to striatally dependent cognitive function was also subsequently obtained in an independent study.

DARPP-32 and drugs of abuse. Drugs of abuse increase dopamine neurotransmission in striatum, a mechanism that has been linked to their action on reward signaling and craving. This suggests involvement of DARPP-32 in the mechanism of action of drugs of abuse. In fact, almost all such drugs, including caffeine, opiates, cocaine, amphetamine, cannabis, LSD, phencyclidine (PCP), nicotine, and ethanol, have been shown to influence the phosphorylation pattern of DARPP-32, despite their varied pharmacology and initial action on quite diverse neurotransmitter systems. In addition, in DARPP-32 knockout mice, the effects of amphetamine, cocaine, cannabis, LSD, PCP, and ethanol are reduced. This evidence indicates that DARPP-32 functions as a key node in a final common pathway of psychotomimetics (psychedelic or hallucinogenic drugs inducing transient states of altered perception resembling or mimicking the symptoms of psychosis) in both the frontal cortex and the striatum. More recently, mice selectively mutated at each of the four phosphorylation "switch" sites discussed above have been generated, which should be useful to further dissect the molecular pathways underlying the effects of drugs of abuse.

DARPP-32 and schizophrenia. The central role of DARPP-32 in molecular mechanisms related to dopaminergic neurotransmission in the striatum, and its diverse pharmacological actions, suggests an involvement in schizophrenia. In postmortem brain, one report found that DARPP-32 protein abundance was selectively reduced in the prefrontal cortex of brain tissue from patients with schizophrenia, but another report found no significant difference in mRNA levels in brains from 16 elderly schizophrenic patients compared to matched controls. Further evidence indicated that the same variants in *PPP1R1B* affecting cognition and frontostriatal function were also associated with risk for schizophrenia in a family sample. However, in a recent study in Japan, some of these variants were not associated with schizophrenia. Thus, some of the data lead to the provocative observation that a frequent haplotype in *PPP1R1B* predicts increased frontostriatal interactions that appeared beneficial (as evidenced by relatively better performance on a wide range of cognitive tasks), yet contributed to risk for schizophrenia. Taken together, these findings raise the question of whether a genetic advantage in normal subjects may translate into a disadvantage in the context of other functional impairments also associated with schizophrenia, such as abnormal function of the prefrontal cortex.

DARPP-32 and depression. Although other neuro-transmitters (for example, acetylcholine and sero-tonin) show stronger links to depression, several lines of evidence support a role for decreased levels of dopamine in this disorder. In rats, chronic administration of antidepressants and lithium (used to combat depression) increases DARPP-32 in the prefrontal cortex, which might therefore be involved in downstream effects of these treatments. In another study, the antidepressant fluoxetine showed an even more extensive pattern of DARPP-32 elevation, additionally increasing DARPP-32 in the striatum and hippocampus, while DARPP-32 knockout mice did not show behavioral reactions to this drug. Electroconvulsive therapy, another effective treatment option for severe depression, acutely reduced DARPP-32, but increased it after long-term treatment. In Japan, a genetic association of variants in *PPP1R1B*, the gene encoding DARPP-32, was found with bipolar disorder, which frequently includes depressive episodes.

Importance. In summary, DARPP-32 is a molecular master switch in neurons that receive the neurotransmitter dopamine, involving it in the mechanisms of normal human cognition, drug addiction, schizophrenia, and depression, among others. Genetic variation in *PPP1R1B*, encoding DARPP-32, has been linked to the function and structure of a circuit encompassing the striatum and prefrontal cortex. These findings highlight DARPP-32 as a promising target for innovative drug treatment in psychiatry.

For background information *see* ADDICTIVE DISORDERS; AFFECTIVE DISORDERS; BRAIN; COGNITION; DOPAMINE; NERVOUS SYSTEM (VERTEBRATE); NEUROBIOLOGY; PSYCHOPHARMACOLOGY; PSYCHOTOMIMETIC DRUG; SCHIZOPHRENIA; SYNAPTIC TRANSMISSION in the McGraw-Hill Encyclopedia of Science & Technology.   Andreas Meyer-Lindenberg; Daniel R. Weinberger

Bibliography. A. Meyer-Lindenberg et al., Genetic evidence implicating DARPP-32 in human frontostriatal structure, function, and cognition, *J. Clin. Invest.*, 117(3):672–682, 2007; P. Svenningsson et al., DARPP-32: An integrator of neurotransmission, *Annu. Rev. Pharmacol. Toxicol.*, 44:269–296, 2004; P. Svenningsson et al., Diverse psychotomimetics act through a common signaling pathway, *Science*, 302(5649):1412–1415, 2003.

Desert dust-storm microbiology

Approximately 34% of Earth's terrestrial cover can be classified as deserts. The two desert areas that contribute the majority of desert dust to Earth's atmosphere are the Sahara and Sahel regions of North Africa and the Takla Makan and Gobi of Asia. The current estimates for the quantity of desert dust that moves some distance in our atmosphere each year ranges from 0.5 to 5 billion tons, with the Sahara and Sahel accounting for 50–75% of the total. Dust

Fig. 1. Continuous transmission of desert dust off the west coast of North Africa forms an atmospheric bridge of dust over the tropical Atlantic Ocean. Image date: August 8, 2001. (*Gene Feldman, SeaWiFS Project, NASA/Goddard Space Flight Center, and GeoEye*)

emanates from North Africa year-round and affects air quality regionally, in the Middle East, Europe, the Caribbean, and the Americas. Movement of dust from North Africa to the Caribbean and the Americas is seasonal, with movement to the northern Caribbean and Central and North Americas occurring during the Northern Hemisphere summer, and to the Southern Caribbean and South America during the Northern Hemisphere winter. **Figure 1** illustrates a period of continuous transmission that has formed an atmospheric bridge extending from the west coast of Africa to the Caribbean and the Americas. Dust clouds moving across the Atlantic typically take 3–5 days to reach the Caribbean. In comparison, the Asian desert dust season occurs between February and April, but large events are capable of global dispersion in the Northern Hemisphere. A large Asian dust event in 1990 affected air quality in the North Pacific, North America, and the North Atlantic, and particles from this event were detected in samples collected in the French Alps. Dust clouds moving off the east coast of Asia take approximately 7 to 9 days to reach the Americas. In addition to these two prime source regions, though, dust storms are common in deserts around the globe, and dust events occurring in other deserts are capable of moving significant quantities of soil vast distances during optimal atmospheric conditions.

Influence of climate change and anthropogenic activity. Since the mid-1960s, a combination of climate flux and anthropogenic activity (development and agriculture) has resulted in a general increase in atmospheric dust transported from the deserts of North Africa and Asia. The North Atlantic Oscillation (NAO) is the position change over the North Atlantic Ocean of the Azores High and the Icelandic Low pressure systems. A more northerly position results in a decrease in precipitation over North Africa. The NAO has been predominantly in a northern position since the late 1960s, and the documented decrease in precipitation has coincided with an increase in dust transport from North Africa to the Caribbean and the Americas. Additionally, data recorded during years of El Niño (the climatic phenomenon associated with sea-surface temperature changes in the Pacific Ocean) demonstrate that this dust transport is further enhanced in those years. Given these conditions and anthropogenic activity along the Saharan perimeter, it is interesting that the overall size

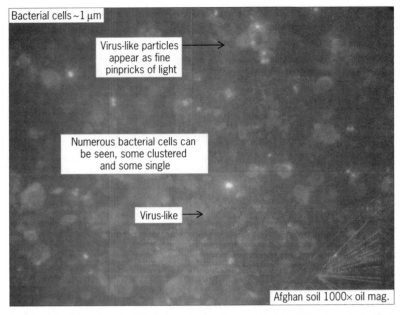

Fig. 2. Afghanistan desert soil sample stained with a nucleic acid stain and imaged using epifluorescent microscopy (1000× magnification and oil). The image shows the presence of bacteria (bright green fluorescence) and virus-like particles. All are attached to particulate matter. (*USGS, Tallahassee, FL*)

of this desert has not changed. However, one example of anthropogenic influence is the diversion of source waters from Lake Chad (bordered by Chad, Cameroon, Nigeria, and Niger) for agricultural and domestic use, which has resulted in drying of the lake [between 1963 and 1997, the water-surface area of the lake declined from 25,000 km^2 to 1350 km^2 (9650 mi^2 to 520 mi^2)]. Due to the fine nature of aquatic sediments, dried lake beds are a dominant source of dust in the atmosphere. Owens Lake in southern California was drained by the city of Los Angeles (for drinking water) by 1926 and is now the primary source of atmospheric dust in North America. The Aral Sea, located within Kazakhstan and Uzbekistan, has suffered a fate similar to that observed with Lake Chad and is another "dust hot spot." A good example of the influence of anthropogenic activity on dust emissions is the loss of perimeter grasslands around the Asian deserts because of detrimental agricultural practices. Because of such practices, China lost ~2100 km^2 (810 mi^2) to desertification each year between 1975 and 1987.

Although humans undoubtedly have contributed to the currently observed dust load rates, annual atmospheric dust loading rates have been much greater during glacial periods throughout Earth's history (more water stored as ice equals more exposed and drier soil). Likewise, during interglacial periods, the atmosphere has been much cleaner because the warming results in less ice cover and more precipitation (wet soils are heavier than dry soils and thus are not as susceptible to suspension in the atmosphere). During an interglacial period that occurred approximately 130,000 years ago, the Sahara was almost completely covered with vegetation.

Desert dust and microbiology. The amount of soil from uncovered ground that can be picked be-

tween thumb and forefinger is about a gram in weight. Regardless of where one is on the planet, that 1 g of soil harbors approximately 1 million to 1 billion bacterial cells. While there are about 10,000 genotypes of bacteria in that community, most of the community is dominated by a few species, with species dominance being dependent on a combination of soil characteristics such as inorganic, organic, and water content. Also present are other microorganisms such as fungi (~1 million per gram in some soils), viruses (~100 million per gram in soils of high moisture content), and protozoa (~10,000 per gram in wet soils). Surprisingly, desert soils harbor the upper range of ~1 billion bacterial cells per gram of soil. **Figure 2** illustrates the presence of bacteria and virus-like particles in a desert soil sample collected in Afghanistan. Although fungi, viruses, and protozoa are present in desert soils, they occur at lower concentrations than in most other soil types because of the lack of moisture. Because of the low water content of desert soils, many of the microorganisms present may be in a state of stasis (low metabolic activity, which is a survival strategy) and thus may not be as readily capable of rapid growth or cell division. The key to an increase in biotic activity in desert soils is the addition of water through precipitation. Film documentaries depicting the rapid and short-lived bursts of small-plant growth in deserts following precipitation events are an excellent reflection of what also occurs on the microbiological scale. From a microbial point of view, there is an alternative to growth induced by precipitation: the soils can be moved to more moist environments. This is precisely what large dust storms do: they move significant quantities of desert topsoil to environments that may favor more rapid microbial growth.

Dust storms: why should we care? It is known that many terrestrial ecosystems benefit from dust transport (for example, plants in the Amazonian and Hawaiian rain forests derive nutrients from African and Asian dust, respectively). Why then should dust storms be of concern, given that dust has always moved through Earth's atmosphere? First, dust clouds moving over industrial, agricultural, and developed lands can pick up toxic compounds that are being emitted into the atmosphere and thus enhance the toxicity potential of a cloud. Emissions originating from the application of pesticides, herbicides, and other sources such as automotive exhaust are examples of the types of compounds that may adhere to dust-cloud particulates. Additionally, these same types of compounds may be present in dust-cloud source-region soils (that is, in the vicinity of dried lake sediments).

With regard to human health risk, toxic compounds may exacerbate the stresses associated with having to breathe particulate-laden air (penetration of particulate matter into the lung environment does not equate to good health) through effects such as immune suppression (active suppression of the immune response). Immune suppression in general is particularly important relative to pathogen exposure and may occur in plants as well as in animals.

Dust-cloud research has shown that 20–30% of the culturable microorganisms are pathogenic (pathogens to some type of plant or animal), and in general the bacterial communities are diverse. One of the first pathogens identified in an African dust sample collected in the northern Caribbean was the sea fan pathogen, *Aspergillus sydowii*, a terrestrial fungus that cannot replicate in the marine environment but that has caused infections throughout the Caribbean Sea. It is possible that *A. sydowii* has been present in dust clouds since this species evolved, but recently, as a result of the introduction of anthropogenic toxins into these clouds, the fungus is now able to cause symptomatic infection. Enhanced dust-cloud toxicity that has an impact an ecosystem-wide health via direct toxicity or through synergy with pathogens is a prime concern.

Second, dust clouds may harbor crop or livestock pathogens through source-area contamination of soils or, similar to the concerns with toxins, may pick up these types of pathogens as the clouds move through agricultural areas. The end result is the potential for the delivery of pathogens to downwind agricultural regions where the same types of crops or livestock exist. Interestingly, in 2003, via experiments conducted aboard the marine-research vessel *JOIDES Resolution* during Ocean Drilling Program Leg 209, two fungal pathogens were isolated from African dust moving over the tropical mid-Atlantic (~15°N, 45°W). One, *Aspergillus dauci*, is a pathogen of Florida carrots, and the other, *Massaria platani*, is the causative agent of Florida sycamore canker. Hence, the direct movement of crop, livestock, native plant or animal, or human pathogens to downwind environments, without regard to synergy, is certainly another concern.

Another big-picture synergistic effect arises from the fact that African dust has been identified as causing harmful algal blooms along Florida's southern and western coastal environments. The deposition of oceanic desert dust, because of its inorganic nutrient content such as iron (which can be utilized for growth in iron-depleted marine waters), has always contributed to the growth of microorganisms in marine waters. However, induced blooms of harmful algae in waters that are currently stressed by other factors, such as inadequate regional sewage-disposal practices, may push a compromised community (that is, the southeast Florida coral reef system) beyond a state of recovery. Of course, there are the more obvious impacts from dust-induced harmful algal blooms: direct loss of marine life, compromised human health via exposure to aerosolized algae toxins, and the resulting economic losses associated with tourism issues.

Obviously, desert dust storms that move around our planet warrant research focused on ecosystem and human health issues. It is clear from a survey of the literature that projects directed toward the identification of pathogen and toxin dissemination are sorely needed. Equally as interesting as the health issues are the questions associated with microbial ecology. What role do these dust clouds play in the distri-

bution of beneficial and harmful microbial species? There are far too few research groups investigating desert dust microbiology. This field is wide open.

For background information *see* ATMOSPHERIC GENERAL CIRCULATION; CLIMATE MODIFICATION; DESERT; DESERTIFICATION; DUST STORM; ECOSYSTEM; EL NIÑO; MICROBIAL ECOLOGY; MICROBIOLOGY; PRECIPITATION (METEOROLOGY); PUBLIC HEALTH in the McGraw-Hill Encyclopedia of Science & Technology.   Dale Warren Griffin

Bibliography. A. S. Goudie and N. J. Middleton, Saharan dust storms: Nature and consequences, *Earth Sci. Rev.*, 56:179–204, 2001; D. W. Griffin, Atmospheric movement of microorganisms in clouds of desert dust and implications for human health, *Clin. Microbiol. Rev.*, 20(3):459–477, 2007; T. D. Jickells et al., Global iron connections between desert dust, ocean biogeochemistry, and climate, *Science*, 308(5718):67–71, 2005; F. Lambert et al., Dust-climate couplings over the past 800,000 years from the EPICA Dome C ice core, *Nature*, 452:616–619, 2008; C. Moulin et al., Control of atmospheric export of dust from North Africa by the North Atlantic Oscillation, *Nature*, 387:691–694, 1997; S. L. O'Hara et al., Exposure to airborne dust contaminated with pesticide in the Aral Sea region, *Lancet*, 355:627–628, 2000; E. A. Shinn et al., African dust and the demise of Caribbean coral reefs, *Geol. Res. Lett.*, 27:3029–3032, 2000.

Diabetes and its causes

Diabetes mellitus is a metabolic disorder that is characterized primarily by chronic hyperglycemia [excessive sugar (glucose) in the blood] resulting from defects in insulin secretion, insulin action, or both. The symptoms of hyperglycemia include those caused by osmotic diuresis (increased urination because of the presence of certain substances, such as glucose, in the kidney tubules) and the resultant loss of calories [including polyuria (passage of copious amounts of urine), polydipsia (excessive thirst), weight loss, and polyphagia (excessive eating)]; blurring of vision caused by fluctuations in the diameter of the ocular lens; and increased susceptibility to infections, particularly urinary tract infections and skin infections. Severe life-threatening acute metabolic complications include diabetic ketoacidosis (excessive amounts of acidic ketones in the blood) and hyperosmolar nonketotic state (a dangerous condition caused by very high levels of blood sugar and dehydration). Chronic effects of poorly controlled diabetes include microvascular complications that affect the eyes (retinopathy, blindness), kidneys (nephropathy, renal failure), and nerves (neuropathy, peripheral and autonomic), as well as macrovascular complications that affect the cardiovascular system, increasing the risk of heart attacks and strokes fourfold compared to the nondiabetic population.

Classification of diabetes. The vast majority of patients with diabetes mellitus can be split into two broad categories (see **table**). Type 1 diabetes is

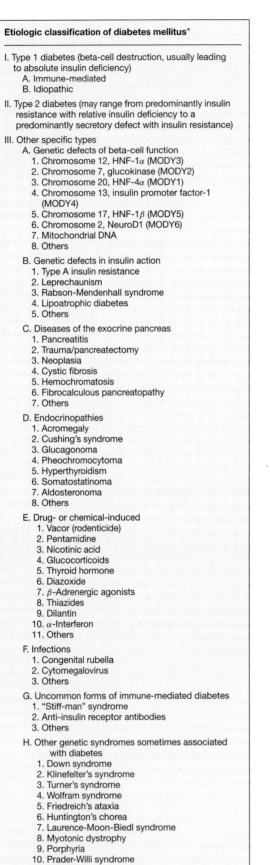

Etiologic classification of diabetes mellitus*

I. Type 1 diabetes (beta-cell destruction, usually leading to absolute insulin deficiency)
  A. Immune-mediated
  B. Idiopathic

II. Type 2 diabetes (may range from predominantly insulin resistance with relative insulin deficiency to a predominantly secretory defect with insulin resistance)

III. Other specific types
  A. Genetic defects of beta-cell function
    1. Chromosome 12, HNF-1α (MODY3)
    2. Chromosome 7, glucokinase (MODY2)
    3. Chromosome 20, HNF-4α (MODY1)
    4. Chromosome 13, insulin promoter factor-1 (MODY4)
    5. Chromosome 17, HNF-1β (MODY5)
    6. Chromosome 2, NeuroD1 (MODY6)
    7. Mitochondrial DNA
    8. Others
  B. Genetic defects in insulin action
    1. Type A insulin resistance
    2. Leprechaunism
    3. Rabson-Mendenhall syndrome
    4. Lipoatrophic diabetes
    5. Others
  C. Diseases of the exocrine pancreas
    1. Pancreatitis
    2. Trauma/pancreatectomy
    3. Neoplasia
    4. Cystic fibrosis
    5. Hemochromatosis
    6. Fibrocalculous pancreatopathy
    7. Others
  D. Endocrinopathies
    1. Acromegaly
    2. Cushing's syndrome
    3. Glucagonoma
    4. Pheochromocytoma
    5. Hyperthyroidism
    6. Somatostinoma
    7. Aldosteronoma
    8. Others
  E. Drug- or chemical-induced
    1. Vacor (rodenticide)
    2. Pentamidine
    3. Nicotinic acid
    4. Glucocorticoids
    5. Thyroid hormone
    6. Diazoxide
    7. β-Adrenergic agonists
    8. Thiazides
    9. Dilantin
    10. α-Interferon
    11. Others
  F. Infections
    1. Congenital rubella
    2. Cytomegalovirus
    3. Others
  G. Uncommon forms of immune-mediated diabetes
    1. "Stiff-man" syndrome
    2. Anti-insulin receptor antibodies
    3. Others
  H. Other genetic syndromes sometimes associated with diabetes
    1. Down syndrome
    2. Klinefelter's syndrome
    3. Turner's syndrome
    4. Wolfram syndrome
    5. Friedreich's ataxia
    6. Huntington's chorea
    7. Laurence-Moon-Biedl syndrome
    8. Myotonic dystrophy
    9. Porphyria
    10. Prader-Willi syndrome
    11. Others

IV. Gestational diabetes mellitus (GDM)

*Patients with any form of diabetes may require insulin treatment at some stage of their disease. Such use of insulin does not, of itself, classify the patient.
SOURCE: Adopted from position statement for diagnosis and classification of diabetes by the American Diabetes Association.

caused by immune destruction of the beta cells in the islets of Langerhans in the pancreas, resulting in an absolute deficiency of insulin. Type 2 diabetes is due to a combination of resistance to the action of insulin compounded by an inability of insulin secretion to compensate for the increased requirement of this hormone. Type 1 diabetes accounts for 5–10% of all cases of diabetes; type 2 diabetes accounts for most of the remaining 90–95%. Type 1 diabetes usually presents acutely, with florid symptoms of hyperglycemia or with emergent diabetic ketoacidosis precipitated by an infection. In contrast, the onset of type 2 diabetes is insidious and gradual, often causing no specific symptoms at all. In most cases of type 2 diabetes, a prolonged prediabetes phase, in which carbohydrate metabolism is abnormal, can be detected by glucose testing in the fasted state (impaired fasting glucose) or after a glucose challenge (impaired glucose tolerance).

Type 1 diabetes. Type 1 diabetes, hitherto known as juvenile-onset diabetes or insulin-dependent diabetes, results from an absolute deficiency of insulin secretion caused by progressive and selective immune-mediated destruction of insulin-producing beta cells in the pancreatic islets of Langerhans. Immunological markers that have been linked to this process and that could help in assessing the risk for type 1 diabetes include islet cell autoantibodies, insulin autoantibodies, autoantibodies to glutamic acid decarboxylase (GAD65), and autoantibodies to tyrosine phosphatases IA2α and IA2β. One or more of these autoantibodies are present in ~90% of all cases of type 1 diabetes. Furthermore, type 1 diabetes is also associated with various human leukocyte antigen (HLA) subtypes (specifically, genotypes DR and DQ) that could be either protective or susceptive to the disease. Children and adolescents with type 1 diabetes usually have a rapid loss of islet viability and can present with diabetic ketoacidosis as an initial manifestation of the disease. Adults, on the other hand, can have a slower onset of the disease that can rapidly intensify with accelerated hyperglycemia or ketoacidosis during infections or other acute illnesses (such as myocardial infarction). Other autoimmune disorders may coexist with type 1 diabetes, including Hashimoto's thyroiditis, Graves' disease, Addison's disease, and myasthenia gravis.

A subtype of type 1 diabetes occurs mostly in individuals of Asian or African descent that is characterized by a lack of endogenous insulin, making them ketosis-prone. These individuals have episodes of ketoacidosis and exhibit varying degrees of insulin dependence between such episodes. Although a strong inherited basis exists, there is no evidence of autoimmunity and the condition is not linked to specific HLA loci.

The model system for classification of type 1 diabetes is still being further defined. The basic concept of this model states that every person starts his or her life with a varying degree of susceptibility to type 1 diabetes. The degree of susceptibility is largely inherited, involving HLA genotypes DR and DQ and, to a lesser extent, other genetic loci termed insulin-dependent diabetes mellitus (IDDM) susceptibility genes. The HLA locus is thought to confer ~50% of the genetic susceptibility, while ~15% is conferred by two other genes—insulin-VNTR (where VNTR stands for variable number of tandem repeats) and CTLA-4 (where CTLA stands for cytotoxic T-lymphocyte antigen)—with minor contributions from other IDDM genes. Environmental factors may also be associated with the immune pathogenesis of type 1 diabetes by initiating beta-cell destruction. These include viruses (enteroviruses, coxsackie, congenital rubella), environmental toxins (nitrosamines), and foods (early exposure to cow's milk proteins, cereals, or gluten). However, a firm and unequivocal link between any of these environmental factors and type 1 diabetes has not been established.

Type 2 diabetes. Type 2 diabetes, hitherto known as maturity/adult-onset diabetes or non-insulin-dependent diabetes, is much more prevalent than type 1 diabetes and typically affects individuals older than 40 years of age. It occurs as a result of resistance to insulin action coupled with an inadequate compensatory insulin secretory response, leading to relative insulin deficiency. This defect is purported to be caused by a loss of the initial insulin secretion in response to a meal, leading to a delayed and flatter peak of postprandial insulin secretory profile. In the early stages of type 2 diabetes, plasma insulin concentrations may actually be higher than in nondiabetic individuals, but are clearly inappropriate with regard to the prevailing glucose concentrations because of impairment in insulin action. Recently, impairment of glucose effectiveness (ability of glucose per se to stimulate its own metabolism) has also been shown to be a prominent factor in the development of this disease. Specific defects in insulin secretion and insulin action have also been described in the prediabetes phases. Moreover, in most people with type 2 diabetes, the progressive decline in beta-cell function leads to such a great reduction in insulin secretory reserves that insulin therapy is necessary for optimum glycemic control.

The incidence and prevalence of type 2 diabetes mellitus are on the rise as a result of increasing obesity (especially in the abdominal region), sedentary lifestyle, and increasing age of the population in both developed and developing countries. Of particular concern is the rapidly increasing prevalence of type 2 diabetes in the adolescent population throughout the world as a result of increasing obesity related to inactivity and poor food habits. Obesity and metabolic syndrome (a cluster of metabolic risk factors, including elevated blood pressure, elevated insulin levels, excess body fat around the waist, and abnormal cholesterol levels) predispose to type 2 diabetes. Although most people with type 2 diabetes are obese, the majority of obese people do not have type 2 diabetes. However, obesity, especially of the upper body, aggravates insulin resistance. Furthermore, frequency of type 2 diabetes is higher in a number of ethnic subpopulations, including Asian Indians, Native Americans, Polynesians, Hispanics, and African Americans.

Heredity. Genetic predisposition has a vital impact on the pathogenesis of type 2 diabetes. There is a 2.4-fold increased risk of type 2 diabetes in individuals with a positive family history of diabetes. Impaired glucose tolerance or diabetes is observed in 15–25% of first-degree relatives of type 2 diabetic patients. The pattern of inheritance in families is complex and multifactorial. If a person has one parent with type 2 diabetes, the lifetime chance of developing diabetes is 38%. If both parents have type 2 diabetes, risk increases to 60% at 60 years of age.

Genetic involvement. Fundamentally, type 2 diabetes is a multifactorial disorder with a pronounced environmental influence on a background of variable genetic predisposition, modifying the natural history and progression of the disease in any given individual. Hence, it is difficult and perhaps impossible with current techniques to quantify risks for each of these components of the disease process. However, discovery of new genes could lead to a better understanding of the various pathophysiological components of the disease and could help achieve homogeneity in clinical trials and physiologic studies. Genomewide association studies (GWAS) using the International HapMap project (a multicountry effort to identify and catalog genetic similarities and differences in humans), among other resources, have much promise as a tool for new gene identification for type 2 diabetes. Although the majority of genes have been linked to regulation of insulin secretion, the FTO (fat mass– and obesity-associated) gene is likely linked to obesity and insulin action.

Maturity-onset diabetes of the young (MODY) and mitochondrial diabetes are the most common monogenic (single-gene) forms of the disease causing defects in beta-cell function, accounting for <10% of cases with type 2 diabetes. MODY variants are inherited in an autosomal dominant pattern and are characterized by impaired insulin secretion with minimal or no defects in insulin action. The most common form, representing ~65% of all MODY cases, is MODY3, with a mutation on chromosome 12q that causes a defect in hepatocyte nuclear factor 1α (HNF-1α). Another form, MODY2, accounts for ~15% of MODY cases and develops from mutation of the glucokinase (GCK) gene on chromosome 7p. Glucokinase plays an important role in insulin secretion (critical for beta-cell glucose sensing) and hepatic glucose uptake. So far, 140 GCK gene mutations have been described. Other causes of MODY are listed in the table.

Point mutations in maternally inherited mitochondrial DNA, most commonly with a distinct adenine (A)–to–guanine (G) substitution, have been found to be involved with diabetes mellitus and deafness, and may also lead to MELAS syndrome (mitochondrial myopathy, encephalopathy, lactic acidosis, and strokelike episodes).

Rarely, diabetes can be caused by severe insulin resistance due to genetic defects. These syndromes are associated with specific clinical conditions such as acanthosis nigricans, lipodystrophy, hyperandrogenism, and other growth disorders (leprechaunism and Rabson-Mendenhall syndrome).

Exocrine pancreatic deficiency. Any acquired process that destroys beta-cell mass can also cause diabetes. These processes include chronic pancreatitis, pancreatectomy, and pancreatic carcinoma. Other rare disorders are specific to certain geographical areas, including the development of diabetes associated with fibrocalculous pancreatopathy related to dietary factors, common to the southwestern coast of India.

Other causes. Excess amounts of several hormones (growth hormone, cortisol, glucagons, epinephrine, etc.) clinically found in various endocrinopathies (see table) can also cause diabetes by downregulating and opposing the action of insulin.

Gestational diabetes mellitus (GDM) is a well-recognized entity and poses a significant risk for the development of type 2 diabetes following the pregnant state.

Drugs (including steroids) can cause or precipitate diabetes in individuals with insulin resistance. Some drugs such as nicotinic acid and glucocorticoids impair hepatic and peripheral insulin action. α-Interferon has been reportedly linked to diabetes associated with islet-cell antibodies. Pentamidine is a beta-cell toxin that causes hyperglycemia as a result of impaired insulin secretion.

For background information *see* CARBOHYDRATE METABOLISM; DIABETES; EPIDEMIOLOGY; GLUCOSE; HUMAN GENETICS; INSULIN; MUTATION; PANCREAS; PANCREAS DISORDERS in the McGraw-Hill Encyclopedia of Science & Technology.

Ananda Basu; Debashis K. Nandy

Bibliography. American Diabetes Association, Clinical practice recommendations, *Diabetes Care*, 28(suppl. 1):S1–S79, 2005; American Diabetes Association, Diagnosis and classification of diabetes, *Diabetes Care*, 20:1183–1197, 1997; American Diabetes Association, Follow-up report on the diagnosis of diabetes mellitus, *Diabetes Care*, 26:3160–3167, 2003; S. Anjos and C. Polychronakos, Mechanisms of genetic susceptibility to type 1 diabetes: Beyond HLA, *Mol. Genet. Metab.*, 81:187–195, 2004; G. Bock et al., Pathogenesis of pre-diabetes: Mechanisms of fasting and postprandial hyperglycemia in people with impaired fasting glucose and/or impaired glucose tolerance, *Diabetes*, 55(12):3536–3549, 2006; D. Devendra, E. Liu, and G. S. Eisenbarth, Type 1 diabetes: Recent developments, *Br. Med. J.*, 328:750–754, 2004; S. S. Fajans, G. I. Bell, and K. S. Polonsky, Molecular mechanisms and clinical pathophysiology of maturity-onset diabetes of the young, *N. Engl. J. Med.*, 345:971–980, 2001; J. A. Maassen et al., Mitochondrial diabetes: Molecular mechanisms and clinical presentation, *Diabetes*, 53(suppl. 1):S103–S109, 2004; K. Owen and A. T. Hattersley, Maturity-onset diabetes of the young: From clinical description to molecular genetic characterization, *Best Pract. Res. Clin. Endocrinol. Metab.*, 15(3):309–323, 2001; M. J. Redondo, P. R. Fain, and G. S. Eisenbarth, Genetics of type 1A diabetes, *Recent Prog. Horm. Res.*, 56:69–89, 2001.

Differential duplicate gene retention

Eukaryotic organisms occasionally double their chromosomes, and this new tetraploid sometimes founds a new lineage whose descendants continue to possess some of the duplicated genes. If the tetraploidy (whole genome duplication) happened recently—within the last approximately 1 million years—then this might be recognized by an increase in chromosome number. This is a result of the fact that the centromere deletions, segmental deletions, and chromosome rearrangements that occur naturally to duplicate genomes have not had enough time to reduce the chromosome number. Recent tetraploidies are especially common among existing flowering plant species, perhaps because they often self-fertilize and have an asexual reproductive option. Ancient tetraploidies, however, are widespread, have a chromosome number similar to that of the ancestor, and are thus more difficult to detect. At the base of the vertebrate phylogenetic tree are two well-supported tetraploidies (wherein the genome of jawed vertebrates underwent two rounds of whole genome duplication that took place between the emergence of urochordates and the radiation of jawed vertebrates); occasional more recent tetraploidies have occurred in fish, amphibians, and yeasts; and multiple, ancient tetraploidies punctuate all plant lineages. These ancient tetraploidies were inferred from the pattern of gene content within genomes that have been sequenced and annotated. **Figure 1** shows an example of data presented as a dot plot supporting the inference of polyploidy from within the genome of a French grape variety, the species useful in wine making. This plot compares the gene content and

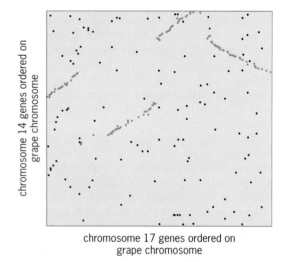

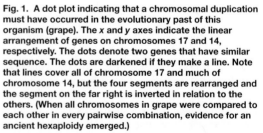

chromosome 17 genes ordered on
grape chromosome

Fig. 1. A dot plot indicating that a chromosomal duplication must have occurred in the evolutionary past of this organism (grape). The *x* and *y* axes indicate the linear arrangement of genes on chromosomes 17 and 14, respectively. The dots denote two genes that have similar sequence. The dots are darkened if they make a line. Note that lines cover all of chromosome 17 and much of chromosome 14, but the four segments are rearranged and the segment on the far right is inverted in relation to the others. (When all chromosomes in grape were compared to each other in every pairwise combination, evidence for an ancient hexaploidy emerged.)

order of grape chromosome 14 (*y* axis) with that of grape chromosome 17 (*x* axis). Each dot indicates where one gene in chromosome 14 matches a gene of similar sequence in chromosome 17. Many such similarities are scattered around the plot, but some form lines. Each line indicates genes of similar sequence and in a similar order that exist on chromosome 14 and most of chromosome 17. Most of these two chromosomes must have been derived from a single ancestral sequence that doubled and then rearranged. Evidence of tetraploidy is observed when these lines of gene colinearity largely cover an entire genome.

Fractionation. The phylogenetic tree of flowering plants is particularly rich in tetraploidy events. Within the model genetic species, *Arabidopsis thaliana*, there have been two sequential tetraploidy events since the origin of the order Brassicales. Recently sequenced papaya, a basal Brassicales, has not had these two tetraploidies (but shares with grape the polyploidy implied by Fig. 1). The most recent tetraploidy in *Arabidopsis* has been studied in detail by at least five research groups from around the world. Initially, all chromosomes doubled and all genes doubled. If particular balances need to be maintained among the genes or their products, these balances are not perturbed by doubling. However, this overall doubling was temporary in the *Arabidopsis* lineage and is more recent than other cases of ancient tetraploidy. The mechanism operating on a tetraploid genome that results in lowering gene content back nearer to that of the ancestor is called fractionation. **Figure 2** illustrates the fractionation mechanism in a case of incomplete fractionation: 25% of the genes in *Arabidopsis* are paired because they were retained from the most recent tetraploidy event. Notice that one or the other of the duplicate genes tends to be lost, but not both. If fractionation had been complete, no genes would have been retained, and there would be no way to prove that the two segments remaining were ever syntenic (located on the same chromosome) without an outgroup genome sequence as a reference. For example, the sequenced papaya genome makes an excellent outgroup for *Arabidopsis*—both being in the same order—because the papaya lineage splits from the *Arabidopsis* lineage just before the last two *Arabidopsis* ancient tetraploidies (thus, papaya is expected to have the ancestral arrangement of genes shown in Fig. 2). Incomplete fractionation is the rule among ancient tetraploidies.

Genes that are resistant to fractionation are biased. The result of incomplete fractionation is that some genes are retained and some are not. In all plant cases studied, this is a highly biased process. For *Arabidopsis*, the best-studied case, four different laboratories came to approximately the same conclusion: Among larger gene categories, genes encoding ribosomal proteins, transcription factors, and protein kinases are significantly overretained (approximately 40%, compared to 25% for the average gene), whereas genes encoding ancient or relatively "simple functions" such as pyrimidine biosynthesis, digestive

protein cleavage, or deoxyribonucleic acid (DNA) repair are underretained (<5%). Hence, proteins with many subunit–subunit interactions or those in complex regulatory networks either tend to be retained or tend to resist fractionation. At the most-retained extreme of 75% ($n = 20$) are genes encoding the proteasome core complex, which is at the center of one of the most complicated machines of life (specifically, the proteasome is the site for degradation of most intracellular proteins). At the least-retained extreme at 0% ($n = 20$) are genes encoding enzymes of pyrimidine base biosynthesis. The results from other plant tetraploidies are similar to the case of *Arabidopsis*.

Gene balance hypothesis. The gene balance hypothesis, the theoretical work of researchers L. Hurst, J. A. Birchler, and R. A. Veitia, derives from experimental genetic work in yeast, plant, *Drosophila* (fruit fly), and human systems. This hypothesis posits that some gene products work best in concentrations balanced with the gene products of other genes. Whole genome duplication does not alter these balances, but fractionation does. Removing single genes alters balance and lowers fitness if strict stoichiometry of products is physiologically optimum. Using the logic of the gene balance hypothesis, genes are retained only because they are costly to remove. Any gene lineage expansions that result are neutral tag-alongs (spandrels) of purifying selection. If the gene balance hypothesis is a primary explanation for biased gene retention, then it follows that tandem (local) duplications of single genes should not be retained for those same sorts of genes that are overretained following tetraploidy. A tandem duplication does change product balance from inception. Many studies with several eukaryotes have investigated gene retention following tandem duplications; genes encoding transcription factors and ribosomal proteins are generally underduplicated. For *Arabidopsis*, a number of studies strongly suggest a reciprocal relationship between tandem duplicate retention and retention following tetraploidy, as expected based on the gene balance hypothesis.

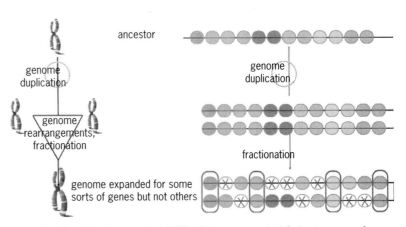

Fig. 2. Fractionation of a new tetraploid back to a gene content, but not a gene order, closer to that of the ancestor. The result is the acquisition of biased gene content. An X denotes fractionation, the biased mechanism. Retained pairs are enclosed in an oval; these are spandrels of purifying selection for the status quo.

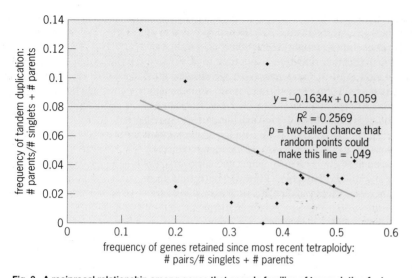

Fig. 3. A reciprocal relationship among genes that encode families of transcription factor proteins exists between the gene retention frequencies following tandem duplication (*y* axis) and following the most recent tetraploidy (*x* axis). This was predicted by the gene balance hypothesis. Only gene families of ≥30 genes were used. Gene number was calculated by condensing each tandem duplicate array to one gene, the parent. Singlet genes + parent genes = total genes. Therefore, the *y* axis monitors the frequency of tandem duplication, not the extent of tandem duplication. When the extent is recorded, the reciprocal relationship is even more significant.

Figure 3 shows a particularly convincing reciprocal relationship, namely, among transcription factor gene families of ≥30 genes in *Arabidopsis*. There is a 4.9% probability that the regression line in Fig. 3 could have been caused by chance; the chance that the relationship is actually positive is very small. A competing and currently popular explanation, called subfunctionalization, predicts a positive relationship between retention of tandem versus post-tetraploidy duplicates. Subfunctionalization posits that each member of a duplicate pair assumes a different part of the gene's original function, and thus both members must be retained. However, subfunctionalization is not the cause of fractionation bias of transcription factor gene content for the plant tetraploidies studied to date.

Implications for the general theory of evolution. There is a tendency among scientists to downplay ignorance with regard to anything to do with evolution. In fact, evolutionary theory, like any other scientific theory, changes with deeper understanding. For example, extrapolating from what is known of tetraploidies and biased gene retention, "connected" genes should increase in their genomic representation with each successive tetraploidy, making a trend of increasing regulatory complexity and perhaps explaining trends in form or behavior. This evolutionary mechanism can be termed balanced gene drive. It seems to have operated at least over the past approximately 150 million years of plant evolution and may still be operating today. A possible method to mitigate this one-way accumulation of regulatory (connected) genes is to prevent tetraploidy, as may have happened in many animal lineages. Another way to mitigate balanced gene drive is to evolve mechanisms that replace the step at which gene dosage is limiting, such as by recruiting proteins to facilitate

optimum subunit–subunit assembly. The prevailing general evolutionary theory, called the *modern synthesis*, codified in the 1940s and 1950s, does not accommodate the concept of balanced gene drive or any other "mutationist" process (in which the mutation mechanism itself—here, fractionation—tends to push evolution in a specific direction). Thus, the topic of differential duplicate gene retention constitutes an important area for future insights into evolution.

For background information *see* CHROMOSOME; DEOXYRIBONUCLEIC ACID (DNA); GENE; GENETIC CODE; GENETICS; GENOMICS; MOLECULAR BIOLOGY; ORGANIC EVOLUTION; PLANT EVOLUTION in the McGraw-Hill Encyclopedia of Science & Technology.

Michael Freeling

Bibliography. K. L. Adams and J. F. Wendel, Polyploidy and genome evolution in plants, *Curr. Opin. Plant Biol.*, 8:135–141, 2005; J. A. Birchler and R. A. Veitia, The gene balance hypothesis: From classical genetics to modern genomics, *Plant Cell*, 19:395–402, 2007; G. Blanc and K. H. Wolfe, Functional divergence of duplicated genes formed by polyploidy during *Arabidopsis* evolution, *Plant Cell*, 16:1679–1691, 2004; M. Freeling, The evolutionary position of subfunctionalization, downgraded, *Genome Dynam.*, 4:25–40, 2008; M. Freeling and B. C. Thomas, Gene-balanced duplications, like tetraploidy, provide predictable drive to increase morphological complexity, *Genome Res.*, 16:805–814, 2006.

Digital elevation models

Grid-based digital elevation models (DEMs) are an important source of terrain data used in analyzing and visualizing surface landforms. In contrast to contour lines used to represent elevations on topographic maps, DEMs are electronic files that store elevation data referenced to cartesian coordinates. Each grid cell is represented by an elevation (z) value, an easting or longitude (x) coordinate, and a northing or latitude coordinate (y). The level of detail, also known as the resolution, in a DEM is a function of the horizontal distance between sampling points (see **table**). DEMs are similar in structure to the digital ground models (DGMs) used in the United Kingdom and the digital height models (DHMs) used in Germany.

DEMs are most often manipulated within geographic information system (GIS) software that

DEM files

Product	Resolution*
USGS 7.5′ (48 contiguous states, Hawaii, Puerto Rico)	30 × 30 m
USGS 7.5′ (Alaska)	1 × 2 arcsec
USGS 15′ (Alaska)	2 × 3 arcsec
USGS 30′ (contiguous states, Hawaii)	2 × 2 arcsec
USGS 1° (U.S.)	3 × 3 arcsec
SRTM (landmasses 60°N to 56°S)	90 × 90 m
USGS GTOPO30 (global)	30 arcsec

*1 arcsec = 30 m.

supports analysis and visualization tools used in resource management, urban planning, navigation, civil engineering, and Earth science. GIS also allows DEM data to be combined with other digital information. For example, a terrain map can be enhanced by superimposing a color aerial photograph over a three-dimensional (3D) perspective diagram. Other geographic information can be added from a digital line graph (DLG), such as a road network.

Most large elevation datasets in the United States are compiled and maintained by federal agencies such as the U.S. Geological Survey (USGS), the National Geospatial-Intelligence Agency (NGA) [formerly the National Imagery and Mapping Agency (NIMA)], and the National Geophysical Data Center (NGDC). For example, the National Elevation Dataset (NED) maintained by the USGS contains elevation data for the contiguous United States and Hawaii, with sampling points separated by 30 m (98 ft). Each NED file corresponds to a USGS 7.5-min topographic quadrangle (topoquad) map at a scale of 1:24,000 (see table). NED files are cropped to the four corners of their corresponding topoquad, making the resulting coverage area quadrilateral with opposite sides that are not parallel (**Fig. 1**).

Global elevation data known as GTOPO30 files are available at resolutions of 30 arc seconds; that is, approximately 1 km (0.6 mi). Completed in 1996, GTOPO30 files were compiled from a variety of vec-

Fig. 2. Shaded relief view of GTOPO30 DEM data. (*USGS*)

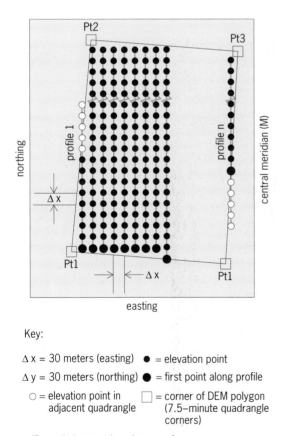

Key:

Δ x = 30 meters (easting)   ● = elevation point

Δ y = 30 meters (northing)   ● = first point along profile

○ = elevation point in   □ = corner of DEM polygon
   adjacent quadrangle   (7.5–minute quadrangle
                                   corners)

(Example is a quadrangle west of central meridian of UTM zone.)

Fig. 1. Structure of a 7.5-min digital elevation model, UTM meter grid.

tor and raster data sources at the USGS Center for Earth Resources Observation and Science (EROS) in Sioux Falls, South Dakota. **Figure 2** shows a GTOPO30 shaded relief view of Central America. Several international organizations contributed data or other support for the development of GTOPO30 files, including the United Nations, Manaaki Whenua Landcare Research Center in New Zealand, Instituto Nacional de Estadistica Geografica e Informatica (INEGI) in Mexico, and the Geographical Survey Institute (GSI) of Japan.

DEM construction. DEMs are constructed using methods ranging from converting data on paper maps to measuring surface elevations with satellite-based sensors. Digitizing information directly from topographic or hydrographic maps involves mathematical algorithms for interpolating elevations at grid points between contour or bathometric lines. Data points can also be collected manually using stereo photos or through automated routines involving the calculation of parallax displacement. The Shuttle Radar Topography Mission (SRTM) has become an important source of DEM data at coarse resolutions. In February 2000, during a mission lasting 11 days, the space shuttle *Endeavor*'s synthetic

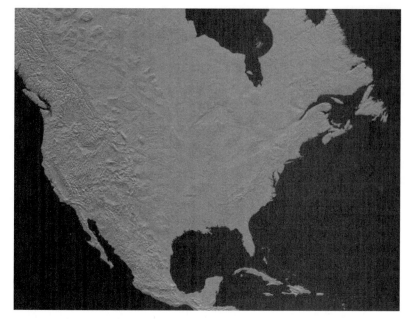

Fig. 3. SRTM view of North America. (NASA/JPL)

aperture radar (SAR) captured elevation data for 80% of the Earth's landmass within a zone extending from 60°N to 56°S latitude. More than one trillion separate measurements of the Earth's topography were captured at a resolution of 92 m (302 ft), measured at the Equator (see table).

Unmanned satellites, such as SPOT-5, QUICKBIRD, IKONOS, and TERRA, have also been used to build DEM datasets using electro-optical scanners. Interferometric techniques are used to measure small differences in height by comparing images taken during successive satellite passes. After comparing the phase information, the images are coregistered to identify differences in phase value for each pixel. Subsequently, altitude values can be determined using phase integration and geometric rectification.

While coarse resolutions are suitable for applications representing large areas, closely spaced grid points are needed for analyzing and visualizing small geographic areas. Light detection and ranging (Lidar) methods have been proven effective for collecting high-resolution DEM data over small areas with vertical accuracies of up to 3 m (10 ft). Lidar sensors emit 5000 to 50,000 pulses per second from aircraft equipped with airborne Global Positioning System (GPS) equipment, producing datasets that can exceed 200,000 data points in 2.6 km² (1 mi²). An inertial measuring unit (IMU) is used to compensate for aircraft pitch, roll, and heading changes. Lidar offers advantages over traditional photogrammetric methods used in compiling elevation data. For example, data can be captured on steep slopes where shadows obscure terrain features. A cooperative program involving the USGS, National Oceanographic and Atmospheric Agency (NOAA), and National Aeronautical and Space Administration (NASA) uses Lidar technology to map coastal areas before and after hurricanes as a means of documenting landscape change.

Accuracy. As with other Earth data, DEMs are subject to errors introduced during measurement, interpolation, and classification. The accuracy of any surface location in a DEM is a function of the distance to the closest sampling (data) point as well as the variation in topography between sampling locations. Vertical DEM accuracy is most often represented as root-mean-square error (RMSE) and can be determined by comparing DEM elevations and corresponding elevations on topographic maps. The USGS requires 90% of DEM data points to have elevation errors of less than 7 RMSE. GPS receivers or high-precision survey equipment can be used for evaluating DEM accuracy.

Research applications. The grid structure of DEMs makes them useful for measuring geomorphic properties of landscapes, such as slope and aspect, and an ideal source of information for delineating surface features, such as drainage basins. DEMs also provide a method for visualizing terrain surfaces. For example, DEMs can be used for creating shaded relief maps that simulate sunlight as it would be seen on a landscape surface (**Fig. 3**). Data contained in DEMs can also be used to construct triangulation irregular networks (TINs). TINS are vector-based (line-based) representations that form networks of nonoverlapping triangles to portray 3D surfaces.

Earth science applications for DEMs include evaluating land subsidence following petroleum extraction and identifying surface inflation prior to volcanic eruptions. DEM data have been used to investigate landslide hazards and the potential for snow avalanches through studies of avalanche pathways. Researchers have used DEM data to evaluate landscape morphology tied to climatic and tectonic factors along a strike-slip fault in northern California. Through an analysis of longitudinal profiles over time, DEM data were useful for demonstrating that stream channels were in equilibrium there. Other projects have used DEM data collected through SRTM to uncover evidence of past tsunami events in areas of coastal Greece, where terrain data helped identify remnant features from catastrophic tsunamis, including irregularly shaped swamps, coastal lagoons, and arc-shaped walls and scarps running parallel to the coast. Data from ASTER (advanced spaceborne thermal emission and reflectance radiometer) have been used to model inland ice-sheet flows in Antarctica's Grove Mountains.

DEMs are also important for solving problems such as viewshed analysis, which enables natural resource managers to create 3D diagrams for evaluating the visual impact of a proposed building or structure on a scenic vista. Algorithms for generating viewsheds from DEM data are based on estimating the elevation difference of grid cells that lie between the viewpoint and the location of interest. Each intervening cell is evaluated to determine if it interferes with the direct line of sight separating the viewpoint from the target location. Civil engineers have used DEMs to identify areas prone to flooding and to estimate the volume of proposed reservoirs. DEM data have many military applications, including guidance systems for cruise missiles and aircraft. Terrain-based navigation systems are equipped with sensors to compare real-time data to onboard terrain databases. Terrain data are also used in telecommunications engineering for predicting radiowave propagation from proposed communications tower and microwave antenna sites.

For background information *see* COORDINATE SYSTEMS; GEOGRAPHIC INFORMATION SYSTEMS; GEOGRAPHY; GEOMORPHOLOGY; INTERFEROMETRY; LIDAR; SPACE SHUTTLE; SYNTHETIC APERTURE RADAR (SAR) in the McGraw-Hill Encyclopedia of Science & Technology.   Thomas A. Wikle

Bibliography. F. Ackerman, Digital elevation models—Techniques and application, quality standards, development, *Int. Arch. Photogramm. Rem. S.*, 30:421–432, 1994; P. Burrough, *Principles of Geographical Information Systems for Land Resources Assessment*, Oxford University Press, 1986; E. Iaasks and R. Srivastava, *Introduction to Applied Geostatistics*, Oxford University Press, 1989; Z. Li, Q. Zhu, and C. Gold, *Digital Terrain Modeling: Principles and Methodology*. CRC Press, 2005.

Disease outbreaks in livestock due to global warming

In September of 2007, bluetongue infection in a cow was reported for the first time in the United Kingdom on a farm near Ipswich, Suffolk. Within a short time, three more cases were identified and the animals destroyed. Soon after, however, the rising number of cases made it obvious that the virus was present in the insect population and was asymptomatic in other cattle. At that point, the government stopped the slaughter of cows while setting up "protection zones," areas where the movement of cattle and sheep are restricted due to potential for disease spread. The bluetongue virus has spread northward into Europe from the Middle East and Africa, yet another warning sign of the effects of global warming.

Bluetongue disease and global warming. Bluetongue disease causes severe symptoms in cattle, sheep, and deer (ruminants), producing ulcers of the mouth, nose, and eyes. Swelling of the head and limbs, lameness, and internal bleeding are followed by trouble in breathing and the tongue turning blue. Pregnant sheep may suffer a spontaneous abortion or stillbirth. Cattle are more likely to suffer early embryonic loss and decreased reproductive efficiency, decreasing their milk and calf production. Death follows in 80% of cases in nonresistant sheep; animals that survive suffer from muscle wastage and their meat is of poor quality. No antiviral drugs are available for treating bluetongue. Therefore, all therapy is strictly supportive. Bluetongue virus (BTV) is transmitted to livestock by biting midges (*Culicoides* species) (**Fig. 1**), resulting in severe hemorrhagic disease in sheep, while remaining asymptomatic in the insect host.

Historically, bluetongue is a disease found in cattle and sheep in Africa and the Middle East. The genus *Culicoides*, the biting midges that transmit the virus, prefer higher winter and summer nighttime temperatures than were the average in Europe previously. The northward migration of the disease matches the areas where winter and night temperatures have increased through Europe. As recently as 1998, the disease had been reported only in the most southern parts of Spain, Italy, Portugal, and Greece. Since that time, the midge has spread through much of these countries into southern France and across the Balkans.

However, the most recent outbreak occurring in northern Europe came from a different source—one that has yet to be identified. Bluetongue suddenly appeared in the Netherlands in August 2006, rapidly spreading into Belgium, Germany, France, and Luxembourg. The virus responsible was different from the strains found in southern Europe—its genetic makeup matched that of viruses from South Africa instead. How this strain traveled to northern Europe is a subject of speculation. One possibility is through air transportation of livestock. As the midges prefer to breed in damp manure, they may have come into the area with a horse shipped in by air from South Africa.

What is really alarming, however, is that the virus survived the cold winter of northern Europe and broke out again in August 2007. For the virus to survive, large numbers of the local midges had to be infected and live long enough to infect many cattle or sheep. Warming allows midges to be active longer during the year, the virus replicates faster at warmer temperatures, and midges that normally cannot carry

Fig. 1. A gravid female *Culicoides dewulfi* collected from a location near bluetongue outbreaks in Belgium in 2006. [Photograph by R. De Deken and M. Madder, *Institute of Tropical Medicine, Antwerp, Belgium; from C. Saegerman, D. Berkvens, and P. S. Mellor, Bluetongue Epidemiology in the European Union, Emerg. Infect. Dis., 14(4):539–544, 2008*]

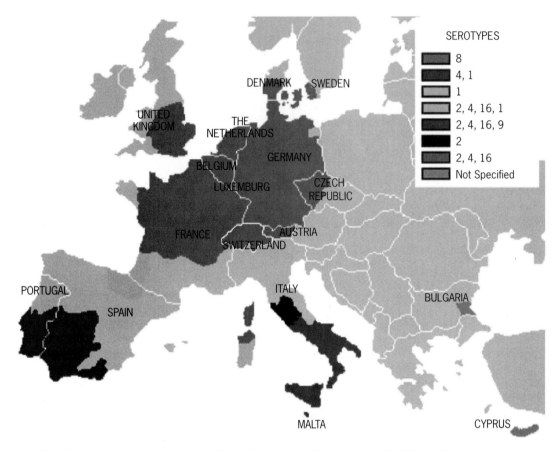

SEROTYPES

8
4, 1
1
2, 4, 16, 1
2, 4, 16, 9
2
2, 4, 16
Not Specified

Fig. 2. Restriction zones in Europe, by serotype. [*From C. Saegerman, D. Berkvens, and P. S. Mellor, Bluetongue epidemiology in the European Union, Emerg. Infect. Dis., 14(4):539–544, 2008*]

the virus become able to carry and transmit it. The paradox is this: The virus cannot persist in the larvae form of the midge, and the adult midges have a life span that is too short to survive through the winter. How, then, did the virus reappear in 2007?

Three possibilities exist to explain how the virus survived through the winter. Some infected midges may have entered barns to survive the colder temperatures. French researchers have examined this possibility and found that this may be the case. A second possible explanation for survival of the virus is that a cow may have carried the disease through the winter. One theory is that T cells may pick up the virus at the site of a midge bite; after the virus enters the cells, it becomes quiescent until a second bite occurs. At this time, it may reactivate and escape the cell into the bloodstream, where it is likely to be picked up again by a biting midge during a blood meal. The final possibility is that it may be carried in the fetus of infected cows. Three cows that had recovered from the disease in the fall of 2007 were exported from Holland to Ireland, where they gave birth to calves infected with the virus. At this point, midges could bite the infected calves and then spread the disease onward to healthy cows. Thus, whatever the method of survival, once the virus reappeared in Belgium in 2007, infected midges were probably blown across the English Channel to the United Kingdom, where they infected their first victims in that country.

Can this disease be prevented or controlled? Killed virus vaccines have been developed to prevent cases of the southern strains of the virus. These have been used since 2004. However, with 24 different strains of the virus, it is difficult to develop vaccines that will protect against all of the viruses simultaneously. A new vaccine for the northern strain (type 8) was put into development in 2006 soon after the outbreak started in northern Europe. In the meantime, governments in northern Europe restricted the movements of animals within areas of infection in the so-called restriction zones (**Fig. 2**), using insecticides to lower midge populations and keeping animals indoors during the most active times for the midges (dawn and dusk). This worked well in Greece; however, as stated previously, it appears that midges in cooler climes will move indoors into barns, whereas they do not seem to do so at warmer temperatures. In May 2008, cattle in the United Kingdom were the first to receive the new vaccine developed against serotype 8. Under European Union law, only cattle and sheep in a restriction zone are allowed to be vaccinated. By the end of the 2008 midge season, approximately 22.5 million doses will be administered to cattle and sheep in Europe.

Spread of disease and link to global warming. Another disease of ruminants, African horsesickness, is caused by a virus closely related to the BTV. Horse advocates in the United Kingdom are already calling for the government to make plans to deal with an outbreak of this illness, which may follow closely the pattern of bluetongue disease. Other diseases that have spread into new geographic regions in recent years, resulting in infection in animal populations, include West Nile virus, chikungunya virus, Rift Valley fever, leishmaniasis, and malaria, one of the world's most deadly diseases. West Nile virus traveled to the United States in 1999, spreading across the country through insects and animals. It may have gotten to North America through infected mosquitoes traveling in a cargo shipment, either by boat or by plane.

Malaria and its vector, the *Anopheles* mosquito, have moved up the slopes of Mount Kilimanjaro in Africa to supposedly malaria-free altitudes. The *Anopheles* mosquito is found worldwide and merely requires the presence of sufficient individuals infected with the *Plasmodium* parasite to result in outbreaks. Three countries in Eurasia that are covered by the World Health Organization—Azerbaijan, Tajikistan, and Turkey—are considered danger zones for mosquito-borne malaria. Leishmaniasis is a protozoan disease spread by sandflies common in many tropical and subtropical climates. As temperatures rise, the sandfly vector may spread into areas previously not endemic for the disease, thereby carrying *Leishmania* with it.

Borrelia burgdorferi, the bacterium responsible for causing Lyme disease, is found in North America, northern Europe, and parts of Asia. Ticks infected with *Borrelia* are being found more frequently in Japan, northwest China, and far eastern Russia. This disease may spread into other countries as temperatures become more temperate in colder locations, allowing the survival of the tick vectors. Other diseases that are vector-borne or transmitted by animals are certain to follow. There is an urgent need to consider the potential effects of global warming on health and disease today and in the very near future. Research and monitoring of the spread of infectious disease urgently require the attention of government health agencies. Coordination between governmental agencies both within a country and between countries will be of overwhelming importance in the control of these spreading infectious diseases. Preparations must be made to deal with the very real problems that global warming will cause in terms of new disease patterns.

For background information *see* AFRICAN HORSE-SICKNESS; ANIMAL VIRUS; BLUETONGUE; CLIMATE MODIFICATION; DIPTERA; DISEASE ECOLOGY; EPIDEMIOLOGY; GLOBAL CLIMATE CHANGE; LYME DISEASE; MALARIA; VACCINATION; WEST NILE VIRUS in the McGraw-Hill Encyclopedia of Science & Technology.
Marcia M. Pierce

Bibliography. A. Bashford (ed.), *Medicine at the Border: Disease, Globalization and Security, 1850 to the Present*, Palgrave Macmillan, New York, 2006; P. J. Hotez, *Forgotten People, Forgotten Diseases: The Neglected Tropical Diseases and Their Impact on Global Health and Development*, ASM Press, New York, 2008; P. R. Murray et al. (eds.), *Manual of Clinical Microbiology*, 9th ed., ASM Press, New York, 2007; B. V. Purse et al., Climate change and the recent emergence of bluetongue in Europe, *Nat. Rev. Microbiol.*, 3(2):171–181, 2005.

Dispersal versus vicariance

Scientists have long been fascinated by the existence of disjunct (geographically discontinuous) distribution patterns such as the one shown in **Fig. 1a**, in which the members of a group of organisms are distributed across the southern continents, now separated by thousands of miles of ocean. How did this type of widely scattered distribution originate? Traditionally, two alternative explanations have been proposed: *dispersal* across a preexisting geographical barrier (for example, a mountain chain); or *vicariance*, the fragmentation of a widespread ancestral distribution by the appearance of a new barrier. Both biogeographical processes result in the isolation of a population by a geographic barrier, followed by differentiation of a new taxon by allopatric (geographically separated) speciation. However, the barrier in the dispersal explanation is older than the geographic disjunction, whereas the appearance of the geographic barrier is responsible for the geographic disjunction in the vicariance explanation, so it cannot be older than the resulting speciation event. Although vicariance and dispersal are not mutually exclusive processes—the opening of the Gibraltar Strait between North Africa and Iberia in the Pliocene was simultaneously a vicariance event for terrestrial organisms and a dispersal event for marine organisms—the history of biogeography as an evolutionary science could be considered until recently as the history of a debate between dispersal and vicariance explanations.

Centers of origin and the vicariance paradigm. Dispersal was for centuries the dominant explanation. Prior to Darwin, every species was assumed to be immutable and unrelated to any other. Darwin's theory of evolution (1859) challenged this immobilist view of species by postulating that organisms originate in an area, the "center of origin," from which some individuals disperse to other areas by chance and later evolve into new species through natural selection. Darwin and his contemporaries, however, still believed in the concept of a stable Earth, in which the size and position of continents had not changed over time.

In the middle of the twentieth century, the discovery of the physical and chemical composition of the Earth finally provided the necessary geological mechanism to explain Earth's changing geography—the theory of plate tectonics. This contributed to the appearance of a new paradigm in historical biogeography—the concept of vicariance. It was no longer necessary to postulate rare and unlikely long-distance dispersal events to explain disjunct geographic distributions: species could be simply carried away as continents split and moved across the surface of the Earth. At the same time, the surge of *cladistics* (a taxonomic theory by which organisms

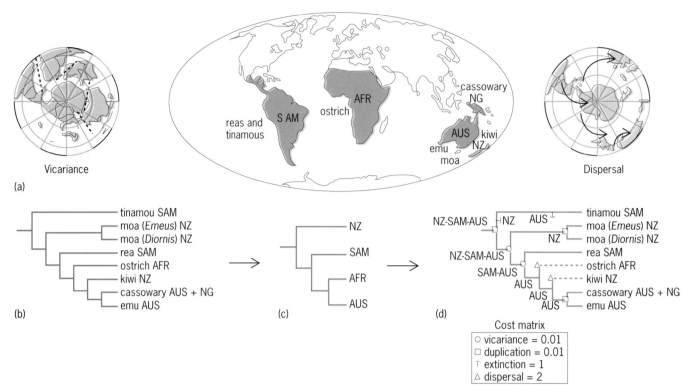

Fig. 1. Distribution patterns of organisms. (*a*) Disjunct geographic distribution of the ratite birds (for example, ostriches) (middle panel) and two alternative historical processes to explain it: (right panel) *dispersal*: the ancestor was originally distributed in one area, the "center of origin," from which it dispersed to the other continents by crossing a geographical barrier (for example, an ocean basin); (left panel) *vicariance*: the ancestor was distributed in a widespread area, the Gondwanan supercontinent, which became fragmented by the sequential opening of ocean basins. (*b*) DNA-based phylogenetic tree and current geographic distributions of ratites. (*c*) "Area cladogram": branching diagram grouping areas of distribution based on their shared endemic taxa, presumably reflecting the history of biotic connections among areas for the group analyzed. (*d*) "Event-based reconstruction": minimum-cost reconstruction of the biogeographical history of ratites (ancestral areas and biogeographical events) necessary to explain the terminal distributions according to the biogeographical cost model. (**Paleomaps from ODSN**)

are grouped and ranked on the basis of the most recent phylogenetic branching point) provided a new methodology to reconstruct phylogenetic relationships among organisms (Fig. 1*b*) without the need to rely on the fossil record. *Cladistic biogeography* was born from the fusion of cladistics with vicariance. It is based on the idea that organisms and areas evolve together. Thus, taxa sharing the same phylogenetic and distributional patterns are assumed to have shared a common biogeographical history, that is, they were part of the same ancestral biota that became divided by a sequence of vicariance-isolating events. For cladistic biogeography, vicariance could be tested by searching for congruence between "area cladograms" (Fig. 1*c*) of multiple codistributed organisms. In contrast, dispersal was considered a rare and random process that was unable to explain shared distribution patterns across organisms with varying ecologies and dispersal capabilities.

Molecular phylogenetics and the resurrection of dispersal. Recent years have witnessed a new paradigm shift in historical biogeography, with dispersal regaining a more prominent, or even primary, role in explaining contemporary distribution patterns. Three factors have contributed to this shift. The first one is the realization that vicariance alone cannot explain all congruent biogeographical patterns.

For example, dispersal is important to explain island biodiversity patterns, especially in isolated oceanic archipelagos such as the Hawaiian Islands that can be colonized only by over-water dispersal. Moreover, if directed by abiotic factors such as prevailing winds and ocean currents, dispersal is also capable of generating nonstochastic (nonrandom), highly concordant distributional patterns such as those expected from vicariance ("concerted dispersal"). For example, dispersal driven by the West Wind Drift, which moves wind and ocean currents eastward around Antarctica, is often argued to explain the close phylogenetic links between the Australian and New Zealand floras.

Second, paleogeographical reconstructions show that many regions present a more complex history than a simple sequence of vicariance/splitting events. In addition to dispersal and vicariance, there is a third process capable of generating congruent distributional patterns across codistributed lineages. "Dispersion" or "geodispersal" is equivalent to range expansion in response to the disappearance of a previous geographic (dispersal) barrier between two areas (**Fig. 2*a***). The collision between two continents that were previously separated by an ocean barrier results in episodes of range expansion occurring simultaneously in different clades. For example, the closing of the Turgai Strait between Europe and Asia in the Oligocene allowed a period of significant biotic exchange between these two continents during the Late Tertiary. Contrary to the strict vicariance model, in which an ancestral area is fragmented by successive splitting events, most regions conform to what has been called a "reticulate" biogeographical scenario, with alternate cycles of area collision (geodispersal) and area splitting events (vicariance) [Fig. 2b].

Probably the most important factor for strengthening the role of dispersal in biogeography has been the advent of molecular phylogenetics. Molecular, deoxyribonucleic acid (DNA)–based analyses are particularly appropriate for biogeographical inference because, when calibrated with fossil evidence, they can provide information on divergence times between lineages (the "molecular clock"). Divergence times can be used to discriminate between dispersal and vicariance scenarios by comparing the estimated time of divergence between the disjunct taxa with the timing of the geographic barrier. Recent molecular studies have used this correlation to show that a variety of patterns are the result of dispersal rather than vicariance. For example, some African chameleons are apparently descendants of lineages that dispersed from Madagascar less than 30 million years ago, long after the separation of both continents. This molecular approach to biogeography has been criticized because fossil calibrations can provide only minimum ages for divergence times and because the rate of mutation generally varies across lineages and over time. Nevertheless, most biogeographers now use several fossils as phylogenetic age constraints and often employ methods that "relax" the clock assumption, allowing DNA sequences to evolve at different rates along the phylogenetic tree.

Dispersal and vicariance: new integrative approaches. In recent years, new methods of biogeographical inference have been developed that seek to integrate both dispersal and vicariance in the biogeographical reconstruction through the use of a model-based approach. *Event-based biogeographic methods*, for example, use a deterministic cost model in which each process receives a cost according to its likelihood to occur (Fig. 1d). The cost of the processes is estimated beforehand using an optimality criterion. To find "phylogenetically conserved distribution patterns," that is, distribution patterns that are somewhat inherited from ancestor to descendants, dispersal is assigned a higher cost than vicariance. This is because dispersal events result in descendants that occur in an area outside the ancestral range, thus breaking the ancestor-descendant geographic association (Fig. 1d). By making explicit the relationship between processes and the expected biogeographical patterns, event-based methods provided some novel insights. However, because of the cost assumptions, these methods generally underestimate dispersal, and they are not truly probabilistic.

Recently, fully probabilistic, *parametric biogeographical methods* have been developed that model range evolution, that is, the change in geographic

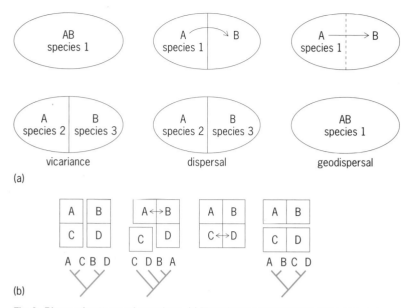

vicariance   dispersal   geodispersal

(a)

(b)

Fig. 2. Dispersal patterns of organisms. (*a*) Three historical biogeographical processes responsible for congruent distributional patterns: vicariance, dispersal, and geodispersal (range expansion in response to the disappearance of a geographic barrier). (*b*) Alternate cycles of area collision (geodispersal) and area splitting (vicariance) can produce *reticulate* biogeographical patterns that cannot be represented by a single hierarchical area cladogram.

range from ancestor to descendants, as a stochastic process that evolves along the branches of the phylogenetic tree according to a probabilistic model for which parameters are biogeographical processes (**Fig. 3a**). Compared to previous approaches, these methods present several advantages: (1) They allow estimation of biogeographical parameters such as dispersal and extinction rates directly from the data.

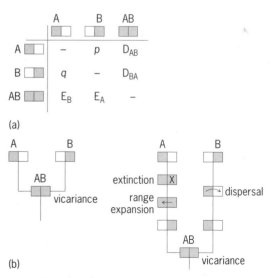

(a)

(b)

Fig. 3. Model-based approach to historical biogeography. (*a*) Range evolution is modeled as a stochastic process with discrete states (geographic ranges) that evolve along the branches of the phylogenetic tree according to a matrix of transition probabilities; its parameters are biogeographical processes (dispersal, p, q; extinction, E_A, E_B; range expansion, D_{AB}, D_{BA}) that determine the probability of range evolution from ancestor to descendants as a function of time. (*b*) The probability of change in geographic range from ancestor to descendants is higher along long phylogenetic branches (here representing time duration) than along shorter branches. Longer branches, or a higher rate of change, also increase uncertainty for the inferred ancestral range.

(2) They incorporate estimates of branch lengths (that is, the evolutionary divergence between lineages or the time since cladogenesis) in the biogeographical inference (Fig. 3b). (3) Relevant biogeographical evidence other than the phylogeny and terminal distributions (for example, ecological and temporal) can be easily integrated into the biogeographical model. For example, to infer colonization patterns in islands, dispersal rates in the transition probability matrix could be made dependent on geographic distance, availability of land connections, or strength of wind currents.

Finally, if dispersal has strengthened its role in current historical biogeography, vicariance has to some extent changed its definition. For the phylogenetic niche conservatism hypothesis, vicariance is the outcome of any environmental change that creates conditions within the species' geographic range that are outside its "ancestral ecological niche" (the range of ecological conditions in which it can maintain viable populations). Individuals are unable to disperse, and the species' range becomes fragmented. Because species tend to retain their ecological niches over evolutionary time, it is this inability to adapt to new environmental conditions, followed by vicariance, that plays a key role in isolating populations and creating new lineages. This concept has recently been used to explain contemporary diversity gradients and for predicting the response of species to climate warming.

For background information *see* BIOGEOGRAPHY; ECOLOGY; ISLAND BIOGEOGRAPHY; PALEOGEOGRAPHY; PHYLOGENY; PLANT GEOGRAPHY; POPULATION DISPERSAL; POPULATION ECOLOGY; SPECIATION; SYSTEMATICS; ZOOGEOGRAPHY in the McGraw-Hill Encyclopedia of Science & Technology.   Isabel Sanmartín

Bibliography. O. Haddrath and A. J. Baker, Complete mitochondrial DNA genome sequences of extinct birds: Ratite phylogenetics and the vicariance biogeography hypothesis, *Proc. Biol. Sci.*, 268(1470):939–945, 2001; J. J. Morrone and J. V. Crisci, Historical biogeography: Introduction to methods, *Annu. Rev. Ecol. Syst.*, 26:373–401, 1995; R. H. Ree and S. A. Smith, Maximum likelihood inference of geographic range evolution by dispersal, local extinction, and cladogenesis, *Syst. Biol.*, 7(1):4–14, 2008; I. Sanmartín, Event-based biogeography: Integrating patterns, processes, and time, in M. C. Ebach and R. S. Tangney (eds.), *Biogeography in a Changing World*, The Systematics Association Special Volume Series 70, CRC Press, Boca Raton, FL, 2007; J. J. Wiens, Speciation and ecology revisited: Phylogenetic niche conservatism and the origin of species, *Evolution*, 58:193–197, 2004.

Dosage compensation of the active X chromosome

Sex chromosomes have evolved from an ordinary pair of autosomes (non-sex chromosomes) in multiple lineages, including flies, worms, and mammals. Lack of recombination between the sex chromosomes has led to loss and differentiation of genes on the Y chromosome, leaving males with a single copy of most X-linked genes. Except for a small region of homology called the pseudoautosomal region, the mammalian sex chromosomes differ significantly in their gene content: For example, the human X chromosome contains about 1300 genes, whereas the Y chromosome contains about 130 genes. This striking divergence results from evolutionary forces that progressively altered the ancestral homologous pair of proto-sex chromosomes. The Y chromosome lost many genes as a result of the suppression of recombination that was needed to avoid the production of abnormal sexual phenotypes. The Y chromosome also accumulated male-advantageous genes around the testis-determining gene. The mammalian sex chromosomes apparently diverged by a stepwise mechanism that progressively suppressed recombination by means of large Y inversions.

The X chromosome is enriched in genes related to sexual reproduction and brain function. Genes that enhance male sexual reproduction are thought to have accumulated because recessive mutations expressed in males as a result of hemizygosity (presence of a single copy of a gene) of the X could give rise to novel functions. Genes involved in female sexual reproduction are also enriched on the X, probably because the X chromosome spends twice as much time in females. Another category of genes enriched on the X are genes expressed in the brain. The high expression of X-linked genes in the brain suggests evolutionary mechanisms for selection of genes that confer enhanced cognitive functions. Such genes may provide a selective advantage to males in sexual reproduction (for example, preferential choice by females), or may have evolved faster because of hemizygosity in males. The special role of the X chromosome in brain function has implication for human diseases; indeed, X-linked mental retardation is common and preferentially affects males.

Dosage compensation. Mechanisms of dosage compensation evolved to protect organisms against deleterious effects of X monosomy (the condition in which one chromosome of a pair is missing in a diploid organism). The regulated dosage of any one gene is not necessarily important for the viability of an organism. However, the gene dosage of a whole chromosome or even a part of a chromosome is vital. In the fruit fly *Drosophila*, being haploid for as little as 1% of the genome reduces viability, and being haploid for more than 3% of the genome is lethal. As in mammals, *Drosophila* females have two X chromosomes, but males have only one X. It has been known for a long time that the single X chromosome in male flies is expressed at twice the level of each X in females, resulting in a balanced expression between the X chromosome and the autosomes.

In mammals, monosomy for even the smallest human chromosome is not viable. Thus, it is not surprising that mammals also evolved a mechanism to ensure balanced expression of the X chromosome and the autosomes. Susumu Ohno was the first to recognize the critical need for X upregulation in mammals when he wrote the following in 1967: "During the course of evolution, an ancestor to placental mammals must have escaped a peril resulting from the hemizygous existence of all X-linked genes in the male by doubling the rate of product of each X-linked gene." The first experimental evidence of X upregulation was found by exploiting a rare evolutionary translocation event that moved the chloride channel gene 4 (*Clcn4*) from the X chromosome in one mouse species to chromosome 7 in another mouse species. *Clcn4* expression was doubled from the single active X as compared to each autosomal copy in mice, resulting from crosses between the two mouse species.

Subsequent microarray experiments to survey the expression of all X-linked and autosomal genes in multiple somatic tissues from several mammalian species (mice, rats, and primates including humans) have shown unequivocally that the active X chromosome is upregulated relative to autosomes in mammals. Indeed, the ratio between the average level of expression of X-linked genes versus that of autosomal genes was calculated to be 1. If there were no upregulation, this ratio would be 0.5. These microarray studies provide concrete evidence that it is indeed important for X expression dosage to be balanced with that of autosomes in mammals (see **illustration**).

Since male and female germ cells contain only one set of autosomes, one might predict that the X chromosome would not need to be upregulated in these haploid cells, which has been confirmed by microarray studies. However, because of a highly complex regulation of the sex chromosomes in germ cells and their precursors, X-linked gene expression varies depending on the stage of germ-cell differentiation. In spermatocytes (precursors of male germ cells), the sex chromosomes are mostly unpaired during meiosis, except at the pseudoautosomal regions. As a result, the X chromosome is mostly silenced by a mechanism that inactivates all unpaired chromosomes. Interestingly, retrogenes [generated by the reverse transcription of ribonucleic acid (RNA) and reintegrated into the genome] that originate from X-linked genes but are located on autosomes supply gene products essential for spermatocyte survival. Following male meiosis, partial reactivation of specific X-linked genes occurs in haploid spermatids. In the female germline, very different events take place, including reactivation of the inactive X in primary oocytes, followed by formation of secondary oocytes in which the X chromosome does not appear to be upregulated. These events suggest either a removal of upregulation marks or a combination of upregulation and repression, as suggested for the *Drosophila* female germline. Further studies are needed to clarify the regulation of X-linked genes in germ cells and early embryos.

Molecular mechanisms. An important question is how X upregulation is established in mammals. The process must have both evolutionary and chromosome-specific components, because a corresponding increase in gene expression does not occur

to correct clinical cases of autosomal monosomies, which are lethal. One possibility is that the basal expression level of each X-linked gene may have increased via deoxyribonucleic acid (DNA) sequence modifications during evolution. Increased levels of steady-state messenger RNA (mRNA) may result from changes in the sequence of promoters and enhancers and/or at the 3′ end of genes. Such mechanisms may increase initiation or elongation rates of transcription, or possibly RNA stability. Another possibility is that X upregulation involves epigenetic changes (that is, changes in the chromatin rather than the DNA sequence), similar to the situation in *Drosophila*. Epigenetic mechanisms could employ both enhancing and repressive marks in order to achieve a perfect balance between the X and the autosomes.

The low transcriptional output observed in haploid germ cells in mammals implies an X-specific partial repression process in these cells, which would be removed in early development to initiate X upregulation after fertilization. Although a balanced expression between the X chromosome and autosomes has been observed at early developmental stages, the interactions between X upregulation and X inactivation, which is also initiated in early development, are not known.

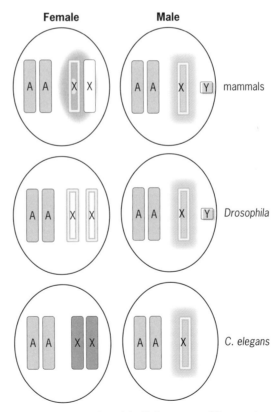

Dosage compensation of the X chromosome (X) occurs in diploid cells of mammals, *Drosophila*, and *C. elegans*. In mammals, the active X chromosome is upregulated in both males (shaded glow) and females (shaded glow), while one X chromosome is silenced by X inactivation in females (white). In *Drosophila*, the X chromosome is only upregulated in males (shaded glow). In *C. elegans*, the X chromosome is upregulated in males (shaded glow) and hermaphrodites, while both X chromosomes are repressed in hermaphrodites (dark gray). Y = Y chromosome; A = autosomes (a single pair represented; gray).

Comparison between species. The findings that X chromosome expression is upregulated in somatic cells of mammals, *Drosophila,* and *Caenorhabditis elegans* (a small nematode worm) suggest that dosage compensation is essential across species (see illustration). In *Drosophila,* gene expression from the X is selectively boosted in male flies, but not in female flies. In contrast, mammals and *C. elegans* increase X-linked gene expression in both males and females (hermaphrodites in *C. elegans*). Then, to avoid functional tetrasomy of X-linked genes in XX individuals, mammalian females silence one X chromosome by X inactivation, whereas *C. elegans* hermaphrodites repress both X chromosomes (see illustration). Thus, the mechanisms of dosage compensation differ among species, probably because they evolved separately in each lineage out of necessity. The selective doubling of gene expression of the male X in *Drosophila* may relate to the use of common pathways between sex determination and dosage compensation in this species. In mammals, the complexity of X chromosome regulation that combines both an enhancing process on the active X chromosome and a silencing process on the inactive X chromosome is perplexing. A potentially beneficial consequence would be to modulate X-linked gene expression between the sexes. For example, some X-linked genes escape X inactivation (that is, are expressed from both alleles) and thus have higher expression in females, which could provide an advantage via a dosage-dependent function.

The molecular mechanisms of upregulation of the mammalian X chromosome may ultimately display common features with those of *Drosophila* and *C. elegans*. Further studies to elucidate how each species achieves dosage compensation between the X chromosome and the autosomes in the germline and soma (the whole of the body of an individual, excluding the germ tract) will provide insight into both the evolution of sex chromosomes and the mechanisms of chromosome-wide gene regulation.

For background information *see* CHROMOSOME; CHROMOSOME ABERRATION; DEOXYRIBONUCLEIC ACID (DNA); GENE; GENE AMPLIFICATION; GENETIC MAPPING; MAMMALIA; RECOMBINATION (GENETICS); RIBONUCLEIC ACID (RNA); SEX DETERMINATION in the McGraw-Hill Encyclopedia of Science & Technology.

Christine M. Disteche; Di Kim Nguyen

Bibliography. D. A. Adler et al., Evidence of evolutionary up-regulation of the single active X chromosome in mammals based on *Clc4* expression levels in *Mus spretus* and *Mus musculus, Proc. Natl. Acad. Sci. USA*, 94:9244–9248, 1997; J. A. Birchler, H. R. Fernandez, and H. H. Kavi, Commonalities in compensation, *Bioessays*, 28:565–568, 2006; V. Gupta et al., Global analysis of X-chromosome dosage compensation, *J. Biol.*, 5:3, 2006; E. Heard and C. M. Disteche, Dosage compensation in mammals: Fine-tuning the expression of the X chromosome, *Genes Dev.*, 20:1848–1867, 2006; D. K. Nguyen and C. M. Disteche, Dosage compensation of the active X chromosome in mammals, *Nat. Genet.*, 38:47–53, 2006; S. Ohno, *Sex Chromosomes and Sex Linked Genes*, Springer-Verlag, Berlin, 1967.

Earliest seafaring

Despite the importance of human seafaring in the history of global colonization, archeologists are uncertain about how, why, and when humans first began to use boats. Physical evidence for boats in the archeological record is sparse and late in time, and our understanding of seafaring is derived from dated archeological remains of earliest human occupation recovered from regions that could have been reached only via sea crossings.

Early history. During the late Pleistocene, when high-latitude glacial conditions reached their maximum, sea levels declined dramatically and allowed populations of modern humans to colonize areas connected by land bridges that were formally isolated islands. However, a number of regions colonized by humans during the Pleistocene were never connected by land bridges, and it can be assumed that humans arrived by boats. The best example of this, and the earliest in time, comes from the continental landmass known as Sahul (Australia, New Guinea, and the Aru Islands), where modern human occupation is dated to at least 55,000 years before present (BP) [**Fig. 1**]. It is known that there have never been land bridges joining the islands of Wallacea to either the Asian (Sunda) or Australian (Sahul) continental landmasses because these two continental regions, and the islands that separate them, had very distinctive faunas prior to human translocation of animals. Sundaland has an Asian placental mammal fauna that includes many species ranging in size from the elephant downwards. The islands of Wallacea, on the other hand, have fewer species and a higher proportion of endemic ones. Australia and New Guinea have a distinctive marsupial fauna. The only placental animals to reach Australia and New Guinea that were not humanly assisted were rats and bats.

While it is usually assumed that seafaring would have been beyond the capabilities of premodern hominins, finds of stone artifacts from the Indonesian island of Flores dated to 800,000 years ago indicate that *Homo erectus* did manage short water crossings. The longest of these, the deep channel separating Bali and Lombok, would have been up to 25 km (15.5 mi) wide. However, these water crossings may not indicate that *H. erectus* engaged in purposive seafaring; rather, it is possible that small groups were blown offshore on floating vegetation and established a founder population. The fact that no remains of *H. erectus* have been found east of Flores, where colonization requires substantially longer and more difficult water crossings, suggests that their arrival on Flores was the result of accidental rather than purposive voyaging.

The water crossings necessary to get seafarers to Sahul were of a vastly different order of magnitude. Even during periods of maximum low sea stand during the last glacial maximum, the water gap to be

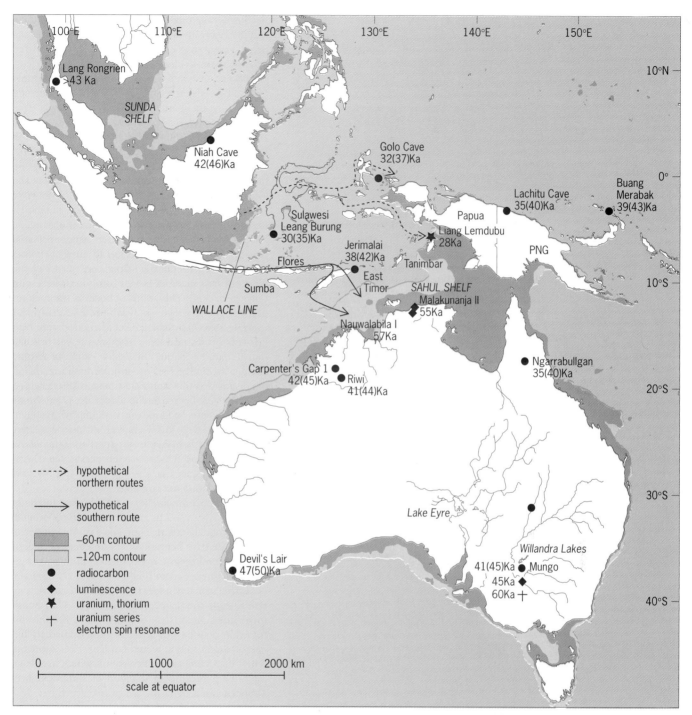

Fig. 1. Map of the region showing the oldest archeological sites in Sunda and Sahul and prospective northern and southern migration routes into Sahul. Radiocarbon dates for sites are shown as uncalibrated (raw) ages; calibrated ages (calendar dates that take into account the variation of production rates of ^{14}C from the present) are given in parentheses. (*Drawn by Carto ANU Section 8-043; Copyright by Susan O'Connor*)

crossed between the south coast of Timor and the expanded northern Australian coastline would still have been on the order of 150 km (93 mi). Between 60,000 and 58,000 BP, sea levels were about 60 m (200 ft) lower than today and the water gap closer to 400 km (250 mi). First landfall sites on the expanded Pleistocene coastline of what is now northern Australia are today submerged, drowned by postglacial rising seas at the end of the Pleistocene. The earliest dated archeological sites are well inland, thus providing no hint as to the type of economy of the

first seafarers to land on this large and unfamiliar continent. However, evidence exists from the Wallacean island of Timor, where the north coast offshore profile drops off steeply to the continental shelf, that provides a window into coastal settlement in the Pleistocene. At the cave site known as Jerimalai, the remains of marine fauna such as fish, turtle, and shellfish date back to 42,000 years BP, demonstrating conclusively that the early colonists were adept at utilizing marine resources. Whether seafaring was facilitated by such skills or whether their

Fig. 2. Double raft known as *kalwa* used in the Kimberley region in historic times. (*Photo by William Jackson, 1917, North-West Scientific Expedition*)

maritime-oriented economy emerged as a response to settlement in this faunally depauperate (impoverished) island is less certain. The water crossings necessary to reach Island Melanesia from the north coast of Papua New Guinea (PNG) were also made by 42,000 years ago, suggesting that voyaging was purposeful rather than accidental (Fig. 1).

A variety of routes to Sahul have been proposed. Based on present evidence, a southern route through the Lesser Sundas (including Flores and Timor) and directly across to the expanded Pleistocene coastline of northern Australia seems most likely due to the minimum distances between island hops and the greatest intervisibility (clear line of sight) between islands. Some researchers have proposed that colonization proceeded via a northern route through Borneo and Sulawesi and thence into northern Maluku and the Bird's Head of Papua (Fig. 1). Excavations on islands on the northern route, such as Sulawesi and Halmahera, have so far failed to establish any settlement as early as on the southern route. An alternative

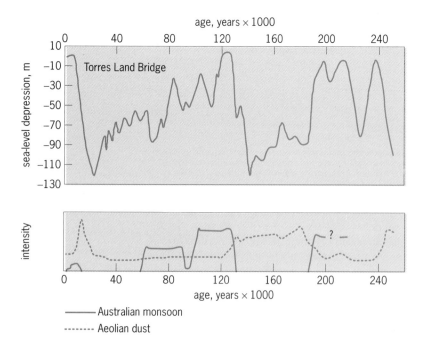

Fig. 3. Variations of sea level and the Australian monsoon for the last 250,000 years. (*Courtesy of John Chappell*)

northern route onto the Sahul Shelf via Buru, Seram, and the Kei (Kai) Islands and onto the shelf near the present-day Aru Islands is also currently unsupported by early dated evidence for human occupation.

Evidence. Archeological evidence of past maritime technology is sparse and patchy in both space and time. Only fragmentary archeological remains of boats have been recovered in Oceania, all less than 1000 years old. The earliest known boats found in the Netherlands, France, and China are dated between 8000 and 10,000 years old. They comprise little more than hollowed-out logs, having no keel and low sides. Such craft may have been suitable for river or lake travel, but would not have been capable of making even narrow sea crossings. In the absence of archeological evidence, researchers have speculated that the lengthy water crossings to Australia would have involved simple but stable craft such as large platform rafts made of lashed bamboo or logs with outriggers. Some depictions of boats in the rock art of the Kimberley region of northern Australia appear to show boats of bundle-reed construction with high upswept prows and sterns. They are crewed by small human figures in the "Bradshaw" style (a unique ancient Australian rock art form of figurative painting discovered by Joseph Bradshaw in 1891), which has elsewhere been dated to older than 17,000 years BP. At the time of European voyaging to Australia, bundle boats were known only from Tasmania. They were made of lashed rolls of paperbark (*Melaleuca* sp.) and were used to cross rivers and for short trips to offshore islands, but were not sufficiently buoyant or stable for long sea crossings.

Large platform rafts akin to those used today on the wide river mouths in southern China provide the best model for early craft, and experimental rafts of this type have been shown to be effective in light weather conditions. Large-diameter varieties such as *Dendrocalamus asper* (giant bamboo) are naturally occurring in mainland Asia, but their natural distribution outside the mainland is uncertain. They occur today throughout Island Southeast Asia, but this distribution may be the result of their spread by humans. Large bamboo would not have been available when the colonists reached Sahul, which may have inhibited a continuing tradition of watercraft use. Timber from a light species of mangrove tree was used for making the rafts used in the Kimberley region of northern Australia in historic times. Capable of carrying up to four people and goods, the rafts were used for journeying to islands up to 15 km (9.3 mi) offshore (**Fig. 2**). A short hardwood paddle could be used to propel them, but most journeys utilized tides and currents. It is doubtful if such rafts would have been sufficiently stable and durable for long sea crossings, and this in itself may be sufficient to explain why there is no evidence of back voyaging to Wallacea.

Once people reached the coastline of Sahul, they appear not to have maintained contact with the populations in the Southeast Asian islands they left behind. Back voyaging would be demonstrated by the translocation of exotic plants, animals, or raw materials such as obsidian from a source area in Sahul. Some useful plants and animals were moved between PNG and Island Melanesia in the Pleistocene, but the earliest evidence postdates first settlement by at least 10,000 years.

Whether or not the rafts or other craft in use over 50,000 years ago could have been equipped with navigational aids, such as sails, has also been much debated. Maritime opportunity, however, may have played a more important role in early seafaring than maritime technology. Although it is usually assumed that maritime migrations would have been facilitated by the reduced distance of the sea crossings between land bodies at times of low sea stand, distance may have been of less significance in the voyaging equation than productive marine environments and wind direction. Rich coastal environments such as mangroves, developed estuaries, backwater swamps, and coral reefs that provide the incentive for maritime activities would have been far less common during falling or low sea stands. Tropical coastal populations are more likely to be equipped for, and engaged in, activities requiring watercraft at times of rising sea levels. Wind regime (the pattern of winds in a given location) is also a significant factor. Navigational aids or knowledge would not have been necessary to make the sea crossing to reach northern Australia via a southern route. During the north Australian wet season, the wind regime at the sea surface is strongly vectored southeastward from Timor, and evidence suggests that this would have been the case between ca. 58,000 and 90,000 BP (**Fig. 3**). Boats drifting off the coast of Timor or Roti during the months of the summer monsoon would inevitably have washed up somewhere on the expanded Kimberley coastline. In favorable conditions, this may not have taken more than a few days. The rising sea level of 62,000 to 59,000 years ago would therefore seem to be a prime time for migration across the Timor Sea. After 58,000 BP, the absence or reduction of the strength of the NW monsoon would have made the voyage considerably more difficult (Fig. 3).

Future studies. While archeological and paleoenvironmental data have allowed a relatively detailed chronological reconstruction of the colonization of the Southeast Asia-Pacific region, there remain many unanswered questions about earliest seafaring. Such questions include the degree to which voyages were intentional, the incentive for migration, and the rate of spread of early seafarers, among others.

For background information see ANTHROPOLOGY; ARCHEOLOGY; EARLY MODERN HUMANS; EAST INDIES; GLACIAL HISTORY; PALEOGEOGRAPHY; PLEISTOCENE; POPULATION DISPERSAL; POPULATION DISPERSAL; SOUTHEAST ASIAN WATERS; ZOOGEOGRAPHY in the McGraw-Hill Encyclopedia of Science & Technology.   Susan O'Connor

Bibliography. A. Anderson, K. Boyle, and J. H. Barrett (eds.), *The Global Origins and Development of Seafaring*, McDonald Institute for Archaeological Research, Cambridge, U.K., in press; M. Balter, In search of the world's most ancient mariners, *Science*, 318:388–389, 2007; S. O'Connor, New evidence from East Timor contributes to our understanding of earliest modern human colonization east of the Sunda Shelf, *Antiquity*, 81:523–535, 2007; S. O'Connor and J. Chappell, Colonisation and coastal subsistence in Australia and Papua New Guinea: Different timing, different modes, pp. 15–32, in C. Sand (ed.), *Pacific Archaeology: Assessments and Prospects*, Les Cahiers de l'Archéologie en Nouvelle-Calédonie, Nouméa, 2002.

Economics of low-tech innovation

As the European Union and the United States are evolving into knowledge societies, the ability to generate, use, diffuse, and absorb new knowledge is increasingly being viewed as critical for economic success. Consequently, conventional wisdom regards high-tech, research-intensive, and science-based industries as the key drivers of future economic prosperity. The policy conclusion is that high-cost industrialized countries should concentrate their efforts on promoting these industries. In this scenario, low- and medium-tech (LMT) industries are deemed to offer very limited prospects for future growth in comparison to high-tech ones and, as a result, receive less explicit policy attention and support. The statistical basis of this perspective is the internationally accepted research and development (R&D) intensity indicator [developed by the Organization for Economic Co-operation and Development (OECD) in the 1960s], which measures the ratio of the R&D expenditure to the turnover of a company or to the output value of a whole sector. Sectors with an R&D intensity of more than 3% are characterized as high-tech or medium-high-tech. Sectors with a R&D intensity below 3% are classified as low- or medium-tech. Mainly "mature" industries, such as the manufacture of household appliances; the food industry; the paper, publishing, and print industry; the wood and furniture industry; and the manufacturing of metal and plastic products, are regarded as low- or medium-tech. In contrast, pharmaceuticals, the electronics industry, medical engineering, vehicle construction, the aerospace industry, large parts of mechanical engineering, and the electrical industries are categorized as high-tech or medium-high-tech.

Surprising stability of LMT sectors. The growing importance of research-intensive and high-tech industries in the context of radically changing social and socioeconomic structures cannot be denied. However, the surprising viability of the nonresearch-intensive industrial sector to this day in the developed economies of the Western countries cannot be ignored either. In relation to the manufacturing industry as a whole, the industrial LMT sector of a large number of industrialized OECD countries has a high employment share of more than 60%; for instance, in 15 EU countries, employment in LMT accounted for about 62% of the total employment in manufacturing in 2005. Thus the question arises: why is this industrial sector so surprisingly stable in high-tech countries?

Overview of low-tech innovation strategies			
	Step-by-step	Customer-oriented	Process specialization
Primary subject area	Incremental product development	Improving the market position; creating new markets	Optimization of process technologies
Examples	Supplier for the automotive industry	Fashion-oriented clothing and furniture industries	Paper manufacturing and food processing
Main conditions	Companies with relatively stable market segments	Broad range of various companies with turbulent market conditions	Companies with highly automated and integrated manufacturing processes

LMT innovation strategies. An answer to this question is that LMT companies are innovative in a specific manner that is generally overlooked in the scientific and public debate. According to recent research findings, the following innovation strategies of LMT companies can be regarded as typical (see **table**).

The first innovation strategy is characterized by the continuous development of given products. This innovation strategy can be referred to as step-by-step product development. Typically, this involves enterprises that manufacture products for relatively stable market segments, such as for special applications in the automotive industry. Generally, this innovation strategy can be regarded as typical for industrial sectors with mature technologies and products, such as fabricated metal products or wood products and furniture. The markets are well defined, and the products are well established and often standardized; production technology is efficient, and the price of products is a main factor in competition.

A second innovation strategy is characterized by innovation measures that are primarily directed at securing and improving the marketing situation of the enterprise. This holds true for the fashion-oriented design of products, the functional and technical upgrading of products, a rapid response to changing customer wishes, taking advantage of market niches, skillful branding strategies, and the expansion of product-related service activities. Unlike those using the first strategy, the enterprises pursuing this strategy belong to a relatively broad range of industrial subsectors. Examples of companies using this strategy are those from the textile and clothing industries and furniture and leather goods manufacturers, whose product development is geared to anticipatable fashion cycles. One can also ascertain a broadening of the spectrum offerings as companies supplement their process functions with services and logistics services tailored to certain customers. This innovation strategy can be referred to as customer-oriented strategy.

A third innovation strategy is primarily directed at the technical and organizational structures of the production process. It is used by companies that employ ultramodern, automated, and capital-intensive process technologies. This innovation strategy may be referred to as process specialization. The enterprises pursuing this strategy belong to industrial subsectors whose products are mostly manufactured at a relatively high level of technological automation. Examples of this strategy are furniture companies that are extensively automated on the basis of a greatly

reduced variety of parts and of simplified processes, sheet-forming companies, firms manufacturing plastic parts or mechanical components or parts made of aluminum, and companies in the food processing industry. These examples show that the term low-technology is basically a misleading designation for the industries and enterprises examined here.

Internal knowledge base. If one looks at the preconditions for the innovation capability of LMT companies, the knowledge base emerges as one of the most fundamental factors. These enterprises pursue virtually no R&D activities, and formalized processes of knowledge generation play an insignificant role. Instead, innovation activities proceed by means of practical and pragmatic doing and using. The knowledge that is relevant for these enterprises shall therefore be regarded as application-oriented practical knowledge. Unlike scientifically and theoretically generated knowledge, which orients itself on criteria such as theoretical relevance and universality, practical knowledge is generated in application contexts of new technologies. It obeys validity criteria such as practicability, functionality, efficiency, and failure-free use of a given technology. The two types of knowledge are, however, difficult to distinguish. To simplify matters, theoretical and scientific knowledge in enterprises (for instance, in the form of systematically acquired engineering knowledge) can primarily be assigned to research, development, and construction processes, while practical knowledge accrues in the context of ongoing operating processes. The term practical knowledge refers to a complex bundle of different knowledge elements. It includes explicit, codified, and formalized elements such as design drawing and requirement specifications for new products and implicit elements such as accumulated experience and well-established, proven, and tested routines for solving technical problems. An example is the innovation strategy of process specialization. Process innovations generally take place in the context of ongoing operations and are mostly initiated and pressed ahead by the staff responsible for the ongoing functions, such as engineers, technicians, master craftsmen, and skilled workers on the shop floor.

External knowledge base. The acquisition and generation of innovation knowledge by no means takes place only within the company. External knowledge sources have proven to be relevant, too. For all innovation strategies, the knowledge of other firms, organizations, and actors and the systematic application of this external knowledge is of decisive importance. This is true for practical knowledge and in particular for scientifically generated knowledge in various forms. Examples of external sources, for instance in the case of the customer-oriented strategy, are the experience of longtime customers concerning new market and demand trends, the expertise of the relevant consultants, or information about foreseeable market trends gained during fair visits. Furthermore, the fashion-oriented design of products by external design agencies often plays an important role in successful sales strategies. Other important external knowledge sources are machine manufacturers and suppliers, who provide theoretically and scientifically generated knowledge in the shape of knowledge incorporated into production technologies and materials, which is often an essential prerequisite for the innovation activities of process specialization.

Knowledge management. Of decisive importance for the innovation strategies of companies is the manner in which they effectively make use of their internally available and externally accessible knowledge. The ability to use knowledge is to a large extent dependent on the professionalism and innovation-related know-how of management. Also, the routines and structures of the company organization, such as the mode of the division of labor, the prevailing communication and cooperation forms, and the connected qualification and personnel structures, play a significant role. In addition, the ability to manage and effectively coordinate network relations across company borders, especially with other companies within the value chain, is a central precondition for successful LMT innovation strategies. An essential requirement for the efficiency of such relations is an organizational structure that is geared to the demands of cross-company cooperation by providing for adequate channels of communication, gateways, and personnel responsibilities oriented toward cooperation. The company management has to be able to harmonize and control the specific competencies and the divergent interests of cooperation partners in such a way that the transfer of the required knowledge is assured.

Development perspectives of LMT. These considerations lead to a new understanding of the restructuring of the economic landscape of knowledge-based countries in the first years of the twenty-first century. The economy does not appear to be undergoing a wholesale structural replacement of old sectors with new ones in the course of the emerging knowledge society. In fact, this process of change is evolving as a restructuring of existing sectoral and technological systems. It is not dominated by industrial activities for which competitive advantage, capability formation, and economic change are generated by front-line technological knowledge. Instead, it is dominated by what are often wrongly termed low- and medium-tech industries. And it is characterized by a specific combination and continuous recombination of high- and low-tech.

LMT industries can play a decisive role in innovations because the involvement of low-tech prod-

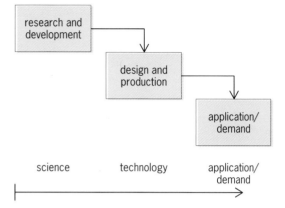

Fig. 1. Schematic representation of the linear model of innovation.

ucts and companies is frequently a core precondition both for the innovativeness of value chains and for the design, fabrication, and use of a range of high-tech products. As research findings on processes of technological change in high-tech economies convincingly show, the relationships between high-tech and non-high-tech sectors are highly symbiotic, and the well-being of high-tech firms and industries depends heavily on their ability to sell their output to other sectors in developed economies.

Innovation models. The research findings on low-tech innovations also prove convincingly that the linear model of innovation, which is widely adhered to in the public debate, inadequately reflects industrial innovation (**Fig. 1**).

The essence of this model is the assumption that research and development activities are the starting point for any kind of innovation and that scientifically generated knowledge gives impulses for the development of new technologies. Research and development are regarded as fundamental and necessary innovation steps that functionally and temporally precede the industrial process. It is assumed that there is a clearly structured course of action, during which the knowledge derived from basic research is transferred, specified, and used step by step via applied research, and is finally employed in the form of concrete technologies in a certain implementation context. According to this model, an effective innovation policy has to focus on research and development in order to encourage industrial innovations. It is then merely necessary to ensure that the new knowledge asserts itself to the point of concrete development projects and field applications by promoting appropriate transfer measures.

However, industrial innovations are mostly not based on scientific knowledge. The relationship may even be the other way around, that is, technology creates the foundation for scientific knowledge. Technological innovations are very often initiated by the requirements of practical problems of the production process and, more importantly, market and application demands. This has been shown by innovation research for many years. In conclusion, an innovation model referred to as the "recursion model of innovation" may have more relevance to

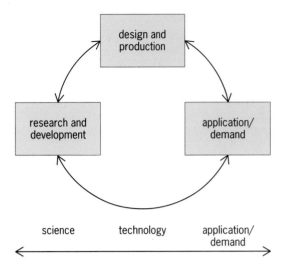

Fig. 2. Schematic representation of the recursion model of innovation.

understanding the processes of technological innovation (**Fig. 2**).

For background information *see* ENGINEERING; ENGINEERING, SOCIAL IMPLICATIONS OF; INDUSTRIAL ENGINEERING; MANUFACTURING ENGINEERING; PROCESS ENGINEERING; PRODUCTION ENGINEERING; TECHNOLOGY in the McGraw-Hill Encyclopedia of Science & Technology.   Hartmut Hirsch-Kreinsen

Bibliography. H. Hirsch-Kreinsen, Low-tech innovations, *Ind. Innovation*, 15:19–43, 2008; P. L. Robertson and R. Patel, New wine in old bottles: technological diffusion in developed economies, *Res. Policy*, 36:708–721, 2007; N. von Tunzelmann and V. Acha, Innovation in "low-tech" industries, in J. Fagerberg, D. Mowery, and R. R. Nelson (eds.), *The Oxford Handbook of Innovation*, Oxford University Press, 2005, pp. 407–432.

Electronic nanopores

Nanometer-diameter pores, known as ion channels, are vital to biology, as they are used to regulate the flow of molecules or ions through the otherwise impermeable nanometer-thick membranes that encompass and isolate cells from their environment. The pores can act like switches, opening and closing in response to chemistry through the binding of a ligand, in response to electricity associated with the membrane potential near the pore, or mechanically through deformation in the membrane. This regulatory motif has been observed in action. Modulation of the ionic current through a single nanopore associated with the salt water in the environment of a cell has been observed using a patch clamp pioneered in Nobel-prize winning work by B. Sakmann and E. Neher. A patch clamp consists of a glass micropipette filled with electrolyte that is in contact with a chlorinated silver wire conducting electricity to a low-noise amplifier. The open tip of the macropipette, which is about 1 micrometer in diameter, is sealed to a portion of a cell membrane containing a pore to record the current through it.

Nanopore sensors. In combination with patch-clamp-type measurements, an open pore can function as a molecular sensor. Some pores are so small in diameter that molecules can only move through them one at a time. Lately, scientists and engineers have been trying to co-opt this feature in order to detect, identify, and count molecules one at a time, making a nanopore the ultimate analytical tool to differentiate molecular types and numbers. The earliest example of a single-molecule measurement used a nanopore derived from the antibiotic peptide, gramicidin, fused in a lipid bilayer. Recently, one proteinaceous pore, in particular, has been a focus of single-molecule measurements: the nanopore derived from α-hemolysin (αHL), a 33-kilodalton protein secreted by *Staphylococcus aureus*. The structure of αHL, represented in cross section in **Fig. 1a**, is known with atomic resolution. The pore self-assembles in a phospholipid layer into seven identical subunits arranged around a central axis; the transmembrane portion is a β-barrel about 5 nm long and 1.5 nm wide at the narrowest point, with two antiparallel strands contributed by subunit; and the extramembraneous domain contains a large vestibule. J. J. Kasianowicz and colleagues were among the first to use αHL to detect and sort single molecules.

Kasianowicz and colleagues discovered that single molecules of polyanionic, single-stranded deoxyribonucleotide acid (ssDNA) in an electrolytic solution could be transported through a pore one at a time by applying an electric field along the axis of the pore. Near room temperature, an αHL pore in a 1M KCl concentration of electrolytic solution remains open and admits an ionic current of about 120 pA in response to a 120-mV potential applied across the membrane. This current can be used to detect a single DNA molecule. When ssDNA moves through the pore it excludes electrolyte from it and the current is blocked. (It is reduced from 120 to about 14 pA.) Individual homopolynucleotides ostensibly have distinct molecular signatures that reflect their composition and how they bind to the pore. For example, the blocking events caused by the translocation of homopolynucleotide poly(adenylic acid) [poly A] reduce the current by 84%, while the translocation of poly(cytidylic acid) [poly C] reduces the current by 90–95%. Both the extent of the blockade and the mean duration of the event can be used to identity an analyte. Since the electric field drives ssDNA through the pore at a rate of 1–10 nucleotide/μs, the duration of the current transient can be used to determine (with some ambiguity) the length of the nucleic acid polymer. Thus, a nanopore can be used as a transducer to ferret out information about the structure associated with a single molecule from an electrical (current) signal.

Using αHL as the prototype, the prospects for single-molecule sensing have been methodically explored. In particular, nanopores have been suggested as a mechanism for sequencing DNA. If it proves feasible, sequencing a single molecule of DNA with a nanopore could eliminate the costly and

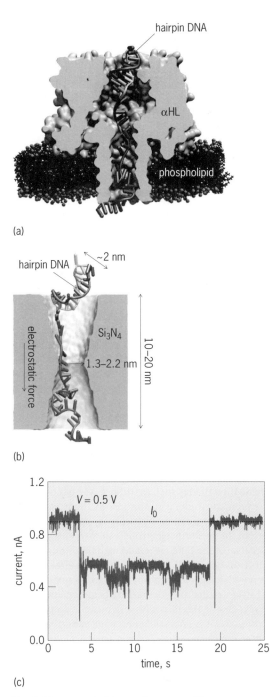

(a)

(b)

(c)

Fig. 1. Nanopores in thin membranes used for single-molecule detection. (*a*) A cross section through the αHL pore taken from a simulation showing a trapped DNA hairpin. The pore diameter is smaller than the stem of the hairpin. (*b*) A cross section through a synthetic pore in a silicon nitride membrane, showing a hairpin trapped in the pore. The diameter of the synthetic pore can be any size. (*c*) A measurement of the electrolytic current through a 1.4-nm-diameter synthetic pore interacting with a DNA hairpin when 0.5 V is applied across the silicon-nitride membrane. The current is blocked, that is, it falls below the open-pore current value, I_0, for an extended duration, indicating that the hairpin is trapped in the pore at low voltage. (*Simulations are a courtesy from A. Aksimentiev*)

error-prone steps in conventional sequencing such as amplification using PCR (polymerase chain reaction) and make it possible to economically and quickly sequence other one-dimensional polymer chains. So far, using αHL, only rudimentary informa-

tion has been recovered about the molecular structure of molecules such as DNA. αHL suffers from some severe limitations. The surface charge on the pore and the limited aperture, which are both associated with the complex protein structure, and the fragility of the lipid bilayer membrane are among the problems. For example, the protein-membrane complex has a limited voltage, temperature, and pH range for operation. This is unfortunate because the secondary structure in ssDNA, which can be unraveled by increasing the temperature or varying pH, hopelessly complicates the interpretation of a current blockade.

Synthetic nanopores. On the other hand, molecular recognition in the pore can be improved or the transport through the pore can be slowed to accommodate the narrow bandwidth and to reduce the noise relative to the signal associated with a blockade current measurement, thus improving the prospects for detection. For example, through protein engineering, H. Bayley and colleagues have successfully created αHL mutants equipped with a molecular adapter that have been used to improve molecular recognition in the pore. In particular, they specifically bind members of a set of nucleoside 5'-monophosphates to cyclodextrin adapters, producing distinct current blockade levels and dramatically improving the blockade current signal relative to the noise. However, the bandwidth and signal-to-noise in measurements of the blockade current, which are irrevocably tied to the membrane capacitance and pore resistance, remain the most severe limitations associated with the detection of single molecules. While it may be possible to differentiate one long strand of homopolynucleotide DNA from another or to discriminate between nucleoside monophosphates by measuring the level of the blocking current, the rate of translocation through the pore, coupled with the bandwidth and noise in the measurement, generally preclude high-fidelity, high-throughput measurements. High throughput is especially desirable for applications such as DNA sequencing for personal medicine or biological computing that require reading literally billions of bases with high fidelity.

For these and other reasons, researchers are seeking alternatives to αHL, its mutants, and the more exotic proteinaceous pores. The feasibility of using a synthetic, inorganic nanopore in silicon-derived or polymeric membranes that mimic biology is being explored. Pioneering efforts by J. A. Golovchenko and D. W. Deamer have demonstrated that a synthetic pore in a silicon nitride membrane, such as the one represented in cross section in Fig. 1*b*, can be used as a stochastic sensor similar to αHL by measuring the blockade current. Figure 1*c* shows a current blockade ostensibly due to a ssDNA molecule with a secondary structure resembling a hairpin trapped in the pore. While it is still a nascent strategy for single-molecule detection, the prospects for synthetic pores go well beyond imitating the operation of αHL and are derived essentially from the flexibility of the fabrication strategy used to manufacture

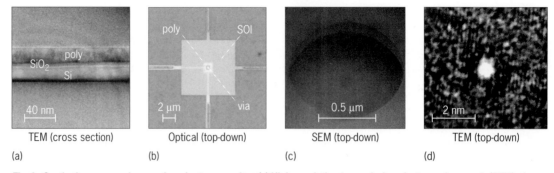

TEM (cross section)   Optical (top-down)   SEM (top-down)   TEM (top-down)
(a)   (b)   (c)   (d)

Fig. 2. Synthetic nanopore in a semiconductor capacitor. (*a*) High-resolution transmission electron micrograph (TEM) of a cross section through the capacitor, showing a thin SiO_2 layer separating two heavily doped electrodes: one polysilicon (poly) and the other crystalline silicon (Si). (*b*) A top-down optical micrograph of the membrane. The polycrystalline (poly) electrode, which is $<2~\mu m$ on edge, is on top of the crystalline silicon (SOI) electrode which is about 10 μm on edge. Two voltage probes are attached to each electrode. The location of the top via shown in *c* is indicated by an arrow. (*c*) A scanning electron micrograph of the top via, which is only about 1 μm in diameter, revealing the underlying membrane. (*d*) A top-down TEM of a nanopore sputtered through the membrane similar to that shown in *a–c*.

them. What is really new about the synthetics is the pore diameter and geometry, and the thickness and composition of the membrane can all be easily controlled with subnanometer precision using semiconductor nanofabrication practices. This precision translates directly into control of the distribution of the electric field in the pore. Such stringent control has already led to the development of the most sensitive device for charge measurement, the single-electron transistor, which has been proposed to be adapted to nanopores.

While they do not self-assemble, synthetic pores have an overwhelming advantage over proteinaceous pores; that is, the manufacturing process is flexible and suitable for integration with silicon microelectronics. Thin, synthetic composite membranes can be fabricated by a variety of means, all of which stem from developments in semiconductor manufacturing. The deposition and metrology of ultrathin films has relentlessly improved over the past 30 years in response to demand from the microelectronics industry to the point where 1-nm-thick films can be uniformly deposited or grown over a 300-mm-diameter wafer with subnanometer precision. In combination with deep ultraviolet lithography and selective etching of vias (vertical pathways), the ultrathin films can be patterned and released to form thin membranes spanning areas that measure only a few micrometers on-edge, as shown in **Fig. 2**. Starting with robust, mass-produced synthetic membranes such as these, a way has been discovered to produce nanopores in them with subnanometer precision by using a tightly focused, bright, high-energy electron beam to sputter atoms. Because of the tight focus and high brightness of the electron beam and subnanometer visualization that is possible, this lithography strategy seems superior to schemes that use focused-ion-beam milling or ion tracks in conjunction with a deposition to produce pores, and it is applicable to a wide variety of membrane materials.

As an example of what is possible, Fig. 2*a–d* shows a nanopore less than 1 nm in diameter through a 40-nm-thick composite membrane formed from a capacitor made from heavily doped silicon electrodes with silicon dioxide (SiO_2) as an insulator between them. Figure 2*a* is a high-resolution transmission electron micrograph (TEM) of a cross section through the capacitor showing the thin SiO_2 layer separating a polysilicon (poly) and crystalline silicon (Si) electrodes. Figure 2*b* is a top-down image of the microstructure containing the membrane, which is formed at the intersection between the polysilicon electrode that is less than 2 μm on edge, and the crystalline silicon electrode that is about 10 μm on edge. It shows the location of the top via through a nitride passivation layer to the membrane below. Figure 2*c* is a scanning electron micrograph (SEM) of a similar via, about 1 μm in diameter. The nanopore is sputtered through the membrane near the center of the via. A TEM of a nanopore through the membrane

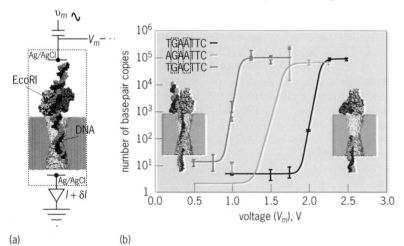

(a)   (b)

Fig. 3. A synthetic nanopore sensor for detecting DNA mutations. (*a*) The complex formed by an EcoRI restriction enzyme bound to a DNA strand trapped in a 3-nm-diameter pore in a silicon-nitride membrane. By applying a large voltage V_m, the electric field in the nanopore can be used to pull on the DNA, introducing a shear force between the enzyme and the cognate sites in DNA that eventually ruptures the bond, producing a signal in the pore current, I. (*b*) Quantitative polymerase chain reaction (qPCR) is used to count the number of copies of DNA that translocate through the nanopore in *a* as a function of the voltage V_m. The results show that there is a threshold in the number of 10^5 base-pair DNA copies originating from EcoRI-DNA complex that permeate the nanopore that depends on the DNA sequence. Superimposed on the data are fits to the curve used to determine the threshold. The cognate sequence GAATTC (black curve) has a threshold voltage of about 2.1 V. In contrast, with a substitution for the third-base, GACTTC (dark gray curve) has a threshold of only 1.1 V. Even a substitution in the flanking sequence, TGAATTC to AGAATTC (light gray curve), affects the threshold appreciably. The left inset shows the DNA-enzyme complex trapped in the pore below threshold. The right inset shows the complex dissociated above threshold. (*Courtesy of A. Aksimentiev*)

is shown in Fig. 2*d*. By leveraging this manufacturing advantage, the shortcomings associated with bandwidth and signal-to-noise encountered in αHL may someday be remedied by either reducing the area of the membrane to reduce the associated capacitance or multiplying the number of pores through it to improve the signal without compromising throughput.

Aside from the convenience of self-assembly, synthetic pores do not really cede anything else to αHL and other proteins. Similar to αHL, synthetic pores have used molecular recognition to control the translocation kinetics and sequence short runs in a strand of DNA. In particular, restriction endonucleases have been used in conjunction with a synthetic pore to identify single mutations in the nucleotides comprising a strand of DNA (a polymorphism).

A single nucleotide polymorphism (SNP) is the most common type of genetic variation. About 9 million SNPs have been identified in the human genome. SNPs can adversely affect protein structure and function, forming a molecular basis for disease. Thus, SNPs are markers in the study of disease genetics. Restriction endonucleases have become synonymous with recombinant-DNA technology due to their high recognition sequence specificity. Restriction endonucleases are used for detecting SNPs; they recognize and bind to (and cleave in the presence of appropriate co-factors such as magnesium ions) a specific sequence of nucleotides. SNPs in the recognition sequence of a restriction endonuclease are known to decrease the affinity of the enzyme for the binding site. This fact, along with the restriction endonuclease, EcoRI, and the electric field in a nanopore has been used to identify SNPs in a recognition sequence of bases: GAATTC, where guanine adenine, thymine, and cytosine are abbreviated G, A, T, and C, respectively. The method is shown in **Fig. 3***a*. When an electric potential, V_m, is applied across the membrane, polyanionic DNA bound to an enzyme and immersed in electrolyte migrates from the cathode toward the anode and becomes trapped in the pore between the enzyme and the electric field. The field in the pore pulls on the trapped DNA introducing a shear force between the enzyme and the target site in the DNA that eventually ruptures the bond and produces a signal in the pore current, δI. A voltage threshold for permeation through the nanopore of DNA bound to a restriction enzyme has been discovered that depends on the sequence and the enzyme, which scales with the dissociation energy. Figure 3*b* illustrates the threshold voltages associated with EcoRI binding to DNA, incorporating the cognate sequence, TGAATTC, and two mutant strands, AGAATTC and TGACTTC. A single mutation in the cognate sequence reduces the threshold by nearly 1 V, but even a change in the flanking sequence has a measurable effect. These observations hold the promise that synthetic nanopores could be used for rapid genotyping of DNA in the future.

Outlook. If a nanopore mechanism can be harnessed for detecting single molecules, it would represent the ultimate analytical tool. The prospects for electrical detection of single molecules, such as DNA, using nanopores seem bright, but remain largely unfulfilled due to shortcomings associated with noise, bandwidth, and the difficulties encountered in creating pores so small. These problems may yet yield to either protein engineering or synthetic pores created in an engineered substrate, but to be practical the solutions will have to provide an economical high-throughput, high-fidelity tool.

For background information *see* BIOPOTENTIALS AND IONIC CURRENTS; CELL MEMBRANES; DEOXYRIBONUCLEIC ACID (DNA); GENE AMPLIFICATION; MICROLITHOGRAPHY; POLYMORPHISM (GENETICS) in the McGraw-Hill Encyclopedia of Science & Technology.

Gregory Timp

Bibliography. M. Akeson, D. Branton, J. J. Kasianowicz, E. Brandin, and D. W. Deamer, Microsecond time-scale discrimination among polycytidylic acids, polyadenylic acid, and polyuridylic acid as homopolymers or as segments within single rna molecules, *Biophysical J.*, 77:3227–3233, 1999; C. Ho, R. Qiao, J. Heng, A. Chatterjee, R. Timp, N. Aluru, and G. Timp, Electrolytic transport through a synthetic nanometer-diameter pore, *Proc. Nat. Acad. Sci.*, 102(30):10445–10450, 2005; J. J. Kasianowicz, E. Brandin, D. Branton, D. W. Deamer, Characterization of individual polynucleotide molecules using a memebrane channel, *Proc. Natl. Acad. Sci.*, 93:13770–13773, 1996; J. Li, M. Gersho, D. Stein, E. Brandin, M. J. Aziz, and J. A. Golovchenko, DNA molecules and configurations in a solid-state nanopore microscope, *Nat. Mater.*, 2:611–615, 2003; B. Rudy et al. (eds.), *Ion Channels, Volume 207 of Methods in Enzymology*, Academic Press, 1992; B. Sakmann and E. Neher, The patch clamp technique, *Sci. Amer.*, p. 28, March 1992; H-C. Wu, Y. Astier, G. Maglia, E. Mikhailova, and H. Bayley, Protein nanopores with covalently attached molecular adapters, *J. Am. Chem. Soc.*, 129:16142–16148, 2007; Q. Zhao, G. Sigalov, V. Dimitrov, B. Dorvel, U. Mirsaidov, S. Sligar, A. Aksimentiev, and G. Timp, Detecting SNPs using a synthetic nanopore, *NanoLetters*, 7(6):1680–1685, 2007.

Electro-optic polymers

Optoelectronics encompasses electro-optic, light-emitting device (LED), photovoltaic (PV), and photorefractive (PR) phenomena. These are quite different phenomena. For polymer materials, each of these derives from the presence of conjugated (interacting) pi orbitals. Of the two types of electron orbitals existing in organic materials, pi orbitals concentrate electron density above and below the plane of the nuclei, while sigma orbitals concentrate electron density between nuclei. Thus, pi electrons are less tightly bound by coulombic (charge) interaction with nuclei than are sigma electrons. While LED, PV, and PR phenomena depend on the absorption/emission of light and the generation and migration of charge (electrons and holes), electro-optic activity does not. Also, the applications of each of these phenomena are quite different; that is, photovoltaics are used for renewable energy, light-emitting devices for displays, photorefractives for real-time holography, and electro-optics for telecommunications, computing, and sensing.

Electro-optic activity is one class of second-order nonlinear optical phenomena, which also include second harmonic generation, optical rectification, and difference frequency generation. Second-order nonlinear optical activity results when the electric component of an electromagnetic field perturbs the charge (for example, electron) distribution of a material, which in turn influences the behavior of a second electromagnetic field. For example, electro-optic activity involves application of a low-frequency (0–15 THz, 1 THz $= 10^{12}$ Hz) field to a material perturbing the charge distribution of the material. An optical-frequency (1–10×10^{14} Hz) field passing through the material senses the perturbed charge distribution; for example, it is slowed down by the increase in charge asymmetry (dipole moment) produced by the first (control) field. Since the velocity of light in a material is specified by the index of refraction of the material (the ratio of the velocity of light in a vacuum to that in the material), electro-optic activity can be considered to be a voltage control of the index of refraction of a material. Note that neither the low-frequency (applied electric field) nor the high-frequency optical field is resonant (ideally, no light is absorbed in the phenomenon of electro-optic activity), and thus the response time is the time required for the electrons to come into phase coherence with the applied electric field. For pi-electron polymeric materials, this time is on the order of tens of femtoseconds (1 fs $= 10^{-15}$ s). Pi-electron materials arguably exhibit the fastest electro-optic response of any material. Other types of electro-optic materials include organic liquid-crystalline materials (where the electric-field-induced charge perturbation is associated with molecular reorientation) and inorganic crystalline materials such as lithium niobate (where charge perturbation is associated with ion displacement under the influence of an applied electric field). In these latter cases, larger masses are moved, and hence slower responses are observed. For liquid-crystalline materials, response times are typically submillisecond (1 ms $= 10^{-3}$ s), while for lithium niobate, response times are subnanosecond (1 ns $= 10^{-9}$ s).

Applications of electro-optic materials include electrical-to-optical signal transduction, optical switching, optical beam steering, active wavelength-division multiplexing (WDM, wavelength or "color" coding of information), radio-frequency signal generation, electromagnetic sensing, radio-frequency beam steering (phased array radar), optical gyroscopes, and analog-to-digital (A/D) conversion.

A brief history of electro-optic polymers. In the late 1970s and early 1980s, considerable interest developed in electro-optic materials, with the goal of taking advantage of the enormous bandwidth available with silica fiber used in telecommunications. Clearly, the bottleneck to higher-speed telecommunications was the rate at which electrical signals could be converted (transduced) to optical signals propagat-ing in silica optical fibers. A high-speed, sensitive electro-optic material was perceived to be the answer to the bandwidth bottleneck, as organic materials certainly possessed the requisite fast response times. Initial interest (in the early 1980s) in organic electro-optic materials focused on crystalline materials such as 2-methyl-4-nitroaniline (MNA). Like crystalline inorganic materials, these materials were difficult to process and integrate with silica fiber optics, and research interest soon turned to polymeric electro-optic materials, which were prepared by dissolving chromophores such as 4-amino-4′-nitrostilbenes and 4-amino-4′-nitroazobenzenes (disperse red dyes) in commercial polymers such as poly(methylmethacrylate), polycarbonate, and polyimides and electrically poling (applying an electric field across) the chromophores near the glass transition temperature of the resulting composite material (see **illustration**). Note that noncentrosymmetric chromophore order is required for electro-optic activity, and this is provided by electric-field poling.

These early materials were deficient in several ways. (1) The electro-optic activity of early polymeric materials was less than that of commercially available lithium niobate (which has a usable electro-optic tensor value of 32 pm/V; 1 pm $= 10^{-12}$ m). (2) The thermal and photochemical stability of these early materials was poor. (3) Processing of early polymeric electro-optic materials and subsequent fabrication of devices was immature, which often led to high device insertion loss. Also, the demand for high modulation (signal transduction) speed was ameliorated by the advent of wavelength-division multiplexing. Thus, in the mid-to-late 1990s, commercial interest in organic electro-optic materials faded.

Since 2000, renewed interest in polymeric electro-optic materials has been driven by increased photonic integration and dramatic improvements in the performance of polymeric electro-optic materials. Materials now exhibit electro-optic activity 10–20 times greater than lithium niobate, permitting devices to operate with millivolt drive voltages. The materials' thermal and photostability exceed telecommunication standards, and devices can be fabricated with optical insertion loss values below those realized using inorganic materials. Polymeric electro-optic materials have been shown to be highly processable (permitting the fabrication of conformal and flexible devices and the mass production of devices by soft and nanoprint lithography techniques) and compatible with disparate materials (for example, polymeric electro-optic materials have been integrated into silicon photonic circuits and devices). Integration with silicon photonics has been particularly important, since the high index of refraction of silicon permits the confinement of light in circuits and devices of smaller dimensions, such as waveguides of several hundred nanometers (1 nm $= 10^{-9}$ m) in width, rather than widths of several micrometers (1 μm $= 10^{-6}$ m). The consequence is the reduction of the size disparity between electronic and photonic circuits. And the use of complementary metal oxide semiconductor (CMOS) foundries to fabricate both electronic and photonic circuits greatly

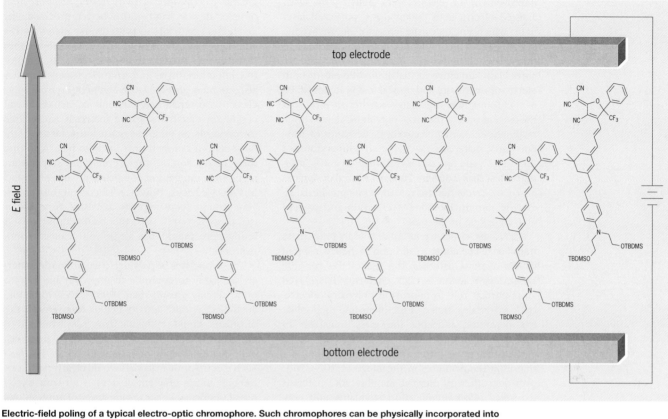

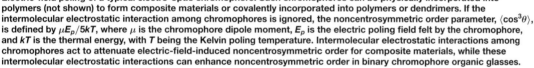

Electric-field poling of a typical electro-optic chromophore. Such chromophores can be physically incorporated into polymers (not shown) to form composite materials or covalently incorporated into polymers or dendrimers. If the intermolecular electrostatic interaction among chromophores is ignored, the noncentrosymmetric order parameter, $\langle \cos^3\theta \rangle$, is defined by $\mu E_p/5kT$, where μ is the chromophore dipole moment, E_p is the electric poling field felt by the chromophore, and kT is the thermal energy, with T being the Kelvin poling temperature. Intermolecular electrostatic interactions among chromophores act to attenuate electric-field-induced noncentrosymmetric order for composite materials, while these intermolecular electrostatic interactions can enhance noncentrosymmetric order in binary chromophore organic glasses.

facilitates the integration of electronics and photonics on the same platform (chip). Such photonic integration permits size, weight, and power savings, together with improvements in performance, reliability, and cost. Critical to the improvement in performance has been the incorporation of polymeric electro-optic materials into 25–150-nm slots in silicon photonic waveguides. Such structures greatly concentrate optical and electric fields, reducing the field strength required for active signal processing. The potential exists for electro-optic modulation and switching with drive voltages in the range 1–10 mV. The improvement in the performance of polymeric electro-optic materials, together with new device concepts, has greatly expanded the range of applications. In addition to telecommunication applications (such as electrical/optical signal transduction, dynamic optical signal routing, and dynamic WDM), applications exist in computing (interconnects between processors and between processors and memory), imaging (THz, IR), and spectroscopy (THz). New materials have also resulted in the advancement of other second-order nonlinear applications (such as optical rectification and difference frequency generation).

Theory-inspired design. Polymeric electro-optic materials have undergone improvement at a rapid rate over the past decade. This is because new quantum-mechanical methods such as real-time, time-dependent density functional theory (RTTDDFT) and statistical mechanical methods such as pseudo-atomistic Monte Carlo molecular dynamics (PAMCMD) have inspired the design of chromophores with dramatically improved molecular hyperpolarizability (ability to exhibit large changes in molecular polarization with the application of electric fields). The incorporation of these chromophores into supramolecular (polymeric and dendritic) architectures results in greater noncentrosymmetric chromophore organization, and hence electro-optic activity. Theory has led to a quantitative understanding of electro-optic activity in chromophore/polymer composite materials (such as those studied in the 1980s and 1990s). In these materials, it was shown to be important to change the chromophore shape from prolate ellipsoid to spherical to optimize electro-optic activity. Theory also led to the introduction and rationalization of two other classes of polymeric electro-optic materials: (1) polymers and dendrimers covalently incorporating chromophores, where covalent bond potentials are used to promote noncentrosymmetric chromophore order, and (2) binary chromophore organic glasses consisting of a chromophore-containing guest material incorporated into a chromophore-containing host material, where the order of the host enhances the order of the guest and vice versa. This latter class of materials also permits improvement of electro-optic activity by laser-assisted electric-field poling using polarized laser light, which enhances the order of photochromic chromophores that undergo photo-driven reorientation. Binary chromophore organic glass materials are particularly attractive in that improvement of both electro-optic activity and optical loss is achieved. Binary chromophore organic glass materials represent the current state of the art of polymeric electro-optic materials, exhibiting electro-optic activities of 300–600 pm/V, material glass transition temperatures of 200–250°C, and optical loss (absorption and scattering) of less than 2 dB/cm. Theory-inspired design has also played an important role in defining material dielectric permittivity, index of refraction, and material glass transition properties.

Diels-Alder chemistry for control of material glass transition temperature. In addition to theory-inspired design of new materials, the use of the versatile Diels-Alder reaction in click chemistry has played an important role in improving polymeric electro-optic materials. Choice of the diene and dienophile has permitted adjusting the glass transition of materials to optimize their processing, including electric-field poling and nanoprint lithography. Glass transition temperatures in the range of 200–250°C provide sufficient thermal stability for materials to satisfy telecommunication standards and contribute to improved photostability. Anthracene diene and acrylate dienophile groups have proven to be particularly useful in achieving the desired properties.

Photostability. As with most organic materials, the photostability of polymeric electro-optic materials is largely a matter of singlet oxygen (excited state) chemistry. Generation of singlet oxygen is the most problematic pathway to the destruction of organic electro-optic chromophores. Photostability of polymeric electro-optic materials is less problematic than for polymeric light-emitting device materials (PLEDs) because electro-optic materials operate at telecommunication wavelengths (1300- and 1550-nm bands), which are far removed from the chromophore absorption wavelengths that activate singlet oxygen. Indeed, a photochemical figure of merit (performance rating) B/σ can be defined for polymeric electro-optic materials, where B is the probability of photodecay from the charge transfer lowest unoccupied molecular orbital and σ is the absorption coefficient for the charge transfer transition. A value of $500 \times 10^{32} \, \text{m}^{-2}$ leads to ten-year stability at normal telecommunication power levels (10–20 mW). Values as high as $5050 \times 10^{32} \, \text{m}^{-2}$ have been observed for polymeric electro-optic materials. Methods for improving photostability include steric protection of the chromophore sites susceptible to singlet oxygen attack, lattice hardening by cross-linking, addition of singlet oxygen quenchers, and packaging to exclude oxygen.

Outlook. Polymeric electro-optic materials are of interest because of the advantages that they afford, relative to their inorganic counterparts. With polymeric electro-optic materials, advantages include

greater electro-optic activity, faster response time (greater device bandwidth), better processability, and better compatibility with diverse materials. Disadvantages include greater optical loss, lower thermal stability, and lower photostability. In addition to competition from inorganic electro-optic materials, one must also consider competition from other classes of materials and phenomena, including semiconductor electroabsorptive materials and charge-injection silicon photonic modulators. Of these, the most serious competitor is likely to be silicon photonic modulators, as they represent a step toward monolithic integration of electronics and photonics in CMOS devices. However, silicon modulators are very energy consumptive and run very hot at the present time. Also, the bandwidths are currently more limited than for polymeric electro-optic materials.

The properties of polymeric electro-optic materials are likely to continue to improve, and electro-optic activity greater than 1000 pm/V is probable. Such electro-optic activity begins to approach that of organic liquid-crystalline materials, while affording improvements in switching speeds from sub-millisecond to subpicosecond (1 ps $= 10^{-12}$ s). Such electro-optic materials would clearly permit extremely energy-efficient devices, with truly exceptional performance for a wide range of applications.

For background information *see* CLICK CHEMISTRY; DENDRITIC MACROMOLECULE; DIELS-ALDER REACTION; MOLECULAR SIMULATION; NONLINEAR OPTICAL DEVICES; NONLINEAR OPTICS; SIGNAL PROCESSING; SUPRAMOLECULAR CHEMISTRY; WAVEGUIDE in the McGraw-Hill Encyclopedia of Science & Technology.   Larry R. Dalton

Bibliography.   T. Baehr-Jones et al., Nonlinear polymer-clad silicon slot waveguide modulator with a half wave voltage of 0.25 V, *Appl. Phys. Lett.*, 92:163303-1, 2008; T. Baehr-Jones et al., Optical modulation and detection in slotted silicon waveguides, *Opt. Exp.*, 13:5216, 2005; L. R. Dalton et al., Polymeric electro-optic modulators: from chromophore design to integration with semiconductor VLSI electronics and silica fiber optics, *Ind. Eng. Chem. Res.*, 38:8, 1999; Y. Huang et al., Fabrication and replication of polymer integrated optical devices using electron-beam lithography and soft lithography, *J. Phys. Chem. B*, 108:8606, 2004; T.-D. Kim et al., Binary chromophore systems in nonlinear optical dendrimers and polymers for large electro-optic activities, *J. Phys. Chem. C*, 112:8091, 2008; B. C. Olbricht et al., Laser-assisted poling of binary chromophore materials, *J. Phys. Chem. C*, 112:7983, 2008; Y. V. Pereverzev et al., Guest-host cooperativity in organic materials greatly enhances the nonlinear optical response, *J. Phys. Chem. C*, 112:4355, 2008; Y. Shi et al., Low (sub-1 volt) halfwave voltage polymeric electrooptic modulators achieved by control of chromophore shape, *Science*, 288:119, 2000; P. A. Sullivan et al., Theory guided design and synthesis of multi-chromophore dendrimers: an analysis of the electro-optic effect, *J. Am. Chem. Soc.*, 129:7523, 2007.

Emerging diseases in marine mammals

The emerging and resurging disease phenomenon over the past 25 years is a growing and serious issue that has implications for human, animal, and ecosystem health on a global basis. Indeed, the complacency of the human and veterinary medical community toward the end of the 1970s in thinking that emerging diseases were under our control is no longer justified. In fact, emerging diseases now have the potential to negatively impact the entire planet because of the closely interconnected global community. The cause of this phenomenon is complex and multifactorial and may reflect an "environmental distress syndrome" whereby ecologic and climatic changes, likely associated with human activities, are encouraging the selection of new and opportunistic pathogens. Similar but less publicized emerging disease trends are now being documented in marine ecosystems and are impacting marine mammals. Emerging and resurging diseases of marine mammals may have zoonotic implications (being transmissible from animals to humans), epizootic potential (affecting many animals of one kind in one region simultaneously), and a complex pathogenesis involving noninfectious cofactors including anthropogenic toxins, immunologic suppression, and other environmental stressors. Advanced diagnostic technologies have greatly enhanced the ability to identify disease etiologies, and marine mammal veterinarians have played a critical role in identifying diseases occurring in marine mammals and the impact that these diseases have on individuals, populations, and the ecosystem as a whole.

Complex infectious and neoplastic diseases. Newly documented multifactorial diseases involving emerging or resurging infectious organisms often with associated neoplasia (tumor formation) are being reported in Atlantic bottlenose dolphins (*Tursiops truncatus*), California sea lions (*Zalophus californianus*), southern sea otters (*Enhydra lutris nereis*), and endangered Florida manatees (*Trichechus manatus latirostris*). In turn, these diseases may provide important information on aquatic ecosystem health. For example, stranded California sea lions have an unusually high incidence of a newly reported urogenital cancer, which is associated with a novel herpesvirus, as well as exposure to anthropogenic contaminants such as polychlorinated biphenyls (PCBs) and dichlorodiphenyltrichloroethane (DDT) that persist in the sea lions' environment. Genetically inbred sea lions, and those with a specific major histocompatibility complex genotype, are more likely to develop this type of cancer. Additionally, benign orogenital neoplasia associated with another novel herpesvirus and a newly sequenced papillomavirus identified as TtPV-2 are now reported in Atlantic bottlenose dolphins in Florida. In some dolphins, the benign oral tumors undergo malignant transformation to cancer (**Fig. 1**). The tumors appear to be sexually transmitted and are now occurring in epidemic proportions in some Atlantic coastal areas. Dolphins with orogenital tumors demonstrate an acute phase inflammatory

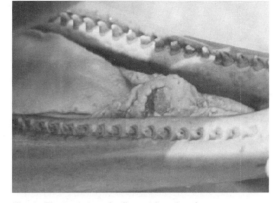

Fig. 1. The tongue and adjacent frenulum (mucous membrane underneath the tongue) of an Atlantic bottlenose dolphin with a squamous cell carcinoma.

response and upregulated innate and humoral immunity, all possible responses to the tumors and/or the viruses associated with the tumors. Additionally, dolphins with tumors have extremely high levels of skin and blood mercury, raising concerns not only for the health of the dolphins but of the coastal human population exposed to this highly toxic metal.

Toxoplasmosis, a protozoal disease caused by *Toxoplasma gondii*, is now reported as a major cause of mortality among southern sea otters in California and is potentially fatal to humans. Although many animal species can serve as intermediate hosts for *T. gondii*, cats are the only animals known to shed the parasite's eggs in their feces. A recent California coastal seroprevalence analysis demonstrated infection in about 52% of stranded sea otters and 38% in free-ranging sea otters. Investigations of the pathogenesis of sea otter *T. gondii* infections provide important information on the changing dynamics of parasite life cycles and infections potentially shared by wildlife, domestic animals, and humans.

Lobomycosis is observed in epidemic proportions in specific geographic regions of Florida's Atlantic coast (**Fig. 2**). Lobomycosis is a rare chronic granulomatous disease of the skin and subcutaneous tissues of dolphins and humans caused by the fungus *Lacazia loboi*. The clinical and pathological manifestations of lobomycosis in humans and dolphins are similar and consist of cutaneous nodules that

Fig. 2. Cutaneous lobomycosis in a bottlenose dolphin from Florida's Atlantic coast.

Fig. 3. Cutaneous papillomas on the lip of an endangered Florida manatee associated with a newly sequenced manatee papillomavirus (TmPV-1).

progress slowly over the course of years, usually without involvement of internal organs or mucous membranes. The reasons for the emergence of this rare disease are unknown, but data now indicate that the disease is associated with profound suppression of adaptive immunity, possibly of environmental origin. Limited evidence exists that lobomycosis may be transmitted from infected animals to humans. However, the high prevalence of dolphin lobomycosis in a Florida coastal region raises concerns for zoonotic or common-source transmission.

The Florida manatee is an endangered marine mammal in coastal waters of the United States, which has a high annual mortality due to human-related factors such as boat impacts. The manatee immune system appears highly developed to protect it against a harsh marine environment, pathogens, and the effects of human-related injury. Nonetheless, the first viral disease associated with benign cutaneous tumors was recently documented in Florida manatees (**Fig. 3**). The tumors are associated with a newly described and sequenced manatee papillomavirus (TmPV-1). Preliminary immunologic data suggest that the manatees with tumors are immunologically suppressed and that the papillomas are caused by activation of latent papillomavirus infections and reinoculation from active infections. The emergence of papillomavirus-induced papillomas in Florida manatees, the possibility of activation of latent infection or transmission of active infection to free-ranging manatees, and the underlying cause of immune suppression predisposing manatees to developing viral papillomatosis are concerns for the future management of this endangered species.

Anthropogenic toxins and biotoxins. Bottlenose dolphins from the eastern coast of the United States and the Gulf of Mexico often have a coastal habitat and many are year-round residents in waters surrounded by human activity. High concentrations of persistent emerging organohalogen contaminants such as polybrominated diphenyl ethers (PBDEs) and perfluoroalkyl compounds (PFCs) and a suite of legacy pollutants such as PCBs, DDT, and trace metals have been found in coastal Atlantic dolphins. For example, levels of PBDEs and perfluorooctane sulfonate (PFOS)

in Charleston, South Carolina, coastal dolphins represent some of the highest measured in marine mammals. Conversely, levels of total and methyl mercury levels in the skin and blood of central Florida coastal dolphins are up to four times higher than those of Charleston dolphins and well above U.S. Environmental Protection Agency (EPA) standards established for fish for human consumption. As apex (top-level) predators, marine mammals, such as dolphins, have unique fat stores and have been known to accumulate high levels of persistent organic lipophilic toxins. During periods of fasting, starvation, lactation, or other physiologic or pathologic demands, stored blubber lipids may be mobilized that may potentially redistribute these toxins, causing health concerns. The high concentrations of these anthropogenic toxins are of particular concern not only for the dolphins but also for the coastal human populations that are exposed to the same toxins.

Harmful algal blooms (HABs) can produce potent biotoxins, which are incriminated in mass mortalities of dolphins, sea lions, and manatees. The HAB problem is growing worldwide and poses a major threat to human and ecosystem health. Recent, often unprecedented Florida manatee and Atlantic bottlenose dolphin epizootics are associated with brevetoxins, which are produced by the "Florida red tide" dinoflagellate *Karenia brevis*. In marine mammals, brevetoxin poisoning is suspected to involve the ingestion of toxins in food sources by manatees and bottlenose dolphins or the direct inhalation of toxins by manatees. The present data suggest that manatee mortality resulting from brevetoxin poisoning may not necessarily be acute, but may occur after chronic inhalation or ingestion and may involve the release of inflammatory mediators that result in fatal toxic shock. Interestingly, the inhalational route of brevetoxin exposure appears to be unique in manatees with regard to other marine mammals, but is shared with humans. Indeed, the inhalational route of brevetoxin exposure in humans is associated with increases in human pulmonary emergency room diagnoses, which are temporally related to Florida red tide occurrences. Additionally, important new data indicate that brevetoxin vectors such as seagrasses and some fish species can result in delayed or remote exposure in manatees and dolphins, respectively. Thus, intoxication may occur in the absence of toxin-producing dinoflagellates, and these unexpected toxin vectors may account for marine mammal deaths long after or remote from a dinoflagellate bloom.

Ocean and human health. As emerging diseases are being increasingly documented and the effects of ecologic and climate change are becoming better understood, concern is being raised about the health of the Earth's oceans. The concept of marine mammal sentinel organisms may provide one way of evaluating aquatic ecosystem health. Such sentinels can provide an early warning system of potential negative environmental trends and permit potential management of the impacts on human and animal health associated with our oceans.

Amidst all other global issues, perhaps we have failed to see the warning signs of possibly larger and more complex health problems to come. This is particularly relevant since the emerging disease data suggest that complex interactions may occur among anthropogenic toxins, infectious agents, and immunologic and genetic factors in many marine mammal species that share coastal environments with humans. Therefore, it is in our own best interest to determine marine mammal health patterns that could potentially impact our own well-being.

[This work was conducted under National Marine Fisheries Permit No. 998-1678-01 issued to the author.]

For background information *see* ANIMAL VIRUS; CETACEA; DISEASE; ECOSYSTEM; ENDANGERED SPECIES; ENVIRONMENTAL TOXICOLOGY; EPIDEMIOLOGY; IMMUNOSUPPRESSION; MAMMALIA; MARINE CONSERVATION; MARINE ECOLOGY; OTTER; PINNIPEDS; SIRENIA; TUMOR; ZOONOSES in the McGraw-Hill Encyclopedia of Science & Technology.
Gregory D. Bossart

Bibliography. G. D. Bossart, Marine mammals as sentinel species for oceans and human health, *Oceanography*, 19(2):44–47, 2006; G. D. Bossart et al., Manatees and brevetoxicosis, in C. Pfeiffer (ed.), *Molecular and Cell Biology of Marine Mammals*, pp. 205–212, Krieger Publishing, Melbourne, FL, 2002; G. D. Bossart et al., Viral papillomatosis in Florida manatees (*Trichechus manatus latirostris*), *Exp. Mol. Pathol.*, 72:37–48, 2002; G. D. Bossart et al., Orogenital neoplasia in Atlantic bottlenose dolphins (*Tursiops truncatus*), *Aquat. Mamm.*, 31(4):473–480, 2005; G. D. Bossart et al., Pathologic findings in Florida manatees (*Trichechus manatus latirostris*), *Aquat. Mamm.*, 30(3):434–440, 2004; J. S. Reif et al., Lobomycosis in Atlantic bottlenose dolphins (*Tursiops truncatus*) from the Indian River Lagoon, Florida, *J. Am. Vet. Med. Assoc.*, 228:104–108, 2006; J. G. Vos et al. (eds.), *Toxicology of Marine Mammals*, Taylor & Francis, London, 2003.

Engineered minichromosomes in plants

Chromosomes are physical carriers of genetic information that all organisms rely on as blueprints. The chromosome consists of deoxyribonucleic acid (DNA) and proteins. The DNA of the chromosome stores genetic information, with the content and the selective expression of which determining the diversity of all living organisms. Chromosomes come in different shapes. Prokaryotic organisms usually have simple circular chromosomes, whereas eukaryotic organisms such as plants and animals have developed highly organized linear chromosome structures. In eukaryotes, DNA and histone proteins are packaged into nucleosomes. Nucleosomes and the "linker" DNA (DNA between nucleosomes) are the basic units of the chromosome fiber called chromatin. The 10-nanometer chromatin is further compacted into other higher-order structures such as the

metaphase chromosomes (that is, chromosomes in the metaphase stage of the cell cycle), which are visible under a light microscope after staining.

The eukaryotic chromosome has three essential parts: centromere, telomeres, and two chromosome arms (see **illustration**). The centromere is seen as the constriction on the metaphase chromosome. During cell division, the cell spindle fibers attach to the chromosome through the centromere to ensure correct chromosome separation to daughter cells. Telomeres are the protection structure at the ends of the linear chromosomes to ensure the stability of linear chromosomes. Although they are structurally important, both centromere and telomere regions

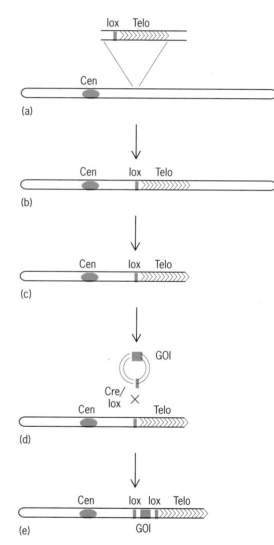

Minichromosome engineering and the manipulation with a Cre/lox site-specific recombination system. (*a*) A plant chromosome is targeted by transforming a telomere-containing construct. (*b*) Telomere repeats and other transgene sequences are integrated into a plant chromosome. (*c*) The integration of telomere sequences at a chromosome site seeds a new telomere and consequently truncates the chromosome to form a minichromosome. (*d*) Genes of interest (GOI) are introduced into minichromosomes by Cre-mediated recombination at the lox site on the minichromosome. (*e*) GOI are integrated into the minichromosome by recombination to confer new traits. Centromere (Cen), lox sequence (lox), Cre recombinase (Cre), telomere sequence (Telo), and genes of interest (GOI) are indicated in the illustration.

are gene-sparse. The genetic information is stored mainly in the euchromatic region (lightly stained chromatins) of the two chromosome arms. In addition, a chromosome needs origins of replication, which are located throughout the length of the chromosome—approximately every 100 kilobases (kb) of DNA in eukaryotes.

Minichromosomes are shortened chromosomes from chromosomal aberrations or engineering. As with normal chromosomes, minichromosomes also have the three essential components—centromere, telomeres, and the replication origins—for stable maintenance in the genome and efficient passage from generation to generation. The chromosome arms in an engineered minichromosome should be minimal in size so that no host genes are expressed from them that can interfere with the host genome. However, minichromosomes should be amenable for genetic manipulation in order to be used as gene expression platforms for genetic engineering. For example, genes can be targeted to minichromosomes; the targeted genes can be expressed in minichromosomes. Subsequently, additional genes can be added to the minichromosomes.

Minichromosome production. There are two approaches for minichromosome construction in eukaryotes: the bottom-up approach (de novo minichromosome construction) and the top-down approach (derivatives of endogenous chromosomes through chromosome breakage).

Bottom-up approach. De novo artificial chromosomes have been reported in yeast and mammals. Yeast has a rather simple centromere sequence requirement. The in vitro assembly of centromeric, telomeric, and replication origin sequence of yeast has led to the successful construction of yeast artificial chromosomes (YACs) that have been widely used as vectors in the cloning of large genomic DNAs. Following the studies in yeast, the first mammalian artificial chromosome was constructed in 1997 from introduced human alpha-satellite DNA (a tandemly repeated DNA family that resides at the centromeres), telomeric DNA, and a large piece of human genomic DNA harboring the origins of replication. However, the de novo artificial chromosome construction in plants has been a challenge for several reasons. First, the current DNA delivery systems in plants present a limit on the introduction of large DNA elements into the plant cell, which is essential for de novo minichromosome construction. Second, there is an epigenetic component of plant centromeres. This means that the DNA sequence alone does not necessarily produce a functional centromere when DNA is introduced into the cell. Other unknown circumstances are also required. Thus, currently, the only successful construction of plant minichromosomes has been through the top-down design.

Top-down approach. Minichromosomes can arise naturally from chromosomal aberration. Systemic production of minichromosomes in mammalian cells was reported in 1992 by the transformation of telomere sequences that truncated the chromosome

arms. By using similar approaches, the first engineered plant minichromosome was constructed in maize by a telomere-mediated chromosomal truncation (see illustration, *a–c*). In this study, two constructs, pWY76 and pWY86, both containing a 2.6-kb direct repeat of the *Arabidopsis* telomeric sequence, a lox site-specific recombination sequence, and a selectable marker gene, were assembled and transformed into immature maize embryos by an *Agrobacterium*-mediated method. By screening initial transgenic events with fluorescence in situ hybridization (FISH), a greater-than-random number of distal transgenes in both pWY76 and pWY86 transgenic plants was recorded, while the number of distal transgenes with the control construct pWY96 (no telomere sequence) was comparable to other transformations. The significantly higher distal loci in telomeric transgenic lines suggested that some of these would be telomere truncation events. Because telomerase adds different copies of telomeric repeats to the newly seeded chromosomal ends in each cell that underwent the truncation process, a Southern hybridization of genomic DNAs digested with restriction enzymes to produce the telomere-containing fragments should display a smear pattern. Twenty truncations in 231 recovered insertions were confirmed by this method, giving an efficiency of 8.7%.

One minichromosome, named R2, was recovered from the truncation of the long arm of chromosome 7. R2 does not pair with normal chromosome 7 and remains stable in any stages of meiosis or mitosis. Thus, R2 could be used as an excellent starting material for minichromosome-based plant genetic engineering.

Engineered miniB chromosomes. B chromosomes are naturally occurring supernumerary chromosomes in many plants and animals. They are mostly inert. For example, there are basically no genes present on the maize B chromosomes. Thus, B chromosomes in maize can be engineered into minichromosome vectors without interfering with the normal genome. B chromosomes have never been found to recombine with standard A chromosomes in the maize genome, and hence the transgenes on the miniB chromosomes will be independent from the normal genome. They can be introduced into any maize variety and eliminated by genetic crossing.

To engineer B-chromosome-based minichromosomes in maize, biolistic-mediated transformation (a method of transfecting cells by bombarding them with microprojectiles coated with DNA) was performed to transform immature maize embryos with 0 to 12 B chromosomes with the telomere-containing constructs. From a total of 281 transgenic events, 32 miniB chromosomes were selected. In addition, 23 transgenic events were found with transgenes on normal B chromosomes that were not truncated. The chromosome truncation rate in B chromosome is 58%, which is much higher than the 8.7% truncation rate recorded for the A chromosome truncations. The lower truncation frequency of A chromosomes compared to B chromosomes may be caused by the lower viability of events with deficient A chromosomes, whereas the missing of parts of the whole B chromosome has no such adverse effects during the tissue-culture and plant-regeneration process. Cells containing truncated A chromosomes were selected against because of the loss of chromosomal parts. MiniB chromosomes are stable during both mitosis and meiosis. The meiotic transmission varies from 12% to 39% through the male parent in different lines.

Maize B chromosomes can accumulate by nondisjunction (failure of homologous chromosomes to separate symmetrically during cell division), during which two B chromosomes will go to one cell during the second pollen mitosis. This results in the two sperm cells with 0 and 2 B chromosomes, respectively. The sperm with the B chromosomes has a greater chance to fertilize the egg during fertilization, so the progeny can have more B chromosomes than their parents. Truncated miniB chromosomes do not undergo nondisjunction because they are missing the long-arm distal region that is required for this process. However, nondisjunction of mini-B chromosomes can be restored by supplying the long-arm distal region. Thus, similar to normal B chromosomes, miniB chromosomes can also accumulate multiple copies in the presence of normal B chromosomes. This can facilitate the studies of gene dosage effects and optimize the yield of interesting gene products.

Although no gene has been found on maize B chromosomes, B or miniB chromosomes can support foreign gene expression. Seventeen events that contained only transgenes on B chromosomes were assayed for beta-glucuronidase (GUS) gene activity. (GUS is used here as a reporter marker gene.) GUS staining was detected in nine events. The other events may be caused by either the incomplete transformation of the gene expression cassette or the inactivation of a transgene normally seen in biolistic-mediated gene transformation. These results showed that transgenes on B chromosomes could be active. Thus, it is possible to use engineered miniB chromosomes in genetic engineering because the genes placed on the engineered miniB will be active.

Manipulation of minichromosomes. A Cre/lox-based recombination strategy was designed for the integration of new genes into minichromosomes (see illus. *d,e*; lox sites are short DNA sequences that are the substrate of a bacterial virus enzyme called Cre). However, a different method was used to test the recombination capability of this system. A maize line containing an R2 minichromosome was crossed with a transgenic line that has a corresponding lox recombination site and expresses the Cre recombinase enzyme. The expected recombination was detected in 10 plants from 120 progeny, indicating the ability of this system for combining new genes into the minichromosomes. Although the recombination was not heritable to the next generation, the recombination in somatic cells demonstrated "proof of concept" that engineered minichromosomes can

be targeted for future needs in the expression of more genes. The Cre/lox recombination efficiency needs to be boosted or other methods need to be developed for future applications of this technology. Other methods include similar site-specific recombination systems (such as FLP/FRT, R/RS, Int/att, or a zinc finger nuclease–mediated recombination system) or simply targeting the minichromosome with a subsequent transformation.

In engineered maize minichromosomes, the meiotic transmission of minichromosomes varied over the range of 12% to 39% through either parent. Thus, methods intended to enhance the transmission should be developed. For example, a fertility-restorer gene can be expressed from engineered minichromosomes in a male sterility system so that only pollen with the minichromosome can be produced to ensure that all progeny inherit the minichromosome.

Applications of engineered minichromosomes. Minichromosome-based genetic engineering has many advantages over traditional genetic transformation methods. In this system, foreign genes can be integrated into an engineered minichromosome, which exists independent of the normal genome. Thus, inactivation of endogenous genes by the integration of foreign genes can be prevented. Second, multiple genes or even a metabolic pathway can be expressed from the minichromosome as a single unit. Third, being an independent chromosome, as opposed to being integrated into the normal genome, linked genetic variants are not transferred when minichromosomes are crossed into new lines. The minichromosome system should be very useful for studies of plant genomics and plant genetic engineering, including gene stacking (combining desired traits into one line) and the use of plants as bioreactors for the production of pharmaceutical proteins and metabolic products. The successful construction of maize engineered minichromosomes can be easily applied to other plants because the key component for minichromosome production, the telomere sequence, is conserved in plants.

For background information *see* AGRICULTURAL SCIENCE (PLANT); CHROMOSOME; DEOXYRIBONUCLEIC ACID (DNA); GENE; GENETIC ENGINEERING; GENETIC MAPPING; GENETICS; MOLECULAR BIOLOGY; NUCLEOSOME; RECOMBINATION (GENETICS) in the McGraw-Hill Encyclopedia of Science & Technology.

Chunhui Xu; Weichang Yu

Bibliography. C. J. Farr et al., Telomere-associated chromosome fragmentation: Applications in genome manipulation and analysis, *Nat. Genet.*, 2:275–282, 1992; F. Han, J. C. Lamb, and J. A. Birchler, High frequency of centromere inactivation resulting in stable dicentric chromosomes of maize, *Proc. Natl. Acad. Sci. USA*, 103:3238-3243, 2006; J. J. Harrington et al., Formation of de novo centromeres and construction of first-generation human artificial microchromosomes, *Nat. Genet.*, 15:345-355, 1997; A. Houben and I. Schubert, Engineered plant minichromosomes: A resurrection of B chromosomes, *Plant Cell*, 20:8-10, 2008; J. M. Vega et al.,

Agrobacterium-mediated transformation of maize (*Zea mays*) with Cre-*lox* site specific recombination cassettes in BIBAC vectors, *Plant Mol. Biol.*, 66:587–598, 2008; W. Yu et al., Construction and behavior of engineered minichromosomes in maize, *Proc. Natl. Acad. Sci. USA*, 104:8924–8929, 2007; W. Yu et al., Telomere-mediated chromosomal truncation in maize, *Proc. Natl. Acad. Sci. USA*, 103:17331–17336, 2006.

Epidemic of obesity

In the United States and across the world, obesity has increased at a startling rate over the past three decades. This sharp increase in obesity rates has led the media and some scientists to label it the "obesity epidemic." Currently, obesity is among the most serious health problems in the United States—the Centers for Disease Control list it as the second leading cause of preventable deaths in America.

Many medical conditions are associated with obesity and, generally, are categorized as being the result of either the high number of fat cells or the fat mass in the body. Diabetes, cardiovascular disease, some types of cancer, and nonalcoholic fatty liver disease are associated with the increased number of fat cells, whereas health conditions such as osteoarthritis and obstructive sleep apnea are due to increased fat mass. Additionally, having a body mass index (BMI) above 32 is associated with a doubled age-adjusted risk of mortality.

Prevalence. Adult obesity is defined by using BMI cutoff points. The BMI is a ratio of weight to height (weight in kg/height in m^2) that is strongly related to body fat and obesity-related health consequences. The World Health Organization has defined "overweight" as having a BMI between 25 and 29.9, "obesity" as having a BMI over 30, and "morbid obesity" as having a BMI over 40. "Normal" weight is a BMI between 18.5 and 24.9. For children and adolescents, the Centers for Disease Control and Prevention Growth Charts define a BMI greater than or equal to the sex- and age-specific 95th percentile as "overweight" and a BMI between the 85th and 95th percentile as "at risk for overweight."

In the United States, epidemiologists have estimated that the prevalence of obesity among adults jumped from 13 to 32% between the 1960s and 2004, while the combined prevalence of obesity and overweight increased by approximately one-third, from 47 to 65%. Certain racial and ethnic minorities and individuals with low socioeconomic status have higher rates of obesity than their counterparts. In 2004, African Americans and Mexican Americans had a 10% greater prevalence rate of overweight than whites (76% versus 64%, respectively). Among women, this difference is even greater: more than 50% of African American women over the age of 40 were obese compared to 35% of white women. Americans who are less educated (having less than a high school education) have a higher prevalence of obesity (with the exception of African American women).

The prevalence of obesity among children and adolescents in the United States has also risen. In 2004, more than a third (35%) of children and adolescents between 6 and 19 years of age were "at risk for overweight," or overweight, and 17% were overweight. This amounts to an annual increase in prevalence of approximately 0.5 percentage point since 1976.

Obesity has increased not just in the United States, but also in countries around the globe. According to the World Health Organization, there are more than 1 billion overweight and 300 million obese adults in the world. Rates of obesity range from less than 5% in Japan, China, and certain African nations to over 75% in urban areas of Samoa. Even in countries that have low rates of obesity, such as China, obesity can be upwards of 20% in urban areas. In most European countries, particularly those in southern, central, and eastern Europe, approximately 20% of all adults are obese.

Causes of the obesity epidemic. The causes of the increasing prevalence of obesity are complex and multifactorial. Genes certainly help determine how vulnerable an individual is to gaining weight: between 25% and 40% of an individual's weight is genetically determined via factors such as basal metabolic rate and the number of fat cells that an individual has. However, genetic factors cannot fully explain why obesity has increased so much in such a short period of time, given that the human genetic makeup cannot change so drastically in one generation.

Scientists have recently begun to document how changes in society are contributing to the obesity epidemic. For one, there has been a major change in the way people live and work that has led to decreased physical activity. Jobs that require physical labor have increasingly been replaced with less physically demanding work. Likewise, improvements in technology and changes in the way that communities are built have led to people increasingly using means of transportation that do not include walking or bicycling, and using home conveniences (for example, washing machines) that do not require the same amounts of physical labor as in previous years.

Perhaps even more importantly, there has also been a major change in the diets of individuals. Over the last three decades, portion sizes in restaurants have increased dramatically, causing people to consume far more calories than they did in previous years. Additionally, the marketing of high-calorie, high-fat, and high-sugar foods, particularly to children, has increased. It is estimated that the average American child will see 21 television food advertisements every day, for a total of 7600 exposures every year. Research shows that this bombardment of food advertisements has the desired effect of causing individuals to consume these high-calorie foods. Finally, high-calorie, low-nutrient-dense, highly processed foods have become cheaper and more readily available over the past three decades. Collectively, these changes have been labeled the "toxic food environment."

Another factor that may be contributing to the obesity epidemic is the idea that some people become "addicted" to certain types of unhealthy, calorically dense foods. This idea is relatively new and not universally accepted among the scientific community. However, researchers are beginning to consider this as one additional factor in explaining the increasing prevalence of obesity.

Solutions and treatment possibilities. The most common way to treat obesity usually involves some type of dietary change. There are hundreds of different diets that individuals can choose from. However, scientific studies are increasingly showing that most diet programs designed for weight loss do not cause people to lose weight over a long period of time. Researchers have shown that, after a 5-year period, over 90% of people typically regain all the weight that they had lost. Individuals may lose weight initially on a given diet, but eventually most people gain back a large percentage—if not all—of the weight they lost. Scientists define a successful diet as one that causes an individual to lose 5–10% of his or her body weight because this amount of weight loss is associated with a number of improved health outcomes. Despite this low success rate of diets, the diet industry in the United States earns approximately $24 billion per year.

Researchers have been working for years to develop drugs that will aid in weight loss. Among the most widely prescribed drugs are orlistat (which reduces intestinal fat absorption) and sibutramine (an appetite suppressant). Thus far, clinical trials have shown that these drugs result in only modest weight loss and, like other prescription medications, can have negative side effects and potential contraindications. Thus, treatment with antiobesity drugs is currently doing little to stem the obesity epidemic.

Weight loss (or bariatric) surgery refers to a variety of surgical procedures that modify the gastrointestinal system to reduce the amount of calories that the body absorbs and/or how many calories an individual consumes. These procedures are supposed to be reserved for individuals with a very high BMI (40 or greater) who have been unable to sustain a significant weight loss in spite of eating a proper diet and exercising, or for patients with a BMI of 35 or greater who also have additional health risk factors as a result of their excess weight. Bariatric surgery typically results in significant weight loss—up to 117 lb (53 kg) on average for some procedures. However, bariatric surgery also frequently causes medical complications in patients. Despite its high success rate, bariatric surgery is not a panacea for stemming the obesity epidemic, given its high cost and that it is not an option for individuals who are not morbidly obese.

Given the limitations of these various treatment options for obesity, many public health and public policy scholars have argued that a prevention-based approach to prevent future weight gain among both adults and children is a more feasible and effective approach for public health. In recent years, public policies that are based on this public health model of disease prevention have been proposed. Obesity prevention policies aim to change the toxic food and lack of physical activity environment that contributes to obesity, such as making healthy foods less expensive and making it easier for people to engage in physical activity (for example, installing bike paths for commuters).

Obesity prevention policies exist both at all levels of government and in the private sector. In the private sector, some workplaces have instituted policies to prevent obesity in their employees, such as reducing the prices of healthy foods in their cafeterias, building exercise facilities, and subsidizing gym memberships. In the United States, the federal government's obesity policies tend to focus more on weight reduction than on prevention of future weight gain. The federal government funds over 300 programs related to obesity, many of which are concerned with public education, such as the MyPyramid program through the United States Department of Agriculture (USDA).

In recent years, public awareness of obesity as a significant health problem and support for obesity prevention policies have increased in the United States. In a 2007 poll, 85% of Americans supported tax breaks for employers who provided exercise space to employees, and 72% said they would support government-sponsored policies that required health insurance companies to cover obesity prevention and treatment programs. It is hoped that these changing attitudes will result in more aggressive measures to stem the obesity epidemic.

For background information *see* ADIPOSE TISSUE; FOOD; FOOD MANUFACTURING; GENETICS; HUNGER; LIPID METABOLISM; NUTRITION; OBESITY; PUBLIC HEALTH in the McGraw-Hill Encyclopedia of Science & Technology.   Victoria L. Brescoll

Bibliography. K. D. Brownell and K. Battle-Horgen, *Food Fight: The Inside Story of the Food Industry, America's Obesity Crisis and What We Can Do about It*, McGraw-Hill, New York, 2003; Institute of Medicine, *Food Marketing to Children: Threat or Opportunity?*, National Academies of Science Press, Washington, D.C., 2006; G. Kolata, *Rethinking Thin: The New Science of Weight Loss—and the Myths and Realities of Dieting*, Farrar, Straus, and Giroux, New York, 2007; M. Pollan, *The Omnivore's Dilemma: A Natural History of Four Meals*, Penguin, New York, 2007.

Epilepsy

Epilepsy refers to the medical condition of recurrent, unprovoked seizures. Epilepsy and seizures are the most common neurologic disorder. Up to 10% of the population will experience a seizure at some point in their life, with most not being diagnosed with epilepsy. A seizure is a sudden change in behavior accompanied by an abnormal discharge of electricity in the brain. A seizure is typically brief and can present in many different ways, including motor changes, sensory phenomena, and alterations in consciousness. Seizures are a symptom of a neurological problem; thus, just as with a symptom such as

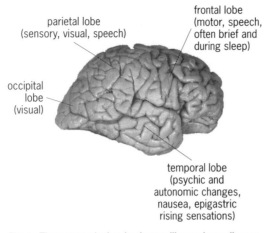

parietal lobe
(sensory, visual, speech)

frontal lobe
(motor, speech,
often brief and
during sleep)

occipital
lobe
(visual)

temporal lobe
(psychic and
autonomic changes,
nausea, epigastric
rising sensations)

Fig. 1. The nature of a focal seizure will vary depending on where in the brain the seizure originates. Some examples are given for seizure types originating from the different lobes. Temporal lobe seizures are the most common.

headache, there are many reasons why a seizure may occur. High fever, meningitis, toxins, chemical imbalances of glucose or electrolytes, stimulant overdose, massive sleep deprivation, and alcoholic withdrawal are all examples of external events that can provoke a seizure.

Types of seizures. A common classification of epilepsy depends on whether seizures start in one brain region and spread, or involve distributed areas of the brain all at once. One class of epilepsy is "localization-related," including focal seizures that clearly start from a specific part of the brain, as happens following stroke and often occurs with brain tumors (**Fig. 1**). This class also includes partial seizures that originate in one part of the brain, usually associated with a specific symptom. The seizure can then spread to the rest of the cerebral cortex, causing

a secondarily generalized seizure that typically appears as repetitive contraction of the arms and legs along with loss of consciousness. Another class of seizures, the primary generalized seizures, involves the brain diffusely from the onset. In a primary generalized seizure, the appearance may be dramatic, with prolonged loss of consciousness and jerking motor movements of all extremities, or can be briefer, as in the case of "absence" seizures where behavior is disrupted very briefly, often with a rapid return to function.

Diagnosis. Seizures can be difficult to diagnose. The nature of some seizures, such as a brief disruption of behavior or involuntary movement, can be mimicked by many other benign conditions. Most seizures do not evolve into epilepsy. However, the risk of this happening is higher after one seizure (especially for an adult) compared to the general population. Although a generalized seizure with motor convulsions is usually obvious, many seizures are subtle and can be unknown even to the person having the event. A careful evaluation will include descriptions of the events by observers, associated medical conditions, any aftereffects when the seizure is over, family history of epilepsy, history of head injury, and any history of repetitive similar events. This evaluation can be supplemented by a scalp recording of brain activity [electroencephalogram (EEG), **Fig. 2**] and brain imaging [magnetic resonance imaging (MRI) or computerized tomography (CT) scan]. In many instances of seizures, these tests will not be necessary. However, several important conditions, such as brain tumors, can first present as epilepsy (**Fig. 3**), so consideration should be given to early imaging when suspicion arises. The EEG is not conclusive, either. Approximately half of individuals with epilepsy will have a normal EEG, unless an epileptic event is captured during the EEG recording.

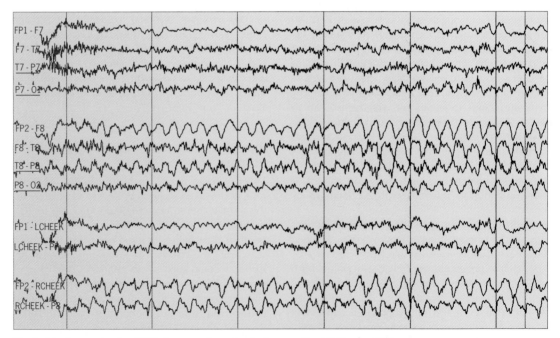

Fig. 2. A brain wave tracing (EEG) shows a seizure developing over the right temporal lobe.

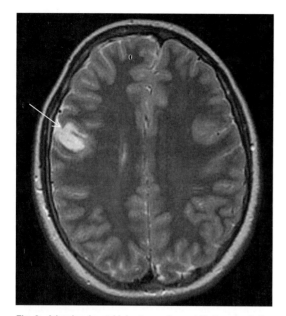

Fig. 3. A benign frontal lobe tumor (arrow) that presented with intractable epilepsy.

Another diagnostic problem occurs with patients who appear to have nonepileptic, psychogenic seizures. In these cases, the seizures are involuntary but psychologically induced. They can occur independently or coexist with other seizures. The diagnosis of psychogenic seizures can be quite difficult. They are often sudden, with dramatic thrashing behavior, but seizures originating from the frontal lobes can appear similar.

Febrile seizures are a special case of provoked events that are common in young children, typically from the ages of 3 months to 5 years old. Children who have longer than 15- to 30-min seizures, more than one seizure in 24 h, and seizures with focal features (for example, involving only one side of the body) are at higher risk for developing epilepsy. In the absence of these features, treatment for one febrile seizure, or even more than one, is often outweighed by the risk of therapies.

As epilepsy is a symptom, the cause can be many different diseases or conditions. Stroke, brain tumors, and metabolic abnormalities are the more common reasons for seizures. Other times, developmental abnormalities can cause the epilepsy. A few genetic syndromes have been found (**Table 1**), although most epilepsies are not inherited. Several genetic syndromes have epilepsy as part of a constellation of associated symptoms. Sturge-Weber syndrome (encephalotrigeminal angiomatosis), tuberous sclerosis, and neurofibromatosis are examples of genetic neurocutaneous syndromes with strong association with epilepsy.

Treatment. The decision to begin medication for epilepsy will depend on several factors. Although epilepsy is a serious condition and repeated seizures can be dangerous, all medications have potential side effects, with some being quite serious. Thus, the decision-making process must achieve a balance between the severity of the seizures, the likelihood of recurrence, the etiology of the seizures, and the proposed therapies.

The majority of epilepsy seizures can be treated with medication (**Table 2**). It is estimated that 20–30% of those first diagnosed with epilepsy will progress to chronic epilepsy, and about two-thirds of these patients can be controlled with continued medicine. The exact medicine to take is a complex decision that depends on the type of seizures, medical and personal factors for a given patient, and a provider's comfort and experience with a given medication. Occasionally, multiple medications are needed. Table 2 gives a partial list of common medications used for seizures. Side effects can occur with any medicine, with sleepiness, rashes, and effects on liver or bone marrow function being common. For most seizure medications that have been on the market for some time, generic substitutions are available. Although sometimes appropriate, many of these substitutions are known to lead to variable blood levels. Thus, generic substitutions (especially for phenytoin and carbamazepine) should not be indiscriminately exchanged for the brand formulation.

Special considerations. Although patients with epilepsy are capable of gainful employment, difficulties in society and the workplace are well documented. Discrimination and misinformation in the workplace occur. Vehicle driving is limited, although the exact rules regarding time after the last seizure,

TABLE 1. Some genetic causes of epilepsy

Name	Gene	Characteristics
Various, including GEFS+ (generalized epilepsy with febrile seizures plus)	SCN1A/SCN1B, GABRG2	Wide spectrum of disorders, including generalized seizures and severe infantile myoclonus (muscle jerking), with variable encephalopathy; genes encode sodium channels and γ-aminobutyric acid (GABA) receptor units.
Benign familial neonatal convulsions	KCNQ2/KCNQ3	Chromosome 20q13.3; seizures commonly outgrown.
Juvenile myoclonic epilepsy	?	Chromosome 6; seizures are brief muscle jerking (myoclonus).
Autosomal dominant nocturnal frontal lobe epilepsy	CHRNA/CHRNB	Gene encodes units of the acetylcholine receptor.
Progressive myoclonic epilepsies	Various	Wide variety of syndromes, including mitochondrial disorders.
Infantile spasms	ARX/CDLK5	X-linked inheritance.
Tuberous sclerosis	TSC1/TSC2	Defined syndrome with skin lesions and benign tumors throughout body; autosomal dominant transmission.

TABLE 2. History of some therapies for epilepsy

Date	Therapies
1886	Surgery to remove epileptic focus
1912	Phenobarbital (Luminal®)
1920	Ketogenic (high-fat, low-carbohydrate) diet
1938	Phenytoin (Dilantin®)
1950–60s	Ethosuximide (Zarontin®), primidone (Mysoline®), valproic acid (Depakote®, Depakene®), tiagabine (Gabatril®), topiramate (Topamax®), diazepam (Valium®)
1970s	Clonazepam (Klonopin®), carbamazepine (Tegretol®, Carbatrol®), lorazepam (Ativan®), midazolam (Versed®)
1990s	Vagus nerve stimulation, felbamate (Felbatol®), gabapentin (Neurontin®), lamotrigine (Lamictal®), levetiracetam (Keppra®), fosphenytoin (Cerebyx®), diazepam rectal gel (Diastat®)
2000s	Oxcarbazepine (Trileptal®), zonisamide (Zonegran®), pregabalin (Lyrica®)

patient and physician reporting requirements, and exceptions vary widely.

The risk of major malformations in the offspring of mothers with epilepsy taking anticonvulsants is about 4–8%, or double that of the general population, although less than the risk of malformations in mothers with other chronic illness, such as diabetes. The use of folic acid has been found to reduce the incidence of fetal malformations. Vitamin K is also recommended in the last six weeks of pregnancy and for the newborn to avoid bleeding complications, which can occur with fetal exposure to various seizure medications. The teratogenicity (capability to cause fetal malformations) of newer drugs has not yet been established. Despite these risks, anticonvulsants should be continued during pregnancy, as the risk of seizures to mother and fetus would be expected to exceed the risk of drugs to the fetus.

Most seizures are self-limiting, but when they persist they become much more dangerous, and this is referred to as *status epilepticus*. The exact amount of time required to establish the diagnosis is unclear and arbitrary, but certainly the risks of permanent brain injury or death become high after 30 minutes of continuous seizure activity. Treatment should begin earlier, and the condition requires prompt medical attention, including cardiopulmonary support, anticonvulsant therapy, and search for the underlying cause. Most cases of status epilepticus will not be in patients with known epilepsy, although occasional exceptions occur.

Unremitting seizures. When two or more medications have failed to control the epilepsy, seizures are considered medically intractable. When a focus of the seizures is suspected (for example, scar tissue or a slow-growing tumor is identified on MRI), surgery is considered and is often (but not always) successful in eliminating seizures. Vagus nerve stimulation (in which short bursts of electrical energy are directed into the brain via the vagus nerve, a large nerve in the neck) is a treatment for partial seizures that is approved by the U.S. Food and Drug Admin-

istration, and on average leads to a 50% reduction in seizure frequency. Injection of a steroid precursor, adrenocorticotropic hormone (ACTH), has been used for intractable infantile spasms; however, this can have serious side effects, being fatal in up to 5% of cases. Some experimental therapies include immunoglobin therapies for a variety of seizure types and stimulation of novel brain targets. Future directions in epilepsy research include a better understanding of the sources of seizures, better detection and localization methods, and better therapies. Many researchers have focused on the abnormal channels that exist in epileptic tissue, including the abnormal sodium and potassium channels seen in some epilepsy models. Some newer imaging methods have used different receptors to image cells that may cause seizure onsets. These can be imaged with positron emission tomography (PET) scanning, and similar information may be seen with newer-generation MRI scanners. Newer detection methods include magnetoencephalography (MEG), but these are very expensive and still uncommon. A higher-density form of scalp EEG shows initial promise in detecting seizure onset and spread. Finally, continued work focuses on when aggressive interventions are best made, with some evidence suggesting that earlier treatment (for example, surgery when indicated) should be done earlier in life for optimal psychosocial outcome.

For background information *see* BRAIN; ELECTROENCEPHALOGRAPHY; MEDICAL IMAGING; MOTOR SYSTEMS; NERVOUS SYSTEM (VERTEBRATE); NEUROBIOLOGY; PSYCHOSIS; SEIZURE DISORDERS; SURGERY in the McGraw-Hill Encyclopedia of Science & Technology.   Jeffrey Ojemann

Bibliography. Commission on Classification and Terminology of the International League against Epilepsy, Proposal for the classification of epilepsy and epileptic syndromes, *Epilepsia*, 30:389–399, 1989; L. B. Holmes et al., The teratogenicity of anticonvulsant drugs, *N. Engl. J. Med.*, 344:1132–1138, 2001; I. E. Leppik, *Contemporary Diagnosis and Management of the Patient with Epilepsy*, 6th ed., Handbooks in Health Care, Newtown, PA, 2006; S. D. Shorvoan et al. (eds.), *The Treatment of Epilepsy*, 2d ed., Blackwell Science, Oxford, 2004; Therapeutics and Technology Assessment Subcommittee of the American Academy of Neurology, Assessment: Generic substitution for antiepileptic medication, *Neurology*, 40:1641–1643, 1990.

Evolution and interrelationships of lizards and snakes

Squamata (snakes and "lizards") is a successful reptile group and one that is important for studying biology. As ectothermic animals (animals whose body temperature is partly dictated by external conditions) that are sensitive to ecological change, squamates aid in the understanding of global climate shifts. As a morphologically and ecologically diverse component of the worldwide fauna today and in the fossil record, squamates represent a key element in

Fig. 1. The Komodo dragon (*Varanus komodoensis*) from southeast Asia is the largest living lizard and may exceed 3 m (9.8 ft) in total length. (*Photo by Rebecca M. Shearman*)

studies of the evolutionary process, evolutionary reconstruction, and Earth history.

Squamate diversity. Squamata includes approximately 8300 extant (living) species inhabiting a wide variety of habitat types and ecosystems. Squamates may burrow, swim through sand, crawl on land, run bipedally across land or water, slither, climb trees, or even glide through the air. The smallest living squamates are among the smallest vertebrates on land at 18 mm (0.7 in.) in total length. In contrast, some living snake species reach 7 m (23 ft) in total length, and Komodo dragons (*Varanus komodoensis*) [**Fig. 1**] commonly exceed 50 kg (110 lb) in mass.

Fossil evidence suggests that squamates and their close relatives, the Rhynchocephalia (including the two living species of tuataras and many fossil species), were present by around 210 million years ago (Ma). Squamate fossils have been found on every continent, including a very large mosasaur from Antarctica. The largest fossil squamate (the mosasaur *Mosasaurus hoffmanni*, known from the eastern United States and from Europe) probably exceeded 17 m (56 ft) in total length, and would have been twice the weight of a *Tyrannosaurus rex*.

Squamata includes seven major groups (**Fig. 2**). Iguania (~1550 extant species) includes iguanas, basilisks, anoles, chameleons, and their relatives. Gekkota (~1200 extant species) is composed of the geckos, scaly-foot lizards, and snake lizards. Scincoidea (~1400 extant species) comprises the skinks and their closest relatives. Lacertoidea (nearly 650 extant species) includes the wall lizards, whiptails, tegus, and the fossil "big-headed" lizards (Polyglyphanodontidae). Anguimorpha (~186 extant species) comprises monitor lizards, Gila monsters, and their relatives, as well as the more obscure glass lizards, alligator lizards, knobby lizards, crocodile lizards, and mosasaurs (fossil). Amphisbaenia (169 named extant species) is a group of mostly limbless lizards that often look like worms. Serpentes includes about 3200 living species of limbless squamates, representing one of many successful lineages of limb-reduced squamates.

Among the most conspicuous identifying characteristics of squamates are the reduction of some skull bones to produce a mobile quadrate and the presence of the hemipenis. The bone connecting the mandible to the rest of the skull in most vertebrates (except mammals and other synapsids) is called the quadrate. The quadrate is a ventrally convex (rounded on the bottom) bone that articulates with a concave cup on the mandible and is usually firmly attached to the skull in most amphibians and reptiles. The squamate quadrate is more loosely attached to the braincase and other skull elements, allowing it to move anteroposteriorly (forward and backward). Snakes and some other squamates may also move the quadrates laterally (to the sides) to help broaden the mouth, allowing these species to swallow prey larger than their heads.

The hemipenis is a paired copulatory organ. Hemipenes come in a variety of shapes and sizes and may be highly ornamented in some cases. The hemipenis is usually deeply divided, with a ridge running along its dorsal surface for the transfer of sperm. Although each male squamate possesses a pair of hemipenes, only one is used at a time.

Snake origins. Modern snakes lack forelimbs and, at most, possess vestiges of hind limbs. However, numerous squamate lineages have lost their limbs. Thus, snakes are squamates lacking conspicuous limbs, but not all limbless squamates are snakes. Pygopodidae (a group of legless lizards including scaly-foot lizards) are limbless geckos. Multiple skink lineages have become limbless. At least two groups of anguid lizards (the group containing alligator lizards

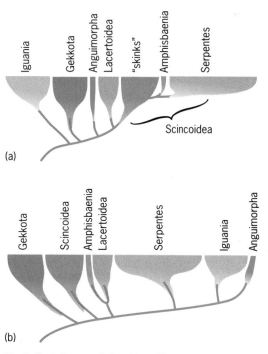

Fig. 2. Evolutionary relationships of Squamata based on (*a*) morphological and (*b*) genetic data. The widths of the shaded areas represent the number of extant species for each group. Note that snakes and amphisbaenians are part of Scincoidea in the top tree, but are separate in the bottom.

and glass lizards) lack external limbs. Snakes are sometimes identified by the absence of movable eyelids and ear openings. However, amphisbaenians, some skinks, and most geckos lack movable eyelids, and amphisbaenians and some skinks may lack external ear openings. Therefore, it may be difficult to tell the difference between snakes and other limbless squamates.

Snakes are very diverse. Some use heat-receptive pits linked with their visual system to find their prey. Snakes may overcome prey with brute strength, by constriction, by injection of venom containing neurotoxin (nervous system disruptors) and/or hemotoxin (which may digest blood cells and blood vessels), or by a combination of these methods. Scolecophidia are tiny snakes sometimes called thread snakes. Some thread snakes have movable upper jaws that are used to sweep prey (for example, insect larvae) into their mouths. Thread snakes often possess degenerate eyes and are sometimes called blindsnakes.

Identifying the squamate group that gave rise to snakes is among the most contentious problems in vertebrate biology. The debate centers on determining which is the basalmost branch on the snake family tree and on what body type is primitive for snakes. The burrowing scolecophidians have been suggested to be the basal snake lineage. However, the oldest known snakes show characteristics of the group containing all snakes except scolecophidians, a group called Oculatophidia ("sighted snakes"). Some of these ancient snakes (for example, *Pachyrhachis problematicus* and *Haasiophis terrasanctus*, from the Cretaceous of the Middle East) possessed tiny, but well-developed, hind limbs with feet

and were adapted for an aquatic environment. This could imply that snakes primitively looked something like modern oculatophidians and were marine. Even so, the snake fossil record is very poor, and numerous other limbed snakes may have been present, but are currently unknown. This is compounded by the nearly complete absence of a blindsnake fossil record.

Most modern evolutionary biologists use cladistic analyses (also called cladistics) to reconstruct the family trees (phylogenies) of organisms. In contrast to older methods that used overall similarity or subjectively chosen characteristics to reconstruct these family trees, cladistics seeks to identify evolutionary changes to isolate branching events on the tree of life. These changes are referred to as "shared, derived, character states" or "synapomorphies." Some scientists use anatomical (morphological) data for this type of analysis, whereas others use genetic (molecular) data. Given the differing strengths of the two data types, the morphological and molecular data are starting to be used in tandem.

Traditionally, snakes have been suggested to be close relatives of monitor lizards (Varanidae) and the giant extinct marine lizards (Mosasauria). However, close examination of the evidence supporting this hypothesis has revealed flaws. A recent morphological cladistic analysis of all squamate groups, including fossils, suggests that amphisbaenians and snakes are closely related to some groups of limbless skinks (Fig. 2a). This would suggest that snakes come from a lineage of already limb-reduced animals. Another hypothesis comes from recent genetic analyses suggesting that snakes are close relatives of iguanians and anguimorphs, and maintains that they are unrelated to amphisbaenians or skinks (Fig. 2b).

Recent fossil discoveries. The Earth has undergone several cycles of warming and cooling over the last 200 million years, and squamate fossils help in the understanding of these cycles. Some fossil anguid lizards from the Middle Eocene (about 45 Ma) are known from northern Canada. Their relatives are known from other parts of North America, Europe, and Asia, demonstrating a warm, circumpolar habitat for these reptiles. Giant marine mosasaurs are known from the central United States, the Sahara Desert, and even Antarctica, helping to reconstruct a tropical world with many continental seas in the Late Cretaceous (99–65 Ma). This may provide important insight in the current era of global warming.

New fossil squamates are found every year. Some recent examples include a gecko relative, a fossil iguanian, and snakes with limbs. A recently described gekkonomorph (a relative of Gekkota) from the Early Cretaceous (nearly 120 Ma) of Mongolia possesses a skull with many primitive features, but with a braincase similar to that of some modern geckos (**Fig. 3**). This discovery was made possible through the use of innovative high-resolution x-ray computed tomography (HXRCT or CT scanning). Herpetologists and paleontologists are increasingly using the HXRCT scans to see hidden details of important specimens without damaging them.

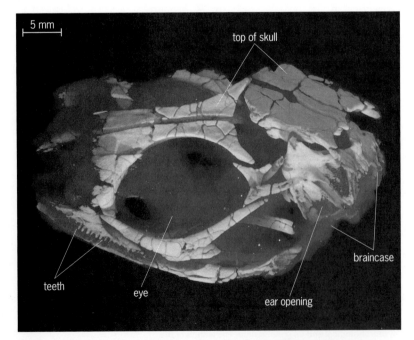

Fig. 3. Cretaceous gecko relative. This image comes from a high-resolution x-ray computed tomograph (HXRCT) of a tiny skull (AMNH FR 21444) that is approximately 120 million years old. Ancient fossils and new technology (like HXRCT) are being used to reconstruct squamate relationships (see Fig. 2).

The discovery of the fossil iguanian *Saichangurvel davidsoni* (~75 Ma) helped to identify an extinct radiation of iguanians endemic to Mongolia. This group, named Gobiguania, may be important for understanding how the earliest iguanians lived and what they looked like. Recent discoveries of limbed snakes such as *Haasiophis terrasanctus* (marine; 99 Ma), *Najash rionegrina* (terrestrial and burrowing; 90 Ma), and others have helped to inform the debate between a marine or terrestrial origin for snakes.

Looking forward. Squamata is among the most successful vertebrate groups in diversity of lifestyle, body form, and number of species. The squamate story adds significant information to the understanding of the world. Further application of innovative techniques and new fossil discoveries continue to help in the comprehension of the diversity of living and fossil squamates. Further exploration of paleontological, behavioral, ecological, and genetic data will be applied to evolutionary investigations to help reconstruct the biological history of this successful and diverse group of reptiles and of our own place in the world.

For background information *see* ANIMAL EVOLUTION; FOSSIL; ORGANIC EVOLUTION; PALEONTOLOGY; PHYLOGENY; REPTILIA; SKINK; SPECIATION; SQUAMATA; SYSTEMATICS in the McGraw-Hill Encyclopedia of Science & Technology.   Jack Conrad

Bibliography. J. L. Conrad, Phylogeny and systematics of Squamata (Reptilia) based on morphology, *B. Am. Mus. Nat. Hist.*, 310:1–182, 2008; J. L. Conrad and M. A. Norell, High–resolution x-ray computed tomography of an Early Cretaceous gekkonomorph (Squamata) from Öösh (Övörkhangai; Mongolia), *Hist. Biol.*, 18:405–431, 2006; R. Estes, K. de Queiroz, and J. Gauthier, Phylogenetic relationships within Squamata, pp. 119–281, in R. Estes and G. Pregill (eds.), *Phylogenetic Relationships of the Lizard Families*, Stanford University Press, Stanford, 1988; S. E. Evans, At the feet of the dinosaurs: The early history and radiation of lizards, *Biol. Rev.*, 78:513–551, 2003; O. Rieppel and H. Zaher, The intramandibular joint in squamates, and the phylogenetic relationships of the fossil snake *Pachyrhachis problematicus* Haas, *Fieldiana (Geology), New Series*, 43:1–69, 2000; T. M. Townsend et al., Molecular phylogenetics of Squamata: The position of snakes, amphisbaenians, and dibamids, and the root of the squamate tree, *Syst. Biol.*, 53:735–757, 2004.

Evolution of African great apes

There are three species of African great apes living today, two chimpanzees and one gorilla, with degrees of subspecies differentiation within the species. Evidence is equivocal in determining the relationships of the African apes with each other and with humans: morphologically, chimpanzees and gorillas are more similar to each other than either is to humans, while molecular evidence indicates that chimpanzees and humans shared a common ances-

tor distinct from that of gorillas. Dating divergence times by the so-called molecular clock (wherein gene mutations occur at a relatively constant rate of molecular evolution) indicates that chimpanzees and humans diverged from each other about 6 million years ago (Ma), and the separation of the chimp-human clade from gorillas is put about 2 million years earlier.

There is considerable uncertainty about the timing and location of the common ancestor of the African apes and humans, although it has been widely assumed, from the time of Darwin and Huxley in the nineteenth century, that the location was somewhere in Africa. The early fossil record of humans is exclusively African. However, the corresponding fossil record for the apes during the past 12 million years is poorly known in Africa, while there is an abundance of fossil apes in Europe and Asia known during the same period. This has led some workers to suggest that the common ancestor of the African apes and humans was derived from European fossil apes such as *Dryopithecus*, which reentered Africa in the late Miocene (~11.6–5.3 Ma). Recently, however, several fossil finds have been made in Africa that begin to bridge the gap, and these will be briefly reviewed here. In addition, predictions have been made about what the common ancestor could be expected to look like, and these will be compared with the new fossil evidence.

New fossil evidence. In August 2007, a fossil ape named *Chororapithecus abyssinicus* was reported from 10.5- to 10-Ma deposits of the Chorora Formation in Ethiopia. This fossil ape is represented by nine isolated teeth (**Fig. 1**), which showed that it had gorilla-sized dentition. The molar teeth combined distinct shearing crests with thickened enamel, an unusual combination not seen in any living ape. The morphology of the shearing crests is described as partly resembling the gorilla condition, and functionally it was probably adapted for feeding on a comparatively fibrous diet. On the other hand, the dentine surface underlying the enamel crown is described

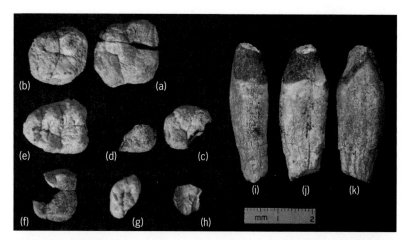

Fig. 1. Isolated teeth of *Chororapithecus abyssinicus*: (*a*) upper second molar, (*b*) upper third molar, (*c*) lower first molar, (*d*) lower molar fragment, (*e*) lower third molar, (*f*) upper third molar, (*g*) upper third molar fragment, (*h*) lower molar fragment, (*i*–*k*) three views of lower canine. (***Photo by Gen Suwa***)

as relatively flat; in combination with the thickened enamel, a functional adaptation to hard and/or abrasive food items is indicated. It is also apparent that *C. abyssinicus* resembles African fossil apes from the period of 17–12 Ma, which are generally hard to distinguish from each other based only on isolated molars and which cannot at present be assigned to any lineage leading to extant African apes. One point of distinction from the earlier fossil apes is that the upper molars are relatively elongated, like the probably thick enameled teeth of *Samburupithecus kiptalami* (see below), but they differ in a number of minor (and extremely variable) characteristics. It is the presence of the shearing crests that mark out *C. abyssinicus* as similar to gorillas, but this characteristic has not been documented for *Samburupithecus* or for most other fossil apes. Thus, it is hard to evaluate the significance of this characteristic as evidence of relationship to gorillas, particularly as it could be a flexible adaptation to diet.

Another recent find, also published in 2007, is *Nakalipithecus nakayamai* from late Miocene deposits in the Nakali valley in northern Kenya. These deposits are dated to between 9.9 and 9.8 Ma, that is, slightly younger than *C. abyssinicus*. The fossils consist of a lower jaw (**Fig. 2**) and 11 isolated teeth. Furthermore, like the Ethiopian fossil, they are large, approximating to the size of gorillas. Of particular significance is the shape and morphology of the upper canine, which is bilaterally broad. This hyper-robusticity of the canine is characteristic of afropithecines, including *Morotopithecus* from earlier in the Miocene, and the elongated upper premolars and relatively deep mandible show similarities with this group of fossil apes. *Nakalipithecus* is slightly larger than *Morotopithecus*, but it would be interesting if the similarities indicate a relationship, since the latter genus appears to have advanced great ape morphological features of its postcranial skeleton. Postcranial remains from Nakali might resolve this issue.

Another fossil ape that falls within this period of 12–8 Ma was described in the late 1990s. This is *Samburupithecus kiptalami*, known by a single upper jaw from the Samburu Hills district in Kenya. The teeth are large, the upper premolars are elongated as in *Nakalipithecus*, and the age of the site, at 9.5 Ma, is close to the age of Nakali. It is difficult to compare

the two fossils at present because *Samburupithecus* is known only from five upper teeth and *Nakalipithecus* mainly from lower teeth, but the upper premolars are similar, and it may be that the two fossils belong to the same species. Finally, four isolated teeth described in 2005 from the late Miocene of East Africa (Ngorora, Lukeino, Kapsomin, and Cheboit) show that fossil apes are there to be found, but they are too fragmentary to provide any information on the origin of the extant African apes.

Another small collection of fossils from Middle Pleistocene deposits at Kapthurin, Kenya, have recently been attributed to an ancestral chimpanzee. Again, the fossils consist only of teeth, and only three have been found so far, but with an age of 0.545 Ma they represent the entire fossil record of chimpanzees. The two upper incisors are particularly convincing, as chimpanzees have distinctive crown morphology, and this is shared by these two fossil teeth. What is particularly important about this find, however, is not that chimpanzees existed at that time (which was inferred already), but their location in East Africa at a time when environments had been fragmented by rift valley formation and most of the forests that had formerly been present had disappeared. By contrast, within the present-day range of chimpanzees, a single femur of unknown context from the Kikorongo Crater in western Uganda may represent a probable late Pleistocene chimpanzee ancestor. The human fossil record from this period and earlier is replete with fossils, and later species were clearly at home in the fragmented rift environments, but the fossil record of the Central and West African forest environments is still unknown, both for chimpanzees and for humans. Either or both may have been present in these western forests, and it is premature to place their environmental dichotomy in time or space in the absence of evidence from these other parts of Africa.

In addition, several recent discoveries of late Miocene fossil apes, some of which have been linked phylogenetically with the orangutan, the other great ape (Asian), are notable. These fossils all fall within the period of 10 to 8 Ma, like the African fossils (that is, *Chororapithecus*, *Nakalipithecus*, and *Samburupithecus*) described above, and the newly discovered fossils come from Bulgaria, Thailand, Turkey, and China, extending the known distribution of apes in Eurasia. The abundance and wide distribution of these fossils from Europe and Asia, together with earlier finds of *Dryopithecus*, have been the basis for suggestions that the line leading to humans bypassed the African apes and originated in Europe, and indeed there is good evidence for hominoid migrations between Africa and Europe during both the Miocene and Pleistocene. However, the absence of a fossil record need not be evidence of the absence of fossils, and thus the suggestions are highly debatable.

Relation to extant African apes. What is of particular interest with these new discoveries is whether any of them can be attributed to lineages leading to the extant African apes. The middle Pleistocene ape from Kapthurin shares significant characters of the

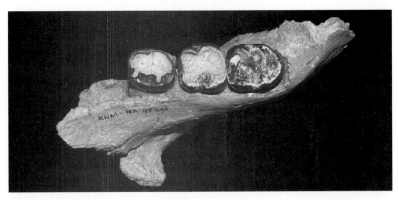

Fig. 2. Fossil jawbone of *Nakalipithecus nakayamai*. (**Photo by Yutaku Kunimatsu, courtesy of the National Museums of Kenya**)

incisors with chimpanzees, and there is a good case for accepting this as evidence of relationship, but these fossils are far too late in date to have any bearing on the divergence of chimpanzees and humans—but what of the late Miocene species? One problem with identifying any fossil close to the common ancestor is that, at so early a stage in the evolution of the descendant groups, few of the later acquired characteristics could be expected to be present, and even those that are present could have been acquired independently. This is particularly a problem when common ancestors are reconstructed from comparison with extant species, what can be termed an evolutionary bird's-eye perspective; this, though, can be counterbalanced by examination of the Miocene precursors from which the possible common ancestor was derived, what we have termed "a worm's-eye view of human evolution." These give an indication of the range of characteristics present in the ape fossil record. For example, we distinguish between two groups of fossil ape: one with robust jaws and enlarged molar teeth with thick enamel, living in seasonal woodland environments, and adapted to some degree of terrestrial positional behaviors; the other had more lightly built jaws, teeth with thinner enamel, and postcrania that were adapted for forelimb suspension and climbing in wetter, less seasonal forests. The former group is known from both Africa and Eurasia, and *Chororapithecus* seems to fit within it; however, the one characteristic that connects this fossil ape to the gorilla lineage is largely unknown for other fossil apes, and hence it is not known if it is phylogenetically significant or not. Some of the characteristics present in *Nakalipithecus* appear to be similar to those found in the much earlier afropithecines, and it may be a later derivative of this lineage.

For background information *see* ANIMAL EVOLUTION; APES; DATING METHODS; DENTAL ANTHROPOLOGY; FOSSIL APES; FOSSIL HUMANS; FOSSIL PRIMATES; MOLECULAR ANTHROPOLOGY; PHYLOGENY; PRIMATES; SPECIATION; TOOTH in the McGraw-Hill Encyclopedia of Science & Technology.   Peter Andrews

Bibliography.   Y. Kunimatsu et al., A new Late Miocene great ape from Kenya and its implications for the origins of African great apes and humans, *Proc. Natl. Acad. Sci. USA*, 104:19220–19225, 2007; D. E. Lieberman, R. J. Smith, and J. Kelley (eds.), *Interpreting the Past: Essays on Human, Primate, and Mammal Evolution in Honor of David Pilbeam*, Brill Academic Publishers, Boston, 2005; S. McBrearty and N. G. Jablonski, First fossil chimpanzee, *Nature*, 437:105–108, 2005; M. Pickford and B. Senut, Hominoid teeth with chimpanzee- and gorilla-like features from the Miocene of Kenya: Implications for the chronology of ape-human divergence and biogeography of Miocene hominoids, *Anthropol. Sci.*, 113:95–102, 2005; C. Stringer and P. Andrews, *The Complete World of Human Evolution*, Thames & Hudson, London, 2005; G. Suwa et al., A new species of great ape from the late Miocene epoch in Ethiopia, *Nature*, 448:921–924, 2007.

Extinction of species

The world is, and always has been, in a state of flux. Over hundreds of millions of years, continents have merged and broken apart, oceans have appeared and disappeared, and mountains have formed and worn away. Together with the geological changes, there have been changes in living things: species, populations, and whole lineages have evolved, while others have disappeared. The evolution of new species and the extinction of others is therefore a natural process. Based on the fossil record, the species present today represent only 2–4% of all species that have ever lived. The remainder are extinct, the vast majority having disappeared long before the arrival of humans.

The extinction of a species usually represents an end point in a long series of population extinctions. During the extinction process, unique evolutionary history is lost at every stage, but the death of the last individual of a species represents the permanent loss of one of life's unique evolutionary and functional forms.

Over geological time, there has been a net excess of speciation over extinction that has resulted in the diversity of life extant today. However, the high number of recent extinctions suggests that the world might now be facing a rapid net loss of biodiversity. Compiling and maintaining an inventory of recent extinctions helps to highlight the growing list of species that are lost forever. Understanding the extent of recent extinctions provides insight into historic extinction rates, which in turn can be compared to the average rates over geological time to determine if current trends are normal or are a cause for concern. An insight into the extinction process helps us to identify species that are at risk of extinction and enables us to highlight taxonomic groups or species from specific regions that are or will be particularly prone to extinction. This article focuses on the extent and causes of recent extinctions and their implications for the future.

Past extinction events.   The fossil record is punctuated by five major mass extinction events. The last and perhaps best known of these occurred 65 million years ago, at the end of the Cretaceous period. That extinction event marked the end of the dinosaurs and opened the way for mammals to become the dominant land vertebrates.

Past mass extinction events took place over many years (for example, the Late Devonian extinction lasted 20 million years), and various theories have been invoked to explain why they happened. These include volcanic eruptions and flood basalt events (basaltic lava flows from fissure eruptions), falls in sea level, impacts involving large asteroids or comets, sustained and significant global cooling or warming, anoxic (oxygen-deficient) events in the oceans, and continental drift. Rather than any one factor being ultimately responsible, it is probable that a number of these factors operated synergistically, resulting in the mass extinctions.

Current extinctions. Many biologists conclude that the widespread and ongoing extinction of species across many families of plants and animals during the current Holocene era is part of a sixth mass extinction—the Holocene extinction event. In broad usage, the Holocene extinction event includes the notable disappearance of large mammals, known as the megafauna (for example, woolly mammoths and saber-toothed cats), by the end of the last glacial period about 9000 to 13,000 years ago. Those extinctions have been attributed to climate change or to the proliferation of modern humans, but they were more likely driven by a combination of both. The Ice Age extinctions are sometimes referred to as the Pleistocene extinction event. However, growing evidence suggests that this extinction event continues through the events of the past several millennia to the present day.

The International Union for Conservation of Nature (IUCN) compiles and publishes the *IUCN Red List of Threatened Species*, which documents extinctions that have occurred globally since the year 1500. A species qualifies for the Red List Category of Extinct when there is no reasonable doubt that the last individual has died. The 2007 edition of the *IUCN Red List* documents 785 extinctions, but this represents only a very small proportion of species that have become extinct during this time period (see **Table 1** for recent extinctions). Many historic extinctions have not been detected or have taken place in taxonomic groups that have not yet been evaluated for the *IUCN Red List*.

Identifying the actual number of historic extinctions is difficult because only ~1.9 million of the world's estimated 5–30 million species have been described. Recent extinctions may be even more prevalent among undescribed species because of their sheer number and the fact that the discovery and description of species tends to be biased toward more broadly distributed and abundant taxa. Furthermore, among the described species, only a few taxonomic groups have undergone thorough conservation assessments to determine whether or not all taxa are still extant. The Global Amphibian Assessment completed in 2004 is an example of a recent assessment of all species that resulted in the listing of an additional 29 extinct species. Similar assessments of such groups as insects, spiders, crustaceans, plants, fungi, and species from poorly studied regions will undoubtedly result in significant increases in the list.

Even where assessments have been conducted, it can take years or decades to prove that a species is truly extinct. The basic paradox of "documenting" extinctions is that absence of evidence is not necessarily evidence of absence. IUCN has begun to highlight species that are believed to be extinct, but have not yet been listed as such because appropriate surveys are still required to confirm that the last individual has died. Systematically flagging these "possibly extinct" species will help to provide a clearer picture of the true extent of recent extinctions. To date, 250 species have been classified as being "Critically Endangered (Possibly Extinct)." Of these, 122 are amphibian species, many of which disappeared relatively rapidly and recently (for example, 18 species of harlequin toad in Central and South America). Other notable examples include Spix's macaw, *Cyanopsitta spixii* (last seen in 2000); the Po'o-uli, *Melamprosops phaeosoma* (this Hawaiian bird was last seen in 2004); and the baiji, *Lipotes vexillifer* (the last confirmed sighting of this Chinese freshwater dolphin was in 2000).

IUCN also records species that are Extinct in the Wild (EW), that is, they exist only in captivity, in

---

TABLE 1. Documented species extinctions from 1982 to 2007*

Group	Species	Common name	Date last recorded	Causes of extinction
Birds	*Myiagra freycineti*	Guam flycatcher	1983	7
	Nesillas aldabrana	Aldabra warbler	1983	1, 7
	Pomarea mira	Ua Pou monarch	1985	1, 7
	Podilymbus gigas	Atitlán grebe	1986	1
	Moho braccatus	Kaua'i 'O'o	1987	1, 2, 7
	Myadestes myadestinus	Kama'o	1989	1, 2, 7
Fish	*Gambusia georgei*	San Marcos gambusia	1983	1, 3, 7
	Telestes ukliva		1988	1, 3, 7
	Alburnus akili	Gokce baligi	1998	5, 7
Frogs	*Eleutherodactylus milesi*		1983	1, 2
	Rheobatrachus vitellinus	Northern gastric brooding frog	1985	1, 2
	Cynops wolterstorffi	Yunnan lake newt	1986	1, 5, 7
	Atelopus ignescens	Jambato toad	1988	2, 3
	Atelopus longirostris		1989	2, 3
	Bufo periglenes	Golden toad	1989	2, 3, 6
	Eleutherodactylus chrysozetetes		1989	1, 2
Snails	*Ohridohauffenia drimica*		1980s	8
	Graecoanatolica macedonica		1992	1
	Pachnodus velutinus		1994	8
Plants	*Oldenlandia adscenionis*		1985	1
	Cyanea dolichopoda	Haha	Post-1990	1, 7
	Crudia zeylanica		1990s	8
	Argyroxiphium virescens	Greensword	1996	1, 7
	Nesiota elliptica	St. Helena olive	2003	1

*Key to causes of extinction: 1 = habitat loss; 2 = disease; 3 = global warming/pollution; 4 = natural disaster; 5 = exploitation/persecution; 6 = restricted range; 7 = invasive species (not disease); 8 = unknown.

cultivation, or as naturalized (introduced) populations. Extinct in the Wild species are in many respects extinct, as they no longer play a functional role in their ecosystems. Also, because successful reintroductions are rare, it cannot be assumed that most of these species will be restored to the wild. The 2007 *IUCN Red List* lists 65 EW species, but this list may grow very rapidly in the next few years with the current amphibian extinction crisis. The growth in the number of EW species is easier to document because these species are often well monitored and conservationists are usually involved in keeping the species alive in captivity or cultivation. However, proving that a species is EW can take years, as it requires confirmation that the last wild individual has died (see **Table 2** for recent records).

Species approaching extinction. The primary function of the *IUCN Red List* is not to document extinctions, but rather to draw attention to those species that are most in need of help, that is, those that are facing a high to extremely high risk of extinction in the wild. The risk of extinction is evaluated using five quantitative criteria, including parameters like the extent of population reduction, small distribution ranges and continuing decline, small populations and continuing decline, very small populations, and very restricted ranges with potential threats that could rapidly result in a species becoming highly threatened or even extinct. Species are assessed against the criteria and placed into one of nine Red List categories. Species classified as Vulnerable, Endangered, or Critically Endangered are termed threatened

---

TABLE 2. Species documented as becoming Extinct in the Wild from 1982 to 2007*

Group	Species	Common name	Date last recorded in wild	Causes of extinction in wild	Possibility of reintroduction
Birds	*Corvus hawaiiensis*	Hawaiian crow	2002	1, 2, 5, 7	Attempt failed because some reintroduced birds died; remainder recaptured and further plans being developed.
	Mitu mitu	Alagoas curassow	Late 1980s	1, 3, 5	Suitable habitat remains, but reintroduction appears difficult.
	Gallirallus owstoni	Guam rail	1987	7	>100 individuals in captivity; reintroduction is under way.
Fish	*Paretroplus menarambo*		Late 1990s	1, 5, 7	Unknown.
	Acanthobrama telavivensis		1999	1, 3	A reintroduction project is under way.
Frogs	*Bufo baxteri*	Wyoming toad	Mid-1990s	1, 2	Captive breeding in place and reintroduction under way, but no self-sustaining population established in the wild.
Mammals	*Oryx dammah*	Scimitar-horned oryx	1996	1, 5	>3000 individuals in captivity; reintroduction is under way.
Plants	*Cyanea truncata*	Punaluu cyanea	1980s	7	Unknown.
	Rhododendron kanehirai		1984	1	Unlikely, as native habitat was completely destroyed.
	Commidendrum rotundifolium	Bastard gumwood	1986	1, 2, 5	Trees have been successfully established.
	Mammillaria glochidiata		Post-1993	5	Unknown.
	Cryosophila williamsii	Lago yojoa palm	1997	1	Plants are in cultivation; reintroduction is possible.
	Mammillaria guillauminiana		1997	4	Unknown.
	Mangifera casturi	Kalimantan mango	1997	1	Unknown; plants are in cultivation.
	Clermontia peleana	'Oha wai	2000	1, 7	Unknown; only one individual remains in cultivation.
	Cyanea superba	Haha	ca. 2000	1, 7	Unknown; plants are in cultivation.
	Cyanea pinnatifida	Haha	2002	1, 7	Unknown; plants are in cultivation.
	Encephalartos nubimontanus		2001	5	In cultivation; could be reintroduced only if area were secure.
	Encephalartos brevifoliolatus		2005	5	In cultivation; could be reintroduced only if area were secure.

*Key to causes of extinction: 1 = habitat loss; 2 = disease; 3 = global warming/pollution; 4 = natural disaster; 5 = exploitation/persecution; 6 = restricted range; 7 = invasive species (not disease); 8 = unknown.

because they face high to extremely high risks of extinction. Listing in a more threatened category implies a higher risk of extinction, but the system recognizes that extinction is a chance process, and that some species in the high-risk categories may persist. This is not because the initial assessment was wrong, but rather because some other factor may delay or prevent the extinction. For example, species with long generation lengths may persist for a long time at very low numbers, or conservation interventions may prevent species from becoming extinct. A recent study has shown that 16 bird species would most likely have gone extinct in the period 1994–2004 if conservation action had not taken place; these include well-known birds like the California condor, *Gymnogyps californicus*, and the Pink Pigeon, *Nesoenas mayeri*.

The 2007 *IUCN Red List* lists over 16,300 species that are threatened with extinction, and this number is rapidly increasing with each update.

Causes of extinction. Humans have played a significant role in the extinction of species prior to historic times, but the true extent of such anthropogenic impacts during the Holocene (the last 11,000 years) remains unclear. However, it is clear that humans are directly or indirectly responsible for most modern extinctions (those over the past 500 years).

Although threats to species are well known, the exact causes of most extinctions are poorly documented. Generally, habitat destruction and degradation, overexploitation, invasive alien species, pollution, and disease have all been major factors. Even when species are relatively well studied, it is often difficult to identify the main cause of extinction, as most species are threatened by more than one process, and these processes often interact in unpredictable ways. Furthermore, the threat process that causes a species to become susceptible to extinction (for example, habitat loss) may be very different from the final process that drives it to extinction (for example, a hurricane).

Threats to species change over time. Invasive species were historically the greatest threat to birds, followed by overexploitation and habitat loss. Today, habitat loss has emerged as the dominant threat to birds, followed by invasive species and overexploitation. This order may change again if predictions about climate change are correct. There are many examples of the effects of climate change on species from around the world (for example, the polar bear, *Ursus maritimus*, was recently listed as Vulnerable because of the impact of predicted losses of Arctic ice), which taken together provide compelling evidence that climate change will be catastrophic for many species.

Why does extinction matter? Biodiversity—the variety of species and their habitats—plays an important role in ecosystem function and in the many services that ecosystems provide. These include nutrient and water cycling, soil formation and retention, resistance against invasive species, plant pollination, climate regulation, and pest and pollution control.

Current rates of species extinction are already significantly higher than background rates and based on current trends are expected to continue rising. This escalating loss of biodiversity therefore has widespread implications for both human and environmental security.

For background information *see* BIODIVERSITY; CLIMATE MODELING; CLIMATE MODIFICATION; ECOSYSTEM; ENDANGERED SPECIES; EXTINCTION (BIOLOGY); GLOBAL CLIMATE CHANGE; MACROEVOLUTION; POPULATION ECOLOGY; SPECIATION in the McGraw-Hill Encyclopedia of Science & Technology.
Craig Hilton-Taylor

Bibliography. J. E. M. Baillie, S. N. Stuart, and C. Hilton-Taylor (eds.), *2004 IUCN Red List of Threatened Species: A Global Species Assessment*, IUCN, Gland, Switzerland/Cambridge, U.K., 2004; D. Jablonski, Background and mass extinctions: The alternation of macroevolutionary regimes, *Science*, 231:129–133, 1986; R. D. E. MacPhee (ed.), *Extinctions in Near Time: Causes, Contexts and Consequences*, Kluwer Academic/Plenum Publishers, New York, 1999; R. Leakey and R. Lewin, *The Sixth Extinction: Patterns of Life and the Future of Mankind*, Doubleday, New York, 1995.

Extratropical cyclone occlusion

The largest fraction of what is called "weather" in the extratropics is delivered by the development and propagation of mid-latitude weather systems. These disturbances, commonly referred to as "lows" and "highs," develop in response to a fundamental instability of the westerlies, known as baroclinic instability, that arises from the permanent but seasonally modulated pole-to-equator temperature contrast. The lows in this endless sequence of disturbances are known as mid-latitude cyclones.

The Norwegian cyclone model. In 1922, the relationship between the three-dimensional (3D) thermal structure and the distribution of clouds and precipitation in these cyclones was first suggested by J. Bjerknes and H. Solberg in their conceptual model of the mid-latitude cyclone, which has come to be known as the Norwegian Cyclone Model (NCM). The genius of this conceptual model was that it placed the instantaneous structure of the cyclone into the context of an identifiable life cycle. According to the NCM, extratropical cyclones developed as infinitesimally small perturbations along a preexisting, globe-girdling discontinuity in temperature known as the polar front, which separated tropical air from polar air. Horizontal temperature advection associated with the initial perturbation converted available potential energy (APE) into eddy kinetic energy (EKE) and led to the amplification of the storm, which, in turn, distorted the polar front into a cold front and a warm front and left a region of homogeneous warm air extending downward to the surface between the two fronts. This region was known as the warm sector, and the air there was known as warm-sector air (**Fig. 1***a*).

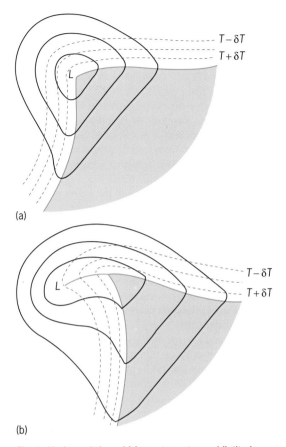

(a)

(b)

Fig. 1. Horizontal view of (*a*) a mature-stage midlatitude cyclone. Solid black lines are sea-level isobars; dashed lines are isotherms; thick solid lines are the cold (dark) and warm (lighter) fronts. The shaded area is the warm sector mentioned in the text. (*b*) An occluded cyclone. The thick solid line connecting the sea-level pressure minimum to the peak of the warm sector is the warm occluded front. Note that the warm occluded front lies in a thermal ridge.

Occlusion. The NCM suggests that as the cyclone reaches its mature phase, with continued intensification and frontal distortion, the fronts "catch up" with one another, instigating a cutting off of the cyclone center from the peak of the warm sector in what was called the "occlusion" of the cyclone. Often a surface boundary remains after occlusion has occurred, and this boundary is referred to as an occluded front. Occluded fronts were thought to occur in two varieties—cold and warm—differentiated based upon whether the air poleward of the warm front was more dense (warm occlusion) or less dense (cold occlusion) than the air behind the cold front. Emphasis on the surface occluded boundary is misplaced and distracting, as consideration of a fully 3D occluded structure offers the most comprehensive insights into the structural and dynamic evolution of the postmature-phase extratropical cyclone. The warm occlusion process described the vertical displacement of warm-sector surface air that resulted from the cold front overtaking, and subsequently ascending, the warm frontal surface. One of the main results of this process was the production of a wedge of warm air aloft, displaced poleward of the surface warm and occluded fronts. The cloudiness and precipitation associated with the development of the

warm occlusion were suggested to result from lifting of warm air ahead of the upper cold front and were consequently distributed to the north and west of the sea-level pressure minimum. As a result of the gradual squeezing of warm air aloft between the two frontal surfaces, the horizontal thermal structure of warm occlusions was characterized by a thermal ridge connecting the peak of the warm sector to the geopotential or sea-level pressure minimum. In nature, this thermal ridge is often manifested as a 1000–500 hPa thickness ridge or as an axis of maximum potential temperature (θ) or equivalent potential temperature (θ_e) in a horizontal cross section and is referred to as the occluded thermal ridge (Fig. 1*b*).

Structure of occluded cyclones. Cyclones are 3D entities. In the development stage, the surface cyclone center, or sea-level pressure minimum, lies to the east of its companion upper-level cyclone center. Such a structure ensures that upward vertical motion, which occurs east of upper-level cyclones, will be located directly above the sea-level pressure minimum. This upward vertical motion not only leads directly to the production of clouds and precipitation, but also evacuates mass from the column of air, thereby lowering the sea-level pressure and intensifying the circulation of the developing storm. As the cyclone approaches the occluded stage, the surface and upper-level cyclones become more vertically stacked. This reduces both the temperature contrasts and vertical wind shear in the vicinity of the surface cyclone. The displacement of the most intense upward vertical motion to the east, away from the sea-level pressure minimum, also inhibits additional development of the surface cyclone center. Hence, occlusion often heralds the beginning of cyclone decay.

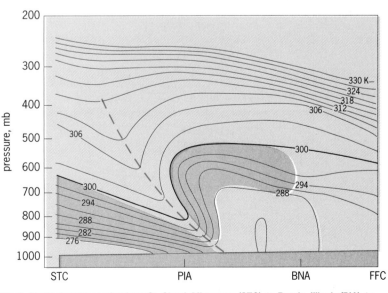

Fig. 2. Vertical cross section from St. Cloud, Minnesota (STC), to Peoria, Illinois (PIA), to Nashville, Tennessee (BNA), to Peachtree, Georgia (FFC), through the occluded thermal ridge of a winter storm over North America on January 19, 1995. Solid lines are the moist isentropes (θ_e), labeled in K and contoured every 3 K. Gray-shaded regions are the warm frontal zone (on the left) and the cold frontal zone (on the right). The thick dashed line is the axis of maximum θ_e mentioned in the text. The thick solid line is the 300-K θ_e isentrope that lies at the warm edge of both the warm and cold frontal zones.

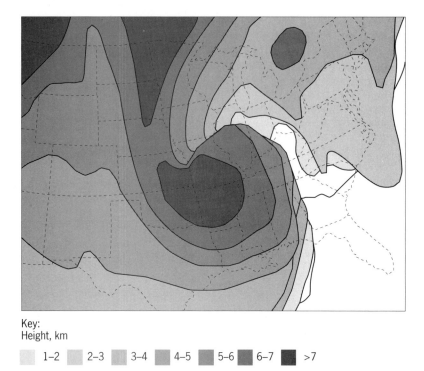

Key:
Height, km

▓ 1–2 ░ 2–3 ▒ 3–4 ▒ 4–5 ▒ 5–6 ▓ 6–7 ■ >7

Fig. 3. Geometric height topography of the 309-K θ_e isosurface from a numerical model simulation of the storm of January 19,1995. Solid lines are height contours, labeled in km and contoured every 1 km, with each region shaded. Darker shading represents a higher elevation. Note the upward-sloping canyon stretching from northern Illinois to northwestern Missouri, as mentioned in the text. Note also the treble-clef shape of the 6- and 7-km elevation contours.

A representative vertical cross section perpendicular to the axis of an occluded thermal ridge reveals the characteristic thermal structure of a warm occlusion, which consists of a poleward-sloping axis of maximum θ_e separating two regions of concentrated temperature contrast, or baroclinicity (**Fig. 2**).

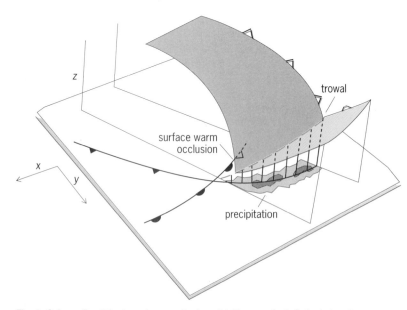

Fig. 4. Schematic of the trowal conceptual model. The gray (color) shaded surface represents the warm edge of the cold (warm) frontal zone. The bold dashed line at the 3D sloping intersection of those two frontal zones lies at the base of the trough of warm air aloft, the trowal. The schematic precipitation in the occluded quadrant of the cyclone lies closer to the projection of the trowal to the surface than to the position of the surface warm occluded front.

The surface warm occluded front is generally analyzed at the location where the axis of maximum θ_e intersects the ground, whereas the base of the warm-air wedge between the two frontal zones (the cold and warm fronts) sits atop their point of intersection.

An alternative means of depicting the warm occluded thermal structure is by considering the topography (geometric or isobaric) of an appropriate θ_e surface selected from a vertical cross section, such as that depicted in Fig. 2. In that section, the 300-K θ_e isentrope lies near the warm edge of both the cold and warm frontal zones of the warm occluded structure. The geometric topography of a similarly selected θ_e surface illustrates the steep slope of the cold frontal zone and the shallower slope of the warm frontal zone (**Fig. 3**). Also evident is a notch in the topography that pokes upward from low levels over central Illinois to northwestern Missouri. This canyon in the 309-K θ_e surface represents the region of overlap of the warm and cold fronts of the warm occluded structure.

With less sophisticated analysis tools at their disposal, scientists at the Canadian Meteorological Service in the 1940s and 1950s observed that the cloudiness and precipitation characteristic of the occluded quadrant of cyclones often occurs in the vicinity of the thermal ridge. This led them to regard the essential structural feature of a warm occlusion to be the wedge of warm air that is lifted aloft ahead of the upper cold front, not the position of the surface occluded front. These studies suggested that the location of a feature called the trowal (*tro*ugh of *w*arm air *al*oft) often bore a closer correspondence to the cloud and precipitation features of occluded cyclones in North America than did the often weak surface warm occluded front. The trowal conceptual model actually represents an extension of the classical occlusion model in that it draws formal attention to the wedge of warm air that is displaced poleward and upward during the occlusion process. The trowal has variously been considered a line connecting the crest of the thermal wave at successive heights and as a "sloping valley of tropical air aloft," or, alternatively, the canyon in the 309-K θ_e surface depicted in Fig. 3. It might best be considered as the 3D sloping intersection of the upper cold frontal portion of the warm occlusion with the warm frontal zone (**Fig. 4**). Compelling evidence supporting these observational findings comes from a number of recent fine-scale numerical modeling studies of occluded cyclones, which have illustrated the structure and thermal evolution of, and airflow through, the occluded quadrant. These studies have identified a coherent airstream, referred to by the author as the "trowal airstream," that originates in the warm-sector boundary layer and ascends cyclonically in the occluded quadrant of the cyclone. This ascent is responsible for the production of the characteristic cloud and precipitation distribution in the occluded quadrant of middle-latitude cyclones.

Dynamics of occluded cyclones. Although extensive investigation of the trowal was made in Canada in the

1940s and 1950s, its relationship to the distribution of clouds and precipitation in the occluded quadrant of cyclones was understood in terms of relative flow along what were considered material frontal surfaces comprising the occluded structure and not as a consequence of some characteristic dynamic process. In 1972, R. M. Morris recognized that the trowal was "a discontinuity in the thermal advection field" and correctly pointed out that such a feature has dynamic significance with respect to the diagnosis of vertical motion.

Another dynamical approach to understanding the forcing for upward vertical motion in the occluded quadrant arises from considering Lagrangian changes in the potential temperature gradient vector. Given the asymmetric, frontal nature of the thermal structure of an extratropical cyclone, its evolution is the integrated result of (1) changes in the vigor of each baroclinic zone and (2) changes in the orientation of the baroclinic zones to one another, both of which occur continuously throughout the cyclone life cycle. Changes in the intensity of baroclinic zones are controlled by a process known as scalar frontogenesis, which modulates the magnitude of the potential temperature gradient and operates on the scale of the individual baroclinic zones. This process is dynamically linked to the production of banded couplets of vertical motion that straddle the baroclinic zone (that is, fronts) and are manifest in the front-parallel bands of clouds and precipitation commonly associated with fronts. Changes in the orientation of the frontal zones with respect to one another are a result of rotation of the potential temperature gradient vector, forced by a process known as rotational frontogenesis. This physical process also has an associated vertical circulation in which couplets of upward and downward vertical motion are distributed along the baroclinic zones.

In 1998, J. Martin showed through analysis of three different cases that the characteristic occluded thermal ridge is produced by differential rotation of the warm and cold frontal baroclinic zones about the cyclone center during the cyclone life cycle. In fact, rotational frontogenesis was found to be the underlying dynamic mechanism responsible for simultaneously creating the characteristic occluded thermal ridge and forcing the majority of the upward vertical air motions within the occluded thermal ridge of the cyclones. In light of this result, the view of the occlusion process, which had been considered to be rooted in traditional, mesoscale frontal dynamics, was reconceptualized as proceeding from the larger synoptic-scale processes that govern the progression of the cyclone through its life cycle.

Latent heat release and the occlusion process. Examination of the tropopause-level distribution of potential vorticity (PV) offers yet another means of identifying a warm occluded structure and is a useful starting point for a discussion of the influence of latent heat release in the evolution of warm occluded thermal structures. The PV distribution at tropopause level often takes on a "treble-clef" shape consisting of an isolated, low-latitude high-PV fea-

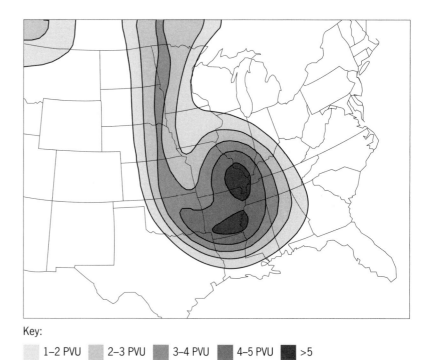

Key:

　1–2 PVU　　2–3 PVU　　3–4 PVU　　4–5 PVU　　>5

Fig. 5. The 9-km potential vorticity from a numerical model simulation of the January 19, 1995, storm. Potential vorticity is labeled, contoured, and shaded in potential vorticity units, PVU. (1 PVU = 10^{-6} m^{2}s^{-1}K kg^{-1}.) Note the notch of low PV that stretches from Wisconsin southward to southeastern Kansas. This is precisely where the trowal is located at this same time (see Fig. 3).

ture that is connected to a high-latitude reservoir of high PV by a rather thin filament of high PV (**Fig. 5**). This same shape appears in the isobaric or geometric height topography of the θ_e surface that lies closest to the warm edges of both the cold and warm frontal zones of the warm occluded thermal structure, particularly above about 6 km (Fig. 3). Such morphological similarity between these two fields is not case specific, but instead is a general consequence of the characteristic thermodynamic structure associated with a positive PV anomaly at the tropopause in which tropospheric isentropes bend upward toward the anomaly, while stratospheric isentropes bow downward toward it (**Fig. 6a**). Thus, regions of large tropopause-level PV sit atop relatively cold columns of air, while relative minima in upper-level PV sit atop relatively warm columns of air. The characteristic thermal structure associated with the horizontal juxtaposition of two upper-level PV anomalies of unequal magnitude, separated by a relative minimum in PV, exhibits an axis of maximum θ beneath the PV minimum that separates two distinct regions of tropospheric baroclinicity (Fig. 6b). This structure precisely depicts a warm occlusion and is a hydrostatic consequence of the treble-clef-shaped upper-level PV structure. Thus, the treble-clef tropopause-level PV distribution, as shown in Fig. 5, is a sufficient condition for asserting the presence of a warm occluded thermal structure in the underlying troposphere.

If one accepts the proposition that the essential structural characteristic of the occluded cyclone is the trowal, then it follows that the occlusion

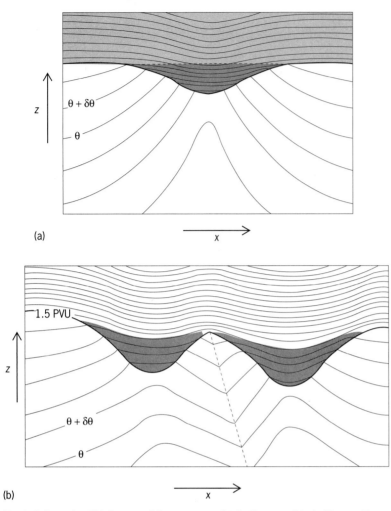

(a)

(b)

Fig. 6. Schematic of (*a*) the potential temperature distribution associated with a positive PV anomaly at the tropopause. Solid lines are isentropes, dark shading is the positive PV anomaly at the tropopause, and light shading represents the stratosphere. Schematic distribution of (*b*) isentropes associated with a horizontal juxtaposition of positive PV anomalies at the troposphere below and the stratosphere above the tropopause, separated by a local minimum in PV. The dashed line is the θ maximum that lies beneath the tropopause PV minimum. Note how it precisely describes the position of the trowal.

tendency to produce an upper tropospheric PV treble clef in that case, the resulting NLHR treble-clef structure was much weaker and slower to develop than the structure present in the FP simulation. They concluded that latent heat release plays a fundamental role in the development of the occluded thermal structure and, therefore, in the occlusion process itself. Taken as a whole, these results served to extend an emerging dynamic and conceptual model of the occlusion process by providing evidence that it depends upon the interaction of a characteristic dynamic forcing for upward vertical motion and the latent heat release that results from that upward vertical motion. In contrast, the cold and warm frontal structures that develop in midlatitude cyclones are a consequence of scalar frontogenesis and are not greatly altered in their essential structural characteristics by the release of latent heat that characterizes them. The fact that the development of an occluded thermal structure appears to depend so intimately on latent heat release further highlights the fundamentally different nature of the occlusion process, as compared to traditional frontogenesis. It appears increasingly clear that further insight into the nature of the occlusion process will arise only when it is no longer viewed as a traditional frontal process.

Occlusion and tropical cyclones. On occasion, upper tropospheric, middle-latitude wave disturbances can migrate sufficiently equatorward to spawn an extratropical surface cyclone in the subtropics. In such a case, the role of latent heat release is enhanced as a consequence of the greater abundance of water vapor, especially over the subtropical ocean. The interaction between enhanced latent heat release and the natural progression toward occlusion can lead to a fairly abrupt erosion of upper-level PV, a reduction in vertical wind shear, and an elimination of baroclinicity. All three of these processes tend to reduce the available potential energy, which, at middle latitudes, results in a gradual but decided weakening of the storm. If the subtropical storm occludes over a warm enough ocean, then the removal of the vertical shear, coupled with enhanced convection, offers the storm an alternative mechanism for continued development of air–sea interaction and wind-induced surface heat exchange (WISHE), an energy transfer mechanism that fuels the growth of tropical cyclones. Thus, the occlusion process likely plays a role in instigating the transformation of extratropical cyclones into tropical cyclones, a process known as tropical transition. Current research seeks to better understand the physical connections between the canonical occlusion process, deep subtropical convection, and tropical transitions. These exciting developments suggest that the process of occlusion, once assumed to be relevant only for mid-latitude storms, probably plays a role in the life cycles of a broader class of cyclonic systems.

For background information *see* BAROCLINIC FIELD; CYCLONE; FRONT; STORM; TROPOSPHERE in the McGraw-Hill Encyclopedia of Science & Technology.

Jonathan E. Martin

process might reasonably be defined as the process(es) by which a cyclone acquires the trowal/treble-clef structure. In 2004, J. Martin and D. Posselt compared companion numerical simulations of the same occluded storm, one run with a full physics (FP) package in the model code, and the other run while withholding the occurrence and associated feedbacks from latent heat release (NLHR). They employed calculations of the diabatic PV tendency and demonstrated that the initial development of the low-PV notch in the treble-clef structure was a direct consequence of the dilution of upper tropospheric PV resulting from latent heat release in the occluded quadrant. As this dilution began to carve out the treble-clef notch, the low-latitude PV feature became progressively more isolated from the higher-latitude reservoir. This cutting-off process, in turn, isolated the tropopause-level cyclonic circulation associated with the low-latitude PV feature. As a consequence, the advection of low PV by the tropopause-level winds contributed to the rapid growth of the PV notch during the late stages of the life cycle. Though they identified a background adiabatic and kinematic

Bibliography. H. B. Bluestein, *Synoptic-Dynamic Meteorology in Midlatitudes*, vol. II, Oxford University Press, 1993; J. R. Holton, *An Introduction to Dynamic Meteorology*, Academic Press, 4th ed., 2004; J. E. Martin, *Mid-Latitude Atmospheric Dynamics: A First Course*, Wiley, 2006.

Femtosecond phenomena

Modern-day technology relies on devices that use new materials and novel structures. Their implementation requires thorough characterization and deep understanding of the underlying physical processes, which involve complex interactions between electrons, magnetic moments, and vibrational degrees of freedom. Very often these processes occur at staggering rates that are beyond the reach of traditional electronic instruments, and approach optical frequencies. Optical spectroscopy, and in particular time-resolved laser spectroscopy, is often the only feasible method to probe them.

Time-resolved laser spectroscopy, in which short laser pulses are used to excite a physical system and probe its response, lends itself to direct investigations of processes that occur on time scales of 10^{-14}–10^{-10} s. It has thus been implemented in studies of diverse phenomena, from molecular motion and chemical reactions, to interactions between quasiparticles and coherent spin dynamics in semiconductors, to magnetization dynamics in magnetic materials. These studies are all driven by a fundamental interest in the highly nonequilibrium conditions that femtosecond laser pulses excite in the different media, as well as by efforts to push the performance of devices to the fundamental limit.

Time-resolved spectroscopy. Various time-resolved techniques have been developed, and they all involve excitation of the sample with several femtosecond pulses at precise time intervals. These techniques are time-resolved adaptations of multibeam experiments with monochromatic lasers, and generally fall into two broad categories: pump-probe spectroscopy and wave-mixing spectroscopy.

In a typical time-resolved pump-probe measurement (**Fig. 1**), the laser beam is split into an intense (pump) beam and a weak (probe) beam. The two beams are then recombined on the sample, after having traversed different path lengths, resulting in a time delay Δt between the arrival of the pump and probe pulses. The incoming pulses are focused on the same spot on the sample, and thus the probe pulse can be used to measure changes in the optical properties that are produced by the pump pulse. (The effect of the probe itself can be safely neglected, since it is very weak.) The time delay between the pulses can be adjusted by simple mechanical means, such as a retroreflecting mirror arrangement on a mobile mirror mount. As the translation stage is moved, the changes of the optical properties can be mapped as a function of Δt, with an accuracy on the order of ~1 femtosecond (determined by the accuracy of the translation stage, which is typically better than

1 micrometer). The temporal resolution of the experiment is therefore limited only by the pulse duration.

The properties that can be measured by the probe include the optical absorption and reflectivity of the sample. Changes of the optical polarization of the probe pulse can also be measured, as a result of being transmitted through the sample or reflected from it. Generally speaking, absorption and reflectivity measurements can provide information on the energy distribution of electronic charge carriers and on vibrational dynamics, while the polarization of the probe pulse is particularly sensitive to the magnetization of the sample. Variations on the basic techniques add frequency resolution and spatial resolution.

In wave-mixing experiments, two or more pump pulses excite the sample at precisely controlled intervals. These pulses interact with the sample and create a nonlinear electronic polarization (that is, a polarization that depends on the electromagnetic fields of all pump pulses), which can in turn be monitored by detecting the faint coherent light that it radiates. Importantly, the rate of decay of the radiated light's intensity provides essential information about decoherence processes.

The measurements rely on sensitive photon detectors and high-quality optical filters and polarizers. Very high sensitivity is essential, since the relative changes of the optical properties induced by the pump pulse (for example, $\Delta R/R$, where R is the sample's reflectivity) are very small, typically on the order of 10^{-3} or less. To reveal these minute changes, a reference measurement (done with the pump beam blocked) must be subtracted from the actual measurement. This is a very challenging task, in view of the multiple sources of noise and background radiation that are practically unavoidable in a typical laboratory environment. However, there is a very elegant solution to this problem, in the form of lock-in detection. A rotating mechanical chopper (Fig. 1) introduces a slow periodic modulation of the pump. The signal received by the detector is also modulated, and the amplitude of this on-off modulation

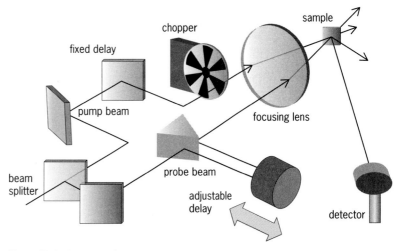

Fig. 1. Typical pump-probe setup.

corresponds to the small differential signal that the researcher is after. The electronics of the lock-in amplifier are designed to detect this small modulation, relying on its stable frequency and phase, which the optical chopper provides as a reference signal.

Ultrafast lasers. Stable femtosecond lasers are the key to femtosecond spectroscopy. A monochromatic laser emits a beam of light with well-defined frequency and wavelength, which correspond to a longitudinal mode of the laser cavity at which the laser gain is highest. (The different longitudinal modes are defined by the condition that the cavity length must be a whole multiple of half the wavelength.) In contrast, a short-pulse laser must have a broad frequency spectrum, a requirement that is related to Heisenberg's uncertainty principle. This implies that a large number of cavity modes (millions of them) must emit at the same time. Not only that, the modes have to be locked together in phase, for if their phases were random, the result would be broadband ("white") noise. To achieve the required bandwidth, femtosecond lasers rely on lasing media that have high gain over a very wide wavelength range, and cavity mirrors with special, broadband coatings. To ensure that all the modes emit with the same phase, a careful balancing of the chromatic dispersion of different components in the laser cavity is required and usually achieved by combinations of adjustable prisms and diffraction gratings or by cleverly designed multilayer cavity mirror coatings ("chirped mirrors").

Early short-pulse lasers used organic dyes as gain media, and mode locking was achieved by synchronously pumping them with "Q-switched" pulsed lasers, which relied on active gain modulation to produce long (typically nanosecond) pulses. The pump pulses induced population inversion in the dye, and once gain was obtained, the dye laser radiated the stored energy in the form of short (typically picosecond or subpicosecond) pulses. Modern lasers take advantage of solid crystal gain media, which are far easier to handle. The most common types are Ti:sapphire (titanium-doped sapphire), which can generate pulses as short as 8 fs at a wavelength of ~800 nm, and erbium- or ytterbium-doped glass fibers, which typically emit 100-fs pulses at ~1500 nm or ~1000 nm. Pumping is done with simpler, monochromatic solid state lasers, such as a frequency-doubled Nd:YAG (neodymium-doped yttrium aluminum garnet) laser or a semiconductor diode laser, and mode locking occurs in response to a slight perturbation, such as tapping on one of the cavity mirrors, or even spontaneously because of noise. The perturbation produces a brief fluctuation of light intensity, which is then strongly amplified as it bounces off the cavity mirrors and goes back through the nonlinear gain medium (or another nonlinear component). This process eventually forms a stable pulse that travels back and forth through the laser cavity, resulting in a periodic pulse train at the output, the repetition rate of which is determined by the cavity length.

The only apparent drawback of these unique lasers is the relatively limited range of wavelengths at which they can operate. This problem can be circumvented by using frequency conversion techniques, such as optical parametric amplification and harmonic generation in nonlinear crystals.

Example. An illustration of the power of ultrafast spectroscopy is provided by a pump-probe experiment on a Ni/FeF$_2$ (nickel/iron fluoride) exchange bias bilayer (**Fig. 2**). Similar bilayers have already found extensive use in magnetic recording devices, but the physics of the interlayer coupling is still poorly understood. The bilayer consists of a thin ferromagnetic layer (Ni) grown on top of an antiferromagnetic layer (FeF$_2$). The atomic contact results in a strong anisotropy of the exchange interaction at the interface between the layers, meaning that the FeF$_2$ defines a preferred direction for the nickel magnetization. When the sample is cooled below the Néel temperature of the FeF$_2$, the antiferromagnetic spins order themselves such that the spins at the interface are in the energetically favorable orientation with respect to the nickel's magnetization. Once this configuration is "frozen," a strong magnetic field in the opposite direction is needed to reverse the nickel magnetization, and when the field is removed, the ferromagnet reverts to the favorable configuration. The result is a shifted magnetic hysteresis loop (Fig. 2a).

When a femtosecond pump pulse excites the bilayer, the temperature of the Ni/FeF$_2$ interface is elevated close to the Néel temperature of the antiferromagnet. Subsequently, the interface cools. Depending on the pump intensity, the excitation may result in a sudden weakening or a complete quenching of the coupling between the layers. The magnetic response is monitored by measuring the polarization rotation of the probe pulse as it is reflected from the sample.

The response of the bilayer to the optical excitation is expected to depend on the relative orientations of the layers, and therefore on the applied magnetic field. If the nickel magnetization and the field initially point in different directions, the pump-induced weakening of the interlayer coupling would be expected to result in a precession of the nickel magnetization around the field vector. Conversely, if the nickel magnetization and the magnetic field are initially parallel, the former would not be expected to react to the pump pulse. The experiment, however, shows a very pronounced precession at high, reverse magnetic fields, when the nickel is supposed to be aligned parallel to the field (Fig. 2b and c). This shows that even when the magnetic field is strong enough to overcome the exchange bias, the magnetization vector is in fact not perfectly aligned with the field (Fig. 2d–f). The experiment also shows that when the pump intensity is high enough to completely quench the interlayer coupling, it is actually the FeF$_2$ which responds (rather than the nickel) and reorders itself, resulting in a reversal of the exchange bias. This type of information is essential for

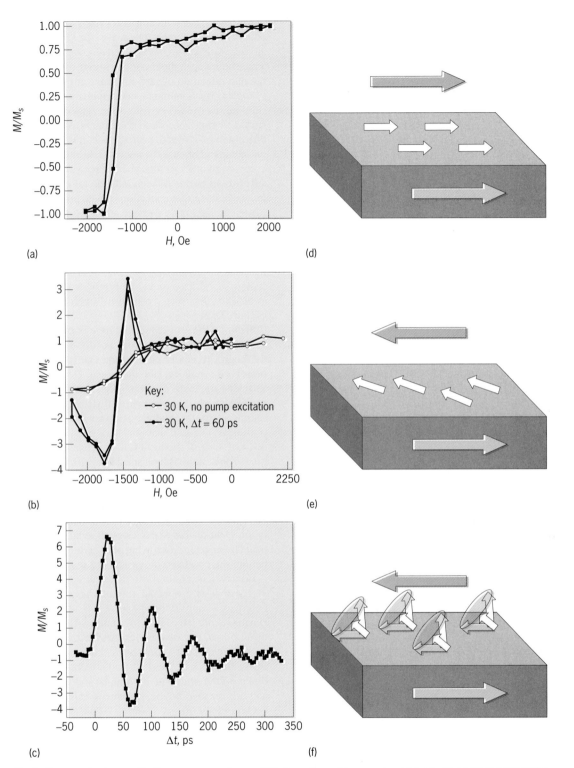

Fig. 2. Time-resolved magnetization measurements on a Ni/FeF$_2$ exchange bias sample, obtained with a typical pump-probe setup, using ~60-fs pulses at 800 nm from an amplified Ti:sapphire laser. (*a*) Typical exchange bias curve, showing a shift of the nickel magnetization reversal to negative fields. (The ratio of magnetization to saturation magnetization, M/M_s, is plotted against the magnetic field H, measured in oersteds; 1 Oe = 79.6 amperes per meter.) (*b*) Comparison of a magnetization measurement as a function of field, taken at a time delay of 60 ps, with a reference measurement taken with the pump beam blocked, showing a strong differential signal at high negative fields. (*c*) The time-resolved signal, measured at a field of −1560 Oe and temperature of 30 K, showing oscillations due to a precession of the nickel magnetization. (*d–f*) Changes in the orientation of the nickel spins as a function of the magnetic field and in response to the laser excitation. The top arrows in *d–f* represent the magnetic field, the front-side arrows show the FeF$_2$ anisotropy, and the smaller white arrows represent the nickel magnetization. (*d*) When the magnetic field is parallel to the FeF$_2$ anisotropy, the nickel spins are also parallel to the field. (*e*) When the magnetic field is strong enough to overcome the FeF$_2$ anisotropy and reverse the nickel spins, they slightly cant towards the FeF$_2$ anisotropy field. (*f*) When heated with a femtosecond pump pulse, the nickel spins in *e* decouple from the FeF$_2$ and start precessing around the magnetic field vector.

understanding the physics of the interlayer coupling, and cannot be inferred from traditional "slow" measurements or from numerical simulations.

For background information *see* ANTIFERRO-MAGNETISM; EXCHANGE INTERACTION; FERROMAG-NETISM; LASER; LASER SPECTROSCOPY; MAGNETIC MATERIALS; NONLINEAR OPTICAL DEVICES; OPTICAL PULSES; ULTRAFAST MOLECULAR PROCESSES in the McGraw-Hill Encyclopedia of Science & Technology.
Shimshon Bar-Ad

Bibliography. P. Corkum et al. (eds.), *Ultrafast Phenomena XV*, Proceedings of the 15th International Conference, Pacific Grove, CA, July 30–August 4, 2006, Springer Ser. Chem. Phys., vol. 88, 2007; M. E. Fermann, A. Galvanauskas, and G. Sucha (eds.), *Ultrafast Lasers: Technology and Applications*, CRC, 2002; J.-M. Hopkins and W. Sibbett, Ultrashort-pulse lasers: big payoffs in a flash, *Sci. Amer.*, 283(3):72, September 2000; H. Kapteyn and M. Murnane, Ultrashort light pulses: life in the fast lane, *Phys. World*, 12(1):31–35, January 1999; A. V. Kimel et al., Femtosecond opto-magnetism: ultrafast laser manipulation of magnetic materials, *Laser & Photon. Rev.*, 1(3):275–287, 2007; C. Rulliere, *Femtosecond Laser Pulses: Principles And Experiments*, Springer, 2004; A. H. Zewail, *Femtochemistry: Atomic-scale dynamics of the chemical bond Using Ultrafast Lasers*, Nobel Lecture, December 8, 1999.

Fossil hominids from Dmanisi

Archeological excavations of the medieval town of Dmanisi, Georgia, starting in the early 1980s revealed the presence of fossilized fauna and stone tools of prehistoric antiquity (**Fig. 1**). These fossils, including the remains of at least five early human (hominid) individuals, have been dated to the Plio-Pleistocene boundary (1.7–1.8 million years ago [Ma]). The discovery of human ancestors outside of Africa at such an early date was an unexpected discovery that challenged previously held notions about the nature of the first migration of early humans out of Africa.

To date, the hominid fossils include five crania (the skull minus the lower jaw), four mandibles (lower jaw), and numerous skeletal elements. Although the fifth cranium has yet to be described, three of the crania and mandibles can be matched, resulting in three complete skulls. Geological evidence supports a relatively brief period of accumulation for the hominid fossils, making this collection of well-preserved fossils, including both a juvenile and a toothless (likely older) individual, an unprecedented glimpse into the paleobiology of a single population of early humans. The simple stone tools associated with these hominids were made from locally available basalt (a type of volcanic rock) and resemble those found at African sites, such as Olduvai Gorge, which date back to 2.4 Ma.

Skull. Of the four crania that have been described to date, the cranial vault (the portion surrounding the brain), face, and parts of the cranial base were preserved for all but one individual, D2280, which is lacking the facial skeleton (**Fig. 2**). D2700 is a juvenile individual, probably less than 13 years of age at time of death, whereas D3444 is an edentulous (toothless) cranium that may have been an older individual. One of the most notable characteristics of the Dmanisi skulls is their small size. The brain size of these individuals (600–775 cm^3) is most comparable to *Homo habilis,* one of the earliest species of our own genus *Homo*. Additionally, some aspects of the Dmanisi cranial morphology are similar to those seen in the earliest *Homo* species, such as strong postorbital constriction (narrowing of the cranium directly behind the eyes) and the profile of the D2700 face.

However, beginning with the earliest descriptions of these fossils, specific resemblances were also noted to a more derived species, *H. erectus*. The Dmanisi skulls have long, low crania with prominent brow ridges, occipital bones that are tightly angled in the midline, bony tori (thickened protuberances) across the back of the skull, and greatest width low down on the skull (Fig. 2). Angular tori are present, and at least two of the specimens exhibit midline keeling of the parietal bones (D2280 and D3444). This particular combination of traits is diagnostic for *H. erectus* and is not found in earlier *Homo* species, including *H. habilis*.

A partial mandible recovered in 1991, D211, was the first evidence of hominid presence at the site of Dmanisi. Although the four crania are similar in size to one another, the four mandibles vary greatly in size, mainly because of the presence of the large D2600 specimen. With the exception of D2600, each of the other three mandibles has been associated with a particular cranium. Many dimensions and characteristics of the mandible are similar in both *H. habilis* and *H. erectus*, but researchers have generally agreed that the presence of certain derived characters, such as the mental trigone (chin)

Fig. 1. Overview of the main excavation areas at the site of Dmanisi, Georgia. Surrounding the site are the remains of the medieval architecture that first attracted scholars to the region. Although many of the fossil hominid remains were located in the area under the tarp (Area 2), fossils have also been recovered from other parts of the site, such as the level area to the left of the tarp (Area 1). (*Photograph by Guram Tsibakhashvili; courtesy of David Lordkipanidze of the Georgian State Museum*)

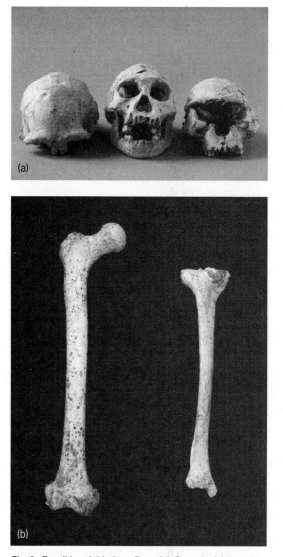

(a)

(b)

Fig. 2. Fossil hominids from Dmanisi, Georgia. (a) Anterior view of three of the Dmanisi crania and one mandible. The specimens are (from left to right) D2280, D2700/D2735 (juvenile cranium and mandible), and D2282. Despite some variation, all crania have marked supraorbital tori (brow ridges) above the orbits, receding foreheads, and marked constriction behind the orbits (postorbital constriction). (b) Anterior view of the D4167 femur (left) and D3901 tibia (right) from the same individual. (*Photograph by Guram Tsibakhashvili; courtesy of David Lordkipanidze of the Georgian State Museum*)

morphology observed in D211, align these specimens more closely with the latter species. If all of the fossil hominids found at Dmanisi are ultimately subsumed into a single species, the size discrepancy among the mandibles may indicate greater levels of sexual dimorphism (diagnostic morphological differences between the sexes) in early *Homo* than previously recognized.

Skeletal morphology. The Dmanisi postcrania (the skeleton minus the skull) have been described as embodying a mix of primitive and more modern humanlike characters. The postcranial sample comprises at least four individuals, one juvenile and three adults (two of which are represented only by ankle and foot bones). One of the adults is larger than the other two and may be associated with the large mandible men-

tioned previously. In addition to small bodies and brains, specific aspects of the shoulder, upper arm, and ankle are more apelike than modern humanlike. However, the body proportions and overall lower limb anatomy are similar to modern humans. Some features, such as the constricted medullary canal (hollow internal space) of the femur shaft, resemble postcranial evidence from *H. erectus* elsewhere. Generally, the small-bodied hominids from Dmanisi appear to have been accomplished bipeds (walking on two legs), although their gait may not have been identical to that of modern humans.

Classification. Although there is widespread consensus that the Dmanisi hominids were part of the *Homo* lineage, there is less agreement regarding their classification at the species level. At least two species are recognized from the period circa 1.7 Ma: the earlier and more primitive *H. habilis* and the more derived *H. erectus*. Both *H. habilis* and *H. erectus* fossils are known from Africa, but *H. erectus* is the earliest hominid species to be distributed outside of Africa—specifically, at sites in China and Java. The oldest *H. erectus* fossils are from sites in East Africa and are approximately the same age as, or slightly older than, the Dmanisi specimens (**Fig. 3**). The features shared in common between the Dmanisi hominids and *H. habilis* appear to be primitive characters and do not necessarily indicate a special relationship between these groups. Most scientists attribute the Dmanisi fossils to *H. erectus* because of the presence of derived features found in *H. erectus* but not in *H. habilis* (as discussed earlier), as well as strong similarities in cranial shape. A separate species, *H. ergaster,* is sometimes recognized for the earlier African *H. erectus* fossils, in which case the Georgian sample may belong in *H. ergaster*. A new species, *H. georgicus,* was also erected to accommodate some or all of the Dmanisi fossils, although the evidence that these fossils are sufficiently distinct to be considered a new species is not convincing.

Evolutionary implications. The recovery of the Dmanisi hominid fossils has reinvigorated debate regarding the initial hominid migration out of Africa and the ability of early *Homo* to adapt to new environments. Before the Dmanisi discoveries, the best evidence for skeletal anatomy in *H. erectus* was the relatively complete juvenile specimen from Kenya nicknamed Turkana Boy (KNM-WT 15000). Turkana Boy is the earliest evidence for modern humanlike body proportions at 1.5 Ma. This suggested that the first hominids that dispersed from Africa may have had larger bodies and relatively longer legs than earlier species of *Homo*. Scientists argued that this body plan may have conferred an advantage for long-distance travel; large body size may also be related to a greater reliance on meat in the diet and a concomitant increase in home range size. This shift in diet and ranging behavior was posited as the impetus for the initial dispersal from Africa. The small-bodied Dmanisi fossils raise the possibility that either smaller body size evolved postdispersal as an adaptation to local conditions in the Caucasus or that a hominid similar in size to *H. habilis* was the first to migrate out

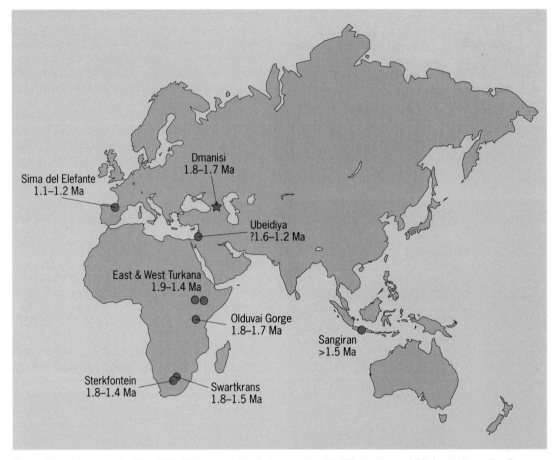

Fig. 3. Map of the most significant Plio-Pleistocene sites between 1.9 and 1.5 Ma that have yielded early *Homo* fossils, including Dmanisi, Georgia. The approximate dates associated with each site are also indicated. Although it is significantly younger, the site of Sima del Elefante in Spain is also shown because it represents the oldest evidence of hominid presence in Europe.

of Africa. The stone tools recovered from the Dmanisi site also argue against the notion that the initial hominid migration out of Africa was mediated by advances in stone tool technology. Furthermore, the fauna found at Dmanisi do not support a broader mammalian migration out of Africa and into this region at the Plio-Pleistocene boundary.

The earliest definitive *H. erectus* fossils from East Africa are 1.8 million years old, although some fragments assigned to *H. erectus* are dated to 1.9 Ma (Fig. 3). The oldest fossils in Java have been dated to >1.5 Ma and possibly 1.8 Ma, but the latter date remains controversial. In contrast, the oldest fossils described from Europe (at the site of Sima del Elefante) are just slightly older than 1 million years old and have been provisionally assigned to *H. antecessor*. Before the discoveries at Dmanisi, *H. erectus* was also not known at higher latitudes until after 1 Ma (in China). This suggests that *H. erectus* migrated out of Africa and into more temperate climates very soon after its initial appearance. This chronology may explain the mosaic of primitive *habilis*-like and derived *erectus*-like characters. If the Dmanisi population was close to the stem of *H. erectus*, it may well have retained some primitive characters present in its (possibly *habilis*-like) ancestor.

Whether the *H. erectus* population at Dmanisi was ancestral to later Asian *H. erectus* or an evo-

lutionary side branch is unclear, although certain detailed resemblances have been noted between the Dmanisi and Asian *H. erectus* populations. Future discoveries of hominid fossils from the Horn of Africa and western Asia regions may further clarify the timing and route of the initial hominid dispersal from Africa. What is clear is that living at higher latitudes would also have entailed adapting to a more temperate environment relative to the more tropical climes to which the ancestors of the Dmanisi hominids were adapted. This is further evidence of the adaptive success of *H. erectus*, a species with a geographic range second only to *H. sapiens* and a time span of nearly 1.8 million years.

For background information *see* ARCHEOLOGY; DATING METHODS; EARLY MODERN HUMANS; FOSSIL HUMANS; FOSSIL PRIMATES; MOLECULAR ANTHROPOLOGY; PHYSICAL ANTHROPOLOGY; PREHISTORIC TECHNOLOGY; PRIMATES; SKELETAL SYSTEM in the McGraw-Hill Encyclopedia of Science & Technology.

Karen Baab

Bibliography. E. Delson et al., *Encyclopedia of Human Evolution and Prehistory*, 2d ed., Garland Publishing, New York, 2000; D. Johanson and B. Edgar, *From Lucy to Language: Revised, Updated and Expanded*, Simon & Schuster, New York, 2006; K. Wong, Stranger in a new land, *Sci. Amer.*, 289(5):74–83, 2003.

Fungal sources for new drug discovery

Fungi play important roles in the environment and have the ability to exploit almost all niches, both natural and human-made. Not surprisingly, their properties are of interest and are being harnessed for use in medicine, food, energy, and industry. Fungi are producers in industrial applications, and are used in pest management programs that harness natural enemies to control pests and diseases. Crop wastes can be degraded and converted into compost or into useful products through the use of specifically formulated microbial communities. They are being used increasingly in bioremediation of industrial waste and are sources of active and marketable molecules. This is not new—fungi, including the yeasts, have been used in Chinese medicine for thousands of years. The best-known examples of the use of fungi are the production of penicillin from *Penicillium* and the use of *Saccharomyces cerevisiae* to brew alcoholic drinks. Humans have used fungi for the production of useful chemicals such as enzymes, biocides, drugs, and dyes; in processes such as biotransformations and biofuel production; and as food. Fungi will provide more solutions to the problems faced by humans with regard to the environment, health (see **table**), agriculture, and the economy at a time when natural resources are being depleted and alternative sources of energy and foods are being sought.

Biopharmaceuticals are valued at $41 billion (U.S. dollars) in the global market, growing at a rate of 21% over the past 5 years. It is considered that the total pharmaceutical market could easily reach $100 billion by the end of the decade. There are huge revenues from fungal-derived drugs—for example, cyclosporin ($1.4 billion) and amoxicillin ($1.7 billion).

Organism potential. More than 100,000 species of fungi have been described, with around 1000 being described annually, but only about 15% of these are available from collections. Unfortunately, many cannot be grown in culture, and others grow poorly, expressing only a limited part of their life cycle—for example, perhaps growing only in the vegetative state, producing mycelia and not developing their anamorph (asexual) or teleomorph (sexual) stages.

More than 1.3 million strains are held in the collections registered with the World Data Centre for Microorganisms, of which more than 400,000 are fungi. There is a long way to go before material will be available for all of the described species and the estimated 1.4 million yet to be described. Targeted isolation programs are needed to make inroads into this enormous task. Additionally, screening for the production of natural products has been done on only a small fraction of the described fungal species. The use of molecular techniques and high-throughput screening increases the capacity to realize fungal potential. Often, screening takes a snapshot of only part of an organism's chemistry, and expression of genes can be strain-specific. Indeed, in some organisms certain genes may be suppressed, whereas others may have multiple gene copies. To ensure that an extract produced for screening can access its full potential, it is essential to have thorough knowledge of the fungus's life cycle, its interaction with its environment, its host, and the organisms in its ecosystem.

Significant effort is being put into ecosystem-based approaches, especially where little-explored areas or extreme environments are involved. The Iwokrama program in central Guyana was established to demonstrate how tropical forest biodiversity may be conserved and sustainably utilized for ecological, social, and economic benefits. Almost 2500 endophytic fungi were isolated from 12 tree species. Of these, 332 were selected and 2800 extracts produced. Eighty-four of the isolates were found to have potent anti-insect activities, 14 exhibited potent antifungal activities, and 13 possessed potent antibacterial activities. This provided a tremendous success rate of 1 in 3, although not all will be new active molecules and quite often activity is a result of a combination of molecules in the extract, thereby making the result not easily exploited. The keys to success in this case were

1. Selection of a unique, unexplored ecosystem.

2. Targeted organisms with unusual properties.

3. Sampling techniques and analysis carried out by specially trained personnel.

4. Selection of characterization techniques as deduplicating tools (that is, tools that identify and remove duplicate entries from a database).

Some examples of natural products from fungi

Fungus	Product	Use
Aspergillus terreus	Mevinolin	Cholesterol lowering
Micromonospora	Dynemicin	Antitumor; antibacterial
Tolypocladium inflatum	Cyclosporin	Immunosuppressant
Penicillium notatum	Penicillin	Antibiotic
Penicillium	Amoxicillin	Antibiotic
Acremonium	Ceftriaxone	Antibiotic
Aspergillus terreus: Monascus ruber	Lovastatin	Cholesterol lowering
Aspergillus terreus: Monascus ruber	Pravastatin	Cholesterol lowering
Claviceps purpurea	Ergotamine	Facilitates childbirth process; postpartum hemorrhage control
Aspergillus terreus: Monascus ruber	Simvastatin	Cholesterol lowering
Cephalosporium	Cephalosporin	Antibiotic
Penicillium griseofulvin	Griseofulvin	Fungistatic, often used to treat ringworm infections

The properties of a fungus are reflected by the habitat in which it grows. Therefore, taking isolates from a wide variety of ecosystems increases the chances of discovering new products. Organism groups that are most likely to provide new compounds include endophytic Basidiomycetes, Ascomycetes, and their anamorphs (asexual state).

The chances of finding active molecules vary enormously, although ensuring a targeted and structured approach, utilizing appropriate technologies, and employing the right skills are imperative for producing rewarding results. From a survey of recently discovered pharmacologically active metabolites from tropical fungi, 13,000 strains of fungi were screened for use as antifungal, antibacterial, antiviral, insecticidal, and antihelminthic agents, or for their ability to fight cancer, diabetes mellitus, inflammation, and other endocrinological conditions. On average, 13% yielded active compounds. However, it is estimated that there is only a 1-in-250,000 chance for an unknown chemical to reach the market; with active compounds, this increases to about 1 in 5000–10,000. Chances of discovering an active molecule are enhanced when fungal metabolism can be manipulated. By linking this to knowledge of ecosystems, targeted isolation programs can give access to the most promising organisms.

Search for active molecules and current activities. Assay techniques are derived from a wide range of different biological disciplines, including biochemistry, bacteriology, microbiology, and ecology, to provide anatomical, physiological, and biochemical data. Many of the methods are simple and can be applied with minimal facilities; others require substantial investment in technology. Media-based tests can be used to determine single activities (for example, urease activity or citric acid production), and methods can be used to investigate multiple activities (for example, chromatography to detect a range of metabolites). Simply inoculating two different microbes on a culture plate to study if they are antagonistic can provide an indication of biological activity. Many tests are commercially available in easy-to-use formats that allow standardization between laboratories. Assays have been designed to identify targets against many common conditions, including neurological diseases, cancers, and infection, and may include utilization of microbial cells, cell lines, nucleic acids, enzymes, and toxicity, as well as protein binding and specific chemical assays.

Metabolic extracts can be readily generated from microbial fermentations. Storage of these in extract libraries will make them easily available for use in different high-throughput screening assays. Chemical libraries are normally built up by preparing hydrophilic and hydrophobic extracts grown in a variety of media in liquid or solid-phase systems. The extracts are created by various means, usually excluding large molecules that may interfere with some high-throughput assays. Further improvement has centered on fractionating of extracts and the purification and characterization of compounds. Techniques such as MALDI-TOF (matrix-assisted laser desorption/ionization time-of-flight) mass spectroscopy are providing comparison data that can deduplicate and help focus on new chemical entities. The development of genomic screening approaches takes discovery to another level. Genomics has enabled the expansion of drug targets, facilitating a fundamental shift from direct screening programs toward rational, target-based strategies. The development of target- and structure-directed screens is now producing higher probabilities of success. Recently, molecular techniques using PCR polymerase chain reaction (PCR), genomic library construction, and heterologous expression have provided an alternative approach to begin exploring the diversity of polyketide biosynthetic pathways in lichens. The techniques can be expanded to cover other pathway types and integrated with conventional culture collection-based screening to provide a comprehensive search for novel chemical entities in these organisms, thereby allowing a paradigm shift in bioprospecting the impact of bioinformatics and the associated technologies. Research into all of these areas is continuing apace, from small university-based research groups and start-up biotech companies to established large pharmaceutical companies. More commonly, partnerships are being created between organizations in order to share expertise, technology, and facilities to improve the likelihood of finding new fungal-derived drugs.

International obligations. Exploitation of biological materials must be in compliance with conventions, treaties, and law, for example, the Convention on Biological Diversity (CBD). The CBD requires that "prior informed consent" (PIC) be obtained in the country where organisms are to be collected. Terms on which any benefits will be shared must be agreed. The benefits may be monetary, but they could be information, technology transfer, or training. If the organism is passed to a third party, it must be under terms agreed by the country of origin. This entails material transfer agreements between supplier and recipient to ensure benefit sharing with, at least, the country of origin. Access and benefit-sharing rules must be followed, and signatory countries to the CBD have agreed on a voluntary code of practice, the Bonn Guidelines.

Outlook. Fungi are found in most ecosystems, in huge varieties and with interesting chemistries. There is a history of their properties being harnessed, and they present great promise. With new technology and tools, as well as better assays to analyze their large numbers and variety, fungi will continue to be an excellent source of natural products.

For background information *see* FUNGAL BIOTECHNOLOGY; FUNGI; GENOMICS; MEDICAL MYCOLOGY; MYCOLOGY; PHARMACEUTICAL CHEMISTRY; PHARMACOGNOSY; PHARMACOLOGY; PHARMACY in the McGraw-Hill Encyclopedia of Science & Technology.

David Smith; Matthew J. Ryan

Bibliography. G. F. Bills et al., Recent and future discoveries of pharmacologically active metabolites from tropical fungi, pp. 165–194, in R. Watling et al. (eds.), *Tropical Mycology*, vol. 2:

Micromycetes, CAB International, Wallingford, UK, 2002; A. T. Bull, *Microbial Diversity and Bioprospecting*, ASM Press, Washington, DC, 2004; P. F. Cannon and P. M. Kirk, *Fungal Families of the World*, CAB International, Wallingford, UK, 2007; D. L. Hawksworth, The magnitude of fungal diversity: The 1.5 million species estimate revisited, *Mycol. Res.*, 105:1422–1432, 2001; J. Kelley et al., The Iwokrama programme: An approach to the sustainable exploitation of genetic resources, pp. 185–195, in N. Lima and D. Smith (eds.), *Biological Resource Centres and the Use of Microbes*, Micotecca da Universidade do Minho, Portugal, 2003; K. ten Kate and S. A. Laird, *The Commercial Use of Biodiversity: Access to Genetic Resources and Benefit Sharing*, Earthscan Publications, London, 1999; M. V. Tejesvi et al., New hopes from endophytic fungal secondary metabolites, *Bol. Soc. Quím. de México*, 1:19–26, 2007; S. K. Wrigley et al., *Biodiversity: New Leads for Pharmaceutical and Agrochemical Industries*, Royal Society for Chemistry, Cambridge, UK, 2000.

Furfurylated wood

Wood is a useful and renewable material. However, it is susceptible to biodegradation (fungal decay and insect and marine borer attack), particularly where the wood is wet. Furfurylated wood is lumber and other wood products modified with a polymer made from furfuryl alcohol ($C_5H_6O_2$). Furfuryl alcohol is a liquid made from plant waste. Furfurylation is accomplished by impregnating wood with a mixture of furfuryl alcohol and catalyst and then heating to cause polymerization. The purpose of furfurylation is to improve physical and biological properties such as resistance to moisture and decay. In addition, furfurylated wood is nontoxic.

Process. Furfurylated wood was first developed in the 1950s. Later technologies for making furfurylated wood were developed in Canada and Sweden in the 1990s, which led to its commercial production in Norway in the early twenty-first century.

Furfural ($C_5H_4O_2$) is an aromatic aldehyde made by hydrolyzing pentoses (5-carbon sugars) in an aqueous, acid solution. The industrial sources of the pentoses used in this process are agricultural residues such as sugarcane bagasse (the fibrous plant residue remaining after the extraction of juice from the crushed stalks of sugarcane) and corncobs. Furfural is converted to furfuryl alcohol by hydrogenation. Furfuryl alcohol is a low-viscosity liquid that can be impregnated into wood. It reacts strongly with itself at low (acidic) pH to form a dark-colored, highly resistant, hard, cross-linked polymer.

Furfurylated wood made during the 1950s used zinc chloride as the polymerization catalyst. This catalyst polymerized the furfuryl alcohol well, but tended to separate from solution as it moved into the wood, producing a nonuniform material in lumber-sized products. Zinc chloride also noticeably reduced the wood's mechanical properties. In the

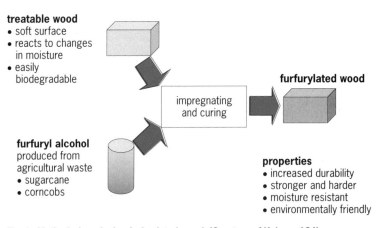

treatable wood
- soft surface
- reacts to changes in moisture
- easily biodegradable

impregnating and curing

furfurylated wood

furfuryl alcohol
produced from agricultural waste
- sugarcane
- corncobs

properties
- increased durability
- stronger and harder
- moisture resistant
- environmentally friendly

Fig. 1. Method of producing furfurylated wood. (*Courtesy of Kebony ASA*)

1990s, alternative catalyst systems using organic anhydrides and acids were developed in Canada and Sweden. Furfuryl alcohol–treating solutions made with these catalysts (in particular, maleic anhydride and citric acid) penetrated wood uniformly and did not degrade the wood.

The new catalyst systems also led to methods for making useful, waterborne furfurylating solutions. This made it possible for the first time to control the polymer loading (the weight percent or concentration of polymer based on the original wood weight). It was found that lower loadings also provided biodeterioration resistance. This led to a whole new opportunity for furfurylated wood, as it could compete in the preservative-treated wood market.

Wood is furfurylated by impregnating wood with monomeric catalyzed furfuryl alcohol either neat (undiluted) or in a water solution and then polymerizing the furfuryl alcohol (**Fig. 1**). To treat lumber, a vacuum and pressure process is used. Curing is done by heating to approximately 100°C (212°F).

When neat furfuryl alcohol polymerizes, about 20% of the furfuryl alcohol is lost in the water of the reaction. Hence, the polymer yield is approximately 80% of the original weight. Neat furfuryl alcohol impregnated into wood produces high polymer loadings. Diluting furfuryl alcohol with water allows the loading to be controlled by the solution concentration. The water distributes the furfuryl alcohol within the wood and then evaporates. The furfuryl alcohol remaining is polymerized, resulting in a uniform, low polymer concentration in the wood.

The polymer made from furfuryl alcohol is dark brown in color, thus darkening the wood. The darkness imparted to the wood depends upon the loading of the wood. Since cell walls are treated, this darkness affects the wood substance as well as any pores filled with polymer. As a result of the polymerized color, furfurylated wood can range in color from slightly browner than untreated wood to very dark brown. Heavily loaded furfurylated wood can even appear to be black.

Furfuryl alcohol swells the wood cell wall because of its small molecular size (allowing impregnation into the wood cell walls) and its hydroxyl group

Fig. 2. Moderately loaded furfurylated maple kitchen countertop. (*Courtesy of Kebony ASA*)

(which swells cellulose). Furfuryl alcohol cures to a hard polymer in the wood, causing the furfurylated wood to remain permanently swollen. As a result, subsequent moisture-induced shrinkage and swelling are greatly reduced. Recent studies have shown that the furfuryl alcohol also covalently bonds to the wood cell wall during polymerization. Therefore, furfurylation is a true chemical modification of the wood.

A polymerizable vinyl monomer such as styrene can also be used as a diluent for furfuryl alcohol. When a styrene–furfuryl alcohol solution is impregnated into wood, the furfuryl alcohol enters and swells the cell wall while the styrene remains in the cell cavity. Upon curing, the styrene and the furfuryl alcohol polymerize. The product made in this way has attributes of both furfurylated wood and vinyl polymer wood composites. It is decay-resistant, highly durable, and strong.

Uses. Because the high loading of furfuryl alcohol polymer in wood produces a very dark, hard material similar to heavy tropical hardwoods, it can be used as a substitute for these woods. This material is highly resistant to biological, thermal, and chemical breakdown. It is nearly as dense as water. Its properties make it useful for high wear or severe conditions. For example, it has been used for knife handles and flooring.

Wood with a moderate loading of furfuryl alcohol polymer still resists biodegradation and is suitable for ground contact or marine use. Its rich, brown color is attractive, making it possible for light-colored woods to simulate dark, expensive woods. Its moisture resistance makes it suitable for applications that can be flooded with water from time to time, such as floors, furniture surfaces, and countertops (**Fig. 2**). Low loading gives sufficient biodegradation protection for aboveground outdoor uses.

Furfurylated wood has greater hardness and elastic and rupture moduli (mechanical properties) than untreated wood. These properties increase with polymer loading of the wood. However, brittleness is the main problem of furfurylated wood. In those instances where the properties of furfurylated wood are needed along with high impact resistance, mixtures of furfuryl alcohol and a vinyl polymer like styrene can be used.

Furfurylated wood's excellent resistance to attack by wood-destroying agents is not caused by toxicity. Instead, chemical changes occur that make it inert to wood-destroying organisms. Furfurylated wood and the raw materials used to make it have been found to be nontoxic to fungi and marine organisms. Combustion tests have shown that furfurylation does not noticeably change the emissions or flammability of the wood.

Wood that can be penetrated with liquids can be furfurylated. This includes sapwood of most species and both heartwood and sapwood of many temperate-zone hardwoods and conifers. Species used commercially include Scots pine, southern yellow pine, Caribbean pine, beech, maple, and birch. Wood density and void content are inversely related. Lower-density woods have more internal pore space available for loading than higher-density species.

Production. Commercial production of moderately loaded furfurylated construction lumber made from Scots pine and southern yellow pine, as well as furniture and flooring lumber from beech, maple, and birch, is currently under way in Norway. Properties needed for this furfurylated wood are improved dimensional stability and biodeterioration resistance at a minimal cost. The appearance and working qualities of moderately loaded furfurylated wood are similar to those features of untreated wood. It is currently being used for outdoor and ground contact applications such as decking, docks and wharves, building siding (**Fig. 3**) and roofs, bridges, waterway liners, highway sound barriers, and outdoor furniture. A dark, hard flooring material from beech using neat furfuryl alcohol–treating solution is being commercially manufactured in Norway as well.

Moderately loaded furfurylated wood is processed and finished similarly to untreated wood and can be worked with normal woodworking equipment. It can be substituted directly for preservative pressure-treated wood in many applications. The main difference from untreated wood or preservative-treated wood is that furfurylated wood has a

Fig. 3. Moderately loaded furfurylated Scots pine siding. (*Courtesy of Kebony ASA*)

greater tendency to split when using mechanical fasteners. Thus, preboring or special screws are necessary.

Heavily loaded furfurylated wood is processed and finished with normal woodworking equipment, although its high density requires slower feed speeds and lighter cuts. Metalworking machines can be used to produce precision, highly polished parts from heavily loaded furfurylated wood, made by using neat furfuryl alcohol or a styrene–furfuryl alcohol mixture as the treating liquid. It can be sanded and buffed to achieve a smooth, hard surface with the wood grain pattern clearly visible. Since it is highly dimensionally stable, the parts retain precise shapes. Material made using neat furfuryl alcohol is very dark, whereas that made with styrene–furfuryl alcohol can be dyed any color. Heavily loaded furfurylated wood is useful for flooring, furniture tops, knife handles and other kitchenware, hobby lathe work, and musical instrument parts.

For background information *see* ALCOHOL; FURFURYL; LUMBER; POLYMERIZATION; STRUCTURAL MATERIALS; WOOD ANATOMY; WOOD COMPOSITES; WOOD DEGRADATION; WOOD ENGINEERING DESIGN; WOOD PROCESSING; WOOD PRODUCTS; WOOD PROPERTIES in the McGraw-Hill Encyclopedia of Science & Technology.   Marc H. Schneider;
Mats Westin; Stig Lande

Bibliography. I. S. Goldstein, The impregnation of wood to impart resistance to alkali and acid, *Forest Prod. J.*, 5:265–267, 1955; S. Lande, M. Eikenes, and M. Westin, Chemistry and ecotoxicology of furfurylated wood, *Scand. J. Forest Res.*, 19(suppl. 5):14–21, 2004; S. Lande, M. Westin, and M. H. Schneider, Properties of furfurylated wood, *Scand. J. Forest Res.*, 19(suppl. 5):22–30, 2004; M. H. Schneider, New cell wall and cell lumen wood polymer composites, *J. Wood Sci. Technol.*, 29:135–158, 1995; A. J. Stamm, Dimensional stabilization of wood with furfuryl alcohol, pp. 141–149, in I. Goldstein (ed.), *Wood Technology: Chemical Aspects*, ACS Symposium Series 43, American Chemical Society, Washington, D.C., 1977.

Gene targeting

It has been recognized for some time that the humble house mouse is an extremely valuable research tool for human medicine. Despite the outward differences in appearance, mice and humans are actually quite closely related with respect to general body plan and physiology, and at the genetic level. Mice and humans share approximately 99% of their genes. That is, if one takes any human gene, there is 99% likelihood that there will be a mouse correlate. Given the high degree of genetic conservation between mice and humans, as well as many ideal laboratory characteristics (such as inexpensive husbandry and rapid life cycle) of mice, it is not surprising that the mouse has come to the forefront of modern medical science as a model for understanding the individual functions of human genes and the genetic disorders that afflict humans.

Early work. The similarities between mice and humans would be almost meaningless if there were no ability to specifically manipulate the mouse genome and generate mice that carry precise mutations. Prior to the mid-1980s, our ability to introduce specific genetic changes to living creatures was limited to single-cell organisms such as bacteria and yeast. This involved making specific alterations in small snippets of deoxyribonucleic acid (DNA) in the test tube and then reintroducing them into the organism. Once in the cell, endogenous cellular machinery catalyzes an exchange between the modified DNA sequence and the endogenous copy in the genome, a process known as homologous recombination. Because these organisms were composed of a single cell, the consequences of a modified genome were immediately apparent. Extending these techniques to the mouse presented two major challenges. First, it had been thought that homologous recombination did not occur in mammalian cells and thus it was impossible to replace normal sequences with modified ones. Second, unlike a bacterium or yeast, mice are composed of hundreds of billions of cells. Therefore, even if one could target modifications into the genome of mouse cells, it would be impossible to modify enough cells to have any effect in the intact organism.

These obstacles were overcome in the 1980s by two independent discoveries. First in 1981, Martin Evans and colleagues reported that inner cell mass cells (ICM) from the mouse blastocyst (**Fig. 1***a*) had the ability to be cultured indefinitely and yet retain the ability to contribute to all tissues including the germline (sperm and eggs). Hence, if one could modify the genome of one of these remarkable cells [called embryonic stem (ES) cells], one would then be able to introduce the cell back into an early mouse embryo. The resulting mouse (called a chimera) would be a composite of normal cells from the host blastocyst and the daughter cells of the modified stem cell (Fig. 1*b*). If some of these daughter cells contribute to the gonad, the gonad (in this case, the testis) would produce modified gametes (in this case, sperm), which through fertilization would transmit the modification to every cell of the offspring (Fig. 1*b*).

The second obstacle, the ability to specifically modify the genome of mammalian cells, was overcome shortly thereafter by two different groups, one led by Mario Capecchi and the other by Oliver Smithies. Contrary to contemporary belief, they both demonstrated that exogenous DNA fragments (obtained from another source) could indeed replace their associated, endogenous (original) sequences via homologous recombination in cultured mouse cells. Capecchi went on to show that homologous recombination is also operational in mouse ES cells (Fig. 1*a*). With the major challenges removed, these groups and others quickly began to introduce specific changes into the genome of mouse ES cells

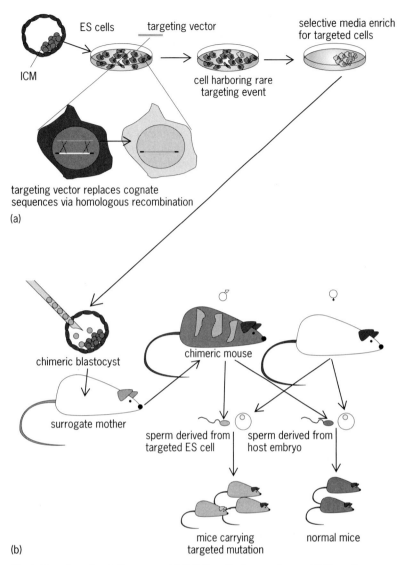

(a)

(b)

Fig. 1. Gene targeting in the mouse. (*a*) Cells from the inner cell mass (ICM) are the source of embryonic stem (ES) cells. Targeting vectors (short bar at top) are introduced into ES cells and on rare occasions line up with cognate sequences in the genome and exchange places through homologous recombination (represented by "X"). After this step, the cells are exposed to selective media that permit only cells that have undergone recombination (cells inside oval at right) to survive. (*b*) Recombined cells are expanded and then injected into blastocysts, where they contribute to the formation of all cell types of the adult mouse (represented by three splotches). Dark gray represents cells derived from the original host blastocyst. The germline (cells that give rise to sperm) likewise can be derived from the modified stem cells (light gray) or the original host (dark gray). Once the modified sperm fertilizes an egg, every cell in the resulting mouse will carry the modification (represented by three mice at bottom).

and then into mice. For their revolutionary discoveries, Capecchi, Evans, and Smithies were awarded the 2007 Nobel Prize in Physiology or Medicine.

Contributions of gene targeting to modern medicine. The sequencing of the mouse and human genomes has revealed that there are some 25,000 genes in both species and, as mentioned above, more than 99% of human genes have a mouse counterpart. Hence, researchers have turned to the mouse for an understanding of how our own genes function. Through gene targeting, researchers have been systematically "knocking out" mouse genes to learn about their individual and collective roles in the mouse. For exam-

ple, it has been known for some time that increasing the activity of genes belonging to the Wnt signaling pathway (one of many cascades that allow cells to communicate with one another) is strongly associated with colon cancer in humans. Using gene targeting, researchers have knocked out members of this pathway in the mouse and have determined that the lining of the gut fails to replenish itself in the resulting mutants, leading to neonatal death. From these findings, one can conclude that one of the normal functions of this pathway is to produce cells to replenish the ever-dying supply of cells in the gut epithelium. Increasing the activity of this pathway would therefore produce too many cellular progenitors, an important preliminary event leading to cancer. Subsequent research has provided very strong evidence that this is very likely to be the case. Performing similar studies, scientists have knocked out the function of some 10,000 genes, nearly one-half of all the genes in the genome. The result has been a literal flood of information regarding the function of individual genes as well as their role in complex molecular circuits.

Another significant contribution of gene targeting to human health has been to generate mouse models of human genetic diseases. In these cases, normal genes are replaced with alleles (alternate versions of genes) that either disrupt gene function or are identical copies of human disease alleles. Mice harboring these mutant alleles often display very similar, if not identical, manifestations of the disease relative to the disorder in humans. These mouse models therefore have proven to be indispensable in understanding the often-complex physiologic and molecular bases for the disease. Furthermore, they provide a front-line testing system for drugs and other therapeutics prior to the initiation of human trials. Through gene targeting, researchers have successfully created models for common genetic diseases such as cystic fibrosis and sickle cell anemia. Further models have been generated for other disorders such as high blood pressure, atherosclerosis, obsessive compulsive disorder, cancer, Alzheimer's disease, and diabetes, to name a few. In short, any disease that has a known genetic component can be recapitulated in the mouse by targeting the causative genetic changes into the mouse genome.

Gene targeting has been typically used to disrupt genes, but researchers have also used this same technology to repair mutant alleles. For example, Rudolf Jaenisch and colleagues repaired a genetic defect in mice that lack the *Rag2* gene (a gene that is crucial for normal immune function). They generated ES cells from the tail of a *Rag2*-deficient mouse. The *Rag2* mutant allele in these cells was replaced with a "normal" copy of the gene via gene targeting. The "repaired" stem cells were then used in two different ways. First, they were used to make chimeras (Fig. 1*b*) that passed the repaired gene to progeny and restored normal immune function. Second, the stem cells were used to repopulate the bone marrow of *Rag2*-deficient mice, which then also restored immune function. This latter approach therefore demonstrates by proof of principle a therapeutic protocol that could be applied to humans.

The first gene-targeting studies all focused on mutant alleles that completely eliminate gene function. Although this yielded much useful information regarding gene function, the consequences of the mutation were exacted the first time the gene was required, usually during embryonic development. Often, the removal of a particular gene caused embryonic lethality, preventing the ability to study its function later in embryogenesis or in the postnatal mouse. To circumvent this difficulty, researchers now often generate what is called a "conditional allele," a version of a gene that is functional until a genetic switch is flipped that renders it inactive. To generate such an allele, one must flank critical regions of the gene with loxP sites (**Fig. 2a**), short sequences of DNA that are the substrate of a bacterial virus enzyme called Cre. This enzyme catalyzes recombination between the loxP sites, which deletes the intervening sequences (Fig. 2b). Hence, prior to Cre exposure, the conditional allele is intact and functions properly. Afterwards, though, critical sequences are deleted, creating a nonfunctional version. Dictating when and where Cre is expressed ultimately controls the time and place where the gene in question will be removed. Using this strategy has yielded vital information regarding gene function in late embryogenesis and even in adulthood.

Future of gene targeting. Despite the avalanche of information provided by previous gene targeting experiments, there is much yet to be done. There are many genes for which knockout alleles have yet to be made. Major funding initiatives are supporting laboratories involved in either improving gene-targeting technology or in systematically knocking out the remaining genes in the mouse genome. The resulting research will provide a framework to understand the basic function of all of the genes in the genome. However, knockout alleles cannot provide the full picture. Other alleles that only partially remove the function or conditional alleles where one has the liberty of knocking out given genes at very specific times and places in the mouse are all very important in expanding the basic scaffold of information provided by the knockout data. The different alleles that one might make for an individual gene are almost limitless. In addition to creating knockout alleles, gene targeting will continue to provide critical disease models that promise to provide vital information of therapeutics and cures.

Despite the unpleasant reputation that the mouse has acquired over the millennia because of its rather tenuous relationship with humans, its utility in providing medical breakthroughs to benefit our species may redeem its status from pest to humankind's best friend.

For background information *see* BIOTECHNOLOGY; DEOXYRIBONUCLEIC ACID (DNA); GENE; GENETIC ENGINEERING; GENETICS; GENOMICS; HUMAN GENETICS;

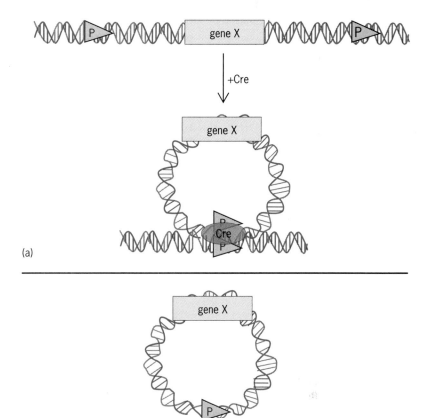

(a)

(b)   gene X deleted from chromosome

Fig. 2. Conditional removal of gene function. (*a*) Critical sequences of a gene are flanked with loxP sites (P, triangles). (*b*) If a particular cell expresses the Cre recombinase enzyme, it will mediate recombination between loxP sites, deleting the intervening sequences and thus eliminating gene function.

MOLECULAR BIOLOGY; MUTATION; RECOMBINATION (GENETICS); STEM CELLS in the McGraw-Hill Encyclopedia of Science & Technology.   Jeffery R. Barrow

Bibliography. M. R. Capecchi, Gene targeting in mice: Functional analysis of the mammalian genome for the twenty-first century, *Nat. Rev. Genet.*, 6:507–512, 2005; M. J. Evans, O. Smithies, and M. R. Capecchi, Gene targeting in mice, *Nat. Med.*, 7(10):1083–1090, 2001; R. Jaenisch et al., Nuclear cloning, stem cells, and genomic reprogramming, *Cloning Stem Cells*, 4:389–396, 2002; Y. Le and B. Sauer, Conditional gene knockout using Cre recombinase, *Mol. Biotechnol.*, 17:269–275, 2001.

Geographic information systems for mine development

In the mining industry, geographic information systems (GIS) have become practical tools in recent years as desktop computing has become both more powerful and affordable. The advanced spatial capabilities of GIS make them particularly well suited for use in mine development, where spatial information is fundamental to the processes of mineral exploration, open-pit mine and underground mine planning and production, and environmental reclamation.

A GIS is a computerized tool designed to manage the relationships between spatial data (maps or aerial photographs) and associated tabular data (records in a database). A GIS was developed as a tool for the effective collection, storage, manipulation, analysis, and display of spatially referenced information. A GIS incorporates image processing, computational geometry, and in some cases spatial statistics and artificial intelligence. A GIS has five essential functions: data acquisition, preliminary data processing, data storage and retrieval, spatial search and analysis, and graphics and interaction.

The spatial search and analysis features, in particular, are what distinguish a GIS from a more traditional computer-aided-design (CAD) database management system interface, which did not have the capability (without extensive user programming) to perform spatial searches and interpolate values.

Spatial data are the basis for nearly all decision making in mineral exploration, mine planning and design, and mine reclamation and environmental planning. The large volume of data involved in the mining cycle and the need to analyze these data have driven the application of GIS to the mining process (**Fig. 1**).

Many GIS have found wide application in mineral deposit exploration for managing ongoing survey, geologic, and environmental data. In addition, most commercially available mine planning software incorporate limited GIS functionality.

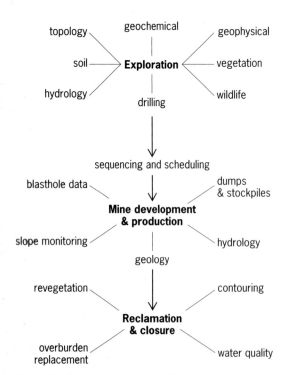

Fig. 1. Data generated during the mining cycle. A variety of data is collected and used during mining activity. This figure does not attempt to represent all of the data generated during the mining cycle.

Mining. One or more GIS can be used at all stages of the mining process and have application in both surface and underground mining. The ability of a GIS to draw from many disparate data sets and graphically display the information has meant that the use of GIS as a mine planning tool offers a more efficient and holistic approach to mine planning.

Exploration. GIS has been identified and widely accepted as a tool for mineral exploration. The ability to query multiple data sets for locations that meet the necessary criteria for exploration, to assign weighting schemes to topologies, and to combine numerous data sets using overlay processing make GIS an invaluable tool for identifying potential mineral deposits. A GIS can also be used to store, manage, and display drilling data. Some packages can perform geostatistical analysis directly.

Land tenure. GIS has also proven to be a valuable tool for managing land leases and claims for mining companies. Traditionally, land-management documents have consisted of stored text-based information regarding land holdings, with separate maps representing claim boundaries. While the maps may have been based on the textual information, they were not automatically generated from this information. Updating the maps to reflect changes in holdings required either manual changes to be made or that the maps be completely regenerated. The creation of multiple maps for the same geographic area resulted in lost efficiency.

With a GIS, professional quality maps can be generated directly from a database that can be used to hold and update all information about land parcels. As a result, current maps are always available to managers and planners, and historic data can be maintained and accessed electronically and with ease. Consequently, decisions are based on the most accurate and timely information. Moreover, managing claims in different jurisdictions (provinces, countries, and so on) is simplified by such a system.

Facilities management. Facilities management is another application of GIS. Dewatering infrastructure (wells, pipe networks, etc.) can be represented visually on a map and related back to tabular data. Using a topological model, the whole system can be simulated, taking into account resistance in the pipe network, valve locations, and so on. The ability to update this information in a timely manner is a major benefit. Electrical infrastructure, tailings (ore processing waste) pipelines, ventilation systems, mill or extraction plant processes, and other systems can be modeled in a GIS in a similar manner.

Haulage road networks also can be planned and optimized using the topological functions of a GIS (**Fig. 2**). Shortest path analysis can be applied to the haul-road topology. The ability to integrate real-time data in a GIS makes equipment monitoring and dispatch a further possibility for integration.

Production. During the production stages of a mine, GIS can be used for ongoing infrastructure management, planning decisions, materials management, volume calculations, and environmental monitoring.

The ability to interface with and simultaneously view multiple datasets, including geologic or block models, drilling data, topographic data, and infrastructure plans enable timely decisions of day-to-day mine planning decisions. These include functions such as shovel moves, where questions of whether sufficient material is exposed at the new location, whether adequate roads and infrastructure (such as electrical supply for shovels) exist at the new location, and whether there are any geologic considerations (faults, soft ground, and complex ore) that need to be considered in advance.

Volume calculations, such as mined volumes or materials placed in waste dumps, can be easily done with updated survey information. Data can also be queried regarding weights, or volumes, of available ore and its properties. The simple interface of a GIS, compared to the relatively specialized background required for using mine planning software, means that data can be accessed and queried by personnel ranging from geologists, surveyors, and mine planners, to mill personnel.

Material tracking, in particular the maintenance of records of material sent to stockpiles and waste dumps, is another GIS use for mine planning. Records of material stored in stockpiles can be used to develop optimal stockpile mining plans, as grade, ore type, lithology, and information on the amount of time passed since mining (and therefore an estimation of the degree of weathering of the material) are all readily available. Similarly, acid-generating waste dump material can be identified and tracked for environmental monitoring, remediation, and closure purposes.

Reclamation. Environmental reclamation is a critical part of the mining process. The onus is now on mining companies to ensure that mining lands are restored to an acceptable state after mining activity ceases. At the minimum, a comprehensive reclamation program should consist of plans for contouring and regrading the land, restoring suitable soil or overburden, revegetation of the area consistent with the natural vegetation and compatible with the altered landscape, watershed management, and a comprehensive monitoring program. The need to store a large amount of data as well as to generate planning maps and visualize what the postmining reclaimed landscape will look like make GIS a natural tool for reclamation planning.

A GIS can be used to develop resloping plans and to generate contours. The ability to model three-dimensional (3D) surfaces in a GIS means cut/fill volumes, seeding requirements, and waterflow paths can be determined. All these are important steps in reclamation planning, and their automation leads to increased efficiency. Surface modeling can also be used to perform "viewshed" analysis; that is, to determine locations from which a mine or its associated infrastructure is visible. This may be critical in receiving approval for mines located near highways, public lands, protected lands, or populated areas, as most people consider mining works a blight on the landscape. The ability to model watersheds and sur-

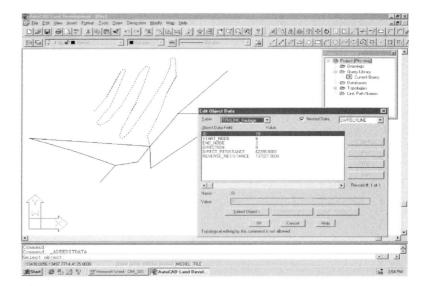

Fig. 2. **Topological modeling of a haulage network.**

face runoff allows a GIS to be used to predict areas potentially affected by acid mine drainage, or by a tailings-dam failure. In this way, appropriate steps can be taken to plan emergency responses to these situations, should they arise. A GIS can also be used in areas of historic mining to identify the worst pollution sources and help plan effective remediation activities.

The map-making facilities of a GIS are exploited in the process of reclamation for producing plans, as well as for producing visually appealing, professional quality maps and illustrations for presentation to both the government and the public. The need for high-quality, 3D representations of the postmining landscape is particularly important for communicating with the public, as it may be difficult to visualize such information on a traditional planimetric map.

Planning. Commercial GIS support the development of end-user applications, which can typically be written in one or more programming languages. In this way, custom routines can be developed that enhance and greatly expand the capabilities of an off-the-shelf GIS software package as a mine planning tool. The advantages of developing planning routines that run inside a GIS are numerous. They include savings on software through a reduction in the number of packages used, better data integration, a potential reduction in software compatibility issues, the ability for the user to program their own routines, and savings in training employees in new software. (It is much easier to learn a new routine within an existing and understood package that to learn an entirely new package.)

Mining generates a large amount of data that must be updated in a timely manner and readily accessible to many groups to be of value. Decisions regarding planning changes are made on a daily, or more frequent, basis. Ease of access to data sources and their timely updating with new geology and survey information ensures the best decisions are being made. With a GIS, access to all data sources is easily achieved without having to regenerate maps or import data from a source to a mine planning or CAD software package, manipulate it, and then generate answers.

Outlook. The benefits of a GIS as part of a mine planning system are numerous. With environmental and social issues increasingly affecting mine planning, a more broad-based approach to design will need to take place. The assimilation of all aspects of the mining process into one integrated system will allow for a broad-based approach to mine design.

For background information *see* DATABASE MANAGEMENT SYSTEM; GEOGRAPHIC INFORMATION SYSTEMS; LAND RECLAMATION; OPEN-PIT MINING; SURFACE MINING in the McGraw-Hill Encyclopedia of Science & Technology.   Ursula Thorley; G. Blackwell

Bibliography. G. F. Bonham-Carter, *Geographic Information Systems for Geoscientists: Modeling with GIS*, 1994; U. Dillon, *The Application of a Geographic Information System to Open Pit Mine Planning and Design* (dissertation), Queen's University at Kingston, 2002; D. Elroi, Applications of geographic information systems to the production and post-production phases of a mine, *National Western Mining Association Conference*, Denver, Colorado, 1993; C. Jones, *Geographical Information Systems and Computer Cartography*, 1997; C. A. Legg, *Remote Sensing and Geographic Information Systems: Geologic Mapping, Mineral Exploration and Mining*, Praxis Publishing Ltd, England, 1994.

Gran Telescopio Canarias

The Gran Telescopio Canarias (GTC, sometimes called GRANTECAN), is a 10.4-m (409-in.) segmented telescope built at the Observatorio del Roque de los Muchachos (ORM), on the island of La Palma in the Canary Islands, Spain. It is considered the largest telescope in the world because of the useful size of its primary mirror. It is also the first segmented telescope built in Europe and is being used as a test bench for the next generation of telescopes, particularly the European Extremely Large Telescope (E-ELT).

The GTC conceptual design was completed during the second half of 1997. The ground-breaking ceremony was in mid-2000, and the first light ceremony was in mid-2007. Telescope and day one instruments were commissioned in 2008. The GTC starts its scientific operation in 2009.

The GTC has been equipped initially with two science instruments: OSIRIS (Optical System for Imaging and Low-and-Intermediate-Resolution Integrated Spectroscopy) and CANARICAM (a shortened form of Canary Camera). Additionally, three second-generation instruments, CIRCE (Canarias Infrared Camera Experiment), EMIR [Espectrógrafo Multiobjeto Infrarrojo (Infrared Multi-Object Spectrograph)], and FRIDA (Infrared Imager and Dissector for Adaptive Optics), are progressing rapidly, and the adaptive optics program is underway to provide the high-resolution images required by FRIDA. With this set of instruments, some of them with unique features for the exploration of the universe, the GTC will be one of the most powerful machines to do astronomy.

Telescope facility. The GTC facility, situated at an altitude of 2267 m (7438 ft), houses the telescope itself and the equipment that provides all the supplies required for it, such as electrical power, cool water, compressed air, liquid nitrogen, and compressed helium (**Fig. 1**).

Environmental features. The telescope is protected from meteorological perturbations with a large hemispherical dome, which can be rotated and has a shutter that opens during the night to see the sky. The telescope chamber, covered by this dome, is ventilated at night through a set of windows (**Fig. 2**). Forced ventilation is used when the outside wind speed is too low to provide that ventilation. In the daytime, this chamber is kept at the expected nighttime temperature with a dedicated air-conditioning system, to avoid heating of the telescope structure. The heat produced by the electrical equipment around the telescope is exhausted away from the facility. All these environmental features are required to keep away any thermal turbulence generated around the telescope. Any thermal turbulence generated inside the dome would destroy locally the high quality of the site for astronomical observations. That quality is a consequence, among other factors, of the very low thermal turbulence in the atmosphere above the site, which allows the light coming from the sky to reach the telescope without perturbations.

Optical configuration. The GTC has an optical configuration called Ritchey-Chrétien. It is formed by two mirrors with hyperbolic polished surfaces that produce very sharp images over a wider field than a single parabolic mirror. The main mirror (M1) is so

Fig. 1. The GTC buildings at the Observatorio del Roque de los Muchachos (ORM).

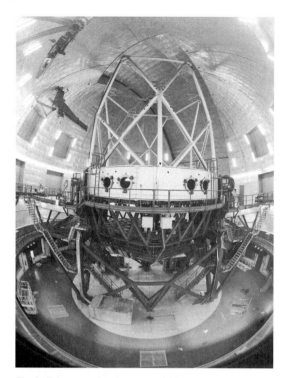

Fig. 2. Telescope chamber of the GTC.

Fig. 3. GTC main mirror with all the segments installed.

large that it cannot be built as a single mirror, and it is formed by 36 hexagonal-shaped mirrors. The secondary mirror (M2) feeds the main focus of the telescope, the Cassegrain focus, behind the main mirror. A flat tertiary mirror (M3), positioned inside a rotating tower, feeds the lateral foci of the telescope: two Nasmyth foci and four folded-Cassegrain foci. Using the GTC Control System, this tertiary mirror can be parked or set in the optical path of the telescope, and its tower can be rotated to point the telescope beam toward one of the various lateral foci that can be used for a particular observation. The science instruments are located at each focus, and they are remotely selected with movement of only the tertiary mirror.

Primary mirror. The GTC primary mirror is formed by 36 hexagonal segments, each 1.87 m (74 in.) between opposite vertexes, 80 mm (3.15 in.) thick, and 450 kg (992 lb) in weight. The total reflecting surface generated is equivalent to a circular mirror with a diameter of 10.4 m (409 in.). These segments are made of a special ceramic material that remains unchanged with any change in temperature, as it has a near-zero expansion coefficient. Because of this characteristic, its optical properties remain unchanged despite any change in the temperature. Polishing of the segments at the level of quality required is a highly specialized and time-consuming task, which took more than six years. The major part of this time was used to set up the production infrastructure and to develop an adequate manufacturing process, but once these were established, the average delivery rate was 22 segments per year (**Fig. 3**).

The positions of the segments are maintained by the GTC Control System through a closed control loop among three position actuators in each segment that move it in height and inclination relative to its neighbors, and relative position sensors located in the adjacent edges between the segments. This control loop corrects in real time the deformation of the supporting structure (gravitational and thermal) and also the global deformations induced by the wind blowing onto the mirror.

Additionally, the reflecting shape of each segment can be modified in real time using moment actuators in the back side, six per segment, installed in the whiffletrees supporting the segments. This capability is used to correct focus and astigmatism of the segment arising mainly from errors in placing the segments in the correct off-axis positions during testing and integration on the telescope, thermal deformation of the structure, and some residual errors in the support of the segment or in the polishing of the mirrors at the factory. This sophisticated control system moves and deforms the 36 segments in real time to produce a primary mirror that operates in the

Fig. 4. GTC secondary mirror installed in the telescope.

Fig. 5. A Nasmyth focus equipped with a slow guiding camera and a high-resolution wavefront sensor in arm 1, and a fast guiding camera in arm 2 for fast corrections with the secondary mirror.

Fig. 6. OSIRIS mounted in the GTC Nasmyth simulator at the laboratory of the Instituto de Astrofísica de Canarias (IAC), before shipment to La Palma.

telescope as a single mirror with the required hyperbolic reflecting surface.

Secondary mirror. The secondary mirror has a serrate contour, following the primary mirror's external contour, and a central hole to achieve an optical design wherein only light coming from the cool sky reaches any of the infrared detectors situated at the various GTC foci (**Fig. 4**). If that careful design were not implemented, parts of the telescope, warm at room temperature, could be seen by the infrared detectors, which would then be saturated. With fast tip-tilt movements of this mirror, the telescope control system corrects any vibration of the telescope produced by the wind blowing on its structure. Those vibrations, if not corrected, would produce image

movements that would destroy any possibility of conducting observations. These fast corrections are possible because this mirror has a very low mass, since it is made from a lightweight (hollowed-out-back) beryllium blank.

Wavefront sensors. Specialized wavefront sensors situated at the telescope focal planes provide information on any deformation of the images resulting from defects of the telescope optics (**Fig. 5**). With this information, the control system reacts on the segments of the primary mirror and on the position of the secondary mirror to correct the measured defects. A closed loop between these wavefront sensors and the actuators on the optics maintains the images generated by the telescope stably despite telescope movements, strong winds, temperature changes, and so forth.

Science instruments. Two science instruments, OSIRIS and CANARICAM, have been built to be operational when the GTC is ready to start observations. They are called the day one instruments, and they determine the science that the GTC will be doing initially.

OSIRIS. OSIRIS provides visible imaging and low-resolution multiobject spectroscopy capabilities over a large field of view (8.4×8.4 arc min). OSIRIS is designed to be a general-purpose instrument at the GTC, but it is conceived as the optimum instrument to do tomography of areas of the universe at different distances from us or at different ages, as distance and age are closely related because of the finite speed of light. This tomographic capability results from the use of tunable filters that make OSIRIS a unique instrument on a 10-m telescope (**Fig. 6**). Tunable filters are optical devices that filter the incoming light in a small wavelength range that can be electronically selected. With OSIRIS, astronomers will get new data to disentangle many of the unknowns concerning the evolution of the galaxies.

CANARICAM. CANARICAM is a thermal infrared imager and spectrograph with capabilities for both polarimetry and coronography. It provides the GTC with novel capabilities for astronomy (**Fig. 7**). CANARICAM and the GTC are the perfect combination to study extrasolar planets and low-mass objects, protoplanetary discs, and active galaxies.

Administration. The GTC is a partnership among Spain, headed by the Instituto de Astrofísica de Canarias (IAC); the main astronomical research institutions in Mexico, the Institute of Astronomy of the National Autonomous University of Mexico (Instituto de Astronomía de la Universidad Autónoma Nacional de México, IA-UNAM) and the National Institute for Astrophysics, Optics, and Electronics (Instituto de Astrofísica, Óptica y Electrónica, INAOE); and the University of Florida.

For background information *see* ADAPTIVE OPTICS; ASTRONOMICAL OBSERVATORY; ASTRONOMICAL SPECTROSCOPY; CONTROL SYSTEMS; CORONAGRAPH; INFRARED ASTRONOMY; POLARIMETRY; TELESCOPE in the McGraw-Hill Encyclopedia of Science & Technology.   Pedro Alvarez

Fig. 7. CANARICAM at the laboratory of the Astronomy Department of the University of Florida, being tested before shipment to La Palma.

Bibliography. J. Cepa et al., OSIRIS: A tunable imager and spectrograph, in *SPIE's International Symposium on Astronomical Telescopes and Instrumentation 2000*, 4008-69, 2000; M. L. Edwards et al., The Canarias Infrared Camera Experiment (CIRCE): Optical and optomechanical design and manufacture, in *SPIE's International Symposium on Astronomical Telescopes and Instrumentation 2008*, 7014-90, 2008; F. Garzón et al., EMIR: The GTC nIR multi-object imager spectrograph, in *SPIE's International Symposium on Astronomical Telescopes and Instrumentation 2006*, 6269-45, 2006; A. López et al., FRIDA: Integral-field spectrograph and imager for the adaptive optics system of the Gran Telescopio Canarias, in *SPIE's International Symposium on Astronomical Telescopes and Instrumentation 2006*, 6269-138, 2006; J. M. Rodríguez Espinosa, P. Alvarez, and F. Sanchez, The GTC: An advanced 10 m telescope for the ORM, *Astrophys. Space Sci.*, 263:355-360, 1999; C. M. Telesco et al., Day one science with CanariCam: The GTC multimode mid-IR camera, in *SPIE's International Symposium on Astronomical Telescopes and Instrumentation 2008*, 7014-25, 2008.

Graphene

Graphene, the basic building block of graphite, is a two-dimensional sheet of carbon atoms arranged in the honeycomb structure (**Fig. 1***a*). In graphite, the graphene sheets are stacked in the Bernal structure (Fig. 1*b*), thereby causing graphite to be a semimetal.

While the covalent chemical bonds between the carbon atoms in a graphene sheet are among the strongest in nature, the inter-sheet van der Waals bonds rank among the weakest. Consequently, microscopically thin layers of graphite are easily mechanically cleaved from a larger crystal by rubbing it against a surface, for example paper, causing pencil traces that consist of microscopic graphite flakes. In fact "graphite" is derived from "graphein" (Greek for "to draw").

Graphene was officially named in 1994, before which it was also called monolayer graphite to denote graphene sheets on various metallic and metal carbide and silicon carbide surfaces. Graphene can be grown on carefully prepared, contamination-free metal samples by heating them to high temperatures in a carbon-containing gas (such as methane, ethylene, or acetylene). It can grow epitaxially (for example, on a nickel surface), whereby the carbon atoms register with the metal atoms on the surface, or as small randomly oriented graphene crystallites (as, for example, on a platinum surface). Bonding to the surface may be so weak that the properties of pristine graphene are retained.

Graphene layers are also grown on the surfaces of single-crystal silicon carbide (SiC). The crystals are heated to high temperatures (greater than 1100°C or 2000°F) in vacuum so that silicon evaporates, and the carbon reconstructs into a stack of one or more epitaxial graphene sheets. Epitaxial graphene on silicon carbide has become recognized as the most promising new material for graphene-based electronics after researchers demonstrated its advantageous properties.

Researchers produced and identified thin graphitic layers on a conducting (degenerately doped) silicon wafer supplied with a thin silicon oxide dielectric coating. The process involved repeatedly "cleaving" a small graphite crystal using Scotch tape, to produce successively thinner graphite flakes on the tape. Pressing the graphite-flake-coated tape against the oxidized silicon wafer transfers exfoliated flakes to the wafer. Flakes as thin as a monolayer could be reliably identified by an optical microscopic method wherein the thin,

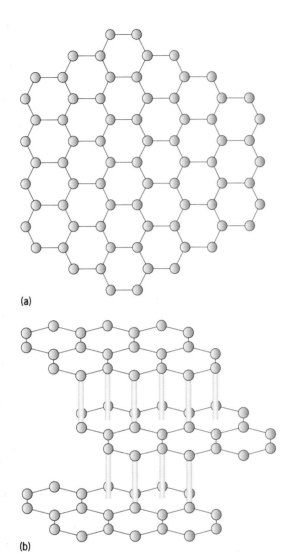

(a)

(b)

Fig. 1. Fig 1. Crystalline structures. (*a*) Graphene sheet consisting of a hexagonal array of covalently bonded carbon atoms. (*b*) Graphite structure consisting of Bernal-stacked graphene sheets. Vertical lines indicate relative positions of atoms in neighboring sheets.

transparent silicon oxide layer served more or less as an interference filter on the silicon so that even a single graphene layer would provide a contrast difference.

A confluence of three circumstances caused recent interest in graphene: the development of epitaxial graphene, with its promise as a new electronics platform; the development of exfoliated graphene for basic research; and the theoretical appeal of its simple but unusual electronic structure (first calculated in 1947).

Electronic structure. The four carbon valence electrons cause the chemical bonds between neighboring carbon atoms. Three of these electrons arrange to form covalent σ bonds to give the graphene sheet its exceptional strength.

The fourth electron oscillates back and forth through the graphene plane, to produce a p_z orbital with lobes above and below the graphene sheet (**Fig. 2***a*). Neighboring p_z orbitals overlap slightly and the energy of a pair decreases (causing bond-

ing) or increases (causing antibonding) depending on their relative electronic motion. Bonding occurs when two neighboring p_z electrons are in phase with each other, antibonding when they are out of phase.

The p_z electrons are mobile and "hop" at a rate $\nu \approx 10^{15}$ times per second from one atom to the next. Quantum mechanically this motion is described as a wave that extends over the entire surface of the graphene layer. These waves come in two varieties: π and π^*: bonding and antibonding states respectively, reflecting the bonding and antibonding nature of the p_z electrons from which the waves are built up. Like light waves (photons), they are traveling waves with a wavelength λ, frequency ν, energy E, and momentum $\boldsymbol{p}$.

The momentum $\boldsymbol{p}$ of the electron in a π (or π^*) state is given by $\boldsymbol{p} = h/\lambda = h\boldsymbol{k}/2\pi$, where h is Planck's constant and $\boldsymbol{k}$ is the wave number, which is a vector that points in the propagation direction. The energy of a π or a π^* electron propagating in the x direction (perpendicular to a hexagon edge) is given by Eq. (1)

$$E = \pm h\nu[1 + 2\cos(kd/2)] \qquad (- \text{ for } \pi, + \text{ for } \pi^*)$$
$$(1)$$

where $d = 0.26$ nm is the graphene lattice constant. This equation describes the π and π^* energy bands as first calculated by P. R. Wallace in 1947 (Fig. 2*b*). The two energy bands cross when $k = K = +4\pi/3d$ and $k = K' = -4\pi/3d$, where $E = 0$, defining the K and K' points. The energy bands have been revealed by angle-resolved photoemission (Fig. 2*c*).

The electronic structure results from filling up the π and π^* bands with all of the electrons in the graphene sheet (allowing only one electron per state), starting with the π electron with the least energy at the bottom of the π band. For neutral graphene the bands fill exactly up to the K and K' points. If the sheet is negatively charged, the additional electrons cause the Fermi level to rise to higher energies; if the sheet is positively charged, the Fermi level is lowered.

Near the K point (and K' point), energy varies linearly with momentum: $E = \pm c^*p^*$, where p^* is the momentum measured from the K point and $c^* \approx 10^6$ cm/s is the Fermi velocity. A complete analysis, including motion in the y direction, shows that $E = \pm c^*|p^*|$ where $|p^*|$ is the magnitude of the momentum. Hence, near the Fermi level the band structure is conical (Fig. 2*b*).

Near the Fermi energy, the electron velocity, given by Eq. (2), is independent of its energy, which is

$$v = dE/dp = \pm c^* \qquad (2)$$

a characteristic property of massless particles such as photons and neutrinos, which travel with the speed of light. In graphene, the electrons are not really massless, and electrons travel with 1/300 of the speed of light. In 1998, T. Ando recognized a formal correspondence between graphene electrons and massless (relativistic) fermions. This recently led to spectacular claims that graphene experiments could replace high-energy experiments.

Some important electronic properties. Important electronic properties of graphene include the possibility of gate doping and quantum confinement, immunity of electrons to backscattering, and the appearance of the quantum Hall effect at very low temperatures.

Gate doping. An electrostatic gate is produced by covering a graphene sheet with a metal-coated insulating layer. Electrostatic forces from an electric potential applied to the gate attract or repel graphene electrons, thereby altering their number. A positive gate potential increases the number and raises the Fermi energy by filling states in the upper cones (vice versa for a negative gate). This affects the conductivity σ, since it is given by Eq. (3) where n is the

$$\sigma = |n|e\mu \qquad (3)$$

surplus of electrons per unit area (the doping density), e is the electronic charge, and μ is the mobility. The wavelength can be tuned since it depends on n through Eq. (4). Gate doping has been demonstrated

$$\lambda_F = \sqrt{4\pi/n} \qquad (4)$$

in prototype field-effect transistors (FETs) and quantum dot devices (**Fig. 3**).

Quantum confinement. As in optics, when electrons are confined in graphene structures with dimensions that are comparable to the wavelength, interference (quantum confinement) becomes important. The wavelength for graphene with $n = 10^{11}$ electrons/cm^2 (obtainable by gate doping) is about $\lambda_F = 100$ nm. Electrons in a graphene ribbon resemble microwaves in a waveguide. Propagation occurs when waves reflected from the walls interfere constructively. Because of quantum confinement, a narrow graphene ribbon with a width of W nm becomes a semiconductor with a bandgap of about 1 eV/W. Quantum confinement effects have been observed both in epitaxial graphene and in exfoliated graphene samples.

Immunity to backscattering. In graphene, electrons can travel relatively large distances without scattering. The reason is as follows. Referring to Fig. 2, a right-moving electron at the K point is in a π orbital, while a left-moving electron is on the π^* orbital ($v = dE/dp$). A charged impurity cannot cause a conversion from π to π^*, hence the scattering is inhibited. This argument applies for all directions of propagation so that the mobilities are large, even in impure samples.

Quantum Hall effect. Moving charges deflect in a magnetic field B. Therefore, electrons in a ribbon that carries a current I accumulate at one side of the ribbon, producing a potential difference V between the two sides (known as the Hall effect). Ordinarily V is proportional to B so that the Hall conductance is $\sigma_H(B) = V/I = CB$, where C is constant. In the quantum Hall effect, observed in high-mobility two-dimensional conductors at very low temperatures, C is not constant, and $\sigma_H(B)$ shows a series of steps at integer multiples of the von Klitzing constant, $g_0 = e^2/h$. Graphene is anomalous: the steps

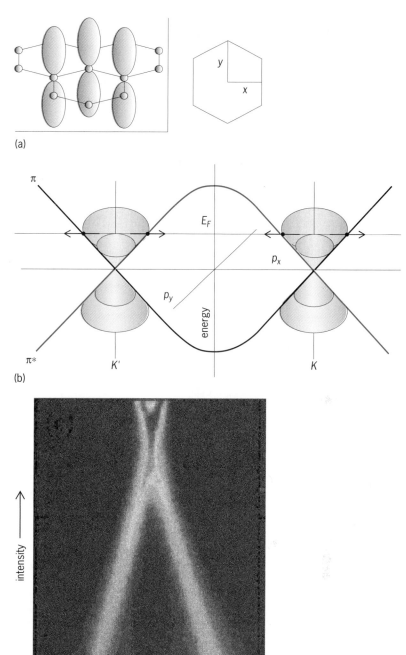

(a)

(b)

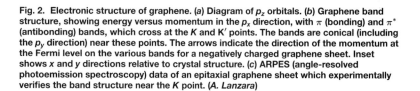

(c)

Fig. 2. Electronic structure of graphene. (*a*) Diagram of p_z orbitals. (*b*) Graphene band structure, showing energy versus momentum in the p_x direction, with π (bonding) and π^* (antibonding) bands, which cross at the K and K' points. The bands are conical (including the p_y direction) near these points. The arrows indicate the direction of the momentum at the Fermi level on the various bands for a negatively charged graphene sheet. Inset shows *x* and *y* directions relative to crystal structure. (*c*) ARPES (angle-resolved photoemission spectroscopy) data of an epitaxial graphene sheet which experimentally verifies the band structure near the K point. (**A. Lanzara**)

are found at half-integer values: $g_n = 4g_0 (n+\frac{1}{2})$, where n is an integer. This phenomenon was predicted in 1998 and observed in exfoliated graphene samples (at cryogenic temperatures in 2005 and at room temperature in very high magnetic fields in 2007).

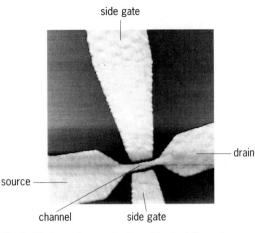

Fig. 3. Electron micrograph of an all (epitaxial) graphene FET consisting of a channel connected to a graphene source and drain. The channel is flanked by two planar graphene side gates. The channel length is about 1 μm. (**W. A. de Heer**)

Graphene types. The two major types of graphene will now be discussed in more detail.

Exfoliated graphene. Exfoliated graphene flakes are well suited for scientific studies. The samples are easy to produce. The combination of the conducting silicon wafer and the thin insulating oxide layer is advantageous for studying and utilizing graphene electronic properties, since the oxide layer forms an insulating barrier that prevents shunting of current through the silicon substrate, and the structure provides a method to adjust the charge density (and Fermi level) by gate doping. The method has been applied to produce devices, including field-effect transistors (gate-doped graphene ribbons), Hall bars (for the quantum Hall effect), single-electron transistors, quantum dots, and *p-n* junctions.

Exfoliated graphene is not suitable for electronics technology. The exfoliated flakes (mostly ultrathin graphite, not graphene) sparsely and randomly cover the silicon oxide surface. Moreover, the graphene sheets are randomly charged, causing localized "puddles" of electrons and holes with relatively large charge densities ($n \sim 10^{11}$-10^{12} charges per cm^2). Nevertheless, mobilities are relatively large ($\mu \sim 10^3$-10^4 cm^2/Vs).

Epitaxial graphene on crystalline silicon carbide. Epitaxial graphene grown on single-crystal silicon carbide crystals covers the entire surface. Silicon carbide crystals have many crystal structures (polytypes) and crystal faces that each graphitize differently. Hexagonal SiC (for example 4H-SiC) can be cut so that one face terminates with carbon atoms (C face) and the opposite face with silicon atoms (Si face). A graphene monolayer grown on the Si face of 4H-SiC has a band gap of about 0.3 eV. This band gap can be exploited in electronic devices.

Multilayered graphene (a stack of up to 100 graphene layers) grows on the C face of 4H-SiC. The first graphene layer (at the SiC interface) is uniformly charged with a charge density of about 5 × 10^{12} electrons per cm^2. It is a good conductor with a mobility $\mu \approx 25,000$ cm^2/Vs. The remaining layers are multilayered epitaxial graphene (MEG) with the electronic properties of a single graphene sheet (in contrast to thin bulk graphitelike exfoliated graphite flakes). MEG is essentially uncharged; it is very pure, and room-temperature mobilities exceeding 250,000 cm^2/Vs have been reported.

Epitaxial graphene-based electronics. Epitaxial graphene is a promising material for graphene-based electronics because high-quality graphene can be grown to uniformly cover the entire crystal. It can be chemically converted to become semiconducting and patterned, using standard microelectronics methods.

In principle, entire integrated circuits can be fashioned from a single epitaxial graphene sheet (as proposed in 2004). For example, patterning allows a narrow ribbon (which serves as the channel of an FET) to be seamlessly connected to a wider ribbon (which serves as the metallic leads to the FET) [Fig. 3].

Epitaxial graphene-based electronics has several advantages over silicon. As with carbon nanotubes, defect-free graphene is a ballistic (essentially resistanceless) conductor that can sustain very high currents. Its properties can be tailored by shape, which includes ribbons and more complex structures. The electronic wavelength is large so that a wide variety of devices (such as quantum dots and quantum interference devices) can be envisioned that could function even at room temperature. Epitaxial graphene electronics promises to operate at much higher speeds (of the order of terahertz) and with less dissipation than is possible with silicon. Moreover, in contrast to silicon, graphene functions well, even at the nanometer scale.

While much research remains to be done, several critical milestones have already been reached. Researchers have patterned arrays of epitaxial graphene field-effect transistors on MEG substrates on a large scale, which is a promising indicator of the feasibility of the technology.

For background information *see* BAND THEORY OF SOLIDS; CHEMICAL BONDING; EPITAXIAL STRUCTURES; GRAPHITE; HALL EFFECT; MOLECULAR ORBITAL THEORY; QUANTIZED ELECTRONIC STRUCTURE (QUEST); TRANSISTOR in the McGraw-Hill Encyclopedia of Science & Technology.   Walter de Heer

Bibliography. C. Berger et al., Ultrathin epitaxial graphite: 2D electron gas properties and a route toward graphene-based nanoelectronics, *J. Phys. Chem. B*, 108:19912–19916, 2004; A. K. Geim and K. S. Novoselov, The rise of graphene, *Nat. Mater.*, 6(3):183–191, 2007; Y. B. Zhang et al., Experimental observation of the quantum Hall effect and Berry's phase in graphene, *Nature*, 438:201–204, 2005.

Gravity Probe B mission

Gravity Probe B (GP-B) is a landmark fundamental physics experiment in space to test Einstein's 1916 general theory of relativity. GP-B uses four spherical gyroscopes and a telescope, housed in a satellite orbiting 642 km (400 mi) above the Earth, to measure, with unprecedented accuracy, two extraordinary effects predicted by the theory: the geodetic effect—the amount by which the Earth warps the local spacetime in which it resides; and the frame-dragging effect—the amount by which the rotating Earth drags spacetime around with it. GP-B tests these two effects by precisely measuring the precession (displacement angles) of the spin axes of the four gyroscopes over the course of a year relative to a distant guide star, and comparing the experimental results with the theoretical predictions. In a polar orbit, the two effects occur at right angles to one another—the geodetic effect in the plane of the spacecraft's orbit, and the frame-dragging effect in the Earth's equatorial plane (**Fig. 1**). Thus, each gyroscope measures both effects simultaneously.

Why perform another test of Einstein's theory? Einstein's general theory of relativity (GR) has passed four important tests since its inception in 1916: (1) the perihelion shift of Mercury's orbit, (2) gravitational deflection of light by a massive body, (3) the analogous radar time delay (Shapiro effect), and (4) the change in orbital frequency of the Taylor-Hulse binary pulsar based on the emission of gravitational radiation. The "equivalence principle," a key assumption underlying Einstein's theory, has also received strong experimental support through NASA's 1976 Gravity Probe A (GP-A) redshift clock experi-

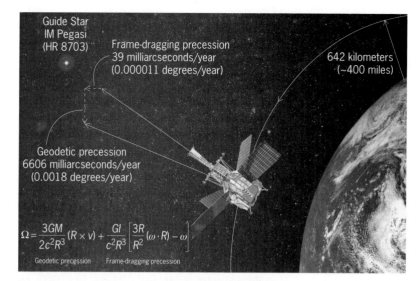

Fig. 1. Conceptual diagram of the GP-B experiment showing the spacecraft in polar orbit, the location of the four gyroscopes at the bottom of the probe in the center of the spacecraft, the geodetic and frame-dragging measurements being made, along with the relativistic gyro precession values derived from Stanford physicist Leonard Schiff's 1960 equation (lower left).

ment and NASA lunar-laser-ranging free-fall measurements. Nevertheless, it is widely believed that our present theories of gravity will eventually be seen as limiting cases of a unified theory in which all four fundamental forces of nature (strong, weak,

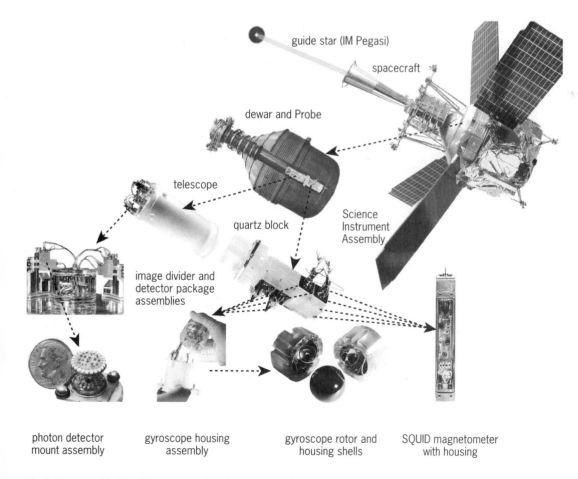

Fig. 2. Scale model of the GP-B spacecraft and an exploded view of the most important components inside.

electromagnetic, and gravity) become comparable in strength at very high energies. But there is no consensus as to whether it is GR, particle physics, or both that must be modified, let alone how.

In this regard, a better understanding of spin, the third of nature's most fundamental mysteries (along with mass and charge) may provide a breakthrough. Frame-dragging, the gravitational analog of magnetism, is central to modern astrophysics as the probable engine for the vast jets of gas and magnetic field ejected from the regions around supermassive black holes as well as keeping those jets aligned across scales of megaparsecs. By focusing on spin, GP-B differs from all the tests of GR previously noted, and it offers by far the best prospect for a trustworthy measurement of frame dragging, the key to understanding the most violent phenomena in astrophysics and the origin of inertia itself.

Experimental design. Conceptually, the GP-B experiment is simple: A near-perfect, superconducting niobium-coated gyroscope is centered along the line of sight of a telescope in a polar-orbiting satellite, 642 km (400 mi) above the Earth. (GP-B actually uses four gyroscopes for redundancy.) At the start of the experiment, both the telescope and the spin axis of the gyroscope are aligned with a guide star (IM Pegasi, also known as HR 8703), whose proper motion has been precisely mapped with respect to a remote quasar. The telescope-guide star alignment is maintained for a year, while the spacecraft orbits over Earth's poles, continuously rolling along the telescope sight axis to average out errors. Throughout the year, the gyroscopes are maintained at a cryogenic temperature of 2.3 K, just above absolute zero. This enables SQUID (superconducting quantum interference device) magnetometers to read out the gyroscope precession over this period in both the plane of the orbit (the geodetic effect) and orthogonally in the plane of the Earth's rotation (frame-dragging effect). Finally, the guide star proper motion data is used to directly relate the gyroscope spin-axis displacement to the reference quasar in the remote universe.

Applying the equation derived by Stanford physicist Leonard Schiff in 1960, the predicted geodetic effect is a tiny angle of 6606 milliarcseconds (0.0018°) in the orbital plane of the spacecraft; the predicted orthogonal frame-dragging effect is a minuscule angle of 39 milliarcseconds (0.000011°)—about the width of a human hair viewed from a quarter mile away. GP-B's measurement of the geodetic effect has an expected accuracy of better than 0.1%, far more accurate than any previous measurements. The frame-dragging effect, which is ~170 times smaller than the geodetic effect, has never directly been measured, but Gravity Probe B is expected to determine its accuracy to better than 5%.

Extraordinary technologies. The GP-B spacecraft is a total system, comprising both the space vehicle and its unique payload—an integrated system dedicated as a single entity to making the measurements of unprecedented precision required by the experiment (**Fig. 2**). The spacecraft and payload are filled with cutting-edge technologies, many of which simply did not exist in the 1960s when the experiment was first proposed.

The spacecraft is built around a 2.7-m (9-ft) tall, 2500-liter (650-gal) dewar (thermos bottle), the largest and the most sophisticated ever flown. Before launch, it was filled with liquid helium, cooled to a superfluid state just above absolute zero. Embedded along its central axis, a cigar-shaped canister called the Probe contains four heat-absorbing windows, a helium-adsorbing cryopump, and the Science Instrument Assembly (SIA)—the pristine spaceborne laboratory for making the GP-B experimental measurements. The SIA is made up of a Cassegrain-style reflecting telescope and a quartz block, optically bonded together. The quartz block contains the four gyroscopes, their centers aligned along the telescope (and spacecraft) central axis. Four digital magnetometers, called SQUIDs (Superconducting Quantum Interference Devices), monitor the spin axis orientations of the gyros. During the 17-month

Fig. 3. GP-B launch on April 20, 2004 at 9:57:24 AM PDT from Vandenberg AFB, CA. (*Photo courtesy of Boeing Corporation*)

mission, the SIA was maintained at 2.3 K (−455.5°F) in a near-total vacuum.

To minimize Newtonian torques on the gyroscopes, the spacecraft is made "drag-free." The Attitude and Translation Control (ATC) system senses any acceleration forces due to solar wind, radiation pressure, and atmospheric drag on one of the four gyroscopes that is designated the "drag-free gyro." The ATC then uses a set of 16 proportional microthrusters to compensate for these drag forces by moving the entire spacecraft. This enables the spacecraft to fly along a near-perfect geodesic. The microthrusters are also used to control the spacecraft's attitude and continuous roll rate.

Helium gas that constantly boils off inside the dewar is "sweated" out of the dewar through a porous plug. The escaping helium gas is captured and reused as the propellant for the 16 microthrusters that keep the spacecraft pointed toward the guide star, and maintain its attitude, position, and roll rate. Outside the dewar, all the systems that provide power, navigation, communication, and control of the spacecraft are mounted to its frame.

GP-B flight mission. On April 20, 2004 at 9:57:24 AM PDT, a crowd of over 2000 on-site observers watched and cheered as the GP-B spacecraft lifted off from Vandenberg Air Force Base on the central California coast (**Fig. 3**). Once in orbit, the spacecraft first underwent a four-month Initialization and Orbit Checkout (IOC), in which all systems and instruments were initialized, tested, and optimized for the data collection to follow. The IOC phase culminated with the spin-up and initial alignment of the four science gyros early in August 2004. On August 28, 2004, the spacecraft began collecting science data. During the ensuing 50 weeks, the spacecraft transmitted over a terabyte of science data to the GP-B Mission Operations Center (MOC) at Stanford, where it was processed and stored in a database for analysis (**Fig. 4**). On August 15, 2005, the MOC team finished collecting science data and began a planned set of calibration tests of the gyros, telescope, and SQUID readouts that lasted 6 weeks, until the liquid helium in the dewar was exhausted at the end of September.

Data analysis and preliminary results. In October 2005, the GP-B science team began the process of analyzing the data. Two intriguing complications, now essentially overcome, have challenged the ingenuity of the data analysis team:

1. The polhode motion of the gyroscope rotors did not remain constant over the life of the experiment. The gyroscope's polhode motion is akin to, though orders of magnitude smaller than, the common "wobble" seen on a poorly thrown (American) football. This complicates the measurement of the relativity effects by putting a small time-varying polhode signal into the data.

2. Instrument calibration tests uncovered much larger than expected classical misalignment torques on the gyroscopes. Careful investigation showed that these torques, and also the time-varying polhode motion, were caused by "patch effect" (contact potential

Fig. 4. GP-B Mission Operations Center at Stanford University during spin-up of the fourth gyroscope to full speed in August 2004. (*Photo courtesy of Bob Kahn, GP-B Public Affairs Coordinator*)

tial difference) interactions between the gyro rotors and their housings.

During 2006–early 2007, the science team made good progress understanding the cause of these effects and developing methodologies for working through them. An announcement of preliminary results was made at the April 2007 meeting of the American Physical Society. The geodetic effect was clearly visible in the raw data, processed at that time to an accuracy level of ∼1.5% (**Fig. 5**). "Glimpses" of the much-smaller frame-dragging effect were noted, but with insufficient accuracy to be counted as a credible scientific result.

Crucial to further progress is devising a method of completely separating the two effects noted above. The difficulty has been the presence of small components of trapped magnetic flux in each superconducting gyro rotor. Members of the team, including several Stanford graduate students, have produced detailed maps of the trapped magnetic flux in the four gyro rotors to predict the time-varying signal in the data. This mapping has resulted in a 500-fold improvement in the determination of the polhode motion throughout the duration of the experiment, essentially solving the first problem.

This, in turn, has enabled the team to implement two complementary methods of removing the disturbing effects of the classical torques. With the classical torques properly handled, rather convincing estimates of frame-dragging are now available, but the team is investigating all possible systematic disturbing effects. As experimentalists, the GP-B science team makes no assumptions about Einstein's theory being right or wrong; rather, the aim is to do everything humanly possible to maximize the precision and accuracy of the final results. It is anticipated that the team will complete the data analysis, publish the final detailed results and

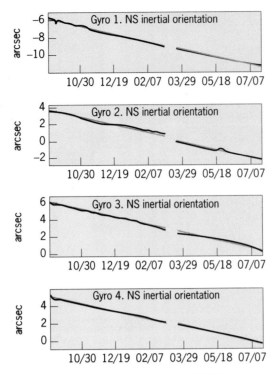

Fig. 5. Seeing Einstein in the raw data. April 2007 preliminary results graph of the geodetic effect measured by the four GP-B gyroscopes. If Newtonian physics were correct, the result would be a horizontal straight line on the top line of each plot. The darker lines represent raw data; the lighter lines represent partially processed data.

bring the experiment to a conclusion by the first half of 2010.

The broader legacy of GP-B. To carry out GP-B, at least a dozen new technologies had to be invented and perfected. For example, the spherical gyros are over 10 million times more stable than the best navigational gyros. The ping-pong-ball-sized rotors had to be so perfectly spherical and homogeneous that it took more than 10 years and a whole new set of manufacturing techniques to produce them. They are now listed in the Guinness Database of Records as the world's roundest human-made objects. The SQUIDs are so sensitive that they can detect a gyro tilt corresponding to 0.1 milliarcsec. Over its 40+ year life span, spin-offs from GP-B have yielded many technological, commercial, and social benefits—for example, GP-B's porous plug for controlling helium in space was essential to several other vital NASA missions. In addition, Global Positioning System (GPS) navigation technology developed for GP-B spacecraft attitude control was recognized in 2006 by the Space Technology Hall of Fame for its contribution to GPS-automated guidance and control systems for many uses, including precision farming and landing aircraft. Most important, GP-B has had a profound effect on the lives and careers of numerous faculty and students—graduate, undergraduate and high school—including 84 Ph.D. dissertations at Stanford and 13 elsewhere. GP-B alumni include the first woman astronaut from the United States, an aerospace CEO, and a Nobel laureate.

For background information *see* GYROSCOPE; RELATIVITY; RIGID-BODY DYNAMICS; SQUID; SUPERCONDUCTING DEVICES; TELESCOPE in the McGraw-Hill Encyclopedia of Science & Technology.   Francis Everitt

Bibliography.   C. W. F. Everitt, et al., Gravity Probe B data analysis status and potential for improved accuracy of scientific results, *Class. Quantum Grav.*, 25:114002, 2008; J. D. Fairbank et al. (eds.), chap. VI: Gravitation and astrophysics, Sec. 2: Gravitomagnetism, jets in quasars, and the Stanford Gyroscope Experiment, and Sec. 3: The Stanford Gyroscope Experiment, in *Near Zero: New Frontiers of Physics*, Freeman, New York, 1998; The GP-B Team, R. Kahn (ed.), *Gravity Probe B: Post-Flight Analysis*, final report to NASA, 2007; R. Ruffini and C. Sigismondi (eds.), Sec. C: The GP-B mission: The orbiting gyroscope experiment around the Earth, in *Nonlinear Gravitodynamics: The Lense-Thirring Effect*, World Scientific, New Jersey, 2003; L. I. Schiff, Motion of a gyroscope according to Einstein's theory of gravitation, *Proc. Natl. Acad. Sci.*, 46:871–882, 1960; J. P. Turneaure, C. W. F. Everitt, and B. W. Parkinson, The Gravity Probe B relativity experiment: Approach to a flight mission, in *Proceedings of the 4th Marcel Grossman Meeting on General Relativity*, pp. 441–464, Elsevier Science B. V., Amsterdam, 1986.

Ground-based measurement of solar ultraviolet radiation

Scientists need to measure solar ultraviolet radiation (UV) because increased exposure to UV has been associated with adverse health impacts to humans. These impacts include skin cancer, especially the most virulent form (melanoma), and diseases of the immune system. There are ecological impacts as well, such as potential harm to amphibians and other sensitive organisms, bleaching of coral reefs, and so on. The amount of UV reaching the Earth's surface increases over time if there are no physical interactions or chemical reactions to absorb it as it passes through the atmosphere. In particular, reduction in stratospheric ozone (O_3) concentration reduces the amount of UV absorbed in the upper atmosphere, which increases the UV intensity at the Earth's surface.

A number of gases, including those associated with global warming, undergo photochemical reactions that contribute to reductions in stratospheric ozone. The ability to track and document long-term trends in UV intensities across the United States helps researchers assess the effectiveness of international agreements such as the Montreal Protocol (signed in 1987 and amended through 1999), in reducing ozone-depleting substances such as chlorofluorocarbons (CFCs). Thus, reliable UV measurements are needed to help researchers learn how human and ecosystem biological response/sensitivity to UV varies with wavelength, and to track how effective domestic laws and international treaties are at protecting

Comparative UV characteristics						
Radiation type	Wavelength, nm	Photon energy	UV absorption by O_3 (stratospheric ozone) and O_2	Amount reaching earth	Surface irradiance	Biological impact
UV-C	200–280	High	High	Low	Low	Low
UV-B	280–320	Moderate	Moderate	Moderate	Moderate	High
UV-A	320–400	How	Low	High	High	Moderate

the upper atmosphere from stratospheric O_3 reductions.

The National Exposure Research Laboratory (NERL) of the U.S. Environmental Protection Agency (EPA) implemented a research program from 1996 to 2004 to measure UV at 21 unique locations throughout the United States to detect trends due to changes in the amount of stratospheric O_3.

Ultraviolet radiation characteristics. Energy reaching the Earth from the Sun facilitates life and also influences weather and climate. UV exists in the portion of the electromagnetic spectrum (in the range 10–400 nanometers) adjacent to the visible wavelengths spanning 400–700 nm.

UV has a higher energy level compared to visible light. It is usually divided into five wavelength ranges based on photon energies and biological effects. The extreme UV band consists of wavelengths between 10 and 120 nm, while the far UV band consists of wavelengths between 120 and 200 nm. The energy in the extreme UV and far UV bands has minimal impact on biological systems. Because biological impacts to living organisms are generally observed at wavelengths beginning at 200 nm, the three UV-wavelength bands that affect biological systems are designated as UV-A, UV-B, and UV-C (see **table**).

Impact of UV on human health and the environment. UV can affect human life, animal life, and the environment. Plants exposed to elevated UV-B levels display physiological changes, which indicate evidence of genetic damage and attempted deoxyribonucleic acid (DNA) repair. It has also been implicated in formation of tropospheric (ground-level) O_3 caused by the photolysis (light-initiated chemical reaction) of nitrogen dioxide (NO_2). UV also plays a role in the formation of photochemical smog [a mixture of NO, NO_2, O_3, volatile organic compounds (VOCs) and peroxy acetyl nitrate (PAN)]. Other effects associated with UV include ocular damage [such as cataracts, cornea damage (via photokeratitis, an ocular analog to sunburn), and skin damage (accelerated aging, skin cancer)].

UV influences the immunosuppressive response in humans, which may play a role in the development of skin cancer. The human skin chromophore (color component) undergoes photoisomerization (light absorption causing change in molecular structure) as a result of UV. *trans*-Urocanic acid, or *trans*-UA (shown below), is the chemical agent that influences the immunosuppression process. *trans*-UA forms in the skin when the amino acid histidine undergoes the process of nitrogen removal in the human liver. *trans*-UA absorbs UV in the same wavelength region as DNA (260–280 nm), so it was originally believed to act as a natural "filter" to protect DNA from UV damage. The photoisomerization process causes the *trans*-UA to change its molecular structure to *cis*-UA. *cis*-UA reduces the human immune response, which is believed to facilitate the formation of skin cancer.

UV instrument types. UV is measured by three major types of instruments, narrowband, broadband, and spectral. The performance and capability of these three instrument types varies depending on which portion of the UV spectrum is measured. Some instruments measure UV at discrete wavelengths, whereas others measure UV in narrow or wide wavelength bands.

Broadband instruments. Broadband instruments can be used to measure the biological effect of UV as a function of wavelength. Broadband instruments give a single number representing the integration of irradiant flux over a selected range of UV wavelengths through the use of weighting functions. This is important when assessing the biological response of living organisms to UV. With broadband instruments, the weighting functions are used to calculate numbers that are specific to different biological responses over various UV-wavelength intervals. Most broadband instruments have their spectral response function/spectral sensitivity function (SRF) instrument response set to approximate the erythemal action spectrum (for human skin reddening/sunburn), which occurs within the 280–320-nm range.

An action spectrum is determined by laboratory and field studies for each biological response of interest. Action spectra are specific for each biological effect [for example, sun burn/skin reddening (erythemal action spectrum)], and they specify the shape (relative magnitude) of the biological response to UV exposure. They are not used to assess the absolute magnitude of the response. These spectral response functions and their associated action spectra are used to indicate the relative impact of UV over a specific wavelength range, with respect to specific biological responses.

Broadband devices convert UV into visible (green) light through the fluorescence of magnesium tungstate. The voltage measured by these instruments is a function of the instrument response and the solar irradiance. The instrument response is a function of the solar zenith angle (SZA), the angle between the Sun/Earth line and the local vertical

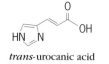

trans-urocanic acid

(zenith) position, and the total column ozone (the total amount of ozone in a column of air stretching from the Earth's surface to space).

Broadband instruments are relatively simple to operate and maintain. The measurements provided by broadband instruments are easy to reproduce. Broadband instrument behavior is consistent because the operation of these systems is based on an observable physical property (the fluorescence of magnesium tungstate). These instruments are the least expensive of the three major types to operate. A disadvantage is that they are sensitive to a wide range of wavelengths, which can lead to invalid results. Some earlier broadband instruments lacked temperature stabilization, which caused the operating temperature and instrument calibration temperature to be different. This flaw led to changes in instrument measurement sensitivity and wavelength range shifts.

Narrowband instruments. Narrowband instruments measure UV in a few, discrete wavelength bands. Their operation requires the use of metallic interference filters. These filters allow transmission of UV in wavelength intervals from 5 to 10 nm wide to as small as 2 nm for the latest generation of narrowband instruments. The use of multiple instruments permits simultaneous measurement of UV in different wavelength ranges. This is the type of instrument currently in use by the U.S. Department of Agriculture's (USDA) UV monitoring network.

Narrowband devices, such as radiometers, provide a measurement of light attenuation due to atmospheric aerosols. These instruments are simple and easy to operate. The metallic interference filters have a tendency to degrade over time, but ion deposition techniques can be used to improve filter stability.

Spectral instruments. Spectral instruments that measure UV are called scanning spectroradiometers. These instruments make continuous, spectrally resolved measurements across either the entire electromagnetic spectrum or specific portions of it. Spectral instruments are designed with various configurations, but most contain photomultiplier detectors with single or double monochromators. Most scanning spectroradiometers have double monochromators, which direct incoming solar radiation passing through an initial diffraction grating (with either 1200 or 2400 lines/nm) and through a middle slit that focuses the light onto a second diffraction grating. The multiple grating configuration minimizes stray light from adjacent wavelengths, which is caused by the rapid shift of UV intensity at wavelengths below 320 nm. Multiple diffraction gratings improve the wavelength resolution of scanning spectroradiometers, which can be as low as 0.5 nm.

It usually takes several minutes for scanning spectroradiometers to make one complete scan of the sky. This introduces a measurement anomaly when clouds pass overhead during a scan period. Spectral instruments with photodiode arrays or charge-coupled devices (CCDs) can record the entire spectrum simultaneously. Maintaining long-term calibration of spectral instruments requires great effort. They are more expensive to operate in comparison to broadband or narrowband instruments, because they require highly skilled operators.

Measuring ultraviolet radiation. Scanning spectroradiometers (for example, Brewer spectroradiometers) measure UV irradiance (spectral irradiance) at each wavelength in the instrument's measurement range (286.5–363 nm) during each scan of the instrument. Spectral irradiance is expressed as power density per unit wavelength (for example, milliwatts per square meter per nanometer [mW/m²/nm]).

An average of 30 instrument scans can be obtained during a typical summer day. The relationship between irradiance and wavelength can be graphed, but the graph alone does not give insight into which UV wavelengths/wavelength range(s) cause the most damage to humans and animals. Action spectra are used to indicate injury as a function of UV wavelength.

A. F. McKinley and B. L. Diffey developed the erythemal action spectrum to determine which wavelength range was most effective in producing the skin-reddening effect (sunburn) in humans. The erythemal action spectrum is used to measure human skin sensitivity to UV. In **Fig. 1**, the spectral overlap function (dotted line) is the product of the action spectrum (solid black line) and the spectral irradiance (dashed line). The spectral overlap function displays the portion of the UV spectrum where the biological response to UV is greatest. The spectral overlap function is normalized to one at 300 nm. This function displays the total spectral response (for the biological effect of interest) in the wavelength range where the action spectrum and the spectral irradiance overlap (300–320 nm for erythema [sunburn]).

The area under the spectral overlap function curve (dotted line) yields the biologically effective dose and is expressed in units of joules per square meter (J/m²). To calculate the biologically effective dose (daily dose), you must integrate the area under the

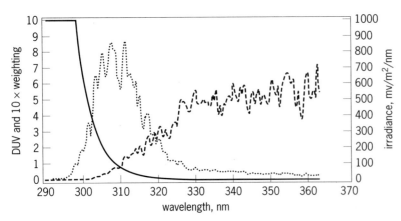

Fig. 1. Erythemal action spectrum illustrating wavelength range affecting human sunburn. The solid black line is the McKinley/Diffey erythemal action spectrum for sunburn, which is used to determine the effectiveness of UV in causing sunburn. The dashed line in is the spectral irradiance at a given location. The dotted line is the spectral overlap function.

spectral overlap function curve with respect to both wavelength (λ) and time as shown in Eq. (1):

$$\text{daily dose} = \int_t \int_\lambda I(\lambda, t) A(\lambda) \, d\lambda \, dt \qquad (1)$$

If the time duration of UV exposure is known with certainty, the biologically effective dose rate (BED) for a given instant in time can be calculated. This is accomplished by integrating the area under the spectral overlap function curve with respect to wavelength only, as shown in Eq. (2). In this case, the bi-

$$\text{BED} = \int_\lambda \text{UV}(\lambda) A(\lambda) \, d\lambda \qquad (2)$$

ologically effective dose rate is expressed in units of milliwatts per square meter (mW/m^2).

EPA's UV measurement data are stored in three types of files. One set of files contains unweighted and weighted (erythemal action spectra) data corresponding to the number of UV scans completed by a monitor during a day. A second set of files contains spectral data in the range 286.5–363 nm (steps of 0.5 nm) with a variable file length, based on the number of monitor UV scans completed in a particular day. The third set of files contains the daily integrated erythemally weighted UV data (DUV).

EPA's UV monitoring network. UV monitoring sites (**Fig. 2**) were located throughout the continental United States, Alaska (Denali National Park), Hawaii (Hawaii Volcanoes National Park), and the U.S. Virgin Islands (Virgin Islands National Park, St. Johns). Fourteen of the sites were located in national parks, and seven sites were located in urban settings.

Brewer spectrophotometer UV measurement equipment tracked the Sun and monitored the variation in solar energy throughout the day. UV was measured at wavelengths from 286.5 to 363 nm in 0.5-nm increments. Therefore, each full scan between the minimum and maximum wavelengths produced 154 separate UV intensity measurement values. The data collected at each network site could be used to calculate both the dose and dose rate of UV received at the Earth's surface at various times throughout the day. The UV data collected by the equipment at each site was processed through a quality assurance protocol to ensure proper characterization of the measured UV intensities.

EPA UV monitoring network results. The 21 sites in the EPA's UV network collected an aggregate of 40,474 days worth of UV data during the operational lifetime of the network. More than over 500,000 individual monitor scans were required to collect the data. Approximately 90% of the raw UV data collected successfully passed a rigorous secondary quality analysis consisting of nine separate screening criteria. Instrument measurements indicate that UV varies seasonally as expected, peaking during the summer months and falling to the lowest levels in winter. Analysis of the UV data indicated that the measured daytime UV levels have increased slightly in the Northeast and Southeast United States dur-

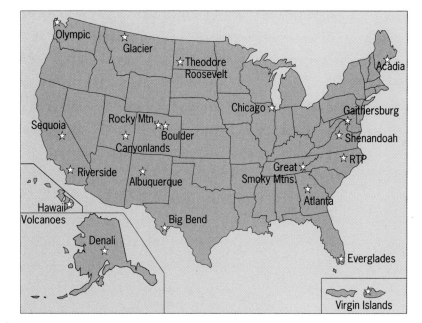

Fig. 2. Location of sites in the EPA's UV monitoring network.

ing the period 1996–2004. The UV measured in the Northwest and Southwest remained relatively constant over the last decade.

Changes in the amount of measured UV over time at each monitor location were analyzed to determine the overall trend (that is, increasing, decreasing, constant). UV measurements were taken at a solar angle of $60°$ and compiled for all monitor locations. By taking measurements at a constant solar angle, the daily and seasonal variations in UV were reduced along with effects of the sun angle. This allowed a consistent basis for comparison between the sites. **Figure 3** shows the trends in measured UV for each region of the United States.

The Northwest sites included Olympic, Glacier, and Theodore Roosevelt National Parks (**Fig. 4**). The Northeast sites included Chicago, Illinois, Gaithersburg, Maryland, and both Acadia and Shenandoah National Parks. The Southwest sites included Riverside, California, Boulder, Colorado, Albuquerque, New Mexico, along with Sequoia, Rocky Mountain, Canyonlands, and Big Bend National Parks. The Southeast sites included Atlanta, Georgia, Research Triangle Park, North Carolina, and Everglades and Great Smoky Mountain National Parks.

UV data from EPA's UV network can be used by researchers to understand the implications of increased UV linked to decreasing stratospheric ozone concentrations. The UV monitors provided accurate UV measurements of known quality for assessing the impact of domestic and international stratospheric ozone policies. The data allow scientists to evaluate the factors affecting UV in the environment. Our improved understanding of the effect of aerosol loading on UV can be incorporated into climate and atmospheric models. The EPA's quality-assured UV data is posted to a publicly accessible. The EPA has a 14-volume analysis of its collected UV data along with peer-reviewer comments on that analysis.

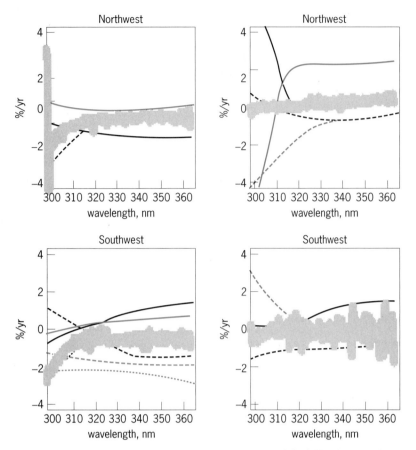

Fig. 3. UV trend measured in percentage per year by wavelength (nm). The thick gray line displays the overall regional trend and the black line displays the trend for individual monitor locations.

The EPA's UV monitoring research program collected, archived, and analyzed a tremendous amount of spectrally resolved UV data. The effect of local factors (such as clouds, haze, etc.) on surface UV at various sites must be adequately determined for various research objectives. EPA's UV measurements were collected in an attempt to answer a myriad of individual questions about UV intensities and their effects on humans, animals, plants, and the environment. Data from EPA's UV network allows characterization of local environment parameters and their effect on surface UV received at any particular site location. Ongoing analysis of the EPA's UV data will continue to provide more clues to how UV trends affect both human and ecosystem exposure.

[Disclaimer: The United States Environmental Protection Agency through its office of Research and Development funded and managed the research described here under the following funding vehicles: contract no. 68-D-04-001 with the University of Georgia at Athens; through Interagency Agreement Number DW13921546-01 with the National Institute for Standards and Technology (NIST); through Interagency Agreement Number DW139392300-01 with the National Oceanic and Atmospheric Administration (NOAA); and through Interagency Agreement no. DW14939466-01-0 with the National Parks Service. It has been subjected to the Agency review and approved for publication. Mention of trade names or commercial products does not constitute endorsement or recommendation for use.]

For background information *see* AEROSOL; AIR POLLUTION; ALBEDO; ATMOSPHERE; ATMOSPHERIC CHEMISTRY; CHARGE-COUPLED DEVICES; DEOXYRIBONUCLEIC ACID (DNA); ELECTROMAGNETIC RADIATION; ENVIRONMENTAL ENGINEERING; IMMUNOSUPPRESSION; NITROGEN OXIDES; PHOTODIODE; RADIATION BIOLOGY; RADIOMETRY; SOLAR RADIATION; STRATOSPHERIC OZONE; TERRESTRIAL RADIATION; ULTRAVIOLET RADIATION; ULTRAVIOLET RADIATION (BIOLOGY) in the McGraw-Hill Encyclopedia of Science & Technology.   Eric S. Hall

Bibliography. W. F. Barnard et al., Daily surface UV exposure and its relationship to surface pollutant measurements, *J. Air Waste Manage. Assoc.*, 53:237–245, 2003; *Brewer MKIV Spectrophotometer Operator's Manual*, SCI-TEC Instruments, Inc., 1503 Fletcher Road, Saskatoon, Saskatchewan, Canada S7M 5S5, OM-BA-C231, REV B, August 15, 1999; J. Brookman, J. N. Chacon, and R. S. Sinclair, Some photophysical studies of *cis*- and *trans*-urocanic acid, *Photochem. Photobiol. Sci.*, 1(5):327–332, 2002; R. W. Ditchburn, *Light*, Dover, New York, 1991; M. Kerker, *The Scattering of Light and Other Electromagnetic Radiation*, Academic Press, New York, 1969; P. S. Koronakis et al., Interrelations of UV-global/global/diffuse solar irradiance components and UV-global attenuation on air pollution episode days in Athens, Greece, *Atmos. Environ.*, 36:3173–3181, 2002; J. Lenoble, *Atmospheric Radiation Transfer*, A. Deepak Publishing, 1993; S. Madronich et al., Changes in biologically active ultraviolet radiation reaching the Earth's surface, *J. Photochem. Photobiol. B*, 46(1–3):5–19, 1998; A. McCulloch, P. M. Midgley, and P. Ashford, Releases of refrigerant gases (CFC-12, HCFC-22, and HFC-134a) to the atmosphere, *Atmos. Environ.*,

Fig. 4. Operational UV monitor located in Theodore Roosevelt National Park, North Dakota.

37(7):889–902, 2003; A. F. McKinley and B. L. Diffey, A reference action spectrum for ultraviolet induced erythema in human skin, *CIE J.*, 6:17–22, 1987; A. V. Parisi, J. Sabburg, and M. G. Kimlin, *Scattered and Filtered Solar UV Measurements*, Advances in Global Change Research, vol. 17, Kluwer Academic Publishers, 2004; G. Ries et al., Elevated UV-B radiation reduces genome stability in plants, *Nature*, 406(6791):98–101, 2000; W. L. Ryan and D. H. Levy, Electronic structure and photoisomerization of *trans*-urocanic acid in a supersonic jet, *J. Am. Chem. Soc.*, 123:961–966, 2001; E. C. Weatherhead and J. E. Frederick, *Report on Geographic and Seasonal Variability of UV Affecting Human and Ecological Health*, Report to U.S. Environmental Protection Agency (14 volumes), Contract no. 4D-5888-WTSA, 2005.

Hallucinations

An hallucination can be defined as a sensory experience that occurs in the absence of corresponding external stimulation of the relevant sensory organ(s). In addition, the experience has a sufficient sense of reality to resemble a veridical (genuine) perception and occurs when one is awake; furthermore, one does not feel that one has direct and voluntary control over it. Thus, an hallucinating person sees things that are not there or hears voices while nobody in the vicinity is speaking, for example. Although visual and auditory hallucinations are most common, hallucinations of taste, smell, and touch can also occur. The exact mechanisms underlying the experience of hallucination remain elusive, but significant strides have been made in recent years to uncover processes in the mind and brain that are associated with hallucinations.

Scope of hallucinations. Hallucinations can occur in the context of a wide variety of neurological and psychiatric disorders. For example, visual hallucinations frequently occur in Lewy body dementia (a progressive brain disease leading to degenerative dementia) and Parkinson's disease. People with acquired blindness may develop Charles Bonnet syndrome, characterized by vivid visual hallucinations. Such a patient may vividly see a dog sitting in his or her living room, but at the same time be perfectly aware that there is no real dog. Patients with Charles Bonnet syndrome are usually not emotionally upset by their hallucinations. Things are different for patients with schizophrenia, however. A large percentage of these patients experience distressing hallucinations when they have psychotic episodes (in which they lose contact with reality). Auditory-verbal hallucinations (hearing "voices") are most common among these patients, although hallucinations in other modalities may also occur. Patients may hear voices speaking with derogatory and abusive content. As well as in neurological and psychiatric illness, hallucinations can occur after ingestion of psychoactive substances, such as LSD (lysergic acid diethylamide) and amphetamines. These drugs affect one or more neurotransmitter systems in the brain associated with hallucination: dopamine, serotonin, acetylcholine, or glutamate. Finally, hallucinations have also been reported in healthy people. As much as 5–10% of the normal population may occasionally experience brief hallucinatory experiences (for example, hearing somebody call your name and discovering there is nobody). An important difference from hallucinations in psychiatric patients is that hallucinations in healthy people are generally not distressing and do not interfere with adequate functioning in daily life.

Hallucinations in the brain. Research using functional magnetic resonance imaging (fMRI) scans to measure brain activation during hallucinations has shown that hallucinations typically activate perceptual areas in the brain. In the case of hearing "voices" in schizophrenia (the most frequently studied types of hallucinations are those in the auditory realm experienced by persons with schizophrenia), this is the speech perception area, the temporoparietal junction in the left hemisphere (**Fig. 1**). However, activation of the same area in the right hemisphere has also been reported, which may have to do with the reduced lateralization of brain areas responsible for speech processing in patients with schizophrenia (in healthy people, speech predominantly activates the left hemisphere). In addition, activation of Broca's area (involved with speech production) has been observed in patients with auditory hallucinations. The lack of activation of the dorsolateral prefrontal cortex (involved in cognitive control) may reflect the fact that patients experience the hallucinations as unintentional, in contrast to mental imagery that does activate this brain area in healthy people. Visual hallucinations have been shown to activate the visual association cortex. Interestingly, a study of patients with Charles Bonnet syndrome showed a striking correspondence between the content of the hallucinations and the activated brain regions.

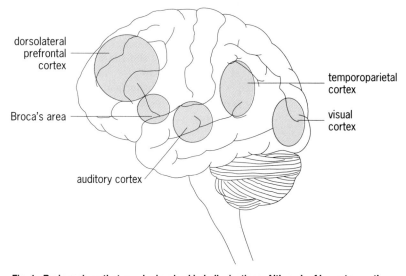

Fig. 1. Brain regions that may be involved in hallucinations. Although of importance, the thalamus is not depicted here, as it is a structure that is buried deep inside the brain. (*Copyright © 2008 by the American Psychological Association; reproduced with permission from A. Aleman and F. Larøi, Hallucinations; the Science of Idiosyncratic Perception, APA Books, Washington, D.C., 2008*)

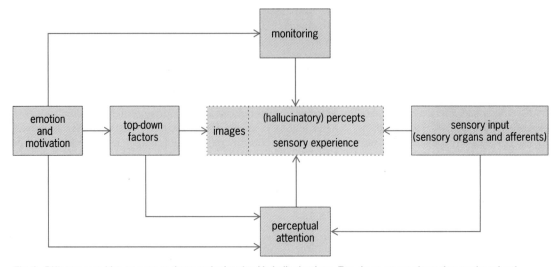

Fig. 2. Different cognitive processes that may be involved in hallucinations. Top-down processing refers to the role of perceptual expectations. (*Copyright © 2008 by the American Psychological Association; adapted with permission from A. Aleman and F. Larøi, Hallucinations; the Science of Idiosyncratic Perception, APA Books, Washington, D.C., 2008*)

Thus, patients with colorful hallucinations activated the color-processing region of the brain, V4. This region was not activated in a patient who hallucinated in black and white. Hallucinations of human faces activated the fusiform face area, which in healthy people is activated when they see real faces.

Top-down processing. It is unclear how these perceptual areas become overactivated, thereby leading to the experience of hallucinations. A recent hypothesis is that overactive perceptual expectations, referred to as "top-down processing," are an important factor in the genesis of some hallucinations. This term refers to activation of attentional and perceptual systems prior to sensory inputs in order to "prepare" the brain to register these inputs as coherent perceptions corresponding to objects or events in the world. If such top-down processing (that is, processing originating in the brain and not from input through the senses) becomes too active, perceptual areas could create images that are as vivid and formed as those arising from incoming information through the senses. **Figure 2** illustrates cognitive processes involved in perception and attention that may contribute to hallucinations. The model suggests different routes toward hallucination. Overactivation and consequently hyperexcitation of perceptual association areas can occur through sensory impairment or aberrant activation from the thalamus, possibly arising from disturbances involving dopamine or acetylcholine systems that regulate interactions between brain regions. This route may be involved in neurological and pharmacological hallucinations. Psychiatric hallucinations may arise from a more "cognitive" route (left part of Fig. 2), in which affective states ("emotion and motivation" in the figure) can trigger internally oriented mental processes and compromise reality monitoring. Reality monitoring refers to the ability to distinguish whether information is internally generated or externally presented. Emotional states can also strongly affect perceptual expectations and bias perception.

Magnetic stimulation. With regard to the treatment of hallucinations, antipsychotic medication may be effective in up to 75% of patients with schizophrenia or other psychoses. However, treatment adherence is poor in a substantial part of the patients, presumably because of unpleasant side effects. In recent years, a novel treatment has been introduced, based on the findings from neuroimaging studies. In this still-experimental treatment, magnetic stimulation of the brain is used to reduce overactivation of speech perception regions. In transcranial magnetic stimulation (TMS), a time-varying magnetic field is generated by a current pulse through a stimulator coil placed over a certain scalp position. The rapid rise and fall of the magnetic field induces a flow of current in the underlying brain tissue and hence influences the activation of neurons. Stimulation at a frequency of one pulse per second has been shown to reduce brain excitability. For treatment of hallucinations in schizophrenia, the targeted scalp position is generally a position that corresponds with the temporoparietal junction. Several studies have now shown that treatment with one pulse per second for at least 15 minutes consecutively, done for at least 5 days, can significantly reduce hallucinations in patients with medication-resistant hallucinations. Patients did not report any major side effects. However, up to now, only small-scale studies have been published, and larger studies are needed to replicate these promising findings and to examine the optimal treatment schemes (that is, in terms of duration, frequency, and intensity of stimulation) in more detail. Researchers are now also using fMRI scans to localize individual brain activity during hallucinations and target these specific areas using TMS.

Conclusions. In summary, hallucinations are puzzling phenomena in which a person gets the impression of perceiving something through the senses, whereas there is actually nothing "out there." In normal perception, signals emanating from the sensory organs play a decisive role in determining what one perceives. In hallucinations, however, internally generated signals emanating from cortical brain centers play a decisive role in determining the sensory experience. Although researchers are still in the process of unravelling the mechanisms that may underlie this, novel treatments are already emerging from neuropsychological approaches.

For background information *see* AFFECTIVE DISORDERS; BRAIN; COGNITION; MEDICAL IMAGING; NEUROBIOLOGY; PARKINSON'S DISEASE; PERCEPTION; PSYCHOPHARMACOLOGY; PSYCHOSIS; PSYCHOTOMIMETIC DRUG; SCHIZOPHRENIA in the McGraw-Hill Encyclopedia of Science & Technology.

André Aleman

Bibliography. A. Aleman and F. Larøi, *Hallucinations: The Science of Idiosyncratic Perception*, APA Books, Washington, D.C., 2008; D. Collerton, E. Perry, and I. McKeith, Why people see things that are not there: A novel perception and attention deficit model for recurrent complex visual hallucinations, *Behav. Brain Sci.*, 28:737–794, 2005; A. S. David, The cognitive neuropsychiatry of auditory verbal hallucinations: An overview, *Cogn. Neuropsychiat.*, 9:107–124, 2004; R. E. Hoffman et al., Temporoparietal transcranial magnetic stimulation for auditory hallucinations: Safety, efficacy and moderators in a fifty patient sample, *Biol. Psychiat.*, 58:97–104, 2005.

Heat-treated wood

It has long been known that heating wood will change its properties. In ancient Africa, natives hardened wood spears by placing a sharpened straight wood stick in the bottom of glowing coals and then pounded the burned end with a rock, repeating this process many times until the end was sharp and hard. The Vikings burned the outside of their ships to make them water-resistant and flame-retardant. In the early part of the twentieth century, it was found that drying wood at high temperature increased dimensional stability (resistance to swelling and shrinkage) and reduced hygroscopicity (ability to take up and retain moisture). Later, it was found that high-temperature drying also increased resistance to microbiological attack. At the same time, the increase in stability and durability increased brittleness and loss in some strength properties, including impact toughness (resistance against breaking after a sharp blow is applied), modulus of rupture (maximum surface stress in a bent beam at the instant of failure), and work to failure (the energy needed to cause fracture). The treatments usually caused a darkening of the wood, and the wood had a tendency to crack and split.

Description. Wood can be heat-treated in several ways: heating wood in the presence of moisture, heating wood in the presence of moisture followed by compression, heating dry wood, and heating dry wood followed by compression. The effect of the heating process on wood properties depends on the process itself. As the wood is heated, the first mass loss is due to the loss of water, followed by a variety of chemical reactions that produce degradation products and volatile gases. As the temperature increases, wood cell wall polymers start to degrade. Pyrolysis (a chemical process in which a compound is converted to one or more products by heat) of the hemicelluloses takes place at about 270°C (518°F), followed closely by that of the cellulose. Lignin is much more stable to high temperature.

Many of the commercial heat-treating processes take place in the absence of air at temperatures in the range 180–260°C (356–500°F) for times ranging from a few minutes to several hours. Temperatures lower than 140°C (284°F) result in less change in physical properties, whereas heating above 300°C (572°F) results in severe wood degradation. Wood has been heated in steam, in inert gases, below molten metal, and in hot oil baths. The mechanism of stability and durability is thought to be due to a loss of hygroscopic hemicellulose sugars and their conversion to furan-based polymers that are much less hygroscopic, and the lost sugars reduce the ability of fungi to attack the heated wood. The mass loss is proportional to the square of the reduction in swelling.

There are a variety of thermal modification processes that have been developed. The results of the process depend on several variables including time and temperature, treatment atmosphere, wood species, moisture content, wood dimensions, and the use of a catalyst. Temperature and time of treatment are the most critical elements, and treatments done in air result in oxidation reactions that do not lead to the desired properties of the treated wood. Generally, mass loss occurs to a higher extent in hardwoods as compared to softwoods.

Several names have been given to the various heat-treated products and heat treatments for wood. These include Staypak® and Staybwood® in the United States, Lignostone® and Lignofol® from Germany, Jicwood® and Jablo® from the United Kingdom, ThermoWood® in Finland, Plato® in the Netherlands, and Perdure® and Retification® in France.

Heating in the presence of moisture. Wood is commonly heated in the presence of moisture for bending and for stability.

Heating for bending. The traditional way of bending wood is to heat dry or wet wood with steam. Heating takes a long time, depending on the thickness of the piece of wood (**Table 1**).

TABLE 1. Mass loss due to heating wood in the presence of moisture

Heating temperature, °C	Time, h	Mass loss, %
120	2	2–3
120	10	5
180	10	10–12
190	2	4–7
190	12	5–10
200	10	10–17
220	1	6–7
235	4	25
240	0.5	7

TABLE 2. Decrease in equilibrium moisture content (EMC) due to heating wood in the presence of moisture

Heating temperature, °C	Time, h	Decrease in EMC, %
190	2	25
190	24	50
220	2	50
235	2	60

Green wood (freshly cut wood that is unseasoned and having a high moisture content) can also be heated using a microwave to produce a softened wood that can easily be shaped into various products such as rocker rails, chair backs, chair legs, etc. Since microwaves heat wet wood from the inside out and steam heats wood from the outside in, microwave heating/bending is much faster and more efficient than bending wood using steam.

Heating wet wood for stability. Wood with a moisture content close to its equilibrium moisture content (EMC; the point where the amount of moisture in the wood balances with that in the atmosphere) can be heated to 180–200°C (356–392°F), resulting in a wood with greatly reduced moisture content (**Table 2**). The high temperature degrades the hemicellulose sugars to furan-based intermediates and volatile gases. The furan intermediates have a lower EMC than the sugars and increase bonding of the wood structure. At a mass loss of approximately 25%, the EMC is lowered by almost the same percentage. Dimensional stability is also increased, but not as much as heating followed by compression (discussed below).

There are two main commercial processes based on heating wet wood for stability and increased biological resistance: ThermoWood® and Plato®. ThermoWood was developed by VTT [Valtion Teknillen Tutkimuskeskus (Finish Wood Research Center)] in Finland and is produced at a plant in Mänttä, Finland. The plant started in the early 1990s with a capacity of 50,000 m³ and has increased to 75,000–80,000 m³ today. The three-stage process is done in the presence of steam. The steam helps protect the wood from oxidative reactions. In the first stage, the wood is heated to 100°C (212°F) for almost 20 h. In the second stage, the wood is heated to 185–230°C (365–446°F) for 10 h followed by the lowering of the temperature in the presence of a water spray. Two types of products are produced: Thermo-S (light brown in color), which is done at 190 ± 3°C (374 ± 5°F) for softwoods and 185 ± 3°C (365 ± 5°F) for hardwoods;

and Thermo-D (dark brown in color), which is done at 212 ± 3°C (414 ± 5°F) for softwoods and 200 ± 3°C (392 ± 5°F) for hardwoods.

Toughness and abrasion resistance of ThermoWood are reduced. ThermoWood has improved durability in pure culture fungi tests, but does not perform well in mixed culture tests (**Table 3**).

Plato (Proving Lasting Advanced Timber Option) wood was developed by Royal Dutch Shell in the Netherlands and involves a four-stage process. The first stage involves heating the wood to 150–180°C (302–356°F) under high-pressure steam for 4–5 h. The wood is then dried to a moisture content of 8–10% and then heated again at 150–190°C (302–374°F) for 12–16 hours, resulting in a drop in moisture content to less than 1%. The wood is then conditioned to 4–6% moisture over a 3-day period.

There is a plant in Arnhem, the Netherlands, producing 12,000 m³ of heat-treated wood per year. Plato wood is dark brown in color, but will weather to the normal gray color in time. It has a 5–20% decrease in modulus of rupture, but a slightly higher modulus (degree) of elasticity.

In addition, the Le Bois Perdure process has been commercialized by PCI Industries based in Quebec, Canada. The process involves drying and heating the wood at 200–230°C (392–446°F) in steam.

All heat-treated woods are glueable and paintable and are used for furniture, flooring, decking, door and window components, and exterior joinery.

Heating in the presence of moisture and compression. When wet wood is heated to 180–220°C (356–428°F) and compressed, the wood structure compresses and remains in this compressed state when dried. The compressed wood is much harder and has a much higher modulus of rupture and elongation. Rewetting the compressed wood reverses the process and the wood swells back to its original thickness.

Heating dry wood. Western white pine has been heated beneath the surface of a molten metal. The metal was a mixture of lead, tin, and cadmium (50:30:20 w/w) and had a melting point of approximately 150°C (302°F). The heating time varied depending on the wood thickness. The heated wood had a greatly reduced EMC and increased dimensional stability, but reduced toughness, hardness, and strength. This process has never been commercialized.

There is one commercial process producing dry heated wood products. Retification is a process developed in France by École des Mines de St. Etienne and involves heating wood in a nitrogen atmosphere

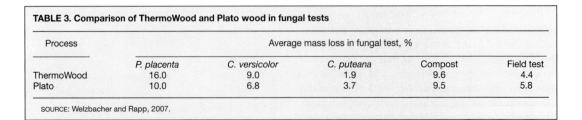

TABLE 3. Comparison of ThermoWood and Plato wood in fungal tests

Process	Average mass loss in fungal test, %				
	P. placenta	*C. versicolor*	*C. puteana*	Compost	Field test
ThermoWood	16.0	9.0	1.9	9.6	4.4
Plato	10.0	6.8	3.7	9.5	5.8

SOURCE: Welzbacher and Rapp, 2007.

to 180–250°C (356–482°F) for several hours. There is a plant near St. Etienne with a capacity of 13,000 m³ per year.

Heating dry wood followed by compression. When dry yellow-poplar is heated and compressed, the thermoplastic lignin flows, giving rise to a product that has increased density, strength properties, and dimensional stability. A temperature range of 150–170°C (302–338°F) was used and the wood was compressed while heated.

Conclusions. Heating wood under a variety of conditions is an environmentally benign process requiring no added chemicals and gives rise to a variety of products with reduced moisture contents and some durability against biological degradation. Most physical properties are reduced, especially abrasion resistance and toughness. While heat-treated wood is not as dimensionally stable or biologically resistant as acetylated wood (Accoya®) or furfurylated wood (Kebony®), it has found applications where some stability and durability are required and where some loss of strength is not critical.

For background information *see* DRYING; HEAT; MOISTURE-CONTENT MEASUREMENT; STRENGTH OF MATERIALS; WOOD ANATOMY; WOOD DEGRADATION; WOOD ENGINEERING DESIGN; WOOD PROCESSING; WOOD PRODUCTS; WOOD PROPERTIES in the McGraw-Hill Encyclopedia of Science & Technology.

Roger M. Rowell

Bibliography. C. A. S. Hill, *Wood Modification: Chemical, Thermal, and Other Processes*, Wiley, Chichester, West Sussex, UK, 2006; M. Inoue and M. Norimoto, Permanent fixation of compressive deformation in wood by heat treatment, *Wood Res. Tech. Notes*, 27:31–40, 1991; M. Inoue et al., Steam or heat fixation of compressed wood, *Wood Fiber Sci.*, 25(3):404–410, 1993; R. M. Rowell, Specialty treatments, in *Wood Handbook*, USDA, Forest Service, Forest Products Laboratory, General Technical Rep. FPL-GTER-113, chap. 19, pp. 1–14, 1999; R. M. Rowell et al., Modification of wood fiber using steam, *Proceedings of the 6th Pacific Rim Bio-Based Composites Symposium*, Portland, OR, 2002; F. Shafizadeh and P. P. S. Chin, Thermal degradation of wood, in I. S. Goldstein (ed.), *Wood Technology: Chemical Aspects*, ACS Symposium Series 43, pp. 57–81, 1977; A. J. Stamm, *Wood and Cellulose Science*, Ronald Press, New York, NY, 1964; USDA, *Wood Handbook*, Forest Service, Forest Products Laboratory, General Technical Rep. FPL-GTER-113, 1999; C. R. Welzbacher and A. O. Rapp, Durability of thermally modified timber from industrial-scale processes in different use classes: Results from laboratory and field tests, *Wood Mater. Sci. Eng.*, 2:4–14, 2007.

History and art of mushrooms for color

While artist Miriam C. Rice was teaching a children's art class about natural dyes at the Mendocino Art Center in Mendocino, California, in the early 1970s, she took a clump of bright yellow *Hypholoma fas-*

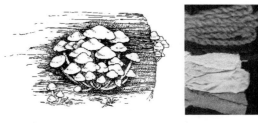

Fig. 1. *Hypholoma fasciculare* and its dyes on wool and silk fibers. (*Illustration courtesy of Dorothy M. Beebee, from M. C. Rice, Mushrooms for Dyes, Paper, Pigments & Myco-Stix™, Mushrooms for Color Press, Forestville, CA, 2007*)

ciculare mushrooms and tossed them into a pot of simmering hot water with a bit of wool yarn. Surprisingly, a clear, bright lemon-yellow dye emerged on the wool—and "mushroom dyeing" was born (**Fig. 1**).

Many fiber artists in the Mendocino area became intrigued with the possibilities revealed in her experimenting. As a result, they continually provided Miriam Rice with new mushrooms to use for color. She experimented with every mushroom that she found for dye potential, attending mushroom fairs and forays, and learning to identify the mushrooms that she was utilizing in her dye experiments. Over a period of several years, she gradually built up a vast collection of labeled samples of mushroom-dyed wool, silk, and cotton. These groundbreaking results quickly spread to textile artists all over the world.

Mushroom dye process. The mushroom dye process is basically the same as for all natural dyes. The fungi are chopped up or mashed and added to water simmered in a nonreactive pot, such as enamel or stainless steel. Usually a 1:1 ratio of mushroom to fiber is used. The dye pot is gently simmered for a specified length of time, usually 30 to 60 min, or until a desired depth of color is achieved in the fibers. Protein fibers, such as wool, silk, mohair, and angora, accept the dyes most readily, although cotton, linen, hemp, and some synthetic fibers may also be successfully used.

For a mushroom to produce a dye, it must contain a water-soluble pigment that will be resistant to sunlight (lightfast) and washing (colorfast, that is, having color that does not fade or run). Many mushrooms contain pigments that will make excellent lightfast and colorfast dyes. These are called "substantive dyes." However, lightfastness and colorfastness are improved by pretreating the fiber with a "mordant." Mordants (metallic salts) are simmered with the fiber in a hot water bath to enable molecular

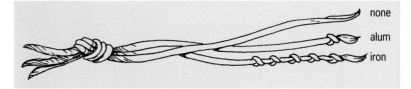

Fig. 2. The current knotting system for mordanted yarns. (*Illustration courtesy of Dorothy M. Beebee, from M. C. Rice, Mushrooms for Dyes, Paper, Pigments & Myco-Stix™, Mushrooms for Color Press, Forestville, CA, 2007*)

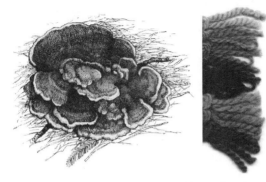

Fig. 3. Polypore *Phaeolus schweinitzii* with samples of premordanted wool fibers. (*Illustration courtesy of Dorothy M. Beebee, from M. C. Rice, Mushrooms for Dyes, Paper, Pigments & Myco-Stix™, Mushrooms for Color Press, Forestville, CA, 2007*)

Fig. 4. *Dermocybe phoenicea* var. *occidentalis* with the top three premordanted yarns dyed only with caps and the bottom three premordanted yarns dyed only with the stipes. (*Illustration courtesy of Dorothy M. Beebee, from M. C. Rice, Mushrooms for Dyes, Paper, Pigments & Myco-Stix™, Mushrooms for Color Press, Forestville, CA, 2007*)

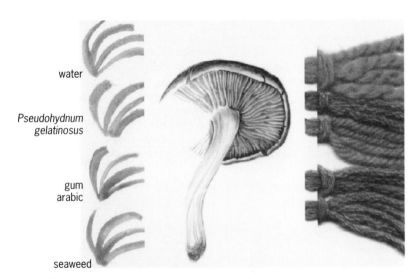

water

Pseudohydnum gelatinosus

gum arabic

seaweed

Fig. 5. Watercolor paints and wool dyes from the mushroom *Dermocybe cinnamomea*. (*Illustration courtesy of D. M. Beebee, from M. C. Rice, Mushrooms for Dyes, Paper, Pigments & Myco-Stix™, Mushrooms for Color Press, Forestville, CA, 2007*)

bonding with the fiber. Then, in a subsequent dye bath, the mushroom pigment, in turn, bonds with the mordant on the fiber. This enables the pigment molecules to make a stronger bond than they would without the mordant, thereby increasing the potential for lightfastness and colorfastness. A mordant also has the potential to change the original dye color.

Miriam Rice originally used five different mordants in her mushroom dye experiments: potassium aluminum sulfate (alum), potassium dichromate, stannous chloride, copper sulfate, and iron sulfate. She developed a knotting system for differentiating different mordanted yarns that is now in universal use among mushroom dyers, enabling them to keep track of which mordanted yarns produce which colors. A series of knots was tied to the end of each skein of yarn:

no knot = no mordant
1 knot = alum mordant
2 knots = potassium dichromate
3 knots = stannous chloride
4 knots = copper sulfate
5 knots = iron sulfate

However, after many years of research, it was found that alum and iron were the only two "safe" mordants that should be employed. The same knotting system remained—with 1 knot for alum mordant and 5 knots for iron mordant—reminding dyers that the missing knots are for those mordants that are considered unsafe for use (**Fig. 2**).

Cream of tartar salt (potassium acid salt of tartaric acid) and Glauber's salt (colorless hydrated sodium sulfate) are often used with mordants to help in distributing the dye evenly in the fiber. Other additives can be added to the dye baths to enhance and alter the color hues by changing the acidity or alkalinity (pH value) of the dye bath. White vinegar is the preferred additive to change the pH of a mushroom dye toward acidic values. Washing soda is used for changing the pH toward alkaline values.

Mushroom dyes. The first successful mushroom dyes were in the range of what are generally called "earth hues"—yellow, gold, orange, burnt sienna, brown, and every shade in between, as exemplified by the dyes from *Phaeolus schweinitzii* (**Fig. 3**). A breakthrough into a new range of color came in the guise of a little brown mushroom that yielded red, rose, and burgundy hues. It was identified by mycologists at the University of Washington as belonging to the Cortinarius family. Many of the species in this group contain extremely lightfast anthraquinone pigments, especially red (**Fig. 4**). Continuing experimentation produced lovely purple, blue, and green hues from other fungi, continuing to enchant and inspire textile artists around the world with this expanded palette of extraordinary dye.

In general, fresh fungi are preferred for dye use, but dried or fresh-frozen mushrooms—and in some cases, mushrooms that are old (bordering on "rotten")—often produce the most intense color. Freezing causes chemical changes to the latent dye pigment—that is, red dyes tend more toward blue tones. The quality of the water used influences the dye as well, especially if chlorine, acids, alkalines, or heavy minerals are present. A weight ratio of 1:1 of mushroom to fiber is recommended at the start. Some fungi have a lot of pigment, whereas others have very little. Thus, with experimentation, careful observation, and note taking, one can decide the best proportions on a case-by-case basis.

Since the First International Mushroom Dyes Textile Show–FUNGI and FIBERS was held in Mendocino in the summer of 1980, interest in mushroom dyes has continued to spread worldwide rapidly, with particular enthusiasm from the Scandinavian countries, Scotland, and Western Australia. In 1985, the International Mushroom Dye Institute (IMDI) was established. The nonprofit research institute was founded to encourage the use of fungal pigments, to further research on their extraction and employment, to encourage research on cultivation of especially desirable fungi, and to provide financial aid to artists and researchers to enable them to participate in the international mushroom dye symposia and exhibitions.

Other developments. In the 1980s, Miriam Rice explored the possibility of making paper out of the fungal detritus that is left over from the dyes. Always a passionate advocate of recycling, she saw this as the natural solution for disposal of the fungal residue from the dye process. She tried many mushrooms and polypores (shelf fungi typically lacking stalks), finding that the polypores that contain chitin as well as cellulose produced the best paper—all colors and textures, and with no additives necessary. Thus, in 1985, the concept of papermaking from fungi was introduced.

Further experiments in the 1990s led in another direction: specifically, making watercolor paints from many of the mushrooms that had also been used for dye. By extracting the fungal pigments and combining them with a variety of media, a variety of watercolor paints have been generated (**Fig. 5**).

In addition, recent research into developing the mushroom pigments into some form of medium for artists to use in drawing and sketching to supplement the watercolor paints has resulted in a drawing medium called "Myco-Stix™." This medium uses pigments extracted from fungi and combines them with a variety of binders to form crayons that can be used as drawing, watercolor, pastel, and encaustic (hot wax) media.

For background information *see* COLOR; DYE; DYEING; FUNGAL ECOLOGY; FUNGI; MORDANT; MUSHROOM; PAPER; PIGMENT (MATERIAL); PRINTING; WOOL in the McGraw-Hill Encyclopedia of Science & Technology.   Dorothy M. Beebee

Bibliography. A. R. Bessette and A. E. Bessette, *The Rainbow beneath My Feet: A Mushroom Dyer's Field Guide*, Syracuse University Press, Syracuse, NY, 2001; K. L. Casselman, *Craft of the Dyer: Colour from Plants and Lichens*, Dover Publications, New York, 1993; M. C. Rice, *Let's Try Mushrooms for Color*, Thresh Publications, Santa Rosa, CA, 1974; M. C. Rice, *Mushrooms for Color*, Mad River Press, Eureka, CA, 1980; M. C. Rice, *Mushrooms for Dyes, Paper, Pigments & Myco-Stix™*, Mushrooms for Color Press, Forestville, CA, 2007.

Human-computer interface design

Advances in computational technology have made computerized devices an integral part of our daily lives. Unfortunately, poorly designed interfaces often make these devices difficult and awkward to use. A conceptual framework for the design of effective human-computer interfaces will be outlined.

Cognitive Systems Engineering (CSE). CSE provides a systematic approach to interface design. The interface is considered a form of decision support: it is used by a person with the goal of completing a task within a work domain. Thus, there are three system components (domain, user, and interface) that contribute a set of mutually interacting "constraints" (**Fig. 1**). How well these components "fit together" will determine the ease of use and the effectiveness of the device.

Domain constraints. Because the interface is viewed as decision support, the logical starting point for interface design is an analysis of the constraints in the work domain. The unfolding events in "law-driven" domains such as process control (for example, a power plant) arise from the physical structure and functionality of the system. In the case of a power plant, the laws of thermodynamics are a fundamental source of regularity. At the opposite end of the

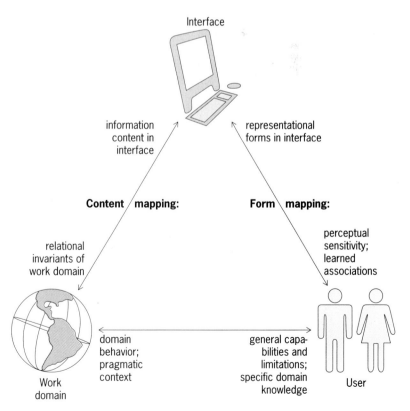

Fig. 1. The Cognitive Systems Engineering (CSE) perspective on interface design: system components and mappings between them.

spectrum are "intent-driven" domains, where the unfolding events arise from the user's intentions, goals, and needs (for example, information retrieval).

User constraints. The capabilities and limitations of users constitute another set of constraints. For example, humans possess extremely powerful skills for perception (obtaining visual information) and action (manipulating objects), but they have extreme limitations in terms of attention and memory (only a few items can be maintained in working memory). Additional constraints are introduced by those who will use the device. Power-plant operators are highly trained and will have homogeneous (uniform) skill sets; this is in contrast to mobile cell phone users, who do not require this level of training and skill.

Interface constraints. A third set of constraints is introduced by the interface. In the early years of computer technology, the interface was designed using an ineffective "conversation" design metaphor: cryptic verbal instructions were communicated through an intermediary (that is, a command line) to the computer. This imposed a severe set of constraints on interaction. Interface technology has evolved considerably since then (high-resolution screens, bit-mapped graphics, precise pointing devices, and so on). This has allowed the use of design metaphors that are predominantly spatial in nature and are far more effective (for example, the desktop metaphor).

Principles of design. The fundamental goal of interface design is to build virtual ecologies of work domains that allow the powerful perception-action skills of the human to be leveraged (in other words, that do not require the use of limited-capacity resources such as working memory). The ease of use and effectiveness of a device will ultimately depend upon the relationships (the "fitting together") between these three sets of constraints (Fig. 1). Three principles of interface design can be used to help these three system components fit together well.

Direct perception. The interface should allow the user to comprehend the current (and perhaps the future) state of the system through powerful visual processes. The first step is to conduct work domain analyses to reveal the domain constraints (physical, functional, and goal-related properties). Effective graphical representations (the virtual ecology) then need to be developed. The visual properties of these representations must match the perceptual and cognitive abilities of the observer (for example, is the user sensitive to variations in the visual features that were chosen?). They must also match the domain constraints (for example, do variations in the visual features accurately reflect variations in, or properties of, the domain?).

Direct manipulation. The interface should allow the user to execute control input through powerful action capabilities applied directly to objects in the interface. A good example is file deletion in the Macintosh OS X operating system. An object in the work domain (a computer file) is represented in the interface (its icon). The user manipulates this representation directly (that is, point, click, drag, drop)

to delete it. In contrast, the deletion of a file via a command-line interface (for example, "rm klunkyinterfaces.doc," where "rm" stands for "remove") or even a pull-down menu is not direct manipulation.

Perception-action loop. There is a potentially symbiotic relationship between direct perception and direct manipulation. The objects in the interface can be designed to support the powerful human skills of both perception and action simultaneously. When this occurs, the display interface and the control interface are merged into one; the perception-action loop (the coupling between a human and the world) is intact.

Examples of interface design. The application of these three general principles will require different design strategies for different work domains. Two examples will be provided.

Law-driven domain. The most effective interface design strategy for law-driven domains is to develop spatial analogs (that is, abstract geometrical forms) that directly reflect the underlying domain constraints (**Fig. 2**). This display combines over 100 individual sensor values into an octagonal form. The octagon is perfectly symmetrical when power-plant conditions are normal (Fig. 2a). Conversely, asymmetrical distortions provide visual evidence that there is a fault (Fig. 2b). The operators learn to associate particular patterns of distortion in the display with particular types of faults. Direct perception has been achieved: the operator can literally see the state of the plant directly, without the need to gather relevant information and perform complicated mental calculations. (Figure 2c illustrates the display in an actual power plant.)

An optimal implementation of direct manipulation in this interface would involve high-level control: the operator would have the capability to point, click, and drag the vertices of the octagon to a desired location, thereby changing the underlying variables. Unfortunately, this is not possible because of the complexity, interconnectedness, and potentially conflicting goals that characterize a power plant. Direct manipulation must therefore be implemented at a lower level. Thus, to change the setting of an individual variable, the user might select a visual indicator and drag it to a new location on a graphical scale (rather than typing it in).

Intent-driven domain. The most effective design strategy for intent-driven domains is to develop spatial metaphors (for example, the desktop metaphor on computers) that relate the requirements for interaction to more familiar objects and activities. The Apple iPhone provides an excellent example.

Direct perception in the iPhone's interface is achieved primarily through static spatial metaphors. These metaphors use physical similarity or symbolic convention (a stylized phone icon) to provide a semantic link between the user's intentions (the need to make a phone call) and the underlying functionality of the iPhone (cell phone). Thus, this design strategy uses graphical representations to leverage preexisting concepts and knowledge, thereby

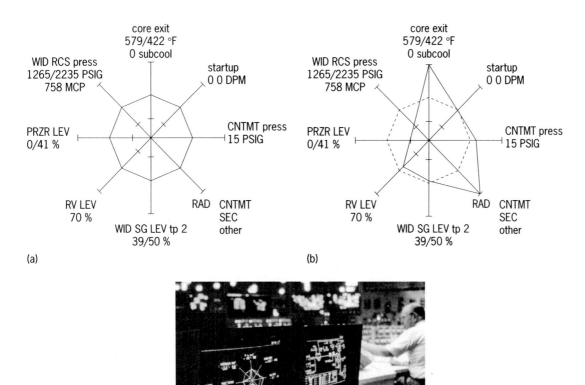

Fig. 2. An abstract geometrical form display used to represent dynamic values of system variables for a process control domain. (*a*) The octagon form in a perfectly symmetrical shape represents the values of variables under normal power-plant conditions. (*b*) The nonsymmetrical distortion of the octagon form characterizes a particular abnormality in the plant (that is, a loss-of-coolant accident). (*c*) The geometrical form display as it appears in an actual power plant. (*Line drawings from W. F. Schaefer et al., Generating an integrated graphic display of the safety status of a complex process plant, United States Patent no. 4,675,147, June 23, 1987; photo courtesy of David D. Woods, Cognitive Systems Engineering Laboratory*)

assisting casual users in understanding and using the iPhone.

It is less obvious that the iPhone uses an overarching spatial metaphor to facilitate navigation among the various applications and modes. This metaphor includes an array of icons (representing applications) in a "home" mode, a row of icons at the bottom of an application interface (representing modes of the application), and a dedicated mechanical button at the bottom of the phone (return to home mode). The spatial dedication of these interface components makes navigation of the iPhone much like navigation in the real world: getting to different applications and modes in the interface is somewhat similar to navigating between rooms in a well-known building.

Direct manipulation in the iPhone's interface represents a true advance in the state of the art. There are no menus, no cursor, no mouse, no stylus; the user's fingers are the only pointing device that is required. The user directly manipulates the objects of interest in the interface in a very natural manner: the touch-sensitive screen recognizes a number of fairly complicated gestures, including taps (single and double), slides, swipes, flicks, and pinches. This gesture-based interaction represents the wave of the future for interface design, even for those devices that are not handheld.

Summary. Designing effective interfaces is a more difficult problem than it initially seems. Cognitive Systems Engineering (CSE) provides a conceptual perspective and principles of design that point to successful solutions. Ultimately, interface design should be driven by the nature of the problems to be solved and by the capabilities and limitations of problem solvers.

For background information *see* COGNITION; COMPUTER; COMPUTER ARCHITECTURE; COMPUTER PROGRAMMING; ENGINEERING DESIGN; HUMAN-COMPUTER INTERACTION; HUMAN FACTORS ENGINEERING; PERCEPTION; PSYCHOLOGY in the McGraw-Hill Encyclopedia of Science & Technology.

Kevin B. Bennett; Shannon M. Posey

Bibliography. K. B. Bennett and J. M. Flach, *Display and Interface Design: Subtle Science, Exact Art*, Taylor & Francis, London, in preparation; K. Mullet and D. Sano, *Designing Visual Interfaces: Communication Oriented Techniques*, SunSoft Press, Englewood Cliffs, NJ, 1995; J. Rasmussen, A. M. Pejtersen, and L. P. Goodstein, *Cognitive Systems Engineering*, Wiley, New York, 1994; K. J. Vicente, *Cognitive Work Analysis: Toward Safe, Productive, and Healthy Computer-Based Work*, Lawrence Erlbaum Associates, Mahwah, NJ, 1999; D. D. Woods, The cognitive engineering of problem representations, pp. 169–188, in G. R. S. Weir and J. L. Alty (eds.), *Human-Computer Interaction and Complex Systems*, Academic Press, New York, 1991.

Human papillomavirus: impact of cervical cancer vaccine

Cervical cancer causes significant morbidity and mortality for women worldwide. Approximately 600,000 cases are diagnosed around the world each year, and it is estimated that 2 million women worldwide are afflicted with this disease at any given time. Because of the lack of cervical cancer (Pap) screening programs in underdeveloped countries, 270,000 women will die each year worldwide, compared to 5000 women in the United States, where screening is much more readily available. After years of research, it is now known that the human papillomavirus (HPV) is responsible for causing cervical cancer through sexual contact. Prior to the introduction of a vaccine against certain types of HPV, mortality from cervical cancer had decreased as a result of Pap screening, but these programs were costly and had yet to be implemented in most countries. This contribution will outline the epidemiology of HPV and its link to cervical cancer, HPV virology, vaccine development, and vaccine indications and effectiveness.

Epidemiology of cervical cancer and HPV. Research studies pointed to an infectious agent spread via sexual contact as a cause of cervical cancer. Early epidemiological studies, in the 1960s, found that there was a low cancer rate among Catholic nuns who were virgins. Women who had multiple sexual partners and who had sexual activity at a young age had increased rates of cervical cancer. Women whose husbands had ex-wives with cervical cancer were also at increased risk, thus suggesting men as carriers. In the 1970s, cells of cervical smears were found to have enlarged irregular nuclei with perinuclear clearing (koilocytes), indicating the presence of HPV infection. Around that same time, Harald zur Hausen (one of the key researchers who demonstrated the role of HPV as the etiological agent of cervical cancer) was able to extract viral particles from plantar warts and isolated several different strains of HPV from cervical cells. After a decade of research, HPV was isolated directly from cervical biopsies.

Currently, there are more than 140 identifiable types of HPV, of which 80 types have been genetically sequenced. The types of HPV are separated into genital and nongenital types. Of the 40 types of genital HPV, low-risk types are responsible for causing genital warts and low-grade cervical lesions. Types 6 and 11 are responsible for 90% of genital warts. Of the high-risk types, types 16 and 18 are responsible for 70% of all cervical cancer cases. Epidemiological studies have shown that these types that cause cervical cancer and genital warts are consistent throughout the world. The current vaccine available in the United States, Gardasil®, contains these four types of HPV.

Virology of HPV and vaccine development. In the early stages of vaccine development, the component to be used in the vaccine that triggers an immune response providing protection against the targeted disease must be identified. HPV is a small, double-stranded deoxyribonucleic acid (DNA) virus that is a member of the papilloma virus family. Its genome produces six early proteins (E proteins) and two late proteins (L proteins). The early proteins, E6 and E7, are essential in the malignant conversion of normal cervical cells to cancer cells and are involved in gene regulation of cellular growth. The late proteins form the icosahedral shape of this virus. These late proteins, when formulated in a vaccine without the infectious DNA, have been shown to produce high levels of neutralizing antibodies that are used by the immune system to destroy infected cells. Only one of the two late proteins (L1) is necessary to produce this effect, versus the effect of using both L1 and L2. The L1 proteins, made by genetically engineered yeast, *Saccharomyces cerevisiae*, or by baculovirus-infected insect cells, can self-assemble into viruslike particles that have an icosahedral shape similar to that of the live virus. These noninfectious particles are then suspended into various formulations to make the different HPV vaccines that are available throughout the world. Future vaccines may include the E proteins to target active infections.

The HPV vaccine has been shown to be effective only in primary prevention; that is, it prevents the initial infection, but it does not treat an existing infection. To determine when the vaccine should be given for the greatest efficacy, research has been carried out to identify when and how HPV is acquired. HPV is transmissible by both sexual and nonsexual routes via bodily secretions. During sexual contact, the cervix is exposed to infectious viral particles. Through disruptions in the cervical mucosa, that is, from abrasions or cervical infections, the virus is able to penetrate to the lower cell layer of the cervix. As the lower cell layer divides, the virus is replicated within these cells and begins the malignant transformation process in these cells.

Initial exposure to HPV can start in adolescents at the age of sexual initiation. By 17 years of age, 40% of adolescents are sexually active and, within 3 years, more than 50% of these are infected with HPV. Clearance of this initial HPV infection can occur by an unknown mechanism in 90% of women. Sloughing of the infected cervical cells and destruction of these cells by the immune system has been postulated as a mechanism for this clearance. Decreased immunity, lack of Pap smear testing, smoking, alcohol consumption, uncircumcised male partners, high parity (the number of children a woman has given birth to), and history of other genital infections increase the likelihood that an exposed patient will progress to a continued infection. Within 5 years, the HPV can lead to precancerous cervical lesions that ultimately could progress to cervical cancer. The current goal of the HPV vaccine is to prevent the primary infection in women. Thus, the vaccine is indicated for young adolescents, 9–26 years of age, preferably prior to their first sexual contact. Research is ongoing to determine whether the vaccine would be effective in preventing infections in females if given to males who can be carriers.

Efficacy of the vaccine. The first studies in the development of the HPV vaccine determined the safety and efficacy in normal human subjects. In 2001, a vaccine using HPV type 16 L1 viruslike particles was given to 72 men and women as a series of three injections (at 0, 1, and 4 months). This vaccine was well tolerated and highly immunogenic in normal individuals. The next step in development was to test the vaccine in women who had no evidence of HPV infection, but who were at risk for acquiring HPV through sexual contact. More than 1500 HPV-negative women, 16–23 years of age, were given either placebo or vaccine in the same three-injection series. Evidence of HPV infection was tested through 48 months by examining serum for HPV antibodies and cervical samples for HPV DNA. This study proved that the HPV 16 vaccine was 100% effective in preventing HPV infection in women at risk compared to placebo.

After this early-phase efficacy trial, a quadrivalent vaccine (a vaccine containing four antigens) was developed against the primary infection of HPV. The vaccine included the top two types that have been shown to cause the most cervical cancer worldwide (HPV 16 and 18) and the top two types that cause genital warts (HPV 6 and 11). More than 12,000 women were enrolled in a further efficacy study and randomized to receive either placebo or quadrivalent vaccine. The vaccine was shown to be 98% effective in preventing HPV-related precancerous or cancerous lesions for up to 3 years, and widespread vaccination of female children and adolescents was recommended. Because the onset of cervical cancer can occur 20 years from the onset of the initial infection, future studies will have to determine if this vaccine's efficacy will last without booster injections. Approval for this vaccine (Gardasil®) was granted by the U.S. Food and Drug Administration (FDA) in June 2006. Some experts estimate that worldwide vaccination of women could potentially decrease the number of cervical cancer deaths by two-thirds.

Indications and recommendations for vaccination. Because of the high percentage of Americans infected with genital HPV, some U.S. public health experts recommend mandatory HPV vaccination. Although federal legislation is pending, no current legislation mandates this vaccination. Currently, the FDA has approved one HPV vaccine (Gardasil®) in the United States, which is currently recommended

for women aged 9–26 years. An additional vaccine (Cervarix®), which includes only HPV 16 and 18, is awaiting approval from the FDA. Both vaccines have been approved in some countries, such as Australia.

The HPV vaccine will not protect against other types of HPV that are not included in the vaccine. Therefore, it is important for future Pap screening to screen for precancerous lesions caused by other HPV types. Sexually active females should receive the vaccine, because it has been shown that the likelihood that they will have all four types is very low. The vaccine falls under pregnancy category B, and thus is not recommended for use in pregnant women. Safety during breastfeeding is not known. The vaccine is not intended for the treatment of active genital warts and cervical cancer, and it will not protect against other sexually transmitted diseases (STDs). Pain at the injection site and fainting have been reported; hence, patients should be monitored for 15 minutes after administration of the vaccine.

Future questions. There are several unresolved questions regarding HPV vaccination. It is to be determined whether this vaccine will be accepted by patients as well as parents. As it is sometimes touted to be "the" STD vaccine, some parents are concerned about increased promiscuity or the stigma that vaccination may impose on their children. Will this vaccine reduce Pap screening and lead to a rise in cervical cancer? Will the vaccine be available in underdeveloped countries, where Pap testing is less readily available? How long will the vaccine last, and are booster shots needed? Should young men also be vaccinated for herd (community) immunity? How about older women? Can this vaccine be safely administered with other vaccines, and are three injections needed? Is it safe and efficacious to give to patients with comorbidities such as human immunodeficiency virus (HIV)? Even with all of these questions, the vaccine appears to be very effective in preventing HPV-related disease. Improvements in this vaccine are currently in the pipeline, and future vaccines may include efficacy against preexisting infection, which would further change the worldwide landscape of cervical cancer.

For background information *see* BIOLOGICALS; CANCER (MEDICINE); EPIDEMIOLOGY; IMMUNITY; IMMUNOLOGY; ONCOLOGY; PUBLIC HEALTH; REPRODUCTIVE SYSTEM DISORDERS; SEXUALLY TRANSMITTED DISEASES; VACCINATION in the McGraw-Hill Encyclopedia of Science & Technology.   Rachel Groff

Bibliography. F. T. Cutts et al., Human papillomavirus and HPV vaccines: A review, *Bull. World Health Org.*, 85:719–726, 2007; Future II Study Group, Quadrivalent vaccine against human papillomavirus to prevent high-grade cervical lesions, *N. Engl. J. Med.*, 356(19):1915–1927, 2007; D. M. Harper, Human papillomavirus vaccines, www.uptodate.com, retrieved January 8, 2008; K. U. Jansen and A. R. Shaw, Human papillomavirus vaccines and prevention of cervical cancer, *Annu. Rev. Med.*, 55:319–331, 2004; H. zur Hausen, Papillomaviruses and cancer: From basic studies to clinical application, *Nat. Rev. Cancer*, 2:342–350, 2002.

Hydrological consequences of global warming

The hydrologic cycle, which includes water fluxes and stores in the atmosphere, oceans, and land regions, is accelerating because of global warming. **Figure 1** shows the primary hydrologic cycle compartments and fluxes. The total volume of water in the Earth system is approximately $1.36 \times 10^9\,km^3$, with over 97% stored in the oceans, 2.9% stored as freshwater in lakes, rivers, ice, snow, and available groundwater, and approximately 0.1% in the atmosphere. The cycling of water consists of atmospheric water vapor transport and precipitation, surface evaporation, transpiration from vegetation, infiltration of water into the ground, overland runoff and riverflow, and water stored in the oceans, snow, glaciers, permafrost, and deep groundwater. Surface water evaporates and is transported by the winds and precipitates back to the surface, where a portion is reevaporated into the atmosphere, another portion infiltrates into the ground, and the remainder runs off into rivers and ultimately into the oceans.

At present, approximately $413 \times 10^3\,km^3$ per year of water is evaporated from the oceans and about $73 \times 10^3\,km^3$ per year evaporates and transpires from land surfaces. The atmosphere annually returns $373 \times 10^3\,km^3$ of precipitation to the oceans and $113 \times 10^3\,km^3$ of precipitation to land. These processes operate at different rates for different locations, depending on the temperature of the lower atmosphere (troposphere) and land and ocean surfaces, as well as the wind speed and direction and the amount of radiative heating. The atmosphere has the shortest residence time, ranging, from seconds to weeks, and is characterized by clouds, storms, and other weather processes. Conversely, the oceans, glaciers, and deep groundwater have much

longer residence times, ranging from months to millennia.

The hydrologic cycle is directly linked to the amount of atmospheric absorption, reflection, and transmission of the Sun's incoming energy (shortwave radiation), as well as the amount of longwave radiation emitted from the land and ocean surfaces. Naturally occurring gases have helped to regulate the atmospheric radiation balance and resulting temperature. However, the increasing rate of human-based emissions of carbon dioxide and other heat-absorbing gases (for example, methane) has changed this balance, leading to an unprecedented warming of the lower atmosphere. The 2007 Intergovernmental Panel for Climate Change (IPCC) reported strong evidence that the atmospheric concentration of carbon dioxide now far exceeds the natural range over the last 650,000 years, and that recent warming of the climate system is unequivocal, resulting in more frequent extreme precipitation events, earlier snowmelt runoff, increased winter flood likelihoods, increased and widespread melting of snow and ice, longer and more widespread droughts, and rising sea level. The effects of recent warming have been well documented, and climate model projections indicate a range of hydrological impacts with likely to very likely probabilities (67–99%) of occurring with significant to severe consequences in response to a warmer lower atmosphere that has an accelerating hydrologic cycle.

Accelerating hydrologic cycle. As the troposphere continues to warm an additional 1–6°C (2–11°F) during this century, the acceleration of the hydrologic cycle will be more evident. One of the most sensitive hydrologic components is the change in atmospheric water vapor, which affects the distribution, type, and timing of precipitation. A warmer air mass is able to hold more water vapor, resulting in more extreme precipitation events in the form of more frequent flooding and more extensive and longer-lasting droughts. This change in water vapor and precipitation roughly translates into dry regions being very likely to become drier and wet regions being very likely to become wetter. **Figure 2a** depicts these trends for a doubling of the 1990 atmospheric concentration of carbon dioxide, with the end-of-the-century precipitation increasing at mid-to-high latitudes and semiarid regions becoming drier and extending northward. The stippled regions are where more than 90% of the IPCC climate model projections agree. Soil moisture is almost certain to decrease as evaporation increases globally with the planet heating up (Fig. 2b and c). Runoff is very likely to increase at high latitudes with increasing precipitation and to decrease at mid-latitudes with decreasing precipitation and increasing evapotranspiration.

Ocean circulation. High-latitude precipitation increases will result in increased surface runoff and freshwater inflow to the high-latitude ocean regions, causing a decrease in salinity, and suggesting a slowdown of the Atlantic Ocean thermohaline circulation of warm tropical water northward and deep cold polar water southward. If global warming continues

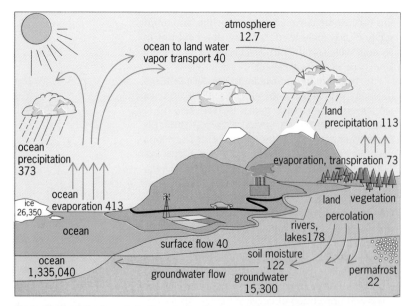

Fig. 1. The hydrologic cycle is characterized by evaporation and transpiration from the surfaces; transport of water vapor in the atmosphere by the wind, clouds, and precipitation; and water storage in snow, glaciers, groundwater, and oceans. Units: 1000 km³ for storage; 1000 km³/yr for exchanges. (*Trenberth et al., 2007*)

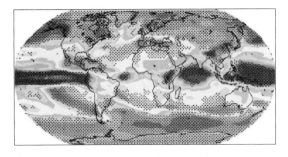

(a) precipitation

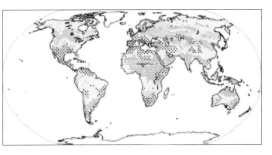

(b) soil moisture

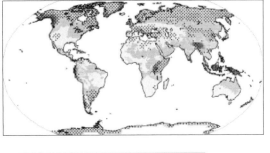

(c) runoff

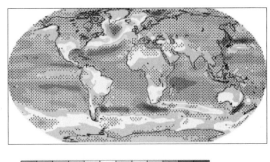

(d) evaporation

Fig. 2. Spatial patterns of climate model projections of (*a*) precipitation, (*b*) soil moisture, (*c*) runoff, and (*d*) evaporation for 2090–2099 under the mid-range A1B emission scenario from the Intergovernmental Panel for Climate Change 2007. (*Climate Change 2007—The Physical Science Basis: Working Group I Contribution to the Fourth Assessment Report of the IPCC, Cambridge University Press, 2007*)

unabated, such a slowdown may tip beyond a stability threshold, resulting in a shutdown of this ocean circulation and a long-term shift in the climate patterns of Europe.

Early snowmelt. Global warming is causing earlier seasonal melting of alpine snowpack, shifting the timing of snowmelt runoff worldwide. Glacial surveys of equatorial and mid-latitude glaciers show retreat of up to 50% in the last 25 years, and evidence is very strong that equatorial glaciers will completely disappear by about 2030. Many regions that rely on snowmelt for domestic and industrial water resources are beginning to feel the effects of this change. Alpine snow reduction is affecting and will continue to affect water resources, including freshwater for domestic use, agricultural and other industrial use, and recreation. An increase in winter and early spring floods is forecast for the future, along with a significant decrease in late spring and summer runoff.

Water resources. Many emerging and developing nations will be affected by decreasing freshwater availability as dry regions expand and drought becomes a more frequent phenomenon. Africa and Southeast Asia are some of the most susceptible regions to drought and are likely to have an increase in famines, water wars, and population declines. Reduced snowpack and early melt will dramatically shift the available water resources for several river systems flowing out of the Himalayas into Southeast

Asia. A combined consequence is the effect on water quality and its impact on human and animal health. Cultures are likely to shift, with mass migrations in response to agriculture failures as a result of freshwater loss.

Alpine glaciers and snowpack provide freshwater to a large segment of society. The western United States is one of the regions most susceptible to drying, and water managers are developing adaptation and coping strategies for less available water. The western United States snowpack has been melting earlier in the spring, the total volume of snowmelt runoff in rivers has been occurring earlier, and summer flows have been below average. Similar trends are being observed worldwide, and major rivers with snowmelt runoff, such as the Colorado River, Ganges River, Mackenzie River, Rhine River, Rio Grande~River, Yangtze River, and Yukon River, will all be subject to reduced flows and reduced hydroelectric power production because of global warming.

Ski resorts in the Alps have been putting down protective covers to slow the retreat of glacial snow in the hope of extending the future of this industry. The sensitivity of reduced runoff is dependent upon the elevation of the snowline and the amount of snow-covered area. Regions with most of the snow cover close to freezing, such as the Cascade Mountains and the Sierra Nevada Mountains, will be the first to experience significant reductions as temperatures continue to rise. The California Climate Change Assessment (2006) has put upper and lower snowpack reduction bounds at 90 and 60% loss, respectively, putting significant pressure on water resource managers to design new coping strategies, with an increased reliance on groundwater storage to meet increasing water demand in California. Groundwater storage is expected to become more important in a warmer world, but a need for deeper wells and increased electricity costs to extract groundwater will result because of groundwater overdraft.

Sea-level rise. Other hydrologic components of the climate system that are changing include Arctic sea-ice decrease, loss of ice from the West Antarctic Ice Shelf, glacial loss from Antarctic and Greenland land masses, and melting of permafrost, resulting in sinking land that damages infrastructure and releases methane into the atmosphere. The recent calving of glaciers from the West Antarctic Ice Shelf does not increase the amount of ocean water mass, because ice shelves do not displace water. However, the land-based ice is undermined from below and can result in a destabilization and ice flow from inland to the sea, with major sea-level changes. The most significant, abrupt hydrologic change is the melting or release of Greenland and/or Antarctic glacial ice, which could result in a sea-level increase of 5 m (16 ft) or more. Such a scenario would drastically change many coastal regions, with large population disruptions in areas such as Florida, Bangladesh, and island nations.

The Global Earth Observing System of Systems (GEOSS) is a United Nations activity for monitoring climate change, including the hydrological cycle, and will lead to improved disaster prevention and management. GEOSS includes satellite, surface, and airborne observations.

Outlook. There is very high confidence that the hydrologic changes happening are man-made and will last for centuries because of the slow response time of the oceans. Society will need to advance coping and adaptation strategies for our warming planet.

For background information *see* ANTARCTICA; CLIMATE MODELING; CLIMATE MODIFICATION; CLIMATIC PREDICTION; DROUGHT; GLACIOLOGY; GLOBAL CLIMATE CHANGE; GREENLAND; HYDROLOGY; OCEAN CIRCULATION; WATER RESOURCES in the McGraw-Hill Encyclopedia of Science & Technology.

Norman L. Miller

Bibliography. California Climate Change Assessment, 2006: Our Changing Climate, Assessing the Risks to California; Intergovernmental Panel on Climate Change, Fourth Assessment Report, 2007: Climate Change 2007; K. E. Trenberth et al., Estimates of the global water budget and its annual cycle using observational and model data, *J. Hydrometeorol.*, 8:758–769, 2007.

Hygiene hypothesis

The "hygiene hypothesis," sometimes referred to as the "old friends hypothesis," is that humans have evolved with certain intestinal bacteria and worms whose presence can "train" the immune system not to overreact against harmless commensal organisms. As people have become more prosperous and are leading "healthier" lifestyles, these organisms have been lost from the intestines, and the immune system is losing its natural balance. This predicts a significant increase in immune diseases, both allergic and autoimmune, which in fact has been observed in the developed world over the last several decades.

Increase in allergies. Statistical surveys in the United States and elsewhere over the past two decades have shown a large increase in the incidence of allergies and asthma that may have been going on since the mid-twentieth century. Much of the increase occurs in rich (developed) rather than poor (undeveloped) countries; and within them, in urban rather than rural communities, and among the urban rich rather than the poor. This has led to the idea that as one eats a more processed, cleaner diet and is less exposed to strangers and to environmental microorganisms, the immune system might not achieve a normal balance and might respond aberrantly to harmless environmental agents such as pollens and molds. In the 1990s, when children in the developing world were put on therapy for intestinal helminths (worms), it was noted that they had a significantly increased incidence of allergies and asthma.

Over the same time period, the model of helper T-cell self-regulation was being developed in many research centers. Two types of helper T cells were recognized. Th1 cells, when activated, produce factors (cytokines) that attract and activate an inflammatory response. This sort of response is valuable for ridding the body of dangerous organisms such as *Mycobacterium tuberculosis*, but it can, if unchecked, lead to chronic inflammatory and autoimmune diseases. Th2 cells are mainly involved in helping B cells to become activated to make antibody. However, if overstimulated, they can drive the B cells to make inordinate amounts of immunoglobulin E (IgE), the antibody class of allergies. The physiological role of IgE, though, is to help the body control helminth infestations. The distinction between responses is important: mouse strains that, for genetic reasons, make only Th1 responses to certain parasites survive; those that make only Th2 responses succumb to the same infections.

The Th1 and Th2 siblings display a sort of molecular rivalry: the cytokines of Th1 cells suppress Th2 development, and vice versa. It is assumed that, in normal people, a balance is achieved that is appropriate for effective immunity.

Formulation of the hygiene hypothesis. Infants are born with an immune system biased toward Th2 responses and gradually develop Th1 and Th2 cells in balance. Thus, when the hygiene hypothesis was first formulated, it suggested that the immune system of children, in the absence of adequate stimulation by intestinal microbes or infections, would remain in an infantile state, biased toward Th2 cells and thus IgE production and allergies.

This hypothesis was very fruitful in stimulating ideas and research about the helper T cells and their mutually antagonistic interactions. In 2005, investigators reported on a small trial of treating very ill patients with Crohn's disease. This condition is caused by T-cell-mediated chronic inflammation (especially of the colon and small intestine). Since this is a characteristic of Th1 cells, the investigators wondered whether a strong Th2 stimulus might override the Th1 cells via the sibling rivalry mechanism, and improve the patients' conditions. Volunteers were fed small quantities of live pig whipworm (*Trichuris suis*) eggs (which would not survive in the human gut long enough to cause disease) since, as noted previously, helminths drive a strong Th2 response. After 24 weeks, 79% of the patients in this small preliminary trial had experienced significant improvement or remission. A similar trial was also successful in patients with ulcerative colitis.

Increase in autoimmunity. While allergists and pulmonologists were developing the hygiene hypothesis to explain the increase in allergies and asthma, rheumatologists, pediatricians, and others were observing a similar increase in incidence of chronic inflammatory diseases, including Crohn's disease and ulcerative colitis, as well as the autoimmune conditions of juvenile (type 1, insulin-dependent) diabetes and multiple sclerosis. Animal and human testing suggested that these are Th1-related diseases. These conditions have the same distribution characteristics as the allergic states: they are prevalent in rich, not poor, countries. In Rwanda, for example, where (usually asymptomatic) helminthic infestation is nearly universal in the rural poor, inflammatory bowel disease is almost unknown.

In another study, researchers in Winnipeg, Manitoba, Canada, have mapped the geographic locations of residences of patients with inflammatory bowel disease to the richest neighborhoods; poor districts have far fewer patients. This pattern is also true for prevalence of asthma.

These findings, though, were not directly compatible with the hygiene hypothesis. If rich (that is, relatively microbe- and helminth-free) people have immune systems dominated by Th2 cells, then that would explain their increased risk for allergies, but not for the Th1 inflammatory diseases.

Old friends hypothesis and regulatory T cells. While all of this information was being gathered and analyzed, a new T-cell category was moving to the forefront of modern immunology: the regulatory T cell, Treg. There are actually several subsets of Treg, but they can all be considered together. Some of their properties were formerly attributed to Th1 or Th2 cells because they have similar surface markers and frequently contaminate other populations. However, Treg make cytokines (interleukin-10, transforming growth factor-β) whose role is to suppress both Th1 and Th2.

Considerable research now shows that the gut, which is full of foreign organisms and molecules, contains a highly specialized set of lining cells and a very organized immune system (there are more lymphocytes associated with the gut than with any other organ). The gut contains commensal microbes, which are harmless or actually helpful in providing nutrients and keeping harmful organisms in check. Commensals must be recognized as such; the mounting of an immune response against them would be futile, as they could never be removed, and endless inflammation would result. Thus, these good commensals are "old friends," and humans respond to their antigens primarily with a strong Treg response. Once activated, Treg keep Th1 and Th2 (and the newly described, highly inflammatory Th17 cell) under control. When harmful organisms invade, the resultant strong innate immune response and tissue damage will result in the Treg giving way to the defensive Th1, Th2, and Th17 response.

Therefore, according to the old friends hypothesis, the presence of many of these helpful commensals causes a gut-associated immune response in which Treg "tell" the other cells to calm down and not respond excessively to minor threats. However, if the commensals have been missing or are scarce, Treg are underrepresented, and minor threats can elicit major, destructive responses. Depending on the nature of the invasion and on the individual's genetic constitution, either Th1/Th17 may dominate, risking chronic inflammation and autoimmunity, or Th2 may dominate, risking allergic diseases.

It has been suggested that the missing "old friends" include helminths, *Lactobacillus* species (which in developed countries are obtained only in yogurt), and many *Mycobacterium* strains, which are harmless relatives of the tuberculosis bacillus.

Future outlook. The hygiene hypothesis will be further refined as new data accumulate. Many researchers are speculating that humans will develop mechanisms to preferentially stimulate Treg, both in normal infants and in people whose immune systems show signs of dysregulation. Perhaps a probiotic supplement may be designed to mimic the missing "old friends." Allergists and others have been of the belief that a child should be exposed to the "germs" of other children via day care or school, thereby giving the immune system a tonic workout and ending up healthier for it. Caution must be used, of course, because there are microorganisms in the environment that can overwhelm any immune system, which is why vaccines are used. Vaccines might be expected to exercise the developing immune system too, but in terms of antigenic load they are minimal compared to a gutful of friendly microorganisms.

For background information *see* ALLERGY; ASTHMA; CELLULAR IMMUNOLOGY; CYTOKINE; IMMUNITY; IMMUNOGLOBULIN; IMMUNOLOGY; INFLAMMATION; INFLAMMATORY BOWEL DISEASE; MEDICAL BACTERIOLOGY; PUBLIC HEALTH in the McGraw-Hill Encyclopedia of Science & Technology.

J. John Cohen

Bibliography. K. Ashenburg, *The Dirt on Clean: An Unsanitized History*, North Point Press, New York, 2007; J. F. Bach, The effect of infections on susceptibility to autoimmune and allergic diseases, *N. Engl. J. Med.*, 347:911–920, 2002; A. DeFranco, R. Locksley, and M. Robertson, *Immunity: The Immune Response in Infectious and Inflammatory Disease*, New Science Press, London, 2007; C. Green et al., A population-based ecologic study of inflammatory bowel disease: Searching for etiologic clues, *Am. J. Epidemiol.*, 164:615–623, 2006; G. A. W. Rook, The hygiene hypothesis and the increasing prevalence of chronic inflammatory disorders, *T. Roy. Soc. Trop. Med. H.*, 101:1072–1074, 2007; R. W. Summers et al., *Trichuris suis* therapy in Crohn's disease, *Gut*, 54:87–90, 2005.

Hypersonic vehicles

Hypersonic speed commonly refers to travel at five times the speed of sound or faster, although there is no sharp boundary between supersonic flight— defined precisely as traveling faster than the speed of sound (Mach 1)—and hypersonic flight. Instead, hypersonic speed is more nebulously defined as traveling fast enough so that strong shock waves and friction in the gas through which a body travels raise the gas temperature high enough that it begins to chemically react (for example, oxygen and nitrogen dissociate from naturally occurring diatomic molecules into atomic species), and at higher speeds even to ionize (when electrons are stripped from gas molecules) and emit significant radiation. At hypersonic speed the heating of vehicle body surfaces due to air friction also becomes significant and can heat those surfaces to extreme temperatures.

Although the extreme thermal environment and performance limits encountered by hypersonic vehicles make their design and fabrication challenging, the effort and investment required to overcome the challenges are worthwhile due to the benefits that hypersonic flight can provide. Efforts have been ongoing since the late 1950s to develop functional hypersonic vehicle designs and associated enabling technologies. Several flight experiment and demonstration programs are ongoing to develop and validate technologies and design concepts that will one day enable practical operational hypersonic flight.

Applications. A hypersonic aircraft that cruises at Mach 6 speed (4000 mph or 6500 km/h) and 95,000 ft (29 km) altitude could travel 7000 nm (13,000 km) in slightly over 2 hours, including the time needed for takeoff, ascent, descent and landing (**Fig. 1*a***). This is significant because almost all major world city-pairs are separated by no more than 7000 nm. Having the ability to travel such great distances in so short a time would be valuable for the delivery of time-critical cargo and packages, and for the relatively small percentage of passengers with the means and desire, or the critical need, to pay a premium ticket price to travel seven times faster than possible by a conventional subsonic airliner. Consider a patient, perhaps a soldier, in critical condition and needing urgent medical care half a world away, or someone in dire need of a donated organ where the only match can be found in another country.

High speed and altitude would also provide advantage to military aircraft and missiles (Fig. 1*b*). The advantage of short flight time would be the same as

for a commercial aircraft, but the important characteristic of vehicle survivability would be enhanced as well, since an aircraft or missile traveling high and fast would be much harder to shoot down than a lower-flying subsonic vehicle. As countermeasures to stealth are developed, reducing the effectiveness of stealth technology, flight at higher speed and altitude could help maintain the advantages currently provided by stealth alone. In the case of a missile, high speed also imparts significant kinetic energy to the vehicle (which increases with the square of velocity) that can be used to replace or supplement the explosive power of a warhead carried by the missile.

(a)

(b)

(c)

Fig. 1. Examples of hypersonic vehicles. (*a*) DARPA/Air Force Blackswift, a hypersonic aircraft (*Popular Science, Nick Kaloterakis*). (*b*) X-51A, an experimental flight vehicle that is representative of hypersonic missiles in size and configuration (*U.S. Air Force/DARPA/Boeing/Pratt & Whitney*). (*c*) FASST (Flexible Aerospace System Solution for Transformation), a hypersonic, air-breathing, two-stage-to-orbit reusable launch vehicle (*Boeing*).

The same technologies that can enable hypersonic atmospheric flight, most notably air-breathing supersonic combustion ramjet (scramjet) engines and high-temperature materials, could also benefit the development of a space transportation system that would allow transit to Low Earth Orbit (LEO) more routinely, affordably, and safely than possible today (Fig. 1c). Scramjet engines are more fuel efficient than chemical rocket engines, allowing the design of reusable launch vehicles employing the former to carry more structure and systems as a fraction of gross weight (in place of unneeded propellant) than those designed to exclusively employ the latter. The greater available mass fraction can be used to accommodate a low-speed propulsion system (turbine engines, for example) that will permit horizontal take-off and landing from a conventional runway, plus increased structure, redundant systems, and vehicle-health-monitoring systems to increase vehicle safety and reliability.

If a reusable launch vehicle (most likely employing two stages) could be developed that flew back and forth from LEO, then launch costs would be driven down dramatically. The result could be the creation of true space commerce, which might include such businesses as orbital space tourism; private resupply of the International Space Station; lower-cost satellite delivery and possibly salvage or repair; commercial launch of space science, research, or manufacturing payloads; and maybe even space-based solar power that could provide electrical power to customers on Earth at a competitive price. *See* SPACE-BEAMED SOLAR POWER.

Critical technologies for hypersonic flight. One of the most critical technologies for enabling hypersonic flight is high-temperature materials. Included in this category are thermal protection systems (TPS) that keep much of the heat generated by air friction from penetrating into vehicle structure and subsystems that must remain relatively cool. An example of a thermal protection system is the low-conductivity silica tiles used on the space shuttle. To illustrate the criticality of high-temperature materials, the leading edge of a Mach 5 aircraft could experience a temperature as high as 1800°F (1250 K) if its radius is on the order of $^1/_{10}$ in. (2.5 mm) to maintain low drag. On the other end of the speed spectrum, the leading edge of a space plane accelerating to orbit under scramjet power would experience a temperature of approximately 5000°F (3030 K) with a 1-in. (25-mm) radius at Mach 15. There is currently no known or developed material that can withstand such a high temperature and remain intact under flight loads. Active leading-edge cooling would therefore be required for a Mach 15 air-breathing hypersonic vehicle. Since leading-edge aerodynamic heating is inversely proportional to the square root of nose radius, very blunt edges such as those found on the space shuttle could be used to reduce material temperature, but high drag prevents their use on an air-breathing vehicle.

Other critical technologies include lightweight reusable tanks that can carry cryogenic propellants such as hydrogen or oxygen for use on hypersonic

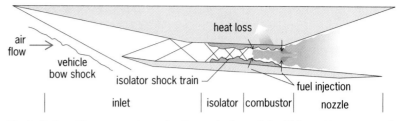
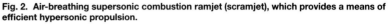

Fig. 2. Air-breathing supersonic combustion ramjet (scramjet), which provides a means of efficient hypersonic propulsion.

space-planes, and advanced vehicle design tools that make possible the design of complex and highly integrated hypersonic vehicles (described below).

Topping the list of critical technologies is the scramjet engine. The scramjet is fairly simple in concept, as jet engines go: It is primarily a shaped duct or tube that contracts on its front end to form an air inlet; has a constant-area or slightly divergent section where fuel is injected, mixed, and burned; and then rapidly diverges on its aft end to expand combustion products to create thrust (**Fig. 2**). The challenge is proper shaping of the duct to provide for efficient air compression, fuel-air mixing and combustion, and gas expansion so that overall scramjet performance is maximized—in concert with the vehicle upon which it is integrated, and operating over a wide range of speed, flight attitude, and atmospheric pressure.

The performance (fuel economy) advantage of scramjets compared to rockets is illustrated in **Fig. 3**, which plots engine specific impulse as a function of Mach number for rockets, turbine engines, ramjets, and scramjets. Specific impulse (I_{sp}) is simply engine thrust divided by propellant mass flow rate. In terms of fuel efficiency, high thrust at low fuel

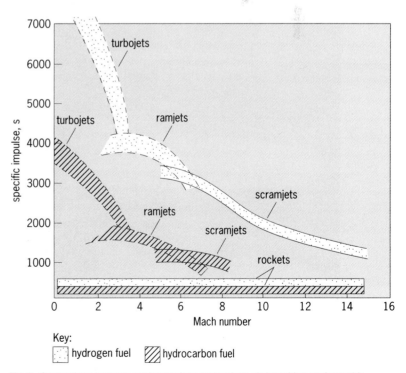

Fig. 3. Comparison between rocket engines and various air-breathing engines with regard to fuel efficiency (evaluated in terms of specific impulse) versus flight speed (evaluated in terms of Mach number).

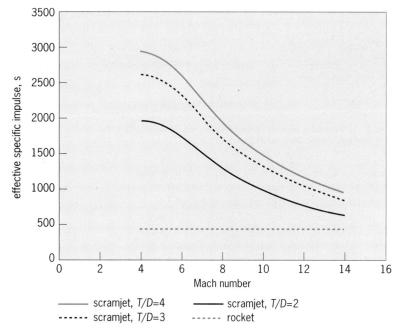

Fig. 4. Comparison of fuel efficiency (evaluated in terms of effective specific impulse) between scramjet-integrated vehicles using hydrogen fuel, with typical thrust-to-drag (*T/D*) values, and a rocket using LOX oxidizer.

flow rate (that is, high I_{sp}) is obviously desirable. Note from the figure that hydrogen fuel provides greater I_{sp} than hydrocarbon fuel, that I_{sp} generally drops with increasing Mach number, and that air-breathing engines have much higher I_{sp} than rocket engines.

The impact of engine I_{sp} on vehicle performance can be evaluated by writing Newton's second law of motion (force equals the time rate of change of momentum) in terms of engine thrust and fuel flow rate (I_{sp}), vehicle drag, and incremental changes in vehicle kinetic and potential energy, yielding the equation below, where W_i is the vehicle initial weight

$$\frac{W_f}{W_i} = \exp\left[\frac{-(\Delta V + g\Delta h/V)}{gI_{sp}(1 - D/T)}\right]$$

and W_f its final weight after being propelled to an incremental increase in energy. Comprising the energy change terms, ΔV is the change in vehicle velocity, Δh the change in vehicle altitude, and g acceleration due to gravity. And finally, V, I_{sp}, and T/D are average values of velocity, engine specific impulse, and vehicle thrust-to-drag ratio, respectively, over the small segment of flight trajectory being analyzed. Inspecting this equation, it can be observed that maximizing the denominator of the exponential, $I_{sp}(1 - D/T)$, often referred to as effective specific impulse, maximizes the vehicle final weight relative to its initial weight, hence minimizes the fuel used to achieve a given change in vehicle energy. In other words,

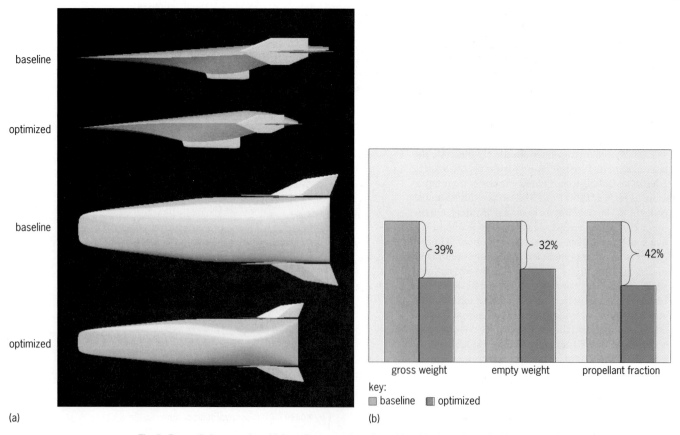

Fig. 5. Dramatic hypersonic vehicle performance benefits achievable through application of multidisciplinary design optimization. (*a*) Comparison of profiles of baseline and optimized hypersonic aircraft and missiles. (*b*) Weight and propellant fraction comparisons of baseline and optimized vehicles.

the larger the engine I_{sp} and vehicle T/D, the more fuel efficient the vehicle system. Along these lines, the fuel efficiency of a hydrogen scramjet-powered hypersonic vehicle can be compared to that of a hydrogen rocket, as is done in **Fig. 4**. Here, the plotted values of effective I_{sp} for a scramjet-powered vehicle were calculated using the mean values of hydrogen scramjet I_{sp} in Fig. 3, and T/D values of two, three and four, which bracket the range of values typical of well-designed air-breathing hypersonic vehicles. The air-breathing engine benefit over a rocket, in terms of vehicle fuel efficiency, is clearly evident.

Design challenge. Hypersonic vehicles that employ air-breathing propulsion systems such as scramjets must meet a large set of demanding, often-competing design requirements. They must also operate over a wide range of environmental conditions during flight, which must be simulated accurately both experimentally and analytically to achieve ultimate design success. To add to the design challenge, the functions of most system components must be highly integrated for vehicle performance and economics to meet requirements. This high degree of component integration is evident in the configurations illustrated in Fig. 1.

The vehicles shown in Fig. 1 are classified as "lifting bodies" because aerodynamic lift is produced by the body itself instead of by a distinct set of wings. When hypersonic lifting bodies employ scramjet engines for propulsion, their entire undersurfaces also contribute to the functions of propulsion and flight control, in addition to aerodynamics. Moreover, the lifting body structure serves as a fuel tank internally and a thermal protection system externally. Such a high degree of functional integration renders the conventional aircraft design process ineffective and conventional design tools inadequate, and advanced methods and tools are therefore required.

The "design technologies" required include the ability to create computer-generated vehicle geometry models that can be "morphed" (that is, their shape modified) by changing numerical values of design parameters; the ability to automate all vehicle discipline analyses (aerodynamics, propulsion, structures, thermal, and so forth); and the ability to employ mathematical optimization algorithms to systematically search for the set of design parameter values that "optimize" the vehicle in terms of performance and economics. When such optimization includes more than one vehicle discipline it is referred to as multidisciplinary design optimization (MDO). MDO is a powerful means of searching design space to discover unique, nonintuitive, and hard-to-find design solutions (**Fig. 5**).

Flight experiment and demonstration programs. The extreme flow energy and thermal environment of hypersonic flight make ground testing in wind tunnels very challenging, though not impossible. This naturally leads to additional differences in the way that hypersonic vehicles are developed as compared to their subsonic or supersonic counterparts. Specifically, some development must be accomplished in flight where air flow conditions can be

(a)

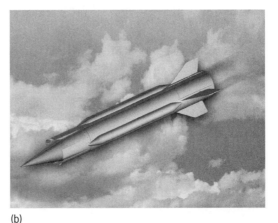

(b)

(c)

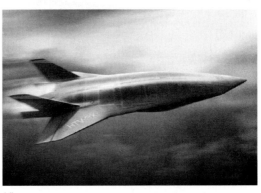

(d)

Fig. 6. Recent and near-term hypersonic flight experiment and flight demonstrations, which are maturing vehicle technology beyond what is achievable through ground testing alone. (*a*) NASA X-43A. (*b*) U.S. Navy/DARPA HyFly. (*c*) U.S. Air Force/Darpa X-51A. (*d*) DARPA/Air Force Blackswift. (*Popular Science, Nick Kaloterakis*)

exactly reproduced, and hardware scale replicated as required.

In the late 1950s and 1960s the first true hypersonic flight testing was conducted using the

X-15 rocket-powered aircraft, which achieved crewed flight at Mach 6.7, and flight exiting and reentering the Earth's atmosphere. The knowledge gained and technology proved by flying the X-15 contributed to development of the space shuttle. In 2004, NASA successfully flight tested the world's first airframe-integrated scramjet engine at Mach 6.8 and then at Mach 9.7 using the X-43A aircraft, proving that scramjets perform as theorized. Several United States hypersonic flight experiments or demonstrations are under development and will begin flight testing sometime between 2008 and 2014, depending upon the program (**Fig. 6**).

For background information *see* ATMOSPHERIC ENTRY; GUIDED MISSILE; HYPERSONIC FLIGHT; SCRAMJET; SPACE TECHNOLOGY; SPECIFIC IMPULSE in the McGraw-Hill Encyclopedia of Science & Technology.
Kevin Bowcutt

Bibliography. T. A., Jackson, Power for a space plane, *Sci. Amer.*, 295(2):56–63, August 2006; D. R. Jenkins, and T., Landis, *Hypersonic: The Story of the North American X-15*, Specialty Press, 2003; C. Peebles, *Road to Mach 10: Lessons Learned from the X-43A Flight Research Program*, Library of Flight Series, AIAA, 2008.

Imaging mechanical defects

Damage detection and location is a problem that is encountered in a broad range of fields, including aircraft and spacecraft integrity, manufacturing and production reliability, structural health monitoring, and infrastructural damage due to aging as well as in response to large earthquakes. A new generation of mechanical damage diagnostics and imaging methods has been developed over the last decade. Nonlinear elastic methods, known collectively as nonlinear elastic wave spectroscopy (NEWS), can be applied to distinguish whether or not a crack exists in an opaque solid. Imaging methods are under development that combine NEWS with the time-reversal mirror (TRM). Time reversal is a remarkable technique for producing large sound amplitudes at the original source location, or locating an anomaly in a solid that scatters waves. Thus, time reversal provides the means to locate an elastic anomaly, and nonlinear elasticity allows one to learn about its nature. The combination of the two techniques is known as time-reverse nonlinear elastic wave spectroscopy (TR NEWS).

Currently, there are no reliable means of determining that a damage anomaly is present or locating it amidst a background of features that may appear as damage but are not. In surveillance, manufacturing, and certification programs, conventional nondestructive inspection does not always work in resolving, locating, and especially characterizing the nature of defects. Development of nonlinear elastic wave damage inspection integrated with the time-reversal mirror may yield a new and powerful approach that provides the means to resolve, locate, and characterize anomalies.

In order to understand how TR NEWS works and to illustrate by example, NEWS will be described. This will be followed by an introduction to time reversal, and then by examples of TR NEWS.

Nonlinear elastic wave spectroscopy (NEWS). In many materials—metals, most ceramics, and individual crystals—Hooke's law, which describes the relation between stress and strain, is linear. But in cracked or otherwise mechanically damaged materials, Hooke's law is nonlinear. This nonlinearity has profound results for dynamic waves: the modulus is strain-amplitude dependent and also pressure-history dependent. Nonlinearity also results in wave distortion, as manifested by harmonic generation, wave modulation (**Fig. 1**), and nonlinear attenuation. These manifestations of nonlinearity are the diagnostics of the presence of damage in a sample.

Linear scattering. A wave traverses a linear elastic material at a speed that is independent of its amplitude. When a wave with central frequency ω_1 encounters a linear elastic scatterer (for example, a void), a scattered wave at the same frequency ω_1 is broadcast from it. Detectors placed around the scattering site on the perimeter of the material can detect the waves. If one has a velocity model of the material, one can formulate what is known as an inverse problem on a computer that uses the scattered wave and the velocity model to find the location of the scatterer. [A velocity model is a numerical model of the velocity structure of a material. It is obtained either by knowledge of the elastic constants and density of a material whose material constants are known and can be found in the literature (a sample of 5180

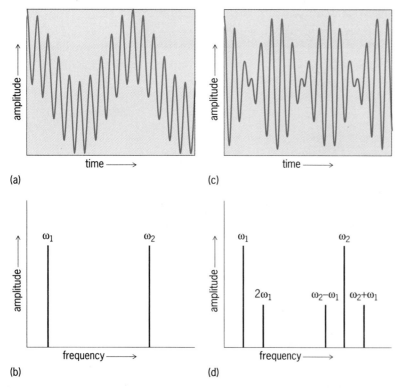

(a)

(c)

(b)

(d)

Fig. 1. Time and frequency response of two waves input into a material containing a crack, a nonlinear scatterer. (a) Two sine waves at frequencies ω_1 and ω_2 broadcast into a cracked solid (where ω is the angular frequency $2\pi f$). (b) Associated frequency spectrum. (c) Detected time signal. (d) Frequency response. A nonlinear scatterer produces both sum and difference frequencies, $\omega_2 + \omega_1$ and $\omega_2 - \omega_1$, respectively, and harmonics such as $2\omega_1$.

steel, for example), or by creating what is known as a tomographic image of the sample. A tomographic image is a numerical velocity model obtained from a suite of velocity measurements.] This is the principle of linear elastic imaging. Extending this idea to two waves, one at frequency ω_1 and the second at frequency ω_2, which are broadcast simultaneously into a linear elastic material and encounter linear elastic scatterers, the principle of linear superposition states that the detected signal is also composed of waves with frequencies ω_1 and ω_2.

Nonlinear scattering. When two wave packets, one at frequency ω_1 and the second at frequency ω_2, with amplitudes A_1 and A_2, are broadcast simultaneously into a linear elastic material and simultaneously encounter a nonlinear elastic scatterer (for example, a crack or other mechanical damage feature), a time- and amplitude-sensitive scattering event is created. At the moment of encounter, the scatterer broadcasts waves with frequencies $\omega_2 \pm \omega_1$ at amplitudes that are proportional to the product of the amplitudes of the wave packets, A_1A_2. Detectors located on the perimeter of the material can detect the waves. Detection of scattered wave energy at frequencies $\omega_2 \pm \omega_1$ is enough to show the presence of one or many nonlinear scatterers. Right away the nonlinear scatterers are picked out from the background of linear scatterers by a frequency signature (Fig. 1). Other geometries (spheres, holes, ellipses, surfaces, and so forth) are linear; they do not create additional frequency components. One can formulate an inverse problem that uses the nonlinear scattered wave and a material velocity model obtained from standard tomography to attempt to find the location of the scatterers. In short, any process that creates "soft" regions in a solid (a crack is viewed as an elastically soft region) will be "seen" by nonlinear wave processes; "linear" anomalies such as voids or interfaces will not exhibit a nonlinear signature. Previous work tells us that some chemical and thermal damage, such as disbonding and delaminations, can be sensed with nonlinear methods. The problem is not detecting the presence of damage; it is locating the damage features. When these methods are used with the time-reversal mirror, the features can be located.

Time-reversal mirror (TRM). The TRM process is simple to understand in an elastically linear medium. Waves are input into a medium by a sound source. They reverberate around the medium and are recorded repeatedly at one or more detectors as time goes on; the reverberant portion of the wave is called the wave coda, a term coined by the seismologist Kei Aki more than 50 years ago. The wave is then literally time reversed and broadcast into the medium from the detectors, which also act as sound sources. If the reversed signals are broadcast in phase with each other, the signals will coalesce at the original source location at an instant in time. The process is illustrated by an experiment conducted in a thin metal plate. **Figure 2a** shows wave emission from the source (a piezoceramic located on the back side of the plate). The wave propagates outward (Fig. 2b and c), analogous to dropping a pebble in a tank of water and watching the waves emanate

outward. Figure 2d shows the complex wave interference that results from reflection off the sample sides. (This is a snapshot of the spatial distribution of the wave coda.) The waves are detected at the eight detectors shown in Fig. 2e. The detected waves are time reversed at these detectors, which are also transmitters—the waves are literally reversed about the time axis—and sent backward into the medium (Fig. 2e). Figure 2f shows the waves focusing on the original source location. After focal time, the waves pass through each other and are reemitted (Fig. 2g and h). The forward and backward propagation are not identical because of effects created primarily by the finite-size transducers. (Energy is redistributed in the broadcast of the time-reversed waves.) This asymmetry is normal in laboratory experiments, and the focusing of waves on the original source location in spite of it is an example of the robust nature of the time-reversal process. The time-reversal mirror has been demonstrated in a solid in its simplest incarnation, and its role in acoustic media has been made widely known by a group at the Laboratoire Ondes et Acoustique (LOA) [Wave and Acoustics Laboratory] in Paris. The concept has existed since at least the 1960s, when what is known as matched filtering, a topic closely related to time reversal, was being explored.

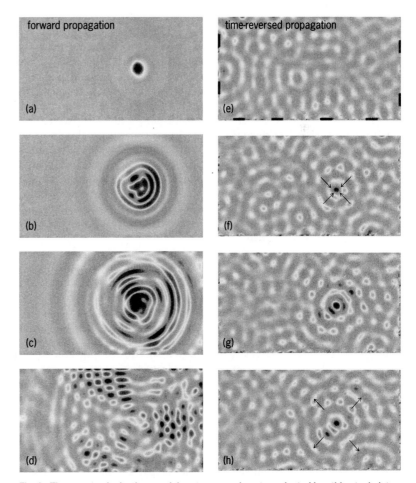

Fig. 2. Time-reversal, elastic wave laboratory experiment conducted in a thin steel plate. Image was created by use of a scanning laser Doppler vibrometer that measures the oscillations perpendicular to the plate. (*a–d*) Forward propagation. (*e–h*) Time-reversed propagation. (*Experimental results courtesy of B. Anderson*)

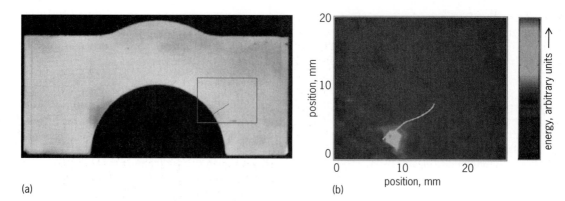

(a)   (b)

Fig. 3. Imaging a surface-breaking crack in a bearing cap, applying TR NEWS. (*a*) Image of bearing cap. The rectangle surrounds the crack location (thin line) and corresponds to part *b*. (*b*) Crack location (line). Lighter shades show the focusing of waves at frequency $\omega_2 - \omega_1$. Energy at frequency $\omega_2 - \omega_1$ is observed to focus only at the crack, the origin of $\omega_2 - \omega_1$ The image was obtained by scanning the area in a stepwise manner using a laser Doppler vibrometer that measures oscillation perpendicular to the sample face. (*Image courtesy of T. J. Ulrich*)

TR NEWS. When two input waves at frequencies ω_1 and ω_2 are scattered by a nonlinear elastic defect in a solid, some of the energy is broadcast directly to a perimeter detector. All of the energy arriving at the detector at frequencies $2\omega_1$, $2\omega_2$, and $\omega_2 \pm \omega_1$ was scattered ("broadcast") from the nonlinear elastic defect at a single moment of time, as described above. The scattered wave arrives at the detector at different times simply because it spent time rattling around in the sample, just as the original sound source did. If the signal arriving at the detector is frequency filtered about one of the nonlinear components, for instance, at one of the modulation frequencies $\omega_2 - \omega_1$, and is time reversed and rebroadcast, it will reassemble itself at the nonlinear scatterer at a single moment of time, that is, it will focus on the nonlinear scatterer. This phenomenon potentially offers a powerful tool for isolating a nonlinear scatterer. If there are several detectors around the perimeter of the sample, carrying out the filtering and time reversal simultaneously for all detectors builds up the strength of the focusing at the scatterer. Thus, with active time reversal, one has the means of dedicated focusing of energy on a nonlinear scatterer, that is, a first step toward interrogating and imaging such a scatterer (**Fig. 3**).

Locating a scattering source. There are many important details left out here, including timing of the rebroadcast so that the wave packets arrive at the nonlinear scatterer simultaneously, the optimal time window of the coda to rebroadcast, the optimal frequencies, and so forth, but the general idea is clear. One employs the nonlinearly generated wave from a crack or other feature for back propagation. A next step is to image inside thick opaque solids, to locate buried cracks or delaminations. In thin solids, this is straightforward if the spatial wavelength used is at least as long as the sample is wide; however, if the feature is hidden well beneath the surface, the problem turns out to be difficult for reasons currently being studied. The buried-feature problem is important in many applications, including laboratory studies, manufacturing, petroleum extraction, and carbon dioxide (CO_2) sequestration.

One approach that is currently in development begins in the same manner. Waves are broadcast into the sample, the nonlinear scatterer produces new waves at other frequencies, and those waves are detected. Then, since one cannot see into the object if it is opaque, the nonlinear components of the wave are back propagated through a velocity model of the object on a computer rather than through the actual object. The computer model must have the appropriate velocity structure for the P and S waves. This often means that standard velocity tomography must first be conducted. Velocity tomography may see the same features, but normally it cannot distinguish a nonlinear scatterer from a linear one.

The protocol to obtain an image is as follows: (1) Forward-propagating waves are collected from the sample, for example, **Fig. 4***a*. The waves scatter off the boundaries and the nonlinear and linear scatterers. (2) A velocity tomography, three-dimensional image is constructed on a computer from velocity

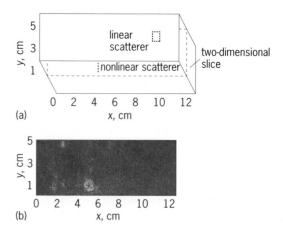

Fig. 4. Imaging in three dimensions, applying TR NEWS. (*a*) Bird's-eye view of a rectangular-shaped solid, showing a linear scatterer, a nonlinear scatterer, and a two-dimensional slice that intersects the two scatterers. (*b*) TR NEWS two-dimensional image obtained for the slice. The nonlinear scatterer is imaged (as seen by the lighter shades at the scatterer location), but the linear one is not. The method provides the means to distinguish between (1) cracks and other damage and (2) interfaces, voids, and so forth. (*Image courtesy of M. Griffa*)

measurements taken from the sample. (3) The laboratory data are filtered about a harmonic, for example, or a sideband if two frequencies are input, and time reversed. (4) The waves are then back propagated through the model on the computer. A computer algorithm conducts a point-by-point search for a focus throughout the volume. The location of a nonlinear scatterer from a two-dimensional slice can be seen in Fig. 4b. This is the general concept of applying TR NEWS to three-dimensional imaging of mechanical damage features.

Considerable work remains for the method to be viable in three dimensions. A number of applications have already come to fruition in solids, especially for damage diagnostics and imaging in thin samples. These range from interrogating potentially cracked solids to landmine detection. Many others are in development, including human bone diagnostics for osteoporosis diagnosis and for fracture healing. Linear time-reversal imaging of earthquake sources is a rapidly evolving subarea of study. In summary, time reversal in solids and TR NEWS are rich areas of research and development that are still in their infancy.

For background information see ELASTICITY; HOOKE'S LAW; MATCHED-FIELD PROCESSING; NONDESTRUCTIVE EVALUATION; NONLINEAR ACOUSTICS; STRESS AND STRAIN; TIME-REVERSED SIGNAL PROCESSING; ULTRASONICS in the McGraw-Hill Encyclopedia of Science & Technology.   Paul A. Johnson

Bibliography.   B. E. Anderson et al., Time reversal, *Acoust. Today*, 4(1):5–16, January 2008; R. A. Guyer and P. A. Johnson, Nonlinear mesoscopic elasticity: evidence for a new class of materials, *Phys. Today*, 52(4):30–35, April 1999; M. Fink et al., Time-reversed acoustics, *Rep. Prog. Phys.*, 63:1933–1995, 2000; L. Larmat et al., Time-reversal imaging of seismic sources and application to the great Sumatra earthquake, *Geophys. Res. Lett.*, 33:L19312, doi:10.1029/2006GLO26336, 2006; A. Parvulescu and C. S. Clay, Reproducibility of signal transmission in the ocean, *Radio Elec. Eng.*, 29:223–228, 1965.

Immune response in inflammatory bowel disease

Crohn's disease (CD) and ulcerative colitis (UC) are inflammatory disorders of the digestive tract that are characterized by abdominal pain, weight loss, malaise, and diarrhea. Although UC affects only the superficial lining of the colon and is continuous from the rectum proximally (toward the beginning of the colon), inflammation in CD can penetrate deep into the bowel wall and can present with multiple discontinuous lesions that affect both the small intestine and the large intestine.

Advances in the understanding of the pathogenesis of these inflammatory bowel diseases (IBDs) have led to a therapeutic revolution. IBD arises from the interaction of genetic and environmental factors that give rise to a dysregulated intestinal immune response.

Intestinal immune system. In the intestine, the mucosal immune system is separated from significant amounts of bacterial and dietary antigens by a layer of cells known as the epithelium. The tone of the mucosal immune system is one of suppression or tolerance toward these nonpathogenic bacteria or food antigens. Defects in this regulation or in epithelial cell barrier function may lead to an aberrant immune response.

There are two facets of the immune responses: innate and adaptive. The innate response is the body's first line of defense against invading microbes. Its objective is to localize and quickly eradicate threats and is mediated by phagocytic cells. However, it is more primitive in that it has neither memory nor specificity for individual pathogens.

The adaptive response evolved as a result of the innate system's inability to augment its response to reinvading pathogens and because large pathogen loads could easily overwhelm this system. Adaptive immunity is characterized by specificity and memory, and it is mediated largely by lymphocytes—T and B cells. These lymphocytes express receptors on their surfaces that recognize microbial antigens, which are molecules specific to individual microbes. T cells also produce many different types of signaling proteins known as cytokines, thereby orchestrating an organized and directed immune response that eradicates infections and gives rise to memory cells.

Recent data have shown that the innate response is essential to initiating adaptive responses. As will be described, this has implications for the disease process. It is not surprising that IBD, a complex immune disorder, involves all these aspects of the immune system.

Innate response. As has been suggested, mucosal barrier defects, as well as inefficient clearance of dead cells and bacteria, may result in excessive inflammation. Previously described environmental factors and recently described genetic susceptibility genes both point to a prominent role for innate immunity and barrier function in the development of disease. Mutations have been described in ATG16L, CARD15/NOD2, and OCTNI/II, which are genes associated with barrier function or innate immunity. In addition, agents that affect the epithelial barrier (nonsteroidal anti-inflammatory drugs, antibiotics, and microbial infections) are common triggers of IBD.

ATG16L is involved in autophagy, whereby cells under stress initiate intracellular programs to undergo orderly cell death. Although the exact effects of the mutation remain unknown, this inability may lead to excessive inflammation. The CARD15/NOD2 mutations were the first susceptibility loci described in Crohn's disease. The mutations affect the innate response and the ability to eradicate bacteria that gain entry to the host. The resulting persistence of bacterial antigens may lead to an adaptive immune response toward an otherwise harmless stimulus. Interestingly, patients with genetic defects in innate immunity [for example, chronic granulomatous disease (an inherited disorder of phagocytic cells) or

Hermansky-Pudlak syndrome (a type of albinism that includes a bleeding tendency)] can have IBD-like disease.

Adaptive immune response. An altered adaptive immune response is most tightly correlated with the disease process. T cells become activated and produce cytokines that affect all other cells in the environment to orchestrate significant inflammatory responses. Studies in rodents and humans suggest that different subsets of T cells become aberrantly activated in CD versus UC.

T-cell responses can be characterized into subsets based on the type of threat they encounter and the different cytokines they produce. The first subset is the T-helper 1 (Th1) response, which is largely a response to intracellular infection (for example, virus or mycobacteria). Generally, this response promotes intracellular killing, the differentiation of cells that mediate direct killing (cytotoxic T cells), and localization of the infectious agent (granuloma formation). Granulomas are clusters of immune cells, inflammatory cells, and scar tissue. Their chief role is to wall off an infecting microbe. Interestingly, many patients with CD have granulomas in their small or large intestine.

Cytokines characteristic of a Th1 response are interferon-γ (IFN-γ), tumor necrosis factor-α (TNF-α), and interleukin-2 (IL-2). These cytokines act in the local environment to enhance intracellular killing (IFN-γ, TNF-α) and inflammatory cell recruitment (TNF-α), and to promote local tissue destruction. The last objective is achieved by the production of enzymes that digest tissue (collagenases, elastases). In an infectious setting, this response walls off microbes while making local tissue inhospitable for bacterial growth. However, when this response is uncontrolled or inappropriate, disease results. Th1 cells are programmed and activated by another proinflammatory cytokine, IL-12. Blocking IL-12 is an effective therapeutic strategy that inhibits the Th1 response and production of T-cell-derived IL-2, TNF-α, and IFN-γ, although other cells can still make these cytokines.

IL-12 is composed of two protein chains: p40 and p35. IL12p40 is shared by another proinflammatory cytokine, IL-23, that affects a recently described T-helper subset, Th17. Th17 cells produce IL-17 and IL-22, proinflammatory cytokines that enhance local tissue destruction. Evidence for a prominent role of Th17 in CD has steadily accumulated. First, IL-17 and IL-22 are increased in inflamed Crohn's mucosa. Second, a recently discovered mutation in the IL-23 receptor protects hosts from developing CD. Last, selective blockade of IL-23, but not IL-12, mitigates IBD-like disease in animal models. This paradigm shift for disease pathogenesis from a Th1 to a Th17 response in CD is also supported by the lack of efficacy of anti-IFN-γ in these patients.

Another T-helper cell subset is Th2. These T-helper cells produce cytokines IL-4, IL-5, and IL-13 and are involved in allergic reactions. UC was initially thought to be a Th2-mediated disease because of increased mucosal IL-13 production. However,

patients with UC exhibit increased production of IFN-γ (Th1) and not IL-4 (Th2). One group has recently shown that another type of T cell, the natural killer T cell, may be responsible for the increased production of IL-13 seen in patients with UC. This cell targets the epithelial cells of the mucosal barrier to become dysfunctional and may explain why UC is a more superficial epithelial injury.

Regulatory cells. Tolerance to nonpathogenic bacteria and food antigens is mediated by regulatory cells and is perhaps the most critical immune pathway in the intestine.

Generally, T cells are divided into two categories: CD4+ and CD8+. There is a series of CD4+ regulatory T cells, including Tr1, TH3, and CD4+CD25+ Treg cells. CD4+ Tr1 cells secrete a potent immunosuppressant cytokine, IL-10, that suppresses local responses to gut bacteria flora. IL-10-deficient mice develop a Crohn's-like inflammatory disease. CD4+ TH3 cells produce transforming growth factor-β (TGF-β), another immunosuppressive cytokine that inhibits T- and B-cell activation. Finally, CD4+CD25+ Treg cells are located more systemically and are dependent on the transcription factor FoxP3. Absence of FoxP3 could lead to autoimmune inflammation of multiple organs. However, these cells are increased in patients with IBD, suggesting that defects in this population are not a cause of disease. Finally, populations of CD8+ regulatory T cells have been described in the gut. Some of these CD8+ cells, TrE cells, have been shown to be deficient in IBD patients.

Regulatory T-cell defects may be involved in the disease process because they hinder a host's ability to tolerate bacterial and food antigens. The presence of antibodies against resident intestinal bacteria in patients with IBD supports this hypothesis. Furthermore, defects in a phenomenon known as oral tolerance (the capacity of the immune system to recognize substances taken orally and to weaken or suppress the immune response to them) have been documented in patients with IBD. Increasing regulatory T-cell activation could be an effective treatment.

Amplification of immune response. Amplification of inflammation results from downstream effects of T-cell activation. Chemoattractant cytokines (chemokines) produced by activated T cells recruit new inflammatory cells to the gut. These cells are prone to producing proinflammatory factors and tissue-destroying enzymes because they have not been programmed in the suppressive environment of the intestine. Additionally, TNF-α induces the up-regulation of adhesion molecules on intestinal blood vessels. This facilitates entry of inflammatory cells into the tissue.

Therapeutic implications. With all the relevant pathways delineated, various therapeutic options are available. To determine efficacy, it is necessary to ask whether the therapy can (1) prevent a cell from getting to its target location, (2) prevent its activation, and (3) prevent the effects of cellular activation if the other two processes cannot be targeted. As products of the immune system have been easier to target, most therapies have focused on the third process. Antibodies to proinflammatory cytokines such as anti-TNF-α and anti-IL12p40 have proven effective in treatment. Antibodies targeting IL-17 are currently in development. Therapies targeting cell homing appear promising as well. Antibodies to adhesion molecules that prevent T cells from gaining access into tissue have been effective in early trials.

Targeting immunoregulation is another promising option. Activating and expanding regulatory T cells may control the excessive inflammation that is characteristic of IBD. However, it is likely that some combinatorial approach targeting defects in the innate and adaptive immune systems as well as in the epithelial cell barrier will have the greatest impact on disease control and possible cure.

For background information *see* ANTIBODY; ANTIGEN; CELLULAR IMMUNOLOGY; CYTOKINE; DIGESTIVE SYSTEM; GASTROINTESTINAL TRACT DISORDERS; IMMUNOLOGY; INFLAMMATORY BOWEL DISEASE; ULCER in the McGraw-Hill Encyclopedia of Science & Technology.   Steven J. Esses; Lloyd F. Mayer

Bibliography. D. C. Baumgart and W. J. Sandborn, Inflammatory bowel disease: Clinical aspects and established and evolving therapies, *Lancet*, 369(9573):1641–1657, 2007; S. J. Brown and L. Mayer, The immune response in inflammatory bowel disease, *Am. J. Gastroenterol.*, 102(9):2058–2069, 2007; C. G. Mathew, New links to the pathogenesis of Crohn disease provided by genome-wide association scans, *Nat. Rev. Genet.*, 9(1):9–14, 2008; R. J. Xavier and D. K. Podolsky, Unravelling the pathogenesis of inflammatory bowel disease, *Nature*, 448(7152):427–434, 2007.

Impedance units from the quantum Hall effect

Quantum effects have opened a new era in metrology for realizing (providing standards for) the physical units. The conventional standards of our macroscopic world (collectively termed "artifacts") alter with time, temperature, and weather; suffer from mechanical shock during transportation; and depend on unwanted side effects and on geometric dimensions that often cannot be measured with sufficient precision. In contrast, quantum effects in the atomic world are unchangeable and can be described with extraordinary precision using only a few fundamental constants.

Today, all commercial measurement instruments, electronic and otherwise, are calibrated through procedures that are traceable to the standards of the national metrology institutes. Their task is to guarantee an internationally consistent system of units by means of comparisons and accreditations, for the benefit of worldwide trade and science. The history of impedance metrology will be briefly recalled, because this best illustrates the advantages of a quantum standard for the national metrology institutes.

Impedance is the generic term for resistance, inductance, and capacitance measured with alternat-

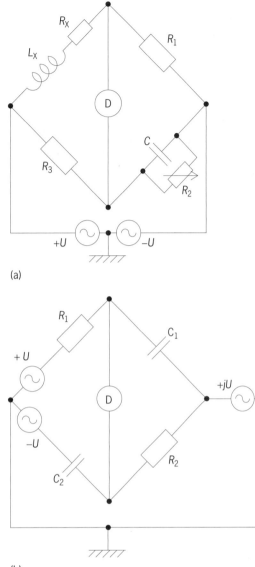

(a)

(b)

Fig. 1. Bridge circuits used in linking impedance standards. (*a*) Maxwell-Wien bridge, an ac Wheatstone bridge that links an inductance L_X to a capacitance C and two resistances, R_1 and R_3. The detector D is nulled when $L_X = R_1 R_3 C$. The resistor R_2 balances R_X, which represents the loss and the lead resistance of the inductor. (*b*) Quadrature bridge, which establishes a precise relation between two capacitances, C_1 and C_2, and two resistances, R_1 and R_2, according to $\omega R_1 C_1 \cdot \omega C_2 R_2 = 1$, where $\omega = 2\pi f$ is the angular frequency of the ac voltages. The voltages $\pm U$ (which are 180° out of phase) are generated by an accurate voltage transformer. The voltage jU is derived from $\pm U$ via an auxiliary R-C network.

ing current (ac), whose units are the ohm, the henry, and the farad, respectively. A Maxwell-Wien bridge (**Fig. 1***a*) can link an inductance standard to capacitance and resistance standards. Capacitance standards (**Fig. 2**) are linked by a quadrature bridge (Fig. 1*b*) to ac resistance standards. These historical, circular links still hold today, and in the past each of the three impedance units was once the starting point for realizing the other two—until experimental limitations and the desire for lower uncertainties necessitated the rotation of their roles.

Fig. 2. Commercial 1-nF capacitance standard. The case has been cut to make the stack of parallel capacitor plates visible. The height of the outer case is about 15 cm (6 in.).

Calculable impedances. Beginning in 1881, a precision glass tube filled with pure mercury realized the ohm with direct current (dc), and the farad and henry were derived from it. The details of this realization sound very antiquated now that modern quantum and atomic standards are available. Later, unacceptable discrepancies became evident, and the mercury ohm could never meet modern requirements for calibrating impedance-measuring instruments as demanded by the electrical and aviation industries and by international agreement for worldwide trade.

To improve this situation, the starting point of the impedance chain was reassigned to a calculable inductor, a coil whose inductance could be calculated from its geometric dimensions as precisely as possible. Even though the accuracy of impedance metrology was improved considerably, the coil dimensions could not all be measured with the desired accuracy, and this circumstance limited the accuracy to an uncertainty of a few parts in 10^6.

In 1956, Australian metrologists invented a precisely calculable capacitor. Its capacitance depends only on its length, and this can be measured by optical interferometry much more accurately than all the complicated dimensions of a calculable coil. The calculable capacitor not only realizes the farad, it is also the starting point of a chain leading to the dc ohm and to the henry. To keep pace with the demand for decreasing uncertainties, it was also necessary to improve the bridge technology, mainly by interconnecting components with coaxial cables. In these cables the currents in the central conductor and the returning currents in the coaxial screen are equal and opposite, so that the total current in every cable is zero. This eliminates stray electromagnetic interactions between different parts of a bridge and between the bridge and external fields.

A calculable capacitor requires extreme machining tolerances, the necessary equipment fills a large laboratory, and it costs an enormous amount of money. At first, only the Australian national metrology institute had a calculable capacitor, and for about 20 years all national metrology institutes working with highest accuracy had to ship their dc resistance standards there for recalibration every few years. On these occasions they realized that their preserved dc resistance units had drifted over the years, and with different drift rates in different countries. Later, a few more calculable capacitors were built, and the uncertainty was reduced to about 5 parts in 10^8, but the large effort required impeded wider dissemination.

Quantum Hall effect. Therefore, the researcher Klaus von Klitzing was immediately aware of the importance of a discovery he made in 1980 concerning the Hall effect under special conditions. The Hall effect is a phenomenon, discovered a century earlier, whereby a conductor carrying an electric current perpendicular to an applied magnetic field develops a voltage gradient that is transverse to both the current and the field. The Hall resistance of a ribbon-shaped conductor whose surface is perpendicular to the field is defined to be the ratio of the voltage across the ribbon to the total current. von Klitzing discovered quantized steps in the Hall resistance whose numerical values depend only on two fundamental constants, the charge e of the electron and Planck's constant h, in the combination h/e^2. This quantum effect occurs only in a special semiconductor layer where the electrons are confined in two dimensions. A strong magnetic field of the order of 10 tesla forces the electrons in the quantum Hall device to—very intuitively speaking—move in closed cyclotron orbits around the magnetic flux lines, similar to an electron in a hydrogen atom moving in a quantized orbit around its nucleus. To avoid a thermal breakdown of the cyclotron orbits, the temperature should not exceed 1.5 K. The quantized Hall resistance can be reproduced with an impressive relative accuracy of about 10^{-10}. Amazingly, and in contrast to all previous impedance standards, this resistance does not depend on any geometric dimension. Fundamental constants do not alter with time and weather, and they do not suffer from mechanical shock during transportation as do classical artifacts. Consequently, if a quantum Hall device is inadvertently destroyed, a new one can be manufactured that has precisely the same resistance value as the old one. This discovery degraded the antiquated attempts to conserve a classical resistance artifact; it brought the quantum era to resistance metrology. Since 1990, the resistance unit has been maintained by the quantum Hall effect. Metrologists now have a fixed point that is precisely the same in every country and to which the slowly drifting resistance standards can be tied as often as desired, without the risk of transporting highly sensitive standards to another country having a calculable capacitor.

The value of the quantum Hall resistance can be measured by means of a calculable capacitor, but this route requires a very fragile resistance artifact to accomplish a calculated transfer from ac to dc resistance. This means that classical artifacts still control the quantum effect.

AC measurements of the quantum Hall effect. In 1993, the PTB (Physikalisch-Technische Bundesanstalt, the German national metrology institute) suggested measuring the quantum Hall resistance with alternating current and using this ac resistance for realizing the unit of capacitance. This proposal shows a way out of reliance on any classical artifacts: using the same quantum effect for the realization of all three units (ohm, farad, and henry), and defining the quantum Hall resistance from the value in the International System (SI) of h/e^2 (assuming that the relation discovered by von Klitzing is exact). The resulting standard for the farad might not be significantly smaller in uncertainty than that derived from the calculable capacitor, but it would constitute a true quantum standard of impedance. Every national metrology institute calibrating impedance standards has the necessary measuring bridges, and if the quantum Hall effect is already used for dc resistance calibrations, the change to ac is only a small step.

Introducing this new quantum standard of impedance has not been without difficulty—proposing a new idea is often easier than the laborious process of learning how to solve the associated problems. Whenever unforeseen effects appear and cannot be understood, scientists like to use the synergy of a collaboration of like-minded partners, combined with the ingenious ideas of other experts. Then, in the end, a problem often disentangles in an unexpectedly simple way. This also happened in the case of ac quantum Hall measurements. Flaws in both the measuring bridges and the calculable resistors caused the discrepancies that were found in earlier measurements made in different national metrology institutes. However, the steps in the ac quantum Hall resistance, when plotted against the applied magnetic field, really are somewhat curved or structured and not as flat as the steps measured with direct current. Furthermore, the steps are slightly shifted with respect to the quantized dc resistance. Both the curvature and the shift are proportional to the measuring frequency, and they are now attributed to capacitances inside the quantum Hall device and to capacitances between the quantum Hall device and its surroundings. In the dc case, these capacitances are charged only during the first nanoseconds after switching on the current, but this transient effect dies away long before the time-consuming dc precision measurement can be started. In the ac case, the quantum Hall device is continuously charged and discharged during each ac cycle, and these charges oscillate in the alternating electric Hall field and sustain polarization losses. Metrologists found that the unwanted ac losses reveal themselves as a measurable voltage drop along the device edges. Either this voltage drop is detected by a separate measurement and then corrected for, or a voltage-controlled electrode is used to eliminate all the electrons that sustain polarization losses. Then, the ac quantum Hall resistance becomes equal to its dc counterpart and can be applied as a precise quantum standard of impedance.

Consequently, at the Conference on Precision Electromagnetic Measurements in June 2008, the PTB reported on a direct link between two 10-nanofarad capacitors and two ac quantum Hall resistances (Fig. 1b and **Fig. 3**), without the need for any classical "calculable" artifact. By means of a 10:1 ratio bridge, the 10-nF capacitors are then stepped down to a 10-picofarad capacitor, the "workhorse" of capacitance metrology, with a relative uncertainty of only 2 parts in 10^8. In the near future, a legal quantum standard of capacitance could well be established in this way, and the time of changing impedance realizations should come to an end.

For background information *see* BRIDGE CIRCUIT; CAPACITANCE MEASUREMENT; CYCLOTRON

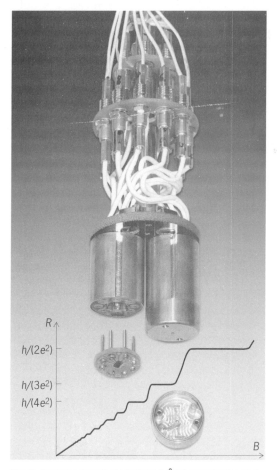

Fig. 3. Innermost part of a helium-3 (^{3}He) cryostat located in the center of a 15-T magnet. The two separately shielded ac quantum Hall resistors are operated at a temperature of 0.3 K. One of the two shields has been opened to show the quantum Hall chip [the small black rectangle on the unplugged, 18-mm-diameter (0.7-in.) printed circuit board]. Background diagram of Hall resistance R versus magnetic field B shows the quantized steps of the Hall resistance as integer fractions of h/e^2.

RESONANCE EXPERIMENTS; ELECTRICAL IMPEDANCE; ELECTRICAL UNITS AND STANDARDS; HALL EFFECT; PHYSICAL MEASUREMENT; RESISTANCE MEASUREMENT; WHEATSTONE BRIDGE in the McGraw-Hill Encyclopedia of Science & Technology.   Jürgen Schurr

Bibliography. M. Kochsiek and M. Gläser (eds.), *Handbook of Metrology*, Wiley VCH, 2008; I. M. Mills et al., Redefinition of the kilogram, ampere, kelvin and mole: a proposed approach to implementing CIPM recommendation 1 (CI-2005), *Metrologia*, 43:227–246, 2006; P. J. Mohr, The fundamental constants and theory, *Phil. Trans. Math. Phys. Eng. Sci.*, 363:2123–2137, 2005; H. L. Störmer and D. C. Tsui, The quantized Hall effect, *Science*, 220:1241–1246, 1983; B. N. Taylor and T. J. Witt, The new international electrical reference standards based on the Josephson and quantum Hall effects, *Metrologia*, 26:47–62, 1989.

In-stream tidal power generation

The prospects of climate change and declining supplies of cheap fossil fuels have led to increasing interest in supplying at least part of humanity's energy needs from renewable sources. Insolation is the largest of these, with a global value of nearly 10^5 terawatts (TW, with 1 TW = 10^{12} W), so that a tiny fraction would provide all of humanity's current usage of approximately 15 TW. The global rate of dissipation of wind energy is close to 1000 TW, suggesting that useful amounts of power could be extracted without major environmental changes. The dissipation rate of ocean tides is only 3.5 TW, and simple arguments show that extracting more than a fraction of this would be impossible and lead to significant tidal changes. Tidal power is not "boundless." It could, nonetheless, provide a significant and economically valuable contribution, possibly of the order of hundreds of megawatts (MW, with 1 MW = 10^6 W) in some locations and perhaps tens of gigawatts (GW, 1 GW = 10^9 W) worldwide, though associated environmental changes are a concern.

Tidal basins. Traditionally, attention has been focused on tidal embayments with large tidal ranges. A simple operating scheme involves trapping water behind a barrage at high tide and releasing it through turbines close to the times of low tide. Generation on the flood rather than ebb, or even two-way generation, is possible. An operating scheme at La Rance in France generates an average of about 70 MW. The installation of the barrage has caused significant (though not necessarily negative) local ecological changes. Concern with such changes and with the possibility of widespread changes in tidal regime have led to barrage schemes being held in some disfavor. The large initial capital cost has also been an obstacle to development.

Tidal streams. Attention has partially shifted to consideration of the exploitation of strong tidal currents, using turbines in much the same way as wind is exploited. It is argued that this can be much less environmentally damaging than a barrage, and also has capital costs that can be incurred and repaid gradually as more turbines are installed.

For a single turbine or a small number of turbines in a channel, the back effect on the tidal regime will be small and the problem is similar to that of wind turbines. The power produced by a turbine is proportional to the water density, the cross-sectional area of the turbine, and the cube of the current speed. Because the density of water is 850 times that of air, a given power may be produced with smaller current than wind speeds. On the other hand, the forces on the turbine and supporting structure are proportional to the fluid density and the square of the fluid speed. Hence the force on a marine turbine is greater than that on a wind turbine of the same power by the ratio of wind to current speed.

Problems associated with fish and marine mammals have parallels with the issue of bird and bat strikes for wind turbines and are likely avoidable, but strikes from neutrally buoyant objects such as deadheads are a concern, and there may be issues associated with corrosion and biofouling. Many turbine designs are being considered and tested, some with shrouds to direct and increase the local flow (**Fig. 1**), but the emphasis of this short review will be on the limitations imposed by ocean dynamics.

Resource assessment. There are two simple situations that serve as a basis for discussion. In the first, the strong tidal current is in a channel connecting two larger basins, with different tidal heights that drive the channel flow but are assumed to be largely unaffected by it or by the deployment of turbines in the channel. In the second situation, the tidal channel connects a semienclosed basin to a larger sea. In both situations, it has been common practice to estimate the available power from the undisturbed kinetic energy flux through the channel, though this is ill-defined for a channel of varying cross section in which the kinetic energy flux varies with position

Fig. 1. Clean Current™ ducted turbine prior to installation in a pilot project at Race Rocks, Vancouver Island, British Columbia, Canada. (*Courtesy of Glen Darou*)

along the channel. Moreover, an obvious consideration is that adding more turbines to the channel would at first lead to increased power generation but would eventually choke the flow and reduce the extracted power. A general problem has been to estimate the maximum power available. Much analysis has considered the use of complete tidal fences across the channel, with all the water forced to pass through turbines. A fence is then really equivalent to a barrage, though with water flow through it all the time.

For the case of a channel connecting two larger regions with prescribed tides, the elevation difference between these regions drives the flow, with the pressure gradient accelerating the flow while opposed by the frictional forces from the bottom and sides of the channel and by the back effect of the force on the turbines. Models also allow for flow separation from the side boundaries at the downstream end of the channel. The power associated with the drag on the flow imposed by the turbines may be regarded as an upper bound to the power available. In many situations, the flow acceleration is a small term, with the flow being in a quasisteady balance. In that case, the power P generated at any instant is given by the equation below, where ρ is the water density, Q the

$$P = \rho Q(g\zeta_0 - \alpha Q^2)$$

volume flux through the channel, g gravity, ζ_0 the instantaneous sea level difference along the channel, and α a coefficient involving channel geometry and friction coefficients. The power P is the product of the mass flux ρQ and the head across the turbines (the term in parentheses in the equation).

In the natural state, the head ζ_0 along the channel is balanced by channel friction and exit flow separation so that the term in parentheses is zero. As turbines are introduced, Q and hence the channel friction are reduced and some of the head is transferred to the turbines. There is a trade-off between the increasing head across the turbines and the decreasing Q. A maximum power of $0.38\rho g\zeta_0 Q_0$ is achieved when Q is reduced to $1/\sqrt{3}$, or 58%, of its original value $Q_0 = (g\zeta_0/\alpha)^{1/2}$, and 2/3 of the original head is transferred to the turbine fence. If the reduction in flow is limited to 10%, then 44% of the maximum power is obtained. If the driving head is sinusoidal, with an amplitude a, then the average maximum power may be written as $0.21\,\rho ga Q_{max}$, with $Q_{max} = (ga/\alpha)^{1/2}$ the maximum flow through the channel in the natural state. The coefficient in the power formula changes a little if local acceleration is important, but in general the maximum average power may be written to within 10% as $0.22\,\rho ga Q_{max}$.

The same formula is generally applicable also, to within about 15%, for the second situation considered, with the channel connecting a large ocean to a small basin, but with a now representing the amplitude of the external tide. The tide within the basin, and hence the elevation difference across the turbine fence, is part of the solution. The simplest case is where the turbines are the only con-

stricting force in the channel, so that in the natural regime the tide in the basin is now equal to that outside. At maximum power, the tidal amplitude in the basin is 74% of the original, with the large elevation difference across the turbine fence being more a consequence of a shift in timing of the tide in the basin than a reduction of its amplitude. The results suggest that a barrage with a suitable operating regime need not be as environmentally damaging as often assumed.

Isolated turbines. The models discussed so far assume that turbines occupy the complete cross-section of the channel. In practice, this is likely to be ruled out by engineering considerations and by the requirements of navigation and the passage of marine mammals. A sufficient number of turbines in different locations can, in principle, produce as much power as would come from a complete fence, but it has been found that one-third or more of the potential power may be lost as the slow water in the wake of an isolated turbine mixes with the fast moving free stream around the turbine.

Applications. For significant amounts of power to be available from tidal streams, the product of the volume flux and the tidal head must be large. This can obviously occur at the entrance to basins with a large tidal range, in which case the potential is much as would be estimated for a barrage scheme. In other regions, large flows can be driven by tidal elevations that are not large but have greatly different timing at different ends of a channel. An example is Johnstone Strait between Vancouver Island and the mainland part of British Columbia, Canada (**Fig. 2**). A peak tidal flow of 311,000 m³/s is driven by a sinusoidal tidal height difference of amplitude 2.1 m (7 ft). The prediction of a maximum available power of 1.3 GW has been supported by more detailed numerical

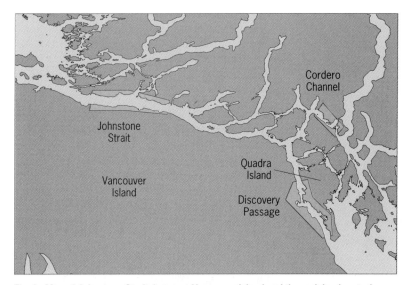

Fig. 2. Map of Johnstone Strait, between Vancouver Island and the mainland part of British Columbia, Canada. A tidal turbine fence in Johnstone Strait could generate a maximum 1.3 GW, before allowing for internal losses, with an associated flow reduction by 42%. Further southeast, the Strait splits into two main channels. Turbines placed in Cordero Channel (or Discovery Passage) alone have a maximum potential of 277 MW (or 401 MW), with the flow partly diverted into the other channel, where the flow is increased by 14% in each case.

modeling. The actual power that might be extracted would be considerably less if significant reductions in flow are to be avoided and also taking into account the wake merging head loss described above, the drag on supporting structures, and internal turbine inefficiencies. Other promising locations include Pentland Firth north of Scotland.

Much of the recent analysis has been for channels, though there are also situations where there are strong tidal currents past headlands in an open sea. Again, there will be a maximum power available, with too many turbines merely diverting the flow. More study is needed.

Further consideration of in-stream tidal power requires recognition that, while the generation of small amounts of power may be environmentally benign, the generation of large amounts would have significant impact on regional flow patterns as well as local conditions. In comparing different options for power generation from the points of view of cost and environmental impact, the use of cold, strong, tidal currents to provide cooling water for nuclear power stations also needs to be considered.

For background information *see* ELECTRIC POWER GENERATION; FLUID-FLOW PRINCIPLES; TIDAL POWER; TIDE in the McGraw-Hill Encyclopedia of Science & Technology.   Chris Garrett

Bibliography. L. B. Bernshtein, E. M. Wilson, and W. O. Song (eds.), *Tidal Power Plants*, rev. English ed., Korean Ocean Research and Development Institute, 1997; L. S. Blunden and A. S. Bahaj, Tidal energy resource assessment for tidal stream generators. *J. Power Energy*, 221:137–146, 2007; C. Garrett and P. Cummins, Limits to tidal current power, *Renew. Energ.*, 33:2485–2490, 2008; G. Sutherland, M. Foreman, and C. Garrett, Tidal current energy assessment for Johnstone Strait, Vancouver Island, *J. Power Energy*, 221:147–157, 2007; Devine Tarbell & Associates, Inc., *Instream Tidal Power in North America: Environmental and Permitting Issues*, Electric Power Research Institute, 2006; D. G. White, Ocean energy in the U.S.: The state of the technology. *Mar. Tech. Soc. J.* 42(1):9–14, 2008.

ITER (International Tokamak Experimental Reactor)

The ITER project is a major step toward realizing the potential of fusion as an essentially limitless, environmentally responsible source of power. ITER is being built in the south of France by a partnership of the European Union, Japan, China, India, Russia, South Korea, and the United States. These partners represent over half the world's population, so ITER is a global response to the global challenge of meeting future energy needs.

Fusion powers the Sun and the stars. The most effective fusion reaction for producing power on Earth uses two isotopes of hydrogen, deuterium (D) and tritium (T). To generate large amounts of power, a gas of deuterium and tritium must be heated to over $10^8\,°C$—ten times hotter than the core of the Sun—in order to allow the nuclei of deuterium and tritium to fuse copiously rather than just bounce off each other's electrical charge. The ensuing fusion reaction converts deuterium and tritium into helium (He) and a neutron, both with huge kinetic energies.

The challenges are to:

1. Heat a large volume of the fusion fuel to over $10^8\,°C$, while preventing it from being cooled (and polluted) by touching the walls of the chamber in which it is confined. This feat is now routinely achieved using a "magnetic bottle" known as a tokamak. ITER will be the world's first tokamak on the scale of a power station.

2. Make a container with walls sufficiently robust to stand up, for several years, to bombardment by the neutrons produced by fusion, which escape the magnetic bottle as they do not feel the magnetic field.

3. Design power stations that will work reliably and can be easily maintained; the reliability of what will be very complex devices will be critical for performance and economic viability, and this is perhaps the biggest challenge of all.

Deuterium is found in natural hydrogen at the level of one atom in 6700; it can easily be extracted from water. Tritium, which is unstable and decays with a half-life of about 12 years, occurs naturally only in small quantities. However, it can be generated in situ in the walls of a fusion reactor by allowing neutrons from the fusion reaction to interact with lithium to produce tritium plus helium. The raw fusion fuels are therefore water and lithium.

Lithium is a common metal that is in daily use (for instance, in mobile phone and laptop batteries); there is enough available in seawater to power the world for millions of years. Used to fuel a fusion power station, the lithium in one laptop battery, complemented by deuterium extracted from 45 L (12 gal) of water, would (allowing for inefficiencies) produce 200,000 kWh—the same as 70 metric tons of coal; this is equal to per capita electricity production in the United States for 15 years (the corresponding figure for the European Union is 30 years). The fact that so much electricity can be produced from so little lithium, without any production of greenhouse gases or long-lived radioactive waste, is the main reason for developing fusion power.

At the center of a fusion power station is a vacuum chamber with a volume of approximately 2000 m³ (70,000 ft³), into which the fusion fuel (D + T) is injected and then heated to over $10^8\,°C$ (**Fig. 1**). The neutrons produced by the fusion reaction are absorbed in the surrounding structure, which is known as the blanket and will be about 1 m (3 ft) thick. The neutrons will heat up the blanket and interact with lithium in the blanket to produce tritium. The heat will be extracted through cooling circuits and used to drive turbines and generate electricity as in a conventional thermal power station, and the tritium will be fed directly back to fuel the plant. The neutrons will also activate the walls, but with suitably chosen materials, the radioactive products will decay with

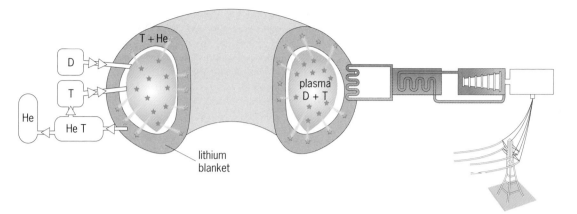

Fig. 1. Conceptual layout of a fusion power station. Fusion power stations are similar conceptually to existing thermal power stations, but with a different fuel and furnace. Some tritium fuel is needed to get started, but subsequently tritium generated in the blanket (as described in the text) will be separated and fed back into the reactor. The small amount of helium generated in the reactor will be released.

half-lives of about 10 years, and all material could be recycled within 100 years.

Studies of fusion power plants suggest that fusion-generated electricity should be able to compete in cost with other low-carbon sources of electricity. (Electricity or heat generated by fusion could also be used to produce hydrogen.)

ITER device. ITER will be an experimental device, not a power station. Although it will not operate continuously and will be slightly smaller than a fusion power plant, it will incorporate the essential components that will make up the core of fusion power stations. Most of these components have been used in small fusion devices, but they have never been used simultaneously or on the scale of ITER. Building ITER is an enormous technical challenge.

The hot fuel is contained in a toroidal chamber, surrounded by magnets, with a volume of 800 m^3 (28,000 ft^3; **Fig. 2**). A combination of magnets and an electric current flowing through the fuel (as in a fluorescent light) holds the hot fuel away from the wall and provides thermal insulation. The current also heats the fuel up to $3 \times 10^7\,°C$, but additional heating methods are required to push the temperatures up to the $10^8\,°C$ required for fusion. The heating systems use either radio-frequency waves, rather like a microwave oven, or banks of accelerators that inject beams of energetic particles. These serve the dual purpose of heating the fuel to the required temperature and maintaining the current flowing round the machine.

The whole ITER machine is enclosed in a cryostat, inside which there is a thermal shield at $-193°C$ ($-316°F$) which helps to insulate the superconducting electromagnetic coils, which are cooled to $-269°C$ ($-452°F$). Diagnostic devices provide real-time data to enable control of the plasma burn and to analyze plasma behavior. The divertor, the main area where plasma contacts the vacuum vessel wall, is a critical component that controls the exhaust of gas and impurities from the reactor and can withstand very high surface heat loads.

Progress in fusion, from T3 to JET to ITER. Experiments in Russia, in a device called T3 with a vol-

ume of about 1 m^3 (35 ft^3), established the tokamak as the most promising basis for a fusion reactor in the late 1960s. Within a few years, a team had started to design the Joint European Torus (JET), with a volume of about 80 m^3 (2800 ft^3), which is still the world's largest tokamak. JET was always seen as an intermediate step, to be followed by something like ITER (which could have been started much earlier if fusion had been funded with a sense of urgency).

Enormous progress has been made in the last few decades, as a result of experiments at JET and other tokamaks around the world. The principal results are summarized in **Fig. 3**, in terms of three key parameters that measure progress:

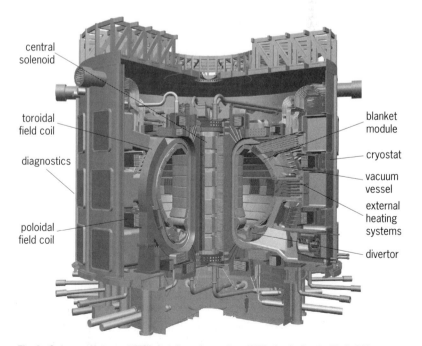

Fig. 2. Cutaway picture of ITER. In tokamaks such as ITER, the fusion fuel is held in a toroidal chamber surrounded by magnets. A current is induced in the fuel by transformer action (the central solenoid is the primary circuit of the transformer and the plasma is the secondary circuit) and, together with the magnets (toroidal and poloidal field coils), creates a helical magnetic structure that holds the hot fuel away from the walls of the chamber. (**Published with permission of ITER**)

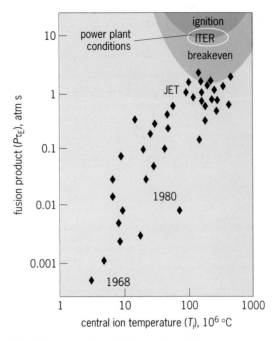

Fig. 3. A sample of results from various tokamaks, which demonstrates substantial progress in fusion over recent decades: from the relatively low-temperature, low-energy-gain points at the bottom left, through results obtained in the 1980s in the middle of the plot, to temperatures above $10^8\,°C$ and conditions close to those needed in a power station that have been achieved by JET and other tokamaks. ITER should reach conditions relevant to power plants.

1. The plasma temperature T, which must be above $10^8\,°C$.

2. The plasma pressure P (at fixed temperature, fusion power grows like P^2).

3. The energy confinement time τ_E, which measures how well the fuel is insulated against heat loss.

It turns out that the fusion product, P (in atmospheres) $\times \tau_E$ (in seconds), is approximately the energy gain of the fusion device, and this must be 10 or more in a fusion power station. The data points in the fusion performance plot (Fig. 3) show results from different tokamaks. Semiempirical scaling laws, which interpolate between results from machines with very different sizes and characteristics, give confidence that ITER will perform as designed.

The expected improved performance of ITER relative to JET (Fig. 3) is easy to understand. ITER will be about twice as big as JET in each dimension, and the ratio of its surface (through which heat escapes) to stored energy (which, for given plasma conditions, is proportional to the volume) will be half that at JET. Furthermore, the heat will on average have twice as far to travel in order to escape. It follows that, all other factors being equal, the energy confinement time at ITER should be four times that at JET, while the fusion product should be more than four times larger, since the magnetic fields will be bigger and the "magnetic bottle" stronger, allowing higher pressures.

ITER organization. Academician Evgeny Velikhov first suggested the construction of a large fusion device (ITER) by an international partnership as a way of building bridges between East and West during the cold war, and of sharing costs. The idea was raised by President Gorbachev of Russia (then the Soviet Union) in summit meetings with President Mitterand of France and President Reagan of the United States in 1985. Europe, Russia, the United States, and Japan established an ITER design team in 1988. The original design (which was twice the size of the device now being built) was rejected in 1998 as being too expensive, and the United States left the project in 1999, rejoining in 2003 together with China and South Korea. India joined in 2005, shortly after the ITER site at Cadarache in the south of France was chosen over another candidate site in Japan. The ITER Agreement was signed in late 2006 and came into force in October 2007.

A design review was carried out in 2007, and final details of the design were to be frozen in 2008. Prototypes of key ITER components have been fabricated by industry and tested. The site has been cleared, and the first substantial agreements for the supply of components, which will be provided largely in-kind by the members, have been signed. It appears likely that ITER will come into operation around 2018.

From ITER to fusion power stations. The aim of ITER is to demonstrate integrated physics and engineering on the scale of a power station. ITER is designed to produce at least 500 MW of fusion power, with an input of about 50 MW. ITER will enter new territory as the first tokamak to produce a burning plasma in which, once it is reached, the operating temperature will be maintained mainly by the fusion process itself, rather than by external heating sources. (The energetic helium nuclei produced in fusion, which will be trapped in the magnetic bottle, will collide with and heat up the D-T fuel.) ITER will also contain test blanket modules that, for the first time, will test the in situ generation and recovery of tritium, and the extraction of the heat.

ITER's success will be a necessary step toward constructing the first real demonstrator fusion power station, known as DEMO. Before DEMO can be built, however, it will be necessary to identify and characterize materials that will be able to survive for long periods in the hostile environment of a fusion power plant, and further develop a range of fusion technologies.

Tests of materials under neutron bombardment show that appropriately chosen materials may be reliable, but it is only by reproducing the real environment of a power station that this reliability can be fully explored. This task can be achieved by constructing an accelerator-based test facility, which has become known as the International Fusion Materials Irradiation Facility (IFMIF). It will consist of two deuteron accelerators that will be focused on a liquid lithium target to produce neutrons with a spectrum and intensity close to those generated in fusion. The final design of IFMIF and the testing of prototypes is being carried out jointly by Europe and Japan. If an early decision is made to proceed directly to construction, IFMIF could be in operation not long after ITER.

Assuming that (1) ITER works as expected, (2) IFMIF is built more or less in parallel with ITER, and (3) work on fusion technologies is increased appropriately, DEMO could be in operation in some 30 years (10 years to build ITER and IFMIF, 10 years to exploit them and feed the results into the design of DEMO, and 10 years to build DEMO). Assuming success, widespread deployment of fusion power might follow 10 to 15 years later. This is disappointingly slow, but by then—as fossil fuels dwindle—the need for new sources of energy will be even greater than today. The world must therefore hope that ITER rapidly succeeds in confirming that fusion can be brought down to Earth as a realistic source of power.

For background information *see* DEUTERIUM; NUCLEAR FUSION; PLASMA (PHYSICS); RADIATION DAMAGE TO MATERIALS; TRITIUM in the McGraw-Hill Encyclopedia of Science & Technology.

Chris Llewellyn Smith

Bibliography. R. Herman, *Fusion: The Search for Endless Energy*, Cambridge University Press, Cambridge, 1990; G. McCracken and P. Stott, *Fusion, the Energy of the Universe*, Elsevier Academic, 2005; National Research Council of the National Academies, *Plasma Science: Advancing Knowledge in the National Interest*, National Academy Press, 2007.

Jumping cirrus above severe storms

Jumping cirrus is a spectacular phenomenon that sometimes is seen at the top of certain severe thunderstorms. It sometimes can be observed by ground-based techniques, if the storm is sufficiently far away and its cloud top can be seen clearly. This phenomenon was first reported in the 1980s by T. Fujita, the atmospheric scientist famous for inventing the Fujita scale of tornado intensity. Fujita often flew jet airplanes above severe storms to observe phenomena associated with thunderstorms. In 1982, Fujita, describing the phenomenon, wrote, "One of the most striking features seen repeatedly above the anvil top is the formation of cirrus cloud which jumps upward from behind the overshooting dome as it collapses violently into the anvil cloud." For the nomenclature of storms mentioned here, please see **Fig. 1**.

In 1989, Fujita provided more detailed descriptions of the jumping cirrus phenomenon, in which he described three more features. (1) Fountain cirrus is a cirrus that splashes up like a fountain, 1 to 2 min after an overshooting dome collapses into an anvil. This appears to be what is mentioned in the quotation above. (2) Flare cirrus is a cirrus that jumps 1–3 km above the anvil surface and moves upwind like a flare. (3) Geyser cirrus is a cirrus that bursts up 3–4 km above the anvil surface like a geyser. These are all associated with obvious vertical motion and appear to be the jumping cirrus he referred to earlier.

Fujita further indicated that the jumping cirrus will drift away from an overshooting area if the above-anvil winds are faster than the translational speed of the overshooting area. If not, the jumping cirrus

moves back toward the overshooting area, which will be covered with a thin or thick veil of stratospheric cirrus. This suggests that the cirrus would jump upstream.

This "upstream jumping" description caused some consternation among storm researchers. How can a thin entity, such as a cirrus, jump against winds that are often easily greater than 10 or 20 knots (nautical miles per hour) at the storm-top level? In the 1980s, there was no known mechanism that could explain how this could happen. While Fujita made fairly detailed descriptions of what he observed about the jumping cirrus, he stopped short of offering any explanation. This lack of explanation, plus the fact that not many aircraft observations were done directly above severe thunderstorms to observe this phenomenon (because of the danger of tremendous turbulence in this region), resulted in reluctance to accept Fujita's observations.

It was not until a few years ago that P. Wang started to study this phenomenon using a three-dimensional thunderstorm model, WISCDYMM (Wisconsin Cloud Dynamical Microphysical Model), that was developed at the University of Wisconsin-Madison by Wang's research group. The group chose a typical midwestern United States severe storm that occurred in Montana in 1981 as the model storm for the simulation study. The model results revealed clearly the physical process of the jumping cirrus phenomenon.

Model simulation of jumping cirrus. **Figure 2** shows a series of 12 snapshots of the RHi (relative humidity with respect to ice) profiles in the central east-west vertical crosssection ($y = 27$ km) of the simulated storm every 120 s from $t = 1320$ to 2640 s. High RHi regions represent locations of high probability of ice crystal formation, and hence they are a reasonable approximation of the cloud boundary, especially the cloud-top region. Since the cloud-top region is

Fig. 1. Nomenclature associated with the storm top environment. Cloud particles at the storm top are mostly ice particles because of the low temperatures (often −40°C or colder) there. The term "cirrus" is used here to represent clouds formed at storm top region.

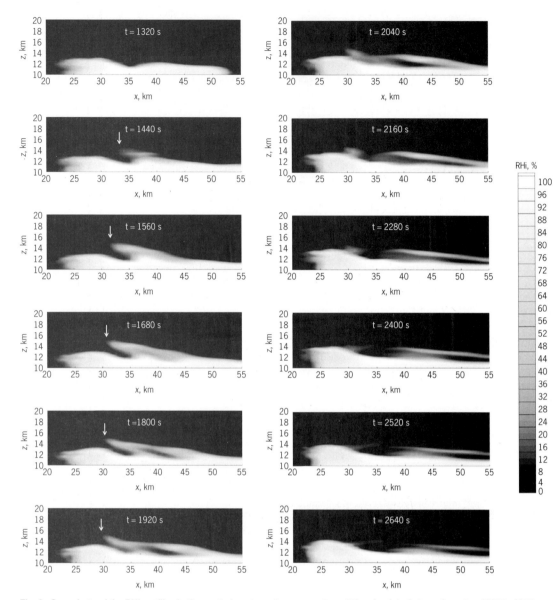

Fig. 2. Snapshots of the RHi profiles in the central east-west cross section of the simulated storm from $t = 1320$ to 2640 s.

the focus here, these snapshots are windowed to 10-20 km vertically and 20-55 km horizontally, with the vertical scale stretched in these views. Note that the range of the vertical axis is from 10-20 km and that the general shear direction is from left to right (west to east).

At $t = 1320$ s, the storm top exhibits a two-wave pattern, with one crest located at the main updraft region ($x \sim 30$ km), and the other at $x \sim 40$ km. At this stage, the overshooting is not yet well developed, and the highest point of the cloud is only slightly higher than the tropopause at 12.5 km. However, the wavy nature of the storm top is already obvious. At $t = 1440$ s, a cloudy patch starts to emanate from the bulge in the cloud top below. This humid patch is the precursor that eventually develops further into full-fledged jumping cirrus. The white arrow pointing at $x \sim 34$ km indicates the approximate position of the left (west) edge of the patch. At the same time, the overshooting top subsides, changing from a height of $\sim$13 km to $\sim$12.5 km, a drop of $\sim$500 m. This

seems to correspond to what Fujita described as the "collapse of the overshooting dome." While the overshooting top is subsiding, the wave crest located at $x \sim 40$ km starts to bulge up and tilt upstream. At $t = 1560$ s, a "jumping cirrus" in the form of a cirrus tongue has developed, with its front edge located at $x \sim 32$ km and reaching an altitude of $\sim$15 km. The cirrus tongue is already higher than the overshooting top and is moving upstream. Note also that a third wave crest appears at $x \sim 48$ km at this time. Thus the average "wavelength" of the waves on the cloud top is approximately 9 km, although the distance between the first two upstream wave crests is only 6-7 km. The "tail" end of the jumping cirrus seems to originate from the detachment from the third wave crest.

As time goes on, the cirrus reaches further west and higher altitude, as can be seen by the locations of the white arrows at the front edge. This upstream and upward motion corresponds to what Fujita described as the "cirrus cloud which jumps upward from behind the overshooting dome." This ascending sequence of the jumping cirrus lasts about 6 min, within which the cirrus rises from $z \sim 12$ to ~ 15 km. The average horizontal and vertical speeds of the jump are about 24 and 8 m s^{-1}, respectively. These are substantial speeds, and certainly justify the term "jumping." The development of the simulated cloud top up to this stage completely verifies Fujita's description of jumping cirrus.

After this time, the cirrus becomes thinner to resemble a plume, and the left half of it is almost detached from storm anvil below. As time goes on, the cirrus plume becomes unstable and breaks into two parts. The western part seems to collapse on, and eventually merge with, the overshooting dome. This could correspond to what Fujita described as the stratospheric cirrus veiling over the overshooting dome.

Mechanism for jumping cirrus formation. What is the mechanism responsible for the formation of jumping cirrus, as described above? Careful analysis of the model results shows that the jumping cirrus forms as a result of cloud-top gravity wave breaking.

Wave breaking at the cloud top. It is well known that severe storms excite gravity waves. Gravity waves are caused by strong updrafts in the core of the thunderstorm, and the stability of the layer above the storm (usually the stratosphere in the case of a severe storm) serves as the restoring factor that results in the wave motion. These waves are called the internal gravity waves because their main energies propagate vertically. Under sufficiently unstable conditions, wave breaking can occur that may result in part of the storm, especially the cloud top, becoming detached and ejected upward into the lower stratosphere. The reason that the jumping cirrus motion points upstream is because of the nature of the waves here. The waves are produced by the blocking of the strong updraft in the storm, which behaves as an obstacle to the ambient flow, similar to the role of a mountain. The waves produced are called the mountain waves whose crests generally point upstream. When these waves break, they will also "jump" upstream as well.

Comparison with photographs. **Figure 3** shows, side by side, a photograph of a jumping cirrus seen on May 24, 1996, in the late afternoon from an airplane above Alabama and Georgia versus a rendered RHi 30% contour surface of the simulated storm top at $t = 1440$ s. The bulge to the west of the overshooting top in the simulated cloud top strikingly resembles the photographed jumping cirrus in the relative location, the upstream-leaning orientation, and the surge shape. This resemblance should lend more weight to the theory of jumping cirrus as described above. In addition, there are now more sightings of this phenomenon, indicating that it is not a rare event, as some thought previously.

The above analysis shows that the behavior of jumping cirrus as observed by Fujita can be explained satisfactorily by the gravity wave breaking mechanism atop thunderstorms. The gravity waves are excited by the strong updrafts in the storm, and

(a)

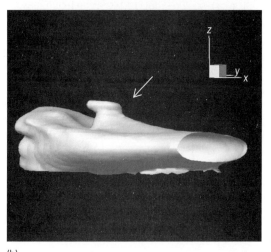

(b)

Fig. 3. Photograph of (*a*) jumping cirrus by Martin Setvak on May 24, 1996, late afternoon from an airplane above Alabama and Georgia (*courtesy of Martin Setvak*). (*b*) RHi 30% contour surface of the simulated storm at $t = 1440$ s. The vertical dimension is enhanced to match the perspective view of the photograph.

the breaking is caused by high instability near the cloud top. Thus, the occurrence of jumping cirrus indicates the presence of such instability, which should imply that the thunderstorm is severe. Such knowledge is potentially useful to meteorological studies.

What does the presence of the jumping cirrus phenomenon imply? In normal gravity wave motions, the shape of the tropopause may be distorted greatly, but there is no net transport of matter across tropopause. In the case of jumping cirrus, on the other hand, wave breaking occurs and there is net mass transfer between the troposphere and stratosphere. Water vapor, ice crystals, and other chemicals may be injected into the stratosphere by this mechanism (similar to the anvil top plume phenomenon; see P. K. Wang, "Atmospheric water vapor," in AccessScience@McGraw-Hill, http://www.accessscience.com, DOI 10.1036/1097-8542.YB041245). Thus the jumping cirrus phenomenon has important implications to atmospheric chemistry and global climate process.

For background information *see* ATMOSPHERE; CLOUD; CLOUD PHYSICS; DYNAMIC METEOROLOGY; STRATOSPHERE; THUNDERSTORM; TROPOPAUSE; TROPOSPHERE in the McGraw-Hill Encyclopedia of Science & Technology.   Pao K. Wang

Bibliography. T. T. Fujita, Principle of stereographic height computations and their application to stratospheric cirrus over severe thunderstorms, *J. Meteoro. Soc. Jpn.*, 60:355–368, 1982; T. T. Fujita, The Teton-Yellowstone tornado of 21 July 1987, *Mon. Weather Rev.*, 117:1913–1940, 1989; P. K. Wang, A cloud model interpretation of jumping cirrus above storm top, *Geophys. Res. Lett.*, 31:L18106, 2004, doi:10.1029/2004GL020787; P. K. Wang, Moisture plumes above thunderstorm anvils and their contributions to cross tropopause transport of water vapor in midlatitudes, *J. Geophys. Res.*, 108[D6]:4194, 2003, doi: 10.1029/2003JD002581.

Koji mold (Aspergillus oryzae)

Aspergillus oryzae, a type of koji mold, is one of the most widely used fungi in industry. Especially with regard to the Japanese way of life, it is quite indispensable, being an essential substance of traditional foods and beverages such as miso (soybean paste), shoyu (soy sauce), and sake (rice wine). Koji, the culture of koji mold on soy, rice, or wheat, is mixed with the substrate and together they are fermented to produce foods. This tradition has been followed for at least 500 years in Japan. Such a long history of food use suggests that koji mold is safe enough to apply in various industrial processes. In addition to its safety, koji mold has the attractive feature of possessing a high ability to produce a variety of natural enzymes. This article describes the application of koji mold for traditional use in food manufacturing as well as its application for industrial enzyme production, which recently has utilized modern gene technology.

Koji mold. Koji is cultured rice grain, wheat grain, or soybean, that is a vital ingredient in the manufacturing of traditional Japanese fermented foods such as sake, miso, and shoyu (**Fig. 1**). Different koji molds are used to produce different foods (**Table 1**). *Aspergillus oryzae* is the most widely used koji mold and therefore is recognized as the most representative, and most well known, koji mold (**Fig. 2**). As with malt for beer production, koji mold provides the enzymes necessary to degrade the substrate for further food fermentation. All these enzymes are capable of being produced by koji mold in large amounts.

The kinds of enzyme activities that are necessary in koji depend on the food to be produced. For instance, koji for sake requires high activity of amylolytic (starch-degrading) enzymes such as amylase or glycoamylase to degrade the rice starch to sugar, which is then fermented by yeast cells to produce alcohol. Also, some protein-degrading enzymes, for example, acid protease and carboxypeptidase, are necessary to convert rice protein to amino acids for use as nutrient for yeast growth, as well as for the aroma in sake. On the other hand, koji for miso or shoyu requires high protease and peptidase activity to degrade the soy protein to amino acids and peptides that contribute favorable taste, flavor, and color to the foods. These koji require lipolytic (lipid-hydrolyzing) activity to bring out aromatic flavor as well. Thus, each koji mold needs to provide the required enzyme activities to the koji. The starter for koji mold, called Tane-koji, is available from various manufacturers. The strains have been bred to give the desired enzyme activity. Tane-koji also can be a mixture of different strains of koji mold.

Koji production: solid-surface fermentation. How does koji production work? First, the substrate for koji, that is, rice, wheat, or soy, is washed, soaked in water, and steamed or roasted to allow the koji mold to grow easily. The cooked substrates are cooled and surface-dried, and then mixed with Tane-koji, which contains spores of koji mold. In the old-fashioned method, the mixed substrate is transferred to woody flamed trays and moved into a small room called the koji-muro (**Fig. 3**). Temperature and humidity are well controlled in the koji-muro so that koji mold can grow appropriately and obtain the necessary enzyme activity to a maximum level. Because koji

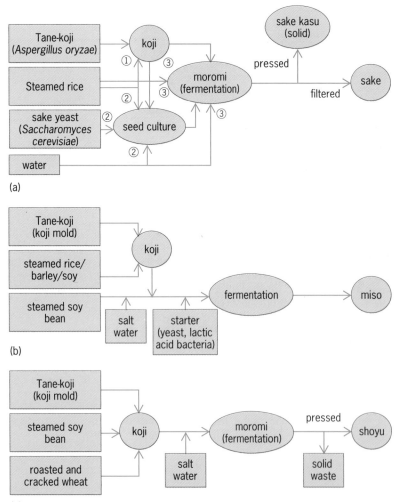

(a)

(b)

(c)

Fig. 1. Food manufacturing processes: (*a*) sake production; (*b*) miso production; (*c*) shoyu production.

TABLE 1. Koji mold used for the production of fermented foods

Food	Koji mold	Substrate for koji	Substrate for food
Sake (rice wine)	*Aspergillus oryzae*	Rice	Rice
Shochu (spirits)	*Aspergillus awamori* *Aspergillus kawachii*	Rice, barley	Rice, sweet potato, wheat, barley, buckwheat
Miso (soybean paste for seasoning)	*Aspergillus oryzae* *Aspergillus sojae*	Rice, wheat, soybean	Soybean
Shoyu (soy sauce)	*Aspergillus oryzae* *Aspergillus sojae* *Aspergillus tamarii*	Soybean, wheat	Soybean

mold is aerobic, the substrates are mixed occasionally to supply enough air to the mold. Too much humidity will lead to contamination by bacteria. Koji mold grows on the surface of the substrate, as well as on the inside of the grain, and produces its enzymes extracellularly. More than 50 different kinds of enzymes have been identified in sake koji. Koji production is usually accomplished in 40–45 h after inoculation.

The old-fashioned manual method requires a great deal of human resources. Today, koji production machines are available and widely used. Koji production is a typical example of solid-surface fermentation, which has been adapted for industrial enzyme production (see below).

Mystery of origin. Taxonomically, *Aspergillus oryzae* belongs to the section *Flavi* of subgenus *Circumdati*, which also contains *A. flavus*, the well-known producer of harmful aflatoxins (mycotoxic metabolites). These two species look very similar morphologically; however, no *A. oryzae* has been found to produce aflatoxins, even after intensive

studies by numerous researchers. *Aspergillus oryzae* probably possesses the whole set of aflatoxin biosynthesis genes, but the genes are not working and thus keep silent. Interestingly, the regulatory factor of these genes, called AflR, is either deficient or nonfunctioning in tested *A. oryzae*. Therefore, the koji mold might be a selected harmless variant of *A. flavus* that developed during its long history.

Taka-diastase. Koji had been used for food fermentation for many centuries without people realizing the presence of enzymes. In the late nineteenth century, Jokichi Takamine, a pioneering Japanese scientist, analyzed the amylolytic enzyme, called diastase at that time, in koji and found that its activity was stronger than that of malt. Takamine's initial goal was to produce whiskey in the United States using the diastase from koji instead of malt. Although he never accomplished this (because of opposition from local malt manufacturers), in 1894 Takamine applied for a U.S. patent for a "Process of Making Diastatic Enzyme" based on his method of growing the koji mold, *A. oryzae*, on wheat bran to obtain the enzyme

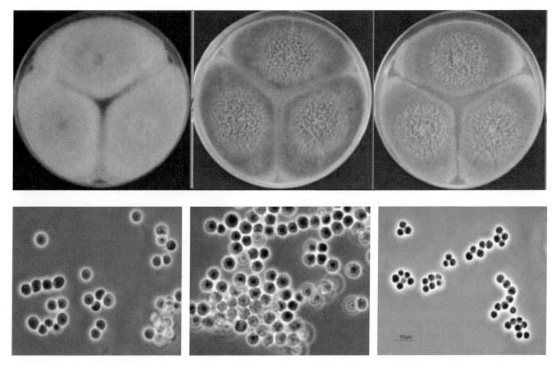

Fig. 2. Koji molds and related fungi. Left to right: *Aspergillus oryzae; A. sojae; A. flavus.* **Top:** Colonies grown on agar plates. **Bottom:** Microscopic images of conidia (spores). (*Courtesy of T. Okuda, Tamagawa University Research Institute, Japan*)

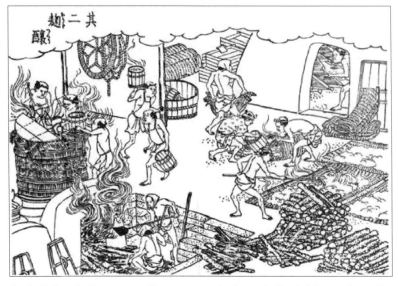

Fig. 3. Koji production as one of the processes of sake production in Edo-era, Japan. The steamed rice is spread on the straw mat for cooling (right) and then mixed with Tane-koji (next place to the left). The room at the back is the koji-muro. (*From Nihon Sankai Meisan Zue, written by K. Shitomi in 1799, which describes traditional foods of Japan in the seventeenth to eighteenth centuries and how people produce or catch these foods*)

extract. He also found that the use of wheat bran generated higher enzymatic activity than the use of rice. Recently, Japanese researchers using DNA array technology have confirmed that wheat bran does indeed produce greater amounts of enzyme compared to rice or soy.

Takamine promoted the enzyme extract, named Taka-diastase, for medical application as a digestive aid in the United States, where it became an enormous success. Taka-diastase was the first commercialized enzyme produced by *A. oryzae*; moreover, it was the first commercialized microbial enzyme ever. The enzyme product is still in use today.

Production of industrial enzymes. Enzymes began to be widely employed in industry in the middle of the twentieth century. Early products were of animal origin, such as bovine trypsin for leather tanning and pancreatic amylase for desizing (removal of size or sizing agents, generally starch, from warp yarns prior to weaving, to protect them against the abrasive action of loom parts). Shortly thereafter, the pancreatic amylase for desizing was replaced by bacterial amylase. The amylase from *A. oryzae* was also applied in the baking industry, where it was added to wheat flour to increase the volume of bread made. In addition, proteases from *A. oryzae* were used for vegetable protein hydrolysis for seasoning. Because of their safe history of food use prior to 1958, carbohydrase (any enzyme that catalyzes the hydrolysis of disaccharides and more complex carbohydrates, such as amylase, cellulase, hemicellulase, pectinase, etc.) and protease from *A. oryzae* have received Generally Regarded as Safe (GRAS) status from the U.S. Food and Drug Administration (FDA). Hence, these enzymes do not require premarket approval.

Solid-surface fermentation has been applied by several enzyme manufacturers following the success of Taka-diastase. Submerged (or liquid) fermentation

(a process used in penicillin fermentation) in closed tanks has also been adapted for enzyme manufacturing. Submerged fermentation is especially capable of producing large volumes of industrial enzymes; it has become the major method utilized today.

Host organism for enzyme production. In the mid-1980s, gene technology began to be applied in industry, including industrial enzyme production. Because of its safe history and high capacity to produce various enzymes, *A. oryzae* was chosen to be a host organism to produce industrial enzymes. The process of gene manipulation is as follows. First, an amylase with high productivity is selected for use. The gene of the desired enzyme is cloned and fused in vitro to the promoter region of the amylase gene from *A. oryzae*. Then, this hybrid gene is transferred into the cell of *A. oryzae* and the gene is allowed to integrate into the chromosome of *A. oryzae*. After it becomes incorporated, the generated *A. oryzae* strain starts to produce the enzyme of interest under the typical conditions for amylase production. By employing this technology, numerous enzymes of even non–*A. oryzae* origin are produced by *A. oryzae* today. This technology has successfully expanded the application of industrial enzymes in many fields (**Table 2**).

During the last two decades, gene expression technology has been significantly improved, and the enzyme production system using *Aspergillus oryzae* has benefited from these advances. For instance, the amylase promoter has been reinforced because of the discovery of a promoter activator called AmyR. Also,

TABLE 2. Industries and applications of enzymes produced by *Aspergillus oryzae*

Industry	Enzyme	Applications
Food	Amylase	Baking, brewing, starch modification
	Hemicellulase	Baking (volume increasing), brewing
	Glucose oxidase	Baking (dough strengthening)
	Protease, peptidase	Seasoning, brewing, peptide formation
	Lipase	Baking (emulsifier substitution, volume increasing), flavor for dairy, products, oil modification, reducing trans fatty acids
	Phospholipase	Oil degumming, lecithin modification
	Lactase	Milk modification, lactose reduction
	Mucor rennet	Cheese production
	Pectin esterase	Fruit and vegetable firming
	Asparaginase	Acrylamide reduction
Detergent	Lipase	Stain removal
	Cellulase	Whiteness, softness
Textile	Cellulase	Denim finishing, texture improvement
	Laccase, catalase	Bleaching, wastewater treatment
Forest product	Lipase	Resin removal, pitch control
	Oxidoreductase	Lignin modification
Feed	Phytase	Feed supplement
	Hemicellulase	Feed supplement

the host strain has been modified so that endogenous amylase and two major enzyme-destroyer proteases are genetically inactivated, allowing greater and more stable production of the desired enzyme. Furthermore, a potential mycotoxin synthetic gene cluster existing in the host strain has been deleted. Together, these provide more advantages for the production of enzymes for food use.

Recent progress on genomics. The genome sequence of *A. oryzae* has been determined by a Japanese research group. The genome is more than 37 megabases (Mb) in size and contains more than 12,000 genes. Of these, 135 are extracellular protease genes, which comprise 1% of the total genes. It has been revealed that *A. oryzae* possesses the genes of carbohydrase enzymes to a greater extent than any other *Aspergillus* species, such as *A. nidulans* or *A. fumigatus*, which supports previous data concerning the characteristic features of koji mold. More genome sequence information will thus enable the compilation of data on different strains or different metabolites at the molecular level, for example, by using DNA array technology, which will help to expand the application of this mold even further.

For background information *see* AFLATOXIN; FERMENTATION; FOOD FERMENTATION; FOOD MANUFACTURING; FOOD MICROBIOLOGY; FUNGAL BIOTECHNOLOGY; FUNGAL GENETICS; FUNGAL GENOMICS; FUNGI; SOYBEAN; YEAST in the McGraw-Hill Encyclopedia of Science & Technology.   Shinobu Takagi

Bibliography. K. Abe et al., Impact of *Aspergillus oryzae* genomics on industrial production of metabolites, *Mycopathologia*, 162:143–153, 2006; K. E. Aidoo, J. E. Smith, and B. Wood, Industrial aspects of soy sauce fermentation using *Aspergillus*, pp. 155–169, in *The Genus Aspergillus—From Taxonomy and Genetics to Industrial Application*, Plenum Press, New York, 1994; J. Bennett and Y. Yamamoto, Dr. Jokichi Takamine: Japanese father of American biotechnology, pp. 54–57, in *Proceedings of the 10th International Congress for Culture Collection (ICCC-10)*, Tsukuba, Japan, 2004; P. H. Nielsen et al., Enzyme applications (industrial), pp. 567–620, in *Kirk-Othmer Encyclopedia of Chemical Technology*, vol. 9, Wiley, New York, 1994; Z. S. Olempska-Beer et al., Food-processing enzymes from recombinant microorganisms—a review, *Regul. Toxicol. Pharmacol.*, 45:144–158, 2006; K. L. Petersen, J. Lehmbeck, and T. Christensen, A new transcriptional activator for amylase genes in Aspergillus, *Mol. Gen. Genet.*, 262:668–676, 1999.

Large Hadron Collider (LHC)

The Large Hadron Collider (LHC) is located at the European Laboratory for Particle Physics (CERN) near Geneva, Switzerland. Currently in the final stage of commissioning (2008), it will be the most powerful particle physics instrument in the world. It consists of two rings in which two counterrotating beams of hadrons (protons) are accelerated to an energy of 7 teraelectronvolts (1 TeV = 10^{12} eV) and brought into collision at four points around the ring containing particle detectors. The center-of-mass energy of 14 TeV will be more than seven times larger than the energy achieved in the currently operating Tevatron at Fermilab near Chicago.

The main objectives of the LHC are reviewed here. Most notable is the search for the Higgs boson, which could explain how other elementary particles acquire properties such as mass. Other fundamental questions to be addressed by the LHC include the origin of the missing mass of the universe and the asymmetry between matter and antimatter that led to the disappearance of antimatter in the universe's early phase of formation. The LHC can also accelerate and collide heavy (lead) ions, in the hope of creating the quark-gluon plasma that is thought to have existed soon after the big bang.

Layout, Technology, and Performance

The collider is housed in a 3.8-m-diameter (12.5-ft) tunnel, 27 km (17 mi) in circumference, at a depth ranging 40–180 m (130–590 ft), straddling the French-Swiss border (**Fig. 1**). The tunnel was formerly used to house the Large Electron-Positron Collider (LEP), which was previously the workhorse of the CERN experimental program. The machine is divided into eight octants, with each octant joined to the next by a long straight section (LSS) approximately 500 m (1600 ft) in length. Four of the long straight sections contain experimental caverns that house the four detectors. The other four long straight sections house machine utilities (**Fig. 2**).

Starting at octant 1, a large cavern specially excavated for the LHC program contains the large ATLAS (A Toroidal LHC Apparatus) detector. At octant 2, a cavern previously existing from the LEP program contains the much smaller ALICE (A Large Ion Collider Experiment at CERN) detector. This detector has been designed to study the primordial

Fig. 1. Large Hadron Collider (LHC) during installation.

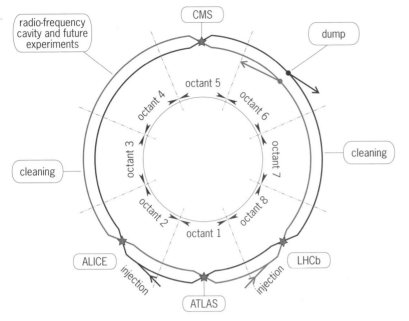

Fig. 2. Schematic layout of the LHC ring.

quark-gluon plasma believed to have existed in the early evolution from the big bang. Octants 3 and 4 contain machine utilities. At octant 3, a collimation system is installed to protect the detectors from unwanted background, and octant 4 contains the radio-frequency cavities that accelerate the beams. Octant 5 contains the other very large detector, the CMS (Compact Muon Solenoid), housed in a cavern specially excavated for the LHC program. It also contains a small experiment (TOTEM) that will make a precision measurement of the total cross section for proton-proton interactions. Octants 6 and 7 contain more machine utilities. At octant 6 is located the beam abort system, which allows the beams to be safely extracted from the machine and deposited on specially designed beam dumps. Octant 7 contains a collimation system similar to that at octant 3. Finally, a cavern previously used for LEP houses the LHCb detector, specially conceived to study the asymmetry between matter and antimatter.

The beams for the LHC are prepared through a tandem of existing accelerators that successively increase the energy of the beams, starting with a linear accelerator at 50 MeV that feeds the Proton Synchrotron Booster (PSB), accelerating to 1.4 GeV. The PSB feeds the Proton Synchrotron (PS), which boosts the energy to 26 GeV before transferring the beam to the Super Proton Synchrotron (SPS), which accelerates it to 450 GeV, finally transferring it to the LHC itself, which accelerates it from 450 GeV to 7 TeV. At this energy the beams rotate in their vacuum chambers for many hours, slowly reducing in intensity as particles are eaten up in collisions.

Technological challenges. The particle trajectories around the 27-km orbit are controlled by very powerful superconducting magnets. Guiding the beams on the quasicircular orbit requires 1232 compact high-field dipole magnets, each weighing 35 metric tons (38.5 short tons), 15 m (49 ft) long, with the two

apertures integrated into a common structure, and with a field of 8.3 teslas, about 3 teslas higher than achieved in previous machines (**Fig. 3**). The vertical magnetic field is oriented in opposite directions in the two apertures, as is required to guide particles of the same charge moving in opposite directions. In order to achieve the unprecedented field strength, advanced superconducting cable had to be developed. Each cable strand of 1 mm (0.04 in.) diameter contains about 9000 fine (6-micrometer) filaments of niobium-titanium alloy in a copper matrix. In all, about 1200 metric tons (1320 short tons) of cable was produced using more than 400 metric tons (440 short tons) of niobium-titanium alloy. Many other superconducting magnets (more than 6000 in total) are needed to periodically focus the beams and to correct for imperfections.

Operation of superconducting magnets built from niobium-titanium superconductor at such high field levels requires that the magnets be cooled to below the normal boiling point of liquid helium at atmospheric pressure (4.2 K). In fact, the LHC operates at 1.9 K, below the transition temperature (2.17 K) at which helium undergoes a phase transition into a quantum liquid, superfluid helium. The production of industrial quantities (130 metric tons or 143 short tons) of superfluid required to cool the 40,000 metric tons (44,000 short tons) of equipment to 1.9 K poses a formidable technical challenge. The only practical method of producing such large quantities of superfluid helium is to reduce the vapor pressure above the helium bath located in a heat exchanger in each magnet unit using enormous pumps (cold compressors). At 50 mbar (5 kPa) pressure, the liquid crosses the lambda point marking the phase transition to superfluid. The pressure must be reduced to 15 mbar (1.5 kPa) to reach the 1.9 K required. The helium inside the magnets themselves is cooled through the heat exchanger until it also reaches 1.9 K, but at atmospheric pressure. Superfluid helium at 1.9 K has a very high specific heat compared with the superconductor and also a very high thermal conductivity; both of these properties help it to absorb unwanted heat input and to transport it away from the coils. The production of such large quantities of liquid helium requires eight refrigerators with a combined installed capacity of more than 140 kW at 4.5 K, making it the largest cryogenic installation in the world. The production of the low (15 mbar) pressure necessary to reduce the temperature to 1.9 K requires eight special cold compressor units in addition to the refrigerators.

The beams will circulate with good (greater than 100 h) lifetime in the LHC only if the beam pipe is under ultrahigh vacuum (less than 10^{-10} mbar or 10^{-8} Pa). Inside the superconducting magnets, the vacuum chamber wall is at 1.9 K, so it acts as a very efficient cryopump provided that the wall is screened from the beam. In the warm sections (the long straight sections), a special nonevaporable getter (NEG) has been developed, consisting of a coating of titanium-zirconium-vanadium (TiZrV) alloy sputtered onto the inside surface of the vacuum pipe.

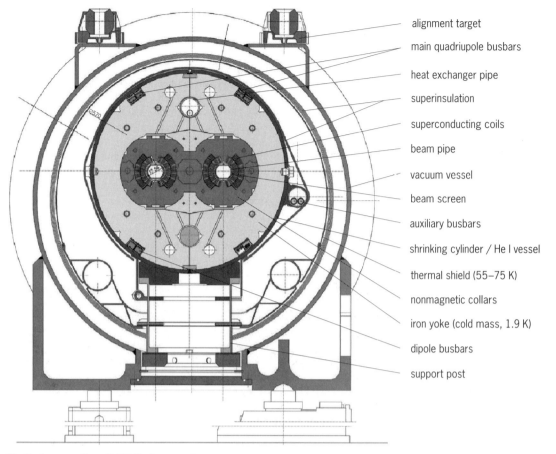

alignment target

main quadriupole busbars

heat exchanger pipe

superinsulation

superconducting coils

beam pipe

vacuum vessel

beam screen

auxiliary busbars

shrinking cylinder / He I vessel

thermal shield (55–75 K)

nonmagnetic collars

iron yoke (cold mass, 1.9 K)

dipole busbars

support post

Fig. 3. Cross section of LHC dipole magnet.

The NEG can be activated at normal bakeout temperature (200°C or 392°F), after which it acts as a distributed pump, reducing the vacuum pressure to the required level.

Performance. The LHC has been designed to produce a very high luminosity (the event rate per unit cross section), about two orders of magnitude higher than achieved in previous hadron colliders. Collisions between particles within each LHC detector will occur at the rate of about 10^9 per second. If data were collected from each collision, they would fill a stack of CD-ROMs 1.6 km (1 mi) high each second. Decisions must be made in real time on what should be discarded and what should be kept for further analysis. Even then, enormous computing resources are needed to further distill the most interesting events from the remaining data. To allow this, a massive distributed computing system (the LHC grid) has been developed so that data can be accessed from anywhere in the world and analyzed locally in one of the institutes collaborating in the experiments.

Another characteristic of the very high luminosity is that the intense beams have a very large stored energy, about 350 MJ, or 80 kg (180 lb) of TNT, two orders of magnitude higher than ever achieved before. If a beam is lost in an uncontrolled way, it can damage machine components. Very sophisticated safety systems are required to monitor beam losses and to abort the beams safely if they become excessive.

It was expected that the first beam would be injected into the machine in September 2008 and that data taking could start soon after. However, during commissioning, an electrical fault was discovered that required warming up a sector of the machine for repair, delaying the start of operation until the Spring of 2009. It was anticipated that initially the luminosity would be low as scientists and engineers learned to handle this very complex scientific instrument. As the luminosity increases and more data are accumulated, there is no doubt that very important discoveries will be made in the new window on the universe opened by the Large Hadron Collider.   Lyndon Evans

Anticipating the LHC

Together with its detectors, the LHC will constitute the world's most powerful microscope, probing the internal structure of matter on the nanonanoscale, below a billionth of a billionth of a meter (10^{-18} m). The LHC will advance the experimental frontier of particle physics to energies above 1 TeV for collisions among the basic constituents of matter. We do not know what the order-of-magnitude leap in energy and resolution will reveal, but we expect exciting results in great profusion, because many of today's outstanding mysteries point to solutions in the new realm that will be opened up by the LHC.

Through a lively interplay between theory and experiment from the early 1970s to the present, particle physicists have crafted a "standard model" of quarks and leptons and their interactions that describes observations throughout the range of energies and distances explored so far. For all its successes, the standard model cannot be the last word: it offers no prediction for the masses of the quarks and leptons, no candidate for the dark matter of the universe, and no explanation of why empty space is nearly weightless.

What hides the electroweak symmetry. Chief among the outstanding mysteries is what distinguishes electromagnetism from the weak interactions. The weak interaction's sphere of influence is restricted to distances smaller than 10^{-17} m, less than 1% of the size of a proton, while electromagnetic radiation can propagate over limitless distances. Differently stated, the weak interactions are transmitted by the massive W and Z bosons, while electromagnetism is carried by the massless photon. The symmetry that relates weak and electromagnetic interactions suggests that all the force carriers, as well as all the quarks and leptons, should have no mass. This is not what we observe, but the symmetries of the laws of nature need not be reflected in the outcome of those laws. In the standard model, a "Higgs field" that permeates all of space engineers a random choice of one out of infinitely many equivalent vacuum states. That choice hides the electroweak symmetry, correctly predicts the heavy W and Z masses, leaves the photon massless, and allows, but does not specify, quark and lepton masses. The satisfying part-per-mille agreement between standard-model calculations and experiment depends on the quantum particle associated with the Higgs field. Lacking that missing piece or a suitable substitute, the theory would come apart at energies around 1 TeV.

The quest for the agent that hides the electroweak symmetry is often expressed in shorthand as a hunt for the Higgs boson, and that high-value target has informed the design and running strategies of the versatile LHC detectors ATLAS and CMS. With sensitivity to many production mechanisms and decay modes, the LHC experiments should be able, after several years of running, to find the (standard-model) Higgs boson all the way up to 1 TeV—the effective upper bound on its mass.

Discovery of the Higgs boson would seem to set the keystone into the electroweak-theory edifice, but many details must be filled in to secure our understanding. Is there a single Higgs boson, or several? Is it an elementary particle, as the standard model prescribes, or might it be composite? Does the Higgs boson give mass to quarks and leptons, or only to the weak-force carriers? If the Higgs boson does generate quark and lepton masses, what sets their values? Just how does the Higgs boson set the stage for the electroweak symmetry to be broken?

Arcane as these questions may seem, the answers will change the way we think about the everyday world. Much more is at stake than finding a fugitive particle: we stand to gain a new appreciation of why we live in a world made of compact atoms linked by chemical bonds into stable structures. An atom's size is inversely proportional to the electron's mass, so

without a mechanism to give mass to the electron, atoms would be infinitely large and lack structural integrity. The universe would be an entirely different place. When we unravel the puzzle of electroweak symmetry breaking, we will understand with new clarity why the world is the way it is.

More new physics on the TeV scale. That may be only the beginning: we have hints of other new phenomena that may be accessible to LHC experiments. Self-consistency of the electroweak theory argues that either the Higgs boson must weigh less than 1 TeV or other new phenomena, such as strong interactions among weak bosons, must appear at TeV energies. If the Higgs boson does materialize below 1 TeV, as circumstantial evidence suggests that it will, that will pose an interesting conundrum in a quantum field theory. Just as the virtual presence of the Higgs boson influences the masses of other particles, including the W and Z, through quantum corrections, other particles act behind the scenes on the Higgs-boson mass. Particles with masses up to the energy at which the standard model yields to a more complete theory can contribute. If the successor is a unified theory of the strong, weak, and electromagnetic interactions or a theory that encompasses gravity, particles with prodigious masses participate in quantum corrections, and tend to tug the Higgs mass far above 1 TeV.

The hierarchy problem refers to the challenge of preserving a Higgs mass below 1 TeV in the face of wide-ranging quantum corrections, or of maintaining the gap between the electroweak scale and the much higher energies at which new physics comes into play. It is a logical possibility that this is simply the way things turn out, the result of a fortuitous cancellation among many large contributions, but experience has taught physicists to be suspicious of such "just-so" stories. If the Higgs-boson mass is (relatively) low for a reason, that reason should show itself in experiments at the LHC. Theorists have examined a number of possible explanations. The extreme sensitivity to quantum corrections evaporates if the Higgs boson is a composite object with a size characteristic of the 1-TeV scale. Accordingly, the LHC might uncover a collection of related composite particles, all with masses on the order of a few TeV.

Supersymmetry offers a reason for the spooky cancellation of quantum corrections. In this picture, every particle has a superpartner that differs by half a unit in spin, but otherwise has identical characteristics. If nature were exactly supersymmetric, each particle and its superpartner would have identical masses, and their influences on the Higgs-boson mass would balance exactly. No superpartners have been found, so we can say with confidence that nature is not exactly supersymmetric, but the net influence on the Higgs mass would still be acceptably small if superpartner masses were less than about 1 TeV, well within the LHC's reach.

Another approach to the hierarchy problem is to imagine that the fault lies not with the electroweak theory but with our conception of gravity. If space has extra dimensions beyond the three that we explore in our daily lives, then perhaps the extra dimensions will lead to new phenomena observable at the LHC. At the distance scales examined until now, gravity is utterly negligible for particle interactions; it is some forty orders of magnitude weaker than the electrostatic force that binds a proton to an electron in the hydrogen atom. Gravity's strength may be comparable to those of other interactions when the extra dimensions come into view. If that happens at the resolution the LHC will provide, experiments might see signs of "missing" energy leaking from the four-dimensional world to excite particles in the extra dimensions, evidence for particle interactions mediated by gravitons, or even highly ephemeral and exquisitely tiny black holes.

None of these possibilities (compositeness, supersymmetry, extra dimensions) ranks as a sure thing, but each would provide leads to a more comprehensive theory. The hierarchy problem does make it plausible that new phenomena, beyond the Higgs boson, await the LHC experiments. When new signals do appear, experimenters and theorists will have a merry time trying to decode what they mean.

Cosmic connection. Evidence from the cosmos also suggests new phenomena on the 1-TeV scale. The rotation curves of spiral galaxies, the large-scale structure of the universe, and characteristics of the cosmic microwave background radiation reveal that most of the matter in the universe (roughly $1/4$ of the total energy density) is unfamiliar stuff called dark matter. The evidence is consistent with the notion that the dark matter is made up of long-lived neutral particles, thermal relics from the early universe. If this dark-matter particle interacts with the strength of the weak force (making it a weakly interacting massive particle, or WIMP), then the requisite numbers could have been produced in the early universe and survived until now, provided that the WIMP mass lies between about 0.1 and 1 TeV. The LHC would thus be poised to help solve one of astronomy's greatest puzzles by creating dark-matter particles in the laboratory. It is tantalizing that nearly every scenario to resolve the hierarchy problem suggests a candidate dark-matter particle.

A Higgs field that fills all of space has cosmological implications because it contributes at least 10^{24} grams per cubic centimeter to the energy density of the universe. Astronomers report that the average energy density of the universe—much of it in the form of the mysterious dark energy—is 54 orders of magnitude smaller. This mismatch is another reminder that something is lacking in our understanding of the electroweak theory.

First things first. As inviting as this menu of possible discoveries is, the first task of the LHC experiments will be to reestablish the standard model by observing W and Z bosons, energetic jets of pions and other particles, and the top quark. The high yield of standard-model particles will help experimenters calibrate the detectors and establish orientation points in the new landscape. First and foremost, the task of the LHC is to explore. Even the shake-down cruise could bring surprises, such as new force particles that decay into electron-positron pairs. The discovery of a new force of nature could provide clues to new symmetries and point toward a unified description of all the interactions. Then it will be on to the central goal—getting to know the new world of the TeV scale.

For background information *see* CRYOGENICS; DARK ENERGY; DARK MATTER; ELECTROWEAK THEORY; HIGGS BOSON; LIQUID HELIUM; PARTICLE ACCELERATOR; QUANTUM CHROMODYNAMICS; STANDARD MODEL; SUPERCONDUCTING DEVICES; SUPERSYMMETRY; WEAKLY INTERACTING MASSIVE PARTICLE (WIMP) in the McGraw-Hill Encyclopedia of Science & Technology.   Chris Quigg

Bibliography. F. Close, *The New Cosmic Onion*, Taylor & Francis, London, 2006; L. Evans, The Large Hadron Collider, *New J. Phys.*, 9:335, 2007; M. E. Peskin, Dark matter and particle physics, *J. Phys. Soc. Jpn.*, 76:111017, 2007; C. Quigg, Particles and the standard model, in G. Fraser (ed.), *The New Physics: For the Twenty-First Century*, chap. 4, pp. 86–118, Cambridge University Press, Cambridge, 2006; C. Quigg, The coming revolutions in particle physics, *Sci. Amer.*, 298(2):46–53, February 2008.

Lusi mud volcano

When sediments are deposited and then buried, they start to compact and undergo chemical changes that turn them into sedimentary rock. Some of the water located between individual grains, along with the fluid and gas produced by the chemical reactions, escapes to the surface. In some cases, this movement of fluid and gas occurs through focused regions a few meters to a couple of kilometers across. During the ascent from depths of up to 5 km (3.1 mi), the fluid and gas can mix with sediment (sand or mud) that has not yet turned into rock. The result is an eruption of sediment, fluid, and gas at the surface. There are thousands of mud volcanoes on Earth, but they are poorly understood phenomena because (1) we cannot witness most of the processes directly as they are occurring underground, (2) little is known about the geological conditions prior to and during eruptions, and (3) unlike igneous systems, there are few fossil mud volcanoes that have been exposed on the surface of the Earth that can be examined in detail.

There are two basic forces that drive the water, gas, and mud mixture to the surface. One is the natural development of abnormally high pressure within sedimentary rock, termed overpressure. The second is the expansion of gas that comes out of solution as the mixture rises to the surface and its pressure decreases. The expansion of gas bubbles gives the mixture sufficient buoyancy to rise. We know little of the detailed structure of the feeder conduits that allow the flow to occur, but they probably consist of a complex system of fractures. In some mud volcano systems, the fluid source does not coexist with the mud source beds. Instead, the fluid comes from

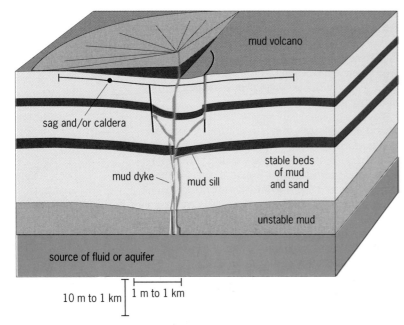

Fig. 1. Cross section through a mud volcano system, showing a source of fluid (aquifer) linked to a source of mud. Mud is eroded and brought to the surface through fractures in the sedimentary rock strata. The mud volcano has a lentoid (lens) shape. (1 m = 0.3 ft; 1 km = 1.61 mi.)

deeper strata and then passes through mud strata that are susceptible to being picked up by the flow underground (**Fig. 1**).

Location and dimensions. The Lusi mud volcano is located in the Porong subdistrict of Sidoarjo in

Fig. 2. Crisp satellite image of the Lusi mud volcano in February 2008. (1 km = 1.61 mi.)

eastern Java, and is the best-known mud volcano on Earth. It started to erupt at about 5 a.m. on May 29, 2006, with the main eruption vent located 150 m (500 ft) from a gas exploration well, Banjar Panji-1. The eruption consists of 79% water and 21% mud and other sedimentary rock fragments. By March 2008, the mud volcano had displaced about ~30,000 people, covered an area of approximately 7.0 km² (2.7 mi²), and was approximately 20 m (66 ft) thick at its center (**Fig. 2**).

Trigger. Despite its unprecedented catastrophic impact on the local population and high media and scientific profile, one of the most fundamental questions about Lusi has yet to be resolved: Was the eruption triggered by the Banjar Panji-1 exploration well or by the Yogyakarta earthquake that occurred on May 27, 2006, two days before the eruption started? Determining the cause is important, as it may determine whether the drilling companies should be compensating those affected and be responsible for remediation work. The destruction caused by the mud volcano has been estimated at hundreds of millions of dollars.

The Yogyakarta earthquake of May 27, 2006 had an epicenter 250 km (155 mi) from the eruption and a moment magnitude of 6.3. It is well known that earthquakes can trigger mud volcano eruptions. Comparison of the magnitude and distance of this earthquake from the mud volcano with historical records of earthquakes that have caused eruptions shows that the Yogyakarta earthquake was either too far away or too small to have been the cause. Also, calculations of the impact that the earthquake would have had on the pressure of underground fluid show that pressure changes caused by the earthquake would have been negligible. Most powerful in this debate are the actual data from the exploration borehole. There are two key facts that point to the exploration borehole as the cause. Firstly, the uppermost 1091 m (3579 ft) of the hole was protected by steel casing and the lowermost 1743 m (5718 ft) was not. Secondly, on May 27, 2006, a decision was made to pull the drill bit out of the borehole. This is commonplace in drilling operations. While this was being done, there was an influx of water and gas into the hole (termed a "kick"). This is also not uncommon, but the kick was not noticed for several hours and potentially dangerous fluids and gas could eventually have reached the surface. Therefore, emergency valves known as blowout preventers were closed at the wellsite. The pressure in the wellbore was measured at the surface, while these valves were shut. These measurements showed that the pressure in the unprotected part of the borehole had passed critical levels and that the pressure in the borehole was slowly dropping. These data demonstrate that the pressure was sufficient to cause the rocks to fracture and crack and for the fluid in the wellbore to start to leak into them. This phenomenon has occurred in other boreholes and is termed a subsurface blowout. In some cases the fluid in the borehole moves toward the surface, working its way up through the cement between the protective casing and the surrounding rock. The wellbore probably provided the connection between water-bearing rock strata at the bottom of the hole (2834 m or 9298 ft) and layers of mud higher up. As the fluid moves from 2834 m depth upward, it passes through mud located at between 1219 to 1828 m (3600 to 5900 ft) depth and picks the mud up in the flow, bringing it the surface. The entrainment of huge volumes of mud is not a common aspect of subsurface blowouts.

Next developmental stages. A region that goes beyond the extent of the mud volcano is now undergoing subsidence. The rate of subsidence is a few centimeters per day. It probably occurs due to the removal of sediment and fluid below the surface and because of the weight of the erupted mud at the surface. New eruptions of fluid and gas are occurring within the area of the mud volcano and around its periphery. These eruptions may be the result of new geological faults forming as the region collapses or because dormant faults are reactivated.

Longevity. The length of this eruption could be estimated, but no calculation has been published to date. The eruption is probably driven by the pressure of the source of the water being higher than the downward pressure of a column of the mud and water that fills the fractures and links to the surface. The water-mud mixture also contains gas, and the expansion of the gas during the ascent of the mixture will be a contributory lift mechanism. Once the pressure of the aquifer has dropped, the main drive for the volcano will stop. But if there are still fractures in the rock providing a pathway to the surface, then the expansion of gas in the fluid will allow the volcano to continue to erupt for many years to come. One could also calculate the volume of the source of the water and also calculate how long it would take for the pressure to deplete to normal levels. If one uses other examples of subsurface blowouts as an indicator as to how long this will continue, then some sort of fluid and gas escape can be expected for decades.

For background information *see* EARTHQUAKE; OIL AND GAS WELL COMPLETION; OIL AND GAS WELL DRILLING; SEDIMENTARY ROCKS; SEDIMENTOLOGY in the McGraw-Hill Encyclopedia of Science & Technology.   Richard J. Davies

Bibliography. D. Cyranoski, Indonesian eruption: Muddy waters, *Nature*, 445:812–815, 2007; R. J. Davies and S. A. Stewart, Emplacement of giant mud volcanoes in the South Caspian Basin: 3D seismic reflection imaging of their root zones, *J. Geol. Soc. London*, 162:1–4, 2005; R. J. Davies et al., Birth of a mud volcano: East Java, 29 May 2006, *GSA Today*, 17(2):4–9, 2007; R. J. Davies et al., The East Java Mud Volcano (2006 to Present): An Earthquake or Drilling Trigger?, *Earth Planet. Sci. Lett.* (in review); M. Manga, Did an earthquake trigger the May 2006 eruption of the Lusi mud volcano?, *Eos Trans. AGU*, 88(18):201, 2007; A. Mazzini et al., Triggering and dynamic evolution of LUSI mud volcano, Indonesia, *Earth Planet. Sci. Lett.*, 261:375–388, 2007; D. Normile, Indonesian mud volcano unleashes a torrent of controversy, *Science*, 315(5812):586, 2007.

Mars Express

Mars Express has completely revamped our understanding of the planet's geological evolution. A great wealth of data has been gathered, allowing us to build a comprehensive and multidisciplinary view of Mars, including the surface geology and mineralogy; the subsurface structure; the state of the interior; the climate's evolution; the atmospheric dynamics, composition, and escape; the aeronomy; and the ionospheric structure. Major advances have been made, such as discovering water ice below the surface, mapping of the various types of ice in the polar regions, tracing the history of water abundance on the surface of Mars in view of the minerals formed at different epochs, discovering the presence of methane in the atmosphere, observing midlatitude auroras above crustal magnetic fields, and establishing much younger time scales for volcanism and glacial processes. Indeed, the presence of methane, independently confirmed by ground measurements, suggests that either volcanism or biological processes are currently active on Mars.

The *Mars Express* spacecraft has been orbiting the Red Planet since December 25, 2003. Its scientific payload includes the High-Resolution Stereo Color Imager (HRSC), the OMEGA infrared mineralogical mapping spectrometer, and the MARSIS subsurface sounding and ionospheric radar to study primarily the solid planet; and the infrared Planetary Fourier Spectrometer (PFS), the SPICAM ultraviolet and infrared atmospheric spectrometer, and the ASPERA energetic neutral atoms analyzer to study the atmosphere and space environment of Mars. Finally, the Mars Radio Science (MaRS) experiment provides information on the surface roughness, the gravity anomalies of the crust, and the ionospheric and atmospheric structure.

Interior and subsurface. MaRS has been probing the Martian interior to study the temporal and spatial variations of the Martian gravity field. Gravity is measured by observing the accelerations of the *Mars Express* spacecraft along its orbit. Flybys above specific targets on the surface of Mars are useful for the determination of the crust density and to ascertain if mantle convection is active at present, in particular in the volcanic Tharsis region for which results point to a high loading density in comparison to the mean density of the Martian crust. The trajectory of *Mars Express* is also disturbed by the mass of the moons Phobos and Deimos, which allows the estimates of each moon's mass to be refined.

The MARSIS radar has provided the first-ever direct subsurface sounding of any planet. This multifrequency synthetic aperture radar is capable of sounding both ionosphere and subsurface. It can detect material discontinuities in the subsurface down to a depth of several kilometers below the surface, allowing the distribution of water, both solid and liquid, in the upper crust to be understood. As a result, we know that the layered deposits near the North Pole are dominated by water ice, with variable amounts of dust. The southern polar cap is more asymmetric,

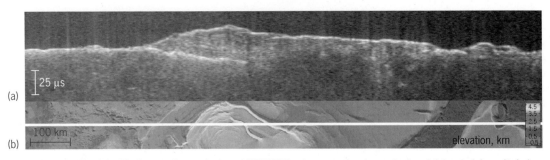

Fig. 1. Observation of the Martian southern polar cap. (*a*) MARSIS radargram, showing water-ice-rich layered deposits below the surface. (*b*) Position of the *Mars Express* ground track (white line) on a topographic map generated by MOLA aboard NASA's *Mars Global Surveyor*. Shades are coded to indicate surface elevation, as shown on the scale. 1 km = 0.62 mi. (*NASA/JPL/ASI/ESA/Univ. Rome/MOLA Science Team*)

with a maximum ice thickness of 3.7 km (2.3 mi) [**Fig. 1**]. Its interior appears to be almost completely water ice, with little dust. The total amount of water in the southern ice cap is equivalent to a global water layer of 11 m (36 ft), constituting thus an extremely large water reservoir. It also appears that the lithosphere under the southern cap is distorted by the heavy ice.

Surface geology and mineralogy. The HRSC imager has provided the first large-footprint high-resolution stereo color images of Mars. It has shed new light on the timing and extent of geological activity (volcanic, fluvial, and glacial) from the very early stages more than 4×10^9 years ago up to the very recent past. Locally geological activity is probably still going on. A key finding is the very recent activity of some of the large volcanoes. The Tharsis province, where Olympus Mons (**Fig. 2**), the largest volcanic edifice in the solar system, is located, is geologically very young, in some places only 5 million years old. In fact, four distinct volcanic episodes have been determined over the past 500 million years. HRSC has allowed surface ages to be precisely determined over large areas: older surfaces have experienced meteoritic bombardment for a longer time, therefore containing a higher number of impact craters (**Fig. 3**). Glacial activity on Mars also appears to be very recent, and is possibly still active in some areas. Ice-related landforms are extremely widespread, over a very large range of latitudes, longitudes, and altitudes. Of particular interest is the discovery of possible glaciers in tropical and equatorial areas, active perhaps only a few million years ago.

OMEGA has provided high-resolution hyperspectral imaging of Mars for the first time, allowing atmospheric and surface constituents, including ices and minerals, to be identified by their spectral fingerprints. It has shown perennial ice deposits of three kinds in the polar regions: water (H_2O) ice mixed with carbon dioxide (CO_2) ice, patches of water ice tens of kilometers wide, and pure water-ice deposits covered by a thin layer of carbon dioxide ice. The discovery of mixed-ice deposits confirms the long-standing hypothesis that CO_2 acts as a cold trap for H_2O ice. The alteration of minerals can be used as a tool to infer the abundance of liquid water on the Martian surface throughout its history. Two main classes of hydrated minerals have been discovered by OMEGA: phyllosilicates (clay minerals) [**Fig. 4**], mainly formed from the aqueous alteration of volcanic rocks in the most ancient outcrops, and hydrated sulfates in younger areas in association with thick layered deposits in Valles Marineris, the largest canyon on Mars (**Fig. 5**). The timing and extent of aqueous alteration on Mars has profound implications for reconstructing its evolution and looking for possible traces of past life. Another important contribution of OMEGA is in complementing the "ground truth" from NASA's rovers investigating the surface composition.

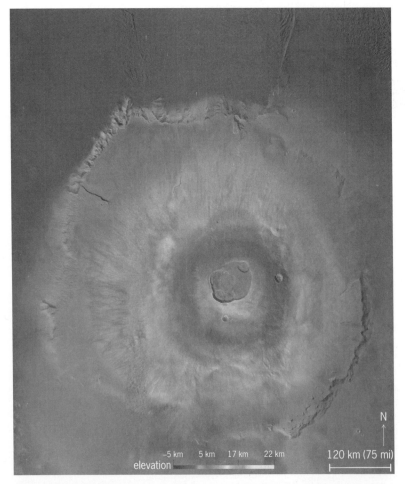

Fig. 2. Olympus Mons as imaged by HRSC, part of a high-resolution Digital Terrain Model (DTM) data set that allows Mars to be visualized in 3D. Shades are coded to indicate surface elevation, as shown on the scale. 1 km = 0.62 mi. (*ESA/DLR/FU Berlin; G. Neukum*)

Atmosphere and climate. For the first time, SPICAM has mapped the atmospheric ozone (O_3), which absorbs the ultraviolet radiation, together with water vapor, making it possible to understand the behavior of both species and how they are linked by chemistry. It also has provided vertical profiles of the density and temperature of CO_2, which are important in planning for future entry probes. SPICAM has also discovered the existence of very high altitude CO_2 ice clouds, giving new information on cloud formation processes and atmospheric dynamics.

Certainly, the most intriguing achievement of PFS is the detection of methane (CH_4) with a global average mixing ratio of 10 ± 5 parts per billion during the first year of observations and a maximum of 30 ppb, indicating that the concentration of this molecule varies with location. These values can be compared to the atmospheric methane concentrations on Earth of 1.75 ppb. This discovery has led to an intense debate in the scientific community on whether the methane is coming from biological or volcanic activity. PFS is also studying global dust storms on Mars, such as the one in July 2007, during which the temperature of the atmosphere increased by 20–40°C (36–72°F) [**Fig. 6**], inflating dramatically the size of the atmosphere around the planet.

Pressure, temperature, and density of the atmosphere are routinely provided by MaRS, from the surface up to 50 km (30 mi) altitude, for a wide range of latitudes. This broad collection of data is useful for comparison with atmospheric models currently being developed.

Ionosphere. SPICAM has detected two new ultraviolet emissions in the upper atmosphere of Mars, one from nitric oxide (NO), which creates a "nightglow" on the night side, and the second from molecular nitrogen (N_2), predicted long ago but never seen. SPICAM has also discovered a new kind of aurora caused by the collisions of electrons, diverted by the crustal magnetic field anomalies at midlatitudes in the southern hemisphere, colliding with atmospheric molecules (**Fig. 7**). Apart from studying the subsurface, MARSIS is also probing the ionosphere. Its active sounding mode gives the electron density as a function of altitude. Very interesting shapes of the ionosphere in the areas of the crustal magnetic field have been discovered. The MaRS experiment routinely derives electron density profiles in the 80–500 km (50–300 mi) altitude range. In addition to a peak in the electron density at around 140 km (87 mi) altitude due to the ionization of the atmosphere by the solar radiation, and the ionopause at about 350 km (220 mi) altitude, a lower layer due to metallic ions from meteors was detected, peaking at about 80–90 km (50–56 mi).

ASPERA has produced new interesting findings on the solar wind interaction all the way down to the lowest point of the spacecraft's orbit (about 250 km or 150 mi altitude) above the day side, which is quite deep in the atmosphere. As a consequence of this interaction, over a period of 3.5×10^9 years some 0.2–4 millibar of carbon dioxide has been

Fig. 3. HRSC view of Maunder crater, an impact structure 90 km (56 mi) in diameter, located in the old southern highlands region of Noachis Terra, showing barchan dunes similar to those in African deserts on its floor. (*ESA/DLR/FU Berlin; G. Neukum*)

Fig. 4. Clay minerals (phyllosilicates) identified by OMEGA in Mawrth Vallis (20°W, 25°N), formed during long-term exposure to water in the oldest regions of Mars, mapped in a color on an HRSC image. (*ESA/DLR/FU Berlin/CNRS/OMEGA Team*)

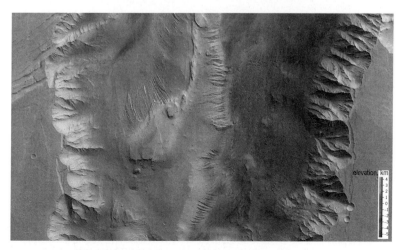

Fig. 5. Candor Chasma (HRSC view), which is situated in the northern part of the Valles Marineris canyon system, and is one of the many radial grabens formed during the Tharsis uplift, which dominated Mars' geological history. Shades are coded to indicate surface elevation, as shown on the scale. 1 km = 0.62 mi. (*ESA/DLR/FU Berlin; G. Neukum*)

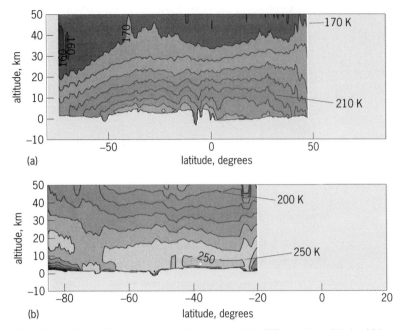

(a)

(b)

Fig. 6. Comparison of temperature variations observed by PFS at various altitudes (*a*) in the absence of a dust storm, and (*b*) subsequently in July 2007 during a global dust storm. 1 km = 0.62 mi. (*ESA/ASI/PFS Team*)

lost due to the solar wind stripping off the high atmosphere; however, this is not an efficient way for the atmosphere to escape. By measuring energetic neutral atoms (ENAs), ASPERA is exploring a new dimension in the solar wind interaction with planets that have no intrinsic magnetic field, as Mars is "shining" in ENAs, partly caused by reflections of inflowing hydrogen ENAs from the solar wind, but mostly due to emissions of plasma, due to the inter-

action of solar and planetary winds with the upper atmosphere.

Scientific output. By mid-2008, the various *Mars Express* experiment teams had published about 300 refereed publications in scientific journals worldwide. The scientific data from the nominal mission is now available in the mission archive for further study by the general public and scientists alike. Together with the principal investigators and their large teams of co-investigators, the tremendous success of this mission is due to a large extent to the various European Space Agency (ESA) teams throughout most of its establishments.

Extension and enlargement of mission. The nominal mission lifetime of one Martian year (January 2004 to November 2005) for the orbiter has already been extended twice, up to May 2009, and further extensions are being requested to fulfill the remaining goals of the mission, such as to complete the global coverage of high-resolution imaging and spectroscopy, as well as subsurface sounding with the radar, to observe atmospheric and variable phenomena, to continue gravity measurements and seasonal coverage, and to revisit areas of discoveries. The scope of cooperation has been enlarged, in particular with NASA's Mars rovers and *Mars Reconnaissance Orbiter*, and with ESA's *Venus Express*, as it is carrying the same instruments to another planet, providing a unique opportunity for comparing our nearest planetary neighbors.

Aid to other current and future missions. Finally, *Mars Express* is providing valuable data for preparing planned missions of ESA's Aurora Exploration Program: first *ExoMars*, including a rover for biological, geophysical, and climatological investigations, and second *Mars-NEXT*, including a network of 3 or 4 surface stations complemented by an orbiter for determining the internal structure, the global atmospheric circulation, and the geology and geochemistry of the landing sites. In particular, *Mars Express* is establishing a surface/subsurface geosciences database and refining the existing atmospheric one to identify potential landing sites of high scientific value and assess risks during atmospheric entry, descent, and landing. Indeed, in May 2008, *Mars Express* contributed to the successful landing of NASA's *Phoenix* mission by providing prelanding characterization of the atmosphere, a procedure which will undoubtedly be useful for future Mars exploration.

For background information *see* MARS; SPACE PROBE; SYNTHETIC APERTURE RADAR in the McGraw-Hill Encyclopedia of Science and Technology.

Agustin F. Chicarro

Bibliography. N. G. Barlow, *Mars: An Introduction to its Interior, Surface and Atmosphere*, Cambridge University Press, 2008; M. H. Carr, *The Surface of Mars*, Cambridge University Press, 2007; A. Chicarro (ed.), *Mars Express: The Scientific Investigations*, ESA SP-1291, in press, 2008; A. Chicarro (ed.), *Mars Express: The Scientific Payload*, ESA SP-1240, 2004; B. Jakosky et al. (eds.), *Mars*, University of Arizona Press, 1993.

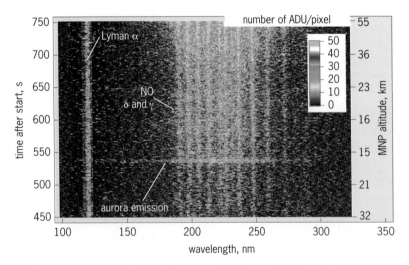

Fig. 7. Auroral emissions observed by SPICAM over midlatitude crustal paleomagnetic anomalies located in the old cratered highlands of the southern hemisphere of Mars. On Earth auroras are always over the poles. 1 km = 0.62 mi. [The scale on the right indicates that the MNP (Mars nearest point) altitude of the region observed at first decreased and then increased to a larger value, reflecting the geometry of the observation. Because the emission was very localized, it was visible in the field of view for only a limited time period (7 s), which accounts for the narrowness of the aurora emission line. ADU (analog digital unit)/pixel is a measure of the light intensity measured by the CCD (charge-coupled-device) camera, based on the voltage read by the analog-to-digital converter circuit of the CCD.] (*ESA/CNRS/SPICAM Team*)

Medicolegal death scene investigation

Medical examiner/coroner's (ME/C) offices investigate about 20% of the approximately 2.5 million deaths that occur in the United States each year. A medicolegal jurisdiction refers to either a county's medical examiner's or coroner's office. There are an estimated 500,000 deaths reported to medicolegal jurisdictions annually in the United States' 3137 counties. All 50 states, four territories, and the District of Colombia differ from one another with regard to their medicolegal system (medical examiner or coroner), statutory authority, resources, and interagency cooperation. Deaths reportable to a medical examiner's or coroner's office usually fall into the categories of violent, suspicious, sudden, and unexpected.

Medical examiner/coroner. In counties that have a coroner system, the coroner is a person who is elected or reelected to a 4-year term. To be elected coroner, the candidate must live in the county and state for a certain number of years but is not required to have any special education, training, or experience. In most of the nation's medical examiner systems, the medical examiner is a physician who is licensed to practice medicine in the state and is appointed by county officials. Usually the medical examiner is a forensic pathologist, a physician who specializes in medicolegal medicine, spending five years in training after medical school performing autopsies, evaluating laboratory test results, and becoming knowledgeable about county, state, federal, and international law.

The medical examiner/coroner is responsible for determining the decedent's cause of death, such as gunshot wound of the head, self-inflicted ligature strangulation, blunt-force trauma in a motor vehicle collision, or cancer of the lung. The medical examiner/coroner certifies many of the deaths that are reported to the office and is also required to determine and certify the decedent's manner of death, including homicide, suicide, accident, natural causes, or undetermined causes.

Personnel from a medical examiner's or coroner's office must work cooperatively with the area's other agencies, such as law enforcement, fire service, and family and children's services, to ensure that a coordinated, thorough, scientific, and unbiased death investigation is conducted. State law defines each agency's authority and responsibilities. Commonly the decedent's body and anything attached to the body are the responsibility of the medical examiner/coroner. The death scene is routinely the responsibility of law enforcement. An established positive multidisciplinary working relationship provides the greatest opportunity to ensure that the death investigation will be professionally handled and will correctly determine the decedent's cause and manner of death.

Medicolegal death investigators. Medical examiner/coroner's offices have investigative staffs that employ medicolegal death investigators (MDI). These are lay professionals who have acquired specialized medical, legal, and forensic science knowledge and expertise. They are experienced in investigating death scenes and identifying evidence that provides critical information for the determination of the decedent's cause and manner of death. Nationwide, more than 1300 of these investigators have been certified by the American Board of Medicolegal Death Investigators (ABMDI).

The medicolegal death investigator is the legal representative of the medical examiner/coroner at the scene and is responsible for receiving the initial death report information from law enforcement, prehospital and medical facility personnel, or other knowledgeable persons; determining if the death is reportable by law and if the office has jurisdiction over this death; and determining if a death-scene investigation or phone investigation will be performed.

Investigation. When it is decided that a death-scene investigation will be performed, the medicolegal death investigator is responsible for ordering transportation support from the decedent's location to the medicolegal office; actualizing the decedent at the death scene, noting injuries, marks, and fragile and trace evidence and developing postmortem interval information; obtaining scene photographs, evidence, and law enforcement information that will be needed by the pathologist to assist in the determination of the cause and manner of death; and notifying the forensic pathologist, morgue staff, ancillary forensic scientists, and office personnel of the death and what procedures need to be performed.

Actualizing the decedent. To complete the death-scene investigation, the medicolegal death investigator determines the decedent's demographic, medical and psychological, occupational, social, and situational history; acquires medical documents, and interviews persons to substantiate the decedent's past history to provide a complete, accurate, and unbiased report of all the information developed regarding the death. In addition, information is provided to other agencies, such as fire service and law enforcement, child protection groups, social service agencies, health departments, local organ and tissue procurement groups, funeral directors, and state and federal agencies that need to be informed of the death, such as the Federal Emergency Management Agency (FEMA), Occupational Safety and Health Administration (OSHA), and transportation departments such as the Federal Aviation Administration (FAA) and National Transportation Safety Board (NTSB).

Following standard operating procedures (SOP) that have been established by the chief medical examiner or coroner, the medicolegal death investigator gathers information from the reporting person (usually a law enforcement officer) to determine if the death was violent, was suspicious, or exhibited other factors that indicate that a scene investigation should be conducted, such as homicide, suicide, infant or child death, motor vehicle fatality, and drowning. Also, it is determined if the decedent was pronounced dead at the scene. If prehospital personnel assessed the patient, found signs of life, and transported him to a medical facility where the person died, then the office's SOP determine whether the investigator will respond to the medical facility or remain in the office until the death is reported by personnel from the medical facility.

In infant deaths, a death-scene investigation re-creation must be done, even if the infant has been removed from the scene and transported to a medical facility. The ME/C investigator responds to the scene to meet the people who were present when the infant was placed in the original position and when the infant was found unresponsive. It is then necessary for the person who placed the infant and the person who discovered that the infant was unresponsive to recreate the scenes, using a child's doll to portray the infant's positions. The medicolegal death investigator then photographically records the placer's positioning and the finder's positioning.

Next, the location of the scene is determined, including street address, type of building, or area, as well as any investigative information that is now available from the reporting party that indicates that a death-scene investigation is warranted.

Once it is determined that the death falls within the criteria set by the ME/C, the medicolegal death investigator contacts the office conveyance company or other transport service provider and relays the scene location, including any known information about the body's condition (such as size, freshness of body, and hazardous circumstances at the scene), notifies the office staff of the known investigation details and the scene location, and proceeds to the death scene.

When the medicolegal death investigator arrives at the death scene, his or her responsibility is to take charge of the decedent and determine who the decedent is (positive identification), what happened that may have contributed to the decedent's cause and manner of death (circumstances of the death), when the death occurred (estimation of time of death), where the death occurred (was the body moved from the location where it was found), and why the decedent died at this time and under these circumstances (cause and manner of death).

This information will then be provided to the medical examiner or coroner who will be certifying the death and assists in the determination of the decedent's cause and manner of death.

Upon arrival at the death scene, the medicolegal death investigator locates the scene commander, explains that he or she has arrived, and pronounces the subject dead, if that has not already been done. Along with other agencies' personnel, the medicolegal death investigator determines if the scene is safe to investigate. If it is not, the fire service, hazmat team, or other safety personnel are requested. When they determine that the scene is safe, the investigators enter and begin the investigation.

The medicolegal death investigator is then briefed by first responders (law enforcement, fire service, and prehospital emergency personnel) as to what information is known at this time. The medicolegal death investigator "walks through the scene" with law enforcement personnel (usually the crime-scene

detectives), who have already scrutinized the scene and determined where the scene evidence (including the decedent) is located. This is done so that no one within the scene contaminates, damages, or loses scene evidence. During this process, it is mutually determined which investigative office is responsible for each particular piece of evidence. A general rule is that the decedent's body and anything physically attached to it is the property of the ME/C. All other evidence within the scene is the property of law enforcement.

It is critical that the scene environment be preserved. An officer is usually placed at the scene entrance so that the entry of unauthorized people can be prevented. It is important that all family, friends, bystanders, and animals be removed from the scene immediately when a dead person has been discovered to preserve the scene's integrity. At appropriate times, the scene should be cordoned off with crime-scene tape to safeguard evidence that is not in the immediate area of the decedent.

Scene photographs. Photographs should be taken by crime-scene investigators immediately after their initial inspection of the scene. Photographs should also be taken by the ME/C investigator so that scene pictures will be available to the forensic pathologist prior to the postmortem examination. Scene photographs should include scene approaches; areas surrounding the body, full-length photos of the body, including the surrounding area; and close-up photographs showing the subject's position, marks or patterns on the body, unique features of the body, the face (for identification purposes), the position of clothing on the subject, and all defects on the body and clothing. After the body is removed from its resting place, a photograph should be taken of the surface on which the subject was resting to discover evidence that may have passed through the body and into the surface. Digital cameras are now used so that the scene photographs can quickly be transferred from the scene, via laptop computer, to the morgue for the forensic pathologist's viewing, prior to the postmortem examination.

Evidence. After photographs of the body have been taken, the ME/C investigator begins to closely inspect the dead person for any marks or injuries that can be clearly seen.

Clothing is not removed or roughly manipulated. Any fragile or trace evidence that will quickly change or deteriorate or that could be easily lost is immediately photographed and secured. Then a methodical body inspection is performed, beginning at the top of the subject's body and moving downward. Not only large obvious trauma but also any marks that are seen whose causative agent is not yet known should be noted and photographed. Postmortem changes are recorded, and the decedent's body temperature is noted to assist the pathologist with issues regarding estimation of the time of death.

Postmortem examination. After the medicolegal investigator's inspection is completed, the decedent's name and case number are printed on the identification tag, and it is attached to the body. The body is then carefully enveloped in an ME/C office's clean white sheet. The wrapped body is placed in a new plastic body bag. On the outside of the body bag is written the decedent's name and ME/C office case number. The subject is then carefully lifted onto a stretcher and placed in the ME/C vehicle that is waiting at the scene.

The medicolegal death investigator briefs the detectives as to the initial observations that were determined during the actualization. The detectives, in turn, update the medicolegal death investigator concerning additional information that has been gathered from witnesses, neighborhood canvas's, and their scene investigation. The medicolegal death investigator contacts the forensic pathologist who will be performing the postmortem examination and provides all the scene information that is now known. The forensic pathologist informs the investigator of the time that the examination will be performed at the morgue, and this information is relayed to the detectives. They can then plan to attend the examination, if necessary. Whether the detectives attend the morgue examination is dependent on their office's SOP and the resources available. The morgue staff is then notified that the decedent is being brought in and the time that the pathologist plans to perform the examination, and x-rays are ordered if needed.

Identification and notification. If the decedent has not been identified, the ME/C investigator will search the subject's personal papers and any other information found on the body for clues to the decedent's identity. Detectives often are able to find next-of-kin information from witness interviews. If the identification is not definite, it will be necessary to establish the identification by scientific means, such as fingerprints, dental comparison, or other forensic science processes, such as deoxyribonucleic acid (DNA) typing.

Once the subject's identification is established, it is necessary to determine who is the subject's next of kin, following the legal chain of hierarchy: spouse, children 18 or older, parents, siblings, and so on. The next of kin will then be located, contacted, and notified of the death. Information developed from the next of kin can provide the name of the subject's dentist, who can be contacted for premortem dental records and x-rays. These can then be compared with postmortem dental x-rays and a scientific identification established.

For background information *see* CRIMINALISTICS; DEATH; FINGERPRINT; FORENSIC BIOLOGY; FORENSIC MEDICINE in the McGraw-Hill Encyclopedia of Science & Technology.   Mary Fran Ernst

Bibliography. S. Clark et al., *Medicolegal Death Investigator: A Systematic Training Program for the Professional Death Investigator*, Occupational Research and Assessment, Inc., Big Rapids, MI, 1996; M. F. Ernst, Medicolegal death investigation and forensic procedures, in R. Froede (ed.), *Handbook of Forensic Pathology*, 2d ed., College of American Pathologists, 2003; M. F. Ernst, Death scene investigation, United States of America, in J. Payne-James et al. (eds.), *Encyclopedia of Foren-sic and Legal Medicine,* Academic Press, 2005; J. Jentzen and M. F. Ernst, Developing medicolegal death investigator systems in forensic pathology, *Clin. Lab. Med.*, 18:279–322, 1998; U.S. Department of Justice, *Research Report: National Guidelines for Death Investigation*, National Institute of Justice, Washington, DC, December 1997.

MEMS sensors

Micro-electro-mechanical systems (MEMS) is a term originally coined in the United States in the 1980s. In Europe the term microsystems technology (MST) is more usual, whereas in Japan the term micromachines is more widespread. These terms all refer to the same technology that is a by-product of the enormously successful semiconductor-microelectronics chip manufacturing industry. In this article the term MEMS will be deployed more generically, as it may be viewed as being a more accurate description of the science and technology underlying these miniature devices whose dimensions are measured in micrometers.

Over the past two decades sensors based on MEMS devices have continued to proliferate in many diverse areas of application, including aerospace, biomedical, defense, space, transport, and telecommunications. In the majority of these applications, the MEMS sensor transduces or converts the mechanical movement of a part of the sensor (controlled by the parameter under observation, such as temperature or pressure) to an electrical signal. The electrical signal can then be amplified, digitized, stored, and manipulated to obtain useful information about the changes in the parameter or measurand under observation. As there are many diverse types of MEMS sensors, covering an equally diverse range of applications, this article will detail only a few key examples to demonstrate the importance of practical MEMS sensors globally. The success of MEMS sensors is also in part attributable to the processing power that microelectronics can offer. Most MEMS sensors are nowadays closely integrated with signal-conditioning circuitry to increase the sensitivity of the sensor, improve its dynamic range, reduce the impact of parasitic effects, and so forth. In addition, due to the small physical size of the MEMS sensors and their associated electronics for excitation and detection, arrays of sensors can be deployed to "map" or monitor the environment of interest to increase the useful information per sensor in comparison with a stand-alone sensor.

Although the exact definition of a sensor remains a matter of debate, the *Oxford English Dictionary* defines a sensor as "a device which detects or measures a physical property and records, indicates, or otherwise responds to it." The debate arises because of the enormously diverse types of "sensing" that occur in academia and industry, ranging from everyday environmental properties such as temperature and pressure to physical properties such as quantized charge and single-photon detection. MEMS sensors are almost universally deployed to measure some physical property in a given operating environment, such as acceleration in automotive vehicles or angular velocity (using gyroscopes) in aircraft. This article will discuss only a few of the main types of MEMS sensors, and highlight some more exotic types whose application, although important, is limited to specific fields.

Accelerometers. These are widely deployed in a range of industries from aerospace and defense to automotive and biomedical. Generally MEMS accelerometers can be categorized into three designs: piezoresistive, capacitive, and resonant. Depending on the application, a particular design may be more advantageous than the other two. Choice of design is often driven by the overall cost and complexity of the sensor as well its functional characteristics, such as sensitivity and dynamic range. Packaging of the MEMS sensor is also a key factor since it accounts for about 75% of its cost. Manufacturers of modern accelerometers offer a choice of analog or digital signal outputs, thereby increasing the number of applications of these sensors. The analog output signal from a sensor can be amplified and made to interact with another circuit as necessary, but a digital sensor output can be easily integrated with modern microprocessor-based systems, without the need for analog-to-digital conversion.

Commercially, Analog Devices was one of the first to launch its ADXL accelerometer in 1991, with $\pm 2\,g$ dynamic range (where g is the acceleration of gravity, equal to 9.8 m/s^2 or 32.2 ft/s^2), a sensitivity of about 0.1 g, and cost of only a few dollars. These single-axis sensors had a significant impact on the automotive industry as they were applied to airbag systems to enhance passenger safety. The ADXL accelerometer is based on a surface-micromachined in-plane proof mass with interdigital capacitive sensors (**Fig. 1**). The motion of the movable frame (or proof mass) caused by acceleration of the whole device is detected as the differential change in capacitance (sensing cell) between the fixed plates and the electrode attached to the movable frame. As the two fixed plates and the electrode attached to the frame form two capacitors in series with a single common point (that of the electrode attached to the frame), it is standard practice to employ these as part of a bridge circuit similar to Wheatstone or Maxwell bridges, thus making measurements immune to supply-voltage variation and giving increased sensitivity. In addition, the output of the bridge circuit applies a compensating signal via the forcing cell, driven by the second electrode attached to the frame, thus forming a feedback loop that maintains the position of the first electrode fixed. The feedback signal is linearly proportional to the applied acceleration. The latest versions of these MEMS sensors can be biaxial or triaxial and can have up to 100 g dynamic range, as well as including on-chip temperature sensors as part of the feedback control to improve the accelerometer sensitivity. A number of other manufacturers offer similar surface-micromachined accelerometers.

Pressure sensors. These are one of the most widely used MEMS sensors, with a host of application areas, ranging from tire- and blood-pressure monitoring to altitude sensing and flow measurements (when combined with the Venturi effect). In general, pressure sensors fall into three categories: absolute pressure sensors, which measure pressure relative to some reference (such as vacuum); gauge pressure sensors, which operate relative to the ambient pressure; and differential pressure sensors. Traditional pressure-sensing techniques, such as manometers, aneroid barometers, and Bourdon tubes, are under increasing competition from MEMS-based sensors, which respond more quickly and have increased dynamic response. MEMS pressure sensors are mostly diaphragm-based, and there are several well-known techniques for detecting the deflection of the diaphragm, such as piezoresistive and capacitive. The material of choice for the diaphragm in MEMS pressure sensors is often silicon due to its elasticity—it does not become plastically deformed and returns to its original dimensions and tension after deflection.

Capacitive MEMS pressure sensors consist of an evacuated cell formed by a fixed electrode and a second diaphragm electrode. They have high sensitivity, low power consumption, and low-temperature sensitivity. However, capacitive pressure sensors can suffer from cross-sensitivity to acceleration, which can be minimized at the expense of added complexity by adding another diaphragm. Capacitive sensors also suffer from nonlinear output, which is often "linearized" by using the sensor over a small dynamic range or by measuring the capacitance at a point offset from the center of the diaphragm (with a slight reduction in sensitivity). A further possibility is to constrain the diaphragm at its center point, which again improves the linearity but reduces the sensitivity. There are also a variety of other types of pressure sensors, such as resonant, microthermopile (employed as thermal conductivity gauges), surface acoustic wave, and optical (where the pressure-sensing diaphragm perturbs an optical path).

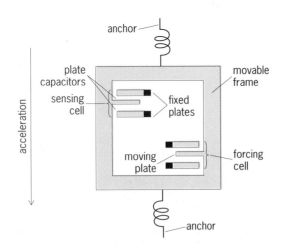

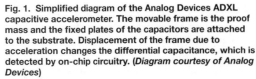

Fig. 1. Simplified diagram of the Analog Devices ADXL capacitive accelerometer. The movable frame is the proof mass and the fixed plates of the capacitors are attached to the substrate. Displacement of the frame due to acceleration changes the differential capacitance, which is detected by on-chip circuitry. (*Diagram courtesy of Analog Devices*)

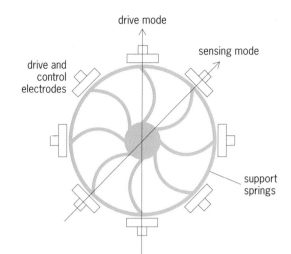

Fig. 2. An outline of a MEMS vibrating-ring gyroscope. *(After S. P. Reilly et al., Overview of MEMS sensors and the metrology requirements for their manufacture, NPL Report, DEPC-EM 008, 2006)*

Gyroscopes. Macroscale gyroscopes normally employ a flywheel of large mass rotating at high speed. The nature of the microdevices and frictional forces prevent this method from being deployed for MEMS-based gyroscopes. Therefore, most micromachined gyroscopes use a mechanical structure that is driven into resonance. Rotation excites a second resonance due to the Coriolis force. A feedback loop is often used to null the oscillation in the second resonant axis. The main difficulty with achieving practical MEMS-based gyroscopes has been the relatively small size of the Coriolis force compared to the driving force that is applied in an orthogonal direction. This limitation has been overcome by using high-Q (quality factor) structures vibrating at the resonance of the sensing axis. Often, active tuning via electrostatic forces is also necessary to fine-tune the resonance due to variation caused by the dimensional tolerances imposed by the manufacturing process. In addition, slight misalignment of the drive mechanism with respect to the axis of freedom can lead to crosstalk.

The earliest surface-micromachined gyroscopes involved two electrostatic comb drives operating perpendicular to each other, one for actuating and one for capacitive sensing. A more universal implementation of a MEMS gyroscope now widely deployed is the vibrating-ring structure (**Fig. 2**). The ring is set to oscillate elliptically while rotational forces couple this vibration to a secondary resonance at $45°$ to the primary mode. Capacitive electrodes placed at $45°$ measure the amplitude of the second mode, which is proportional to the angular velocity of the ring structure. MEMS gyroscopes are increasingly finding consumer applications in addition to their traditional aerospace and military applications.

Microwave power sensors. Radio-frequency (RF) MEMS is a large and rapidly growing field where micro- and nanomachined devices are finding extensive applications. These applications include switches, voltage dividers, ac/dc voltage references, antennae, phase shifters, and couplers, as well as microwave power sensors. A microwave power sensor is one example of an electromagnetic field transducer whose fundamental principle could be deployed in a range of other sensing applications such as sensing force and acoustic vibration, and thermal and mechanical, as well as biological and chemical, activities.

For microwave power detection, a thin metallic membrane is suspended above a metallic coplanar waveguide (CPW), which carries the microwave signal (**Fig. 3**). The central conductor of the waveguide and the membrane, which is held at shield potential (0 V), form a capacitor. Since the force between the plates of any capacitor is proportional to the square of the potential difference across the plates, an attractive electrostatic force is generated by the electric field component of the microwave between the central conductor of the coplanar waveguide and the membrane. Since the power flow in a coplanar waveguide is given by the voltage squared divided by the impedance of the waveguide, this force is also proportional to the power flow through the waveguide. The microwave-induced attractive force displaces the membrane until it is balanced by the elastic spring force of the membrane. This displacement is detected by separate capacitive electrodes placed on either side of the waveguide using a precision 1-kHz capacitance meter, since the change in capacitance is only of the order of femtofarads.

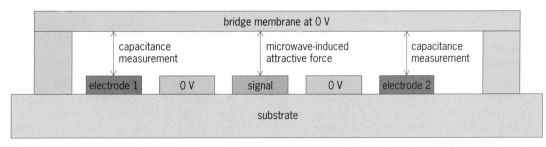

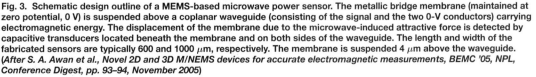

Fig. 3. Schematic design outline of a MEMS-based microwave power sensor. The metallic bridge membrane (maintained at zero potential, 0 V) is suspended above a coplanar waveguide (consisting of the signal and the two 0-V conductors) carrying electromagnetic energy. The displacement of the membrane due to the microwave-induced attractive force is detected by capacitive transducers located beneath the membrane and on both sides of the waveguide. The length and width of the fabricated sensors are typically 600 and 1000 μm, respectively. The membrane is suspended 4 μm above the waveguide. *(After S. A. Awan et al., Novel 2D and 3D M/NEMS devices for accurate electromagnetic measurements, BEMC '05, NPL, Conference Digest, pp. 93–94, November 2005)*

The spring constant and the sensitivity of the microwave power sensor are designed for a specific dynamic range of power, which, if exceeded, can lead to membrane stiction against the waveguide. The membrane is designed to have negligible influence on the microwave power flow in the coplanar waveguide through accurate impedance matching to the characteristic impedance of the waveguide. The main advantages of this type of power sensor over others (such as thermistor- and diode-based sensors) is that it does not dissipate the microwave power, can be designed for a variety of dynamic ranges, and offers greater linearity. The limitations of this type of sensor are mostly related to its design and operational complexity.

The sensor shown in Fig. 3 was designed at the National Physical Laboratory, United Kingdom, and fabricated by Microfabica using its EFAB® process. This process allows novel three-dimensional (3D) structures (with heights up to a few hundred micrometers) to be realized, in contrast to conventional MEMS devices, which are often planar and have a height limitation of only a few micrometers due to their fabrication process. Research in the field of 3D MEMS is only just beginning and it is likely to be another decade before other practical applications of 3D MEMS devices emerge.

For background information *see* ACCELEROMETER; BRIDGE CIRCUIT; CAPACITANCE MEASUREMENT; CORIOLIS ACCELERATION; GYROSCOPE; MICROELECTRO-MECHANICAL SYSTEMS (MEMS); MICROSENSOR; MICROWAVE POWER MEASUREMENT; PRESSURE MEASUREMENT; Q (ELECTRICITY) in the McGraw-Hill Encyclopedia of Science & Technology.

Shakil A. Awan

Bibliography. S. A. Awan et al., Novel 2D and 3D M/NEMS devices for accurate electromagnetic measurements, *BEMC '05, NPL, Conference Digest*, pp. 93–94, November 2005; S. Beeby et al., *MEMS Mechanical Sensors*, Artech House Publishers, 2004; S. Beeby, M. Stuttle, and N. M. White, Design and fabrication of a low cost microengineered silicon pressure sensor with linearized output, *IEE Proc. Sci. Meas. Technol.*, 147:127–130, 2000; P. Grieff, Silicon monolithic micro-mechanical gyroscope, *Proc. Transducers '91*, pp. 966–968, 1998; S. P. Reilly et al., *Overview of MEMS sensors and the metrology requirements for their manufacture*, NPL rep. DEPC-EM 008 2006.

MESSENGER mission

On January 14, 2008, the MErcury Surface, Space ENvironment, GEochemistry, and Ranging (*MESSENGER*) spacecraft flew by the planet Mercury. No other space probe had visited Mercury since *Mariner 10* flew by the innermost planet three times in 1974–1975. All the instruments in *MESSENGER*'s payload operated successfully during the 55-hour period centered on closest approach at 201.4 km (125.1 mi) above the planet's surface. About 21% of Mercury's surface (**Fig. 1**) was imaged at close range for the first time, and observations were made of Mercury's exosphere, magnetosphere, and surface. *MESSENGER* is scheduled to fly past Mercury two more times (October 2008 and September 2009) before becoming the first spacecraft to orbit Mercury in March 2011.

Objectives. The mission is focused on answering six key questions about Mercury:

1. What planetary formational processes led to the high ratio of metal to silicate in Mercury?

2. What is the geological history of Mercury?

3. What are the nature and origin of Mercury's magnetic field?

4. What are the structure and state of Mercury's core?

5. What are the radar-reflective materials at Mercury's poles?

6. What are the important volatile species and their sources and sinks on and near Mercury?

These questions are addressable by observations that can be made from an orbiting spacecraft, and their answers bear not only on the planet Mercury but more generally on the comparative formation and evolution of all of the terrestrial planets.

Instruments. The *MESSENGER* payload consists of seven instruments plus radio science. The instruments (**Fig. 2**) include the Mercury Dual Imaging System (MDIS), which has a wide-angle camera with 11 filter channels and a higher-resolution, monochrome narrow-angle camera; the Gamma-Ray and Neutron Spectrometer (GRNS), which incorporates two sensors, a Gamma-Ray Spectrometer (GRS) and a Neutron Spectrometer (NS); the X-Ray Spectrometer (XRS); the Magnetometer (MAG); the Mercury Laser Altimeter (MLA); the Mercury Atmospheric and Surface Composition Spectrometer (MASCS), which consists of a moving-grating Ultraviolet-Visible Spectrometer (UVVS) and a Visible-Infrared Spectrograph (VIRS); and the Energetic Particle and Plasma Spectrometer (EPPS), which consists of an Energetic Particle Spectrometer (EPS) and a Fast Imaging Plasma Spectrometer (FIPS). The instruments communicate to the spacecraft through fully redundant Data Processing Units (DPUs).

Spacecraft. The *MESSENGER* spacecraft body is 1.42 m (4.7 ft) tall, 1.85 m (6.1 ft) wide, and 1.27 m (4.2 ft) deep, and constructed of a lightweight graphite-epoxy composite. The front-mounted ceramic-fabric sunshade, 2.5 m (8.2 ft) tall and 2 m (6.6 ft) across, is mounted on a titanium-tube structure (**Fig. 3**).

Power system. Power is provided by two rotatable solar arrays that extend about 6 m (20 ft) from end to end across the spacecraft. The two 2.6-m^2 (28-ft^2) panels are one-third gallium arsenide solar cells and two-thirds solar reflectors (mirrors), included to moderate thermal input. The arrays are used to charge a 22-cell, 23-A-h nickel-hydrogen battery that provides spacecraft electrical power during eclipses. The power-system electronics regulate the power output of the system; they provided about 390 W near Earth's distance from the Sun and will provide 640 W in Mercury orbit.

Propulsion system. A key characteristic of *MESSENGER* is its initially large mass fraction of propellant and low-mass (81.7-kg or 180-lb) propulsion system that enables a velocity-change capability of 2250 m/s (7380 ft/s). Features include three custom-built titanium tanks that feed a dual-mode propulsion system having one bipropellant (hydrazine and nitrogen tetroxide) thruster for large maneuvers and 16 monopropellant thrusters for small trajectory adjustments and attitude control. This system, with 599 kg (1320 lb) of propellant at launch (54% by mass), is used to perform five large deep-space maneuvers (three accomplished by mid-2008) that, along with the six planetary gravity assists (four performed by mid-2008) and a final large Mercury-orbit-insertion burn, permit the spacecraft to be placed into Mercury orbit. Smaller trajectory-correction maneuvers employ hydrazine burns, and the three-axis-stable flight configuration is maintained through a combination of the monopropellant thrusters, reaction wheels, and solar radiation pressure on the steered solar arrays. The main tanks have no internal propellant management devices other than a vortex suppressor at the outlet, and under zero acceleration between bipropellant maneuvers the fuel and oxidizer are not localized in the tanks. Thrusters therefore rely on an initial low-acceleration "settling"

Fig. 1. *MESSENGER* wide-angle camera image of the hemisphere of Mercury seen as the spacecraft departed the planet. The image, at a resolution of approximately 2.5 km/pixel (1.6 mi/pixel), includes all of the Caloris basin (upper right) and 21% of the planet's surface never before seen at close range. (*NASA/Johns Hopkins University Applied Physics Laboratory/Carnegie Institution of Washington*)

burn, executed with hydrazine in a bladder-containing auxiliary tank, to position fuel and oxidizer at the exit ports of the main tanks before each bipropellant maneuver. This management approach places burdens on burn design and flight

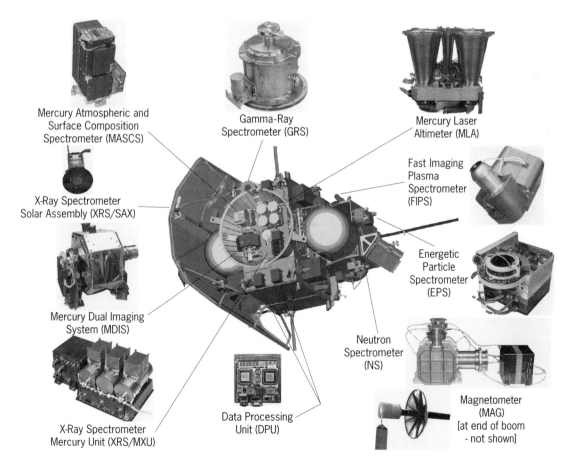

Mercury Atmospheric and Surface Composition Spectrometer (MASCS)

X-Ray Spectrometer Solar Assembly (XRS/SAX)

Mercury Dual Imaging System (MDIS)

X-Ray Spectrometer Mercury Unit (XRS/MXU)

Gamma-Ray Spectrometer (GRS)

Data Processing Unit (DPU)

Neutron Spectrometer (NS)

Mercury Laser Altimeter (MLA)

Fast Imaging Plasma Spectrometer (FIPS)

Energetic Particle Spectrometer (EPS)

Magnetometer (MAG) [at end of boom - not shown]

Fig. 2. Montage of the *MESSENGER* instruments showing images of flight units and their locations on the spacecraft. (*NASA/Johns Hopkins University Applied Physics Laboratory/Carnegie Institution of Washington*)

Fig. 3. Completed *MESSENGER* spacecraft being prepared for vibration testing at the Johns Hopkins University Applied Physics Laboratory prior to shipment to Cape Canaveral and subsequent launch. (*NASA/Johns Hopkins University Applied Physics Laboratory/Carnegie Institution of Washington*)

software but helped to minimize mass and hardware cost.

Avionics. A redundant processor performs all main spacecraft functions, while other processors use rule-based engines to provide for spacecraft health and safety. Command and data handling, guidance and control, and fault-protection functions are implemented in (redundant) Integrated Electronics Modules (IEMs). As with all *MESSENGER* systems, significant development and manufacturing design effort went into minimizing the mass of these systems while retaining a high level of robustness for the mission. The system is built around RAD6000 processors and includes redundant solid-state recorders with 8 gigabits of memory, implemented with a VxWorks-based, file-system protocol. The RAD6000 processor is based on the IBM reduced instruction set computer (RISC) single-chip central processing unit (CPU) and is radiation-hardened to function reliably in Mercury's environment over the long duration of the mission. VxWorks, one of the real-time operating systems that run on these units, is "UNIX-like" and can be used in a multitasking environment. The telecommunications system runs at X-band and employs redun-

Fig. 4. Launch of *MESSENGER* from Cape Canaveral Air Force Station Space Launch Complex 17B on a Delta II 7925H-9.5 at 06:15:56.537 UTC on August 3, 2004. (*NASA/Johns Hopkins University Applied Physics Laboratory/Carnegie Institution of Washington*)

dant transponders, solid-state power amplifiers, and a suite of antennas that includes two phased-array antennas, the first electronically steered antennas designed for use in deep space.

Mission timeline. The seventh competed Discovery Program mission flown by the National Aeronautics and Space Administration (NASA), *MESSENGER* was launched on August 3, 2004 (**Fig. 4**). The spacecraft subsequently executed gravity-assist maneuvers at Earth (August 2, 2005) and twice at Venus (October 24, 2006, and June 5, 2007).

Prior to insertion into orbit about Mercury in 2011, *MESSENGER* will have traveled 7.9×10^9 km (4.9×10^9 mi) in 15 revolutions around the Sun. The Mercury orbital phase of the mission as currently planned will last one Earth year (equivalent to just over four Mercury years or two Mercury solar days).

Mercury flyby observations. During *MESSENGER*'s first flyby of Mercury, MDIS obtained 1213 images of the surface as well as an additional 104 images for optical navigation and the production of approach and departure movies. The Caloris basin—the largest, best-preserved impact basin on Mercury—was imaged in its entirety for the first time. A number of large monochrome mosaics were taken at a range of resolutions, some suitable for stereogrammetry, and a series of color frames was acquired for photometric analysis. MASCS obtained the first high-resolution spectral reflectance measurements (at ultraviolet to near-infrared wavelengths) of surface composition, conducted night-side and dayside limb scans of species in Mercury's tenuous atmosphere, and mapped the structure of Mercury's cometlike tail of neutral sodium atoms that extends from the planet in the direction opposite that of the Sun. MAG measured Mercury's intrinsic magnetic field near the equator and documented the major plasma boundaries of Mercury's magnetosphere. (Mercury's intrinsic magnetic field is of internal origin, the product of either a magnetic dynamo within the fluid outer core or a permanently magnetized upper crustal layer, in contrast to fields of external origin, such as those generated by currents in the magnetosphere.) EPPS made the first measurements of low-energy ions in Mercury's magnetosphere and its heliospheric environment. MLA carried out the first laser altimetric profile of the planet, and GRNS and XRS provided a first look at surface elemental composition. The radio science experiment reduced uncertainties in the long-wavelength components (that is, lowest spherical harmonics) of Mercury's gravity field.

Surface observations. Although Mercury's surface is densely populated with impact-craters, *MESSENGER* images have revealed substantial new information on the geological history of the planet. From temporal constraints on the emplacement of smooth plains indicated by superposition relationships and measurements of impact-crater density, MDIS images provided evidence for widespread volcanism on Mercury, and candidate sites for volcanic centers were identified. MDIS also revealed newly imaged lobate scarps and other tectonic landforms supportive of the hypothesis that Mercury contracted globally in response to interior cooling and growth of a solid inner core (**Fig. 5**). From the density of fault structures now recognized, the magnitude of global contraction is at least one-third greater than appreciated from *Mariner 10* observations. Reflectance spectra of Mercury show no evidence for ferrous iron (FeO) in surface silicates, and the Neutron Spectrometer sensor on GRNS indicates an upper bound of 6% on the surface elemental abundance of iron. The reflectance and color imaging observations support earlier inferences that Mercury's surface material consists dominantly of iron-poor, calcium-magnesium silicates with an admixture of spectrally neutral opaque minerals. The Caloris basin, one of the few areas of Mercury that experienced horizontal extensional faulting, has at its center a radial pattern of extensional troughs unlike anything seen elsewhere on the planet (**Fig. 6**). MLA demonstrated that the equatorial topography of Mercury has at least 5 km (3 mi) of relief.

Space environment observations. *MESSENGER* also confirmed that Mercury's internal magnetic field is primarily dipolar, inventoried the heavy ions that fill the magnetosphere, and detected two current-sheet boundaries on the outbound leg of its trajectory. A current sheet is a surface that separates regions of differing magnetic field orientation or intensity. One of the current sheets was located at Mercury's magnetopause (the boundary to the region of space

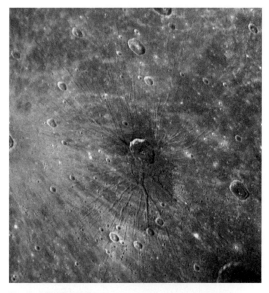

Fig. 6. *MESSENGER* narrow-angle camera image of Pantheon Fossae, a set of more than 100 narrow troughs that radiate outward from the 41-km-diameter (25-mi) Apollodorus crater and occupy the central region of the interior floor of the 1550-km-diameter (960-mi) Caloris basin. (*NASA/Johns Hopkins University Applied Physics Laboratory/Carnegie Institution of Washington*)

dominated by Mercury's magnetic field rather than that of the solar wind field), as expected, while the other, about 1000 km (600 mi) closer to the planet, may mark the edge of a boundary layer for planetary ions that have been ionized and accelerated by the solar wind or possibly signals that solar wind protons have much greater access to Mercury's magnetosphere than is the case at Earth. MASCS mapped a north–south asymmetry in the sodium tail and determined the ratio of sodium to calcium near the tail and also near the dawn terminator.

Administration. The Johns Hopkins University Applied Physics Laboratory built and operates the *MESSENGER* spacecraft and manages the mission for NASA.

For background information *see* MAGNETOSPHERE; MERCURY; OPERATING SYSTEM; PLANET; SPACE COMMUNICATIONS; SPACE POWER SYSTEMS; SPACE PROBE; SPACE TECHNOLOGY; SPACECRAFT PROPULSION in the McGraw-Hill Encyclopedia of Science & Technology.

Sean C. Solomon; Ralph L. McNutt, Jr.

Bibliography. J. C. Leary et al., The MESSENGER spacecraft, *Space Sci. Rev.*, 131:187–217, 2007; J. V. McAdams et al., MESSENGER mission design and navigation, *Space Sci. Rev.*, 131:219–246, 2007; S. C. Solomon et al., MESSENGER mission overview, *Space Sci. Rev.*, 131:3–39, 2007.

Metal-free hydrogen activation and catalysis

Hydrogen (H_2) is the simplest diatomic molecule. Despite this simplicity, H_2 plays a very significant role in today's economy and is destined to become even more important in the future. Hydrogenation—the chemical addition of H_2 across double or triple

Fig. 5. *MESSENGER* narrow-angle camera image, about 500 km (300 mi) across, illustrating the interplay of cratering, volcanism, and deformation on Mercury. A peak-ring basin (left) has been nearly filled with smooth plains material. The basin was later disrupted by the formation of a prominent scarp or cliff, the surface expression of a major thrust fault, which runs alongside part of the basin's northern rim and may have led to the uplift seen across a portion of the basin floor. A smaller crater in the lower right of the image has also been deformed by the scarp, showing that the fault system was active after both impact features had formed. Smooth plains material abutting and ponding against the lower, northeastern edge of the scarp suggests that volcanism continued in this area after this fault system ceased to be active. (*NASA/Johns Hopkins University Applied Physics Laboratory/Carnegie Institution of Washington*)

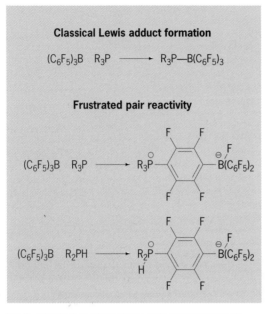

Fig. 1. Classical and frustrated Lewis-pair reactivity of phosphine donors with a boron-based Lewis acid.

bonds, mediated by transition-metal catalysts—is used in the production of a wide range of products, including foodstuffs, plastics, rubbers, commodity and specialty chemicals, and pharmaceuticals. Research efforts targeting more efficient hydrogenation processes have been prompted by concerns over the environmental and health impact of residual metals as well as the high cost of precious-metal catalysts. Perhaps more familiar is the impending role for H_2 as a major energy source. This prospect

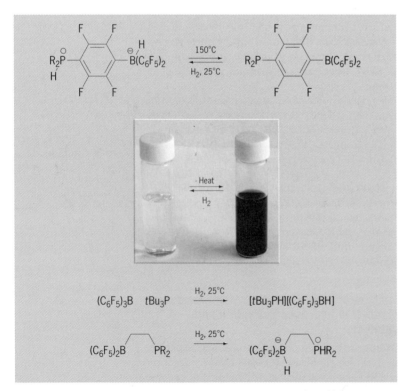

Fig. 2. Activation of H_2 by frustrated Lewis pairs. Insert: Left, solution of PH/BH; right solution of P/B.

promises the environmental "holy grail" for a fuel whose consumption generates only water. For this to be implemented on a large scale, the scientific and engineering research communities must find new ways to provide hydrogen on demand, quickly and cheaply, as well as to recharge spent material. Regardless of the application, the development of new technologies for storing or using hydrogen will require the discovery of fundamentally new H_2 chemistry.

One new approach to the activation of H_2 recently reported stems from an old concept. In 1923, G. N. Lewis put forth a description of acids and bases, categorizing molecules as electron-pair donors or acceptors. This concept provides an understanding of much of the molecular interactions in main-group and transition-metal chemistry. A twist on this long-standing axiom of inorganic chemistry was to ask: What is the result of the combining electron-rich and electron-poor species where the molecular structures preclude direct interactions of the electron acceptor and donor atoms? Further, in such sterically "frustrated" Lewis pairs (FLPs), what is the result in terms of reactivity of the unquenched Lewis acidity and basicity?

To probe these questions, reactions were examined using the commercially available Lewis acid, tris(pentafluorophenyl)borane [$B(C_6F_5)_3$] with phosphorus-based donor molecules. It is well known that small phosphines such as PMe_3 ($Me = CH_3$) form stable P–B bonds, affording what is described as a donor-acceptor adduct of the formula $Me_3PB(C_6F_5)_3$. In contrast, reactions of $B(C_6F_5)_3$ with tertiary or secondary phosphines (such as R_2PH), which have sterically bulky substituents, did not yield such simple Lewis adducts. Instead, these reactions gave white, air- and moisture-stable solids, such as [$R_2PH(C_6F_4)BF(C_6F_5)_2$] (**Fig. 1**). Further reaction of these salts with Me_2SiHCl resulted in the replacement of the B-bound fluoride with a hydride (H$^-$) anion. These phosphonium-borate salts [$R_2PH(C_6F_4)BH(C_6F_5)_2$], or (PH/BH), are among a rare class of compounds containing both protic (positive) and hydridic (negative) centers, known as zwitterionic salts.

Activation of H_2. Given the nature of the species, it is perhaps not surprising to find that heating the colorless solution of the PH/BH zwitterion in solution to 150°C resulted in the quantitative liberation of H_2 and generation of an orange-red solution of the phosphino-borane, $R_2P(C_6F_4)B(C_6F_5)_2$ [P/B] (see insert, **Fig. 2**). This P/B product is also a unimolecular FLP, as there is no evidence of intermolecular coordination of P to B either in solution or the solid state. In a remarkable and unprecedented finding, exposure of a solution of this neutral phosphino-borane to H_2 at 25°C led to the reformation of the zwitterionic salt (Fig. 2). This facile interconversion of the PH/BH and P/B species represents the first non-transition-metal system that reversibly releases and takes up hydrogen.

In a similar fashion, the H_2 was split to give H$^+$ and H$^-$ by astonishingly simple reactions with

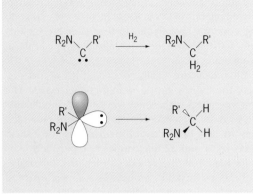

Fig. 3. Reaction of a carbene with H₂.

combinations of sterically frustrated phosphine/borane pairs. Thus, while crowded phosphines R_3P ($R = t$Bu, $C_6H_2Me_3$) do not react with $B(C_6F_5)_3$, exposure of mixtures of these FLPs to 1 atm (101 kPa) H_2 resulted in the immediate formation of the salts $[R_3PH][HB(C_6F_5)_3]$. However, these salts do not liberate H_2, even when heated above 150°C. More recently, the G. Erker research group in Germany has shown that related compounds in which P and B are linked by a two-carbon chain also activate H_2 to give the corresponding salts $[R_2PH(C_2H_4)BH(C_6F_5)_2]$ (Fig. 2). The mechanistic details of reactions of FLPs with H_2 continue to be the subject of study. Preliminary experimental and computational studies support the view that the most likely course is polarization of H_2 by the Lewis acid, followed by protonation of the Lewis base. However, the possibility that the Lewis base interacts with H_2 first cannot be dismissed, as low-temperature matrix isolation experiments have indicated direct interactions of phosphine with H_2.

In a related recent study, G. Bertrand and coworkers at the University of California, Riverside, described the reaction of some carbene (divalent carbon with two nonbonded electrons; for example, R_2C) derivatives with H_2. Here as well, the hydrogen molecule is split to give formally an H⁺ and H⁻, transforming the carbene carbon irreversibly to a methylene (R_2CH_2) group. Since the carbene carbon possesses an electron pair orthogonal to a vacant electron acceptor p-orbital on carbon, it can be viewed as a unique FLP where both the donor and acceptor sites reside on the same atom (**Fig. 3**).

Hydrogenation catalysis. Based on the facile heterolytic cleavage of H_2 by FLPs, a catalytic cycle for metal-free hydrogenation (reduction) was envisioned where the H_2 is activated with the proton and hydride transferred to an unsaturated organic molecule, thus regenerating the FLP for reaction with more H_2. In this way, the FLP would become a catalyst for reduction. Indeed, this notion was demonstrated (**Fig. 4**). The zwitterionic PH/BH salts have been shown to effect the transfer of a proton and hydride to the C–N double bonds of imines ($R_2C\!=\!NR'$) to give the corresponding amines cleanly and in high yield at temperatures of 80–120°C and H_2 pressures of 1–5 atm (101–507 kPa). In addition, FLP catalysts

effect the reductive ring opening of aziridines (NC_2 three-membered rings). While this is the first metal-free catalyst system to reduce organic substrates with H_2, it is only effective for imines with sterically demanding substituents on N. However, coordination of imines with less bulky N-substituents and nitriles ($C\!\equiv\!N$) to $B(C_6F_5)_3$ does allow for the reduction of these species as well. This limitation arises as less-hindered imines or nitriles bind tightly to the borane center of the P/B phosphino-borane catalyst, thus quenching further H_2 activation. In the case of imine reduction, the mechanism involves initial protonation of the imine by the phosphonium center, followed by BH attack of the iminium salt. For $B(C_6F_5)_3$-bound nitriles these steps are reversed, with initial hydride transfer being followed by protonation to give the product amine-borane adduct. In very recent work, it has been shown that sterically hindered imines themselves can act as the Lewis-base partner of an FLP, permitting reduction of imines with simply $B(C_6F_5)_3$ and H_2.

Outlook. The unique H_2 chemistry unveiled by using FLPs offers the potential of significant new technology on both shorter- and longer-term

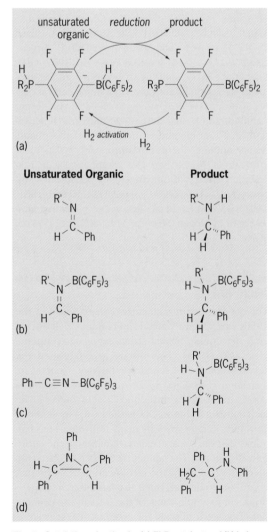

Fig. 4. Catalytic reduction by (a) FLP catalysts of (b) imines, (c) nitriles and (d) aziridines.

timescales. The use of FLP hydrogenation catalysts seems viable. The application of catalyst design and optimization strategies is likely to significantly broaden the range of reducible substrates, thus providing both the environmental and monetary advantages of eliminating trace metals in products and replacing traditional precious-metal catalysts. In the longer term, the finding of the reversible binding of H_2 by FLPs prompts much interest in the potential for hydrogen storage materials. While the present phosphine/borane systems based on bulky Lewis acid-base combinations clearly suffer from an extremely low capacity by weight percent for hydrogen capture and release, these findings do provide new insights and strategies for activating H_2, which is a key consideration in the development of materials that have higher storage capacity.

For background information *see* ACID AND BASE; ALTERNATIVE FUELS FOR VEHICLES; BORON; CATALYSIS; HYDROGEN; HYDROGENATION; MATRIX ISOLATION; NITRILE; OXIDATION-REDUCTION; PHOSPHORUS; REACTIVE INTERMEDIATES in the McGraw-Hill Encyclopedia of Science & Technology.

Douglas W. Stephan

Bibliography. P. A. Chase et al., Lewis acid catalyzed hydrogenation: $B(C_6F_5)_3$ mediated reduction of imines and nitriles with H_2, *Chem. Commun.*, March 6, 2008, DOI: 10.1039/b718598g; P. A. Chase et al., Metal-free catalytic hydrogenation, *Angew. Chem. Int'l Ed. Engl.*, 46(42):8050–8053, 2007; G. D. Frey et al., Facile splitting of hydrogen and ammonia by nucleophilic activation at a single carbon center, *Science*, 316(5823):439–441, 2007; P. Spies et al., Rapid intramolecular heterolytic dihydrogen activation by a four-membered heterocyclic phosphane-borane adduct, *Chem. Commun.*, (47):5072–5074, 2007; G. C. Welch and D.W. Stephan, Facile heterolytic cleavage of dihydrogen by phosphines and boranes, *J. Am. Chem. Soc.*, 129(7):1880–1881, 2007; G. C. Welch et al., Reversible metal-free activation of hydrogen, *Science*, 314(5802):1124–1126, 2006.

MicroRNA

The era of modern molecular biology began in the 1950s when James Watson and Francis Crick discovered the double helical structure of deoxyribonucleic acid (DNA). This discovery established the "central dogma" of molecular biology. Genetic information is codified in DNA. This information is copied to a messenger ribonucleic acid (mRNA). The genetic code in the mRNA is then used to generate a protein. Recent work by many individuals has culminated in the sequence determination of the entire human genome. The genome sequence has been immensely valuable for the identification of genes that contribute to disease, but it also has raised new questions. For example, most of the DNA in the genome does not seem to code for proteins. These regions were sometimes termed "junk DNA" because they did not have an obvious purpose. It is now known that important genes are hidden in these uncharacterized regions of DNA. These include genes that do not code for protein but instead generate biologically active RNAs. One very important type of these "noncoding RNAs" is the microRNA.

MicroRNAs are small, noncoding RNAs that regulate the expression of protein coding genes. The first microRNA discovered was *lin-4*. Groundbreaking work by Victor Ambros and colleagues revealed that the only products of the *lin-4* gene were short RNAs of 61 and 22 nucleotides (a typical mRNA is more than 1000 nucleotides in length). In parallel work, Gary Ruvkun's group was studying *lin-14*, a conventional protein-coding gene that is regulated by *lin-4*. Surprisingly, they found that the *lin-14* mRNA could bind to the 22-nucleotide *lin-4* RNA. This binding of the microRNA (*lin-4*) to its mRNA target (*lin-14*) led to repression of *lin-14* expression. These initial studies uncovered a novel paradigm for gene regulation. There are now more than 800 microRNA genes that have been discovered in the human genome, with many others probably remaining to be identified. In whole, this class of small, noncoding RNAs is believed to regulate one-third of all genes in the human genome.

How microRNAs are made. A microRNA begins its life as a long RNA that is called the primary transcript, or pri-miRNA (see **illustration**). The important feature of all pri-miRNAs is the presence of a stem-loop structure. This is a region of RNA that can bind to a neighboring region, forming a short double helix, with a loop at the end. The biologically active part of the microRNA is the "mature" region, which is one strand of the helix. This region must be removed from the remaining pri-miRNA before it can regulate the expression of target genes. This is accomplished by two enzymatic reactions. The first step is mediated by the nuclease enzyme Drosha in

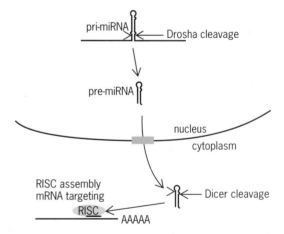

Overview of the microRNA biogenesis and effector pathway. MicroRNA genes are encoded in the genome. Transcription of the gene leads to the generation of the pri-miRNA, which is processed by the enzyme Drosha. The liberated stem-loop pre-miRNA is transported to the cytosol, where the enzyme Dicer removes the loop. One strand, termed the "mature" microRNA, is bound to the protein complex RISC. The mature microRNA guides the RISC to complementary sequences in mRNA targets. This results in a block in protein production from targeted mRNAs.

cooperation with its cofactor DGCR8 (also known as Pasha). Drosha cuts the stem-loop away from the flanking ends of the pri-miRNA. This RNA intermediate is typically 60 to 80 nucleotides in length and is termed the precursor, or pre-miRNA. The pre-miRNA is exported out of the nucleus into the cytoplasm, where the second processing step occurs. The nuclease enzyme Dicer, in cooperation with its cofactor TRBP (also known as Loquacious), cuts the loop away from the stem region. The product of this reaction is an approximately 21-nucleotide, double-stranded RNA helix. One or both strands of this helix are the final product of the microRNA gene and are referred to as the "mature" microRNA.

How microRNAs work. The ultimate goal of a microRNA is to regulate the expression of its target mRNA. The target sequence is recognized by the microRNA itself. The microRNA binds to complementary sites in the target mRNA, essentially forming a small double-helix region. The microRNA does not act alone, but acts as part of a large protein complex termed RISC (RNA-induced silencing complex). This complex is very similar to the complex that mediates the gene-silencing phenomenon of RNA interference (RNAi; in this process, foreign, double-stranded RNA is recognized and degraded by specialized protein complexes, believed to be an evolutionarily conserved defense mechanism against RNA viruses and transposable elements). In fact, microRNAs are the naturally occurring triggers of the RNAi pathway. The core protein in RISC is termed Argonaute. This multifunctional protein binds the mature microRNA and also contains a nuclease domain. This allows Argonaute to cut the mRNA in half, preventing protein production from that mRNA. This cleavage activity is possible only for a microRNA/mRNA pair that is perfectly matched. Many microRNAs do not match their corresponding target sequences perfectly. Therefore, Argonaute cannot cut and the RISC switches to an alternate mode—it reduces protein expression by preventing efficient translation of the mRNA. The exact way that RISC does this is not yet clear. Translation of an mRNA to protein is a complex event with a plethora of protein factors. Most models suggest that RISC prevents initiation of translation, although other models have been put forth. However, the end result is less protein generated from the targeted mRNA.

Roles of microRNAs in biology. More than 800 microRNAs have been identified in the human genome.

Some of these microRNAs are expressed in specific tissues, and some are widely expressed. All tissues, however, express some combination of microRNAs. Recent work has demonstrated that individual microRNAs have important roles in development of certain cell types. For example, the microRNAs miR-155, miR-223, and miR-150 control the production of certain populations of white blood cells. As a result, mice lacking these microRNAs have impaired immune response. Another microRNA, miR-208, is specifically expressed in heart muscle. This microRNA regulates the response of the heart to stress. Mice lacking this microRNA cannot respond properly to cardiac injury resulting from high blood pressure. In addition to these tissue-specific microRNAs, other microRNAs are expressed in many tissues. In some cases, these microRNAs regulate basic cell functions such as proliferation and differentiation, often in opposing directions. Cells that have high levels of the Let-7 microRNA often have low levels of the microRNA cluster miR-17-92, and vice versa. It is believed that Let-7 promotes differentiation and miR-17-92 promotes cell proliferation. Therefore, the balance of these sets of microRNAs is important for avoiding excessive proliferation that could result in cancer. In addition to human microRNA genes, these small RNAs have been found in almost every organism that has been studied. MicroRNAs have even been found in viruses. It should be noted, though, that microRNAs were only very recently discovered. Most microRNAs have not been functionally characterized, and their biological roles are unknown.

How microRNAs contribute to disease. Because microRNAs regulate a diverse set of biological pathways, it might be expected that aberrant expression of specific microRNAs would contribute to disease. This is now known to be true. The disease most strongly connected to microRNAs is cancer. A multitude of studies have shown that every type of cancer has altered expression of microRNAs. One common theme is the elevated expression of miR-17-92 and the reduction of Let-7, as described above. Studies have linked these microRNAs to established oncogenes and tumor suppressors. Furthermore, enforced expression of miR-17-92 in the mouse promoted the development of several cancers, whereas enforced expression of Let-7 inhibited tumor growth. Other microRNAs have been linked to more specific cancers. For example,

MicroRNAs that are associated with cancer

MicroRNA	Cancer type	Known mRNA targets
miR-15, miR-16	Chronic lymphocytic leukemia	Bcl2
miR-155	B-cell lymphoma	Bach1, Jarid2, Cutl1
miR-17-92	B-cell lymphoma, many others	Bim, E2F1, PTEN
miR-143, miR-145	Colorectal carcinoma	ERK5
miR-21	Glioblastoma, many others	PTEN, PCD4
Let-7	Many	Ras, Myc, HGMA2
miR-371-373	Teratocarcinoma	LATS2
miR-34	Lung, others	Many cell cycle genes
miR-10b	Metastatic breast carcinoma	HoxD10
miR-126, miR-335	Metastatic breast carcinoma	Sox4, Tenascin-C

miR-15 and miR-16 have been linked to chronic lymphocytic leukemia, and miR-371-373 microRNAs have been associated with teratocarcinoma (see **table**). In addition to cancer, other diseases have been linked to microRNA function, including cardiac disease, schizophrenia, and viral pathologies.

MicroRNAs as therapeutic targets. With the mounting evidence for a role of microRNAs in disease, there has been significant interest in therapeutics directed against some of them. It is very likely that inhibitors of miR-17-92 would be an effective therapy against tumors that depend on high-level expression of these microRNAs. Several "anti-miR" or "antagomiR" strategies are being developed. They are all based on small synthetic RNAs that bind to the offending microRNA and prevent access to the target mRNAs. Although this approach would work for disease-promoting microRNAs, they would not be applicable for microRNAs whose loss leads to disease. For such cases, microRNA mimics would restore the function of these microRNAs. This is being tested for Let-7 as a cancer therapeutic. The challenge for all these approaches is the ability to deliver the therapeutic to cells. RNA and related molecules are not taken up efficiently by cells, so extensive modification must be done to improve therapeutic qualities. The other challenge for microRNA therapeutics is that much still needs to be learned about microRNA biology. Although an anti-miR to a cancer-causing microRNA sounds beneficial, that microRNA does serve some important function, otherwise it would not be present in the human genome. For this reason, the inhibition of microRNAs may lead to unexpected side effects. Nevertheless, these difficulties have not dissuaded pharmaceutical companies from embarking on microRNA research and studying its potential value in therapeutics.

Outlook. The microRNA pathway, though only recently discovered, has energized investigators because of its important role in biology. Researchers are only beginning to understand this pathway, its roles in normal biology, and its contribution to disease. It is also worth noting that many other noncoding RNAs appear to exist in the human genome, with functions unrelated to microRNAs but probably as important. The findings so far offer only a small glimpse into the complexity of the human genome.

For background information *see* CANCER (MEDICINE); DEOXYRIBONUCLEIC ACID (DNA); GENE; HUMAN GENOME; MOLECULAR BIOLOGY; NUCLEOPROTEIN; NUCLEOTIDE; ONCOGENES; PROTEIN; RIBONUCLEIC ACID (RNA); TUMOR SUPPRESSOR GENE in the McGraw-Hill Encyclopedia of Science & Technology.   Scott M. Hammond

Bibliography. D. P. Bartel, MicroRNAs: Genomics, biogenesis, mechanism, and function, *Cell*, 116:281–297, 2004; A. Esquela-Kerscher and F. J. Slack, Oncomirs—microRNAs with a role in cancer, *Nat. Rev. Cancer*, 6:259–269, 2006; S. M. Hammond, MicroRNA therapeutics: A new niche for antisense nucleic acids, *Trends Mol. Med.*, 12:99–101, 2006; R. C. Lee, R. L. Feinbaum, and V. Ambros, The *C. elegans* heterochronic gene *lin-4* encodes small RNAs with antisense complementarity to *lin-14*, *Cell*, 75:843–854, 1993; B. Wightman, I. Ha, and G. Ruvkun, Post-transcriptional regulation of the heterochronic gene *lin-14* by *lin-4* mediates temporal pattern formation in *C. elegans*, *Cell*, 75:855–862, 1993.

Midrange wireless power transfer

At the turn of the twentieth century, Nikola Tesla devoted his extraordinary abilities to pioneering research on radio-frequency electromagnetism, with the ultimate goal of perfecting a system for delivering electric power wirelessly over continental distances. Although Tesla's dream was not fully realized, this work led directly to the development of the radio. In recent years, there has been a proliferation of portable electronic devices, most of which rely on modern developments of radio technology to transfer data wirelessly. These devices are frequently marketed as wireless, although the consumer must still plug them into a power outlet regularly, at least to recharge their batteries. A method for supplying energy without cables, particularly if the source of wireless power and the receiving object are separated by midrange distances of at least several feet, could make these devices completely wireless.

Transferring power through electromagnetic fields. Transferring power from a source to a device using electromagnetic fields involves driving the source so that it generates a time-varying electromagnetic field, which in turn excites a current in the receiver. The energy contained in this induced current can then be extracted and used to power an electric device.

The nature of the electromagnetic field involved depends on its frequency of oscillation. At sufficiently high frequencies, the field will be essentially composed of electromagnetic waves with wavelength $\lambda = c/f$, c being the speed of light and f the frequency. At lower frequencies, the wavelength of the associated waves will become longer. And a device placed at a distance from the source that is much shorter than the wavelength will be predominantly subjected to a near-field, which, while oscillating in time, has the spatial profile of a combination of a magnetic field created by a constant current and an electric field created by a static distribution of charges.

Wireless communication is essentially wireless power transfer, although the power transferred to receivers need only be enough to convey information. A main obstacle to using antennas designed for communication for efficient power transfer is that the radiation flows out from the source in nearly all directions, and whatever is not intercepted by the receiver is lost. One solution to this problem is to focus the electromagnetic waves and aim the resulting beam at the receiver, a method developed by William Brown at Raytheon in the 1960s. Although this technique can be used to transfer power over long distances, it requires a mechanism for tracking the location of the receiver. Furthermore, one needs to operate at fairly high frequencies (of the order of GHz or more) to properly collimate the radiation, and at these frequencies the beam is very sensitive to objects obstructing the line of sight between the source and the receiver.

While electromagnetic waves are fundamentally composed of equal parts magnetic and electric field, in the sense that the time-averaged power contained in each type of field is the same, near-fields can be predominantly electric or predominantly magnetic. Tesla did a significant amount of work on power transfer through oscillating electric fields, and his findings are often reproduced in science demonstrations featuring Tesla coils. The main drawback to using electric fields is that they interact much more strongly than magnetic fields do with extraneous objects, in particular metallic ones. These electric fields also give rise to large voltages, which create a hazard for anyone who accidentally closes the circuit between two surfaces at significantly different voltages.

We are therefore led to consider near-field magnetic fields. The transfer of power through magnetic fields relies on the principle of magnetic induction, which is the basis for many ubiquitous electrical devices, including voltage transformers. Magnetic fields interact weakly with nonmetallic objects, and the power transfer is largely unaffected if the line of sight between transmitter and the receiver is obstructed by most types of material. Moreover, the scheme is generally safer for human beings than the alternatives. Even metallic objects do not ruin the transfer, so long as they are not too large and too close to either the source or the device.

Proper design. Magnetic induction is typically used to transfer power when the distance between the source and the receiver is much smaller than the characteristic size of either. For midrange wireless power transfer, we would like to reverse this ratio, making the distance significantly larger than the source and device size, but trying to increase this separation without a suitable redesign leads to low efficiency. A good starting point is to exploit the phenomenon of resonance.

Perhaps one of the more popular and dramatic demonstrations of the power of resonance features a glass of wine and an opera singer. The glass of wine has a discrete set of mechanical resonances because of its shape and material, and these resonances manifest themselves when one taps the glass gently with a utensil and the resulting vibrations gradually die out over several seconds. When the opera singer starts singing a particular note, the sound waves carrying her voice will drive the glass at that particular frequency. At generic frequencies, the glass will not vibrate significantly more than other objects, but if the singer hits a note corresponding to a resonant frequency of the glass, the amplitude of the vibrations will increase drastically, possibly causing the glass to shatter as a result of mechanical overstress.

In the same vein, one can build electromagnetic resonators, with the simplest design being an LC circuit composed of an inductor, L (which can be a simple loop of wire), and a capacitor, C. If the receiver is designed to resonate at the frequency of the mag-

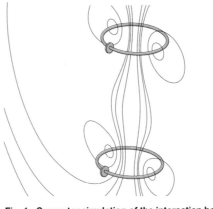

Fig. 1. Computer simulation of the interaction between two simplified electromagnetic resonators, with magnetic flux lines shown.

netic field generated by the source, it will capture the power contained in the field much more effectively (**Fig. 1**), leading to an improved efficiency of power transfer.

Although resonance is a key idea in making power transfer over midrange distances viable, it is not the whole story. When the electronic circuits responsible for driving the source and extracting power from the receiver are optimally designed (a nontrivial but doable engineering challenge), the efficiency depends only on three dimensionless parameters. Two of these are the quality factors, Q, of the source and the device (respectively, Q_S and Q_D), which are inversely proportional to the fraction of energy dissipated (because of electrical resistance and, to a smaller degree, radiation) per cycle of oscillation. The remaining parameter, k, represents the strength of the interaction between the source and the device, and is proportional to the fraction of energy transferred between the two objects per cycle. It depends on the geometry of each object as well as on their position relative to each other, with its magnitude decreasing with increasing separation.

The efficiency of the transfer depends on the combination $k\sqrt{Q_S Q_D}$ (**Fig. 2**), and if the value of this variable is greater than about 1, the efficiency becomes sufficiently high for many practical applications. From the above descriptions of Q_S, Q_D, and k,

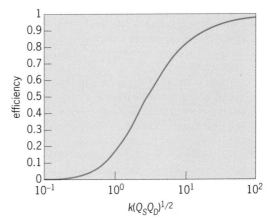

Fig. 2. Semilogarithmic plot of the efficiency as a function of key system parameters.

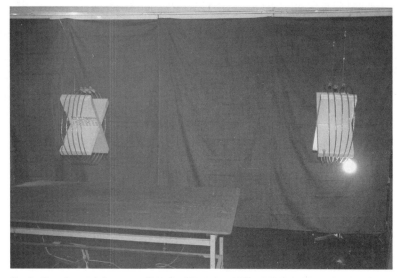

Fig. 3. Experimental demonstration of midrange power transfer between two identical electromagnetic resonators, each 60 cm (approximately 2 ft) in diameter, separated by 2 m (6.5 ft), to light a 60-W bulb. The source resonator, on the left, is inductively coupled to the driving circuit (not shown). (*A. Kurs et al., Wireless power transfer via strongly coupled magnetic resonances, Science, 317:83, 2007*)

we see that $k\sqrt{Q_S Q_D} > 1$ means that energy is transferred from the source to the device before much of it gets wasted, thereby enabling efficient power transfer. One must be careful when trying to maximize $k\sqrt{Q_S Q_D}$, however, since improving one variable may have detrimental effects on another.

Our research team recently set out to show that it was possible to build an apparatus where the condition $k\sqrt{Q_S Q_D} > 1$ held at midrange distances (**Fig. 3**). We built two identical electromagnetic resonators, 60 cm (2 ft) in diameter and operating at 10 MHz with quality factors of approximately 1000. The coupling between the resonators was such that at a separation of 2 m (6.5 ft), the efficiency of the power transfer was still 50%, while at a distance of 1.25 m (4 ft), still more than twice the diameter of the source and the device, efficiency was above 80%. This performance may be improved significantly with further refinements to the design. The amount of power transferred was up to 100 W, enough to power three laptop computers, and was limited mostly by the circuit driving the source.

Outlook. Although there is still room for improvement in performance, and further work must be done to reduce the size of the receiver so that it fits into portable electronic devices, it seems that, at least from a technical perspective, reasonably efficient midrange power transfer is feasible. Because no line of sight between the source and the devices is required, a source may be placed in an inconspicuous location (for example, embedded in a wall), from which it can transmit power to all the devices within range.

This scheme may be used wirelessly to recharge the batteries of portable electronic gadgets, and potentially to power them directly, reducing battery use. Batteries contain toxic chemicals and need to be replaced periodically, as their performance degrades over time. It is therefore possible that wireless power transfer would be advantageous for the envi-

ronment and cheaper overall to the consumer for a variety of applications. Other technologies that may benefit include sensors in inaccessible locations and biomedical devices, such as artificial hearts, which could be recharged noninvasively and efficiently.

For background information *see* ELECTRIC FIELD; ELECTROMAGNETIC FIELD; ELECTROMAGNETIC INDUCTION; ELECTROMAGNETISM in the McGraw-Hill Encyclopedia of Science & Technology.   André Kurs

Bibliography. W. C. Brown, The history of power transmission by radio waves, *IEEE. T. Microw. Theory*, MTT-32:1230, 1984; A. Karalis et al., Efficient wireless non-radiative mid-range energy transfer, *Ann. Phys.*, 323:34, 2008; A. Kurs et al., Wireless power transfer via strongly coupled magnetic resonances, *Science*, 317:83, 2007; N. Tesla, *Nikola Tesla: Lectures, Patents, Articles,* Health Research, 1973.

Mineral textures in lavas

To better understand volcanic phenomena, researchers are interested in how magma evolves while stored in the Earth's crust and how it subsequently ascends and erupts at the surface. The textures and compositional variations that minerals develop in response to magmatic changes can be used to interpret the rates and mechanics of how these processes operate. Textural analysis of lava flows and lava domes, along with compositional analysis of minerals and surrounding melt (glass), can be particularly effective in studying these processes and ultimately aid in volcanic hazards assessment.

A crystallizing magma produces a mineral assemblage characteristic of its pressure, temperature, and composition. Magma will experience changes in pressure or temperature as a result of the ascent to the surface during eruption or mixing with a hotter magma. When this happens, previously stable minerals will respond by developing textural and compositional variations that record information about the decompression or heating event. This article discusses examples of how textures in lavas (flows and domes), as well as textures of minerals themselves, are used to understand some of the most important magmatic processes that occur during magma storage in the shallow crust and magma ascent during lava-producing volcanic eruptions.

Lava textures. Magma, or molten rock, is the complex solution of crystals, melt, and volatiles (such as H_2O, CO_2, and SO_2). Texture is a term that geologists use to describe the noncompositional properties of magma when it solidifies to rock. Although one lava flow can possess several different textures, in a broad sense, lava texture is based on (1) the proportion of glass (solidified melt) relative to crystals, (2) the size of crystals, and (3) the shape and/or character of individual crystals that are observed on the scale of a hand sample or thin section under a petrographic microscope. The crystallinity of igneous rocks, or the percentage of mineral grains relative to glass in rocks formed through the crystallization of magma, can range from 0 to 100%. Most lavas have crystallinities ranging from <1 to 40%. Magmas with crystallinities of more than 40% are usually too viscous to flow to the surface. Grain sizes of minerals in lavas range widely from <0.01 to >20 mm. The shapes and textures of individual minerals in lavas vary from crystals with well-formed crystal faces to crystals that are broken, rounded, embayed, or enclosed.

What controls the proportion of glass to crystals in lava? To answer this, we must first understand what volcanic glass is, how it relates to magma, and how it forms. Glassy texture forms as a result of magma cooling at a sufficiently fast rate to prevent randomly dispersed ions from migrating and organizing themselves into an orderly crystalline structure. Thus, glass does not occur in igneous rocks that had enough time to fully crystallize underground. Field observations of glassy rocks in volcanic regions support the hypothesis that fast cooling rates produce glass. Rising magma batches that erupt and cool at the Earth's surface harden to form glass-rich lavas. Some lava flows, such as those that erupt under water or adjacent to a glacier, develop a thick glassy crust around their surface in response to very rapid cooling. And small pieces of magma that are blown violently from a volcanic vent into the much cooler atmosphere harden to form jagged fragments of glassy rock known as pumice or scoria. An example of volcanic glass is shown in **Fig. 1**a. The hand specimen

Hand sample view

Microscope view
(plane-polarized light and cross-polarized light)

Fig. 1. General textures of lava flows and domes record information about the pressure, temperature, and composition of the melt from which they crystallized. This information can be used to interpret the rates and mechanics of how these processes operate, which ultimately aid in the assessment of volcanic hazards. In the examples presented here, the rocks have roughly the same chemical composition but different textures. Pictures of hand samples are shown on the left with a penny for scale, with corresponding photomicrographs in plane-polarized light (middle) and cross-polarized light (right), showing (a) rhyolite obsidian with glassy texture, (b) rhyolite lava dome sample with aphanitic texture, consisting of mineral grains visible in microscopic view, (c) dacite lava flow sample with porphyritic texture, where two separate rates of cooling are recorded as two distinct populations of crystals (phenocrysts and microlites), and (d dacite lava dome sample with coarsely porphyritic texture, with plagioclase crystals up to 1 cm in diameter.

displays a conchoidal fracture, with the sharp edges typical of broken glass. No distinct grains are visible, but when viewed under a microscope, distinct flow bands composed of innumerable minute crystals can be observed.

If crystal growth from a melt requires time for the ions to collect and organize themselves, then the crystallinity of a volcanic rock should increase as the cooling rate decreases. The term aphanitic is a textural classification that describes volcanic rocks with very fine-grained crystals surrounded by glass. The crystals are so small that a microscope is needed to observe them. This texture indicates relatively rapid cooling, but not nearly as rapid as the quenching of magma to form a rock entirely composed of glass. An example of aphanitic texture is shown in Fig. 1b. Many aphanitic rocks contain numerous small spherical or ellipsoidal cavities, referred to as vesicles. Vesicles are produced by gas bubbles trapped in the solidifying rock. As hot magma rises toward the Earth's surface, the confining pressure diminishes, and volatiles that were once dissolved at greater depth exsolve into a separate bubble phase. In lavas, vesicles tend to develop in the upper part of a lava flow, just below the solid crust, where the upward-migrating bubbles are trapped.

What controls the size of the crystals in lavas? Like crystallinity, the size of the crystals observed in lavas is also related to the cooling rate. One of the most common lava textures is known as porphyritic, which is a textural classification used to describe volcanic rocks with two distinct sizes of crystals. Examples of porphyritic texture are shown in Fig. 1c and d. The larger crystals are referred to as phenocrysts, whereas the smaller crystals that compose the groundmass, or the matrix of the rock, are referred to as microlites. This texture is particularly interesting because it requires the parent magma to have experienced two distinct cooling regimes during solidification, with information about each regime being recorded in the separate crystal groups. Phenocrysts record a cooling regime that favors the slow growth of a few crystals at temperatures only slightly cooler than the temperature at which the magma becomes entirely molten. A cooling regime such as this is likely to occur in a magma storage region (magma chamber) located in the upper 5–15 km of the Earth's crust. In contrast, microlites record a cooling regime favoring the rapid formation (that is, nucleation) of many new crystals, coupled with extremely limited growth. Interestingly, experimental studies combined with investigations of magmatic fluid mechanics and observations of active eruptions show that heat loss is not the main driving force for microlite crystallization. Instead, crystallization induced by isothermal rapid decompression, which facilitates volatile exsolution, is a more likely mechanism in the generation of microlites. Thus, porphyritic texture records the crystallization of phenocrysts dominated by slow crystal growth resulting from small degrees of prolonged heat loss of the shallow magma body to the surrounding crust in addition to rapid crystallization of micro-

lites resulting from virtually isothermal ascent to the surface during eruption.

Mineral-melt textures. In addition to the general lava textures just described, minerals themselves may develop other textures in response to chemical reactions with the surrounding melt. These mineral textures record a variety of different processes that chemically and thermally modify the magma. Oscillatory zoning of plagioclase feldspar (**Fig. 2a** and b) is one of the most common mineral-melt textures observed in lavas. Other minerals also display this texture, such as amphiboles, pyroxenes, and zircon. Like tree rings, oscillatory zones record changing conditions in the surrounding environment as the phenocrysts grow. The oldest zones are at the core of the crystal, and the youngest are at the crystal rim. The magnitudes of the amplitude and wavelength pattern of the oscillatory zoning (that is, how close together the compositional bands occur) can be used to determine its origin. Smaller variations in amplitude and wavelength are related to a self-organization process related to the substitution of Ca + Al and Na + Si during relatively slow crystallization of plagioclase feldspar. Larger variations may be caused by several processes, including the episodic injection of hot, more mafic magma into a magma chamber or the transport of a growing plagioclase crystal through a compositionally and/or thermally stratified magma by convective stirring.

Another widely observed mineral texture group is known as sieved textures, which are particularly common in plagioclase crystals and may be superimposed over previously formed oscillatory zoning (Fig. 2c and d). Sieved textures are often divided into two types based on the size and morphology of the sievelike cavities in the crystal. "Spongy" sieve texture refers to the type characterized by a coarser-grained (20–200-μm diameter) interconnected network of cavities and channels, whereas "dusty" sieve texture refers to the densely packed and finer-grained type (1–10-μm diameter). Results from hydrothermal experimental studies suggest that the spongy-type texture represents a dissolution process that develops in response to rapid (~10 m/s) decompression of magma from the lower crust (~40 km in depth) to shallow crustal magma storage regions (~8 km in depth). Dusty-type sieved texture in plagioclase also represents a dissolution process. It forms in response to rapid heating (200°C over 0.5–1.5 h) of magma, such as might occur during the replenishment of a magma storage region by a primitive magma of contrasting composition and higher temperature. During heating, plagioclase rims develop a tightly packed network (dusty texture) of micrometer-size glass inclusions and high-anorthite plagioclase that migrates inward toward the crystal core over time. Dusty textures commonly truncate previously formed spongy or oscillatory zoning textures (Fig. 2e and f) or may be enclosed by a clear euhedral to subhedral rim of elevated anorthite plagioclase, compared to the plagioclase core or oscillatory zoning texture. ⁻ ‛hickness of dusty texture zones has been us   estimate the amount of

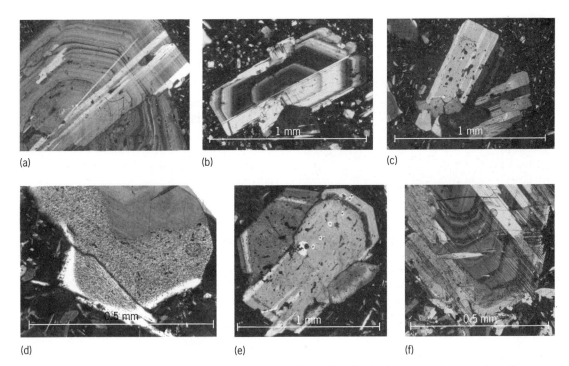

Fig. 2. Common mineral-melt reaction textures observed in plagioclase that develop in response to chemical reactions between plagioclase phenocrysts and the surrounding melt during processes that chemically and thermally modified the magma. All photomicrographs are from cross-polarized light. Examples of (*a*) and (*b*) oscillatory zoning, (*c*) spongy-sieved texture, (*d*) dusty-sieved texture, (*e*) dusty-sieved texture encasing spongy plagioclase, and (*f*) dusty-sieved texture superimposed over previously developed oscillatory zoning.

time required for mixing of an intruding hot primitive magma with a cooler host magma, from which the plagioclase originally crystallized. Such estimates are based on experimental results showing that dusty texture develops at a rate of approximately 4μm/h, given a surrounding temperature of 950–1050°C.

A third group of common mineral textures is known as reaction rims. They are crystalline coronas that are commonly observed to encase a wide variety of minerals, including quartz, olivine, and amphiboles. The rims form as a result of reactions between a crystal and the surrounding melt in response to changes in the composition, temperature, or pressure of the surrounding melt. Unlike oscillatory zoning, reaction rims progress from the mineral-melt boundary inward toward the core of the reacting mineral. Thus, the oldest portion is on the outside of the rim and the youngest portion is on the inside of the rim. The minerals formed within the reaction rims are useful indicators of the magmatic processes responsible for the change in the surrounding melt. For example, amphibole phenocrysts in porphyritic andesite and dacite lavas commonly display fine-grained reaction rims, predominantly consisting of plagioclase and orthopyroxene with lesser magnetite (**Fig. 3**). Experimental studies have demonstrated that these rims develop around amphiboles as magma slowly (0.001–0.02 m/s) rises from storage regions in the upper crust (5–15 km), which causes volatiles once dissolved in the melt at depth to eventually exsolve because of their lower solubility in melt at lower pressures. Amphiboles, which contain 4–5 wt.% H_2O in their structure, destabilize as a result of this dry-

ing and react with the surrounding melt to form a reaction rim composed of microlite-sized anhydrous minerals. Interestingly, magma that ascends to the surface quickly and erupts explosively does not contain amphiboles with fine-grained reaction rims, suggesting that reaction rims develop only in slowly ascending magma that erupts as lava flows or lava domes.

Another example of reaction rims is those surrounding quartz crystals in porphyritic dacite lavas, which often display coarse-grained reaction rims composed of plagioclase, clinopyroxene, and magnetite (Fig. 3*b* and *c*). Not only are all of the rim minerals stable at higher temperatures and more primitive melt compositions than quartz, but also the quartz crystal faces are irregularly embayed, which suggests that the quartz crystal was dissolving as the reaction rim formed. Thus, these reaction rims are often interpreted to represent an abrupt change in the temperature, and possibly the composition, of the melt surrounding the original quartz crystal caused by the introduction of new, hot, primitive magma while stored in the shallow crust. A similar mechanism has been suggested for coarse-grained reaction rims of plagioclase, clinopyroxene, and magnetite surrounding irregularly embayed amphibole crystals (Fig. 3*d*). Note how this reaction rim contrasts with the amphibole rim that develops in response to decompression, in that it is coarser-grained and surrounds an embayed amphibole.

For background information *see* AMPHIBOLE; ANDESITE; ANORTHITE; DACITE; IGNEOUS ROCKS; LAVA; MAGMA; MAGNETITE; MINERALOGY; PETROLOGY; PYROXENE; RHYOLITE; VOLCANIC GLASS; VOLCANO in

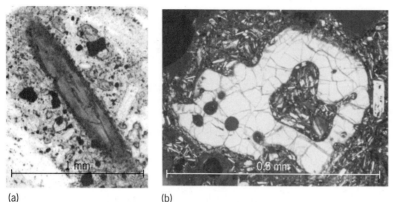

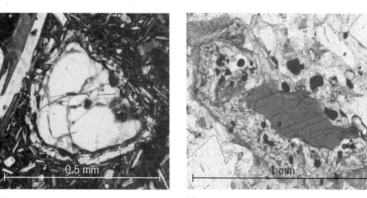

(a)

(b)

(c)

(d)

Fig. 3. Examples of common mineral-melt reaction rims observed in amphibole and quartz that develop in response to chemical reactions between the phenocrysts and the surrounding melt during processes that chemically and thermally modified the magma. Photomicrograph in plane-polarized light of (*a*) hornblende amphibole surrounded by a fine-grained reaction rim, formed in response to ascent (decompression) to the surface, composed of plagioclase and orthopyroxene with minor amounts of Fe-Ti oxides; photomicrograph in cross-polarized light of (*b* and *c*) an embayed quartz crystal enclosed by a reaction rim of clinopyroxene, plagioclase, and Fe-Ti oxides, formed in response to a sudden thermal and compositional change of the surrounding melt during a magma-mixing event; and photomicrograph in cross-polarized light of (*d*) an embayed hornblende crystal enclosed by a coarse-grained reaction rim of clinopyroxene, plagioclase, and Fe-Ti oxides, also formed in response to a sudden thermal and compositional change of the surrounding melt during a magma-mixing event.

the McGraw-Hill Encyclopedia of Science & Technology.   Brandon L. Browne

Bibliography. J. Blundy, K. Cashman, and M. Humphreys, Magma heating by decompression-driven crystallization beneath andesite volcanoes, *Nature*, 44, 76–80, 2006; B. L. Browne and J. E. Gardner, The influence of magma ascent path on the texture, mineralogy, and formation of hornblende reaction rims, *Earth Planet Sci. Lett.*, 246, 161–176, 2006; K. V. Cashman and S. M. McConnell, Multiple levels of magma storage during the 1980 summer eruptions of Mount St. Helens, WA, *B. Volcanol.*, 68, 57–75, 2005; J. E. Hammer and M. J. Rutherford, Kinetics of decompression-induced crystallization in silicic melt, *J. Geophys. Res.*, 107, 10,029–10,053, 2002; B. D. Marsh, Crystal size distributions in rocks and the kinetics and dynamics of crystallization I. Theory, *Contrib. Mineral. Petr.*, 99, 277–291, 1988; M. J. Rutherford and J. Devine, Magmatic conditions and processes in the storage zone of the 2004-06 Mount St. Helens eruption: The record in amphibole and plagioclase phenocrysts, in D. R. Sherrod, W. E. Scott, and P. H. Stauffer (eds.), *A Volcano Rekindled: The First Year of Renewed Eruption at Mount St. Helens, 2004–2006,* U.S. Geological Survey Professional Paper 1750, chap. 31, 2008.

Miniaturized ion traps

The quadrupole ion trap is an instrument designed to store and separate charged particles using a combination of radio-frequency (RF) and direct-current (dc) electric fields. The ion trap has many important applications in biology, geology, physics, and chemistry, particularly in the areas of mass spectrometry, precision atomic and molecular spectroscopy, and quantum information science. Two ion-trap geometries are commonly used: the Paul trap, which uses a three-dimensional quadrupole field, and the linear trap, which uses a two-dimensional quadrupole field (**Fig. 1**). Depending on the operating conditions, a charged particle can form either a stable or an unstable trajectory inside such a quadrupole potential. By varying the operating voltages, frequency, and ambient pressure, it is possible to trap ions and particles with a large range of mass-to-charge ratios in a controlled environment over a period of days. The mass-to-charge ratio is a dimensionless ratio between the mass number of an ion and its charge number, and has been demonstrated to fall in the range 10^2–10^9, using a single Paul trap with a 1-mm radius.

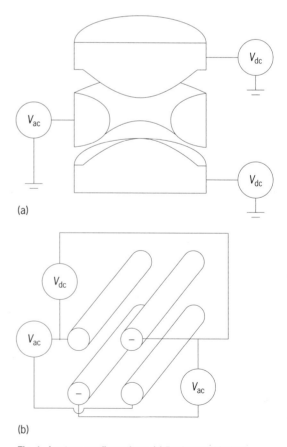

Fig. 1. Ion trap configurations. (*a*) Paul trap with three electrodes. (*b*) Linear quadrupole trap with four rod electrodes

Recently, there have been multiple efforts in miniaturizing ion traps, using silicon integrated circuit fabrication techniques. Most of the research has concentrated on experiments in quantum computing and the development of small mass spectrometers. Modern microfabrication and nanofabrication techniques have been applied to scale the dimensions of the quadrupole ion trap from centimeters and millimeters to micrometers. The motivation of these efforts is to reduce the size of the device to lower its cost, power consumption, and weight, without sacrificing performance. In particular, the ability of fabricating a large array of ion traps in a dense and closely coupled configuration provides a scalable technical path to increase the complexity and functionality in ion-trap quantum information processing and to extend the sensitivity and operating pressure in ion-trap mass spectrometry.

Application in quantum computation. Trapped ions have been used to demonstrate many of the required operations for quantum information processing. The "qubit" used for quantum processing can be one of the nuclear hyperfine levels shown in the electronic energy-level diagram for single-electron ions in **Fig. 2**. The qubit is the basic unit of information in quantum computing and is different from the classical bit in that it can be zero, one, or a superposition of both.

Single-qubit quantum gates are implemented using two laser beams incident on the trapped ion. The difference frequency between the two laser beams is tuned to the hyperfine splitting. To avoid spontaneous emission noise, the laser beams are detuned from the excited P state by a frequency Δ, as indi-

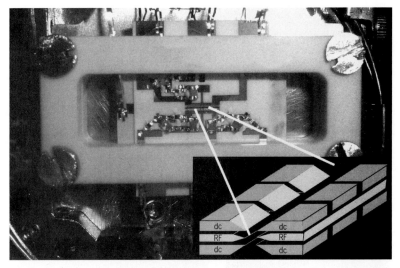

Fig. 3. A multilevel electrode ion trap mounted in an ultrahigh vacuum (UHV) chamber. The 3D inset shows the orientation of the RF and dc electrodes. The ion is trapped along the electrode axis of symmetry. The lateral separation between the RF electrodes is 200 μm. This trap was built by researchers in C. Monroe's group, then at the University of Michigan.

cated by the broken line in Fig. 2. Two-qubit gates are implemented using the shared vibrational motion of two ions in a common ion trap. These two types of gates are sufficient for universal quantum computation and have been used to demonstrate many quantum computational processes, including qubit teleportation, quantum error correction (3-qubit), quantum Fourier transform, and entangled-state purification.

Ions are trapped for these operations in multilevel electrode traps mounted in ultrahigh vacuum chambers, as shown in **Fig. 3**. The multilevel electrode traps are similar to the four-rod trap shown in Fig. 1b. There is a trapping potential well between the symmetric RF and control electrodes.

Planar electrode traps have recently been introduced and demonstrated by J. Chiaverini and colleagues. They are much simpler to fabricate than the multilevel traps. **Figure 4a** shows a scanning-electron micrograph of a planar electrode trap fabricated on a silicon wafer, using silicon microfabrication techniques. The well depth of the planar trap, determined by the electric fields above the ion trapping position (Fig. 4b), is much smaller than the trap depth for the multilevel traps, typically only a few percent of the multilevel trap depth. However, the trap depth is sufficient to capture ions that are photoionized and then rapidly cooled by the laser Doppler effect to temperatures of a few millikelvin, much less than the well depths in the range above 200 K.

Both the RF rails and the control electrodes are aluminum metallic films approximately 1 micrometer thick. The RF rails are separated by approximately 100–150 μm and the RF rail width is 20 μm. Silicon fabrication technology can be applied to make chips with multiple ion traps and large arrays of trapped ions that can be used for future quantum information processors.

The silicon fabrication techniques also allow for much smaller trap dimensions. For quantum gates,

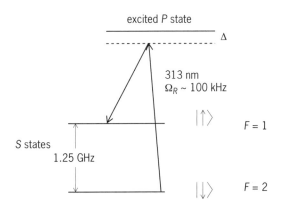

Fig. 2. Energy levels (not drawn to scale) for a single-electron ion (for example, Be$^+$) with a nucleus with spin. The electron's magnetic field at the nucleus splits the nuclear spin states by a hyperfine splitting of 1.25 GHz for Be$^+$. The qubit used for quantum processing is the spin state of the ion's outermost electron and nucleus, symbolized by up and down spins (arrows). F is the total spin of electron and nucleus. These states can be set to an arbitrary superposition of up and down spin states using two laser beams tuned near the excited P state electronic transition with a difference frequency equal to the hyperfine splitting. The laser detuning, Δ, from the P-state level prevents spontaneous emission noise from the P state to S state. The single-qubit gate time is determined by the Rabi frequency Ω_R of the induced laser transition between the qubit states, typically in the 100-kHz range. This Rabi frequency is the rate at which the spin states transition from one to the other.

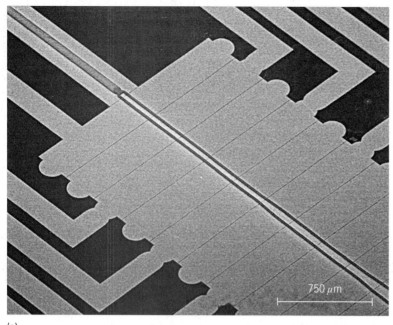

(a)

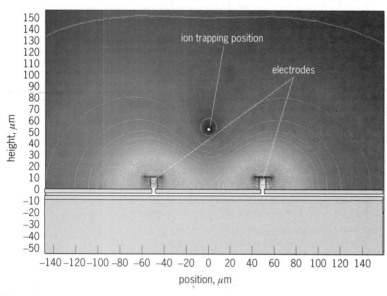

(b)

Fig. 4. Planar ion trap. (*a*) Scanning electron micrograph. The segmented control electrodes move the ions along the RF trap line. (*b*) Schematic drawing of the electric fields above the planar electrodes. The RF voltage is applied to the electrodes elevated on silicon dioxide (SiO₂) rails. The ion trapping position is determined by a null in the RF field between and above the RF electrodes, located at the white spot. This trap was fabricated at the New Jersey Nanotechnology Consortium, until recently located at Bell Laboratories.

magnitude, the inverse-fourth-power dependence of the heating rate makes this rate far too large. It has recently been shown by Isaac Chuang and coworkers that cryogenically cooled traps have heating rates reduced by over two orders of magnitude. This opens the way for ion traps with much smaller dimensions.

It is hoped to scale the number of trapped ions to at least 40–100 ions. Ion counts in this range would allow quantum simulation of condensed matter systems that cannot be attained by classical computational techniques. Up to 50 or 100 ions can be trapped in a single trap; however, the ion vibrational and spatial structure becomes much too complex for useful quantum processing. Spatial multiplexing of the ion traps is one solution that is presently being explored. In this multiplexing scheme, only a few ions are trapped at a particular location, and the ions are shuttled between various trap locations for quantum gate operations. The shuttling process has been shown to be rapid (~10 μs) and efficient (efficiencies near 100%) for relocating the ion along a linear trap structure (Fig. 4*a*) to a trap at a distance of a few millimeters. This performance is sufficient for many quantum processing tasks. Shuttling through trap junctions that fork into various directions also seems possible; however, the study of junction transport is in the early stages and it is not clear whether ions will be heated as they transit through junctions.

Application in mass spectrometry. A conventional quadrupole ion trap mass spectrometer has three main components: (1) an ionization source that ionizes the incoming unknown analyte, (2) an ion trap that stores and separates the analyte by mass-to-charge ratio, and (3) a detector that detects the ions as they are ejected from the trap. Ionization of molecules can be achieved by application of a high electric field by optical or electron-beam irradiation or by electrospray ionization. In the mass-selective axial instability mode, the mass spectrometer operates by changing the trap potential well such that ions of ascending mass-to-charge ratio become unstable and are ejected. The ejected ions are detected by a Faraday cup or an electron multiplier. By scanning the operating voltage or frequency, a mass histogram of the unknown analyte can be measured with a limit of detection of up to femtogram (10^{-15} g) range. The entire instrument operates in high vacuum, typically at 10^{-8} torr (10^{-6} Pa), and is not portable.

The ion trap requires at least two electrodes oriented in a planar or three-dimensional geometry (**Fig. 5**). There are two advantages in making the ion trap smaller. The first advantage is to increase the number of trapped ions per unit area. The number of ions inside a trap is roughly proportional to the radius of the trap. It is possible to replace a single trap with an array of N small traps with the same total area and potentially gain a factor of $\sqrt{N}$ in the number of trapped ions. The second advantage is in the reduction of operating pressure. The operating pressure is related to the mean free path of a gas, that is, the average distance between collisions. In general, the smaller the ion trap, the smaller the required mean free path for the operation of the mass

the ions must be cooled to near their vibrational ground states. It has been found that ions heat at an anomalous rate in all traps tested to date. At present, the cause of the anomalous heating is not known. The ion heating rate is measured to vary as the inverse fourth power of the distance between the ion and the nearest metal electrodes. Ion heating rates at room temperature for electrode-ion spacings near 60 μm are at the level of one vibrational quantum per millisecond, tolerable for many quantum operations that typically occur on a time scale near 10 μs. If the trap dimensions are reduced by an order of

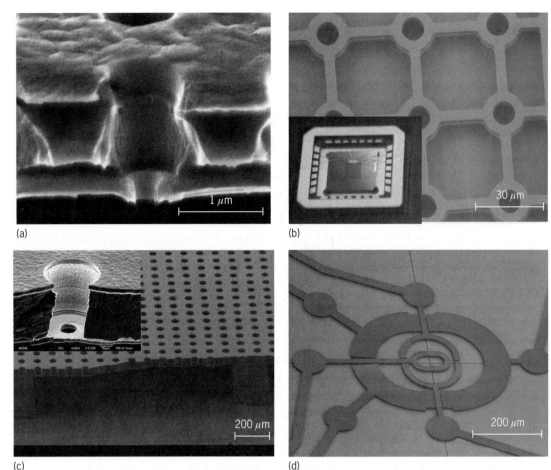

Fig. 5. Scanning electron micrographs of quadrupole ions traps designed for mass spectrometer applications. (*a*) Cylindrical traps with aluminum electrodes and diameter of 0.9 μm. (*b*) Cylindrical traps with silicon electrodes and diameter of 10 μm. Inset shows a packaged chip. (*c*) Cylindrical traps with silicon electrodes and diameter of 40 μm. Inset shows an enlarged view of a single trap. (*d*) Planar trap with silicon electrodes. All traps were fabricated at the New Jersey Nanotechnology Consortium.

spectrometer. A smaller mean free path means higher operating pressure. A trap radius of 20 μm is estimated to operate at a pressure of 1 torr (133 Pa), which can be achieved using a small pump. Using the same assumption, operation at atmospheric pressure requires a trap with a radius of 100 nm, a dimension that is achievable using current fabrication technology. Thus, miniaturization of ion traps can lead to a portable mass spectrometer that does not need a pump, at least not an expensive one.

Several technical challenges are being actively investigated. The RF voltage frequency must be increased as the trap size is reduced; this requires the device design to be impedance matched to the driving circuit. The operating electric field and geometry must also be engineered such that there is no gaseous breakdown. And while ion detectors such as the Faraday cup can operate at high pressure, a sensitive ion detector with amplification that operates at atmospheric pressure is needed. The goal of current research is to integrate the entire mass spectrometer on a chip that operates at high pressure with high resolution and sensitivity.

For background information *see* DOPPLER EFFECT; LASER COOLING; MASS SPECTROMETRY; MASS SPECTROSCOPE; MICROLITHOGRAPHY; PARTICLE TRAP; QUANTUM COMPUTATION in the McGraw-Hill Encyclopedia of Science & Technology.

Stanley Pau; Richart E. Slusher

Bibliography. E. R. Badman and R. G. Cooks, Special feature: Perspective miniature mass analyzers, *J. Mass Spectrometry*, 35:659–671, 2000; J. Chiaverini et al., Implementation of the semiclassical quantum Fourier transform in a scalable system, *Science*, 308:997–1000, 2005; J. Labaziewicz et al., Suppression of heating rates in cryogenic surface-electrode ion traps, *Phys. Rev. Lett.*, 100:013001, 2008; S. Pau et al., Microfabricated quadrupole ion trap for mass spectrometer applications, *Phys. Rev. Lett.*, 96:120801, 2006; R. Reichle et al., Experimental purification of two-atom entanglement, *Nature*, 443:838–841, 2006.

Mixed layers (oceanography)

The oceanic surface mixed layer is the layer of almost uniform density resulting from the interaction between stratifying and destratifying processes. The mixed layer extends from the surface of the ocean to the top of the pycnocline (see **illustration**). The pycnocline is the region where seawater density

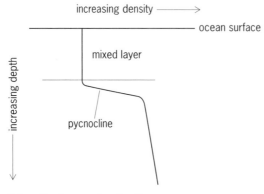

Schematic diagram of a mixed layer.

changes rapidly with depth. In the ocean, density is not directly measured but calculated using temperature, salinity, and pressure measurements from a variety of data collection platforms, such as ship-board surveys, moored instruments, and profiling floats. At shallow depths, as in the vicinity of the surface mixed layer, the pressure can be assumed to be constant, and density depends on temperature (inversely) and salinity (directly) through a very complex, nonlinear relationship which is quantified using the equation of state. Stratification results from an increase in temperature, a decrease in salinity, or both. Conversely, destratification results from a decrease in temperature, an increase in salinity, or both. Common processes that cause stratification include surface heating by the Sun and freshwater addition by precipitation (rain or snow) or by ice melting. Destratifying processes include surface cooling, increase of salinity due to evaporation or brine rejection during formation of ice, and mechanical mixing by winds, tides, and currents. Flow of water over a rough bottom can increase mixing.

Mixed layers and air-ocean interaction. The sea surface is in continuous interaction with the atmosphere above it. The ocean gains heat from the atmosphere primarily by absorption of short-wave radiation as well as by heat conduction and long-wave back radiation. It loses heat to the atmosphere by short-wave reflection, by turbulent transfer, by latent heat loss due to evaporation, and by long-wave radiation. Net absorption of short-wave solar radiation increases the temperature of the water. The increase is greatest in surface waters and diminishes rapidly with depth. This distribution is altered by convection, advection, and stirring by wind waves. In a stably stratified water column, lighter water overlies denser water. For mixing to take place in a stably stratified water column, denser water that is colder or saltier must be displaced upward against gravity and lighter water that is warmer or fresher is displaced downward against buoyancy forces. The energy required to accomplish this is reflected as a change in the potential energy of the water column. Thus mixed layer depths (MLDs) redistribute mass, momentum, and energy in the water column and participate in their exchange between the ocean and atmosphere.

Calculating mixed layers. There are various ways to quantify MLDs. The simplest way is to obtain a vertical profile of ocean density and choose the top of the pycnocline. The lack of density data in most cases makes this method unsuitable. Often, in low latitudes and away from coastal boundaries, density follows temperature, and temperature profiles are substituted for density profiles. In this case, the top of the thermocline (the region of rapid temperature change with depth) can be chosen to define the MLD. But this method does not work in high-latitude areas or in areas with heavy precipitation or freshwater discharge. In such areas (for example, fjords), density tends to follow salinity and the halocline (region of rapid salinity change) may be substituted for the pycnocline. In addition, sometimes profiles have multiple pycnoclines or thermoclines, which are relics of previous mixing events, and it is important to choose the correct one, according to the application. For example, if the deepest mixed layer of the year is desired, it is necessary to select the bottom-most pycnocline. Conversely, one may want to choose the most recent mixed layer, and to do this the shallowest pycnocline must be selected. MLDs can also be calculated using more complicated algorithms that are commonly available.

Variations in mixed layers. Mixed layers vary in time and space. The MLD can vary regionally from a few centimeters, as in a daytime, tropical, freshwater system, to thousands of meters in depth, as in the winter, in the Labrador Sea. MLDs also vary on different time scales. Ocean water density at any location is affected by both temperature and salinity. Therefore changes in temperature, salinity, or both will change the depth of the mixed layer. A diel (24-h) pattern of diurnal shoaling (becoming shallower) and nocturnal deepening follows the diel pattern of solar insolation. There is also an annual pattern of summer shoaling and winter deepening as the amount and length of daylight changes with the seasons. An annual pattern in wind velocity also plays a role in the annual pattern of MLDs. For example, in the northern Gulf of Alaska, severe winter storms help to cause deep winter mixed layers. In other cases, episodic storms, such as hurricanes and typhoons, may cause deviations from the annual patterns. Near coasts and in high latitudes, where there may be high freshwater discharge and melting ice, mixed layers tend to be shallower. Conversely, in areas and times of ice formation, mixed layers underlying the ice tend to be deeper due to brine rejection.

Mixed layers and primary productivity. Mixed layers are significant because they influence primary production in the ocean as well as climate. Sunlight is available at the surface of the ocean and it decreases exponentially with depth. Most of the primary production in the ocean occurs in the euphotic zone, or the thin layer of the surface ocean, where sunlight penetrates into the water column. Just as with plants on land, photosynthetic organisms in the ocean require both sunlight and nutrients for primary production. Nutrients are mainly available from the deep ocean by remineralization of organic matter that die and sink. Vertical mixing is an important process by which nutrients are made available to the euphotic zone. At low latitudes, where light is abundant throughout the year, deepening mixed layers can bring additional nutrients into the euphotic zone and increase biological production. At high latitudes, where light levels change seasonally and light angles are more oblique than at lower latitudes, deepening of the mixed layer can decrease new production since it mixes water and organisms into darker, light-limited zones. Thus at higher latitudes, mixed layers can increase production through nutrient enhancement or decrease production by removing primary producers from optimal light conditions. Shallow mixed layers in all systems may also decrease net production by allowing higher rates of predation. By controlling the productivity in the ocean, MLDs significantly affect the entire food web, including higher trophic levels such as fish, and marine mammals and birds.

Mixed layers, climate, and climate change. The upper surface of the mixed layer is also the atmosphere-ocean boundary layer. The mixed layer is actively involved in the connection between the atmosphere and the oceanic interior. Mixing modifies properties, such as sea-surface temperature and partial pressure of gases, which directly control the exchange of heat, moisture, and gases (such as carbon dioxide and water vapor, both of which are greenhouse gases) between the atmosphere and the ocean. Moreover, the high density and specific heat of water compared to air makes the ocean a greater reservoir of heat than the atmosphere (approximately 1000:1). Oceans mostly absorb heat in their mixed layers at lower latitudes and transport it to higher latitudes, where there is a heat deficit. For this reason mixed layers can play a significant role in climate and climate change.

In recent years, MLDs have played another role in controlling the Earth's climate. The oceans are natural sinks of anthropogenic or human-induced carbon dioxide. Primary production in the ocean, as on land, incorporates carbon into organisms. However, unlike on land, when these organisms die, the carbon sinks to the bottom of the ocean. It takes hundreds of years for the carbon to remineralize and be released back into the earth-ocean-atmosphere system. Thus, essentially, this process traps carbon and makes it unavailable for global warming. In recent years, scientists have tried to enhance this process in selected regions of the ocean by fertilization experiments that add nutrients, specifically iron, to increase primary production and "fix" more anthropogenic carbon to reduce greenhouse gases in the atmosphere.

According to the Intergovernmental Panel on Climate Change (IPCC) report, the current warming trend of the global ocean is $0.13°C$ ($0.23°F$) per decade. This is accompanied by greater precipitation in the mid and high latitudes and higher evaporation in the lower latitudes. Both the warming and the freshening trends in mid to high latitudes serve to increase stratification of the water column, thus reducing MLDs and consequently reducing nutrient re-plenishment into the euphotic zone. However, since mid and high latitudes may be light limited, this may increase primary production by increasing light levels, as mentioned earlier. In lower latitudes, the outcome is even less certain because the temperature trend is toward higher stratification, leading to shallower MLDs, while the salinity trend is toward lower stratification, which leads to deeper MLDs. It is possible that the temperature effects may dominate because of the nonlinearity of the effects of temperature and salinity on ocean density. In this case, the MLDs here would shoal too, reducing nutrient flux into the euphotic zone and decreasing primary productivity. Additional light levels from the shoaling MLDs may further decrease productivity by photoinhibition or enhanced predation.

Even though MLDs are so crucial to the understanding of climate, the weakest part of most ocean-atmosphere coupled models is the parameterization of the mixing. This is mainly because measuring turbulence is difficult with present instrumentation. Understanding, measuring, and parameterizing turbulence and mixing accurately is a major focus of current oceanographic research.

For background information *see* ATMOSPHERIC GENERAL CIRCULATION; CLIMATE MODIFICATION; INSOLATION; OCEAN; OCEAN CIRCULATION; OCEANOGRAPHY; SEAWATER; SEAWATER FERTILITY; TERRESTRIAL RADIATION in the McGraw-Hill Encyclopedia of Science & Technology.   Nandita Sarkar

Bibliography. L. H. Kantha and C. A. Clayson, *Small Scale Processes in Geophysical Fluid Flows*, International Geophysics Series, vol. 67, Academic Press, 2000; K. H. Mann and J. R. N. Lazier, *Dynamics of Marine Ecosystems*, 2d ed., Blackwell Science Inc., 1996.

Moisture in houses

Water is vital to our survival. Without water, we die of thirst, but too much water will drown us. It is not all that different with our homes. If the indoor air is too dry, we are uncomfortable in our homes, and our skin as well as the wood furniture begins to crack. However, too much humidity or water creates a host of different problems, including mold growth, wood decay, buckling wood floors, and rusting of metals. Thus, it is clear that moisture needs to be controlled, both for human health and for the health of the building.

Moisture problems. Mold fungi are abundant in nature, and are found inside and outside the home at any time. Mold growing on interior surfaces inside the home is unsightly and may produce an unpleasant odor. In addition, mold spores may cause respiratory problems and other allergic reactions. Hypersensitive persons or persons with an impaired immune system may have a very severe reaction to mold. Unlike decay fungi, mold generally does not destroy the material it is growing on. Mold needs air, a source of food, and moisture. The first two factors are nearly impossible to control, so moisture control is our principal weapon against excessive mold growth. Mold will grow on most surfaces when the relative humidity at that surface exceeds a critical value for a long enough time to allow the mold to establish itself. The longer that conditions are favorable for mold growth, the more likely it is to occur. These conditions vary with mold species, the nature of the material surface, and temperature. Xerophilic ("dry-loving") mold species can grow at surface relative humidities as low as 65%. However, for the species that are dominant in the indoor environment, more humid surface conditions are needed, somewhere above 80% relative humidity. The temperature at the surface also has to be conducive for mold growth, somewhere between 5°C (41°F) and 40°C (104°F).

Dust mites also can trigger allergies and asthma. As the name suggests, dust mites can be found in house dust and in places such as mattresses, carpets, and soft furniture. Dust mites thrive at high relative humidity (over 70%) at room temperature, but will not survive sustained relative humidities below 50%. As with mold, these relative humidities relate to local conditions in the carpet, mattress, or other material, which may be quite different from the relative humidity in the middle of the room.

If moisture accumulates between paint film and substrate, the paint may blister or peel. Water condensing on the back of siding may result in streaks and stains, as well as buckling or warping of wood-based siding. Excessive water in masonry may cause efflorescence (white powdery stains) or may cause spalling (chipping) in freezing weather.

In rare cases, structural damage may occur because of wood decay or corrosion of metal fasteners (see **illustration**). Usually water entry is the cause for such structural damage. Wood decay requires that the wood remain above 30% moisture content for a long enough period of time, and with temperatures somewhere between 5°C (41°F) and 40°C (104°F). When the wood dries out, the decay fungi go dormant, but will continue their destruction as soon as favorable conditions return.

Moisture movement in buildings. The goal of moisture control is to maintain moisture and humidity conditions within appropriate limits for the sake of human comfort and health, and for the survival of the home itself. Long-term accumulation of mois-

ture inside materials or building cavities should be avoided. To understand how effective control can be achieved, it is useful to recognize how moisture moves through buildings, through building components, and in and out of building materials.

Moisture can enter and move through a building as water or as water vapor. Most moisture problems are caused by liquid water entering the home in such amounts that it cannot be safely absorbed, evaporated, and removed. Water may come in through the foundation, through leaks in the roof or walls, or from leaks inside the home. Porous materials such as brick and concrete can wick up (absorb) moisture through capillary suction. Taking showers, bathing, dishwashing, and cooking also contribute moisture to the home, with lesser amounts coming from plants, people, and pets. The contributions from these sources are highly variable, and vary from household to household.

Water in the form of water vapor is carried by air flows or moves through materials by diffusion. Usually, air movement dominates the flow of water vapor. Thus, controlling air movement is the key to controlling the flow of water vapor. Vapor diffusion, which is driven by water vapor concentration differences rather than air pressure, is much slower and usually plays a much less important role.

Hygroscopic (moisture-retaining) materials such as wood, masonry, paper, and fabrics can safely store large amounts of water. The amount of moisture contained in these materials varies with the season and from region to region. They "adsorb" or "desorb" water vapor as a function of the relative humidity of the air in contact with the material: When the relative humidity drops, the material dries and contributes water vapor to the indoor air; when the relative humidity rises, the material adsorbs water vapor, removing it from the air. Interior hygroscopic materials therefore provide a moisture "buffer" that evens out large fluctuations in indoor humidity.

General moisture control strategies. Some typical moisture control strategies include rain water management, air and vapor control, and attic and crawl space venting.

Rainwater management. It is obvious that limiting water entry into the building should be the first priority. Moisture control begins at the design and construction stage. For instance, complicated roof designs sometimes lead to details that are difficult to flash (weatherproof) effectively. Insufficient elevation of the concrete foundation may make it impossible to provide proper site drainage. During construction, materials should be kept dry. If materials do get wet, they should be dried out before the building is fully closed in. This is best done with special drying equipment.

Here are some important water management strategies to consider:

1. Landscape so that water flows away from the building rather than into the building.

2. Provide gutters with working downspouts that deposit the water away from the foundation.

Structural wall damage caused by water penetration from a bathroom shower.

3. Install proper flashing around windows and around anything that attaches to and/or penetrates the exterior wall or roof.

4. Avoid roofs that drain into walls.

5. If possible, provide roof overhangs.

Air and vapor control. After water entry prevention has been taken care of, water vapor flow may need to be managed as well. This becomes especially important in cold climates (for example, Minnesota) or hot and humid climates (for example, the Gulf coast). Because air flow distributes water vapor far more efficiently than vapor diffusion, controlling air leakage through building components is the first priority. Airtight construction requires a continuous air barrier system that completely envelops the building enclosure. This means that there cannot be any major interruptions or holes in the air barrier anywhere. Simply installing an air barrier material on the exterior will not make the wall, or the home, airtight. The air barrier has to connect with air barriers in the other walls and ceilings, as well as with air seals around windows, doors, and any other penetration. In homes, much air leakage occurs through the top of the wood frame walls, so the top of the wall cavity needs to be sealed before the wall is enclosed, including partition walls. Building an airtight home requires much attention to detail, but trying to make the home airtight after it has been built is far more difficult to do. It should be noted that air barriers are currently not required in residential building codes and have not yet been widely adopted in residential housing. One reason is that many builders believe that airtight construction leads to moisture problems. In reality, airtight construction has rarely been associated with moisture problems, whereas air leaks have frequently been shown to lead to condensation and mold. Airtight houses do need a ventilation system to deliver fresh air to the interior and remove moisture from the house to avoid excessive indoor humidity in winter and poor indoor air quality. These systems can range in complexity from fresh air intakes connected to the return plenum of the furnace and air-conditioning system to separately ducted air-to-air heat exchanger systems. In humid climates, it is recommended that the ventilation air be dehumidified as well.

After attending to air tightness, there may still be a need for vapor diffusion control with vapor retarders (also called vapor barriers). In cold climates, an interior vapor retarder is usually required on the interior "warm side" of the walls and ceiling. There is a misconception that only polyethylene can be used for vapor retarders, but other materials such as some types of kraft paper (high-strength paper made from wood pulp) and some vapor barrier paints also qualify. Special "smart" vapor retarders have recently appeared on the market, whose water vapor transfer rate varies with relative humidity conditions. This provides for increased drying in case the wall or ceiling gets wet. In hot, humid climates, interior vapor retarders are not recommended because the vapor drive in these environments is almost always from the outside toward the inside. Especially low-permeance vapor retarders such as polyethylene or vinyl can pose condensation hazards in such climates. The same holds true for milder climates if the exterior walls have exterior brick. There is a misconception that vapor retarders need to be installed without gaps or holes. However, if an air barrier system is in place, continuity of the vapor retarder is not necessary.

Attic and crawl space venting. It has long been an established principle that attic ventilation is a critical part of moisture management in homes, and attic ventilation has been universally adopted in the residential building code. In addition, attic ventilation is credited with protecting the roofing shingles, providing energy savings during the cooling season, and preventing ice dams during winter. A review of the history and technical merits of attic ventilation has revealed that those code requirements do not have a sound technical basis. It is also obvious that attic ventilation can be counterproductive in warm, humid climates, especially when ducts and air-conditioning equipment are located in the attic. While attic vents are potentially effective in cold climates for moisture control and prevention of ice dams, there are more effective alternative strategies for achieving the same objectives. During the winter, indoor humidity control has consistently been found to be the most effective way to prevent condensation in the attic as well as condensation and mold elsewhere in the home. Air sealing and insulating the attic floor is the best way to prevent ice dams. Only in areas with extreme snowfall is attic venting necessary in order to prevent ice dams. In all other regions, it is possible to build a well-performing home with a roof that is not vented. Venting is recommended in cold climates, but should be considered optional and should be viewed as one of several moisture control strategies.

Similarly, venting is not always effective in lowering humidity in crawl spaces. If the outdoor air is warm and humid, bringing it into the cooler crawl space area will raise, not lower, the humidity there. Again, other strategies, such as a vapor barrier on the floor, have been shown to be more reliable in limiting humidity. Of course, ensuring through water management that rainwater does not accumulate under the house is the most important action.

For background information *see* AIR POLLUTION, INDOOR; ARCHITECTURAL ENGINEERING; BUILDINGS; FUNGI; HUMIDITY; MOISTURE-CONTENT MEASUREMENT; VAPOR PRESSURE; VENTILATION; WOOD PROPERTIES in the McGraw-Hill Encyclopedia of Science & Technology.   Anton TenWolde

Bibliography. ASHRAE, *Handbook of Fundamentals*, Chapters 23 and 24, American Society of Heating, Refrigerating, and Air-Conditioning Engineers, Atlanta, 2005; J. Lstiburek, *Builder's Guide to Cold Climates*, Building Science Press, Somerville, MA, 2006; J. Lstiburek, *Builder's Guide to Hot/Humid Climates*, Building Science Press, Somerville, MA, 2005; W. B. Rose, *Water in Buildings, An Architect's Guide to Moisture and Mold*, Wiley, New York, 2005; W. B. Rose and A. TenWolde, Venting of attics and cathedral ceilings, *ASHRAE J.*, 44(10):26–34, 2002.

Motor vehicle event data recorder

Every day, thousands of people are killed and injured on the roads. In 2006, in the United States alone, there were an estimated 5,973,000 police-reported traffic crashes, in which 42,642 people were killed and 2,575,000 people were injured, with 4,189,000 crashes involving property damage only. An average of 117 people died each day in motor vehicle crashes in 2006; that is, about 1 every 12 minutes. Motor vehicle crashes are the leading cause of major brain and spinal injuries and the leading cause of death for every age from 2 through 34. The personal, social, and economic costs of motor vehicle crashes include pain and suffering; direct costs sustained by the injured persons and their insurers; indirect costs to taxpayers for health care and public assistance; and for many victims, a lower standard of living and quality of life. During the past 20 years, motor vehicles accounted for over 90% of all transportation fatalities, and an even larger percentage of accidents and injuries. Our increasingly mobile society exposes all age groups to the risks of crashes, as passengers, drivers, and pedestrians. In contemporary society, automobiles play an indispensable role in transporting people and goods, and yet motor vehicle crashes remain a major public health problem. The health care cost of motor vehicle crashes is a national financial burden that must and can be reduced. Overall, the impact of motor vehicle crashes on a society is enormous. The U.S. National Highway Traffic Safety Administration (NHTSA) estimates the cost of crashes at well over $500 million a day, totaling $230 billion annually. These costs include property damage, medical care, insurance administration, emergency services, legal and court costs, travel delay, productivity losses, and costs to employers.

According to the European Transport Safety Council, the numbers in European Union (EU) countries are quite similar, where 43,000 die and over 3.5 million are injured annually, with motor vehicle injuries being the leading cause of death and hospital admission for citizens under the age of 45. Globally, vehicle crashes are the single largest cause of death of men between the ages of 15 and 44 and the ninth disease burden worldwide. Currently, worldwide, one person dies every minute of the day in a motor vehicle crash. Every day, 3000 people are killed in crashes on unsafe roads. Many of these deaths are preventable. The World Heath Organization (WHO) projects that by the year 2020, road traffic injuries will be the third highest disease burden in the world. There has yet to be a highway safety countermeasure that has resulted in significant reductions of deaths, injuries, and crashes.

Crash information. Worldwide, research and development is underway into systems that link highway infrastructure and telecommunications using emerging technologies via computers, electronics, and advanced sensing systems. Crash information is critical to understanding causes leading up to the crash, occupant kinematics and vehicle performance during a crash, and postcrash events. Manufacturers, engineers, policy makers, researchers, and others rely on crash information to improve vehicle design, shape regulatory policy, develop injury criteria, detect vehicle defects, and resolve investigations and litigation.

Motor vehicles have markedly transitioned from mechanical machines with mechanical controls to highly technological vehicles with integrated electronic systems and sensors (**Fig. 1**). Modern automobiles generate, use, and analyze electronic data to improve vehicle performance, safety, security, comfort, and emissions.

The NHTSA states in Federal Regulation 49 Code (CFR) 563: Event Data Recorders (EDR) that an EDR is a device or function in a vehicle that records the vehicle's dynamic time-series data during the time period just prior to a crash event (for example, vehicle speed versus time) or during a crash event (for example, total change in vehicle velocity [delta-V] versus time), intended for retrieval after the crash event. For the purpose of this definition, the event data do not include audio or video data. Thus, unlike a flight data recorder on an aircraft, an EDR does not record cabin conversations. It only records very brief data from the computer network and sensors in the vehicle.

Capture of a subset of vehicle data surrounding a crash on an EDR makes important information readily available for medical responders, crash investigators, and researchers. The degree of societal benefit from EDRs is related directly to the number of vehicles operating with an EDR and the ability to retrieve and use these data. Having standardized data definitions and formats allows the capture of vehicle crash information.

An EDR records and stores data, as measured by various sensors on an automobile, for a few seconds prior to and after a crash. If the forces are great enough, the EDR will record the data of near-crash events as well. The EDR is often a part of the occupant restraint (airbag and seatbelt tensioner)

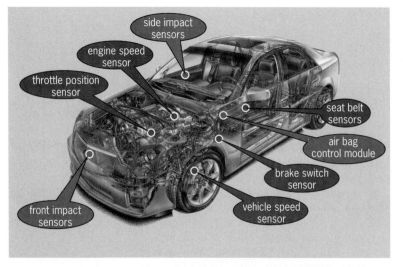

Fig. 1. Various sensors on an automobile that send information to the EDR, which is located on the airbag control module. (*Courtesy of Harris Technical Services*)

control module (**Fig. 2**), which is usually located in the passenger compartment (**Fig. 3**).

During operation of a vehicle, data from its various sensors are continuously recorded and overwritten by the EDR. In the event of a crash, typically involving airbag deployment, the NHTSA has specified 15 types of data to be recorded and stored in memory, including delta-V, vehicle speed, accelerator depression, braking, and seatbelt use. In the event of a crash, for example, the vehicle speed is stored for 5 s prior to the crash and the change in velocity after the crash, delta-V, is measured and stored for 0.25 s. All commercial EDRs have retrieval tools to recover crash data, typically possessed by vehicle manufacturers and law enforcement.

Data. Individual crash events and aggregate data have value for end users, depending on the application and data used.

Automotive industry. These include data-driven design of vehicles, using larger numbers of crashes across a continuum of severity; early evaluation of system and vehicle design performance; and international harmonization of safety standards.

Insurance industry. Data help to identify fraudulent claims, which cost more than $20 billion annually; improve risk management; expedite claims; and decrease administrative costs. Insurers require accurate crash data for subrogation of claims and recovery of expenses.

Government. Uses include standards development and revision, identifying problem injuries and mechanisms, stipulating injury criteria, and investigating defects. State and local officials require crash information to identify problem intersections and road lengths, to determine hazard countermeasures, and to evaluate the effectiveness of safety interventions.

Researchers. In human factors research, data are needed for studying the human–machine interface; crash causation, the effects of aging and medical conditions, and fatigue; and biomechanics research on human response to crashes, harmonized dummy development, and injury causation.

Medical providers. Crash data are useful for developing on-scene field triage of motor vehicle crash victims; improved diagnostic and therapeutic decisions; automatic notification of emergency providers; and better organization of trauma and EMS system resources.

Public. Data are used in policy decisions, vehicle design, emergency response, and roadway design.

Controversy. The implementation of EDRs has not been without controversy. The U.S. Department of Transportation (USDOT) docket management system contains over 1000 submissions. These reflect a decade-long debate among automakers, government regulators, safety and privacy advocates, and the public.

Issues such as the ownership of EDR data, how EDR data can be used/discovered in civil litigation, how EDR data may be used in criminal proceedings, whether EDR data may be obtained by the police without a warrant, whether EDR data may be developed into a driver-monitoring tool, and the nature

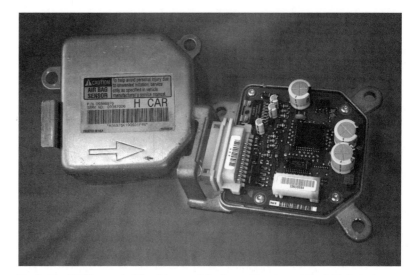

Fig. 2. The EDR is a part of the airbag module shown. (*Courtesy of Harris Technical Services*)

and extent that private parties will have or may contract for access to EDR data are generally within the realm of state law and are being addressed by state legislatures. State legislation varies but, generally speaking, the major issues are ownership, privacy, recoverability, access, use/misuse/mischief, corruption, credibility, disclosure, tampering, consumer protection, and consumer acceptance. Eleven states have enacted laws covering EDRs since 2004, including Arkansas Code 27-37-103, California Code 9950-9953, Colorado Statutes 12-6-4, Connecticut Public Act 07-235, Mane Statutes 29A-1-17-3, New Hampshire Statutes 357-G, New York Laws 4A16 416-B, Nevada Statutes 484.638, North Dakota Code 51-07-28, Oregon House Bill 2568, Texas Statutes 547.615, and Virginia Code 46.2-1088.6.

Consumer protection and acceptance. Privacy is the most important issue in terms of the success or failure of implementing EDRs in surface transportation. In the absence of federal regulation or state legislation, the consumer should have the ability to control access to collected crash data. This is important because data from automotive "black

Fig. 3. The EDR, contained within the airbag module, is bolted to the floor inside the vehicle. (*Courtesy of Harris Technical Services*)

boxes" increasingly have been used in court cases nationwide, including ones dealing with vehicular homicide in which speed was an issue. To achieve consumer protection, a vehicle connector lockout apparatus prevents data tampering. The motorist/owner simply plugs a mechanical lockout into the diagnostic connector port under the dash and turns the key lock to prevent data tampering. Then, as needed, the lockout can be unlocked for vehicle maintenance or inspection. This helps to assure consumer acceptance.

For background information *see* AUTOMOBILE; HIGHWAY ENGINEERING; TRAFFIC-CONTROL SYSTEMS; TRANSPORTATION ENGINEERING in the McGraw-Hill Encyclopedia of Science & Technology.

Thomas M. Kowalick

Bibliography.   CFR 49 563: Event Data Recorders, *Federal Register*, 73 (9), 2168–2184, and corrections 73 (30), 8408–8409, 2008; H. C. Gabler, J. Hinch, and J. Steiner, Event Data Recorders: A Decade of Innovation, SAE International, 2008; T. M. Kowalick, *Black Boxes: Event Data Recorder Rulemaking for Automobiles*, Micah Publications, 2006.

mTOR

Cell growth and proliferation are regulated by various intracellular and environmental cues, including nutrients, growth factors, energy, and stress. Recent studies have suggested that the mammalian target of rapamycin (mTOR) plays a key role in sensing and integrating these diverse signals, thus placing mTOR as a central regulator of cell growth and proliferation. Interest in mTOR is also spawned by clinical applications of mTOR inhibitors in transplantation, prevention of coronary restenosis (reconstruction or renarrowing of a coronary artery after it has been treated with angioplasty or stenting), and treatment of cancer.

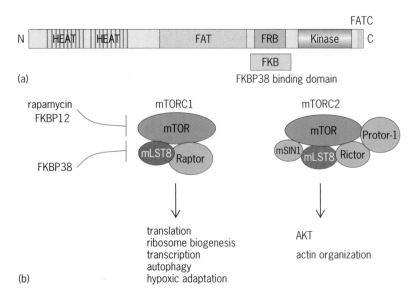

(a)

(b)

Fig. 1. Schematic presentation of (a) mTOR structure and (b) the mTOR complexes. See text for details.

Rapamycin. Rapamycin, the drug that targets mTOR, is a macrocyclic lactone antibiotic that was originally isolated as an antifungal agent from soil bacteria collected from Easter Island in the South Pacific. Although it was never developed into an antifungal drug, its fungicidal activity allowed the use of yeast as a model to study its action mechanism, which led to the identification of its intracellular target (TOR) in yeast and subsequently in mammals and other eukaryotes.

While rapamycin inhibits mTOR, it does not bind to mTOR directly. Upon entering cells, rapamycin forms a complex with a cytosol protein called FK506-binding protein 12 (FKBP12). The resulting FKBP12-rapamycin complex, in turn, binds specifically and irreversibly to mTOR and interferes with its function, presumably by blocking its access to substrates. In most eukaryotic cells, inhibition of mTOR by rapamycin results in growth arrest. However, the sensitivity of each type of cell to the drug varies. The cell type–dependent response to rapamycin permits applications of this drug for treating certain human diseases in which the disease-causing cells are more sensitive to the drug than other cells.

mTOR and the mTOR complexes. mTOR is a large protein with more than 2500 amino acids and a molecular weight of ~280 kDa. It belongs to the phosphoinositide-3-kinase-related kinase family (PIKK). Members of this family of kinases are characterized by their large sizes (2500–5000 amino acids in length) and possession of a kinase domain that resembles the catalytic subunit of phosphoinositide-3-kinase (PI-3K). Despite the sequence similarity with PI-3K, mTOR does not possess the ability to modify phospholipids. Instead, it phosphorylates serine (Ser) or threonine (Thr) residues in protein substrates, and hence functions as a serine/threonine protein kinase.

In addition to the kinase domain, mTOR contains several other regions that can be distinguished based on their sequence features and function (**Fig. 1a**). Immediately N-terminal to the kinase domain is the region where the FKBP12-rapamycin complex binds, commonly referred to as the FRB domain. Certain mutations within this domain disrupt the binding of mTOR with the FKBP12-rapamycin complex, rendering it resistant to the drug. The N-terminal half of the protein contains as many as 20 tandem sequence repeats called the HEAT motifs. These motifs comprise 39–43 imperfectly matched amino acids and are involved in mediating the association of mTOR with other proteins. In addition, mTOR also possesses a FAT domain located in its middle region and a FATC at the C-terminus. These domains are commonly found in members of the PIKK family. However, their roles in mTOR function are not clear.

mTOR elicits its function in the context of two structurally and functionally distinct complexes, termed mTOR complex 1 (mTORC1) and complex 2 (mTORC2). The two complexes share mTOR and an mTOR-interacting protein termed mLST8 (also called GβL). While mTORC1 contains a single unique component, Raptor, mTORC2 has several, including

Rictor, mSIN1, and Protor-1 (Fig. 1*b*). Despite the presence of mTOR in both complexes, the FKBP12-rapamycin complex binds only to mTORC1. It is believed that the presence of the unique components in mTORC2 shields the FRB domain and prevents the FKBP12-rapamycin complex from binding to mTOR. However, prolonged presence of rapamycin is able to affect mTORC2 function by binding to newly synthesized mTOR before its assembly into the complex.

Functions of the mTOR complexes. Rapamycin has been widely used as a chemical probe to study the function of mTORC1 in mammalian cells. Many growth-related processes, including translation, transcription, autophagy (the cellular process of self-digestion), and hypoxic (oxygen-deficient) adaptation, are found to be sensitive to the drug. Among these mTORC1-dependent processes, translation is the one that receives the most attention, owing to the identification of two direct targets of mTORC1 in this process, including the eukaryotic initiation factor 4E (eIF-4E) binding protein 1 (4E-BP1) and S6 kinase (S6K). 4E-BP1 is a translation repressor that binds to and inhibits eIF-4E, a translation initiation factor that recognizes the cap structure at the 5' end of most messenger ribonucleic acids (mRNAs). Phosphorylation of 4E-BP1 by mTORC1 prevents the binding and releases eIF-4E for translation initiation. S6K is responsible for phosphorylation of S6 ribosomal protein, which is required for translation of mRNA containing tracks of polypyrimidine (5'-TOP) in their 5' sequence. A majority of these transcripts encode ribosomal proteins that are required for protein synthesis in cells. mTORC1 phosphorylates S6K at Thr389, located in the activation loop of the kinase, leading to its activation. Both S6K and 4E-BP1 contain a sequence motif termed TOR signaling motif (TOS) that is recognized by Raptor, which functions as an adaptor protein to recruit substrates to mTOR. mTORC1 has been implicated in many other cellular processes, although the underlying mechanisms remain elusive.

In comparison with mTORC1, the function of mTORC2 has not been well characterized. The major known function ascribed to mTORC2 is control of the oncogenic kinase, AKT (also known as protein kinase B). mTORC2 has been found to phosphorylate AKT at Ser473, a site in a C-terminal hydrophobic region outside the catalytic domain. Phosphorylation at this site is not required for AKT activity, but is important for maximal activation of the kinase in response to mitogenic stimuli. It may also be involved in targeting AKT to different substrates. In addition, mTORC2 is involved in control of actin organization and cell morphology through several members of the Rho family of small guanine triphosphatases (GTPases), including Rho, Rac, and Cdc42.

Regulation of the mTOR complexes. The proximate regulator of mTORC1 (**Fig. 2**) is its endogenous inhibitor FKBP38, a protein that is structurally related to FKBP12. However, unlike FKBP12, which binds and inhibits mTORC1 only in the presence of rapamycin, FKBP38 is capable of binding directly

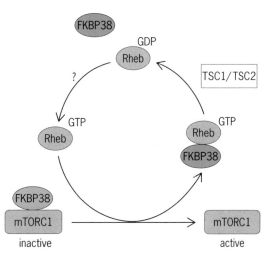

Fig. 2. Regulation of mTORC1 by Rheb GTPase. Rheb activates mTORC1 by preventing FKBP38 from binding to mTOR. In its GTP-bound form, Rheb interacts with FKBP38 and releases mTORC1 for activation. Following hydrolysis of the bound GTP to GDP with the facilitation of the TSC1/TSC2 complex, Rheb loses the ability to interact with FKBP38 and allows it to inhibit mTOR. Rheb is reactivated by a putative nucleotide exchange factor that catalyzes the exchange of GDP for GTP.

to mTOR and inhibiting mTORC1 activity in the absence of the drug. FKBP38 binds to a region in mTOR embracing the FRB domain (Fig. 1*a*). Therefore, its effect on mTOR function is expected to be similar to that of the FKBP12-rapamycin complex. The binding of FKBP38 with mTORC1 is regulated by Rheb, a small GTPase that is structurally related to Ras. Like Ras and other small GTPases, the activity of Rheb is dictated by its guanine nucleotide binding states. It is active when bound to GTP, but becomes inactive upon hydrolysis of guanosine triphosphate (GTP) to guanosine diphosphate (GDP). In the GTP-bound state, Rheb interacts directly with FKBP38 and prevents its association with mTORC1, hence releasing the complex for activation. The rate of GTP hydrolysis of Rheb is controlled by a GTPase activating protein (GAP) complex formed by two tumor suppressors, harmatin (TSC1) and tuberin (TSC2). The TSC1/TSC2 complex promotes hydrolysis of GTP to GDP by stimulating the intrinsic GTPase activity of Rheb. In doing so, it keeps Rheb in the GDP-bound inactive state, and consequently impedes mTORC1 activity.

The TSC1/TSC2 complex is regulated by an intricate signaling network that receives and propagates diverse signals (**Fig. 3**). A common mechanism for the regulation involves altering the formation or function of the complex through phosphorylation of TSC2. Several kinases have been shown to carry out this function, among which are AKT and an adenosine monophosphate (AMP)-activated protein kinase, AMPK. AKT is a key component in the PI-3K pathway. This pathway is activated by receptor tyrosine kinases (RTK) residing on the cell surface in response to growth factor stimulation. Once activated, PI-3K catalyzes the conversion of phosphatidylinositol-4,5-bisphosphate (PIP$_2$) to phosphatidylinositol-3,4,5-triphosphate (PIP$_3$), a

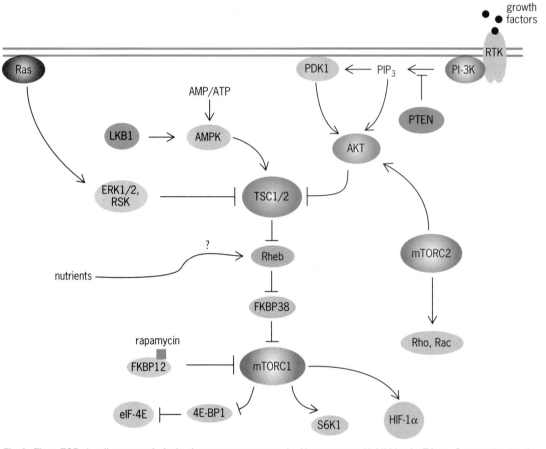

Fig. 3. The mTOR signaling network. Activation processes are marked by arrows and inhibition by T bars. See text for details.

bioactive messenger. The elevated PIP_3 levels in the plasma membrane recruit PDK1 and AKT, two serine/threonine protein kinases containing a pleckstrin homology (PH) domain that binds PIP_3. (PH domains are unique modules of around 100 amino acids found in several proteins involved in signal transduction; they were first detected in pleckstrin, a cytoskeletal protein.) AKT is subsequently phosphorylated by PDK1 and mTORC2, respectively, at residues Thr308 and Ser473, leading to its full activation. Phosphorylation of TSC2 by AKT disrupts the TSC1/TSC2 complex, thus reducing its activity toward Rheb.

AMPK is activated in response to low energy levels when the intracellular ratio of adenosine monophosphate/adenosine triphosphate (AMP/ATP) rises. Phosphorylation of TSC2 by AMPK leads to an increase in the GAP activity of the TSC1/TSC2 complex, and consequently downregulation of mTORC1. This mechanism couples energy signals to mTORC1 and causes its activity to fall when intracellular energy is limited. In addition, extracellular signal-regulated kinases (ERK1 and ERK2) and ribosomal S6 kinase (RSK) have been shown to phosphorylate and inhibit TSC2. These kinases are activated by Ras in response to mitogen stimulation, thus relaying mitogenic signals to mTORC1. Therefore, by serving as a target for multiple kinase pathways, the TSC1/TSC2 complex acts as a signaling hub to channel signals from different sources to mTORC1. However, how

this complex integrates both positive and negative signals from multiple pathways is still a mystery.

mTORC1 activity is also regulated by mechanisms independent of the TSC1/TSC2 complex. An endosome-located PI-3K has been shown to activate mTORC1 in response to amino acid availability. Phosphatidic acid (PA), another bioactive messenger, is able to bind to mTOR and enhance mTORC1 activity. In addition, Rheb and mTORC1 itself may be regulated directly by amino acid conditions. These diverse mechanisms highlight the complexity in mTORC1 regulation and are consistent with the role of mTORC1 as a sensor and integrator for signals of different origins. In contrast, mechanisms that regulate mTORC2 are largely unexplored.

mTOR and cancer. Several upstream regulators of mTORC1 are tumor suppressors, including PTEN, LKB1 (also named STK11), TSC1, and TSC2. Inactivation of these suppressors is often associated with tumorigenesis. PTEN, a phosphoinositide phosphatase, counteracts the activity of PI-3K by converting PIP_3 back to PIP_2. Its inactivation results in accumulation of PIP_3—hence enhancement in activity of the PI-3K–AKT pathway. Loss of PTEN has been found in vast numbers of cancers. LKB1 is a Ser/Thr kinase that phosphorylates and stimulates AMPK activity. Inactivating mutations in LKB1 are associated with Peutz-Jeghers syndrome, which is characterized by benign tumors (hamartomas) in the intestine. On the other hand, inactivating mutations in the TSC1 and TSC2 genes give rise to tuberous sclerosis complex syndrome, which is manifested by the occurrence of hamartomas in brain, heart, kidney, and skin. All these tumor suppressors are negative regulators of mTORC1, and their inactivation causes hyperactivation of mTORC1, which contributes to tumor formation. Two major downstream targets of mTORC1, including HIF-1α and eIF-4E, are believed to play critical roles in the process. HIF-1α is a transcription factor that is required for expression of many genes involved in tumor angiogenesis. eIF-4E enhances the production of several growth-promoting factors, such as c-Myc and cyclin D1. Activation of these two factors, as a result of mTORC1 hyperactivation, drives tumor growth and angiogenesis, and hence promotes tumorigenesis.

mTOR inhibitors as therapeutic agents. Given the central role of mTOR in cell growth and proliferation, it is not surprising that the specific mTOR inhibitor rapamycin has found many clinical applications. Rapamycin is currently used as an immunosuppressive drug for preventing acute rejection in renal transplantation and as a stent-coating drug to block coronary artery restenosis in balloon angioplasty. These clinical applications are based on the ability of the drug to inhibit the proliferation of T lymphocytes and vascular smooth muscle cells, respectively.

Several rapamycin derivatives (rapalogs), including temsirolimus, everolimus, and AP23573, have been developed as anticancer agents and are currently in clinical trials. Temsirolimus has been recently approved by the Food and Drug Administration for treatment of patients with advanced renal cell carcinoma (RCC). However, not all types of tumors are sensitive to rapalogs. The mechanisms underlying rapalog sensitivity are poorly understood. While abnormal signaling activity in mTORC1 and its upstream regulators, such as PI-3K and AKT, may predispose cancer cells to rapalogs, many other factors may modify or alter the sensitivity. As a result, accurate prediction of tumor responsiveness to rapalogs remains a challenge at the current time. However, lessons learned from both basic research and clinical use of these types of drugs will provide insights to guide better strategies for their use in cancer therapies.

For background information *see* AMINO ACIDS; CANCER (MEDICINE); CELL (BIOLOGY); CELL DIVISION; ENZYME; ONCOLOGY; PROTEIN; PROTEIN KINASE; SECOND MESSENGERS; SIGNAL TRANSDUCTION in the McGraw-Hill Encyclopedia of Science & Technology.
Yu Jiang

Bibliography. X. Bai et al., Rheb activates mTOR by antagonizing its endogenous inhibitor, FKBP38, *Science*, 318:977–980, 2007; G. G. Chiang and R. T. Abraham, Targeting the mTOR signaling network in cancer, *Trends Mol. Med.*, 13:432–442, 2007; J. B. Easton and P. J. Houghton, mTOR and cancer therapy, *Oncogene*, 25:6436–6446, 2006; S. Faivre, G. Kroemer, and E. Raymond, Current development of mTOR inhibitors as anticancer agents, *Nat. Rev. Drug Discov.*, 5:671–688, 2006; K. Inoki and K-L. Guan, Complexity of the TOR signaling network, *Trends Cell Biol.*, 16:206–212, 2006; S. Wullschleger, R. Loewith, and M. N. Hall, TOR signaling in growth and metabolism, *Cell*, 124:471–484, 2006.

Multiferroics

Our technological world is built on functional materials. Such materials are essential to the operation of devices ranging from cell phones to laptops and beyond, which have long utilized many different types of functional materials, including magnets, ferroelectrics, and more. Today, however, the focus of researchers is on the future—on multifunctional and smart materials that will enable the next generation of computing, memory, logic, and more. One example of such materials is the multiferroics—materials that possess two or more of the properties ferromagnetism, ferroelectricity, and ferroelasticity in a single phase. The presence of multiple types of order (that is, ferromagnetic, ferroelectric, and so forth) means that multiferroics can have a spontaneous magnetization that can be changed by an applied magnetic field, a spontaneous polarization that can be changed by an applied electric field, and a spontaneous deformation that can be changed by an applied stress. Furthermore, such multiferroic materials can have cross-coupling between properties, thus giving rise to the ability to tune magnetic order with electric fields, polarization with magnetic fields, and so forth. It is the combination of these intriguing properties that has made multiferroics the focus of much research over the past decade and has poised them to make a great impact on the technological world.

Order in multiferroics. Fundamental to multiferroics are the properties that make these materials of great interest. The first type of order that is important in multiferroics is ferroelectric order. A material is said to be ferroelectric when it has a spontaneous electric polarization, has two or more orientational states in the absence of an electric field, and can be shifted from one to another of these states by an electric field. Any two of the orientational states are identical in crystal structure and differ only in electric polarization at zero applied field. Ferroelectric materials are invariant under time-reversal symmetry, but must violate spatial inversion symmetry. (A physical system is said to have time-reversal symmetry when one can replace time *t* in the physical equations representing it with $-t$ and obtain the same answer. To violate spatial inversion symmetry, a material must not possess a center of symmetry. A center of symmetry is a point in space such that, if a line is drawn from any point or atom of the unit cell to the center of symmetry and extended an equal distance beyond it, an analogous point or atom will be encountered.) Ferroelectrics are materials that undergo a phase transition from a high-temperature phase that behaves as an ordinary dielectric to a low-temperature phase that has a spontaneous polarization whose direction can be switched by an applied electric field. Any lattice of oppositely signed atomic species is inherently unstable and relies on short-range interactions between adjacent electron clouds in the material to stabilize the structure. In ferroelectric materials, these interactions result in the formation of a double-well energy potential that stabilizes a distorted (noncentrosymmetric) structure in preference to the centrosymmetric, nondistorted structure.

In the case of classic perovskite (materials with the chemical formula ABO_3) ferroelectrics like lead titanate ($PbTiO_3$) and barium titanate ($BaTiO_3$), the Ti $3d$–O $2p$ orbital hybridization is essential for stabilizing the ferroelectric distortion. (Orbital hybridization is the concept of mixing atomic orbitals to form new hybrid orbitals suitable for the qualitative description of the atomic bonding properties of a crystal.) It has also been found that most perovskite ferroelectrics have B-site ions that are formally d^0 in nature. Thus, the lowest unoccupied energy levels are the d states, and they tend to hybridize with the O $2p$ orbitals, resulting in the double-well potential.

A material is said to be ferroelastic when it has two or more orientation states in the absence of mechanical stress (and electric field) and can be shifted from one to another of these states by mechanical stress. It is imperative that two of the orientational states are identical or enantiomorphous in crystal structure and different in mechanical strain tensor at null mechanical stress (and electric field).

Magnetic materials, on the other hand, must violate time-reversal symmetry (meaning that when one replaces time t in the physical equations representing magnetic order with $-t$, one will not obtain the same answer), but are invariant under spatial inversion; in other words, when magnetic moments are present in a crystal, the antisymmetry operator may also be present. The antisymmetry operator changes the sign of the object on which it operates, thereby creating a much more complex set of symmetry conditions for magnetic materials. A material is said to be a ferromagnet when there is long-range, parallel alignment of the atomic moments, resulting in a spontaneous net magnetization even in the absence of an external field. Ferromagnetic materials undergo a phase transition from a high-temperature phase that does not have a macroscopic magnetization (wherein atomic moments are randomly aligned, resulting in a paramagnetic phase) to a low-temperature phase that does. There are other types of magnetism, including antiferromagnetism (wherein atomic moments are aligned antiparallel) and ferrimagnetism (wherein dipoles are aligned antiparallel, but one subset of dipoles is larger than the other, resulting in a net moment). The theory of magnetism is a rich field and beyond the scope of this treatment, but is explained by the existence of the quantum-mechanical exchange energy, which causes electrons with parallel spins and therefore parallel moments to have lower energy than electrons with antiparallel spins. Inherent to this concept is the presence of unpaired electrons in a material. Thus, a requirement for magnetism in transition metals is a partially filled d orbital.

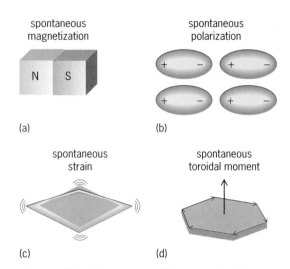

Fig. 1. Multiferroics possess multiple types of ordering, including (a) magnetism, (b) ferroelectricity, (c) ferroelasticity, and (d) ferrotoroidicity, and often have strong coupling between the order parameters.

Multiferroism and magnetoelectricity. Multiferroism describes materials in which two or all three of the properties ferromagnetism, ferroelectricity, and ferroelasticity occur in the same phase (**Fig. 1**). Additionally, the current trend is to extend the definition of multiferroics to include materials possessing two or more of the ferroic or corresponding antiferroic properties such as antiferroelectricity and antiferromagnetism. Recently, the idea of ferrotoroidic materials has been added into consideration (Fig. 1d). These are materials that possess a stable and spontaneous order that is taken to be the curl of the magnetization or polarization. (The curl of a vector field is the amount of rotation or angular momentum of the contents of a given region of space.)

Only small subgroups of all magnetically and electrically polarizable materials are either ferromagnetic or ferroelectric, and fewer still exhibit both properties simultaneously. In these select materials where multiple properties exist, however, not only is there the possibility that electric fields can reorient the polarization and magnetic fields the magnetization, but there is a possibility of coupling an electric field to magnetization and a magnetic field to electric polarization. This functionality offers an extra degree of freedom in these materials, and hence they are referred to as magnetoelectric materials. Magnetoelectricity is an independent phenomenon that can arise in any material with both magnetic and electronic polarizability, regardless of whether it is a multiferroic or not. Creating such materials, however, is rather difficult. Investigating a number of factors including symmetry, electronic properties, and chemistry leads to an understanding of the scarcity of magnetoelectric multiferroics: this scarcity is a result of the seeming contradiction between the conventional mechanism for the formation of a spontaneous polarization in ferroelectrics (which requires an empty d orbital) and the formation of magnetic order in materials (which results from partially filled d orbitals). The focus of many researchers, in

Pathways to multiferroism in materials

Pathway to multiferroism	Mechanism for multiferroism	Examples
A-site driven	Stereochemical activity of A-site lone pair gives rise to ferroelectricity, and magnetism arises from B-site cation.	$BiFeO_3$, $BiMnO_3$
Geometrically driven	Long-range dipole-dipole interactions and oxygen rotations drive the system toward a ferroelectric state.	$YMnO_3$, $BaNiF_4$
Charge ordering	Noncentrosymmetric charge ordering arrangements result in ferroelectricity in magnetic materials.	$LuFe_2O_4$
Magnetic ordering	Ferroelectricity is induced by the formation of a symmetry-lowering magnetic ground state that lacks inversion symmetry.	$TbMnO_3$, $DyMnO_3$, $TbMn_2O_4$

turn, has been on designing and identifying new multiferroic materials. A number of pathways to create single-phase thin-film and bulk multiferroics have been identified (see **table**). One of the major challenges to the field of multiferroics is the fact that many phases possess only one strong type of ordering (for instance, a phase that has strong ferroelectricity and weak magnetism—too weak to be useful in applications). This is the result of the fact that the mechanism driving the formation of the secondary order is typically very unstable in nature; however, this does not preclude strong coupling between those properties. Additionally, many multiferroics possess low ordering temperatures [that is, less than 150 K ($-123°C$ or $-190°F$)], and thus, to date, no room-temperature single-phase, ferromagnetic-ferroelectric multiferroic has been identified.

Designing multiferroics and magnetoelectrics. A recent resurgence of research on multiferroics has been aided by the advent of new growth and characterization techniques as well as theoretical approaches to materials discovery. Work on the growth of these materials has accelerated as a consequence of better thin-film deposition techniques, including pulsed-laser deposition, sputtering, metal-organic chemical vapor deposition, and molecular beam epitaxy. Recent advances in the in situ characterization of thin-film growth have also enabled researchers to fabricate extremely high quality heterostructures of these complicated materials in hopes of artificially creating novel multiferroic phases. These advances in growth have, in turn, been combined with new synchrotron-based, optical, and scanning-probe characterization techniques that better enable researchers to characterize the multiple properties of these materials. Finally, this work continues in parallel with first-principles theoretical modeling of these systems, which has led to predictions of new metastable multiferroic phases as well as interface phenomena in artificial heterostructures.

State-of-the-art multiferroics. One area that is receiving much interest today is the pursuit of room-temperature functionality with multiferroics. A number of devices that would take advantage of multiferroic materials have been proposed, including devices designed to enable electric-field control of ferromagnetism. As noted above, no room-temperature single-phase, ferromagnetic-ferroelectric multiferroic has been discovered so far. Nonetheless, researchers have identified a number of intriguing

pathways to achieve such room-temperature functionality. Much of this work has been built on the ability to create novel heterostructures of functional materials. One example is composite nanostructures of magnetic nanopillars embedded in a ferroelectric matrix (**Fig. 2**a). Such nanostructures utilize an interface-mediated coupling resulting from the three-dimensional heteroepitaxial nature of these structures to couple ferroelectricity and ferromagnetism at room temperature. These structures have been shown to possess very large magnetoelectric coupling coefficients and offer an exciting possibility for a novel nonvolatile electrically written magnetic storage. *See* NONVOLATILE MEMORY DEVICES.

Another manifestation of exciting multiferroics research is the creation of layered heterostructures (Fig. 2b). The idea is to combine traditional functional materials—such as ferromagnets—with multiferroic materials to create new functionalities at room temperature. One example of this has been demonstrated using the multiferroic bismuth ferrite ($BiFeO_3$)—which is known to be both ferroelectric and antiferromagnetic at room temperature—where

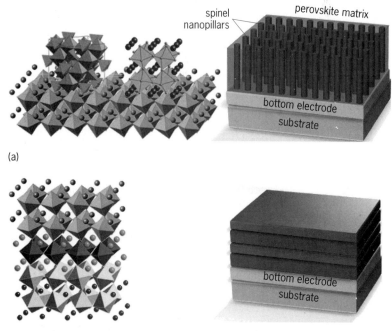

(a)

(b)

Fig. 2. Atomic (left) and macroscopic (right) structures of model multiferroic heterostructures. (a) Three-dimensional nanostructure of magnetic nanopillars embedded in a ferroelectric matrix. (b) Layered heterostructure.

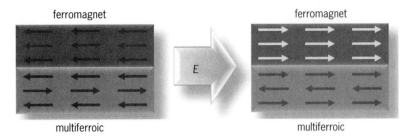

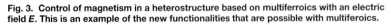

Fig. 3. Control of magnetism in a heterostructure based on multiferroics with an electric field *E*. This is an example of the new functionalities that are possible with multiferoics.

researchers have devised a way to control ferromagnetism with an electric field at room temperature. Utilizing the magnetoelectric coupling between the ferroelectricity and antiferromagnetism in a $BiFeO_3$ thin film (100–200 nm thick), electric fields can be used to control antiferromagnetism. This antiferromagnetic order is, in turn, coupled via an exchange coupling interaction to a thin ferromagnetic layer (2.5–10 nm thick) on top of the $BiFeO_3$ (**Fig. 3**). [For an antiferromagnet and a ferromagnet that are in contact, exchange coupling refers to the interaction between magnetic moments in the antiferromagnet and the ferromagnet that can give rise to new and different properties for the ferromagnetic material.] Thus, upon application of an electric field, there is a corresponding change in the nature of antiferromagnetism and, in turn, a change in the ferromagnetic material as well (Fig. 3).

Multiferroics are poised to have an impact on the world by enabling new functionalities that so far have been unachievable. Such materials could make possible a new generation of memories, logic, computers, and sensors that utilize electric-field control of ferromagnetism. Researchers continue to search for new materials, taking advantage of the significant advances in techniques of thin-film and single-crystal growth over the past decade, and at the same time have engineered new types of materials physics by combining these intriguing materials into novel structures and with traditional functional materials such as ferromagnets. In the end, the future for multiferroics will rely on continued fundamental research that will enable a better understanding of the fundamental physics underlying the coupling in these materials, and continued attempts to integrate multiferroics into device structures that utilize the novel functionalities made possible by these intriguing materials.

For background information *see* ANTIFERROMAGNETISM; CALCULUS OF VECTORS; CRYSTAL GROWTH; EXCHANGE INTERACTION; FERRIMAGNETISM; FERROELECTRICS; FERROMAGNETISM; MAGNETIC SUSCEPTIBILITY; NANOSTRUCTURE; PEROVSKITE; PHASE TRANSITIONS; POLARIZATION OF DIELECTRICS; SEMICONDUCTOR HETEROSTRUCTURES; SYNCHROTRON RADIATION in the McGraw-Hill Encyclopedia of Science & Technology.   Lane W. Martin; Ramamoorthy Ramesh

Bibliography. M. Fiebig, Revival of the magnetoelectric effect, *J. Phys. D*, 38:R123152, 2005; N. A. Hill, Why are there so few magnetic ferro-

electrics, *J. Phys. Chem. B*, 104:6694–6709, 2000; M. E. Lines and A. M. Glass, *Principles and Applications of Ferroelectrics and Related Materials*, Clarendon Press, Oxford, 1977, reprint 2001; R. C. O'Handley, *Modern Magnetic Materials: Principles and Applications,* Wiley, New York, 2000; R. Ramesh and N. A. Spaldin, Multiferroics: progress and prospects in thin films, *Nat. Mater.*, 6:21–29, 2007.

Multiple-proton radioactivity

The frontiers of our knowledge of the nuclear world advance continuously following progress in experimental techniques. The quest to reach very unstable nuclei at both the neutron-rich and the neutron-deficient limits of the chart of nuclei is motivated by expectations that the structure of an atomic nucleus changes when the imbalance between the number of neutrons and the number of protons increases. New features and new phenomena are expected. These generally pose a challenge to theoretical models that were developed to describe, and thoroughly tested with, nuclei in the vicinity of stable ones.

Proton emission. The atomic nucleus, or the nuclide, is a system composed of protons and neutrons (nucleons), held together by strong nuclear forces. Such a system is stable when, for a given number of protons, a specific number of neutrons is present. When a system is formed away from a stable configuration, radioactive decays will change it until stability is reached. For example, when a nucleus is neutron-deficient, a beta positron (β^+) decay may occur, in which a proton transforms into a neutron (which is accompanied by the emission of a positron and an electron neutrino). Radioactive processes, such as beta (β) decays, alpha (α) emission, or fission, are generally slow enough to consider such an unstable system as a well-defined nuclear object whose properties can be measured. When a nuclide is sufficiently far from stability, however, a limit is encountered beyond which nucleons are no longer bound by nuclear forces. In a nuclide beyond the neutron-deficient edge of stability (the proton drip line), protons are not bound and thus may be ejected. This process—proton radioactivity—does not always occur immediately, in contrast to emission of a neutron from a system beyond the neutron drip line, because the Coulomb barrier hampers the emission of charged particles. To escape, the unbound proton has to undergo quantum tunneling through this barrier (an analogous phenomenon occurs in alpha decay). The probability of tunneling depends strongly on the thickness of the barrier, which in turn depends on the energy of the unbound proton and on its angular momentum. Thus, measurement of the proton energy and half-life provides important information on the initial quantum state, which can be compared with predictions of theoretical nuclear models. This method proved to be a very fruitful tool of nuclear spectroscopy, yielding a wealth of data on the nuclear structure at the edge of nuclear stability.

Proton emission may also occur when a nucleus is excited to an energy higher than the proton binding energy. Such a situation happens commonly in beta positron decays of nuclei far from stability. The decay may proceed to highly excited, unbound states of the daughter nucleus from which protons are emitted. Since the proton emission is generally much faster than the beta decay, such a process is called beta-delayed (β-delayed) proton emission. The first proton emission was observed from the long-lived excited state of cobalt-53 (^{53}Co) in 1970. The first proton radioactivity from the ground state was observed in 1982 for lutetium-151 (^{151}Lu) and for thulium-147 (^{147}Tm).

Two-proton emission. Two protons can also be emitted from an excited nuclear state. Such a process was discovered in 1983 when two protons were detected following the beta decay of aluminum-22 (^{22}Al). Later, many other two-proton emissions were observed from states populated in beta decay or in nuclear reactions. For all these states, however, the single-proton emission channel was open, and in almost all cases two-proton emission could be described as a sequence of two single-proton emissions. There was an intriguing possibility, predicted by Vitaly Goldansky in 1960, that a nucleus with an even number of protons (an even-Z nucleus) in the ground state may be unstable against two-proton emission while being stable against one-proton radioactivity. The latter would require extra energy to break the pairing interaction between protons and thus cannot proceed spontaneously. In contrast, the unbound pair of protons can be ejected. Both particles have to come out simultaneously, which makes it a three-body process. Such two-proton radioactivity (2p) provides more observables than single-proton radioactivity; in addition to the decay energy and the half-life, the angular and energy correlations between emitted protons contain information on the state of the initial nucleus, on the interactions between protons, and on the mechanism of the decay.

Ground-state two-proton radioactivity was discovered in 2002 in the decay of iron-45 (^{45}Fe). Later, two-proton decay was also identified in zinc-54 (^{54}Zn). These extremely neutron-deficient nuclei are very difficult to synthesize. Presently, the only practical method of doing so is based on the fragmentation reaction of a high-energy heavy-ion beam impinging on a relatively thin production target. Ions of interest emerging from the target with high energy are purified by a magnetic separator and are unambiguously identified in flight by means of transmission detectors (thin detectors which the ions pass through), ion by ion. The single-ion sensitivity is crucial, for the production rates are very small; first results were deduced from the statistics of about 10 ions. Selected and identified ions were implanted into a silicon detector where only the decay time and the decay energy were measured. The interpretation relied on comparison with model predictions; no hypothesis other than two-proton decay could explain the data.

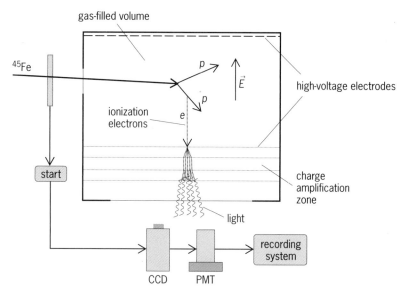

Fig. 1. Schematic, simplified view of the optical time-projection chamber. Primary ionization electrons (e) drift in a uniform electric field ($\vec{E}$).

Optical time-projection chamber. To meet the challenge of recording both protons separately and to establish correlations between them, a different method of detection must be used. A novel technique, developed at the University of Warsaw, Poland, combines a gaseous time-projection chamber with optical digital imaging. The resulting optical time-projection chamber (OTPC) allows the tracks of charged particles to be recorded. The selected ions and their charged decay products are stopped in a volume filled with a counting gas (**Fig. 1**). The primary ionization electrons drift in a uniform electric field, with a velocity of about 1 cm/μs, toward a double-stage amplification structure formed by parallel-mesh flat electrodes. In the second multiplication stage, emission of ultraviolet photons occurs. After conversion of their wavelength to the visual range by a thin luminescent foil, these photons are recorded by a digital charge-coupled device (CCD) camera and by a photomultiplier tube (PMT). The camera image represents the projection of particles' tracks on the luminescent foil. The signals from the PMT are digitized with a 50-MHz sampling frequency, providing information on the drift time, which is related to the position along the axis perpendicular to the image plane. The recording sequence of the OTPC, lasting about 25 ms, is selectively triggered by signals from the identification detectors. The combination of the CCD image and the drift-time profile allows the tracks to be reconstructed in three dimensions.

Study of ^{45}Fe decay. A detailed decay study of ^{45}Fe with the help of the OTPC was performed in 2007 at the National Superconducting Cyclotron Laboratory at Michigan State University in East Lansing. About 120 decays of ^{45}Fe were recorded, yielding the results discussed below. Another detection method for two-proton radioactivity, also based on a time-projection chamber, is being developed at the CEN in Bordeaux, France (Centre d'Etudes Nucléaires de

energy, MeV

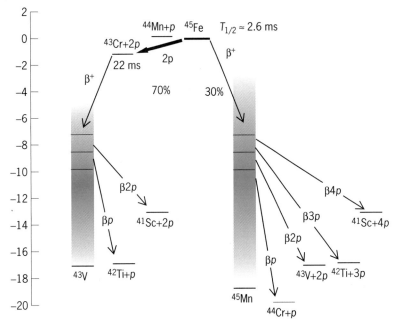

Fig. 2. Partial decay scheme of ^{45}Fe, with half-life $T_{1/2} \approx 2.6$ ms. The energy scale is shown relative to the ground state of ^{45}Fe.

which the three-body dynamics is the crucial ingredient. Moreover, this pattern suggests that the emitted pair of protons occupied p and f orbitals (with orbital angular momenta $1\hbar$ and $3\hbar$, respectively, where $\hbar$ is Planck's constant divided by 2π) in the ^{45}Fe nucleus. In addition, this conclusion turns out to be consistent with the measured decay energy and the half-life. All the observables together on the one hand provide a rigorous test of models describing two-proton radioactivity, and on the other hand shed light on the physical properties of a very exotic nuclear system.

Prospects. This new experimental tool opens new possibilities in nuclear spectroscopy. According to the nuclear shell model, a proton shell closes at proton number $Z = 28$. In addition to ^{45}Fe ($Z = 26$), nickel-48 (^{48}Ni, $Z = 28$) is suspected to be a two-proton emitter, and ^{54}Zn ($Z = 30$) is known to be a two-proton emitter. It is anticipated that detailed investigations of the proton-proton correlations in these three cases will yield important information on the nuclear structure around the $Z = 28$ shell closure at the limit of nuclear stability. The search for other cases of two-proton radioactivity continues with the

Bordeaux Gradignan), but so far has not provided data on the mechanism of two-proton decay.

The decay scheme of ^{45}Fe is complex and consists of two branches (**Fig. 2**). The main branch (70%) is the two-proton emission to chromium-43 (^{43}Cr), which subsequently decays by beta positron transitions to excited states of vanadium-43 (^{43}V), from which beta-delayed one- and two-proton emission are known to occur. The second decay mode of ^{45}Fe (30%) is the beta positron decay to excited, unbound states of manganese-45 (^{45}Mn). Here, many beta-delayed particle-emission channels are energetically open, including beta-delayed three-($\beta 3p$) and four-proton ($\beta 4p$) transitions. The beta-delayed protons have much larger energies (and thus longer tracks) than protons from the two-proton radioactivity, which clearly distinguishes the two decay modes. In the experiment, about 90 events of the two-proton decay type were found (**Fig. 3**). Others were classified as beta positron decay of ^{45}Fe followed by one-, two-, and even three-proton emission (**Fig. 4**). The beta-delayed three-proton emission events represent the first experimental observation of this decay mode.

The detailed analysis made it possible to establish the distribution of the opening angle between proton tracks from the two-proton radioactivity of ^{45}Fe, which provided the first information on the mechanism of this decay. The distribution exhibits two broad maxima: one centered around $50°$, and a second centered around $145°$. Such a pattern is at variance with the hypothesis that the two-proton decay could be a two-body process mediated by a virtual helium-2 (^{2}He) particle, a diproton. It evidences the truly three-body character of the decay, as it agrees with a model of L. V. Grigorenko and his colleagues in

(a)

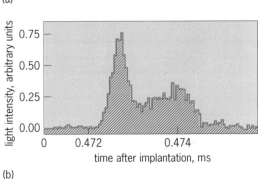

(b)

Fig. 3. Example of two-proton decay of ^{45}Fe. (*a*) Image recorded by the CCD camera. A track of a ^{45}Fe ion entering the chamber from the left is seen. Two short tracks are protons emitted about 0.47 ms after stopping of the ^{45}Fe ion in the gas. (*b*) Part of the time profile of the light intensity measured by the PMT, showing in detail the two-proton emission.

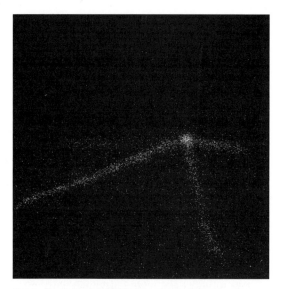

Fig. 4. Example of a CCD image of beta-delayed three-proton emission from ^{45}Fe. A track of a ^{45}Fe ion entering the detector horizontally from the left is seen. The three protons emitted some milliseconds later escape from the chamber.

hope that this phenomenon will reveal more secrets of nuclear structure.

For background information *see* CHARGE-COUPLED DEVICES; NUCLEAR STRUCTURE; PHOTOMULTIPLIER; RADIOACTIVITY; TIME-PROJECTION CHAMBER in the McGraw-Hill Encyclopedia of Science & Technology.
Marek Pfützner

Bibliography.   J. S. Al-Khalili and E. Roeckl (eds.), *The Euroschool Lectures on Physics with Exotic Beams*, vol. I, Lecture Notes in Physics, vol. 651, Springer, Berlin, New York, 2004; S. K. Blau, New experiments view details of nuclear two-proton emission, *Phys. Today*, 61(1):25–27, January 2008; D. J. Dean, Beyond the nuclear shell model, *Phys. Today*, 60(11):48–53, November 2007.

Naegleria fowleri infections due to drought

During the summer months of 2007, seven fatal infections caused by *Naegleria fowleri* occurred in Florida, Texas, and Arizona. All cases were in boys and young men who participated in recreational water sports at freshwater lakes during the summer. At the time, Florida was suffering from drought conditions through much of the state, leading to lowered water levels and increased temperature of the water. Arizona was in moderate to severe drought conditions during most of 2007, while Texas experienced typical high temperatures and had drought conditions present in 2005, 2006, and 2007. What connection, if any, exists between drought and these tragic fatalities?

Naegleria fowleri and primary amebic meningoencephalitis. *Naegleria fowleri* is a small ameba with an unusual appearance under the microscope (**Fig. 1**). This organism is a protozoan. Protozoans are grouped in the Kingdom Protista and consist of single-celled eukaryotic cells with typical organelles such as nuclei, mitochondria, and so on. *Naegleria fowleri* is motile through the use of pseudopodia ("false feet"); it extends its membrane outward and its cytoplasm flows to follow, allowing it to ooze along a surface or in water. The structure and microscopic appearance of a typical ameba is shown in **Fig. 2**.

Naegleria fowleri has two structural forms in its life cycle: a trophozoite form, which is the amebic cell that infects humans; and a rounded, thick-walled cyst, which is resistant to temperature extremes and mild chlorination. Most cases of *Naegleria* infection that have been reported occurred in people who have recently been swimming in warm, natural bodies of freshwater. *Naegleria fowleri* is a free-living ameba that is an accidental invader of humans, causing these infections only when victims inadvertently become exposed to the ameba. This ameba feeds on bacteria and algae that live in the sediment of freshwater lakes and ponds. However, when individuals swimming in these waters stir up the sediments and allow water to enter their nose, the ameba can enter and attach to the nasal epithelial cells. At this point, the ameba burrows into the olfactory mucosa, starts to multiply, and eventually travels into the brain.

Once in the brain, multiplication of the ameba occurs in the meninges (the membranes surrounding

Fig. 1. Scanning electron micrograph of *Naegleria fowleri*. The "eyes" and "mouth" of its facelike appearance are its attachment and feeding structures. [*From M. K. Cowan and K. P. Talaro, Microbiology: A Systems Approach, 2d ed., McGraw-Hill, New York, 2009*]

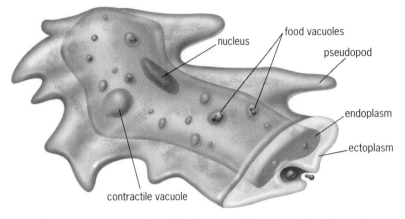

nucleus

food vacuoles

pseudopod

endoplasm

ectoplasm

contractile vacuole

Fig. 2. Structure of an ameba. [*From M. K. Cowan and K. P. Talaro, Microbiology: A Systems Approach, 2d ed., McGraw-Hill, New York, 2009*]

the brain) and in the neural tissue. Eventually, primary amebic meningoencephalitis (PAM) develops. Hemorrhage and edema occur; biopsies of brain tissue exhibit trophozoite forms of the ameba when examined by microscopy. Tissue necrosis is found in response to the infections. However, little can be done clinically to distinguish PAM from acute bacterial meningitis, which is why many clinicians do not consider PAM when diagnosing patients with severe meningitis. Another factor leading to the fatalities associated with this pathogen is that treatments for PAM and for infectious bacterial meningitis are extremely different, and what works for one infection will do nothing for the other.

The onset of symptoms may occur any time from 2 to 14 days after exposure to *N. fowleri*. Symptoms may include changes in the patient's sense of taste or smell, followed by fever, headache, anorexia (loss of appetite), nausea, and vomiting.

Most patients with PAM exhibit confusion, progressing to coma, and the infection is rapidly fatal within a week. PAM should be considered whenever acute meningitis occurs in a patient with a history of swimming in freshwater that is potentially contaminated with ameba. Cerebrospinal fluid (CSF), when taken from these patients, frequently has an increased number of erythrocytes (red blood cells), giving it a bloody appearance. The most important diagnostic tool involving CSF is examination of the fluid using a wet-mount slide preparation. This allows the detection of motile trophozoites, which are usually not recognized by Gram stain due to disruption of the amebas when the slide is fixed. Treatment is usually with amphotericin B, an antifungal agent that is given both intravenously and directly into the spinal canal (intrathecally). This drug has been used in all cases where patients survived the infection. However, its use is not always successful, even when administered within 24 h of patient admission and diagnosis.

Reasons for infections. *Naegleria fowleri* is a thermophilic organism, meaning it prefers warm temperatures. It can tolerate water as warm as $45°C$ ($113°F$). It is found worldwide in naturally warm waters, as well as in waters polluted with heat from energy pro-

duction or other sources. The concentration of the ameba in warm water can be greater than one cell for every 25 mL of water. Consequently, it is obvious that millions of people have been exposed to this pathogen. Despite this apparent high level of exposure, only 33 cases have been reported in the United States during 1996–2007.

Does immunity to this ameba exist? Antibodies specific for *N. fowleri* have been demonstrated in human populations, but the role they play in protection is uncertain. Children often can carry the ameba as normal biota, or microbes, present in the body, especially during the summer months. Whether this leads to specific antibody production is unknown.

In 2007, seven boys and young men died from infection with *N. fowleri*. Why only male victims? And why are most cases found in boys? The tendency of boys to undertake more boisterous activities in water may be a possibility. People become infected when they wade through shallow water, stirring up the bottom. If water is allowed to enter and shoot up the nose, perhaps by someone doing a somersault in shallow water, the ameba can enter the nasal passages and latch onto the host cells.

How is infection with this ameba connected to drought conditions? As drought continues in a geographic area, standing water levels in bodies of freshwater drop, thereby concentrating the number of ameba found in the water. Therefore, where once the ameba was found in low concentrations during a period of adequate rainfall, it is now found in much higher levels—and exposure rate can increase significantly. High temperatures in the summer only exacerbate the situation because bacteria and algae are multiplying exponentially. Thus, conditions are conducive for amebic disease: food, warmth, and swimmers.

In order to avoid *N. fowleri* exposure, the Centers for Disease Control and Prevention (CDC) in the United States recommends staying out of bodies of warm freshwater, hot springs, and thermally polluted water (such as found around power plants). The CDC also advises avoiding water-related activities in warm freshwater during periods of high water temperature and low water levels, such as those seen during drought conditions. If individuals are taking part in water-related activities in these kinds of freshwater bodies, they are advised to hold their nose shut or use nose clips to close the nose and prevent exposure of the nasal passages. Swimming pools should be properly maintained and appropriately treated to prevent multiplication of the ameba. Finally, the CDC recommends that recreational swimmers avoid digging in or stirring up the sediment of these lakes or ponds, so as to avoid increasing the concentration of ameba in the water.

In the future, more cases of infection with *N. fowleri* are likely to be seen. El Niño conditions, changes in weather patterns, increases in human populations, and the effects of global warming may all play a role in whether this disease will become a more common occurrence in the United States and worldwide.

Infections due to this ameba have occurred in 15 of the southern tier states, including the ones mentioned in this article: Texas, Florida, and Arizona. This disease is of concern to health authorities and to the general population. The ameba prefers warmer temperatures. As worldwide temperatures increase, this ameba is bound to thrive—and cause more cases of infection in recreation seekers.

For background information *see* AMEBA; AMOEBIDA; CLIMATIC PREDICTION; DROUGHT; FRESH-WATER ECOSYSTEM; MEDICAL PARASITOLOGY; MENINGES; MENINGITIS; PARASITOLOGY; PATHOLOGY; PROTOZOA; PUBLIC HEALTH in the McGraw-Hill Encyclopedia of Science & Technology.   Marcia M. Pierce

Bibliography. M. K. Cowan and K. P. Talaro, *Microbiology: A Systems Approach,* 2d ed., McGraw-Hill, New York, 2009; P. R. Murray, K. S. Rosenthal, and M. A. Pfaller, *Medical Microbiology,* 5th ed., Mosby, St. Louis, 2005; D. Schlossberg (ed.), *Infections of Leisure,* ASM Press, Washington, D.C., 1999.

Neanderthal DNA

Neanderthals are the closest evolutionary relatives of modern humans. Based on fossil and archeological records, they are large-brained people with a distinct body plan and skull morphology compared to modern humans, associated with Mousterian (Middle Paleolithic) stone tools. Neanderthals had dominated in Europe and western and central Asia for 200,000 years, until their population declined for unknown reasons. After retreating to isolated refuges in southern Europe, they finally died out approximately 28,000 years ago.

Neanderthal bones were first recovered in Engis Cave, Belgium, in 1829, but were neglected by scientists for the next 100 years. The first scientific description of the bones belonging to the skeleton of a Neanderthal was conducted by Hermann Schaaffhausen at the University of Bonn in 1861. He analyzed bones with unusual morphological characteristics that were found in a rock quarry in the Neander valley in Germany, now known worldwide as the Feldhofer Grotto. In 1863, William King from Queen's University in Ireland declared the species name to be *Homo neanderthalensis.* Neanderthals became the first extinct people to be recognized as a species distinct from modern humans.

Theories of Neanderthal and modern human evolution. Since the first recognition of Neanderthals as a separate species, the origin, evolutionary history, and extinction of the Neanderthals, as well as the relationship between Neanderthals and modern humans, have been among the most strenuously debated issues in human evolution.

Several competing models have been proposed that explore the potential genetic contributions of Neanderthals to modern humans. According to the regional continuity model, Neanderthals are the direct ancestors of the modern Europeans, and the Neanderthal genome has directly evolved into the genome of modern European people, with gene flow

occurring between different populations around the world. The other main competing model—namely, the replacement or out of Africa model—hypothesizes that all modern humans stemmed from an ancestral population in Africa that had evolved there about 100,000–200,000 years ago. This ancestral population subsequently dispersed throughout the world, completely replacing the Neanderthals in Eurasia. The out of Africa model suggests that gene flow between modern humans and Neanderthals was absent, with their genomes diverging at a point in the past.

Between these two extreme models of total genetic continuity and total replacement lie several intermediate models, advancing the possibility of medium to low levels of genetic exchange between early modern human colonizers of Asia and Europe and the aboriginal population of Neanderthals, possibly resulting in hybrid populations.

Neanderthal mitochondrial DNA. The mitochondrial genome is a circular deoxyribonucleic acid (DNA) molecule found in the mitochondria of the cell, known as mitochondrial DNA (mtDNA), with a size of approximately 16,500 base pairs. The most extensive mtDNA variation in humans is concentrated in the noncoding control region. Within the control region, the hypervariable segment I (HVS-I), consisting of 400 nucleotide base pairs, can be sequenced to detect mtDNA diversity between closely related species, populations of the same species, and individuals. Because of multiple copies (around 1000 copies per cell), small segments of mtDNA molecules can be recovered from biological remains, under certain circumstances, even after tens of thousands of years.

The first successful analysis of Neanderthal mtDNA, published in 1997, involved the retrieval of the HVS-I from the Feldhofer Neanderthal 1 by polymerase chain reaction (PCR, a technique for copying and amplifying the complementary strands of a target DNA molecule). This was the first attempt to assess the genetic relationship of Neanderthals and modern humans directly by analyzing Neanderthal DNA. The study demonstrated that the Neanderthal mtDNA falls outside the variation of modern humans. However, since this conclusion was based on a single ancient specimen, it was greeted with caution by some scholars who had doubts about the authenticity and interpretation of the results.

DNA samples from other Neanderthals were needed to verify the original conclusion. In 2000, a second mtDNA sequence from a Neanderthal specimen, found in Mezmaiskaya cave in the northern Caucasian mountains in Russia, was successfully analyzed by an independent team of researchers. This second sequence provided corroboration of the uniqueness of Neanderthal mtDNA, as well as new insight into Neanderthal mtDNA diversity and the age of the earliest Neanderthal divergence. These initial results opened the gate to the study of Neanderthal population genetics.

Neanderthal mtDNA diversity. These early studies were crucial in pointing out the possibilities and

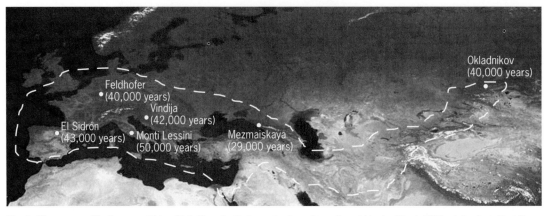

Fig. 1. The geographical range within which Neanderthal remains have been found (dashed line). White dots show the sites of the discoveries used to estimate the almost complete HVS-I of the Neanderthal mtDNA. The approximate ages of the Neanderthal fossils are shown in parentheses. (*Satellite imagery: NASA's Earth Observatory*)

limitations of the analysis of Neanderthal mtDNA. Subsequent studies of additional specimens led to the successful analysis of HVS-I sequences from six other Neanderthals: Feldhofer 2, Vindija 75 and 80 from Croatia, El Sidrón from Spain, Monti Lessini from Italy, and Okladnikov from Russia (**Fig. 1**). These Neanderthal HVS-I sequences have been compared with the Cambridge Reference Sequence (CRS), which serves as a reference for modern human mtDNA (**Fig. 2**). The results show that the Neanderthal mtDNA pool falls outside the mtDNA variation of modern humans (**Fig. 3a**).

In addition, the Neanderthal mtDNA sequences can be investigated with regard to their phylogenetic relationship within the Neanderthal mtDNA pool. The Feldhofer 1 and 2, Vindija 75 and 80, and El Sidrón sequences are very closely related, whereas the Mezmaiskaya, Okladnikov, and Monti Lessini are more divergent, and may represent different lineages of the Neanderthal phylogenetic tree (Fig. 3b). More sequences from different geographic regions are required in order to understand the relationships between different Neanderthal populations within Europe and Asia. Seven other Neanderthal specimens—El Sidrón 77, Vindija 77, Engis and Scladina from Belgium, La Chapelle-aux-Saints and Les Rochers-de-Villeneuve from France, and Teshik-Tash from Uzbekistan—have also yielded very short fragments of 30–190 base pairs that fall within the Neanderthal mtDNA pool.

Neanderthal nuclear genes. The analysis of Neanderthal nuclear genes is essential for understanding the genetic differentiation between modern humans and Neanderthals. However, such an analysis is extremely problematic because of the very low amounts of nuclear DNA fragments in ancient bones. It is also complicated by the predicted high similarity between Neanderthal and modern human DNA, making possible contamination by modern human DNA difficult to identify.

Two nuclear genes, *FOXP2* and *MC1R*, have been analyzed from Neanderthal specimens. The human *FOXP2*, one of a complex of genes involved in the development of speech and language, has two nucleotide substitutions in exon 7, changing two amino acids and making the human FOXP2 protein different from the chimpanzee FOXP2 protein. Exon 7 from the Neanderthal *FOXP2* gene, sequenced from two individuals from the El Sidrón cave, shares these two substitutions, as well as six other substitutions found in the intronic region, with modern humans.

	1	1	1	1	1	1	1	1	1	1	1	1	1	1	1	1	1	1	1	1	1	1	1	1	1	1	1	1	1	1	1	1
	6	6	6	6	6	6	6	6	6	6	6	6	6	6	6	6	6	6	6	6	6	6	6	6	6	6	6	6	6	6	6	6
	0	0	0	1	1	1	1	1	1	1	1	1	1	1	1	2	2	2	2	2	2	2	2	2	2	2	3	3	3	3	3	3
	7	8	9	2	3	4	5	5	6	7	7	8	8	8	0	2	3	3	4	5	5	6	6	6	7	9	1	2	4	6	6	6
	8	6	3	9	9	8	4	6	9	2	8	2	3	9	9	3	0	4	4	6	8	2	3		3	8	1	0	4	1	2	9
																								a								
CRS	A	T	T	G	A	C	T	G	C	T	T	A	A	T	T	C	A	C	G	C	A	C	T	-	C	A	T	C	C	G	T	G
Feldhofer 1	G	T	C	A	T	T	C	G	T	T	T	A	C	C	C	T	G	T	A	A	G	T	T	A	T	G	C	T	C	G	C	G
Feldhofer 2	G	T	T	A	T	T	C	G	T	T	T	C	C	C	C	T	G	T	A	A	A	T	T	A	T	G	C	T	C	G	C	G
Vindija 75 and 80	G	T	T	A	T	T	C	G	T	T	T	C	C	C	C	T	G	T	A	A	G	T	T	A	T	G	C	T	C	G	C	G
Mezmaiskaya	A	C	T	A	T	T	T	A	T	T	T	C	C	C	C	T	G	T	A	A	A	T	T	A	T	G	C	T	T	G	C	G
Monti Lessini	A	T	T	A	T	T	T	G	T	T	T	C	C	C	C	T	G	T	A	A	G	T	T	A	C	A	C	T	C	A	T	A
El Sidrón	G	T	T	A	T	T	C	G	T	T	C	C	C	C	C	T	G	T	A	A	G	T	T	A	T	G	C	T	C	G	C	G
Okladnikov	A	T	C	A	T	T	T	G	T	C	T	A	C	T	C	T	G	T	A	A	A	T	C	A	T	G	C	T	C	G	C	G

Fig. 2. The mtDNA HVS-I diversity found in eight Neanderthals. The nucleotide substitutions and insertions relative to the Cambridge Reference Sequence (CRS) are shaded.

The *MC1R* gene encodes the melanocortin 1 receptor that regulates skin and hair pigmentation in humans. Loss-of-function mutations in this gene result in pale skin and red hair in modern Europeans. A small exonic fragment of 128 base pairs was sequenced from two Neanderthals recovered from Monti Lessini and El Sidrón. Both carry a new substitution at the *MC1R* gene that is absent in modern humans. This mutation changes an amino acid in the melanocortin 1 receptor, causing a partial loss of its function and reducing the pigmentation level of skin and hair.

Neanderthal genome. The sequencing of the Neanderthal genome has proved to be one of the most challenging whole-genome projects to date. It became possible only after the invention of the next-generation high-throughput sequencing technology, which allows sequencing of each DNA molecule directly from a complex DNA mixture. This technology makes it theoretically possible to sequence ancient single-copy nuclear genomes, including the restoration of the genomes of Neanderthals.

To date, approximately 1,065,000 base pairs of the Neanderthal nuclear genome have been sequenced from the Vindija 80 Neanderthal using high-throughput sequencing approaches. However, these impressive results represent a set of short DNA sequences of up to 120 base pairs, dispersed over the Neanderthal genome, and overall account for less than 0.04% of the nuclear genome. No Neanderthal-specific DNA changes were discovered, and 80% of the studied sequences appeared to be due to modern human DNA contamination. With the development of next-generation sequencing technologies, the analysis of additional Neanderthal remains, and the study of Neanderthal sequences by independent research teams, the potential exists to sequence significant parts of the Neanderthal genome.

Admixture between Neanderthals and modern humans. The Neanderthal and modern human mtDNA pools form distinct evolutionary groups that diverged significantly earlier than when modern human mtDNA began evolving in the ancestral population of modern humans (Fig. 3*a*). These data, as well as the absence of Neanderthal mtDNA lineages among the early modern humans of Europe, suggest either that Neanderthals have not contributed their mtDNA to modern humans at all or that such admixture was so low that it can be neglected. However, mtDNA is a single locus with maternal inheritance and can provide only a partial insight into the pattern of Neanderthal and modern human gene flow.

Variation in the nuclear genome has also failed to support any detectable gene flow between Neanderthals and modern humans. The five Y chromosome loci, as well as 65,000 base pairs of the Neanderthal genome sequenced from a metagenomic library, do not demonstrate any signs of admixture. The light skin and hair phenotype in Neanderthals has a different genetic background in the *MC1R* gene found in modern Europeans, supporting independent evolution of this phenotype.

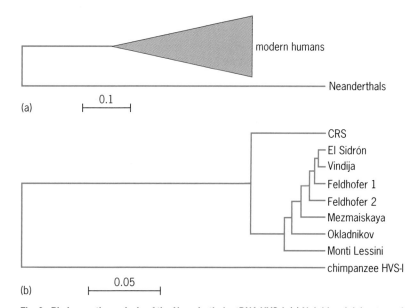

Fig. 3. Phylogenetic analysis of the Neanderthal mtDNA HVS-I. (*a*) Neighbor-joining tree of the Neanderthal mtDNA sequences and 2500 modern human HVS-I sequences from 43 modern populations. (*b*) Neighbor-joining tree of the Neanderthal mtDNA HVS-I along with the Cambridge Reference Sequence (CRS) and one chimpanzee mtDNA sequence, which was randomly selected. The scale bars show the genetic distances between populations or DNA sequences on the phylogenetic trees. The genetic distance ranges from 0 for populations with no differentiation and DNA sequences with no differences to 1 for highly differentiated populations and DNA sequences.

Age of divergence of Neanderthals and modern humans. The rate of nucleotide substitutions in Neanderthal and modern human sequences has been used to estimate the time of divergence between different segments of their genomes. The Neanderthal mtDNA, the best molecular clock, dates the split between the Neanderthal and modern human lineages to approximately 600,000 years ago, and the time of the most recent common ancestor between the western and eastern Neanderthals at about 250,000 years ago. The age of the most recent common ancestor of the modern human mtDNA pool is 120,000–150,000 years.

The Neanderthal genome sequences, without evidence of admixture, provide an estimated time of divergence of 700,000 years, with a 95% confidence interval of 470,000 to 1,000,000 years. This estimate is in good agreement with the age of divergence of Neanderthals and modern humans based on the mtDNA data. These dates are generally consistent with evidence from the fossil record.

For background information *see* DEOXYRIBONUCLEIC ACID (DNA); EARLY MODERN HUMANS; EXTINCTION (BIOLOGY); FOSSIL HUMANS; GENE; GENOMICS; HUMAN GENOME; MITOCHONDRIA; MOLECULAR ANTHROPOLOGY; NEANDERTALS; PHYSICAL ANTHROPOLOGY in the McGraw-Hill Encyclopedia of Science & Technology.   Igor Ovchinnikov; William Goodwin

Bibliography. E. Culotta, Ancient DNA reveals Neanderthals with red hair, fair complexions, *Science*, 318:546–547, 2007; R. E. Green et al., A complete Neandertal mitochondrial genome sequence determined by high-throughput sequencing, *Cell*, 134:416–426, 2008; M. Höss, Neanderthal population genetics, *Nature*, 404:453–454, 2000; M. Krings et al., Neanderthal DNA sequences and the origin of modern humans, *Cell*, 90:19–30, 1997; D. M. Lambert and C. D. Millar, Ancient genomics is born, *Nature*, 444:275–276, 2006; I. V. Ovchinnikov et al., Molecular analysis of Neanderthal DNA from the northern Caucasus, *Nature*, 404:490–493, 2000.

Negative thermal expansion materials

Substances showing negative thermal expansion (NTE) over small temperature ranges have long been known. For example, water exhibits negative thermal expansion over a range of several degrees close to its freezing point. The world we live in would be a very different place if it were not for that NTE, which causes ice to float on water. NTE is also known to occur in many materials, such as silicon (Si), at very low temperatures, but this NTE disappears well before approaching ambient temperature. NTE also occurs well above ambient temperature over limited temperature ranges in materials such as some crystalline forms of silicon dioxide (SiO_2). The discovery in 1995 that cubic zirconium tungstate (ZrW_2O_8) exhibits strong NTE over a temperature range exceeding $1000°C$ led to a surge of interest in materials exhibiting NTE.

The primary application of NTE materials is as components in composites to lower the overall thermal expansion of the composites. NTE materials have demonstrated the ability to reduce the thermal expansion of composites in which the matrix is a metal, polymer, or ceramic. Reducing thermal expansion to very small values has obvious applications of improving dimensional stability and reducing thermal stress fractures, but there are also applications in which a particular value of thermal expansion is needed to avoid thermal expansion mismatch. In electronics, matching the thermal expansion of Si is desired to avoid circuit failures. Another very different application is the use of NTE materials in composites for filling teeth, where failure of the filling can occur as a result of thermal mismatch between the tooth and the composite filling material.

Transverse thermal motion. Although several mechanisms can give rise to NTE, the dominant mechanism is thermal vibration of the central atom in an M–X–M linkage. The vibration of this atom perpendicular to the linkage can cause the M–M distance to decrease with increasing temperature. For example, in CO_2 at low temperatures, the average O–O

distance contracts as temperature increases. Most commonly, NTE occurs in oxides, where the relevant linkage is M–O–M and M is a metal cation typically with either a tetrahedral or octahedral coordination to oxygen. For optimum NTE, the M–O bond should be a strong covalent bond, because such bonds show very low thermal expansion. Thus, the thermally excited vibration of O perpendicular to the M–O–M linkage can dominate the thermal expansion. It is not necessary for the M–O–M linkage to be linear. NTE will occur with a bent linkage as a result of vibration of O perpendicular to the plane of the bent $\overset{O}{\underset{M\ M}{\diagdown\diagup}}$ linkage. In an extended structure, one expects that there will be correlations of the vibrations of the various oxygen atoms. The simplest model for such correlations assumes that the MO_x polyhedra are rigid; that is, the M–O distances and the O–M–O angles do not change as the MO_4 tetrahedra or MO_6 octahedra rock back and forth, giving O vibrations that result in NTE. Such vibrations are rigid unit modes (RUMs). This is shown schematically for the cubic ReO_3 structure in **Fig. 1**. For many materials that exhibit NTE, there is a structure collapse transition as the temperature is lowered, and NTE behavior is not observed below this transition.

The cubic ReO_3 structure is an example of NTE in which the MO_x polyhedra are all octahedra. The various common forms of SiO_2 are examples of NTE in which all the polyhedra are tetrahedra. We may mix tetrahedra and octahedra in various ratios. The primary mixtures of interest are AMO_5, $A_2M_3O_{12}$, and AM_2O_7, where the A cations are octahedrally coordinated and the M cations are tetrahedrally coordinated. The important condition met for all of these examples is the twofold coordination of oxygen.

The common forms of SiO_2, such as quartz and cristobalite, contain SiO_4 tetrahedra and O bound to two Si atoms. The SiO_4 tetrahedra can be considered to be rigid units. Nonetheless, NTE behavior is very limited in these common forms of SiO_2. They all have structure collapse transitions well above ambient temperature. However, one has only to open up the structure and NTE behavior becomes very pronounced. A synthetic form of SiO_2 having the faujasite structure has a structure based on corner-sharing tetrahedra as in the common forms of SiO_2. However, the faujasite form of SiO_2 has a very open structure and strong isotropic NTE over the measured temperature range of 10 to 580 K (see **table**). The strongest NTE behavior yet found over an extended temperature range is found for $AlPO_4$-17 (see table). This is a very open structure based on corner sharing of SiO_4 and AlO_4 tetrahedra. Structure collapse transitions do not occur for either faujasite SiO_2 or $AlPO_4$-17.

Combining tetrahedra and octahedra in equal amounts gives the formula AMO_5, where the octahedral A cation can be Nb or Ta and the tetrahedral M cation can be P or V. There are different ways of connecting these polyhedra, and all exhibit NTE behavior, but all show a structure collapse transition with decreasing temperature. This transition is usually above room temperature. However, in $TaVO_5$, the transition is below room temperature. The $TaVO_5$

Fig. 1. The ReO_3 structure depicted as corner-sharing ReO_6 octahedra. Rocking of the octahedra back and forth corresponds to correlated transverse motion of O in the Re–O–Re linkages. The tilted structures to the left and right are contracted relative to the center structure even though there is no change in size or shape of the octahedra.

Thermal expansion coefficients at room temperature, °C	
Material	α/K^{-1}*
Polypropylene	90×10^{-6}
Copper	16.6×10^{-6}
Al_2O_3	6.5×10^{-6}
Si	3×10^{-6}
SiO_2 (amorphous)	0.5×10^{-6}
SiO_2 (faujasite)	-4.2×10^{-6}
$PbTiO_3$	-3.5×10^{-6}
$TaOVO_4$	-4.4×10^{-6}
$Lu_2W_3O_{12}$	-6.8×10^{-6}
ZrW_2O_8	-8.7×10^{-6}
$AlPO_4$-17	-11.7×10^{-6}

* α = (change in length divided by length) per degree of temperature change.

structure has orthorhombic symmetry with NTE behavior along all three crystallographic axes.

The $A_2M_3O_{12}$ formula is the only formula for which all tetrahedra share corners with only octahedra and all octahedra share corners with tetrahedra only. Again, there are different ways to connect these polyhedra, and NTE behavior is observed for the different connectivities, even though RUMs are not present. Many, but not all, compounds in this $A_2M_3O_{12}$ family have a structure collapse transition with decreasing temperature. One compound without a structure collapse transition is $Y_2W_3O_{12}$. This compound shows strong NTE behavior over the measured temperature range of nearly 1400°C. The symmetry of the structure is orthorhombic, with NTE behavior along all three crystallographic axes. Detailed structural studies of this compound using neutron diffraction were used to obtain accurate values for thermal motion as a function of temperature. This information was used to calculate the degree of NTE behavior expected based on the perpendicular thermal motion of O in the Y-O-W linkages. Good agreement was found with the observed NTE behavior.

There are many compounds of the AM_2O_7 type, where A is octahedrally coordinated and M is tetrahedrally coordinated. There are no RUMs in this structure, but NTE behavior is found in several cases. For example, strong NTE behavior is observed for ZrV_2O_7 from about 100 to at least 800°C. A structure collapse structure occurs on cooling below 100°C, with positive thermal expansion below this transition. The structure of ZrW_2O_8 is very similar to that of ZrV_2O_7, but the bond between tetrahedral groups is missing. This makes the structure more flexible, and it now contains RUMs. Strong NTE behavior is observed for ZrW_2O_8 over a temperature range of 1400°C (**Fig. 2**). The structure possesses cubic symmetry. Thus, the NTE behavior is isotropic, which is an advantage for applications.

The ReO_3 structure is the simplest structure composed of octahedra only in which we might expect NTE (Fig. 1), but strong NTE behavior over extended temperatures has not been observed in this structure. There are very few compounds with this structure. For ReO_3 itself, very low thermal expansion

is reported, which is apparently negative over some range of temperature. The ReO_3 structure is found for TaO_2F, which has extremely low thermal expansion over a wide temperature range. We can conclude that the transverse thermal motion of the anions in ReO_3 and TaO_2F is important in lowering thermal expansion, even though strong NTE behavior is not observed.

Oxygen is frequently coordinated to two metal atoms in oxides. Thus, the emphasis for NTE in oxides is on M-O-M linkages. However, O-M-O linkages do occur in a few cases. Both Cu_2O and Ag_2O have such linkages. NTE behavior is found in Cu_2O below room temperature, and this behavior extends through room temperature for Ag_2O. The O-Cu-O linkage also exists in some $CuMO_2$ compounds, such as $CuScO_2$, and NTE behavior results.

The anion in M-X-M linkages need not be oxygen. Strong NTE behavior is frequently found for M-CN-M linkages, and again this behavior is due to the transverse thermal motion of the CN group. Transverse thermal motion can give rise to NTE in polymers along the strongly bonded chains. However, the many weak bonds in polymers lead ultimately to a strong overall positive thermal expansion. The transverse thermal motion mechanism for NTE does not depend critically on twofold coordination. For example, in graphite, the carbon atoms in the strongly bonded sheets all are in threefold coordination. Thermal motion of the carbon atoms perpendicular to the sheet causes the sheets to contract on heating. However, the weak bonds between sheets give a strong positive thermal expansion perpendicular to sheets, giving an overall positive thermal expansion.

Other mechanisms. Strong NTE behavior is observed in $PbTiO_3$ below 480°C, and this is not related to the mechanism described above. Instead, it is related to polyhedra becoming more regular with increasing temperature. Polyhedra are expected to shrink as they become more regular. Some compounds contain a rare earth element in mixed electron configurations. The relative amounts of these configurations can be temperature-dependent, and

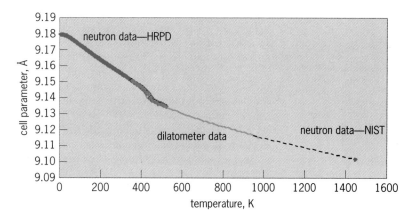

Fig. 2. Thermal contraction in zirconium tungstate (ZrW_2O_8). The cubic unit cell edge is plotted versus temperature. A direct measurement of a ceramic body is superimposed (dilatometer data).

this can give rise to NTE behavior, as it does in $Sm_{0.75}Y_{0.25}S$. The larger $(4f^5)(5d^1)$ configuration gives way to the smaller $(4f^6)$ configuration as temperature increases. There is also an extrinsic mechanism for NTE that is frequently observed in ceramics based on anisotropic particles. Ceramics are generally sintered (fired) at high temperature. On cooling a ceramic with grains that have strongly anisotropic thermal expansion, microcracks usually occur. These can cause expansion of the ceramic body, even if the grain themselves are contracting. On heating such a material, the cracks can close, giving NTE.

For background information *see* COMPOSITE MATERIAL; COORDINATION CHEMISTRY; CRYSTAL STRUCTURE; OXIDE; OXYGEN; STRUCTURAL CHEMISTRY; THERMAL EXPANSION in the McGraw-Hill Encyclopedia of Science & Technology.   Arthur W. Sleight

Bibliography. G. D. Barrera et al., Negative thermal expansion, *J. Phys. Condens. Matter*, 17:R217, 2005; J. S. O. Evans, Negative thermal expansion materials, *J. Chem. Soc. Dalton Trans.*, 3317, 1999; A. W. Sleight, Compounds that contract on heating, *Inorg. Chem.*, 37:2854, 1998.

Neuroeconomics

Neuroeconomics is a recent consilient discipline (that is, a discipline that combines the principles of other disciplines to produce a comprehensive analysis) that measures brain activity while experimental subjects make decisions. Because the brains of all animals are "economic," that is, they have limited resources to achieve necessary goals, neuroeconomics experiments are not limited to studies of human beings, but have also employed apes, monkeys, and rodents. Economics is the study of constrained decision making, and it uses both mathematical and statistical models of the decision goals and outcomes without considering the mechanisms leading to decisions. Neuroscience has focused primarily on cataloging mechanisms without considering the purpose of decisions. For this reason, neuroeconomics is a natural combination that draws from the best of, and extends, both fields.

Decisions can be modeled mathematically with three components: a decision maker's preferences, beliefs, and constraints. Such models produce empirically testable predictions. Each of these three factors can be measured using the methods of neuroscience. Many decision models in economics predict choices quite well—for example, individuals purchasing things in competitive markets. In other cases, standard models do not predict behavior well—for instance, as in some models of strategic decisions involving other people.

Neuroeconomists have investigated both individual and social decisions in order to understand the processes behind the models that predict behavior accurately, as well as to improve the models that do not predict behavior well. Because many economic models are specified mathematically and have been studied in both the laboratory

and the field, they provide sharp predictions when seeking to find the brain mechanisms involved in decisions.

Utility functions. One of the most fundamental ideas in economics is that a person's preferences are represented by a utility function. Such a function relates an individual's experience with things to that individual's own valuation of those things. Vanilla and chocolate ice cream may cost the same amount, but one buys chocolate because one gets more utility from it. One's consumption of ice cream is constrained, or limited, by a variety of factors, for example, how much money one has. If the price of chocolate ice cream is substantially higher than that of vanilla, one may switch to vanilla. How are such things decided? Direct measurement of brain activity in monkeys has shown that brain cells (neurons) calculate utility. Brain imaging experiments have replicated this work in humans, revealing a network of regions that appear to calculate the value of different choices. Utility calculations draw on both evolutionarily old regions in the midbrain and newer cortical regions on the outer surface of the brain. The older regions appear to get the individual to focus on finding options, while the cortical areas integrate this information with prices to guide the individual toward the "best" choices. "Best" in this case means the choices that were most advantageous for producing progeny over the evolutionary history of *Homo sapiens*. Some of these choices, though, may be maladaptive in the modern environment. An example of a maladaptive choice is the preference for high-fat foods. During the long history of the human species, such foods were rare and were greatly valued for their high caloric content. In today's developed societies, this preference for high-fat foods (and their low cost) is producing high rates of obesity.

Standard utility maximization models also predict that people are risk averse; that is, they typically prefer a sure thing to a risky choice, even if the risky choice has a larger average payoff. Risk aversion has been localized by a number of laboratory analyses to an area of the brain called the anterior insula. This brain region makes you feel queasy when you smell rotten food, and makes your palms sweat when you are riding a roller coaster. Knowing the brain region that causes risk aversion allows scientists to understand why people vary in their responses to risk, as well as to help treat those who are pathologically risk averse or who take excessive risk, like compulsive gamblers.

Game theory. Game theory is a branch of mathematics that describes how to make choices involving other people who are also making decisions. Game theory can describe how best to make chess moves, how to negotiate an employment contract, and how to make myriad other decisions involving other people. Many game theoretic models have choices that are cooperative (sharing benefits) and choices that are selfish (hoarding benefits). Understanding why people choose to cooperate or to be selfish is vitally important because it is not possible to live in a free society unless people choose to behave cooperatively with others most of the time, even when they are not being monitored by the government.

Unfortunately, many game theoretic models do not predict behavior very accurately. For example, consider a set of choices known as the "Ultimatum Game." Suppose you were given $100 and asked to propose some split of it to another person in a different room. No communication with this person is allowed, and you will never meet him or her. The other person knows that you were given $100 and that you have to propose a split of the money. Here's the catch: if the other person accepts your proposal, you are both paid the money; but if your proposal is rejected, you both get nothing. What would you do? Standard game theoretic models predict that any offer, no matter how small, will be accepted, since some money is always preferable to nothing. However, in most developed countries, offers of $20 or less are nearly always rejected. Neuroeconomics experiments have shown why. Stingy offers produce strong activation in the interior insula, suggesting that low offers are rejected because people are disgusted by them. Human brains have evolved for social interactions, and it was typically better to lose some resources to punish a stingy person than to build a reputation for being exploitable. On the other hand, why would anyone ever make an offer in the Ultimatum Game that is generous, that is, larger than needed to be accepted? Neuroeconomists thought that empathy toward others might drive people to be generous. They tested this by giving people more of a brain chemical called oxytocin that increases empathic behaviors. Infusing oxytocin into people's brains using a nasal spray increased generosity to a stranger in the Ultimatum Game by 80%. This shows that people cooperate with them because they emotionally identify with them and do not want to hurt them.

Trust. The role of oxytocin in decisions to trust a stranger with one's money has also been studied by neuroeconomists. Any transaction that occurs over time, like a financial investment, has a degree of trust embedded in it since there are no perfectly enforceable contracts. Indeed, the general level of trust among people in a country is among the strongest predictors of which countries will have rising standards of living: high-trust countries see rapid increases in incomes. But, an open question is: Why would you ever trust a stranger with your hard-earned money? If someone shows that he or she trusts you by investing money with you, neuroeconomics studies have found that the receiver's brain releases oxytocin. In addition, the more oxytocin released by people's brains, the more they returned some of the invested money (which typically earns a large return) to the trustee. This is surprising because, in these experiments, there is no obligation to return any money at all. To prove that brains use oxytocin to help determine whom to trust, neuroeconomists have infused oxytocin into the human brain. When this is done, more than twice as many people show maximal trust in a stranger by sending that stranger all their money.

The neuroeconomists' findings on generosity and trust present a conundrum for traditional economics: trustworthy people (typically more than 90% of people studied) could have kept all the money they controlled for themselves. Instead, these people freely chose to return often a large proportion of the money to the person who initially trusted them. Why? Recent brain imaging experiments have shown that monetary transfers to another person indicating trust activate regions in the brain that reinforce behaviors by making them pleasurable. This brain reward circuit prominently uses the neurotransmitter dopamine. Because humans are social creatures, our brains have evolved to make cooperative behaviors, including trust, rewarding. Brain imaging studies have shown that even donating money to charity appears to activate brain regions associated with empathy (through oxytocin) and reward (through dopamine). These studies also reveal the importance of emotions when making economic decisions.

Punishment. What happens when someone betrays your trust? If you are like most people, you don't like this at all, and you want to let the other person know it. When people are given the chance to spend some of their own money to punish another person for betrayal, they readily do so. Costly punishment occurs even if the individuals involved will not interact with each other again. This has been called moralistic punishment. Physiologically, when one is betrayed, testosterone, a hormone associated with aggression, spikes. The act of punishment also activates dopaminergic reward regions of the brain. Individuals punish because they are angry, and they find it rewarding to punish betrayers—even at a cost to themselves. The threat of punishment is an important mechanism that sustains cooperative behaviors, even among those who might consider being selfish.

Outlook. Rather than the classical view of humans as "*homo economicus*" (purely rational and self-interested), research in neuroeconomics suggests that humans could more appropriately be called "*homo reciprocans*"—reciprocating creatures who are influenced by emotion. These early but important neuroeconomics studies indicate that the human brain is wired to evaluate the utility of options and to extract economic value from social interactions. While neuroeconomics is a new field, it holds the promise to improve the ability to understand one's own choices, to better predict the choices of friends and customers, and to guide government policy. Neuroeconomics studies also allow scientists to help those who make poor choices, including criminals, those with psychiatric disorders, and those under extreme stress, such as soldiers.

For background information *see* BRAIN; COGNITION; DECISION ANALYSIS; DECISION THEORY; DOPAMINE; GAME THEORY; INFORMATION PROCESSING (PSYCHOLOGY); INSTRUMENTAL CONDITIONING; LEARNING MECHANISMS; MEDICAL IMAGING; NEUROBIOLOGY in the McGraw-Hill Encyclopedia of Science & Technology.   Paul J. Zak

Bibliography. C. F. Camerer and E. Fehr, When does "economic man" dominate social behavior?, *Science,*

311:47–52, 2006; C. F. Camerer, G. Loewenstein, and D. Prelec, Neuroeconomics: how neuroscience can inform economics, *J. Econ. Lit.*, 43:9–64, 2005; M. Kosfeld et al., Oxytocin increases trust in humans, *Nature*, 435(2):673–676, 2005; P. J. Zak, Neuroeconomics, *Philos. T. Roy. Soc. B*, 359:1737–1748, 2004; P. J. Zak, The neurobiology of trust, *Sci. Amer.*, pp. 88–95, June 2008.

Neuromorphic and biomorphic engineering systems

Many biological systems, from the molecular scale to the macroscale and from the body to the brain, display remarkable efficiency and robustness. For example, a single mammalian cell, approximately 10 micrometers in size, performs complex biochemical signal processing on its mechanical and chemical input signals with highly noisy and imprecise parts, using approximately 1 picowatt (10^{-12} W) of power. Such signal processing enables the cell to sense and amplify minute changes in the concentrations of specific molecules amid a background of confoundingly similar molecules, to harvest and metabolize energy contained in molecules in its environment, to detoxify poisonous molecules, to sense if it has been infected by a virus, to communicate with other cells in its neighborhood, to move, to maintain its structure, to regulate its growth in response to signals in its surroundings, to speed up chemical reactions via sophisticated enzymes, and to replicate itself when it is appropriate to do so. The approximately 20,000-node gene-protein and protein-protein molecular network within a cell makes even the most advanced nano-engineering of today look crude and primitive.

The brain is made of approximately 22×10^9 neurons that form a densely connected network of approximately 240×10^{12} synaptic connections. This network performs approximately 10^{15} synaptic operations per second at approximately 14 W of power, several orders of magnitude more energy efficient than the most advanced computers. The brain can perform real-time, reliable, complex tasks with unreliable and noisy devices. It uses remarkably compact hardware built with a rich array of biochemical and biophysical devices and is architected with a 3D interconnect technology that allows three orders of magnitude more connectivity than the most advanced engineering systems of today. The brain is adaptive and plastic with rapid learning and generalization capabilities that outperform the most sophisticated machine-learning algorithms.

Can we learn from nature to build better engineering systems that are equally impressive, robust, and efficient? The goal of neuromorphic engineering, a term coined by Carver Mead, is to take inspiration from neurobiological architectures to build better engineering systems, "morphing" them with insight from their natural neurobiological domains to be useful in artificial engineering domains. More generally, we can define a biomorphic system as one that takes inspiration from any architecture in biology, for example, the architecture of cells, to create a morphed version that is useful in an engineering context. Thus, airplanes are biomorphic architectures that are inspired by the winged flight of birds. A neuromorphic silicon cochlea or silicon retina is inspired by the architecture of the ear or the eye and performs highly parallel nonlinear filtering, gain control, and compressive computations on an audio or image input respectively.

Relation of engineering to biological systems. Biomorphic solutions have sometimes been reinvented by engineers without their even knowing that they are biomorphic or that they already exist in nature: The use of chirp signals for accurate range sensing in radars was invented by engineers around World War II, but bats had already been using ultrasonic chirps for range sensing in their biosonar systems for millions of years. Positive-feedback circuits were invented about 100 years ago but have been present in sodium ion channels for more than 100 million years. Thus, knowledge of systems in nature can provide useful ideas for engineering. Several biomorphic architectures, such as machine-learning and pattern-recognition systems inspired by the operation of neurons in the brain, are already widely used in artificial systems.

In biomorphic systems, it is important to keep the insightful "baby" and throw out the cluttering "bathwater" details. Certain architectures in biology may be accidents of evolution, may be more suited to the constraints of a biological organism, and may serve or may have served a purpose that we do not yet understand. Consequently, their relevance to a different engineering context where the constraints are different may be questionable. Birds are not airplanes and airplanes are not birds, although the study of one can shed insight into the study of the other. Hence, it is important to evaluate a biomorphic engineering system by traditional engineering metrics to insightfully understand where value can be added.

Types of biomorphic systems. From an engineering point of view, where do biomorphic systems add value? They clearly have the potential to shine in the following kinds of systems:

1. Ultralow-power and highly energy efficient sensing, actuating, and information-processing systems.

2. Signal processing and pattern-recognition systems that need to operate in noisy environments and over a wide dynamic range of inputs.

3. Robust and efficient computation with noisy and unpredictable devices.

4. Systems with feedback, adaptation, and learning at multiple spatial and temporal scales.

5. Systems that integrate technologies from diverse domains.

6. Self-repairing systems.

7. Self-assembling systems.

8. Energy-harvesting systems.

9. Robotic systems.

Features of biomorphic systems. How do biomorphic systems appear to accomplish these feats?

These systems have many common features, some of which will be discussed.

1. The exploitation of analog physical basis functions for computation in biochemical, biomechanical, or bioelectronic technologies, rather than mere digital logic functions, as in traditional computation. The increase in efficiency then arises out of exploiting the graded analog degrees of freedom in each signal and device rather than treating signals as being merely "on" or "off" or treating devices as just being switches, as in digital systems. However, the robustness of such analog computing systems is established in a different manner from digital systems. Clever feedback-and-regulation systems and inherently noise-robust topologies are built to ensure that the overall output and overall system is robust even though each device and signal in it is not. It is this combination of feedback and analog system design that allows biological systems to operate both robustly and efficiently. Rather than operate in a collective fashion, with several low-precision digital elements that collectively interact to implement a high-precision or complex operation, biological systems operate with many low-precision analog elements that collectively interact to implement a high-precision or complex operation. The interactions and processing can have a hybrid analog-digital nature with both all-or-none digital and graded analog processes being present. All-or-none interactions are often useful in performing digital signal restoration in analog systems that are prone to the effects of noise and in making decisions. For example, after extensive analog processing has been performed on the multiple inputs of a neuron, the final output of a neuron is often an all-or-none "spike" or voltage pulse. This spike is fired by the neuron as soon as the voltage near a particular somatic region of the neuron exceeds a threshold.

Biomorphic systems have used programmable analog processing with feedback-and-feedforward calibration to construct ultralow-power cochlear-implant processors for the deaf. Such processors have lowered processing power by more than an order of magnitude while being robust to several sources of noise and while maintaining high levels of flexibility.

2. The use of highly sophisticated technologies: The ear is an example of a highly sophisticated technology that integrates microfluidics, micromechanics, piezoelectrics, and microelectronics into a system to perform more than 10^9 arithmetic operations per second of spectrum-analysis computations with 14 μW of power in a volume not much larger than the size of a pea. Biomorphic systems have mimicked the architecture of the cochlea in radio-frequency (RF) technologies to construct efficient ultrawide-band spectrum analyzers, that is, RF cochleas. As another example, the brain's 3D interconnect technology allows a fanout of approximately 10,000 per neuron rather than just approximately 5 per logic gate in electronic technologies today.

3. The use of nonlinear and adaptive processing: During development, cells implement ingenious nonlinear diffusion-and-degradation partial differential equations that ensure that cell differentiation is robust to variations in the concentrations of "morphogen" molecules. These concentration values are important in deciding whether the cell develops or "differentiates" into one cell type versus another, for example, a liver cell versus a kidney cell. Learning is an adaptive or feedback process that alters the parameters or topology of a system over slow time scales such that it more efficiently processes signals in its environment. Learning is ubiquitous in biological systems. Nonlinear and adaptive signal processing in the ear has led to a bio-inspired companding algorithm for improving the perception of speech recognition in noise in both deaf patients and in artificial speech-recognition systems.

4. The use of scalable cellular architectures: The use of scalable cellular architectures, with strong integration of processing, memory, sensing, actuation, and communication functions in each local cellular unit, rather than only in specialized regions as in traditional architectures, is common in several systems in biology. Such architectures range from a network of skin cells to a network of neurons in the brain. John von Neumann, the inventor of traditional digital architectures named after him, was himself aware of the limitations of his inventions. He engaged in research on cellular automata and analog computation in his later years, inspired by computation in the brain.

5. The use of ingeniously clever algorithms: The control of eye movement appears to use sophisticated feedback-control loops that function with good speed, accuracy, and stability in spite of large delays in the system, a challenging task in engineering. Biological systems appear to use knowledge and learning to constantly tune parameters in predictive architectures that can compensate for such delays. In general, biological systems are adept at tuning themselves to respond optimally to the signal statistics in their environment such that they are both robust and efficient. Increasingly, advanced circuits in engineering are using predictive digital compensation of errors in analog systems to enable better robustness and efficiency.

Historical and evolutionary perspectives. The use of nature to inspire better engineering is as old as it is new. The most advanced aircraft are beginning to explore the use of turbulence just as birds do, even though the Wright brothers gave up on such strategies because they were too difficult to implement at the time. Nevertheless, airplanes are still very far from achieving energy efficiencies that can compete with that of a bird.

Humans have a long way to go before their architectures will successfully compete with those in nature, especially in situations where ultra-energy-efficient or ultralow-power operation are paramount. In evolutionary environments, food was always a scarce resource, so biological systems that needed to harvest their energy and raw materials from food needed to be very efficient. The result has been the creation of incredibly energy-efficient architectures.

Biomedical applications. Several biomorphic and neuromorphic architectures have led to useful circuits for biomedical applications, where learning from biology is only natural in helping to fix systems when they do not work. Ultralow-power operation is also important for these applications. For example, ultra-low-power analog-to-digital conversion inspired by the operation of spiking neurons in biology has led to a very energy efficient converter for biomedical applications. Another bio-inspired algorithm has helped create an algorithm that has benefits for helping the deaf perceive music.

Learning architectures that mimic the connectivity and adaptation in connection strengths of synapses, that is, the junctions that connect neurons to each other, have been very important in artificial machine learning systems. Many learning systems exploit local learning rules at artificial synapses to automatically create self-organizing systems at a more global network level. Such systems automatically teach themselves to function from mere examples or by extracting patterns in their input data. Ultralow-power analog learning circuits are being applied to create architectures for learning and decoding the movement intentions of paralyzed patients from their brain signals.

The field of biomorphic design suggests that we can mine the intellectual resources of nature to create devices useful to humans, just as we have mined her physical resources in the past. Such mining will require us to combine inspiration with perspiration and to understand how nature works with insight.

For background information *see* ADAPTIVE CONTROL; ADAPTIVE WINGS; ANALOG COMPUTER; ARTIFICIAL INTELLIGENCE; BIOMEDICAL ENGINEERING; BRAIN; CELL ORGANIZATION; CELLULAR AUTOMATA; CONTROL SYSTEMS; CYBERNETICS; HEARING IMPAIRMENT; NEURAL NETWORK; NEUROBIOLOGY; NONLINEAR CONTROL THEORY; SIGNAL PROCESSING in the McGraw-Hill Encyclopedia of Science & Technology.

Rahul Sarpeshkar

Bibliography. L. C. Aiello et al., The expensive-tissue hypothesis: the brain and digestive system in human and primate evolution, *Curr. Anthropol.*, 36(2):199–221, 1995; G. Cauwenberghs and A. Bayoumi, *Learning in Silicon: Adaptive VLSI Neural Systems*, Kluwer Academic Publishers, 1999; A. Eldar et al., Robustness of the BMP morphogen gradient in *Drosophila* embryonic patterning, *Nature*, 419:304–308, 2002; C. Mead, Neuromorphic electronic systems, *Proc. IEEE*, 78:1629–1636, 1990; D. A. Robinson et al, A model of the smooth-pursuit eye-movement system, *Biol. Cybern.*, 55(1):43–57, 1986; R. Sarpeshkar, Analog versus digital, *Neural Comput.*, 10:1601–1638, 1998; R. Sarpeshkar et al., An ultra-low-power programmable analog bionic ear processor, *IEEE Trans. Biomed. Circuits Syst.*, 52(4):711–727, 2005; R. Sarpeshkar et al., Low-power circuits for brain-machine interfaces, *IEEE Trans. Biomed. Circuits Syst.*, in press, 2008; L. Turicchia et al., A bio-inspired companding strategy for spectral enhancement, *IEEE Trans. Speech Audio Process.*, 13:243–253, 2005.

Neutron scattering

Neutron scattering is a widely used technique for determining both the atomic and magnetic structures of materials and the way in which these structures evolve in time. It can provide remarkably precise and unique information about the structure and dynamics of a variety of materials that are often difficult to interrogate by other means. As scientific interest moves to increasingly complex structures and the goals of nanoscience are pursued, new and more powerful neutron sources currently coming on line will be able to contribute new insights, building on a tradition that started 60 years ago with experiments by scientists such as Clifford Shull and Bertram Brockhouse.

The success of neutron scattering, and its ability to complement other techniques for probing materials structure, such as x-ray diffraction, electron microscopy, and nuclear magnetic resonance, derive from several unique aspects of the interaction between neutrons and matter. Paradoxically it is the weakness of this interaction that gives the technique one of its primary advantages, its ability to obtain quantitative information nondestructively. Because the interaction is weak, the number of neutrons scattered with particular changes of momentum, energy, and spin can be calculated very accurately, allowing precise amplitudes for structural and magnetic fluctuations to be obtained. However, this weak interaction also means that the percentage of neutrons scattered by a sample is small. Sometimes, the scattered neutron signal is too weak to be measured accurately. Accordingly, practitioners of the method are always seeking more powerful sources of neutrons as well as new experimental methods that allow them to make better use of available neutrons.

New and upgraded neutron sources. The next few years promise to be exciting because two major new neutron sources are poised to begin contributing scientific results and several expansions of existing facilities are under way. In the first category, the United States and Japan have recently completed the construction of high-power spallation neutron sources that produce neutrons from nuclear reactions that occur in a heavy metal target when it is bombarded with high-energy protons from an accelerator. While the two sources use different accelerator technologies to produce their protons, both employ streams of ~1 microsecond long proton pulses and both will deposit 1 MW or more of power into their mercury neutron-production targets when they reach full design performance around 2010. By mid-2008, the U.S. version, the Spallation Neutron Source (SNS) at Oak Ridge National Laboratory, had already operated at 500 kW as part of its ramp-up to full power (**Fig. 1**). At the design power of 1.4 MW for its first phase, the SNS will produce roughly an order of magnitude more neutrons than the current world-leading spallation source, ISIS in the United Kingdom, which has a proton beam power of about 130 kW.

Not to be outdone by new developments, ISIS has almost completed its own upgrade that will add an additional neutron production target together with a suite of instruments in a new experimental hall. In the United States, the Department of Commerce is adding a new experimental hall and neutron instruments to its facility at the National Institute of Standards and Technology in Gaithersberg, Maryland. Australia has recently commissioned its own reactor-based facility, and China is building both a pulsed spallation source similar in power to ISIS and a 60-MW nuclear reactor that matches the power of the world's current leading reactor-based neutron scattering facility, the Institut Laue-Langevin in Grenoble, France.

Types of neutron sources. Since the early 1980s, the worldwide neutron scattering community has debated the relative merits of producing neutrons by fission in a nuclear reactor and by spallation using pulsed proton beams. With current technologies, the preferred neutron source depends on the nature of neutron scattering experiment to be performed and three different types of source have to be compared for each application: continuous neutron sources, such as nuclear reactors, and pulsed spallation neutron sources with either long (1–2 ms) or short (~1 μs) pulses. Both the United States and Japan chose to build short-pulse spallation sources as their next-generation facilities because this type of source offers potential advantages for many scientific studies and because existing accelerator technologies allowed significant gains in performance over current facilities of this type. In the United States and Europe, scientists are planning long-pulsed neutron sources for future projects, recognizing that the availability of high-power versions of each type of neutron source (continuous, long-pulse, and short-pulse) allows the optimum source to be chosen for each scientific application.

Applications of neutron scattering. Neutron sources are expensive to build, so their expected scientific impact must be commensurably large. Among the scientific areas that new and upgraded neutron sources are expected to impact are fluid transport in porous media; electrochemical processes at surfaces; improved functional materials such as composites, ceramics, coatings, and lubricants; surfactants at interfaces; ion-conducting glasses; molecular motions of chemical groups in catalysts and other important organic molecules; nanoparticles and low-dimensional magnetic and nonmagnetic systems; sensors of various types; high-temperature superconductors and correlated-electron materials; macromolecular complexes of biological significance; and enzyme activity. Several of the recent experiments described below support the notion that new and more powerful neutron sources can be expected to produce unexpected and significant scientific advances in many of these areas.

Fig. 1. Spallation Neutron Source at Oak Ridge National Laboratory in Tennessee.

Contrast variation. An important technique used in neutron scattering, called contrast variation, is based on the fact that different isotopes of the same element often have different scattering powers for neutrons. Hydrogen is a case in point. One of the early contributions of neutron scattering to polymer science occurred in 1972 when scientists realized that if one hydrogenated polymer chain is mixed with many otherwise equivalent, but deuterated, polymer chains, the hydrogenated chain stands out for neutrons much like a single strand of green fettuccine in a bowl of white noodles. The result was one of the first quantitative demonstrations of the way polymer chains organize themselves in a melt. The use of contrast variation has continued to impact polymer science since those early experiments. Among other results, this has led to a better understanding of the microstructures of layered and bulk block copolymers as well as models for the interdiffusion of different polymer types across an interface between them. But as knowledge is accumulated, more sophisticated scientific questions are asked, often requiring smaller and smaller sample volumes to be probed and therefore providing less signal in neutron scattering experiments. For example, recent experiments measured the radius of gyration of a polymer cast as a layer whose thickness could be chosen to be either greater or less than the radius of gyration of the polymer chain in the bulk (that is, layers thinner than 10 nanometers). Other studies investigated the microstructure of block copolymers within grooves only 50 nm wide and 200 nm deep etched into a silicon wafer. In both cases, the scientific questions concerned changes in the structure caused by strong spatial confinement, an important topic in both nanoscience and biology.

Studies of polymer dynamics. In addition to exhibiting an average conformation, polymer chains also change shape dynamically in the molten state, imposing limits on technologies for polymer processing, for example. The chain dynamics is dominated by the effects of chain entanglement, which give

one of the strongest
hydrogen-bonding sites

Key:

H

C

O

Zn

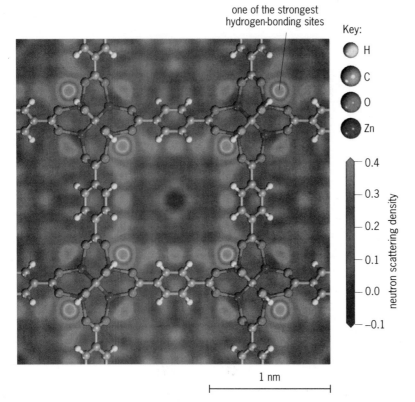

neutron scattering density

— 0.4
— 0.3
— 0.2
— 0.1
— 0.0
— –0.1

1 nm

Fig. 2. Neutron scattering results for a metal-organic cage structure. The location of one of the strongest hydrogen-bonding sites is labeled; the other sites with the same appearance are also strong binding sites. A ball-and-stick model of the molecular structure is superimposed on a contour plot of the measured Fourier difference neutron scattering density. The neutron diffraction data on which the image is based were taken at the NIST Center for Neutron Research. The strength of the hydrogen binding at various sites was also calculated. (*Taner Yildirim*)

rise to topological constraints restricting motion. Although Pierre-Gilles de Gennes proposed his reptation model to explain how individual chains snake their way through the entangled mess of neighbors in 1971, a verification of his model at the molecular level took 30 years and several generations of neutron scattering instrumentation. Eventually a technique called neutron spin echo demonstrated that reptation provided the best description of the way in which a single chain loses memory of its past configurations. To unambiguously confirm a particular model for chain movement, researchers had to make measurements at sufficiently long correlation times. Both the power of the neutron source and the quality of the scattering instrumentation impose limits on accessible values of this time. The current generation of neutron sources and instrumentation can reach correlation times of $\sim$350 ns, and with these machines scientists have now found slight deviations from the reptation model, ascribable to fluctuations of the contour length of the chains, that show up for polymer chains of smaller molecular weight. At the SNS, the same group that made these polymer measurements has constructed a new spin echo instrument capable of probing correlation times of up to a microsecond. Because the SNS is a national user facility, open for peer-reviewed access to any scientist, this new instrument will allow multiple groups to search

for new understanding of the dynamics of polymers and other complex fluids.

Studies of MOFs. The future success of hydrogen fuel cells depends on the discovery of new materials that can store large amounts of hydrogen. Metal-organic framework compounds (MOFs) are a class of nanoporous materials which may be useful in this respect. To improve these materials it is important to understand the way in which hydrogen is absorbed. Recent experiments, again making use of isotopic contrast and neutron scattering, have shown that the metal-oxide clusters in the MOF are primarily responsible for absorption of hydrogen and that the organic linkers between the clusters are relatively unimportant (**Fig. 2**). As part of the discovery process, it is likely that a large number of measurements like those presented in Fig. 2 but for different compounds under a variety of thermodynamic conditions will be needed. The rate of discovery is clearly linked to the number of experimental resources that can be used and, in this sense, the recent expansions of neutron scattering capability both in the United States and overseas can be expected to have a significant impact.

Studies of magnetic structures. In addition to being scattered by nuclei in materials, neutrons are also affected by the fluctuation of magnetic fields, and there is a long history of discovery based on neutron scattering in the area of magnetism, starting with confirmation of the Néel state of antiferromagnets in 1949 and extending to present-day revelations of magnetic fluctuations in high-temperature superconductors. The phenomenon of giant magnetoresistance (GMR) has been incorporated surprisingly rapidly into modern-day technologies such as computer disk drives. Many of the recent fundamental questions about artificially layered magnetic structures, such as those that display GMR, have been answered by neutron scattering using a technique called polarized neutron reflectometry. Just as in the case of the polymer films described earlier, the scientific issues involve the effects on magnetism of confinement, neighboring materials, and the interfaces between them. Starting with simple layered systems, for example, scientists found that, in some circumstances, magnetization reversal occurs differently on one side of the hysteresis loop of so-called exchange-biased systems than on the other: on one side of the loop, reversal of sample magnetization occurs by domain wall motion, while on the other, magnetization rotation occurs. The next step was to examine the magnetization reversal process in even more tightly confined systems comprising magnetic islands and stripes. This, and the study of phenomena such as vortices in arrays of magnetic nanodots less than 100 nm in diameter, are examples of experiments that severely stretch the limits of present-day neutron scattering capability but which will become much easier with more powerful neutron sources.

For background information *see* ANTIFERROMAGNETISM; ARTIFICIALLY LAYERED STRUCTURES; COPOLYMER; MAGNETORESISTANCE; NEUTRON SCATTERING; NANOSTRUCTURE; POLYMER; SPALLATION REACTION in the McGraw-Hill Encyclopedia of Science & Technology.

Roger Pynn

Bibliography. P. G. de Gennes, Reptation of a polymer chain in the presence of fixed obstacles, *J. Chem. Phys.*, 55:572–579, 1971; M. R. Fitzsimmons et al., Neutron scattering studies of nanomagnetism and artificially structured materials, *J. Magn. Magn. Mater.*, 271:103–146, 2004; R. L. Jones, et al, Chain conformation in ultrathin polymer films, *Nature*, 400:146–149, 1999; A. Wischnewski et al., Molecular observation of contour-length fluctuations limiting topological confinement in polymer melts, *Phys. Rev. Lett.*, 88:058301, 2002; T. Yildirim and M. R. Hartman, Direct observation of hydrogen adsorption sites and nanocage formation in metal-organic frameworks, *Phys. Rev. Lett.*, 95:215504, 2005.

Nitrogen fixation

Nitrogen fixation is the process of turning free nitrogen gas (N_2) into chemical compounds, usually ammonia (NH_3), which can be used by growing plants as a source of nitrogen. Gaseous N_2 makes up the majority of the atmosphere, but higher organisms cannot use it because the N_2 molecule is extremely stable and unreactive. Nitrogen-fixing reactions are vital for life on Earth because all organisms contain nitrogen atoms in essential molecules such as DNA, RNA, and proteins. Natural nitrogen fixation is done by microorganisms through the action of the nitrogenase enzymes. Chemical N_2 fixation is done through the Haber-Bosch process, which is used to produce about 120 million tons (110 million metric tons) of ammonia per year. World agricultural production is highly dependent on nitrogen fixation. It is because of the need for fertilizers that chemical nitrogen fixation is done on such a gigantic scale. The amount of ammonia manufactured by the chemical industry is greater than the amount of natural nitrogen fixation, and the world's population could not be supported without this input of fixed nitrogen. However, the additional fixed nitrogen in the environment has led to pollution by nitrate (NO_3^-) and nitrite (NO_2^-) in natural waters.

Chemical fixation. Three N_2-fixing reactions have been used commercially. Two of these, the cyanamide process and the arc process, are much less prevalent. The majority of nitrogen fixation today uses the Haber-Bosch process, as shown in reaction (1) of N_2 with hydrogen gas (H_2) to give ammonia (NH_3).

$$N_2(g) + 3H_2(g) \rightarrow 2NH_3(g) \qquad (1)$$

This reaction does not occur at a feasible rate without a catalyst, and the catalysts for nitrogen fixation invariably contain metals. Fritz Haber discovered near the beginning of the twentieth century that ammonia production can be catalyzed by several different metals, including iron, rhenium, ruthenium, osmium, cobalt, and uranium. Iron catalysts have the most desirable balance of activity, cost, and durability. The iron catalyst is produced under special conditions, with small amounts of potassium and silica, to give it high surface area and texture. This special iron catalyst was developed by Carl Bosch, who shared the Nobel Prize in Chemistry in 1931 for his work on high-pressure ammonia production. His catalyst is still used today on a scale that makes it one of the five largest-volume chemical processes.

Why is high pressure necessary? The equilibrium between the left and right sides of reaction (1) favors the left side more at higher temperature and favors the right side more at higher pressure. So, it would be ideal to work at low temperatures, but the Haber-Bosch process is extremely slow at room temperature and requires high temperatures (around 500°C or 930°F) to proceed at a useful speed. At this temperature, the equilibrium is very unfavorable, and so pressures of greater than 100 times atmospheric pressure are needed to produce a reasonable amount of ammonia. The high temperatures and pressures in the Haber-Bosch process makes ammonia synthesis very energy-intensive, with more than 1% of the world's total energy output going for Haber-Bosch ammonia synthesis, primarily for use in fertilizers.

Attempts to study the mechanism of this reaction led to many advances in surface science at the end of the twentieth century. The nature of N_2 binding to the iron surface was demonstrated by Gabor Somorjai and Gerhard Ertl (the latter was recognized with the Nobel Prize in Chemistry in 2007). N_2 binds most strongly to iron surfaces that are rough on the atomic scale. The bound N_2 is nearly parallel to the surface, suggesting that the N_2 may lie in a surface defect, interacting with multiple iron atoms. N_2 is subsequently broken down to nitrogen atoms on the surface, and finally, reaction with H_2 gives the product, NH_3.

Chemists who study solution reactions have searched for metal-containing molecules that can do the same transformation at lower temperature. One way of overcoming this challenge was reported by Richard Schrock, who showed that a molybdenum complex can transform N_2 into NH_3 multiple times at room temperature (**Fig. 1**). This system only functions correctly with a specific combination of reducing agent, acid, and a bulky hydrocarbon group (R) and further development is needed to it make a practical solution catalyst for NH_3 synthesis.

Nitrogenase enzymes. Evolution led to bacteria that fix N_2 at room temperature. These bacteria are known as azatrophs, from the Greek for "eats nitrogen." Azatrophs were responsible for most of the

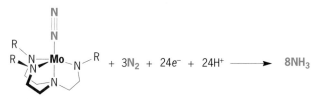

Fig. 1. Catalytic ammonia formation by a molybdenum-containing compound. R represents a large hydrocarbon group that protects the metal center. Electrons (e^- from a reducing agent and protons (H^+) from an acid are added to N_2.

fixed nitrogen on Earth until the invention of the Haber-Bosch process. The most famous azatrophs infect the root nodules of legumes in a symbiotic relationship in which the plant provides sugar to the bacteria and the bacteria fix nitrogen for the plant. The increased amount of fixed nitrogen in the soil is one reason why farmers rotate crops between legumes and grains to make their farms more productive. Azatrophic microorganisms are also found in various soils, in gut bacteria in animals, and in the oceans.

The enzymes (catalytic proteins) in azatrophs that react with N_2 are called nitrogenases. Under the best conditions, the nitrogenase enzymes catalyze reaction (2).

$$N_2(g) + 8H^+(aq) + 8e^- + 16ATP^{3-}(aq)$$
$$+ 16H_2O(l) \rightarrow 2NH_3(aq) + H_2(g)$$
$$+ 16ADP^{2-}(aq) + 16H_2PO_4^-(aq) \qquad (2)$$

Protons (H^+) are provided by water, electrons (e^-) are provided by biological electron donors, and ATP is adenosine triphosphate, a molecule present in all living organisms that releases energy when combined with H_2O to give ADP (adenosine diphosphate) and free phosphate.

Four types of nitrogenases are known, and all require dissolved iron for their formation. The most common and most active is molybdenum nitrogenase, which is encoded by *nif* genes in microorganisms such as *Azotobacter vinelandii*, *Clostridium pasteurianum*, and *Klebsiella pneumoniae*. When there is not enough molybdenum in the environment, these organisms have a "plan B," whereby they use *vnf* genes to express a vanadium (V) nitrogenase. When both molybdenum (Mo) and vanadium are scarce, there is another backup system, and *anf* genes express an iron-only nitrogenase. These three enzyme systems (*nif*, *vnf*, and *anf*) work in a roughly similar fashion. The activity and efficiency decrease in the order Mo > V > Fe. A fourth type of nitrogenase, from *Streptomyces thermoautotrophicus*, has been studied more recently. This fourth type differs from the others in a few important ways. Whereas the first three types of nitrogenase are air-sensitive and are poisoned by carbon monoxide, this new nitrogenase is air-stable and is not affected by carbon monoxide. Also, *S. thermoautotrophicus* lives at higher temperatures (50–65°C or 120-150°F) than the other organisms. However, the environment of the metal(s) in this enzyme is not yet known. It is possible that greater understanding of this enzyme could lead chemists to discover more robust catalysts for fertilizer production.

Molybdenum nitrogenase. The most active nitrogenase has catalytically essential iron and molybdenum. The molybdenum nitrogenase system is composed of two components, the Fe protein and the MoFe protein. The Fe protein adds water to ATP and uses the energy from this reaction to push electrons into the FeMoco, a metal-containing site within the protein where N_2 is transformed into ammonia.

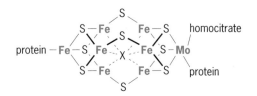

Fig. 2. FeMoco, where molybdenum nitrogenase fixes N_2. X is an unidentified atom (C, N, or O).

The FeMoco is an eight-metal cluster (**Fig. 2**), in which the metals are held together by nine sulfur (S) atoms and by an atom (X) at the center. The identity of X is currently the source of intense controversy. Also debated is whether N_2 binds at the molybdenum atom (like Schrock's solution catalysts) or at an iron atom (like the Haber-Bosch catalyst). Researchers continue to use many synthetic, biochemical, and physical techniques to understand how natural enzymes fix N_2 under ambient conditions.

Concerns of nitrogen fixation as part of the nitrogen cycle. Ammonia produced by nitrogen fixation (chemical or biological) is converted into nitrite (NO_2^-) and nitrate (NO_3^-) by soil bacteria, and it is these forms of fixed nitrogen that plants can use. However, some of the nitrite and nitrate are lost. Metabolic enzymes in bacteria reduce some of the nitrite and nitrate to N_2 through denitrification, in order to get energy. Also, nitrite and nitrate are very soluble in water, and they can be washed out of the soil and into lakes, aquifers, streams, and rivers. This problem is exacerbated because chemical fertilizers are typically added in batches rather than the rate at which they are needed by the crops.

Nitrite and nitrate pollution in groundwater has a number of ill effects. First, nitrite deactivates hemoglobin by converting it to the "met" form that cannot bind oxygen (O_2). Infants are especially susceptible to this effect if they ingest nitrate-containing water, and met-hemoglobin anemia is also called "blue baby syndrome." Second, the nitrite and nitrate in rivers and lakes is a nutrient for algae and bacteria in the water. Therefore, bodies of water downstream from many farms are prone to algae blooms and other problems of water tainted by microorganisms. In extreme cases, this leads to eutrophication, where the microorganisms take up all the dissolved O_2 in the water, and other marine life cannot survive. For example, a "dead zone" appears most summers in the Gulf of Mexico (off the coast of Louisiana) that is 5000 to 8000 mi^2 (13,000 to 21,000 km^3). Refinement of agricultural practices may reduce the extent of this problem.

For background information *see* AMMONIA; CATALYSIS; ENZYME; EUTROPHICATION; HEMOGLOBIN; NITROGEN; NITROGEN CYCLE; NITROGEN FIXATION in the McGraw-Hill Encyclopedia of Science & Technology.   Patrick L. Holland

Bibliography. G. J. Leigh, *The World's Greatest Fix*, Oxford University Press, 2004; J. R. Postgate, *Nitrogen Fixation*, Cambridge University Press, 2000; B. E. Smith et al. (eds.), *Catalysts for Nitrogen Fixation*, Kluwer, 2004.

Nonmetallocene single-site catalysts for polyolefins

Polyolefins, exemplified by polyethylene (PE) and polypropylene (PP), exhibit many useful properties. They are lightweight and possess excellent mechanical strength, flexibility, processability, chemical inertness, and recyclability. Polyolefins, therefore, are cost effective. They are used in the manufacture of numerous products, such as toys, plastic shopping bags, food packages, shampoo and detergent bottles, storage boxes, disposable diapers, sneakers, and car bumpers.

Although heterogeneous Ziegler-Natta catalysts, such as $MgCl_2$-supported $TiCl_4$, which have multiple active sites, currently dominate the polyolefin production market, catalysts with one active site (single-site catalysts) based on group 4 metallocenes (**Fig. 1a**) are increasingly becoming viable in industry. Single-site group 4 metallocene catalysts possess a well-defined active site that permits control over polymer molecular weight, molecular-weight distribution, and polymer stereochemistry. These active sites also allow for the commercial production of linear low-density PEs, isotactic PPs (stereoregular polymer with methyl groups on one side of the polymer chain), syndiotactic PPs (stereoregular polymer with methyl groups on alternate sides of the polymer chain), plus uniform comonomer incorporation for producing amorphous ethylene/1-butene copolymers and others.

Following the tremendous advances made by the single-site group 4 metallocene catalysts in the syntheses of polyolefins with controlled microstructures, a new generation of catalysts is being developed that provides higher catalyst productivity and greater control over polymer microstructures. As a result of a formidable amount of academic and industrial research, a diverse number of highly active nonmetallocene single-site catalysts based on both early and late transition metals have now been discovered. Prominent examples include diimine-ligated Ni and Pd complexes and pyridinediimine-ligated Fe complexes for the late transition-metal catalysts, and pyridylamine-ligated Hf complexes and phenoxyimine-ligated group 4 transition-metal complexes (FI catalysts) for the early transition-metal catalysts (Fig. 1b–e). The development of highly active nonmetallocene single-site catalysts for olefin polymerization has had a significant impact on recent polymerization catalysis, and this has led to the production of new polyolefin-based materials with unique properties.

Diimine-ligated Ni and Pd complexes. In 1995, M. Brookhart reported on bulky diimine-ligated complexes of Ni and Pd, which represented the first examples of late transition-metal catalysts capable of polymerizing propylene as well as ethylene to high-molecular-weight polymers. The steric protection of the vacant axial coordination site by the bulky diimine ligand (**Fig. 2a**), which reduces chain-transfer rates, is responsible for the production of high-molecular-weight polymers, with molecular weights up to 1,000,000 g/mol being possible. At lower temperatures, the polymerization proceeds in a living (nonterminating) fashion, and di- and triblock copolymers can be prepared from ethylene and α-olefins.

A crucial feature of the diimine-ligated complexes is their versatility for controlling branching of the resulting PE chain by changing the ethylene pressure and temperature. With Ni and Pd complexes, PE can be synthesized with branching ranging from linear (~0) up to 100 branches per 1000 carbon atoms. Under the same polymerization conditions, Ni complexes generally form less branched PE. The

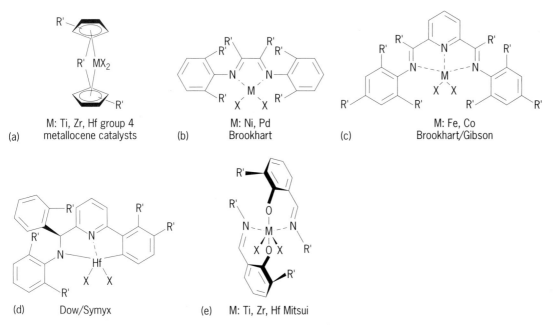

(a) M: Ti, Zr, Hf group 4 metallocene catalysts

(b) M: Ni, Pd Brookhart

(c) M: Fe, Co Brookhart/Gibson

(d) Dow/Symyx

(e) M: Ti, Zr, Hf Mitsui

Fig. 1. General structures of (a) group 4 metallocene catalysts and (b–e) high-performance nonmetallocene single-site catalysts.

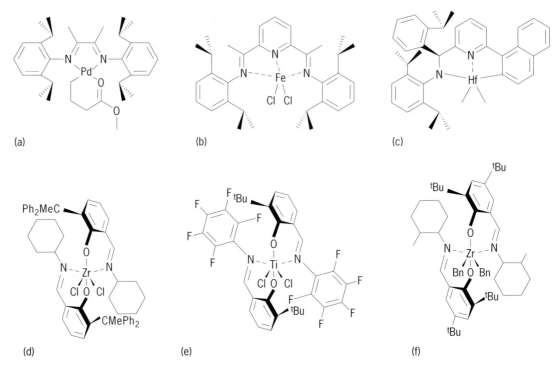

Fig. 2. Structures (a–f) of nonmetallocene single-site catalysts.

proposed mechanism for the formation of branched polymers contains the reversible steps involving β-H transfer to the metal, bond rotation, and hydride insertion, as depicted in **Fig. 3**, to which the name "chain walking polymerization" was given, since the central metal "walks" on the polymer chain.

Another important feature is that the diimine-ligated complexes display good functional-group tolerance. Importantly, the Pd complexes, such as Fig. 2a, can copolymerize ethylene with methyl acrylate (MA) to form highly branched random copoly-

mers with ester groups attached to the ends of the branches, representing the first examples of ethylene and MA copolymerization through coordination polymerization.

Pyridinediimine-ligated Fe complexes. In the late 1990s, M. Brookhart and V. C. Gibson independently developed pyridinediimine-ligated Fe complexes (Fig. 1c) for ethylene polymerization, representing the first reports of Fe-based catalysts capable of polymerizing ethylene with high efficiency. After activation with a cocatalyst such as methylalumoxane

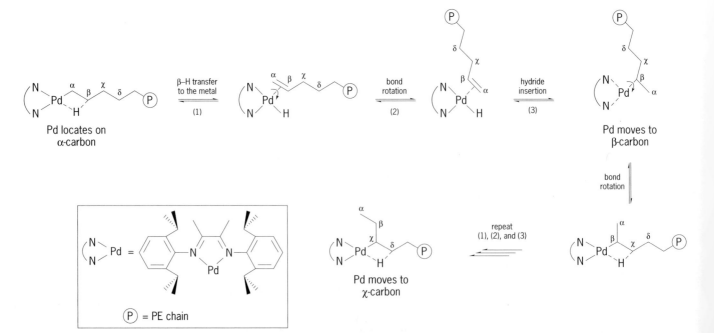

Fig. 3. Proposed mechanism for the formation of branched PE.

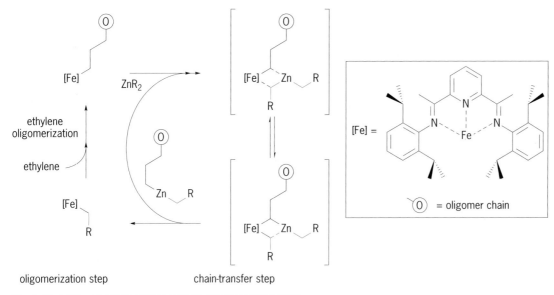

Fig. 4. Fe-catalyzed reaction showing chain-growth reaction on Zn.

(MAO), the Fe complexes produce linear PE at very high activity, which compares favorably with the activity of the group 4 metallocene catalysts. The product molecular weight displays a marked dependency upon the catalyst's aryl substituents, resulting in α-olefins to high-molecular-weight PE. Increasing the size of the substituent, which decreases chain-transfer rates, results in the formation of a higher-molecular-weight product. In addition to β-H transfer, chain transfer to Al is operative for ethylene polymerization with the Fe complexes (Fig. 2b), leading to PE with high polydispersity (molecular-weight distribution), $M_w/M_n = 19.2$, as calculated by the weight-average molecular weight (M_w) divided by the number-average molecular weight (M_n), where $M_w/M_n = 1$ is monodisperse. The Fe complexes may function as good catalysts for the production of high-density PE.

Interestingly, ethylene oligomers with a low polydispersity (for example, $M_n = 700$, $M_w/M_n = 1.1$) can be obtained when the reaction is done in the presence of $Zn(C_2H_5)_2$. The low polydispersity is reminiscent of a living polymerization, in which no chain transfer is involved. The molar ratio of the oligomers produced and the Zn added is practically 2, suggesting that long-chain dialkyl Zn is formed by an Fe-catalyzed chain-growth reaction on Zn, which after hydrolysis affords ethylene oligomers (**Fig. 4**). These results show that Zn facilitates highly efficient chain transfer to and from the Fe center, representing the first example of reversible chain transfer for olefin polymerization catalyst systems.

Pyridylamine-ligated Hf complexes. Researchers at Dow Chemical and at Symyx recently developed pyridylamine-ligated Hf complexes for olefin polymerization, using high-throughput screening (HTS) methodology. The Hf complexes possess unique structures featuring the presence of Hf-aryl sigma bonds. It is difficult to imagine the development of such a catalyst without using HTS methodology, indicating the usefulness of HTS methodology for the dis-

covery of new high-performance catalysts for olefin polymerization.

After activation, the Hf complexes exhibit unprecedented catalytic behavior for the polymerization of propylene. For example, Hf complex (Fig. 2c) with borate can form highly isotactic, high-molecular-weight PP with a peak melting temperature (T_m) of about 150°C in a high-temperature (>100°C) solution process. This is a rare example of a catalyst that displays molecular-weight capability and stereoselectivity at high polymerization temperature. Also of note is that modification of the ligand structure by monomer insertion into the Hf-aryl bond results in catalytically active species that possess two diastereotopic coordination sites, in which 1,2-monomer insertions occur at one face of the catalyst.

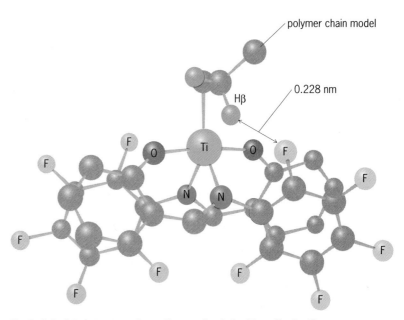

Fig. 5. Calculated structure of an active species derived from Fig. 2e. tBu groups are omitted for clarity.

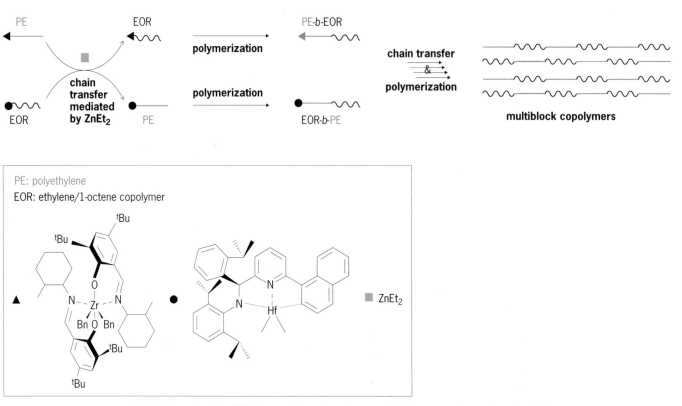

Fig. 6. Formation of multiblock copolymers through a reversible chain transfer, mediated by ZnR₂.

Bis(phenoxyimine)-ligated Ti, Zr, and Hf complexes.
Bis(phenoxyimine) group 4 transition-metal complexes, also known as FI catalysts, were developed by researchers at Mitsui Chemicals in the late 1990s. After activation, FI catalysts display strikingly high ethylene polymerization activity, which often exceeds that of the group 4 metallocene catalysts. The highest catalyst turnover frequency exhibited by Zr-FI catalyst (Fig. 2d) reached 64,900/s per atom, which is two orders of magnitude greater than that seen with common group 4 metallocene catalysts.

The very high catalyst efficiency and readily varied steric and electronic nature of the phenoxy-imine ligands have yielded a wide variety of unique polyolefins, some of which were previously unavailable via other means of polymerization. The polyolefins include ultrahigh-molecular-weight PEs, selective vinyl-terminated PEs, and ultrafine noncoherent particle PEs that are smaller than 10 μm in size. The vinyl-terminated PE can readily be transformed to epoxy- and diol-terminated PE, which is a valuable material for block and graft copolymers containing a PE and a polar polymer segment.

Interestingly, fluorinated Ti-FI catalysts, such as Fig. 2e, can induce thermally robust living ethylene and highly syndiospecific living propylene polymerizations. The highly controlled living nature is proposed to originate from an attractive interaction between the *ortho*-F and a β-H on the growing polymer chain, which suppresses chain transfers (**Fig. 5**).

The fluorinated Ti-FI catalysts allow access to di- and multiblock copolymers containing crystalline and amorphous segments from ethylene, propylene, and higher α-olefins. The diblock copolymer composed of PE and amorphous ethylene/propylene copolymer segments possesses a good combination of extensibility and toughness.

Recently, researchers at Dow Chemical reported the production of multiblock copolymers of PE and amorphous ethylene/1-octene copolymer segments, using a catalyst system consisting of Zr-FI catalyst (Fig. 2f), pyridylamine-ligated Hf complex (Fig. 2c), and ZnEt₂. The Zr-FI catalyst selectively generates PE even in the copresence of ethylene and 1-octene, while the Hf complex forms amorphous copolymers, resulting in the formation of multiblock copolymers through a reversible chain transfer mediated by ZnR₂ (**Fig. 6**). Unlike the random copolymer of similar density, the block copolymer possesses a 40°C higher melting temperature ($T_m \sim 120°C$), while maintaining excellent elastic properties. The development of an FI catalyst with extremely high ethylene selectivity as well as the reversible chain-transfer nature has made it possible to produce these unique polymers.

For background information *see* CATALYSIS; LIGAND; METALLOCENES; POLYMER; POLYMER STEREOCHEMISTRY AND PROPERTIES; POLYMERIZATION; POLYOLEFIN RESINS; SINGLE-SITE CATALYSTS (POLYMER SCIENCE) in the McGraw-Hill Encyclopedia of Science & Technology.

Hiromu Kaneyoshi; Terunori Fujita

Bibliography. D. J. Arriola et al., Catalytic production of olefin block copolymers via chain shuttling polymerization, *Science*, 312:714, 2006; M. Brookhart et al., Late-metal catalysts for ethylene homo- and copolymerization, *Chem. Rev.*, 100:1169, 2000; T. Fujita et al., FI catalysts: new olefin polymerization catalysts for the creation of value-added polymers, *Chem. Rec.*, 4:137, 2004; V. C. Gibson et al., Advances in non-metallocene olefin polymerization catalysis, *Chem. Rev.*, 103:283, 2003; V. C. Gibson et al., Surprisingly reactive ligands and a gateway to new families of catalysts, *Chem. Rev.*, 107:1745, 2007; J. C. Stevens et al., Nonconventional catalysts for isotactic propene polymerization in solution developed by using high-throughput-screening technologies, *Angew. Chem. Int. Edit.*, 45:3278, 2006.

Nonvolatile memory devices

Nonvolatile memories are semiconductor devices that retain their memory state when power is removed. These memories were first proposed in 1967 in the form of metal-nitride-oxide-semiconductor (MNOS) and floating gate concepts, with the floating gate structure ultimately becoming the most prevalent. First commercialized as the EPROM, Erasable Programmable Read Only Memory, almost 40 years ago, nonvolatile memory has evolved today to the form of flash memory as the fastest-growing memory segment, driven by the rapid growth of portable devices such as digital cameras, cellular phones, and music players. Flash memory allows for a subarray within the chip to be erased at one time, typically of 256 Kbytes. The most prevalent version of flash, NAND flash, is orientated toward data-block storage applications. [The term NAND actually refers to a logic gate "NAND" (Not AND), since the memory cells within NAND flash are configured similar to the transistors of a NAND logic gate.] The other main version of flash is called NOR, oriented toward embedded code execution. The nonvolatile NAND flash memory stores code and user files (music files, photos, videos, and so forth) in these products. This growth has been fueled by ever lower cost per memory bit achieved by the reduction of the memory cell area to smaller dimensions, a process called scaling. By the end of 2008, NAND flash manufacturing was expected to cross below the 50-nanometer node (where 50 nm is the width and length dimensions of the memory cell, along with the spaces between adjacent cells), and memory capacities were expected to reach 32 Gbit per chip. Scaling allows for more memory bits to occupy the same area from generation to generation and thus higher-capacity memories and lower cost per bit are attained. Availability of lower cost and higher-capacity memory, in turn, drives more application growth, with flash solid-state drives now poised to compete in computer applications against traditional magnetic hard disk drives.

The continual scaling to smaller dimensions, however, poses many technological challenges. These

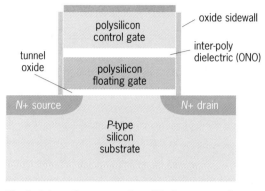

Fig. 1. Schematic cross section of flash memory cell.

challenges spark research into alternative storage approaches from the charge storage of flash memory, in hopes of achieving better scalability, and thus lower-cost memory, in the future.

Flash cell basic operation. A floating-gate nonvolatile flash memory cell is shown schematically (**Fig. 1**). It is a metal-oxide-semiconductor (MOS) transistor with two gates: a floating gate and a control gate. The memory cell consists of an *N*-channel transistor with the addition of an electrically isolated polysilicon floating gate. Electrical access to the floating gate is only through a capacitor network of surrounding SiO_2 oxide layers and source, drain, transistor channel, and polysilicon control gate terminals. Any charge present on the floating gate is retained due to the inherent $Si\text{-}SiO_2$ energy barrier height, leading to the nonvolatile nature of the memory cell. A typical structure includes a thin tunneling oxide ($\sim$8 nm), an oxide-nitride-oxide (ONO) inter-poly dielectric (IPD) about two or three times thicker than the tunneling oxide, and a short electrical channel length. Because the only electrical connection to the floating gate is through capacitors, the flash cell can be thought of as a linear capacitor network with an *N*-channel transistor attached. The threshold voltage of the device can be changed by modifying the charge on the floating gate, which can retain this charge for many years. Data can be stored in the memory cell by adding or removing charge (electrons). In its simplest form, two threshold levels (high and low) can store one binary bit of information in each memory cell. This concept can be extended to store more levels, a technique commonly called MLC (multilevel cell). Four levels of charge allow two binary bits of data per physical memory cell to be stored. The number of levels is theoretically quantized by the number of electrons stored on the floating gate, but in practice is limited to far less (typically to two or three logical binary bits per physical memory cell), since it is difficult to differentiate the exact number of electrons given manufacturing variations. Adding and removing charge from the floating gate is normally achieved by Fowler-Nordheim tunneling (tunneling through a potential barrier, as is observed in field emission).

Capacitor network scaling challenges. In addition to basic structural and mechanical integrity of the

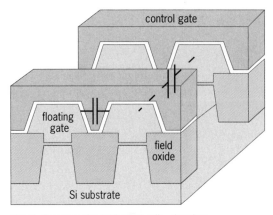

Fig. 2. Diagram of neighboring cells, showing representative parasitic capacitances.

memory cell at nanometer dimensions, the primary limitations to scaling are electrical considerations.

Floating-gate interference and charge retention are important attributes for cell scaling. As feature size scales, parasitic capacitive coupling from the floating gate of one cell to that of its neighbor causes an unwanted shift in threshold voltage (**Fig. 2**). This parasitic interference is predicted to become as large an effect on the memory state as is the actual stored charge. If this prediction were to occur, it would not be possible to discern the state of the memory cell from that of the neighbors. The other key limitation for scaling is the inability to significantly scale the thickness of the tunnel oxide, due to charge retention considerations (required for nonvolatile data retention). Stress damage to the tunnel oxide through program and erase cycles (writing and erasing data in the memory) causes data retention loss, cell disturbs, and erratic behavior of the memory cells. This is exacerbated by the limited number of stored charges on the floating gate and MLC operation requiring discrete levels of charge (**Fig. 3**).

Potential scaling solutions. Replacement of the continuous floating gate structure with charge-trap-flash (CTF) that has discrete trapping layers (that is, nanodots or a nitride film) is viewed as the most likely scaling solution. The advantage of CTF is that the charge is stored in discrete locations, whereas on the continuous floating gate, the charge is free to move within the floating gate itself. Therefore, with CTF, a tunnel oxide defect will only cause loss of electrons local to the trap, versus loss of the charge state of the entire memory cell, as would occur in the floating gate structure. Another advantage is the reduced parasitic interference, due to reduced capacitive coupling of traps of one cell to those of another. The first attempts at semiconductor nonvolatile memories, in the 1960s, were based on CTF concepts. Major tradeoffs between data retention time and write speed left little room for a successful engineering compromise in those early attempts. Generally, traps of sufficiently low energy to capture electrons quickly (required for fast memory writing speed) also release the electrons quickly, leading to data-retention problems. With modern material deposi-

tion techniques, engineered barrier heights and profiles of trapping layers are possible, leading credence to a potential solution.

Maintaining adequate capacitive coupling of the control gate to the floating gate and the silicon channel is another key aspect to scaling of the memory cell area. This is required to keep the voltages and electric fields as low as possible and to provide control of the channel for reading and programing. As the memory cell dimensions scale, maintaining control of the channel requires a thinner IPD between the control gate and the floating gate, approaching the thickness of the tunneling dielectric. The requirement of IPD is to provide good capacitive coupling to the floating gate from the control gate, while minimizing any leakage through the dielectric. Unlike the tunnel oxide, this dielectric is not expected to support charge transfer during programming and erase operations. For the past decade, the IPD has been composed of an ONO sandwich film. ONO scaling has been achieved by thinning the dielectrics while maintaining control of the thickness. Moves toward CTF storage approaches also impose increased electrical constraints on the present ONO films.

An alternative approach would be to replace the IPD with a higher dielectric constant material (high-k), as compared to SiO_2. Since this film is not expected to transport charge (as is the case with the

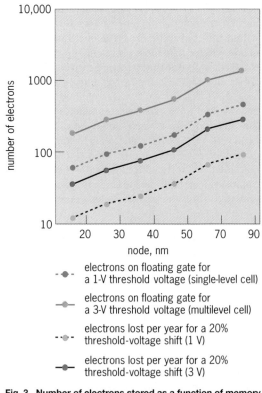

Fig. 3. Number of electrons stored as a function of memory dimension and requirements for data retention. Node nm is the width and length dimensions of the memory cell in nanometers, along with the spaces between adjacent cells. (*After K. Prall, Scaling Non-Volatile Memory Below 30 nm, IEEE Non-Volatile Semiconductor Memory Workshop, 2007*)

tunnel oxide), it has fewer constraints and it may be easier to find a candidate to meet the memory cell requirements, as compared to replacing the tunnel oxide. Possible options include Al_2O_3 or HfO_2, different from logic devices or DRAM capacitors, due to differing requirements. The gate electrode should be chosen consistent with the high-k film to make sure the energy barrier is engineered to reduce unwanted charge injection.

Innovative methods will continue to be required to meet the challenges of memory cell scaling. As the planar dimensions start reaching fundamental limits of electric fields, the vertical dimension, or scaling into 3D, is becoming an additional area of future research for decreasing memory cell size. These and other innovations are expected to be deployed over the coming years, continuing memory scaling, albeit with more difficulty each generation. Lastly, new memory concepts now in the research phase, such as phase-change memory or magnetic memories, not relying on charge storage as a basis of the memory cell, provide new opportunities for future memories.

For background information *see* LOGIC CIRCUITS; SEMICONDUCTOR MEMORIES; TRANSISTOR in the McGraw-Hill Encyclopedia of Science & Technology.
                                                    Albert Fazio

Bibliography. M. Bauer et al., A multilevel-cell 32Mb flash memory, in *Technical Digest, IEEE International Solid State Circuits Conference*, pp. 132–133, February 16, 1995; J. Brewer and M. Gill, *Nonvolatile Memory Technologies with Emphasis on Flash*, Wiley-IEEE Press, 2008; W. Brown and J. Brewer, *Nonvolatile Semiconductor Memory Technology: A Comprehensive Guide to Understanding and Using NVMS Devices*, IEEE Press, 1997; G. W. Burr et al., Overview of candidate device technologies for storage-class memory, *IBM J. Res. Devel.*, 52:1288–1295, 2008; J. De Blauwe, Nanocrystal nonvolatile memory devices, *IEEE Trans. Nanotechnology*, 1(1):72–77, 2002; K. Prall, Scaling, Non-Volatile Memory Below 30nm, in *IEEE Non-Volatile Semiconductor Memory Workshop*, 2007.

Nuclear RNA polymerases IV and V

In all living cells, genetic information stored as deoxyribonucleic acid (DNA) is copied into ribonucleic acid (RNA) polymers that mediate the chemical transactions of life. For instance, messenger RNAs (mRNAs) carry the codes for proteins to the ribosomes, where proteins are made. So-called noncoding RNAs (ncRNAs) that do not encode proteins also carry out essential functions. For instance, ribosomal RNAs (rRNAs) and transfer RNAs (tRNAs) are essential for protein synthesis. Other ncRNAs guide RNA processing events. Because of the importance of RNA, the enzymes that transcribe DNA into RNA, known as DNA-dependent RNA polymerases, are critical to life. Bacteria and archaea have a single DNA-dependent RNA polymerase. However, in eukaryotic organisms, including fungi, plants, and animals, three different DNA-dependent RNA polymerases, abbreviated as Pol I, Pol II, and Pol III, are essential for cell viability (see **table**). It was long assumed that eukaryotes use only Pol I, II, and III. However, the sequencing of the *Arabidopsis thaliana* genome revealed that plants possess two additional nuclear RNA polymerases, Pol IV and Pol V (formerly known as Pol IVa and Pol IVb, respectively) that are not strictly required for cell survival but that specialize in the production of ncRNAs that mediate gene silencing.

Pol IV, Pol V, and transcriptional gene silencing. RNA-mediated gene silencing can occur by preventing transcription (the synthesis of RNA by DNA-dependent RNA polymerases), which is therefore known as transcriptional gene silencing (TGS), or by guiding the cleavage or inactivation of RNAs after transcription has occurred, which is known as posttranscriptional gene silencing (PTGS). Both types of silencing are used in defense of the genome. For instance, PTGS helps combat the proliferation and spread of viruses by degrading viral gene transcripts or, in some cases, the viral RNA genome. Likewise, TGS is important for taming retrovirus-like transposable elements, known as retrotransposons, which multiply by being transcribed into RNAs that are then copied into DNA elements that can insert into new chromosomal locations. Retrotransposons are common throughout the genomes of most multicellular eukaryotes. If they are not transcriptionally silenced, they can proliferate and do harm by integrating into the middle of essential genes, thereby abolishing their functions. Engineered transgenes that are generated through recombinant DNA technology and integrated into the plant genome are frequently silenced by PTGS or TGS, presumably because the introduction of the foreign genes can trigger the defenses that have evolved to combat invading viruses and retrotransposons.

Functions for Pol IV and Pol V are best known with regard to the transcriptional gene silencing of retrotransposons and transgenes via the so-called RNA-directed DNA methylation pathway, a pathway that has been genetically defined in *Arabidopsis thaliana*. In this pathway, long double-stranded RNAs (dsRNAs) generated by the collaboration of Pol IV and RNA-dependent RNA polymerase 2 (RDR2) are cleaved by the enzyme Dicer-like 3 (DCL3) to generate short dsRNA molecules whose component strands are 24 nucleotides (nt) in length (see **illustration**). Individual 24-nt RNAs, known as short-interfering RNAs (siRNAs), become bound to argonaute4 (AGO4) and are thought to guide AGO4 to loci whose sequences can pair with the siRNAs. In a subsequent step, the DNA sequences that match the siRNAs are modified by adding methyl groups (one carbon and three hydrogen atoms) onto the cytosines (shown as ^{me}C in the illustration), one of the four chemical building blocks of DNA. The cytosine methylation reaction is catalyzed by a specific DNA methyltransferase enzyme, DRM2 (domains rearranged methyltransferase 2). Whereas Pol IV is

Multisubunit nuclear RNA polymerases of *Arabidopsis thaliana*		
Enzyme	Abbreviation	Transcripts or functions
RNA polymerase I	Pol I	18S, 5.8S, and 25S rRNAs
RNA polymerase II	Pol II	mRNA, miRNA, numerous small nuclear RNAs
RNA polymerase III	Pol III	tRNA, 5S rRNA, several small nuclear RNAs
RNA polymerase IV	Pol IV (also Pol IVa)	RNA precursors of siRNAs
RNA polymerase V	Pol V (also Pol IVb)	Targeting of RNA-directed DNA methylation

Note: Specificities of the five *Arabidopsis* multisubunit nuclear RNA polymerases are shown. Pol I transcribes the three largest RNAs inherent to the structure and function of ribosomes (rRNAs). Pol II transcribes messenger RNAs (mRNAs) that encode proteins, microRNAs (miRNAs) that direct the cleavage of mRNAs or that block the mRNAs from being translated into proteins, and small nuclear RNAs that guide various RNA processing events. Pol III transcribes the smallest of the four rRNAs, transfer RNAs (tRNAs) that carry amino acids for protein synthesis, and several small nuclear RNAs. Pol IV is somehow involved in the production of siRNAs. Pol V generates transcripts that play a role in the cytosine methylation of the transcribed locus.

required near the beginning of the pathway leading to the production of siRNAs, Pol V acts at the target loci to make them receptive to modification by DNA methylation. The details of how this occurs are not fully understood. One model suggests that siRNAs associated with AGO4 bind to Pol V transcripts while they are still being synthesized, thereby bringing the DNA methylation and gene silencing machinery to the vicinity of the affected loci (see illustration).

Pol IV and Pol V in flowering and stress responses. In addition to playing genome defense roles, such as the silencing of transposons, Pol IV and Pol V play roles in regulating genes that control flowering time, responses to environmental stresses, and responses to pathogen attack. Under short-day conditions (8 h light, 16 h dark), flowering is significantly delayed in Pol IV and Pol V mutants, which is also true in other mutants that disrupt the RNA-directed DNA methylation pathway. In fact, several genes that were initially identified based on their effects on flowering time in *Arabidopsis* when they suffered mutations

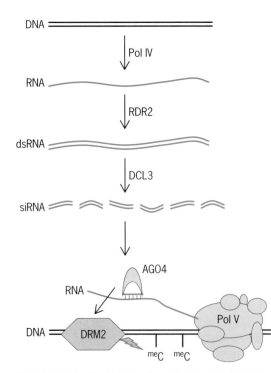

Model depicting the distinct roles of Pol IV and Pol V in the RNA-directed DNA methylation pathway in *Arabidopsis*. See text for terms and details.

appear to be components of the RNA-directed DNA methylation pathway or to be target genes that are regulated by the pathway.

Pol IV and Pol V do not always work as part of the same pathways. For instance, Pol IV plays an important role in responses to environmental stresses or pathogen attacks that are regulated by a class of siRNAs known as nat-siRNAs because they are generated from overlapping Natural Antisense Transcripts. The dsRNAs that give rise to nat-siRNAs stem from adjacent genes whose distal ends overlap. In the examples studied thus far, one of the genes is transcribed continuously, but the other gene is turned on as a response to the stress (for example, drought, nutrient deficiency, bacterial infection). Production of RNAs from each of the paired genes results in the formation of dsRNA in the region of overlap between their mRNAs. The activity of the nuclease enzyme Dicer generates siRNAs that direct the posttranscriptional degradation of the constitutively transcribed gene, thereby bringing about a physiological response that is beneficial for surviving the challenge. Although it is not entirely clear what Pol IV does within the pathway(s), Pol IV mutants are defective for stress-responsive nat-siRNA production. However, in many cases studied thus far, Pol V mutants have no effect on the production of the nat-siRNAs.

Pol IV and mobile silencing signals. Upon turning the PTGS machinery against an invading virus or highly expressed foreign transgene, plant cells generate signaling molecules, thought to be RNAs, that can prepare distant cells to fight the invader. This is a particularly useful strategy for combating the systemic spread of viruses throughout the plant. Pol IV has been shown to be required for production of molecules responsible for the short-range spread of RNA silencing from one cell to another via intercellular connections known as plasmadesmata. In this case, the signaling molecules are thought to be siRNAs. Pol IV is also required in the long-range transmission of silencing signals from the bottom of the plant to the top of the plant via the vascular system. In the case of long-range silencing, Pol IV is dispensable for the production of the mobile signals, which are apparently not siRNAs but may be longer RNAs derived from the genes triggering the silencing response. However, Pol IV is required for the perception of the long-distance silencing signal and the subsequent siRNA-mediated silencing of the target RNA at the distant location once the signal has been perceived. Pol V does not appear to be involved in either short-range or long-range silencing. Collectively, these observations are consistent with the hypothesis that Pol IV functions in the production of siRNAs, whereas Pol V plays a different role in the cell that is limited mostly to loci that are subject to RNA-directed DNA methylation.

Origins of Pol IV and Pol V. Eukaryotic DNA-dependent RNA polymerases I, II, and III are each composed of 12–17 protein subunits that interact to form the fully functional enzymes. Although the full subunit structures of Pol IV and Pol V are not known, their largest and second largest subunits clearly evolved from duplicated genes for the corresponding subunits of Pol II, and there is evidence that both enzymes utilize small subunits that are also present in Pol II. The evolutionary relationship among Pol II, Pol IV, and Pol V may also explain the observation that siRNA-mediated TGS and PTGS mechanisms occur in diverse organisms including fungi, worms, flies, and mammals, as well as plants, yet only plants have Pol IV and Pol V class polymerases. Presumably, Pol II fulfills the roles of Pol IV and Pol V in these other organisms. This hypothesis is supported in fission yeast by evidence indicating that mutations in at least two Pol II subunits can disrupt siRNA-mediated transcriptional gene silencing.

Priorities for future studies. The biological functions for Pol IV and Pol V have only been recently defined, so there is still a great deal to be learned about these enzymes. In particular, it will be important to learn how Pol IV and Pol V are recruited to their sites of action, the characteristics of the RNAs that they produce, the complete subunit structures of the functional Pol IV and Pol V enzymes, and the proteins that interact with Pol IV and Pol V in order to regulate or coordinate their activities.

For background information see DEOXYRIBONUCLEIC ACID (DNA); ENZYME; GENE; MOLECULAR BIOLOGY; NUCLEOPROTEIN; PROTEIN; RETROVIRUS; RIBONUCLEIC ACID (RNA); RIBOSOMES; TRANSCRIPTION; TRANSPOSONS; VIRUS in the McGraw-Hill Encyclopedia of Science & Technology.   Craig S. Pikaard

Bibliography. P. Brodersen and O. Voinnet, The diversity of RNA silencing pathways in plants, *Trends Genet.*, 22:268–280, 2006; P. Cramer, Multisubunit RNA polymerases, *Curr. Opin. Struct. Biol.*, 12:89–97, 2002; C. S. Pikaard et al., Roles of RNA polymerase IV in gene silencing [Epub ahead of print], *Trends Plant Sci.*, 13:390–397, 2008; M. Zaratiegui, D. V. Irvine, and R. A. Martienssen, Noncoding RNAs and gene silencing, *Cell*, 128:763–776, 2007.

Organic-inorganic interfaces

Organic electronic and magnetic materials are being incorporated into electronic devices as alternatives to purely inorganic systems. Most of the physical properties of these materials, such as the ability to conduct electrons, can be custom tailored via the modification of the chemical repeat unit or the addition of one or more functional groups. This has led to an expanding library of materials, phenomena, and innovative devices. In addition, organic-based devices offer mechanical flexibility, lightweight design, and low-cost fabrication, all versatile characteristics uncommon in their inorganic counterparts. The science of the interaction of inorganic materials with organic electronic and magnetic materials is a growing field in its own right, and of particular importance to designing and optimizing electronic devices.

Some of the devices that are affected by the interfaces of organic electronic and magnetic materials are organic solar cells (OSCs), organic field-effect transistors (OFETs), organic light-emitting diodes (OLEDs), and organic spin valves (OSVs). These devices are fabricated using layer-by-layer depositions (via evaporation, sublimation, spin-coating, and so forth) of organics and inorganics. For instance, a simple bilayer OSC can be fabricated by sandwiching two organic semiconductor layers, one electron-accepting layer (acceptor) and one electron-donating layer (donor), between an inorganic metal anode and an inorganic metal cathode (**Fig. 1**). It is essential that the two organic semiconductors be electronically different (for example, acceptor-donor) so that the internal potential energy gradients at the organic-organic (o-o) interface will be strong enough to assist in the separation of photo-generated bound states (excitons) of a negative electron and a positive quasiparticle hole. For example, the o-o interface of a bilayer OSC between copper phthalocyanine–based donor and perylene tetracarboxylic derivative–based acceptor splits the excitons to achieve efficiencies over 1%. The area of the interface, which limits the number of separated charge carriers, can be increased dramatically by using dispersed acceptor-donor blends [such as buckminsterfullerene and poly(*p*-phenylene vinylene) blends] in a bulk-heterojunction device. These separated charge carriers can then be collected at organic-inorganic (o-i) interfaces with the metal electrodes. These interfaces can have nonohmic electronic behavior that increases the series resistance of the device, leading to decreased efficiency. In OFETs and OLEDS, charge carriers are injected across similar o-i interfaces from metal electrodes into an organic transporting material. The electrons and holes experience potential energy barriers to injection at interfaces that dramatically affect device performance; large injection barriers require high operating voltages. In OSVs symmetry is broken at an o-i interface, which enables interconversion of up spin and down spin and reduces spin lifetime. Indeed, a fundamental understanding of the degree to which dissimilar materials at an interface react to form new bonds, exchange charge, redistribute charge, transfer spin, and transfer photons, as well as the physical phenomena associated with these interactions, is vitally important to improving organic electronic devices.

Organics. Polymers (often termed "plastics") are long chains of carbon-based monomers linked

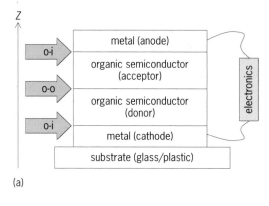

(a)

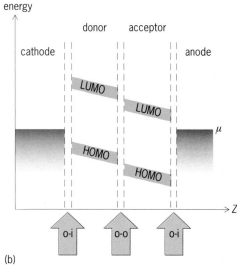

(b)

Fig. 1. Organic solar cell (OSC). (*a*) Side view of the layers in an OSC. (*b*) Schematic energy-level diagram of the OSC including the organic semiconductor's highest occupied molecular orbital (HOMO) and lowest unoccupied molecular orbital (LUMO) levels. Solid arrows indicate interfaces between dissimilar materials, while broken lines emphasize the need to understand how key energy levels behave at those interfaces.

dered, there is periodic, quantum-mechanical overlap of molecular orbitals. This causes the distinct energy levels to broaden into continuous bands of energy states as found in the band theory of inorganic metals and semiconductors. However, for disordered organic materials the intermolecule interaction is generally weak and the organic electronic structure is usually best described by distinct, localized energy states. In these systems, a maximum of two electrons (with opposite spin) can occupy each distinct energy state. The two principal energy states for organics are the highest occupied molecular orbital (HOMO) and the lowest unoccupied molecular orbital (LUMO), and are analogous to the valence band and conduction band in band theory, respectively. Two other important parameters considered at interfaces are the chemical potential (μ), which is a thermodynamic quantity related to the energetic position of electrons in a system, and the vacuum level, which is related to the energy needed to free an electron from the solid to just outside the surface of the solid. The relative positions of these and other energy levels, with respect to the energy levels of adjacent materials, dictate how charge carriers are transported across interfaces.

Interfaces. One of the most important interfaces in organic devices is between an organic material and an inorganic metal electrode. This o-i interface provides electrical connection between an organic device and an external circuit. Commonly, at least for vertically constructed devices, an organic layer is deposited on top of an already existing metal electrode by either thermal sublimation or spin-coating. In this case the interaction between the metal and the organic is largely physical without much chemical reaction due to thin native oxide layers and water monolayers on the metal surface that act as barriers to reaction. Chemical reaction can occur more readily, however, when the organic is deposited on an especially clean metal surface under ultra-high-vacuum conditions where these surface layers are not present. Subsequent metal electrodes are often deposited in reverse by thermally evaporating metal onto an organic layer. In this case, the kinetic energy of the metal atoms can provide the activation energy for reaction, and these interfaces can be dominated by chemical reactions including the formation of new organometallic bonds. In addition, the high kinetic energy of the metal atoms (or clusters of atoms) can lead to deep penetration into the organic layer, complicating any treatment of the interface as ideally flat. Organic interfaces on metals are being used to control corrosion. While some interfaces, such as paint, only serve as a barrier to corrosion, differences in chemical potentials at polymer-metal interfaces can lead to substantially reduced corrosion rates for iron and aluminum. For the case of iron, this is effective even for remote surfaces away from the polymer coating.

Other important interfaces include those between an organic and an inorganic semiconductor and those between two organic layers. One common example of an o-i interface is between a metal-oxide

together. Although the everyday polymers used for such things as water bottles and trash bags are nonmagnetic electrical insulators, there exists a wealth of polymers, as well as smaller molecule organics, that exhibit fascinating electronic and magnetic properties. When fabricating organic devices that exploit these properties, small-molecule organic films are typically thermally sublimated while polymers are typically wet processed using techniques such as spin coating. The method of deposition and the deposition environment significantly impact how materials interact at interfaces. These interactions can affect the physical morphology of the interface as well as the energies of electrons near the interface.

Understanding the electronic structure of organics, and how electronic structures change near interfaces, is essential to modeling the behavior of electrons at junctions between materials. The valence atomic orbitals of atoms in molecules overlap with one another and form molecular orbitals with distinct energy levels. When organic monomers are closely spaced, such as along polymer chains, and when small-molecule organic solids are highly or-

inorganic semiconductor such as titanium dioxide (titania) and an organic dye for use in dye-sensitized solar cells (including Grätzel cells). At this interface the organic is covalently bonded with the inorganic semiconductor to promote charge transfer; the formation of new bonds is exploited to harvest photogenerated charge carriers and improve device efficiency. In addition, stacked (tandem) hybrid solar cells have utilized an o-i interface as a tunnel junction. Organic-organic interfaces are of particular interest for their use in OSCs and OLEDs. The interface area of bulk-heterojunction OSCs is increased by controlling the interpenetration of organic layers to improve device efficiency. The broken symmetry at the o-o interface of an OLED promotes the formation of exciplex states, which are bound states of excited acceptor-donor pairs at the o-o interface, which improves radiative recombination and controls emission wavelengths.

Phenomena. A simple method of describing the energetics between two materials is to assume that their chemical potentials and vacuum levels align at the interface (**Fig. 2**). The main idea behind this method is to use basic electronic properties of bulk materials to predict how electrons behave near interfaces. The first assumption is that an offset in chemical potentials between two materials at an interface leads to charge-carrier exchange across the interface (from the bulk) toward thermodynamic equilibrium. This assumption leads to a shifting of energy levels (energy-level bending) as a function of thickness, but is valid only when a significant number of mobile charge carriers are present near the interface. Heavily doped polymers exhibit modest shifts, and undoped organic semiconductors have essentially flat energy levels. The second assumption of vacuum level alignment, also known as the Mott-Schottky limit, is largely invalid for organic interfaces due to the formation of interfacial dipoles. When charge carriers transfer across the interface from one material to the other (from the first few atomic layers) or redistribute in response to another layer, an electronic dipole shifts the energy levels of the organic material. These two assumptions are insufficient in describing the complex interactions at interfaces; more sophisticated models are currently being developed.

Interfacial electronic dipoles that shift energy levels are generally strongest for metal-organic interfaces and weaker for organic-organic interfaces. Interfacial dipoles are generated by a number of phenomena including charge redistribution outside metal surfaces, molecular polarization due to image charges at metal interfaces, transfer of charge from one layer to another, mixing of electron densities, and permanent dipoles of organic molecules. The magnitude of dipoles at metal-organic interfaces is often associated with the redistribution of electrons that extend past the metal surface. This "pillow effect" is created when electrons are pushed back into the metal by the Coulombic repulsion from an organic layer (**Fig. 3**). Interfacial dipoles shift the organic's energy levels, which can increase or decrease the potential energy barriers to injection and

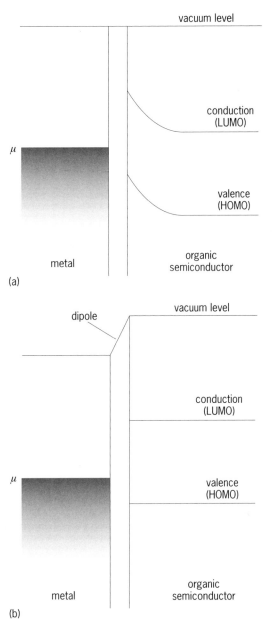

(a)

(b)

Fig. 2. Schematic energy-level diagrams of a key interface between a metal and an organic semiconductor assuming (*a*) energy-level shifting (bending) due to alignment of the chemical potential μ as well as vacuum-level alignment and (*b*) no energy level shift (flat levels) and an interfacial dipole.

greatly affect device performance. Interfacial dipoles for o-o interfaces are smaller because the closed-shell nature of molecules mitigates charge redistribution, transfer, and mixing. The formation of electric dipoles at interfaces is more the rule than the exception and a proper understanding of their role is essential to controlling charge transport in electronic devices.

The nature of the interface also controls the ability to inject spin-polarized charge. Studies have been performed that look at spin-injection across Schottky contacts. Efficient spin injection requires that the semiconductor be driven far out of local thermal equilibrium. Incorporation of Schottky or tunnel barriers at the o-i interface is key to spin injection.

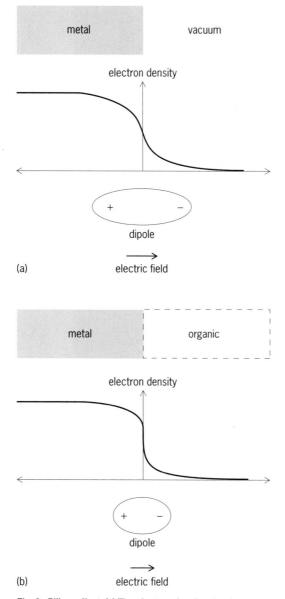

Fig. 3. Pillow effect. (a) The electron density of a clean metal surface extends into a vacuum and generates a electric dipole. (b) The interfacial dipole resulting from the pillow effect, when the organic pushes some of the metal's electron density back into the solid.

Single-molecule o-i interfaces between an organic acceptor with π-orbitals and a transition-metal atom with partially filled d-orbitals can enable the formation of a semiconductor with a valence band that has all electrons with one sign of spin and a conduction band with opposite spin, and is a magnet even at temperatures above the boiling point of water.

Controlling interfaces. The future of organic electronic devices lies in the ability to predict and control the science of interfaces. One method used to accomplish this is to control the interfacial dipoles by depositing a thin interfacial layer between two materials. This interfacial layer can prevent metals from chemically reacting with organics and can shift the potential energy barriers to charge injection. For instance, self-assembled monolayers with permanent dipoles are well-suited for this task because of the ease of deposition and the ability to chemically tailor the dipole moment of the molecule. Also, insulating alkali halide (for example, LiF) or conducting polymer interfacial layers have been shown to be extremely useful for modifying the charge transport across the interface. It should be noted that the mechanism by which interfacial LiF layers improve injection is not well understood. Light emission at o-o interfaces can be governed by charge-transfer complexes (exciplexes), while, for example, a redox polymer between layers can be used to make a symmetrically configured ac light-emitting (SCALE) device. Oxygen and water contamination at interfaces, typically introduced during deposition or from exposure, can destroy an organic device by breaking bonds. However, sometimes a small native oxide layer on the surface of a metal is required to prevent an organic from chemically reacting with the metal. The biocompatibility of electronic devices will no doubt be improved with a better understanding of interfaces. It indeed appears as though interface science and technology will lead to highly structured interfaces that precisely control physical phenomena inside organic electronic devices.

For background information *see* BAND THEORY OF SOLIDS; BIOELECTRONICS; CHEMICAL EQUILIBRIUM; INTERFACE OF PHASES; MOLECULAR ELECTRONICS; MOLECULAR ORBITAL THEORY; MONOMOLECULAR FILM; SEMICONDUCTOR in the McGraw-Hill Encyclopedia of Science & Technology.   Austin R. Carter;
Arthur Epstein

Bibliography. J. Albrecht and D. Smith, Electron spin injection at a Schottky contact, *Phys. Rev. B*, 66:113303, 2002; P. Blom and D. Markov, Device physics of polymer: Fullerene bulk heterojunction solar cells, *Adv. Mater.*, 19:1551, 2007; M. Fahlman and W. Salaneck, Electronic structure of hybrid interfaces for polymer-based systems, *J. Phys.: Condens. Matter*, 19:183202, 2007; H. Ishii, and K. Seki, Energy level alignment and interfacial electronic structures at organic/metal and organic/organic interfaces., *Adv. Mater.*, 11 no. 8, 1999; N. Koch, Organic electronic devices and their functional interfaces., *Chem. Phys. Chem.*, 8:1438, 2007; J. Kortright and A. Epstein, Bonding, backbonding, and spin-polarized molecular orbitals: basis for magnetism and semiconducting transport in V[TCNE]$_{x\sim2}$, *Phys. Rev. Lett.*, 100:257204, 2008; V. Prigodin and A. Epstein, Spin-driven resistance in organic-based magnetic semiconductor V[TCNE]$_x$, *Adv. Mater.*, 14:1230, 2002; W. Salaneck et al., *Conjugated Polymer and Molecular Interfaces: Science and Technology for Photonic and Optoelectronic Applications*, Marcel Dekker, New York, 2002; J. Veinot and T. Marks, Toward the ideal organic light-emitting diode. The versatility and utility of interfacial tailoring by cross-linked siloxane interlayers, *Acc. Chem. Res.*, 38:632, 2005; Y. Wang and A. Epstein, Interface control of light-emitting devices based on pyridine-containing conjugated polymers, *Acc. Chem. Res.*, 32:217, 1999.

Origins of modern amphibians

The evolutionary origin of modern-day amphibians [Lissamphibia: frogs and toads (Anura), salamanders and newts (Urodela or Caudata), and the limbless caecilians (Apoda; "rubber eels"); **Fig. 1**] has been one of the more vexing problems confronting biologists for many years. In the past decade in particular, this issue has been contentious, with three different hypotheses offered to solve the problem. This controversy has, in turn, motivated new research from several different angles, including paleontology, anatomical- and molecular-based analyses of relationship, and molecular-based estimates of divergence timing ("molecular clocks"). The evidence has been mounting, especially in recent years, and a resolution appears to be in sight. This article briefly examines the controversy and some of the new evidence from fossils and molecular studies that is helping to establish the origins of modern amphibians.

Problem of origin. Modern approaches to delineating the relationships among species (a practice called phylogenetic systematics) utilize shared

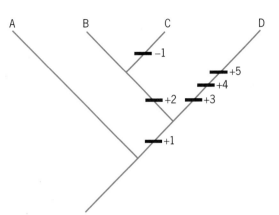

Fig. 2. Shared, derived features over evolutionary time. Time works from the bottom to the top. As features are acquired (black bars with + signs), they are passed on to all descendant lineages. Species C shows a reversal in feature 1 (black bar with − sign) to the condition in A, an example of homoplasy. Species D exhibits a large number of unique features acquired through a long period of independent evolution. These independently acquired features are not evidence of common ancestry. If all taxa evolve independently for long enough, these independent features might converge to the same state, which can make it difficult to reveal true relationships.

evolutionary novelties, either a new anatomical structure (for example, feathers in advanced theropod dinosaurs) or a substitution of one of the base pairs in the sequence of deoxyribonucleic acid (DNA) in either the nucleus or mitochondria of a cell. These novelties, if beneficial, are passed on to the offspring of the organism that possesses them. Over time, these shared, unique features (called synapomorphies) accumulate, and can be read back by computer programs as evidence of the pattern of evolutionary branching over time (**Fig. 2**). However, evolutionary processes can also wipe out these important pieces of evidence; for example, base pairs can switch back and anatomical features can be lost (species C in Fig. 2). Instances when evolutionary information is lost through reversal or when the same state in two species is acquired convergently (together called homoplasy) can lead to clustering of organisms into incorrect arrangements by the computer programs; therefore, it is critical that as many synapomorphies as possible be considered by the programs to counteract the disturbances created by homoplasy.

Multiple factors create difficulty in inferring the relationships among modern amphibians, affecting both anatomical and molecular techniques. First, skeletons, which provide the majority of potential synapomorphies that can link modern organisms to fossil amphibians, are very highly specialized, and the numbers of bones in the skull (a very important feature for studies of relationship) are greatly reduced. The specializations overprint (overwrite) the phylogenetic signal, whereas the reduction in ossification reduces the potential number of synapomorphies, which is frequently based on the interrelationships of one bone to another. Molecular studies similarly suffer from overprinting because of the change over time in fast-evolving sets of DNA

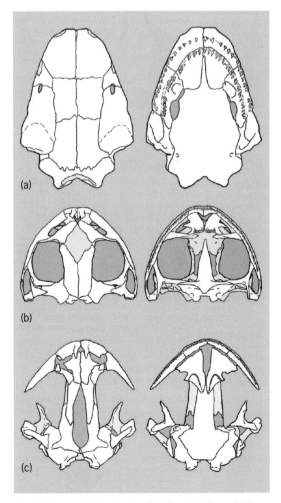

Fig. 1. Lissamphibians in dorsal and palatal views: (a) the caecilian *Dermophis mexicanus*; (b) the anuran *Gastrotheca walkeri*; (c) the caudatan *Salamandrella keyserlingii*. [*From J. S. Anderson, Focal review: The origin(s) of modern amphibians, Evol. Biol., in press, December 2008; used with permission*]

(such as mitochondrial DNA), whereas more slowly evolving DNA might not capture rapid evolutionary splits. As a result of extinctions, it is only possible to sample the most recent "tips" of the evolutionary tree, which allows more overprinting to take place as species evolve independently and which creates possible statistical errors in clustering.

Extinction, which creates gaps in the molecular sample, is mirrored by the unevenness of geologic preservation, which creates gaps in the anatomical record. In the fossil record of amphibians, there are major gaps between the very first definitive fossils of the earliest lissamphibians (Triassic for frogs, Jurassic for salamanders and caecilians) and the fossils from the Pennsylvanian and Permian. Because of these gaps, it has historically been easy to suggest various patterns of relationship, and even the modern computer programs suffer from a scarcity of evidence.

Despite these difficulties, three main hypotheses have been offered in recent years for the origins of modern amphibians. The first, called the temnospondyl hypothesis (TH), suggests that all modern amphibians share a single evolutionary origin among amphibamid temnospondyls (an early group of extinct amphibians) (**Fig. 3***a*). These temnospondyl fossils share a number of features with lissamphibians, such as large openings on the palate and a special condition of the teeth, called *pedicely*, in which the tooth cusp is separated from its base by a zone of weak ossification, creating an inwardly swinging hinge. The TH has been the overwhelming opinion of most scientists since the late 1960s. The second, the lepospondyl hypothesis (LH), posits that all lissamphibians share a single evolutionary origin among a different group of fossil amphibians. Lepospondyls are a diverse assemblage of small animals that share a single vertebral centrum (most temnospondyls have multiple parts in their vertebrae). The LH suggests specifically that lissamphibians are related to a highly specialized group of elongate lepospondyls called lysorophians (Fig. 3*b*). This relationship is supported not by shared anatomical features, but by shared anatomical "losses." Because these "loss features" frequently occur in unrelated organisms, they are not considered as strong evidence of relationship; however, the LH has served as a motivation for much of the research conducted on amphibian origins (see below). The third hypothesis is the polyphyly hypothesis (PH), which takes a little from both of the preceding ideas: Frogs and salamanders are related to temnospondyls, according to the TH, but the highly specialized caecilians are related to another group of lepospondyls—the recumbirostran ("recurved-nose") microsaurs (Fig. 3*c*). This idea is supported by the same synapomorphies for frogs and salamanders (together termed Batrachia) as the TH; however, it considers the differences seen between caecilians and batrachians not as unique specializations for burrowing, but as synapomorphies with the similar looking microsaurs.

New fossils. In recent years, some of the gaps in the fossil record have begun to be filled. Much work has gone into better understanding of the anatomy

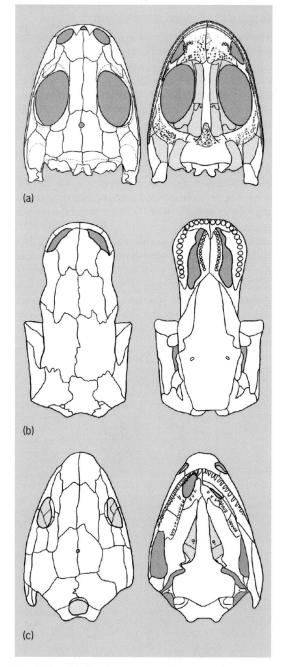

Fig. 3. Possible fossil progenitors of lissamphibians in dorsal and palatal views: (*a*) the amphibamid temnospondyl *Doleserpeton*; (*b*) the lysorophian lepospondyl *Brachydectes*; (*c*) the recumbirostran ("microsaur") lepospondyl *Rhynchonkos*. [*From J. S. Anderson, Focal review: The origin(s) of modern amphibians, Evol. Biol., in press, December 2008; used with permission*]

and relationships of amphibamid temnospondyls, thereby clarifying features of note that should be looked for in the lissamphibian groups. Most significantly, a new "missing link" between temnospondyls and the earliest frogs and salamanders was discovered. Named *Gerobatrachus* (literally, "elderly frog"), this fossil is intermediate to batrachians and temnospondyls in a number of features (**Fig. 4**). The skull is both broader and more lightly built than in amphibamids, with some foreshortening in the rear as seen in batrachians. The palate is more strutlike, the big temnospondyl palatal fangs are replaced with rows of small teeth on raised patches, and the marginal teeth are tiny, very numerous, and pedicellate. The number of vertebrae (17) is halfway between amphibamids (21) and the first batrachians (14). *Gerobatrachus* also shares a unique combination of two ankle bones (the wrist is unknown) called the basale commune, previously seen only in salamanders. The anatomy suggests that *Gerobatrachus* is from just prior to the split between the lineages that would become frogs and salamanders.

The other major fossil find related to this subject is the very earliest known caecilian, named *Eocaecilia* ("dawn caecilian"), which has been recently thoroughly described. This fossil is clearly a caecilian because it has numerous features unique to that group: a highly co-ossified skull, a unique jaw-opening mechanism, and possibly the first traces of the sensory organ known as the tentacle (**Fig. 5**). However, it retains a number of features from more distant ancestors, including limbs, and numerous bones remain free from co-ossification or have not been lost. *Eocaecilia* is most similar in its appearance to the recumbirostran microsaurs, but similarity alone is not sufficient evidence of relationships. Because there still remains a sufficiently large anatomical gap between *Eocaecilia* and the alternate fossil amphibians, caecilian origins will probably remain an area of debate for the near future.

Molecular studies. As technology has made it easier to sequence DNA, genes have played an ever-increasing role in inferring evolutionary relationships because each of the thousands of base pairs in a single gene is potentially informative. Since 2003 there have been at least 11 studies at the level of all modern amphibians, with many more studies looking individually at each group. These studies are converging on two important results. First, they unanimously support a single evolutionary origin, with caecilians branching off earliest. At initial glance, this would appear to support the TH (because *Gerobatrachus* makes the LH highly unlikely). However, the early splits between the amphibian groups occurred rapidly, with subsequent long periods of independent evolution, which is a scenario in which "long branch attraction" might be present, falsely clustering lissamphibians together. This is especially of concern with the studies that relied heavily on the fast-evolving mitochondrial genome.

The second finding of note relates to "molecular clock" estimates of divergence. The fossil record, when read literally, seems to suggest that lissamphibians had arisen by the Triassic. However, molecular estimates push this well back into the latest Devonian and early Carboniferous. This discrepancy can be alleviated under the PH. When the fossil record of all temnospondyls and lepospondyls is considered, the divergence timing implied by the molecular genomes is matched, and the age of *Gerobatrachus* generally matches the estimated dates for the batrachian divergence. These paradoxical outcomes require further research to unravel how both the TH and PH might be simultaneously supported.

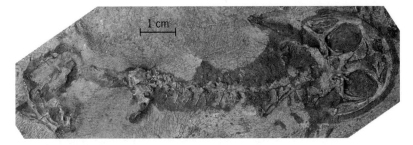

Fig. 4. *Gerobatrachus*, a stem batrachian from the Early Permian of Texas. [*From J. S. Anderson, Focal review: The origin(s) of modern amphibians, Evol. Biol., in press, December 2008; used with permission*]

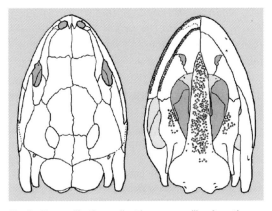

Fig. 5. *Eocaecilia*, the earliest known caecilian from the Early Jurassic, in dorsal (left) and palatal (right) views. [*Redrawn from F. A. Jenkins, D. M. Walsh, and R. L. Carroll, Anatomy of Eocaecilia micropodia, a limbed caecilian of the Early Jurassic, Bull. Museum Comp. Zool., 158(6):285–365, 2007; used with permission*]

Development. Study of the timing of onset of developmental events is revolutionizing the understanding of mechanisms for evolutionary change, and the lissamphibian problem has benefited from "evo-devo" (evolutionary developmental biology) studies. Modern amphibians are receiving intensive new attention, which is clarifying long-standing questions such as, for instance, the identity of the individual elements that comprise the highly co-ossified caecilian skull. Data on the sequence of ossification of skull bones have received special attention because of an extraordinary collection of larval temnospondyls from Germany. The sequence seen closely matches that of the most primitive extant salamanders, and some have argued that this is evidence for common ancestry. However, further work suggests that the pattern is much more generalized, being present in other tetrapods, including lepospondyls and mammals, and even in distantly related species such as ray-finned fish. Clearly, these sorts of studies are in the early stages, and much more remains to be discovered before the question of modern amphibians reaches a consensus opinion among biologists.

For background information *see* AMPHIBIA; ANIMAL EVOLUTION; ANURA; APODA; LEPOSPONDYLI; LISSAMPHIBIA; LYSOROPHIA; PHYLOGENY; SYSTEMATICS; TEMNOSPONDYLI; URODELA in the McGraw-Hill Encyclopedia of Science & Technology.   Jason S. Anderson

Bibliography. J. S. Anderson, Incorporating ontogeny into the matrix: A phylogenetic evaluation of developmental evidence for the origins of modern amphibians, pp. 182–227, in J. S. Anderson and H.-D. Sues (eds.), *Major Transitions in Vertebrate Evolution*, Indiana University Press, Bloomington, IN, 2007; M. J. Benton, *Vertebrate Palaeontology*, Blackwell, Malden, MA, 2005; R. L. Carroll, *Vertebrate Paleontology and Evolution*, W. H. Freeman, New York, 1988; W. E. Duellman and L. Trueb, *Biology of Amphibians*, 2d ed., The Johns Hopkins University Press, Baltimore, MD, 1994; R. R. Schoch and A. R. Milner, Structure and implications of theories on the origin of lissamphibians, pp. 345–377, in G. Arratia, M. V. H. Wilson, and R. Cloutier (eds.), *Recent Advances in the Origin and Early Radiation of Vertebrates*, Verlag Dr. Fredrich Pfeil, Munich, Germany, 2004.

Oyster mushrooms

Oyster mushrooms are an excellent source of many nutrients, antioxidants, and dietary fiber. More than nine species of oyster mushrooms are cultivated worldwide, making the genus *Pleurotus* one of the most diverse groups of the cultivated edible mushrooms. Their common name is derived from the shape of their fruiting body (a specialized, spore-producing organ), which resembles an oyster shell, not from their taste. The fruiting bodies of oyster mushrooms occur naturally in many colors, ranging from white to pink, blue, yellow, brown, and gray (**Fig. 1***a–d*).

Oyster mushroom production worldwide has increased dramatically over the last several years and now ranks third behind the common cultivated mushroom (*Agaricus bisporus*) and the shiitake (*Lentinula edodes*). Total world production of edible and medicinal mushrooms in 2003 was estimated at over 14 million metric tons. Of this amount, oyster mushroom production accounted for about 20% of the total, or 2.8 million metric tons. China is the main producer and consumer of oyster mushrooms.

Production technology. A number of factors are involved in the production of a crop of oyster mushrooms.

Spawn. Pleurotus spp. are grown from mycelia (threadlike filaments that become interwoven) that are propagated on a base of sterilized cereal grain, usually millet or rye. This mycelium/grain mixture is called spawn and is used to "seed" mushroom substrate. Most spawn is made from a stored culture rather than from spores because spores are likely to yield a new genotype and performance would be difficult to predict. Spawn making is a rather technical task and is not feasible for the common mushroom grower. Most commercial spawn companies produce

Fig. 1. (*a*) King oyster mushrooms (*Pleurotus eryngii*) fruiting from supplemented cottonseed hull substrate. (*b*) *Pleurotus ostreatus* fruiting from straw/cottonseed hull substrate contained in plastic bags. (*c*) *Pleurotus cornucopiae* fruiting from cottonseed hull/wheat straw substrate contained in black bags. (*d*) Fruiting and harvesting of *P. eryngii* from substrate contained in polypropylene bottles.

spawn from grain and mycelial cultures that have met strict quality-control standards. These standards include use of nondamaged whole-kernel grain, evaluation of inoculum production performance, strict control of environmental parameters during incubation, and verification of the spawn's biological purity and vigor.

Substrates. Oyster mushrooms may be produced on a variety of raw materials. These include cereal and peanut straw, sawdust, cottonseed hulls, olive mill waste, coffee pulp, grass, "spent" mushroom substrate from cultivation of other species, wastepaper and cardboard, spent brewer's grains, and corncobs. Nutrient supplements in the form of commercial delayed-release nutrients (specially formulated nutrients encapsulated in a denatured protein coat that have a delayed availability until the mushroom fungus has thoroughly invaded the substrate, thereby minimizing early utilization by competitive microorganisms within the substrate), oilseed meals, wheat or rice bran, and grains such as millet, rye, or corn may be added to the basal ingredients to improve the yield, quality, or medicinal properties of the mushrooms. Farmers are perpetually searching for alternative substrates that may be more readily available or cost-effective, or that may provide higher yield and better mushroom quality.

Substrate preparation. There are almost as many methods for substrate preparation as there are substrates. In general, substrate preparation requires formulation, mixing, and watering of the ingredients followed by pasteurization or sterilization to reduce or eliminate competitive organisms that may interfere with mushroom production. These pests include bacteria, fungi, nematodes, fly eggs and larvae, and mites. Pasteurization or sterilization may occur in bulk or in individual bags or bottles. On some commercial farms where bag production is the method of choice, ingredients are fed into revolving mixers, water is added to the desired level, and live steam (steam direct from a boiler and under full pressure) is injected into the mixer while it is in operation. The substrate then is cooled and spawn is added aseptically to the mixer in bulk or broadcast onto the substrate as it moves along on a conveyor belt. The air supplied to the spawning area usually is filtered [using a high-efficiency particulate air (HEPA) filter, 99.9% efficiency] to help minimize the reintroduction of competitive organisms during the bagging operation.

Spawning and spawn rate. Increasing the amount of spawn used to inoculate substrate (up to 5% of the wet weight of the substrate) may result in increased mushroom yields of up to 50%. This is due to several factors. First, the increased level of nutrients available in higher levels of spawn provides more energy for mycelial growth and development. Second, more inoculum points, available from increased levels of spawn, provide a faster spawn run and thus faster completion of the production cycle. Finally, a more rapid spawn run decreases the time that noncolonized substrate is exposed to weed molds and harmful bacteria.

Delayed-release nutrients. At the time of spawning, a commercial delayed-release nutrient may be added to the substrate to increase mushroom yield and reduce the time to first harvest. Delayed-release nutrients were developed in the early 1970s to stimulate yield in the common cultivated mushroom. Later, it was found that delayed-release nutrients also were effective for oyster mushrooms. Yield increases of up to 90% or more may be possible from substrates supplemented with 6% (dry wt basis) nutrient. Delayed-release nutrients also may reduce the time from spawning to harvest by 2–3 days. Use of supplements, however, may result in overheating of the substrate if growers are not able to anticipate the resultant heat surge (since the released nutrients cause the substrate to heat up excessively) and maintain spawn-run temperatures near the optimum of 23–29°C (73–84°F). Additional cooling capacity is required when higher levels of supplement are used.

Production in bottles. In many parts of Japan, Korea, and China, and to a lesser extent in the United States, growers have chosen a bottle-based production system. Substrate is filled into polypropylene bottles contained in trays (usually 16 bottles per tray), sterilized, cooled, and then inoculated with spawn. After the spawn run, the bottle lids are removed and a 1–2-mm (0.04–0.08-in.) layer of the substrate surface is removed to stimulate uniform production of mushroom primordia on the surface. After the mushrooms are harvested, they are weighed and packaged for shipment to market.

Postharvest handling and marketing. Picking and packaging methods often vary from grower to grower. Freshly harvested mushrooms must be kept refrigerated at 2–5°C (35.6–41°F) to prolong the shelf life. Oyster mushrooms typically are packaged and sold at retail in units of 100–150 g (3.5–5.3 oz). Often, oyster mushrooms and other specialty types are used to highlight the common cultivated mushrooms, which may be sold whole, sliced, or in bulk. Some growers also have considerable success marketing their product at farmers' markets.

In recent years, the trend for oyster mushroom sales has been toward the retail market. This trend is driven partly by an increased interest in oyster mushrooms and by the convenience packaged products offered to the consumer. In some retail markets, 90% of the oyster mushrooms are purchased by only 10% of the customers.

Nutrients. Oyster mushrooms are a fine source of many nutrients (**Fig. 2**). They are an excellent source [containing >20% of the recommended daily allowance (RDA) per serving] of niacin (vitamin B_3) and are a good source (containing >10% of the RDA) for riboflavin (vitamin B_2), pantothenic acid (vitamin B_5), copper, and potassium. Oyster mushrooms also contain rich amounts of dietary fiber, thiamine (vitamin B_1), protein, folic acid, zinc, vitamin B_6, manganese, magnesium, and selenium. On the other hand, oyster mushrooms are low in fat, sodium, and calories.

Dietary fiber. Oyster mushrooms contain numerous complex carbohydrates including polysaccharides

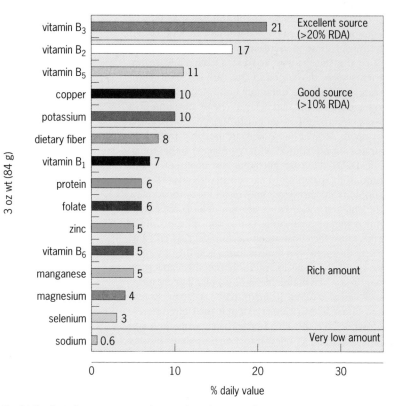

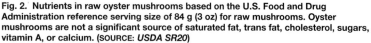

3 oz wt (84 g)

% daily value

vitamin B₃ — 21 — Excellent source (>20% RDA)
vitamin B₂ — 17
vitamin B₅ — 11
copper — 10 — Good source (>10% RDA)
potassium — 10
dietary fiber — 8
vitamin B₁ — 7
protein — 6
folate — 6
zinc — 5
vitamin B₆ — 5
manganese — 5 — Rich amount
magnesium — 4
selenium — 3
sodium — 0.6 — Very low amount

Fig. 2. Nutrients in raw oyster mushrooms based on the U.S. Food and Drug Administration reference serving size of 84 g (3 oz) for raw mushrooms. Oyster mushrooms are not a significant source of saturated fat, trans fat, cholesterol, sugars, vitamin A, or calcium. (SOURCE: *USDA SR20*)

such as glycogen and glucans, monosaccharides, disaccharides, sugar alcohols, and chitin. Most polysaccharides are structural components of cell walls and are indigestible by humans, so they may be considered dietary fiber. Dietary fiber may help prevent many diseases prevalent in affluent societies.

Potassium. Oyster mushrooms are a good source of potassium, an element that is important in the regulation of blood pressure, maintenance of water in fat and muscle, and proper functioning of cells. An 84-g (3-oz) oyster mushroom contains more potassium than a banana or an orange.

Antioxidants. Oyster mushrooms are good sources of antioxidants and rank with carrots, green beans, red peppers, and broccoli as good sources of dietary antioxidants. They are rich sources of polyphenols, which are the primary antioxidants in vegetables, and are the best source of L-ergothioneine (ERG), a potent antioxidant that is produced in nature only in fungi. Oyster mushrooms contain over 40 times more ERG than any other known (except other mushrooms) dietary source of ERG.

Outlook. It takes about six to eight weeks to produce a complete crop of oyster mushrooms—from substrate preparation to spawning to crop termination. The development of improved technology to cultivate oyster mushrooms more efficiently should allow the consumer price to decline. At the same time, product quality should increase, thus furthering demand. It is anticipated that oyster mushroom production will continue to increase worldwide because of these mushrooms' unique culinary characteristics and relative ease of production.

For background information *see* ANTIOXIDANT; FOOD MANUFACTURING; FOOD MICROBIOLOGY; FUNGAL BIOTECHNOLOGY; FUNGAL ECOLOGY; FUNGI; MUSHROOM; MYXOMYCOTA; NUTRITION in the McGraw-Hill Encyclopedia of Science & Technology.
Daniel J. Royse

Bibliography. N. J. Dubost, B. Ou, and R. B. Beelman, Quantification of polyphenols and ergothioneine in cultivated mushrooms and correlation to total antioxidant capacity, *Food Chem.*, 105:727–735, 2007; A. Gregori, M. Svagelj, and J. Pohleven, Cultivation techniques and medicinal properties of *Pleurotus* spp., *Food Technol. Biotechnol.*, 45:238–249, 2007; MushWorld, *Mushroom Growers Handbook 1: Oyster Mushroom Cultivation*, MushWorld, Seoul, Korea, 2004; P. Oei, *Mushroom Cultivation: Appropriate Technology for Mushroom Growers*, 3d ed., Backhuys Publishers, Leiden, the Netherlands, 2003; D. J. Royse, T. W. Rhodes, S. Ohga, and J. E. Sanchez, Yield mushroom size and time to production of *Pleurotus cornucopiae* (oyster mushroom) grown on switch grass substrate spawned and supplemented at various rates, *Bioresource Technol.*, 91:85–91, 2004.

Paleogenomics and biocalcification

One of the main events in the evolution of animals was the transition from soft-bodied organisms to those that possessed mineralized hard parts for protection and support. This major evolutionary hallmark supported the rapid diversification of animals and their occupation of a diverse range of novel ecological niches at the dawn of the Phanerozoic, between 560 and 530 million years ago (mya). The understanding of the evolution of genes and gene regulatory networks that enable and control the construction of mineralized body parts (biominerals) is therefore vital to the understanding of how animals, including humans, evolved and will also provide clues for the fabrication of biomimetic materials (that is, artificial materials designed to mimic natural forms).

Until recently, studies of biomineralization processes were mainly restricted to describing the (ultra) structure of the minerals and their interaction with the organic material of the organism. However, recent methodological progress in molecular biology has opened up a new dimension, allowing the elaboration of the molecular mechanisms that initiate and control biomineralization.

Biomineralization. Biomineralization refers to the processes by which organisms form minerals (such as calcium carbonate, calcium phosphate, silica, and iron oxides) under strict biological control. These processes are therefore defined as biologically controlled mineralization, with the skeletons formed being integral, functional parts of the organisms. There are, however, several other serendipitous processes linked to living organisms (for example, microbes) or organic substances that result in mineral (mainly calcium carbonate) formation that are not under the strict control of organisms. Biologically induced mineralization applies to processes where inorganic minerals are precipitated by adventitious precipitation—for example, when biologically produced metabolic end products cause relatively minor perturbations of microenvironments, when structural templates such as cell walls are provided, or when particular cations are released from the cell. Often the organism itself does not derive any benefit from the process, and many forms of pathologic mineralization occur as biologically induced mineralization. Organomineralization refers to a nonbiologically supported precipitation in which organic substrates provide templates for calcium carbonate mineral nucleation, often linked to degrading organic matter.

While both biologically induced mineralization and organomineralization play important roles in the formation of carbonate systems in the geological record (microbialites, stromatolites, ooids, etc.), it is the biologically controlled mineralization that is responsible for the production of sophisticated biomaterials such as bones, shells, and teeth, some of which show exceptional strength and beauty, as in the case of pearls.

In biologically controlled mineralization, organic macromolecules are usually associated with the process of mineral formation, which can comprise up to 5% of the total mass of a biomineral and fulfill important functions during its initial nucleation, growth, and termination, as well as contributing to the biomechanical properties of the mature product. Such organic macromolecules are subclassified as "soluble organic matrices" (SOM), that is, those that are water soluble after dissolving the mineral by EDTA (ethylenediaminetetraacetate) or acetic acid, and "insoluble organic matrices" (IOM). IOM are more or less neutrally charged, polymerized to a high degree, and important as frame-building matrices (such as collagen, chitin, or silk proteins). SOM molecules are often highly acidic (due to negatively charged carboxylate groups), are typically rich in aspartic acid (Asp) and/or glutamic acid (Glu), and usually have mean molecular weights of 10–30 kilodaltons. While organic macromolecules appear ubiquitous and pivotal in biologically controlled biomineralization, an increasing number of studies now show that a transient amorphous (noncrystalline) phase of calcium carbonate (ACC) is frequently involved in the formation of the earliest stages of calcium carbonate biominerals. However, little is known about the ACC or the organic macromolecules involved in calcification in most animal taxa.

Several SOM proteins have been sequenced and characterized, mainly in sea urchins, mollusks, and vertebrates, but the precise functions of any of these are mostly unknown, and their evolutionary origin and relationships remain largely unresolved. Knowledge about the genes and regulatory mechanisms involved in biocalcification of the most basal calcifying invertebrate taxa, the sponges and the cnidarians (such as corals), is also deficient, although it is indispensable for understanding the evolution of metazoan biocalcification. Understanding the genetic underpinning of biocalcification is also of timely importance to assess how resilient important calcifying organisms, such as the frame-building scleractinian corals, are to global change and ocean acidification.

Evolution of biocalcification. In the geological record, major animal phyla began biomineralizing shortly before, or during, the so-called Cambrian Explosion, about 530 mya, where a high diversity of animal groups suddenly appeared in the fossil record. Many causes have been postulated for this biomineralization event, including the evolutionary driving force of predation or changes in ocean chemistry that resulted in greater calcium levels than before.

The question of whether the ability to biocalcify evolved in animals several times independently, possibly to produce armor in response to increased pressure from predators, or whether preexisting calcium regulation and transport systems in late Precambrian metazoans provided the evolutionary prerequisites for their eventual use in biomineralization still remains a conundrum. If calcium carbonate biomineralization has evolved as a physiological response (mutual to many taxa) to increasing calcium levels in the late Precambrian ocean, then an evolutionary link between the proteins and the regulatory elements involved in skeletogenesis among metazoan groups can be postulated. This leads to the hypothesis that the last common ancestor of the Metazoa provided a "genetic biomineralization toolkit," a core set of biomineralization genes that has been elaborated upon to produce the range of biomineralized structures that are observed in the Earth's history.

Paleogenomics. Paleogenomics is an emerging field in paleontology that combines modern techniques of molecular biology and genetics with phylogeny as reconstructed from the fossil record and extant genomes. The key aim of paleogenomics is the reconstruction of ancient genomes and the function of ancestral genes. This is primarily accomplished by the comparative analysis of genomes of key species of extant organisms, for which genomic information is already available, and then deducing the organization and inferring the function of ancestral genomes and the scenario that has led to the subsequent evolution of present-day genomes (see **illustration**). Paleogenomics adds a deep-time perspective to the field of genomics, and information is gained not only about the genes that built long-extinct creatures, but also on the origin and evolution of the gene regulatory networks that are in operation to specify the construction of the body parts of modern-day organisms.

Paleogenomics can also involve ancient deoxyribonucleic acid (DNA) and/or proteins of extinct organisms to aid in the reconstruction of their genomes and evolutionary relationships. However, nucleic acid recovery is possible only in rare circumstances from exceptionally well preserved and geologically

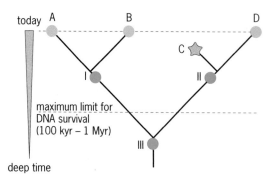

today

A   B   D

C

I   II

maximum limit for
DNA survival
(100 kyr – 1 Myr)

III

deep time

Conceptual sketch of paleogenomics. Genomes of extant taxa A and B can be used to reconstruct genomic features and function of genes of their common ancestor (I). Similarly, ancient DNA from well-preserved specimens of extinct organisms (C) might contribute to revealing their phylogenetic position or, in concert with closely related extant organisms (D), allow the reconstruction of genes and their function in their common ancestor (II). However, because of the fragile nature of nucleic acids, the maximum limit for DNA survival in exceptionally well preserved samples (for example, permafrost) is between 100,000 and 1 million years. Hence, genomes, genomic features, and gene functions in ancestral organisms that are much older than this limit (III) can be reconstructed only by comparing genes and genomes of extant organisms (for example, here, A and D).

relatively young organic material (maximum age: 100,000 to 1 million years).

Paleogenomics of biocalcification. If, as outlined above, calcium carbonate biomineralization systems in the major metazoan phyla evolved from a core skeletogenic toolkit of the last common ancestor of animals, then ancestral metazoan systems should be used as an anchor for a paleogenomic approach to reconstruct this toolkit and explore the evolutionary road map of diversification.

The first metazoans that produced a biologically controlled calcium carbonate skeleton during the late Vendian (end of the Precambrian) were small shelly fossils (for example, *Cloudina*). Archaeocyathida, however, were the first metazoans that produced massively calcified skeletons during the Tommotian (early Cambrian), and are now assigned to the Porifera (sponges), the most ancestral metazoans still extant today. Archaeocyathida were reef-building organisms and produced a calcareous basal skeleton. Similar rigid hypercalcified basal skeletons are known throughout the Earth's history from the so-called sclerosponges or coralline sponges, a polyphyletic grouping of sponges that contributed to the development of reefs during various periods—for example, the so-called Chaetetids, Stromatoporoids, and Sphinctozoans in the Paleozoic and Mesozoic. About 15 sclerosponge taxa are still extant, mainly dwelling in cryptic habitats of coral reefs. Investigation of biomineralization processes and the genetic machinery involved in calcification in these living fossils (living species belonging to an ancient stock otherwise known only as fossils) proved to be a successful avenue to acquire novel knowledge about ancestral metazoan biologically controlled calcium carbonate biomineralization systems and to comparatively reconstruct genomic features and function of the last common ancestor of animals.

Recent paleogenomic studies have examined the common inheritances of gene toolkits that were involved in skeletonization. By using a calcifying coralline sponge, the living fossil *Astrosclera willeyana*, a key component of an ancestral skeletogenic toolkit inherited from the last common ancestor of all animals—the *alpha*-carbonic anhydrase gene—that is instrumental in biocalcification was recognized. Duplication of this gene allowed skeletonization to evolve in many clades of metazoans. The recently sequenced genome of the sea urchin *Strongylocentrotus purpuratus* revealed that key genes for the echinoderm skeleton (the stereome) are echinoderm-specific and are likely to be the same as in the earliest echinoderms and their direct ancestor of more than 520 mya. However, in certain gastropods (*Haliotis asinina* and *Lottia scutum*), an unexpected evolutionary complexity of the skeletogenic genes in the tissues responsible for secreting the shell (the "secretome" of the mantle) suggests that there are significant molecular differences among shell-building gastropods. This contributes, along with the modular design of the molluscan mantle, to the spectacular variety observed among the architecture of mollusk shells.

This research area is still in its early stages, and a much more detailed characterization of the core ancestral skeletogenic toolkit is still needed, along with the untangling of lineage-specific diversifications and innovations, to understand how the impressive diversity and beauty of skeletal structures originated and evolved, and how this information can be used as a template for the invention of new biomimetic high-performance materials.

For background information *see* ANIMAL EVOLUTION; ARCHAEOCYATHA; CAMBRIAN; CARBONATE MINERALS; GENE; GENOMICS; LIVING FOSSILS; METAZOA; PALEONTOLOGY; PHYLOGENY; PORIFERA; SCLEROSPONGE; SKELETAL SYSTEM in the McGraw-Hill Encyclopedia of Science & Technology.   Gert Wörheide

Bibliography. D. J. Bottjer et al., Paleogenomics of echinoderms, *Science*, 314:956–960, 2006; S. T. Brennan, T. K. Lowenstein, and J. Horita, Seawater chemistry and the advent of biocalcification, *Geology*, 32:473–476, 2004; D. J. Jackson et al., Sponge paleogenomics reveals an ancient role for carbonic anhydrase in skeletogenesis, *Science*, 316:1893–1895, 2007; D. J. Jackson et al., A rapidly evolving secretome builds and patterns a sea shell, *BMC Biol.*, 4:40–49, 2006; H. A. Lowenstam and S. Weiner, *On Biomineralization*, Oxford University Press, Oxford, 1989.

Pandemic of 1918

The Spanish influenza pandemic of 1918–1919 is considered the most devastating outbreak of infectious disease in recorded history. In fact, the Spanish flu killed more people in a period of 25 weeks than acquired immune deficiency syndrome (AIDS) has killed in 25 years. Conservative estimates place the number of fatalities at 20 million; others believe that the number may be as many as 50 million within a very short time frame. This pandemic is sometimes called the "forgotten" pandemic because of the way it has passed out of the collective consciousness. The pandemic was caused by an unusually severe and virulent strain of the influenza A virus subtype H1N1. Recently, scientists bioengineered genes from the original chromosome of the 1918 influenza virus into template influenza viruses that had very limited capability of causing disease. Genes for two of the major proteins in influenza virus were used: hemagglutinin (HA) and neuraminidase (NA). It was found that the HA protein made these viruses highly lethal in mice, which previously had not developed disease after being infected with the template virus. Further studies in macaques found that much of the damage results from the host's immune response, which causes a "cytokine storm" in the patient. These results suggest that the primary reason for the pandemic's severity was the virulent proteins of the virus and not the conditions of the era in which it occurred.

Spanish flu pandemic. The Spanish flu appeared in two locations in March of 1918: Fort Riley in Kansas and Queens, New York, with cases appearing within one week of each other. Within a short period, the disease spread to France and Sierra Leone, from which it then traveled to other regions of the world. It was labeled the "Spanish flu" only after it traveled from France to Spain in November of 1918. By the time the pandemic was considered to be "over," it had spread worldwide as far as the Arctic regions and remote islands in the Pacific.

Symptoms of the Spanish flu were so severe that it was frequently misdiagnosed as dengue, cholera, or typhoid fever. Patients hemorrhaged from their mucus membranes, including the nose, stomach, and intestine. Bleeding from the ears occurred in some patients, along with petechial hemorrhaging in the skin (petechiae are capillary hemorrhages no larger than the head of a pin). Bacterial pneumonia as a secondary infection caused most fatalities; however, many died from the influenza directly, suffering from massive hemorrhaging and fluid buildup in the lungs. What was most striking about this pandemic was that it caused the most severe disease in young adults, and less severe disease in the elderly and young children. Suddenly stricken victims would be too weak to walk within hours. Many would die the next day. Patients would often appear to have a blue tint to their skin; some would die coughing up massive amounts of blood from the hemorrhaging in the lungs. Others hemorrhaged from their intestines, dying of the resultant blood loss. Autopsies revealed lungs up to six times their normal weight.

Seven times as many people died of the 1918 influenza than did in World War I. In Alaska, 60% of the native Inuit population died. South Pacific islands lost about 20% of their population. How then did it spread to the rest of the world from Kansas and other places in the United States? It is believed that soldiers in the 15th U.S. Cavalry carried the influenza strain with them to Europe on their way for service. From there, the epidemic spread outward across Europe and then across the rest of the world.

Why was this particular strain of influenza virus so deadly? Until 1997, little was known about the genetics of the particular viral strain. Jeffery Taubenberger, a scientist working at the Armed Forces Institute of Pathology in Washington, D.C., analyzed lung tissue from U.S. servicemen who had died of the Spanish flu during the pandemic. The tissue had been fixed in formalin and embedded in paraffin to preserve it before being stored in the Army's medical collection. Further lung tissue containing the virus from frozen bodies buried in a mass grave dug in permafrost in 1918 in Brevig Mission, Alaska, was also provided to Taubenberger's research team. Between these two sources of tissue, Taubenberger was able to sequence the coding regions of nine of the ribonucleic acid (RNA) fragments that make up the viral genome. These included the coding for the HA and NA proteins produced by the virus. Based on the sequence of the genome, it is now believed that the 1918 virus originated in birds rather than in humans or in swine, which was the original hypothesis.

Bioengineering a newly virulent influenza virus. In 2004, a research group led by Yoshihiro Kawaoka at the University of Wisconsin–Madison placed the genes for HA and NA directly into viruses of low virulence for mice. When these recombined viruses were used to infect mice, HA caused a highly virulent infection in the mice. These viruses could infect the entire lung in the mice, inducing high levels of macrophage-derived cytokines. This in turn led to a massive influx of inflammatory cells and severe hemorrhage in the lungs, leading to death in the mice.

In 2007, Kawaoka's group reconstructed the Spanish influenza virus using the sequenced genome and reverse genetics. Using the reconstructed virus in mouse studies, it was found that 39,000 times more virus particles were present in the lungs of mice than when they were infected with any other H1N1 strain. Death in mice would then occur within six days of infection. Animal studies in macaques using the reconstructed virus led researchers to believe that death was due to what is known as a "cytokine storm," an overreaction of the body's immune system to the virus.

Infection in the macaques caused acute respiratory distress, typified by increased levels of cytokines in the lungs, which was followed by death. This would explain why the virus killed young adults more often than the young or the elderly—the immune system in adults between 20 and 40 years of age is more powerful than in other age groups. A decrease in lung function in the macaques was demonstrated by a drop of up to 36% in blood oxygen saturation. When autopsied, the lung tissue of the 1918 virus-infected macaques was filled with bloody liquid, greatly reducing their ability to breathe. By day 8 of the study, the macaques infected with the control influenza virus showed healing taking place in the lungs. This was in contrast to the lungs of the 1918 virus-infected macaques, where damage was increasing and lungs were becoming less and less oxygenated. This decrease in ability to oxygenate the blood accounts for the "blue tint" in the faces of victims of the original pandemic; insufficient oxygen was getting into their bloodstream.

A cytokine storm or systemic inflammatory response syndrome (SIRS) can be defined as an immune system that is overreacting to the pathogen. Macrophages and T cells travel to the site of infection, producing massive amounts of cytokines. Cytokines are signaling proteins that induce immune responses. The release of these cytokines in turn leads to inflammatory responses in the lungs, producing many of the symptoms experienced in the macaques, as well as in the victims of the 1918 pandemic. The cytokines produced included interleukin-6, interleukin-8, monocyte chemotactic protein-1 (CCL2), and CCL5 (RANTES). Interestingly, the avian H5N1 "bird" influenza strain of current concern also causes a cytokine storm or SIRS in its victims.

Future outlook. What do these results indicate in terms of modern-day disease? It is clear from the above data that an influenza virus that can cause severe overreaction of the immune response is frequently lethal, in both animals and humans. It is also clear that another pandemic is almost certain should such a virus become highly contagious between humans. The only reason that there have not been more fatalities caused by the present-day avian influenza virus is that it is not believed to be readily transmissible between humans. At this time, all cases appear to have occurred through exposure to birds. However, if the H5N1 strain should infect a human who is also infected with a human form of influenza, there is a high probability of recombination of the two viruses into a more contagious and lethal human form. If this happens, a pandemic of epic scale is likely to be seen.

Fortunately for humankind, influenza can be prevented through the development of vaccines. At this time, candidate vaccines are being prepared in a number of research facilities that should immunize against the current strain of avian influenza. In addition, researchers are attempting to develop what is known as a "universal" vaccine, one that is capable of protecting against all strains of influenza viruses affecting humans. Current vaccines are being examined using animal models to test their effectiveness. When these vaccines become available, the chances of suffering through a pandemic like that of 1918 will dwindle and hopefully fade away.

For background information *see* CYTOKINE; DISEASE; EPIDEMIC; EPIDEMIOLOGY; GENE; GENETIC ENGINEERING; INFECTIOUS DISEASE; INFLUENZA; PUBLIC HEALTH; VACCINATION; VIRULENCE; VIRUS in the McGraw-Hill Encyclopedia of Science & Technology.

Marcia M. Pierce

Bibliography. A. W. Crosby, *America's Forgotten Pandemic: The Influenza of 1918*, University of Texas Press, Austin, 2003; D. Kobasa et al., Aberrant innate immune response in lethal infection of macaques with the 1918 influenza virus, *Nature*, 445:319–323, 2007; D. Kobasa et al., Enhanced virulence of influenza A viruses with the haemagglutinin of the 1918 pandemic virus, *Nature*, 431:703–707, 2004; T. Shors, *Understanding Viruses*, Jones and Bartlett, Sudbury, MA, 2009; T. M. Tumpey et al., Pathogenicity of influenza viruses with genes from the 1918 pandemic virus: Functional roles of alveolar macrophages and neutrophils in limiting virus replication and mortality in mice, *J. Virol.*, 79(23):14933–14944, 2005.

Phanerozoic predation intensity and diversity

The idea that species interactions, such as predation, parasitism, or competition, helped shape the history of biodiversity is not a new one. Indeed, Charles Darwin compared the biotic world to a surface of tightly packed wedges, and the only way for a new species—in his metaphor, a wedge—to gain a foothold was to drive out another wedge. By observing species interactions in the modern living world, it seems apparent that organisms' struggles for food, habitat, and reproduction were the fundamental processes responsible for the biodiversity that we see today. Predation, the killing and consumption of an organism (prey) by another organism (predator), has been shown to be particularly important in determining species diversity in modern environments. The classic experiments of ecologist Robert Paine in intertidal environments in the Pacific Northwest illustrated the role of predators in increasing the number of species in a local environment. Paine removed the top invertebrate predator, the sea star *Pisaster*, from a shoreline environment, and the number of species in the experimental plot dropped from 15 to 8. The explanation is that by feeding—also termed "cropping"—predators prevent other species from monopolizing the environment and outcompeting their competitors. Steven M. Stanley extended the cropping hypothesis to explain the Cambrian explosion, a geologically sudden macroevolutionary event during which nearly all animal phyla appeared in the fossil record.

Theoretical and empirical work by Stephen Jay Gould and colleagues suggested that organismal interactions might not be as important in long-term evolutionary trends as was previously thought; factors producing change on short ecological time scales did not scale up to long-term evolutionary time scales. Gould envisioned evolution as a process working through a hierarchy of three tiers. Functioning at the first tier are the everyday ecological processes with which evolution is typically associated: predation, competition, parasitism, and so on. The process of punctuated equilibrium, the geologically rapid evolution of new species from isolated populations, occupies the second tier. Mass extinction, the geologically rapid extinction of a major portion of the biota, occupies the third tier. Processes that occur on higher tiers can counteract trends that accumulate at lower tiers. From this framework, Gould predicted that trends produced at the first tier would proceed rapidly and would not scale up to long-term evolutionary trends; therefore, ecological interactions are not important in the macroevolutionary theater.

Hence, understanding the role of ecology in macroevolution is a major theme in evolutionary paleoecology. The goal of this study is to investigate the role of predation in the history of biodiversity.

Previous studies of predation in the fossil record. In many geological settings, it is rare to preserve direct evidence of predation in the fossil record. There are examples of exceptional preservation where a predator was preserved with the contents of its last meal in its gut; however, these are rare indeed. There are two special cases in which reasonable evidence of predation, or attempted predation, is readily preserved in the fossil record. As they grow by accretion (or the incremental growth of layers), many shelled invertebrates record the events of their lives, including interactions with their own predators. Many predators gain access to the soft, nutritious flesh inside by either drilling through or breaking the armored shells. Predatory drill holes are typically round in outline and penetrate straight through the prey's valve (**Fig. 1**). Drill holes can be drilled by a diverse array of predators, including many types of snails and worms, and complete drill holes are interpreted as successful kills. Many jawed and clawed predators (such as crabs, fish, and birds) break the shells of their prey. Occasionally, the prey survive an attack and live to repair their broken shells (Fig. 1). These drill holes and repair scars are readily preserved on shells in the fossil record; thus, a fossil record of predator-prey interactions has been preserved as well.

Many paleontologists and biologists have documented predatory drill holes and repair scars in shelled organisms ranging in age from the first biomineralizing organisms 549 million years ago to living organisms today. The frequency of predation intensity has fluctuated through geologic time and may at times be affected by mass extinctions. Previous studies addressing predation frequency over long periods of geologic time tended to focus on broader classification levels (for example, the drilling frequency of all bivalves or gastropods in a sample) rather than reporting species-level predation frequencies. There is a significant loss of information when species-level data are pooled together; fortunately, many studies have reported predation frequencies of individual species over shorter intervals of geologic time. It is from these studies that one is able to more fully understand the history of predator-prey interactions and its relation to the history of biodiversity.

Phanerozoic histories. Elucidating the history of predation and diversity and examining the relationship between the two is a tall order indeed and certainly cannot be completed by a small number of investigators collecting the raw data on their own. Fortunately, there is a rich data source available in the published literature produced by hundreds of paleontologists from the last two centuries. In order

Fig. 1. Examples of predatory traces. The clam on top was killed by a drilling predatory snail. The drill hole is visible at the top. The snail on the bottom was attacked by an unknown predator that peeled the shell back from the opening, similar to the action of a can opener. For what could be attributed to a number of reasons, the predator was not able to gain access to the fleshy part of the prey and ceased the attack. Subsequently, the surviving snail continued to grow as usual by depositing new layers. The snail's shell grew larger, but evidence of the failed attack remained as a jagged edge (arrow).

to understand the history of predation on marine invertebrates, it was necessary to compile a database of species-level predation frequencies spanning the geologic history of biomineralizing animals. These data came from 196 peer-reviewed papers, representing 2292 occurrences of predation traces (that is, drill holes and repair scars). Biodiversity data were taken from the work of the late Jack Sepkoski, who compiled a diversity database for fossil marine animals and protists for the Phanerozoic fossil record (the last 540 million years).

The histories of predation intensity and biodiversity (genus-level diversity) are remarkably congruent (**Fig. 2**). When the data are binned by the geological periods of the Phanerozoic Eon (including only periods with at least 10 occurrences of species-level predation frequency data), average predation frequency is significantly correlated with average genus-level diversity. The significant correlation persists regardless

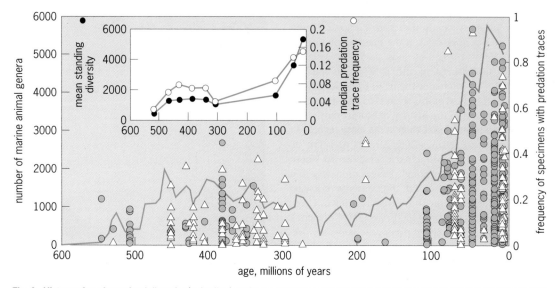

Fig. 2. History of marine animal diversity (color line) and predation trace frequencies for marine invertebrate prey (circle: drill hole; triangle: repair scar). Inset: Mean diversity values (black) and median predation frequency (white) values. (*Modified from J. W. Huntley and M. Kowalewski, Strong coupling of predation intensity and diversity in the Phanerozoic fossil record, Proc. Natl. Acad. Sci. USA, 104(38):15006–15010, 2007*)

of which measures (mean, median, etc.) of predation intensity and diversity are used. The correlation is also significant when the data are corrected for the tendency of two time series to autocorrelate. (Autocorrelation occurs when residual error terms from observations of the same variable at different times are related. Thus, in computing the correlation of the time series, an error is correlated with the error immediately before it, distorting the true nature of the correlation.)

These results thus document the relationship between predation and biodiversity through the geologic history of animal life. What does this correspondence mean? Caution must be used when assigning a causal relationship to a correlation. There are three possible causal relationships to consider when two variables are correlated. Imagine that variables X and Y are correlated through time: a change in X could have driven a change in Y, a change in Y could have driven a change in X, or changes in X and Y could both have been driven by an external factor(s). One of the following scenarios could have produced the correlation between predation intensity and biodiversity through geologic time.

1. Increasing predation intensity drove the amplification of biodiversity over geologic time. In this case, the cropping explanation would indeed scale up from a short-term ecological process to a macroevolutionary process; ecological processes would thereby be a driving force in macroevolutionary patterns.

2. Rising diversity increased the likelihood of the evolution of shell-drilling or shell-crushing predators. In this case, independently increasing biodiversity is responsible for the stochastic (random) evolution of predators with relatively rare predatory strategies (that is, drilling and crushing); macroevolutionary patterns therefore influence ecological processes.

3. Uneven preservation and sampling of the fossil record produced the trends in predation intensity and biodiversity. It has been noted that trends in biodiversity mimic trends in the amount of rock preserved from various geologic periods, the implication being that the likelihood of finding a fossil increases when there is more rock preserved from a given time. Therefore, Sepkoski's genus-level diversity curve may reflect the changing quality of the fossil record through time more than the actual history of biodiversity. Perhaps it is easier to find fossil samples that record the behavior of predators during times when preservation and sampling are more complete. However, the plausibility of this third scenario is reduced when we consider the nature of our two types of data, namely, diversity and predation intensity.

Our measures of diversity are ultimately based on the number of species that have been discovered. The predation frequency data, however, are not solely based on the number of predation events found, but are calculated as the frequencies of predation events from within single samples. Inequality in preservation or sampling over the geological time scale should have little effect on the proportion of shells within a single sample that were attacked.

Regardless of which scenario explains the correlation between predation intensity and biodiversity, the result is significant because it suggests that a common causative mechanism exists between macroevolutionary and macroecological patterns. Perhaps ecological interactions are intimately related to the macroevolutionary process after all.

For background information *see* ANIMAL EVOLUTION; BIODIVERSITY; ECOLOGICAL COMPETITION; EXTINCTION (BIOLOGY); GEOLOGIC TIME SCALE; MACROEVOLUTION; PALEOECOLOGY; PALEONTOLOGY; PHANEROZOIC; POPULATION ECOLOGY;

PREDATOR-PREY INTERACTIONS; TAPHONOMY in the McGraw-Hill Encyclopedia of Science & Technology.
John Warren Huntley

Bibliography. S. J. Gould, The paradox of the first tier: An agenda for paleobiology, *Paleobiology*, 11(1):2–12, 1985; J. W. Huntley and M. Kowalewski, Strong coupling of predation intensity and diversity in the Phanerozoic fossil record, *Proc. Natl. Acad. Sci. USA*, 104(38):15006–15010, 2007; P. H. Kelley and T. A. Hansen, The fossil record of drilling predation on bivalves and gastropods, in P. H. Kelley, M. Kowalewski, and T. A. Hansen, (eds.), *Predator-Prey Interactions in the Fossil Record*, Kluwer Academic/Plenum Publishers, New York, 2003; M. Kowalewski, A. Dulai, and F. Fürsich, A fossil record full of holes: The Phanerozoic history of drilling predation, *Geology*, 26(12):1091–1094, 1998; R. T. Paine, Food web complexity and species diversity, *Am. Nat.*, 100(910):65–75, 1966; J. J. Sepkoski, A compendium of fossil marine animal genera, *Bull. Am. Paleontol.*, 363:1–560, 2002; S. M. Stanley, An ecological theory for the sudden origin of multicellular life in the Late Precambrian, *Proc. Natl. Acad. Sci. USA*, 70(5):1486–1489, 1973.

Phytochemicals: human disease prevention agents

Plant-based foods, such as fruit, vegetables, and whole grains, which contain significant amounts of bioactive phytochemicals and have potent antioxidant activity, may provide desirable health benefits beyond basic nutrition and reduce the risk of developing chronic diseases. The beneficial effects associated with plant-based food consumption are in part due to the existence of phytochemicals.

Phytochemicals (from the Greek *phyto*, pertaining to plants) are plant chemicals. They are defined as bioactive non-nutrient plant compounds in fruits, vegetables, grains, and other plant foods and have been linked to the reduction of risk for major chronic diseases. It is estimated that more than 8000 individual phytochemicals have been identified, but a large percentage still remain unknown and need to be identified and studied before we can fully understand the health benefits of phytochemicals in whole foods. However, more and more convincing evidence suggests that the benefits of phytochemicals in fruit, vegetables, and whole grains may be even greater than are currently understood. They are believed to counteract the oxidative stress induced by free radicals involved in the etiology of a wide range of chronic diseases or disorders such as cancer, heart diseases, stroke, diabetes, Alzheimer's disease, and cataracts, as well as some of the functional declines associated with aging. Because phytochemicals differ widely in composition and ratio from fruits to vegetables to grains, and often have complementary mechanisms to one another, it is suggested that one consume a wide variety of these plant-based foods.

Phytochemicals can be classified as phenolics, alkaloids, nitrogen-containing compounds, organosulfur compounds, phytosterols, and carotenoids (**Fig. 1**). The most studied of the phytochemicals are the phenolics and carotenoids.

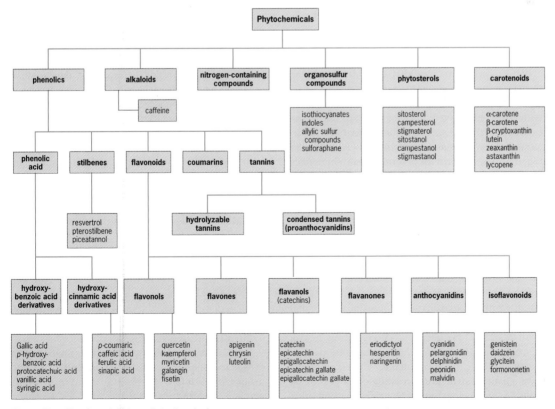

Fig. 1. Classification of dietary phytochemicals.

Phenolics. Phenolics are compounds possessing one or more aromatic rings with one or more hydroxyl groups. They generally are categorized as phenolic acids, flavonoids [structure (**1**)], stilbenes,

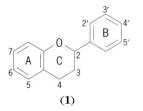

(**1**)

coumarins, and tannins. Phenolics are the products of secondary metabolism in plants, providing essential functions in the reproduction and growth of the plants, acting as defense mechanisms against pathogens, parasites, and predators, as well as contributing to the color of plants. In addition to their roles in plants, phenolic compounds in our diet may provide health benefits associated with reduced risk of chronic diseases. Among the 11 common fruits consumed in the United States, cranberry has the highest total phenolic content, followed by apple, red grape, strawberry, pineapple, banana, peach, lemon, orange, pear, and grapefruit. Among the 10 common vegetables consumed in the United States, broccoli possesses the highest total phenolic content, followed by spinach, yellow onion, red pepper, carrot, cabbage, potato, lettuce, celery, and cucumber. Phytochemicals in whole grains have not received as much attention as the phytochemicals in fruits and vegetables. Recent research has found that total phytochemical content and antioxidant activity of whole grains have been commonly underestimated in the literature because bound phytochemicals of whole grains were not included. Most whole-grain phenolics are in bound form: 85% in corn, 76% in wheat, and 75% in oats. Therefore, whole grains contain more phytochemicals than previously reported and are good sources of dietary phenolics on a per-serving basis. The beneficial effects associated with whole-grain consumption are in part due to the existence of the unique phytochemicals that complement those in fruits and vegetables when consumed together. The majority of potential health-beneficial phytochemicals of whole grains are present in the bran/germ fraction. Refined-wheat flour loses 83% of total phenolics, 79% of total flavonoids, 78% of zeaxanthin, 51% of lutein, and 42% of β-cryptoxanthin when compared to whole-wheat flour.

It is estimated that flavonoids account for approximately two-thirds of the phenolics in our diet and that the remaining one-third are from phenolic acids.

Flavonoids. Flavonoids are a group of phenolic compounds with antioxidant and biological activity that have been identified in fruits, vegetables, and other plant foods and have been linked to reducing the risk of major chronic diseases. More than 4000 distinct flavonoids have been identified. They commonly have a generic structure consisting of two aromatic rings (A and B rings) linked by three carbons that are usually in an oxygenated heterocyclic ring, or

C ring [structure (**1**)]. Differences in the generic structure of the heterocyclic C ring classify them as flavonols, flavones, flavanols (catechins), flavanones, anthocyanidins, or isoflavonoids (isoflavones) [Fig. 1 and structures (**2**)–(**7**)]. Flavonols (quercetin,

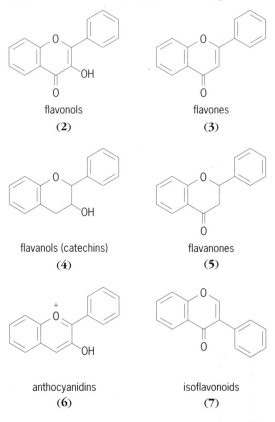

flavonols
(**2**)

flavones
(**3**)

flavanols (catechins)
(**4**)

flavanones
(**5**)

anthocyanidins
(**6**)

isoflavonoids
(**7**)

kaempferol, and myricetin), flavones (luteolin and apigenin), flavanols [catechin, epicatechin, epigallocatechin (EGC), epicatechin gallate (ECG), and epigallocatechin gallate (EGCG)], flavanones (naringenin), anthocyanidins (cyanidin and malvidin), and isoflavonoids (genistein and daidzein) are common flavonoids in the diet [Fig. 1 and structures (**8**)–(**22**)]. Flavonoids are most frequently found in nature as conjugates in glycosylated or esterified forms, but can occur as aglycones (the nonsugar components of a glycoside molecule resulting from hydrolysis of the molecule), especially as a result of the effects of food processing. Many different glycosides can be found in nature because more than 80 different sugars have been discovered bound to flavonoids. Anthocyanidins give the red and blue colors to some fruits and vegetables.

Human intake of all flavonoids is estimated at a few hundred milligrams to 650 mg/day. The total average intake of flavonols (quercetin, myricetin, and kaempferol) and flavones (luteolin and apigenin) is estimated as 23 mg/day, of which quercetin contributes ~70%, kaempferol 17%, myricetin 6%, luteolin 4%, and apigenin 3%.

Phenolic acids. Phenolic acids, another major source of dietary phenolics, can be subdivided into two major groups: hydroxybenzoic acid and hydroxycinnamic acid derivatives (**Tables 1** and **2**). Hydroxybenzoic acid derivatives include *p*-hydroxybenzoic,

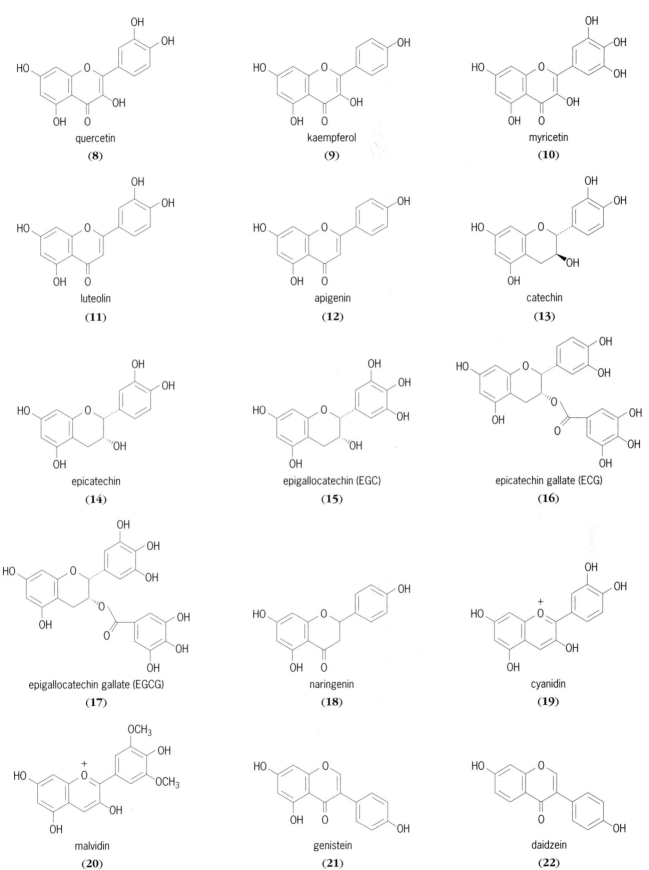

quercetin
(8)

kaempferol
(9)

myricetin
(10)

luteolin
(11)

apigenin
(12)

catechin
(13)

epicatechin
(14)

epigallocatechin (EGC)
(15)

epicatechin gallate (ECG)
(16)

epigallocatechin gallate (EGCG)
(17)

naringenin
(18)

cyanidin
(19)

malvidin
(20)

genistein
(21)

daidzein
(22)

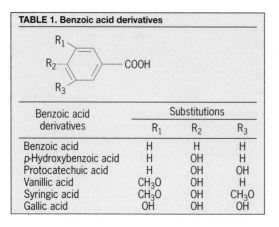

TABLE 1. Benzoic acid derivatives

Benzoic acid derivatives	Substitutions		
	R_1	R_2	R_3
Benzoic acid	H	H	H
p-Hydroxybenzoic acid	H	OH	H
Protocatechuic acid	H	OH	OH
Vanillic acid	CH_3O	OH	H
Syringic acid	CH_3O	OH	CH_3O
Gallic acid	OH	OH	OH

TABLE 2. Cinnamic acid derivatives

Cinnamic acid derivatives	Substitutions		
	R_1	R_2	R_3
Cinnamic acid	H	H	H
p-Coumaric acid	H	OH	H
Caffeic acid	OH	OH	H
Ferulic acid	CH_3O	OH	H
Sinapic acid	CH_3O	OH	CH_3O

protocatechuic, vanillic, syringic, and gallic acids. They are commonly present in the bound form and are typically components of complex structures such as lignins and hydrolyzable tannins. They can also be found in the form of sugar derivatives and organic acids in plant foods.

Hydroxycinnamic acid derivatives include p-coumaric, caffeic, ferulic, and sinapic acids. They are present mainly in the bound form, linked to cell-wall structural components such as cellulose, lignin, and proteins through ester bonds. Ferulic acids occur primarily in the seeds and leaves of plants, mainly covalently conjugated to mono- and disaccharides, plant cell-wall polysaccharides, glycoproteins, polyamines, lignin, and insoluble carbohydrate biopolymers. Wheat bran is a good source of ferulic acids, which are esterified to hemicellulose of the cell walls. Free, soluble-conjugated, and bound ferulic acids in grains (corn and wheat) are present in the ratio of 0.1:1:100. Food processing, such as thermal processing, pasteurization, fermentation, and freezing, contributes to the release of these bound phenolic acids.

Caffeic, ferulic, p-coumaric, protocatechuic, and vanillic acids are present in almost all plants. Chlorogenic acids and curcumin are also major derivatives of hydroxycinnamic acids present in plants. Chlorogenic acids are the ester of caffeic acids and are the substrate for enzymatic oxidation leading to browning, particularly in apples and potatoes. Curcumin is made of two ferulic acids linked by a methylene in a diketone structure and is the major yellow pigment of mustard.

Carotenoids. Carotenoids are nature's most widespread pigments, with yellow, orange, and red colors, and have also received substantial attention because of both their provitamin A and antioxidant roles. More than 600 different carotenoids have been identified in nature. They occur widely in plants, microorganisms, and animals. Carotenoids have a 40-carbon skeleton of isoprene units [structures (23)–(29)]. The structure may be cyclized at one or both ends, may have various hydrogenation levels, or may possess oxygen-containing functional groups. Lycopene and β-carotene are examples of acyclized and cyclized carotenoids, respectively. Carotenoid compounds most commonly occur in nature in the all-trans form. The most characteristic feature of carotenoids is the long series of conjugated double bonds forming the central part of the molecule. This gives them their shape, chemical reactivity, and light-absorbing properties. β-Carotene, α-carotene, and β-cryptoxanthin function as provitamin A (which is converted to vitamin A in the liver). Zeaxanthin and lutein are the major carotenoids in the macular region (yellow spot) of the retina in humans.

Orange vegetables and fruits, including carrots, sweet potatoes, winter squash, pumpkin, papaya, mango, and cantaloupe, are rich sources of the carotenoid β-carotene. Tomatoes, watermelons, pink grapefruits, apricots, and pink guavas are the most common sources of lycopene; 85% of the lycopene intake in the United States comes from processed tomato products such as ketchup, tomato paste, and tomato soup.

Carotenoid pigments play important functions in photosynthesis and photoprotection in plant tissues. The photoprotection role of carotenoids originates from their ability to quench and inactivate reactive oxygen species such as singlet oxygen formed from exposure to light and air. This photoprotection role is also associated with its antioxidant activity in human health. Carotenoids can react with free radicals and become radicals themselves. Their reactivity depends on the length of the chain of conjugated double bonds and the characteristics of the end groups. Carotenoid radicals are stable by virtue of the delocalization of the unpaired electron over the conjugated polyene chain of the molecules. This delocalization allows addition reactions to occur at many sites on the radical. Astaxanthin, zeaxanthin, and lutein, excellent lipid-soluble antioxidants, scavenge free radicals, especially in a lipid-soluble environment. Carotenoids at sufficient concentrations can prevent lipid oxidation and related oxidative stress.

Role in prevention of cancer. Cells in humans and other organisms are constantly exposed to a variety of oxidizing agents, some of which are necessary for life. The key factor is to maintain a balance between oxidants and antioxidants to sustain optimal physiological conditions. Overproduction of oxidants can cause an imbalance leading to oxidative stress, especially in chronic bacterial, viral, and parasitic infections. Oxidative stress can cause oxidative

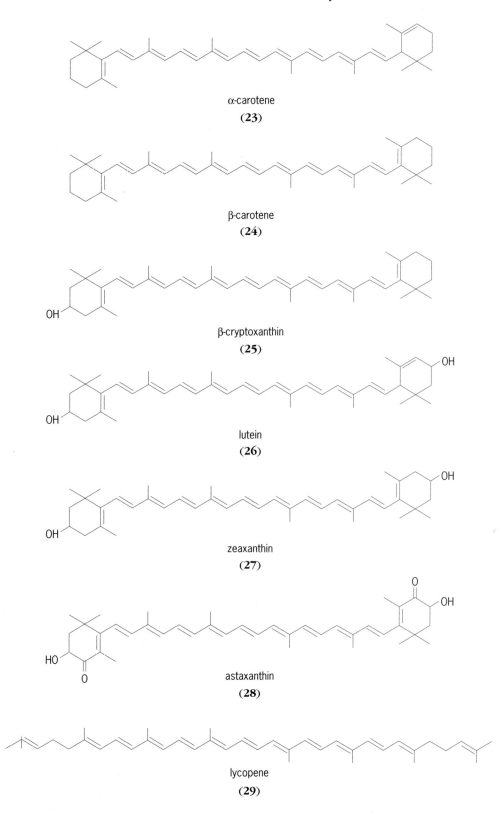

α-carotene
(**23**)

β-carotene
(**24**)

β-cryptoxanthin
(**25**)

lutein
(**26**)

zeaxanthin
(**27**)

astaxanthin
(**28**)

lycopene
(**29**)

damage to large biomolecules such as lipids, proteins, and deoxyribonucleic acid (DNA), resulting in an increased risk for cancer.

Strong epidemiological evidence suggests that regular consumption of fruit and vegetables can reduce cancer risk. The risk of cancer was twofold higher in persons with a low intake of fruit and vegetables than in those with a high intake. Intake of flavonoids from fruits and vegetables was inversely associated with all-cause cancer risk and cancer of the alimentary and respiratory tracts. Lung cancer risk has been inversely associated with flavonoid and quercetin intake.

Dietary phytochemicals can act to prevent cancer or interfere with its progression at virtually every stage of cancer development. Studies to date have demonstrated that phytochemicals in common fruit, vegetables, and whole grains can have complementary and overlapping mechanisms of action, including (1) antioxidant activity and scavenging of free radicals; (2) regulation of gene expression in cell proliferation, cell differentiation, oncogenes, and tumor-suppressor genes; (3) induction of cell cycle arrest and apoptosis (programmed cell death); (4) modulation of enzyme activities in detoxification, oxidation, and reduction; (5) stimulation of the immune system; (6) regulation of hormone-dependent carcinogenesis; and (7) antibacterial and antiviral effects. Therefore, a recommendation that consumers eat a wide variety of fruit, vegetables, and whole grains daily is an appropriate strategy to reduce the risks of chronic diseases and to meet their nutrient requirements for optimum health.

For background information *see* ANTIOXIDANT; CANCER (MEDICINE); CAROTENOID; DISEASE; FLAVONOID; FOOD; FOOD SCIENCE; FREE RADICAL; OXYGEN TOXICITY; PATHOLOGY; PLANT METABOLISM in the McGraw-Hill Encyclopedia of Science & Technology.   Rui Hai Liu

Bibliography. K. K. Adom and R. H. Liu, Antioxidant activity of grains, *J. Agric. Food Chem.*, 50:6182–6187, 2002; G. Block, B. Patterson, and A. Subar, Fruit, vegetables, and cancer prevention: A review of the epidemiological evidence, *Nutr. Cancer*, 18:1-29, 1992; Y.-F. Chu et al., Antioxidant and antiproliferative activities of vegetables, *J. Agric. Food Chem.*, 50:6910-6916, 2002; P. Knekt et al., Dietary flavonoids and the risk of lung cancer and other malignant neoplasms, *Am. J. Epidemiol.*, 146:223-230, 1997; L. Le Marchand et al., Intake of flavonoids and lung cancer, *J. Natl. Cancer Inst.*, 92:154-160, 2000; R. H. Liu, Health benefits of fruits and vegetables are from additive and synergistic combination of phytochemicals, *Am. J. Clin. Nutr.*, 78:517S-520S, 2003; R. H. Liu, Potential synergy of phytochemicals in cancer prevention: Mechanism of action, *J. Nutr.*, 134:3479S-3485S, 2004; J. Sun et al., Antioxidant and antiproliferative activities of fruits, *J. Agric. Food Chem.*, 50:7449-7454, 2002.

Plant defense against herbivorous insects

Plants and their insect herbivores comprise a major proportion of the Earth's biodiversity, with roughly 22% of all macroscopic described species being plants and 26% being herbivorous insects. In the process of finding causal links between these interacting groups of organisms, ecologists and evolutionary biologists were driven by two major questions: Why are there so many species of plants and insects? Why is the world still green, given the diversity and abundance of plant-eating organisms? Research on these questions quickly led scientists to a third question: Why do plants produce so many chemical compounds that are seemingly not important for life-sustaining processes? Many of these chemicals, called secondary metabolites to differentiate them

from primary metabolites necessary to sustain fundamental vital functions (for example, proteins, carbohydrates, and lipids), have long been used as active ingredients in medicinal and spiritual tinctures, but their ecological functions were not immediately obvious to scientists. In an influential paper in 1964, biologists P. R. Ehrlich and P. H. Raven proposed that secondary metabolites function as toxic plant defenses. They argued that insects are under strong natural selection to cope with toxic plant secondary metabolites and that this pressure would favor insects that are resistant to chemical defenses. The resistant insects would in turn put plants under natural selection to evolve new defensive metabolites in order to keep pace with the adapting herbivores. This continuing cycle of coevolving plant defenses and insect counterdefenses could thus contribute to the formation of new plant and animal species.

In answering the question of why plants produce secondary chemicals, this theory also appeared to answer (in part) the questions of why there are so many plants and herbivorous insects (coevolution), and why the world is green (plants are defended). The study of plant defenses against insects was the basis for a whole new field of research: the chemical ecology of plant-insect interactions.

How do plants defend themselves? Plants have four principal ways of coping with attacking insects:

1. Plants can tolerate damage; that is, they can regrow the lost tissue and thereby avoid any negative effects of herbivory.

2. Plants can express physical defenses that make them less approachable (e.g., spines, thorns, prickles, or hairs) or less palatable (e.g., silicic acid and silicate crystals, or lignified cell walls).

3. Plants can produce secondary metabolites that function as direct, toxic chemical defenses against insects.

4. Plants can attract the natural enemies of the herbivores (e.g., predators) using chemical signals or rewards.

All of these plant strategies can be further characterized as constitutive (always present) or induced (produced only when attacked by herbivores) (**Fig. 1**). Moreover, individual plant species vary considerably in how many strategies they deploy and in the specific traits (e.g., chemicals) that characterize each strategy. Chemical direct and indirect plant defenses received a lot of scientific attention because they were identified as major factors driving organismal interactions and influencing community dynamics.

Defensive secondary metabolites. A number of secondary metabolites and defensive proteins, including alkaloids, terpenoids, polyphenols, glucosinolates, proteinase inhibitors, and polyphenol oxidases, among others, have been shown to be toxic or antidigestive to insects, but only a few have been tested for their ability to actually provide a fitness benefit to plants (Fig. 1).

The limited proof for a defensive function (an increase in plant survival and/or reproduction) of secondary chemicals is caused partially by experimental difficulties in "producing" plants that synthesize a particular compound and those that don't for comparative experiments. New genetic and molecular tools can be used now to better identify and evaluate natural variation among plants or to genetically engineer plants for such functional comparisons.

More interesting, two rather scientific problems of the functional analysis of chemical plant traits illustrate the complexity of chemistry-mediated interactions. First, a single compound alone may not function as an efficient chemical defense, but might do so only in concert with other compounds in the plant. For example, leaves of plants in the mustard family (Brassicaceae) contain nontoxic glucosinolates (mustard oils), which are converted into toxic thiocyanates, isothiocyanates, and nitriles by another compound, the enzyme myrosinase, when tissue has been damaged; this synergistic function of plant metabolites was termed "mustard oil bomb" for its potency as a defense.

Second, most plants are attacked by many different insect species and frequently have to cope with a diversity of herbivores, as well as pathogens and diseases at the same time. Different species of herbivores may have different responses and sensitivities to different chemicals, resulting in a weaker overall relationship between a particular chemical and the plant's resistance to attacking insects. This, in turn, may lead to a more diffuse coevolutionary process between plants and insects, the direction of which is determined by the relative impact of each of the attacking herbivore species. However, it is generally accepted that secondary compounds that mediate resistance to insects are also likely to benefit the plant's survival and reproduction (Fig. 1).

Defensive compounds are categorized by their modes of action and can be toxic, antidigestive, and/or antinutritive. Toxins (e.g., alkaloids, terpenoids, phenolics, and glucosinolates) poison generalist herbivores, forcing specialists to invest resources in detoxification mechanisms that in turn incur growth and development costs. Antidigestive compounds such as proteinase inhibitors are inducible by herbivory and influence herbivore performance by inhibiting digestive enzymes (proteinases) in the insect gut. Antinutritive enzymes, such as polyphenol oxidases, decrease the nutritive value of wounded plants by cross-linking proteins or catalyzing the oxidation of phenolic secondary metabolites to reactive and polymerizing quinones.

Costs of plant defenses. In the absence of herbivory, plants get no benefit from the production of defensive compounds, and thus these compounds can represent a net cost. Such costs arise, for example, from the allocation of plant resources (energy, nutrients) to defenses, when they could have been allocated to growth or the production of offspring. Production of toxic compounds can also harm the plant itself, by negatively affecting its primary metabolism. In addition to these physiological costs, there may also be ecological costs. For example, specialized herbivores can be attracted to their host plants and may additionally sequester toxic plant metabolites

Constitutive	**Induced response**
Produced at all times (may vary during ontogenesis)	Produced after herbivore/pathogen damage (or other environmental changes)

Plant resistance
Plant traits that reduce herbivore/pathogen survival, reproductive output, or preference for a plant—*herbivore/pathogen perspective*

Plant defense
Plant traits that decrease the negative Darwinian fitness consequences of an herbivore attack for the plant—*plant perspective*

Direct defense	**Indirect defense**
Physical (for example, thorns, spines, prickles, leaf toughness, silicates) or chemical traits (for example, toxic, antidigestive, and antinutritive compounds) that protect plant tissue from herbivores/pathogens	Plant traits that facilitate presence (for example, domatia, extrafloral nectar, food bodies) or foraging behavior (for example, induced volatile organic compound emission) of the natural enemies of the herbivores (third trophic level; for example, parasitoids, predators)

Fig. 1. Plant defense nomenclature.

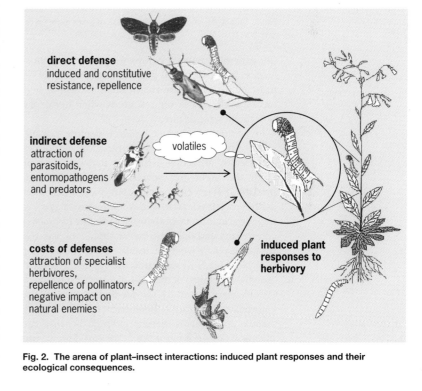

direct defense
induced and constitutive resistance, repellence

indirect defense
attraction of parasitoids, entomopathogens and predators

volatiles

induced plant responses to herbivory

costs of defenses
attraction of specialist herbivores, repellence of pollinators, negative impact on natural enemies

Fig. 2. The arena of plant–insect interactions: induced plant responses and their ecological consequences.

for their own defenses against natural enemies and so compromise the plant's interactions with beneficial insects, such as parasitoids and predators. Alternatively, defensive compounds can end up in floral tissues (e.g., pollen or nectar) and thereby have negative effects on mutualistic pollinators (**Fig. 2**).

Induced direct and indirect defenses. To avoid these costs, many plant species produce defenses only when actually attacked, a strategy known as induced defenses. All attacked plants undergo dramatic reconfiguration of their metabolism (compared to unattacked plants). However, in many species, these metabolic changes result in a higher resistance

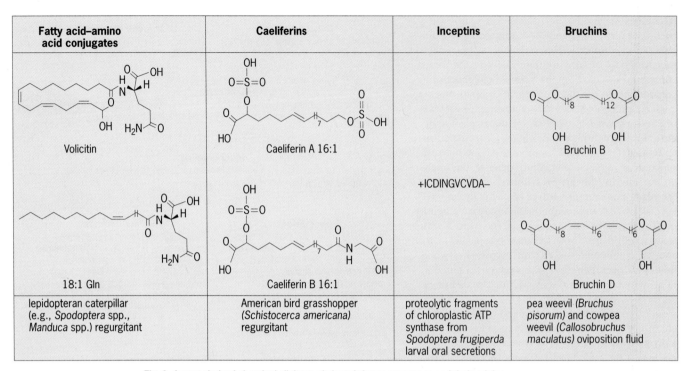

Fatty acid–amino acid conjugates	Caeliferins	Inceptins	Bruchins
Volicitin	Caeliferin A 16:1	+ICDINGVCVDA–	Bruchin B
18:1 Gln	Caeliferin B 16:1		Bruchin D
lepidopteran caterpillar (e.g., *Spodoptera* spp., *Manduca* spp.) regurgitant	American bird grasshopper (*Schistocerca americana*) regurgitant	proteolytic fragments of chloroplastic ATP synthase from *Spodoptera frugiperda* larval oral secretions	pea weevil (*Bruchus pisorum*) and cowpea weevil (*Callosobruchus maculatus*) oviposition fluid

Fig. 3. **Insect-derived chemical elicitors of plant defense responses and their origins.**

(for example, reduced herbivore survival or performance) of the damaged plant to attackers. Induced resistance can alleviate the negative effects of herbivory while also saving resources when herbivory is unpredictable or rare.

In addition to induced direct defenses, some metabolic plant responses can function as indirect defenses by attracting organisms that consume or harm the herbivores (parasites, parasitoids, and predators) (Fig. 2). For example, parasitoids such as the parasitoid wasp *Cardiochiles nigriceps* lay their eggs in tobacco budworm (*Heliothis virescens*) caterpillars, and the wasp larvae then consume the host caterpillars from the inside. Wasps locate their caterpillar hosts by detecting the organic chemical "smells" (volatiles) of the host plants that have been specifically induced by the feeding activity of the caterpillar. The plant's "defensive" release of these volatile chemicals is thereby indirect, because the chemical does not harm the herbivore directly, but instead acts indirectly through the wasp. Induced volatiles have also been shown to be used by predators and nematode pathogens of herbivores as attractive signals. Moreover, herbivores may use those same signals in order to avoid plants that have been damaged and might already be defended. Induced volatiles can even induce neighboring plants to ready their defenses, before the enemy arrives. Such eavesdropping on the neighbor's signal can give the receiver plant a competitive advantage over the sender plant. Thus, in how far and through what mechanism induced volatile compound emissions increase the relative fitness of the plant depends strongly on the interacting community and the interplay between direct and indirect plant defenses (e.g., synergistic or antagonistic).

Elicitation of plant responses. The changes in the induced direct and indirect defenses of plants are often very specific to the particular insect species, or even to insect age. This specificity arises from a process of signaling that integrates information from the type of damage (for example, whether the insect is a chewing caterpillar or a sap-sucking aphid) and herbivore-derived chemical stimuli. Collectively, these signals (termed elicitors) can indicate the identity and age of the herbivore species and elicit the specific defensive response of the plant. Some chemical elicitors are partially derived from compounds in the plant tissue and are altered in the insect's digestive tract, whereas others are insect digestive enzymes such as β-glucosidase (**Fig. 3**). Recent studies have even identified chemical elicitors in oviposition fluids of female bean weevils (*Bruchus pisorum*), which elicit specific plant responses during oviposition (Fig. 3). It remains largely unknown how the plants perceive these elicitor compounds, but they are both necessary and sufficient to explain the specificity of plant responses.

Plants integrate the complex information provided by herbivore feeding style and the various chemical elicitors through the differential activation of hormonal signaling pathways, a process that has been referred to as signaling pathway cross-talk. Following damage and/or the deposition of chemical elicitors, plant hormones (for example, jasmonic acid, salicylic acid, and ethylene) are mobilized and turn on specific genes associated with both primary and secondary metabolic pathways, resulting in changes in the production of defensive compounds. Typically, hormone signaling is not restricted to the damage site but is propagated throughout the plant, resulting in a plantwide (systemic) resistance to herbivores or a systemic emission of volatile compounds to attract natural enemies of the herbivores.

Further ramifications. The primary and secondary metabolic pathways of plants are not independent from each other, and the changes that occur in these pathways following herbivory are interconnected. This interconnectedness means that responses to multiple insect species are integrated in the attacked plant. Consequentially, these defensive responses affect other organisms, including pathogens, pollinators, predators, and other herbivores, in a concerted manner. Indeed, induced plant defenses can mediate complex interactions between any organisms that are directly or indirectly associated with the plant, and therefore have an extended sphere of influence that is much larger than the plant itself (Fig. 2). These include interactions on the cellular level, known from plant–pathogen interactions, and reaches up to the whole plant and community levels. For example, plant defenses in general and induced defenses in particular can have significant effects on the growth of entire herbivore populations as well as on the composition of the insect community that relies on the plant. The emerging picture no longer shows plants as passive consumable organisms but rather as active, dynamic, and influential participants in the interaction with the surrounding ecosystem. Utilizing the plants' natural defenses and considering the various plant strategies and consequential multitrophic interactions in modern breeding programs and agricultural practices has a high potential to lead to a more sustainable agriculture with fewer negative effects on natural resources and biodiversity.

For background information *see* AGRICULTURAL SCIENCE (PLANT); CHEMICAL ECOLOGY; ECOLOGY; INSECT CONTROL, BIOLOGICAL; INSECTA; PLANT; PLANT PHYSIOLOGY; PREDATOR-PREY INTERACTIONS in the McGraw-Hill Encyclopedia of Science & Technology.
André Kessler

Bibliography. P. R. Ehrlich and P. H. Raven, Butterflies and plants: A study in coevolution, *Evolution*, 18:586–608, 1964; R. Karban and I. T. Baldwin, *Induced Responses to Herbivory*, Chicago University Press, Chicago, 1997; A. Kessler and I. T. Baldwin, Plant responses to insect herbivory: The emerging molecular analysis, *Annu. Rev. Plant Biol.*, 53:299–328, 2002; L. M. Schoonhoven, J. J. A. van Loon, and M. Dicke, *Insect-Plant Biology*, Oxford University Press, Oxford/New York, 2005.

Plant defense against pathogens

Land plants diversified over 360 million years ago. This occurred well after the bacteria and fungi, which have a much more ancient origin. Although few fossils record the coexistence of plants and pathogens, early Devonian (400 million-year-old) Rhynie Chert (a fossil bed of hard, dense, sedimentary rock composed of fine-grained silica located near the village of Rhynie, Scotland) contains plant fossils that display symptoms of disease and harbor ascomycete fungal reproductive structures, indicating an ancient and potentially parasitic association between these organisms. Additional evidence of a long coevolution between plants and their pathogens is found in the highly specialized and elaborate infection and invasion structures produced by pathogens, as well as the elaborate and specialized defense systems that plants possess. In addition, recent plant and microbial genome analyses have revealed the existence of large families of plant genes that play roles in pathogen detection and defense, as well as pathogen genes that are important in virulence, providing additional evidence for the coevolution of plants and pathogens.

Coevolution of host resistance and pathogen virulence traits. The great reduction in host fitness caused by disease provides strong selective pressure for plant variants that have enhanced disease-resistance traits. Under pathogen pressure, such plants will outcompete more susceptible plants in the population, thus preserving their defense-related traits and fueling the evolution of resistance. However, at the same time, pathogens have also been subjected to selection for variants that can evade or overcome plant resistance, thus enabling the microbe to gather nutrition from the parasitized host, reproduce, and preserve its genetic contribution to the population. This coevolutionary landscape, and the long association between plants and their pathogens, has led to a diverse array of host disease-resistance mechanisms that make up the plant defense network, together with pathogen countermeasures that can defeat host defenses. Thus, a standoff usually exists, with most plants exhibiting strong resistance to most potentially pathogenic microbes that they encounter and some adapted or specialist pathogens having the ability to infect and cause disease on plants by use of specific mechanisms for evading, suppressing, neutralizing, or outpacing host defenses.

Passive defenses. Plant disease resistance is composed of both passive defense mechanisms and active processes that are elicited by pathogen exposure. The first lines of defense are those that a pathogen encounters when it comes into physical proximity or contact with a potential host. These defenses may take the form of physical barriers to infection, such as a tough plant cell wall that can resist pathogen invasion and a waxy cuticle that prevents water accumulation (which usually aids pathogen invasion). However, many pathogens have evolved mechanisms to defeat these physical barriers by secretion of cell wall- or cuticle-degrading enzymes, use of chemical or physical methods for overcoming these defenses, or invasion tactics that evade physical barriers by exploiting wounds or natural openings into plant cells, thus permitting the pathogen ingress into plant tissue where it can establish infection.

Pathogens may encounter a second passive line of defense in the form of preformed antimicrobial chemicals that accumulate in plants, which may provide a toxic or harsh environment for pathogen growth and reproduction. Such chemicals have been called phytoanticipins, in reference to a plant "anticipating" pathogen attack and prophylactically producing defense compounds. Examples of preformed antimicrobial chemicals include cyanogenic glycosides, glucosinolates, and detergent-like compounds called saponins. Many of these compounds are produced as nontoxic precursors that require chemical cleavage to activate their antimicrobial properties, and may also be sequestered in plant cell compartments, protecting the plant cell from toxicity. For example, cyanogenic glycosides and the enzymes that process them are isolated from each other within a plant cell; cell damage can allow the combination of these components, thereby liberating the potent respiratory poison cyanide. Cell damage and ensuing cyanide production may occur after infection of the plant or herbivore damage. This defense is countered by some pathogen strains that are largely resistant to cyanide, due to detoxification systems or by employing a cyanide-insensitive respiration pathway. The preformed defense chemicals known as saponins are distributed widely across the plant kingdom and are believed to disrupt pathogen cell membranes by binding to membrane sterols. Saponins usually require activation by pathogen attack before their toxic properties are exposed. Well-known examples of saponins include oat avenacin and avenacosides, and α-tomatine (prior to tomato saponin). Pathogens also have evolved defenses against these preformed defense compounds. For example, the oat root-infecting pathogen *Gaeumannomyces graminis* var. *avenae* infects roots to produce "take-all" (a severe root disease) and relies upon the production of an avenacin-degrading enzyme for its virulence on oats that produce avenacin. Similarly, many tomato fungal pathogens exhibit tolerance to the leaf- and fruit-localized α-tomatine, usually through deployment of enzymes that detoxify this saponin. Fungal variants that lack the saponin-detoxifying activity show much less virulence on α-tomatine-rich tissues.

Active defenses. In addition to passive defense mechanisms, plants have highly effective pathogen-induced defenses. These responses are induced upon pathogen detection and require specific biochemical signal transduction pathways that induce a diverse array of responses, including activating new gene expression, altering plant cell physiology, or inducing programmed plant cell death. To activate these defense responses, the plant must be able to detect the presence of pathogen, either through recognition of specific pathogen-derived molecules or through detection of secondary effects of infection, such as plant cell wall fragments that result from the cell wall–degrading enzymes that some fungal pathogens produce. Plant cells may also be able to indirectly sense infection by monitoring a disruption in cellular homeostasis caused by a parasite that is actively harvesting nutrients from the cell, or by monitoring the status of plant proteins that are specifically targeted by pathogen-delivered toxins. These diverse indicators of pathogen attack must be sensed by plant cell receptors that, upon activation, stimulate signal transduction pathways that regulate responses to infection, including defense.

Phytoalexins. One of the responses displayed by plants following pathogen challenge is the production of phytoalexins, which are low-molecular-weight compounds that have antimicrobial activity. These are a chemically diverse group of compounds, usually having a core structure composed of multiple aromatic rings, to which side groups are attached that are often plant species-specific. In many cases, plants within the same family produce a characteristic set of phytoalexins that share the same core structure, which is assembled by one or more primary biochemical pathways that also produce a myriad of other chemical structures essential for life.

Phytoalexin levels increase following exposure to a diverse range of pathogens, and other stressors such as ultraviolet (UV) light or oxidative stress can also be effective inducers of their accumulation. Although phytoalexins are by definition antibiotics, it has been difficult for investigators to demonstrate a role in defense for all phytoalexins, as the antimicrobial activity in various cases may not be sufficiently concentrated or produced quickly enough to thwart infection. In addition, as nonbiotic stressors may induce accumulation of some phytoalexins, they are not just specific responses to infection. Despite these considerations, strong evidence supports an important role for some phytoalexins in defense. To be effective in preventing disease, a phytoalexin must be produced at a sufficiently high concentration to interfere with pathogen fitness and must be localized to the site of fungal or bacterial attack. This has been demonstrated for some well-studied phytoalexins.

Like other resistance mechanisms, pathogens have also evolved counterdefenses, such as that of the pea pathogen, *Nectria haematococca*, which produces a virulence factor in the form of an enzyme that inactivates, by demethylation, the host phytoalexin, pisatin. Most *N. haematococca* strains that lack the virulence factor are sensitive to pisatin and are less virulent on pea. Similarly, transfer of the pisatin demethylase gene into fungal strains that lack this activity enhances their virulence on pea. Together, these data show that pisatin contributes to, but is not the sole source of, resistance in pea to fungal pathogens.

Induction of phytoalexin biosynthesis in plants may also be triggered by a highly specific pathogen-detection system encoded by resistance genes (discussed below). Pathogen perception by the encoded resistance proteins not only activates phytoalexin production in some plants, but importantly induces the expression of other highly effective defense mechanisms that together can provide complete resistance to the detected parasite.

Resistance proteins. The best-known pathogen recognition receptors in plants are encoded by "resistance" genes, which were identified genetically through research done in the 1940s by Harold Flor on flax and the flax rust pathogen *Melampsora lini*. In genetic crosses, Flor identified flax *Resistance* (*R*) genes that were essential to prevent

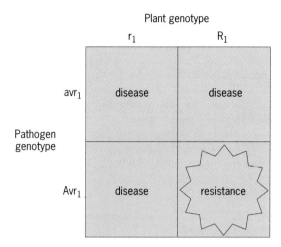

Fig. 1. Gene-for-gene resistance occurs when a host possesses an *R* gene (R_1) that corresponds to a matching pathogen *Avr* gene (Avr_1). In the absence of either (r_1 or avr_1), the plant fails to mount a defense response, and disease ensues.

infection by specific rust isolates. In genetic analyses of the rust pathogen, he identified fungal *Avirulence (Avr)* genes and showed that rust resistance in flax occurred only if the host possessed an *R* gene that corresponded to a matching rust *Avr* gene. The requirement that matched *R* and *Avr* genes be present to elicit a host resistance response led Flor to propose his "gene-for-gene" resistance hypothesis (**Fig. 1**). Since Flor's work with flax rust, many instances of gene-for-gene (or race-specific) resistance have been documented in many plant species, functioning against a diverse variety of parasites including bacteria, fungi, viruses, and oomycete phytopathogens.

Since the 1990s, many plant *R* genes have been cloned, allowing investigations into how the encoded R proteins function in the perception of pathogen attack and inactivation of host defense responses. Most R proteins have a conserved structure composed of a nucleotide-binding site (NBS) and leucine-rich repeat (LRR) domain. NBS-LRR proteins may be localized in the cytoplasm or anchored in the plant cell membrane, and constitute large gene families in plants; in the fully sequenced genomes of *Arabidopsis* and rice, approximately 150 and 400 NBS-LRR proteins are estimated to exist, respectively. Several models have been proposed to explain how pathogens are detected in plants by resistance proteins, with pathogen detection mechanisms depending upon the identity of the plant pathogen and its host. For example, molecules derived from the rice blast fungus appear to be directly detected by their interaction with the rice resistance protein Pi-ta, initiating a signaling pathway in rice that leads to blast resistance. However, in several cases, plant resistance proteins appear to sense pathogen attack only indirectly by monitoring the status of important plant proteins that are "attacked" or targeted by pathogen-produced protein toxins. According to this "guard" model, pathogens deploy virulence proteins that attack a limited number of key host cellular targets.

Consequently, plants have evolved mechanisms for "guarding" or detecting attack upon these targets using R proteins that, upon sensing an attack, activate host defenses that combat the pathogen. Uncertainty still exists with regard to the function of the different regions of R proteins in pathogen detection, guard function, and signal transduction.

When activated, gene-for-gene resistance responses can produce total or partial resistance to an attacking pathogen. While the actual mechanisms that provide resistance are poorly understood, one of the hallmarks of this kind of resistance is the localized death of plant cells at the site of infection, called the hypersensitive response (HR). The HR is a form of programmed cell death (apoptosis) and is thus under the genetic control of the plant, and results in a microscopic or macroscopic cluster of dead plant cells that surround the invading pathogen. The HR may play a role in defense by isolating the parasite from water and nutrients, generating toxic compounds that inhibit microbial growth, and generating signals that induce nearby healthy plant cells to bolster defenses and thus constrain the infection.

Systemic acquired resistance. Pathogen-induced defense responses not only function at the site of pathogen attack, but also may occur systemically to produce whole-plant resistance. To activate a systemic response, some kind of systemic or long-distance signal must be produced at the site of the initial pathogen-recognition event, which then travels to distant plant parts to activate resistance (**Fig. 2**).

The best-understood form of whole-plant pathogen-induced resistance response is called systemic acquired resistance (SAR), which is frequently but not always activated in response to a localized infection that triggers the HR. SAR leads to broad-spectrum resistance, meaning that it is effective against a diverse array of pathogen groups (bacteria, fungi, oomycetes, and viruses), and has been documented in monocot and dicot plants, suggesting that it is widespread in the plant kingdom. Induction of SAR in tobacco and *Arabidopsis* has

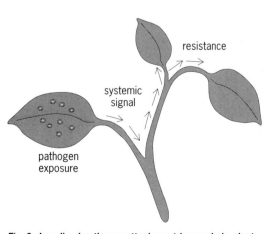

Fig. 2. Localized pathogen attack can trigger whole-plant resistance. Arrows represent a systemic signal that induces resistance in distant plant parts, such as occurs in systemic acquired resistance.

pathogen $\xrightarrow{\text{\quad * \quad}}$ SA $\longrightarrow$ NIM1 NPR1 $\longrightarrow$ PR genes $\longrightarrow$ SAR
detection

Fig. 3. Signal transduction pathway model for induction of systemic acquired resistance (SAR). Pathogen detection by the plant triggers the accumulation of salicylic acid (SA), which signals through a pathway defined by the *NIM1/NPR1* gene, leading to induction of *Pathogenesis-Related* (*PR*) genes and resistance. The long-distance signal in SAR acts upstream of SA in this pathway, indicated by an asterisk.

been shown to require accumulation of salicylic acid (SA), which is a chemical produced by plants in response to pathogen attack. Strong evidence for the requirement of SA accumulation in the induction of SAR came from studies of plants unable to accumulate SA due to expression of the bacterial *nahG* gene that encodes salicylate hydroxylase (which breaks down SA); nahG plants were unable to express SAR and had a general hypersusceptibility to pathogens. Interestingly, grafts between nahG rootstocks and normal tobacco scions were still able to develop SAR in scion leaves after a primary infection on the rootstock, whereas the reciprocal grafts could not. This indicated that the long-distance signal is not SA, although systemic tissue requires SA to activate SAR. The SA signal transduction pathway that activates SAR has been identified in genetic studies of *Arabidopsis*, which has revealed a single gene called *NIM1* or *NPR1* that is required for plants to respond to the SA signal and activate transcription of downstream *Pathogenesis-Related* (*PR*) genes (**Fig. 3**). Expression of *PR* genes is likely to be responsible for the robust disease resistance observed in plants expressing SAR.

Outlook. Genetic, biochemical, and genomic approaches have been essential for identifying plant genes required for pathogen detection, signal transduction, and defense. Important goals of this research have been to define the molecular mechanisms and signaling pathways by which plants detect pathogens or their attack, and to describe how plants combat these parasites using chemical and protein weapons, programmed cell death, or other defenses. Greater understanding of the molecular mechanisms that govern plant defense will enable novel strategies to be devised for manipulating endogenous plant defenses in agriculture, enhancing food and plant production, and reducing the use of chemical disease control agents. In addition, greater understanding of the mechanisms by which plants detect pathogens, such as through R proteins, may enable the engineering of pathogen detection capability into formerly naïve plants, permitting them to recognize and respond defensively to new pathogen races or species.

For background information *see* AGRICULTURAL SCIENCE (PLANT); DISEASE ECOLOGY; IMMUNOLOGY; PARASITOLOGY; PATHOGEN; PHYTOALEXINS; PLANT; PLANT PATHOLOGY; PLANT VIRUSES AND VIROIDS; SIGNAL TRANSDUCTION in the McGraw-Hill Encyclopedia of Science & Technology.　　Terrence P. Delaney

Bibliography. G. N. Agrios, *Plant Pathology*, 5th ed., Elsevier Academic Press, San Diego, 2005; J. L. Dangl and J. D. Jones, Plant pathogens and integrated defence responses to infection, *Nature*, 411:826–833, 2001; T. P. Delaney, Salicylic acid, pp. 635–653, in P. J. Davies (ed.), *Plant Hormones: Biosynthesis, Signal Transduction, Action!*, Kluwer Academic Publishers, Dordrecht, the Netherlands, 2004; W. E. Durrant and X. Dong, Systemic acquired resistance, *Annu. Rev. Phytopathol.*, 42:185–209, 2004; R. Hammerschmidt, Phytoalexins: What have we learned after 60 years?, *Annu. Rev. Phytopathol.*, 37:285–306, 1999; J. A. Lucas, *Plant Pathology and Plant Pathogens*, 3d ed., Blackwell Science, Oxford, 1998.

Plant juvenility

The development of the aboveground portion of seed plants, also called the shoot, is characterized by four phases: embryonic, juvenile vegetative, adult vegetative, and reproductive. The embryonic stage includes the period between the fertilization of an egg and the formation of a mature embryo within a seed. In contrast to animal embryos, plant embryos do not contain precursors of all of the organs found in an adult plant. Instead, the embryonic phase establishes the basic axes of growth, and new organs are continually formed throughout the life of the plant from a pool of stem cells at each end of the shoot-root axis. These stores of stem cells are called the shoot apical meristem (SAM) and the root apical meristem (RAM), respectively. Thus, while the apical end of an angiosperm embryo includes only one or two cotyledons, one or more immature true leaves, and the SAM, upon germination the plant repeatedly initiates new organs from the SAM—first vegetative organs (leaves, stems, and branches) and then reproductive ones (flowers).

Juvenile and adult vegetative phases. Vegetative development itself is subdivided into two phases—juvenile vegetative and adult vegetative—and the developmental process whereby a plant goes from the juvenile to the adult state is called vegetative phase change. The juvenile and adult phases of vegetative development are formally defined by their difference in reproductive competency—juvenile plants and tissues are reproductively incompetent, and adult plants and tissues are reproductively competent. In addition to changes in reproductive ability, a number of other morphological and physiological characteristics also change during vegetative growth. These traits are unique to each species and include differences in leaf morphology, disease resistance, phytochemical production, leaf retention, wood quality, and so on. Rarely do these traits strictly correlate with one vegetative phase or the other. Most of these changes are gradual and vary on a continuum. For example, in *Arabidopsis thaliana*, which has been used as a model of vegetative phase change, the only clear-cut morphological marker of vegetative phase change is production of trichomes (hairlike epidermal structures in plants) on the abaxial (lower) surface of the leaves: juvenile leaves possess trichomes only on the adaxial (upper) surface, while adult leaves have trichomes on both surfaces. At the same time, several other characteristics, including leaf shape, flatness, and serration, vary continuously: juvenile leaves are rounder and flatter with smoother margins, while adult leaves are more oblong and curled with more serrated margins.

Interestingly, adult plants are developmental chimeras of juvenile and adult tissues. This is most obvious in mature specimens of some woody plants (trees, shrubs, and woody vines), which have juvenile branches originating from the lower part of the trunk and adult branches above. This is particularly obvious in many oak species, such as the sawtooth oak (*Quercus acutissima*). In this species, the juvenile branches do not produce flowers and retain their leaves from the previous season until the spring, while the adult branches on the same plant make flowers and lose their leaves earlier. This chimeric habit is in clear contrast to the juvenile-to-adult transition in animals, where there is a global change throughout the individual during puberty, such that the entire body becomes adult.

Relationship between vegetative phase change and reproductive onset. The strict definition of the juvenile and adult phases of vegetative development by their reproductive competency may lead to the misconception that vegetative phase change and the transition to flowering are one and the same. In fact, certain factors can coregulate both of these phase transitions. By using nonreproductive traits as markers of phase identity, such as abaxial trichomes in *Arabidopsis*, however, scientists have found mutants that functionally separate these two transitions. These mutants include those that allow for an earlier onset of adult vegetative characteristics without any effect on reproductive timing, and those that have a normal vegetative phase transition but delayed flowering. This means that a plant must be in an adult phase of growth in order to flower, but that an adult phase is not sufficient for reproduction. Thus, when a plant makes the transition from the juvenile to the adult state, it gains the competency to sense floral-inducing signals, which are signals whose expression is separately regulated by reproduction-inducing pathways.

Small RNAs regulate the juvenile-to-adult transition. Based on these results, it has generally been thought that overlapping signaling pathways control the postembryonic developmental transitions. Much is understood about the pathways regulating flowering. Until recently, however, little was known about what controls the switch from juvenile to adult development. In order to identify these factors, genetic screens were performed in *Arabidopsis* and maize to find developmental timing mutants with an early or late juvenile-to-adult onset. In the last few years, several of these mutants have been shown to have lesions in genes involved in the biogenesis or activity of small ribonucleic acids (sRNAs).

sRNAs are 19 to 24 nucleotides in length and are processed from longer RNA precursors. These sRNAs bind to complementary sequences in messenger RNAs (mRNAs, which encode proteins) and direct their cleavage or inhibit their translation into proteins. The net result in both situations is a reduction in the protein expression of the target genes. Some sRNAs subsequently induce transcriptional silencing of the target genes, as well.

There are two major types of sRNAs—microRNAs (miRNAs) and small interfering RNAs (siRNAs)—and these differ in their origin and their target genes. As seen in **Fig. 1**, miRNAs are formed from the primary transcripts of longer, nonprotein-coding RNAs that have self-complementarity, enabling the RNA to form a double-stranded hairpin structure. The miRNA is found within this hairpin, duplexed with a partially complementary miRNA (miRNA*). While still in the nucleus, a Dicer family protein cleaves the hairpin twice, releasing the duplex. The duplex or the single-stranded miRNA is transported to the cytoplasm by an miRNA nuclear exporter. In an RNA-induced silencing complex (RISC), an Argonaute family protein unwinds any duplexes, allowing the single-stranded miRNA to bind to an mRNA to which it has partial complementarity. The Argonaute protein then cleaves the mRNA within the miRNA target site or inhibits its translation. Importantly, the primary miRNA transcript is complementary to its target genes only within the short miRNA target site. Many miRNA targets are genes that are important in various plant developmental processes.

By contrast, siRNAs are formed from long, double-stranded RNA precursors (Fig. 1). These precursors are polymerized from single-stranded mRNAs by RNA-dependent RNA polymerases (RdRPs or RDRs). Dicers cleave the double-stranded mRNA into many siRNAs. These siRNAs then generally target the cleavage of perfectly complementary transcripts that are homologous to the original mRNA throughout its entire length. The action of siRNAs in this manner is called posttranscriptional gene silencing (PTGS) or RNAi. The normal functions of siRNAs in plants include defense against RNA viruses and transposon repression.

Recently, several *Arabidopsis* mutants have been identified that display an early juvenile-to-adult transition due to mutations in genes that are important for the biogenesis or activity of sRNAs. These genes include *HST* (a gene encoding an miRNA nuclear exporter), *SGS3* (a plant-specific gene involved in PTGS), *RDR6* (an RNA-dependent RNA polymerase gene), *DCL4* (a Dicer-like gene), and *AGO7/ZIP* (an Argonaute family gene). *SGS3* and *RDR6* had previously been discovered in screens for mutants with defects in PTGS, but no developmental roles had been assigned to them. A microarray analysis of the *sgs3*, *rdr6*, and *ago7* mutants identified three transcription factors (*ARF3*, *ARF4*, and *SPL3*) that are expressed more highly in the mutants than in wild-type plants. Further experimentation showed that SGS3, RDR6, DCL4, and AGO7 normally repress the expression of *ARF3* and *ARF4* via their involvement in the biogenesis of a new class of siRNAs called *trans*-acting small interfering RNAs (ta-siRNAs). The ta-siRNAs are formed from long, double-stranded RNA precursors like siRNAs, but they are

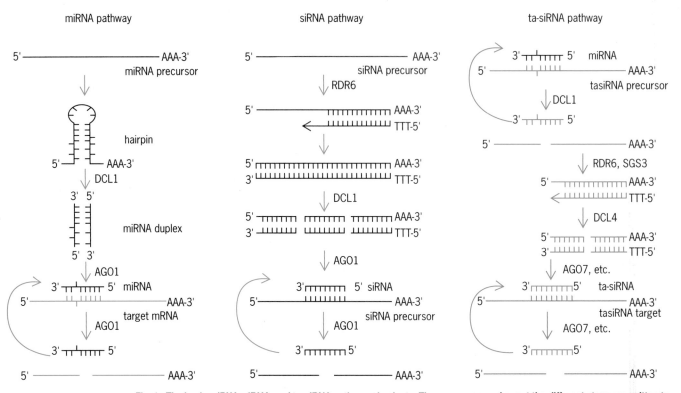

miRNA pathway   siRNA pathway   ta-siRNA pathway

Fig. 1. The basic miRNA, siRNA, and ta-siRNA pathways in plants. The enzymes carrying out the different steps are written to the right of the downward arrows. The miRNA pathway schematic shows processing of the hairpin into an miRNA:miRNA* duplex by the Dicer-like protein DCL1. The Argonaute protein AGO1 can cleave the miRNA target mRNA, as shown, or it can inhibit its translation. In the siRNA pathway, the siRNAs target other siRNA precursors like the original transcript, while in the ta-siRNA pathway the ta-siRNAs target non-self transcripts. For simplicity, RDR6 is shown polymerizing the entire second strand of the siRNA precursor, starting from the polyA tail at the 3′ end, but RDR6 can actually start at any point along the transcript. In the ta-siRNA pathway, the ta-siRNA-directed cleavage is carried out by AGO7 in the case of the *tasiARF* and its targets *ARF3* and *ARF4*, but the action of other ta-siRNAs is mediated by different Argonaute proteins. In all three pathways, the active sRNAs are able to direct the cleavage or translational inhibition of multiple targets, as shown by the curved arrows.

unique in that they originate from nonprotein-coding RNAs and target (Fig. 1). The biogenesis of ta-siRNAs is also more complicated because their mRNA precursors are targets of miRNAs. In the case of the ta-siRNAs that silence *ARF3* and *ARF4* (*tasiARF*), the miRNA *miR390* directs the cleavage of the ta-siRNA precursor at a specific target site. SGS3 stabilizes the cleavage fragments, allowing the 3′ cleavage fragment to be double-stranded by RDR6 and cleaved by DCL4 into ta-siRNAs, some of which go on to target *ARF3* and *ARF4* transcripts for silencing. Overexpressing *ARF3* in transgenic plants was found to reconstitute the early adult onset phenotype seen in *sgs3*, *rdr6*, *dcl4*, and *ago7* mutants, suggesting that the misregulation of *ARF3* and/or *ARF4* largely explains the phenotype of the mutants. Interestingly, while *ARF3* and *ARF4* seem to promote vegetative phase change, their mRNA expression does not increase during the juvenile-to-adult transition in wild-type plants. However, there could be a change in the protein levels or activity of these genes that has not been discovered to date.

In contrast, *SPL3*, the third gene that is significantly upregulated in the ta-siRNA biogenesis mutants, is expressed in a temporal pattern: low levels in juvenile tissue and high levels in adult tissues (**Fig. 2**). This is due to the complementary expres-

sion pattern of the miRNA *miR156*, which targets *SPL3* and 10 more of the 17 *SPL* genes for cleavage and/or translational inhibition. Overexpression of *miR156* significantly delays vegetative phase change, and overexpression of an *miR156*-insensitive version of *SPL3* hastens it. Orthologs of the *SPL3* and *miR156* genes are also found in maize, where a similar relationship is seen. This *SPL-miR156* module is actually evolutionarily conserved throughout the

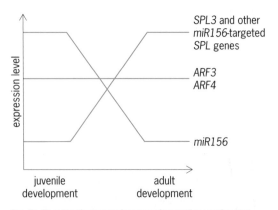

Fig. 2. The developmental expression patterns of genes that regulate vegetative phase change. This graph is meant to show only the changes in expression for each gene during development rather than actual expression levels or relative levels between genes.

plant kingdom, further suggesting its importance. It is still not clear why *SPL3* is upregulated in *sgs3*, *rdr6*, and *ago7*, however, because these genes do not regulate miRNAs. Therefore, it is likely that the effect of these genes on *SPL3* expression is indirect. The early adult onset phenotype of the miRNA exporter *hst*, however, is probably due to the lower levels of both *miR156* and *miR390* in this mutant.

Outlook. The *miR156* and the *SPL* genes that it regulates are the only genes known to have temporal expression patterns that affect vegetative phase change, and work is currently under way to discover the basis for the temporal expression pattern of *miR156*. The future of the field of vegetative phase change, however, involves going from the study of individual genes to a genome-wide characterization of the gene expression patterns associated with the juvenile and adult phases of development. Along with this, it will be interesting to see if there are any previously unknown global, whole-organism effects at the level of gene expression as a result of the juvenile-to-adult transition, which is the typical case in animals. The initial studies will focus on *Arabidopsis* and maize, the major laboratory models for vegetative phase change, but it will be important to relate this information to what happens in woody plants, where phase change significantly affects the timing of reproduction and, thereby, the speed of breeding programs. In white oak trees (*Quercus alba*), for instance, it can take 30–40 years for a tree to enter the adult phase of growth and begin to reproduce. It will also be vital to understand how vegetative phase change and disease resistance can be uncoupled in the instances where they are coregulated.

For background information *see* APICAL MERISTEM; LEAF; PLANT; PLANT GROWTH; PLANT MORPHOGENESIS; PLANT ORGANS; PLANT PHYSIOLOGY; PLANT REPRODUCTION; RIBONUCLEIC ACID (RNA); STEM in the McGraw-Hill Encyclopedia of Science & Technology.
                                        Matthew R. Willmann

Bibliography. I. Bäurle and C. Dean, The timing of developmental transitions in plants, *Cell*, 125:655–664, 2006; R. S. Poethig, Phase change and the regulation of developmental timing in plants, *Science*, 301:334–336, 2003; R. S. Poethig et al., The function of RNAi in plant development, *Cold Spring Harb. Sym.*, 71:165–170, 2006; M. R. Willmann and R. S. Poethig, Conservation and evolution of miRNA regulatory programs in plant development, *Curr. Opin. Plant Biol.*, 10:503–511, 2007.

Plumes and banded iron formations

Banded iron formation (BIF) is a chemical sedimentary rock containing >15 wt% iron (Fe). Most commonly, Fe-rich layers alternate with chert, although they may also be interbedded with layers made up of carbonate or silicate minerals, or sulphidic shale. As the major source of iron ore, BIF is of crucial economic importance. Sedimentologic and geochemical investigations of BIF continue to yield significant insights into the chemistry of the Precambrian atmosphere-hydrosphere system and the evolution of life.

Algoma-type sequences. Algoma-type BIFs, common in Archean greenstone belts, are associated with mafic and ultramafic, magnesium- and iron-rich volcanic units. Many exhibit an unequivocal association with localized submarine magmatism and high-temperature hydrothermal processes. These BIFs are often interbedded with mafic units containing pillow lavas. They may grade into volcanogenic massive sulfide deposits, contain remnant hydrothermal chimneys, and display geochemistries consistent with deposition from hydrothermal fluids. Similar Fe-rich deposits occur in Phanerozoic ophiolites, and form today along mid-ocean ridge crests and in back-arc basins, where high-temperature hydrothermal activity occurs.

Superior-type sequences. Depositional environments associated with Superior-type BIFs are more variable. Some precipitated on the continental shelf (such as the Mozaan Group, Pongola Supergroup, and the Cauê Formation, Minas Supergroup); others, in foreland basins (for example, BIFs of the Labrador Trough) or back-arc basins (for example, the Brockman Supersequence of the Hamersley Basin). Many of the largest units were deposited during a marine transgression, or at a sea-level highstand. While some Superior-type BIFs display granular or oolitic textures, indicative of deposition above the wave base, most accumulated in deeper water, on the outer shelf or slope. Units that accumulated below the wave base display banding on a variety of scales, ranging from macrobands tens of meters thick to millimeter-thick microbands. There are no modern analogs for these BIFs. In fact, with only a few notable exceptions, deposition of Superior-type BIFs was confined to the period between about 2.9 and 1.8 billion years ago (Ga). Deposition of most Superior-type BIFs occurred prior to and during the Great Oxidation Event of 2.47–2.40 Ga, although those of the Animike Group and Labrador Trough were deposited some ~500 My later.

The degree to which bacteria moderated the precipitation of Fe in BIFs remains largely unresolved. Recent work has implicated anoxygenic photoautotrophic bacteria and bacteria performing dissimilatory iron reduction in BIF accumulation (the former in the 3.7 Ga Isua Supracrustal Sequence, and the latter in the ~2.45 Ga BIFs in the Hamersley Supergroup, West Australia, and the Transvaal Supergroup, South Africa).

Source of iron for banded iron formation. The volcanogenic depositional environment of Algoma-type BIFs virtually demands that they accumulated by incorporating a hydrothermal source of Fe, and their geochemistries support that inference. Similarly, whole-rock rare earth element (REE) geochemistry, or the geochemistry of Fe-rich layers in Superior-type BIFs, uniformly demonstrates positive europium anomalies (Eu/Eu* >1). Coupled with depleted REE patterns and neodymium (Nd) isotopic signatures, the data support precipitation of Superior-type BIFs from seawater containing a

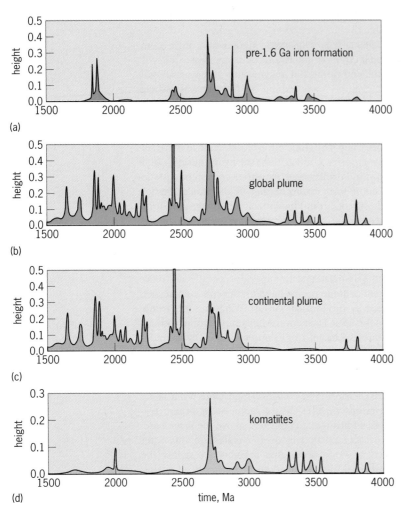

(a)

(b)

(c)

(d)

Time series of mantle plume activity and iron formations. All time series were constructed from radiometric ages by summing each age and its error as a Gaussian distribution. (a) Iron formations. (b) Global plumes: massive dike swarms, komatiites, flood basalts, and layered intrusions. (c) Continental plumes: massive dike swarms, flood basalts, and layered intrusions. (d) Komatiites. Correlations between global plumes and komatiites versus iron formation are significant at the 99% confidence level. The ages of continental plumes correlate more weakly at an 85% confidence level.

cesses and ambient seawater, whose chemistry reflected riverine input of weathering residues. Eruption and subsequent weathering of mafic-ultramafic flood basalt sequences with a comparatively high burden of Fe would have increased to some extent the riverine flux of Fe to the oceans. Increased mafic-ultramafic submarine volcanism along ridge crests or ocean plateaus would have elevated hydrothermal fluxes of Fe to the oceans. By inference, deposition of both Algoma- and Superior-type BIFs should be correlated with pulses of mafic-ultramafic magmatic activity commonly attributed to mantle superplume events.

Banded iron formation linked to mantle plume magmatism. There is a strong statistical correlation between the ages of both Algoma- and Superior-type BIFs and ultramafic-mafic magmatic events, documented by eruption of komatiites (high-magnesium oxide content volcanic rocks) and large igneous provinces (LIPs), and the emplacement of large groups of dikes and layered ultramafic igneous intrusions (see **illustration**). Similar Phanerozoic igneous units are generally considered to have been manifestations of mantle plume activity, although other robust models have been developed recently for several important LIP sequences.

A number of processes may contribute to the apparent causal link between important mafic-ultramafic volcanic events and the accumulation of BIF. First, weathering of continental flood basalt sequences, or oceanic plateaus that attain a subaerial expression, would increase the flux of iron to the global oceans. Second, the production of ocean-centered LIPs would elevate the hydrothermal flux of iron from high-temperature water-rock reactions. This effect might be exacerbated, as ocean plateaus have a shallower average topographic expression than mid-ocean ridges or back-arc basins, and geochemical studies suggest that as pressure is reduced, more Fe is released per unit of circulating seawater. Hydrothermal plumes emanating from oceanic plateaus might also be released in a shallower portion of the water column, facilitating Fe transport to shelf-slope environments. As the production of oceanic plateaus promotes marine transgressions, it may be no coincidence that many of the largest Superior-type BIFs were deposited during relatively high stands of sea level, including units in the Hamersley, Transvaal, and Minas Supergroups. Finally, mantle-plume volcanism promotes development of oceanic plateaus and seamounts, and may provoke continental rifting, thereby increasing the environments that appear to have favored BIF deposition.

Archean and Paleoproterozoic mantle superplumes. A significant period of mantle plume volcanism occurred around 2.7 Ga. This event may have been associated with mantle overturning, or resonance between outer core nutation and the luni-solar tides. As M. E. Barley and others have noted, widespread mafic and ultramafic volcanism, a hallmark of this period, was accompanied by unprecedented metallogenesis, including the deposition of BIF. Flood basalts

component of hydrothermal effluent. The hydrothermal signature may have been introduced via upwelling of deep, reducing seawater bearing a hydrothermal constituent. While it is commonly accepted that Superior-type BIFs received Fe from a hydrothermal source, the source of other constituents, notably silicon (Si), has been more hotly debated.

In fact, it is clear that the chemical sedimentary rocks in some Algoma- and Superior-type BIFs also received constituents from weathered terranes. Samarium-neodymium (Sm-Nd) isotopic signatures of the earliest Algoma-type BIFs in the Isua Supracrustal Sequence are best explained by mixing between a hydrothermal source and an ambient seawater source that received the weathering products of a basaltic continental crust. The largest Superior-type sequences also received some fraction (and perhaps a substantial fraction) of their major constituents from a continental source.

BIF deposition therefore both recorded and was controlled by Fe fluxes from hydrothermal proin the Kaapvaal's Ventersdorp Supergroup and the Pilbara's Fortescue Group have ages constrained within a range of 2.8–2.7 Ga. The Superior and Yilgarn Cratons host numerous greenstone belts of this age that contain both Algoma-type iron formations and komatiites. Some of these have been interpreted as accreted oceanic plateaus. The peak in numbers of BIFs deposited during this time is driven by deposition of Algoma-type units (see illustration).

A second significant period of mantle plume volcanism occurred during the period 2.5–2.4 Ga. Flood basalts in the Karelian Supergroup and the Imandra-Varzuga Supergroup were erupted during this period. The Metachewan and Hearst dike swarms (Superior Craton), the Great Dike (Kaapvaal Craton), and the Widgiemooltha dike swarm (Yilgarn Craton) were emplaced during this time. The bulk of the Fe in Superior-type BIFs (which contain about 85% of the Fe in all BIFs) accumulated during this period, including that deposited in the Hamersley and Transvaal Supergroups. The Great Oxidation Event also occurred during this time, which abrogated BIF deposition for at least 500 My. By inference, the flux of reducing gases associated with these volcanic events, and the deposition of Fe in BIF, would have countered the rise of atmospheric oxygen (O_2). Global oxygenation therefore may have occurred in fits and starts as volcanism and attendant BIF deposition waxed and waned.

Impacts, LIP production, and banded iron formation. Several statistical studies have documented notable strength at a period of about 30–35 My in time series of the ages of large igneous provinces and of high-Mg lavas. This periodicity is analogous to the frequency of terrestrial impact events. Geophysical models support extensive decompression melting in the wake of the impact of a large (diameter > 10 km) extraterrestrial body, resulting in development of an LIP.

Archean and Paleoproterozoic units containing evidence of impact are known from the Pilbara and Kaapvaal Cratons. A. Glikson has shown that seven of these units are overlain by iron-rich sedimentary rock, including some combination of BIF, jaspilite, and ferruginous shale. According to Glikson, the impactors were large enough that they may have promoted flood basalt volcanism and/or the uplift and exposure of mafic terranes, increasing riverine and/or hydrothermal fluxes of Fe to the oceans. Glikson and his colleagues have hypothesized that enhanced Fe availability then prompted deposition of the iron-rich units. While they caution that this hypothesis requires extensive testing, and that any putative correlation between impact ejecta and the deposition of Fe-rich sediment may be spurious, the hypothesis provides a robust explanation for the statistical correlations observed between impacts and mafic-ultramafic volcanism, and between such volcanism and the deposition of BIF.

For background information *see* ARCHEAN; BANDED IRON FORMATION; BASALT; CONTINENTS, EVOLUTION OF; CRATON; EARTH INTERIOR; HOT SPOTS (GEOLOGY); HYDROTHERMAL VENT; LAVA; MAGMA; MID-OCEANIC RIDGE; OOLITE; OPHIOLITE; PHANEROZOIC; PRECAMBRIAN; PROTEROZOIC; SEDIMENTARY ROCKS; SUBDUCTION ZONES; VOLCANO in the McGraw-Hill Encyclopedia of Science & Technology.   Ann E. Isley; Dallas H. Abbott

Bibliography. M. E. Barley et al., The Late Archean bonanza: Metallogenic and environmental consequences of the interaction between mantle plumes, lithospheric tectonics and global cyclicity. *Precambrian Res.*, 91:65–90, 1998; M. E. Barley, A. L. Pickard, and P. J. Sylvester, Emplacement of a large igneous province as a possible cause of banded iron formation 2.45 billion years ago, *Nature*, 385:55–58, 1997; M. E. Barley, A. Bekker, and B. Krapez, Late Archean to Early Paleoproterozoic global tectonics, environmental change and the rise of atmospheric oxygen, *Earth Planet. Sci. Lett.*, 238:156–171, 2005; L. A. Derry and S. B. Jacobson, The chemical evolution of Precambrian seawater; evidence from REEs in banded iron formations, *Geochim. Cosmochim. Act.*, 54:2965–2977, 1990; P. G. Erikkson et al., Late Archean superplume events: a Kappvaal-Pilbara perspective. *J. Geodyn.*, 34:207–247, 2002; R. Frei and A. Polat, Source heterogeneity for the major components of ~3.7 Ga banded iron formations (Isua Greenstone Belt, Western Greenland): tracing the nature of interacting water masses in BIF formation, *Earth Planet. Sci. Lett.*, 253:266–281, 2007; A. Glikson, Asteroid ejecta units overlain by iron-rich sediments in 3.5–2.4 Ga terrains, Pilbara and Kaapvaal cratons: accidental or cause-effect relationships?, *Earth Planet. Sci. Lett.*, 246:149–160, 2006; A. Glikson and J. Vickers, Asteroid mega-impacts and Precambrian banded iron formations: 2.63 and 2.56 Ma impact ejecta/fallout at the base of BIF/argillite units, Hamersley Basin, Pilbara Craton, Western Australia, *Earth Planet. Sci. Lett.*, 254:214–226, 2007; A. E. Isley and D. H. Abbott, Plume-related mafic volcanism and the deposition of banded iron formation, *J. Geophys. Res.*, 104:15,461–15,478, 1999; A. E. Isley and D. H. Abbott, Implications of the temporal distribution of high-Mg magmas for mantle plume volcanism through time. *J. Geol.*, 110:141–158, 2002; C. M. Johnson et al., Iron isotopes constrain biologic and abiologic processes in banded iron formation genesis. *Geochim. Cosmochim. Ac.*, 72:151–169, 2008; C. A. Spier et al., Geochemistry and genesis of the banded iron formations of the Cauê Formation, Quadrilátero Ferrífero, Minas Gerais, Brazil, *Precambrian Res.*, 152:170–206, 2007.

Polymer nanocomposites with carbon nanotubes

Over the last decade, academic and industrial scientists and engineers have become increasingly interested in tuning the properties of polymers by adding small quantities of nanoparticles. Polymers include plastics, rubbers, adhesives, and epoxies, and when nanoparticles are incorporated in them, these materials are called polymer nanocomposites. There

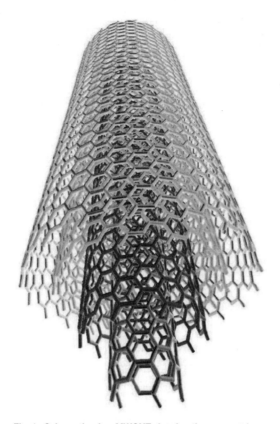

Fig. 1. Schematic of an MWCNT showing the concentric SWCNTs with hexagonal arrangement of *sp²* hybridized carbon atoms. (*Reproduced with permission from Eindhoven University of Technology, The Wondrous World of Carbon Nanotubes: A Review of Current Nanotube Technologies, 2003*)

is a vast array of nanoparticles available that can be sorted by shape into plates, spheres, and rods, where their smallest dimension is 1–100 nanometers.

Among the various nanoparticles, materials scientists and engineers have identified carbon nanotubes (CNTs) as having great potential to revolutionize the technological landscape. CNTs come in two common classes: single-walled (SWCNTs) and multiwalled (MWCNTs). SWCNTs are graphene sheets (hexagonal arrangements of sp^2 hybridized carbon atoms) rolled into a seamless cylinder with diameters of ∼1 nm and lengths that can be several hundreds or thousands of times the diameter. MWCNTs form under slightly different synthetic conditions and consist of concentric SWCNTs and have larger diameters, in the range of 10–50 nm (**Fig. 1**). CNTs are particularly attractive as fillers for polymer nanocomposites because they possess a unique combination of high mechanical strength, high electrical and thermal conductivity, and low density.

The first commercial success for polymer nanocomposites containing CNTs came from Hyperion Catalysis International. Electrically conductive MWCNT/polymer nanocomposites provide advantages over other filled polymer systems during electropainting, a method by which car bumpers are painted. Polymer nanocomposites with CNTs continue to enter niche markets for applications that require multifunctional materials. Advances in surface functionalization, nanocomposite fabrication,

and characterization tools are making significant contributions to this field, as mechanical, electrical, and thermal property enhancements continue to be reported.

The challenge of controlling the interfacial zone. One of the striking features in polymer nanocomposites is the enormous interfacial area per unit volume between the nanofillers and the polymer matrix. This is simply the result of the small size of the nanoparticles, as shown qualitatively in **Fig. 2a** and *b*. Consider, for example, two polymer composites with the same volume fraction of particles, but the spherical fillers in one are microparticles (radius of 3 μm) and in the other are nanoparticles (radius 3 nm). As the size of the spherical particles drops from the microscale to the nanoscale, the number of particles increases by 1 billion times, the mean particle-particle separation decreases by 1000 times, and the total interfacial area increases by 1000 times.

The increase in interfacial area is particularly important, because recent work in polymer physics has uncovered intriguing effects on polymer properties, particularly how the polymers move when they are near surfaces or otherwise spatially constrained. The region over which the polymer properties are altered is estimated to be on the order of the polymer molecules in the matrix, that is, 1–10 nm. To first approximation, the thickness of the interfacial region (Fig. 2) is independent of the particle size. Consequently, as particles decrease in size and the interfacial area increases, the volume of the polymer near the particle/polymer interfaces ($V_{\text{interface}}$) increases dramatically.

Figure 2c shows the calculated quantities for the interfacial volume of plate, sphere, and rod particles. The vertical axis gives the ratio of the interfacial volume to the particle volume ($V_{\text{interface}}/V_{\text{particle}}$), so that for a specific volume fraction of filler particles, the plot provides the volume of the interfacial region adjacent to the particles. The relative interfacial volume depends on the shape and aspect ratio of the particles, where the aspect ratios are defined as height/diameter for plates (aspect ratio <1) and length/diameter for rods (aspect ratio >1). Each curve represents a different relative interfacial thickness δ, which, for spheres, is defined as the ratio of the thickness of the interfacial zone to the radius of the sphere. Microparticles have $\delta \sim 0.01$, so that at any aspect ratio, the interfacial volume is negligible relative to the volume of the particles, ($V_{\text{interface}}/V_{\text{particle}} < 0.03$). In contrast, nanoparticles have $\delta \sim 1$–10, and the volume of the interfacial zone in nanocomposites exceeds the volume of the particle. For example, by dispersing 0.5 vol% of spherical nanoparticles (radius ∼2 nm) in a polymer (interfacial thickness ∼8 nm and $\delta = t/R = 4$), the volume fraction occupied by polymers near the particle/polymer interfaces is ∼32 vol%. Thus, the role of this interfacial zone is obviously paramount in controlling and designing the properties of polymer nanocomposites.

To illustrate the dramatic effect of the interfacial zone in polymer nanocomposites, we provide

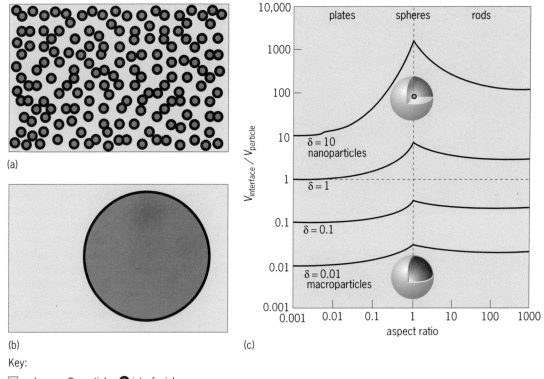

(a)

(b)   (c)

Key:

☐ polymer   ● particle   ○ interfacial zone

Fig. 2. Schematics (not to scale) showing the significant difference in the interfacial volume relative to the particle volume in polymer composites with (a) nanoparticles and (b) microparticles. (c) The ratio of the polymer volume near the particle/polymer interface to the particle volume ($V_{interface}/V_{particle}$) as a function of the particle aspect ratio for four relative interfacial thicknesses (δ). The interfacial thickness is assumed to be independent of the particle size and related to the size of the polymer molecule. As particles shrink from the macroscale ($\delta \sim 0.01$) to the nanoscale ($\delta > 1$), the interfacial volume exceeds the particle volume and might dominate the properties of the polymer nanocomposites. (*Adapted from MRS B., 32:314, 2007*)

examples for nanocomposites with CNTs. R. Haggen-mueller and coworkers observed that the presence of SWCNTs in high-density polyethylene (HDPE) induces a dramatic increase in the number of nucleation sites and thereby decreases the average HDPE crystallite size (**Fig. 3**). While the neat HDPE forms large spherulites (~30-50 μm) with twisted lamellae, the SWCNT/HDPE composites show smaller crystallites (~1-5 μm), even at a SWCNT loading as low as ~0.25 wt%. Furthermore, the HDPE crystals grow perpendicular to the nanotubes, and this altered polymer morphology in the interfacial zone significantly influences the composite material properties. J. N. Coleman and coworkers observed an unmistakable correlation between increasing MWCNT loading in poly(vinyl alcohol) and increasing crystallinity and mechanical properties. A. H. Barber and coworkers reported that in experiments to pull individual MWCNTs from an epoxy resin, the polymer in the interfacial zone resisted shear deformation to a significantly greater extent. A. Eitan and coworkers characterized MWCNT/polycarbonate composites and concluded that the improved mechanical and flow properties were due to the presence of immobilized polymers in the interfacial zone around the nanoparticles.

From these and other reports, it is evident that predicting the properties of polymer nanocompos-

ites by simply adding the properties of the polymer matrix and the filler is inadequate. For mechanical properties, in particular, the contribution of the polymer near the particle/polymer interface is distinct from the contribution of the polymer matrix farther away from the particles. And as the particles become smaller, the fraction of polymer near the filler can dominate the properties. The ultimate performance of polymer nanocomposites, specifically polymer nanocomposites with carbon nanotubes, requires that the specific design of particle/polymer interfaces optimize the mechanical properties for desired applications. Furthermore, the electrical transport properties of polymer nanocomposites are nonadditive with respect to particle concentration, because a percolating filler network dominates the composite properties.

Tuning the interface between polymers and nanotubes. While nanotube dispersion in polymers remains an active area of study, many have turned their attention to tuning the nanotube/polymer interface as a means of improving a wide variety of properties. The majority of the work to control the interface has focused on mechanical properties, because electrical conductivity is dominated by the CNT-CNT interactions rather than the CNT-polymer interactions. Interfacial modifications can be covalent or noncovalent, with the polymer matrix

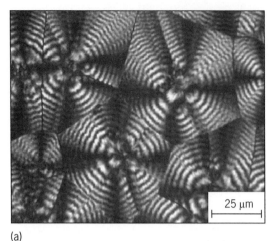

(a)

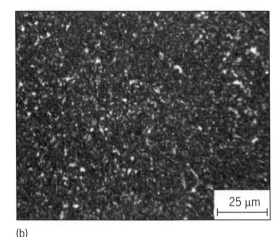

(b)

Fig. 3. Transmission optical micrographs with crossed polarizers of (*a*) high-density polyethylene (HDPE) and (*b*) HDPE with 0.25 wt% SWCNT, where the HDPE crystals are dramatically smaller in the presence of SWCNT. This and related results demonstrate that the particle/polymer interface nucleates and templates crystallization, thereby illustrating how the polymer matrix is altered in the vicinity of the nanotubes. (*Reproduced with permission from Macromolecules, 39:2964, 2006*)

interacting with the CNTs either through strong covalent bonds or through weaker secondary bonds, which might include hydrogen bonding or simply van der Waals interactions. Note that many of these strategies used to manipulate the interface will also enhance CNT dispersion.

The covalent bonding approach to controlling the interfacial properties builds upon recent developments in CNT functionalization and is particularly important for mechanical properties. For example, M. Moniruzzaman and coworkers combined nylon-6,10 and SWCNTs that were covalently functionalized with carboxylic acid chloride groups, using an in-situ polymerization method. The mechanical properties of these composites were much improved relative to those of nylon-6,10 with nonfunctionalized SWCNTs (**Fig. 4**). Similar mechanical property enhancements were reported by grafting short molecules, such as polymer brushes, on to the surface of the CNTs.

The noncovalent bonding approach involves a wide range of strategies derived from designing mis-

cible polymer blends. Polymer miscibility is promoted when the two components have a specific, favorable interaction, such as hydrogen bonding or dipole-dipole interactions. Work by M. Mu and K. I. Winey highlights the importance of weak van der Waals interactions, where they demonstrate the importance of polymer molecular weight when the polymer matrix is glassy. When the molecular weight is high, there are, on average, more points of contact between the polymer and the CNT per polymer chain. Thus, load transfer is more efficient when the polymer molecule is large relative to the particle size (**Fig. 5**).

Outlook. Significant strides have been taken in the past few years toward realizing the applications of polymer nanocomposites containing CNTs. As the quantity of commercial CNTs increases in the coming years, the practicality and properties of these nanocomposites will continue to improve. Reaching their full potential requires a fundamental understanding of the mechanisms that drive the properties at various length scales. Controlling the polymer-CNT interface is critical for dispersion and

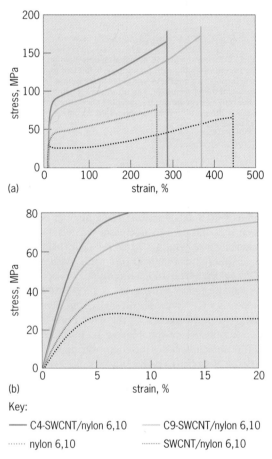

Fig. 4. Mechanical response (stress versus strain) of (*a*) mechanical response (stress versus strain) of fibers prepared by melt spinning three 1 wt% SWCNT/nylon 6,10 nanocomposites. (*b*) Enlarged low strain regime of part *a*. Two different composites with long alkane (C9) and short alkane (C4) linkages between the SWCNTs and nylon 6,10 were compared with neat nylon 6,10 and the composite made from nonfunctionalized SWCNTs. Carefully tuning the chemistry at the particle/polymer interfaces improves the elastic modulus, strength, and fracture toughness. (*Reproduced with permission from Nano Lett., 7:1178, 2007*)

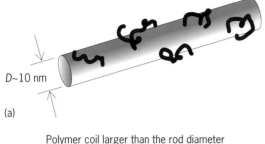

Polymer coil smaller than the rod diameter
mechanical properties—no improvement

D~10 nm

(a)

Polymer coil larger than the rod diameter
mechanical properties—good improvement

D~10 nm

(b)

Fig. 5. Another recently identified method of controlling the interfacial properties, (particularly the ability to transfer mechanical load from the polymer matrix to nanotubes) is the relative size of rod-shaped fillers and polymers. When the filler is nanoscale, a large polymer is able to entangle the rodlike fillers and thereby increase the resistance to mechanical deformation. Note that the lengths of the CNT fillers are considerably longer than shown here.

mechanical properties. And as demonstrated above, recent advances in tuning polymer/CNT interfaces have been successful. Pressing questions for the future include establishing the thickness of the interfacial zone, measuring its properties, evaluating the impact of this interfacial zone, and optimizing the interfacial zone for particular materials and applications. As an increasing number of scientists and engineers contribute to the understanding of these materials, we predict an exciting future for the innovation and commercialization of polymer nanocomposites using CNTs.

For background information *see* CARBON NANOTUBES; COMPOSITE MATERIAL; INTERMOLECULAR FORCES; NANOPARTICLES; NANOTECHNOLOGY; POLYMER; SHEAR; STRESS AND STRAIN in the McGraw-Hill Encyclopedia of Science & Technology.

Arun K. Kota; Karen I. Winey

Bibliography. A. H. Barber et al., Fracture transitions at a carbon-nanotube/polymer interface, *Adv. Mater.*, 18(1):83–87, 2006; J. N. Coleman et al., Reinforcement of polymers with carbon nanotubes. The role of an ordered polymer interfacial region. Experiment and modeling, *Polymer*, 47:8556, 2006; M. Daenen et al., *The Wondrous World of Carbon Nanotubes: A Review of Current Nanotube Technologies*, Eindhoven University of Technology, Eindhoven, Netherlands, 2003; A. Eitan et al., Reinforcement mechanisms in MWCNT-filled polycarbonate, *Compos. Sci. Technol.*, 66:1162, 2006; R. Haggenmueller et al., Single wall carbon nanotube/polyethylene nanocomposites: nucleating and templating polyethylene crystallites, *Macromolecules*, 39:2964, 2006; M. Moniruzzaman et al., Tuning the mechanical properties of SWCNT/nylon 6,10 composites with flexible spacers at the interface, *Nano Lett.*, 7:1178, 2007; M. Mu and K. I. Winey, Improved load transfer in nanotube/polymer composites with increased polymer molecular weight, *J. Phys. Chem. C*, 111:17923, 2007; K. I. Winey and R. A. Vaia, Polymer nanocomposites, *MRS Bull.*, 32:314, 2007.

Polymers through noncovalent bonding

Natural polymers (such as wool, silk, rubber, and cotton) have been used for many centuries, and artificial polymers (such as vulcanized rubber and cellulose nitrate) have been obtained from them since the end of nineteenth century. Synthetic polymers (such as Bakelite or polystyrene) were discovered soon afterward. These materials with remarkable mechanical properties were believed to be made from many small molecules, which somehow aggregated into large structures: colloidal aggregates. It took several decades and the work of pioneers such as H. Staudinger and W. Carothers to prove that polymers are in fact made of extremely large molecules whose repeat units are covalently bonded together. This understanding established an approach to polymer chemistry that has enabled the synthesis and development of a huge variety of polymers, which have effectively changed our lives. At the end of the twentieth century, the new concept emerged of supramolecular polymers, which are self-assembled small molecules (**Fig. 1**) held together by reversible noncovalent interactions, such as hydrogen bonds, metal-ligand complexation, π– π stacking, and host-guest interactions. It is noteworthy that supramolecular polymers do not comply

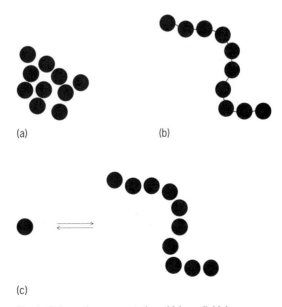

(a)　　　　　　　　(b)

(c)

Fig. 1. Schematic representation of (*a*) a colloidal aggregate (noncovalently and randomly interacting small molecules), (*b*) a polymer (covalent chain of monomers), and (*c*) a supramolecular polymer (noncovalent and reversible chain of monomers).

with Carothers' classical definition of a polymer as a single macromolecule. In fact, they are reminiscent of the nineteenth-century idea of polymers as colloidal aggregates. Nevertheless, supramolecular polymers have a lot in common with their covalent counterparts, and are now accepted as a special type of polymer.

Potential advantages. Because of their macromolecular architecture, supramolecular polymers can display polymerlike rheological or mechanical properties. The fact that noncovalent interactions are involved means that the assembly can be reversibly broken. This reversibility brings additional features, compared to usual polymers, which can potentially lead to new properties, such as improved processing and recyclability, self-healing behavior, or stimuli responsiveness. Indeed, the noncovalent nature of the backbone means that the length of the chains can be tuned by suitable stimuli. For example, heating usually weakens noncovalent interactions, which then leads to a shortening of the chains and thus a strong decrease in viscosity. This may be useful for fabricating materials that are easily processable at elevated temperatures (if the chains are short), but display good properties at low temperatures (if the chains are long). Moreover, the dynamic nature of the chains (that is, the chains break and recombine continuously) should lead to spontaneous self-healing of defects or cracks appearing during the lifetime of the material.

Design features. A key point is the selection of the interacting unit expected to drive the self-assembly process. First, the choice of a particular kind of noncovalent interaction has consequences on the final properties of the material, such as solvent resistance or color. For example, hydrogen bonding can lead to strong interactions in a nonpolar environment, but can be very weak in aqueous or alcoholic solvents because of hydrogen bonding of the solvent. On the other hand, metal-ligand complexes are usually difficult to manipulate in nonpolar media because of their limited solubility. The second point concerns the thermodynamic stability of a particular interacting unit. The strength of the interaction directly controls the length of the supramolecular polymer chains. In the case of a monomer bearing two identical self-complementary associating units, the number average degree of polymerization of the final supramolecular polymer is $DP_n \approx (KC)^{0.5}$, where C is the total concentration in repeat units (in mol/L) and K is their association constant (in L/mol). This means that if a degree of polymerization of about 1000 is sought for a bulk supramolecular polymer ($C \approx 1$ mol/L), then an association constant of $K \approx 10^6$ L/mol is required. Moreover, if the material is intended to be used in dilute solution, or as an additive, then the decrease in concentration has to be compensated by a corresponding increase in association constant. Thirdly, the kinetic characteristics of a particular interacting unit control the dynamic properties of the material. If a mechanical stress is applied for a time shorter than the time needed to break a chain, then the material shows an elastic response in the same manner as a covalently bonded polymer. In contrast, if a mechanical stress is applied for a long time, then the material starts to flow or creep because the constraints in the material can be released by breaking the chains, releasing entanglements, and then reforming the chains. Finally, the design of the monomer's backbone connecting the two (or more) interacting units is also important, as it will influence the solubility, rigidity, crystallinity, and thermal stability of the material.

Significant examples. Because of its directionality and versatility, hydrogen bonding is the most widely used interaction in the field of supramolecular polymers. **Figure 2** shows three typical systems. The quadruple hydrogen bonding motif of ureidopyrimidone (Fig. 2a) has been designed to form very strong dimers ($K_{dim} = 2 \times 10^7$ L/mol in chloroform at 25°C). Consequently, difunctional monomers form long chains even in dilute solutions: from the value of the equilibrium constant, a degree of polymerization of $DP_n = 1800$ ($M_n = 1.3 \times 10^6$ g mol^{-1}) at 0.04 mol L^{-1} (30 g L^{-1}) in chloroform can be estimated. These solutions show a high viscosity ($\eta/\eta_0 = 12$ at a concentration $C = 0.04$ mol L^{-1}) and a high concentration dependence of the viscosity ($\eta/\eta_0 \sim C^{3.7}$). Moreover, a wide range of di- or trifunctional oligomers, such as poly(ethylene-butylene), poly(tetramethyleneoxide), and poly(ε-caprolactone), have been functionalized with ureidopyrimidone. These supramolecular polymers show bulk mechanical properties at room temperature that are comparable to those of the corresponding high-molar-mass covalent polymers, but much reduced melt viscosity.

Recently, L. Leibler and coworkers have developed a supramolecular polymer based on fatty-acid dimers and trimers bearing amidoethylimidazolidone groups (Fig. 2b). The absence of crystallization of the fatty acid platform, together with a subtle control of the thermodynamics and dynamics of association of the amidoethylimidazolidone groups are responsible for the impressive properties of these self-healing elastomers. The system shows recoverable extensibility up to several hundred percent and little creep under load. In striking contrast to conventional cross-linked or thermoreversible rubbers made of macromolecules, these systems, when broken or cut, can be simply repaired by bringing together their fractured surfaces to self-heal at room temperature. Moreover, repaired samples recover their extensibility, and the process of breaking and healing can be repeated many times.

Bis-ureas have also been shown to form interesting supramolecular architectures. If a parallel or antiparallel orientation of the two ureas is enforced by the spacer connecting them, then long one-dimensional supramolecular assemblies can be expected. Depending on the exact nature of the spacer and the lateral substituents, it is possible to tune both the structure and the dynamic character of the assemblies. With symmetrical spacers and regular substituents, crystallization of the bis-urea is favored. However, using an unsymmetrical spacer and/or branched

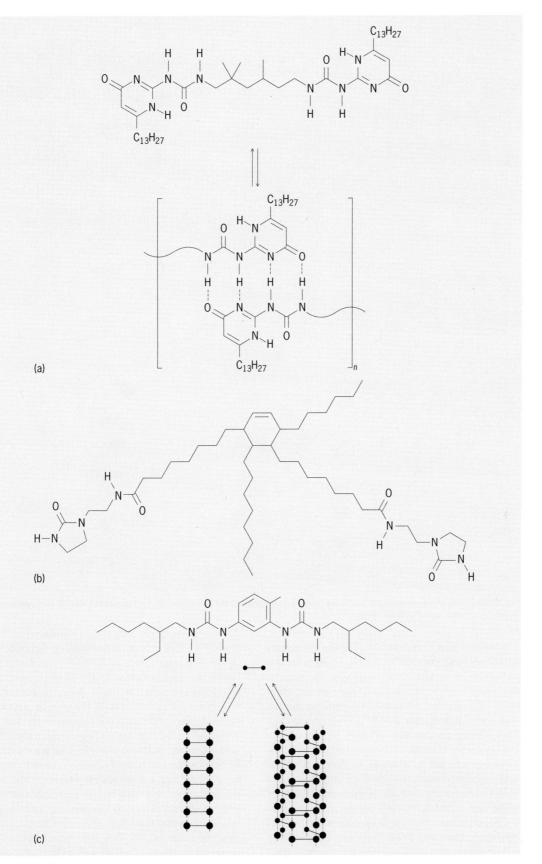

(a)

(b)

(c)

Fig. 2. Examples of supramolecular polymers based on hydrogen bonding self-associating units: (*a*) ureidopyrimidone (*E. W. Meijer et al.*), (*b*) amidoethylimidazolidone (*L. Leibler et al.*), and (*c*) bis-urea (*L. Bouteiller et al.*).

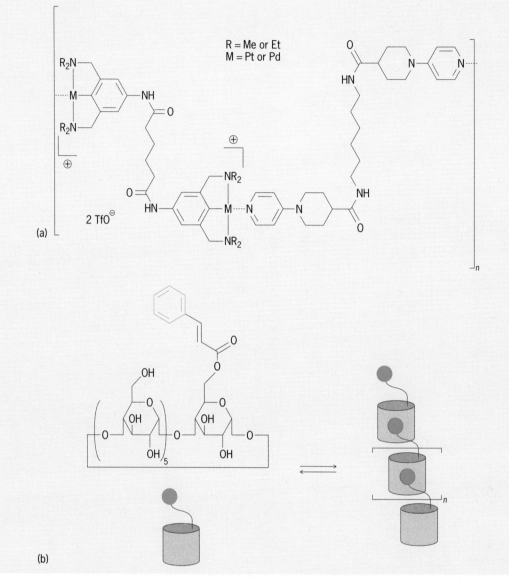

Fig. 3. Examples of supramolecular polymers based on (*a*) metal-ligand complexes (*S. L. Craig et al.*), and (*b*) hydrophobically driven host-guest complexes (*A. Harada et al.*).

substituents can destabilize competing crystalline structures and stabilize dynamic supramolecular polymers. Bis-ureas with a 2,4-toluene spacer (Fig. 2c) indeed form dynamic supramolecular polymers in nonpolar solvents. The remarkable feature about this system is that it displays two distinct supramolecular architectures, which are stable over a wide range of concentrations and temperatures and are in dynamic exchange with the monomer. Small-angle neutron scattering shows that both structures are long and fibrillar, with the high-temperature structure being thinner than the low-temperature structure. A ladderlike supramolecular arrangement has been proposed for the high-temperature, thin-filament structure and a thick tubular arrangement has been proposed for the low-temperature structure. The bis-urea thin filaments can be very long, and consequently the solutions show a high viscosity ($\eta/\eta_0 = 8$ at a concentration $C = 0.04$ mol/L in 1,3,5-trimethylbenzene, at

$T = 20°C$) and a high concentration dependence of the viscosity ($\eta/\eta_0 \sim C^{3.5}$). However, the solutions are not viscoelastic, probably because the relaxation of entanglements by chain scission is very fast ($\tau < 0.01$ s). In contrast, the tubular structure yields strongly viscoelastic solutions in the semidilute regime. The rheological properties of these bis-urea solutions can thus be switched from a viscoelastic behavior (at low temperatures) to a purely viscous behavior (at high temperatures). Moreover, the transition has been shown to be fast, reversible (without hysteresis), and extremely cooperative. The conversion of tubes into thin filaments occurs within a temperature range of 5°C only. This transition can be triggered by temperature as well as by a change in the solvent or monomer composition.

Other interactions can also be used to form supramolecular polymers. In particular, metal-ligand interactions have the significant advantage of covering a wide range of stability, from weak complexes to covalent interactions. Moreover, S. L. Craig showed that the rate of association can be tuned independently from the thermodynamic stability of the complexes. For example, replacing R = methyl groups by R = ethyl groups on the pincer ligands, as shown on **Fig. 3a**, results in two orders of magnitude slower scission and recombination rates, without affecting the association constant. Additionally, the presence of metal ions in metal-ligand based supramolecular polymers makes it possible to introduce optical, redox, or catalytic properties. In particular, the assembly can be triggered by changing the oxidation state of the metal ion.

Hydrophobic interactions are potentially the most suitable interaction for designing water-soluble supramolecular polymers. However, they are difficult to control because hydrophobic interactions are not directional. To get around this problem, it is possible to use hydrophobic interactions in combination with another interaction, such as hydrogen bonding or a host-guest complex (Fig. 3b).

For background information *see* CHEMICAL BONDING; COORDINATION CHEMISTRY; COORDINATION COMPLEXES; HYDROGEN BOND; INTERMOLECULAR FORCES; LIGAND; POLYMER THERMODYNAMICS; RHEOLOGY; SUPRAMOLECULAR CHEMISTRY; VISCOSITY in the McGraw-Hill Encyclopedia of Science & Technology.   Laurent Bouteiller

Bibliography. L. Bouteiller, Assembly via hydrogen bonds of low molar mass compounds into supramolecular polymers, *Adv. Polym. Sci.*, 207:79–112, 2007; L. Bouteiller et al., Thickness transition of a rigid supramolecular polymer, *J. Am. Chem. Soc.*, 127:8893–8898, 2005; A. Ciferri (ed.), *Supramolecular Polymers*, Marcel Dekker, Inc., 2005; S. L. Craig et al., Orthogonal control of dissociation dynamics relative to thermodynamics in a main-chain reversible polymer, *J. Am. Chem. Soc.*, 125:15302–15303, 2003; A. Harada et al., Cyclodextrin-based supramolecular polymers, *Adv. Polym. Sci.*, 201:1–43, 2006; L. Leibler et al., Self-healing and thermoreversible rubber from supramolecular assembly, *Nature*, 451:977–980, 2008; E. W. Meijer et al., Reversible polymers formed from self-complementary monomers using quadruple hydrogen bonding, *Science*, 278:1601–1604, 1997.

Porous alumina

Porous materials have been studied for years and are widely used in many industrial applications. More recently, they have become relevant in the field of nanotechnology as well as in fundamental studies of the new physics occurring in confined geometries. In particular, anodized porous alumina has a broad range of applications as a filter, gas and biological sensor, photonic crystal, and nanotemplate for the synthesis of various functional nanostructures, among others. Porous alumina has also been proposed as a model system for the understanding of capillary condensation and related effects. This broad interest originated from the discovery of self-organization of the porous structure in anodized alumina under certain conditions, leading to a hexagonal-close-packed array of nearly ideal cylindrical pores.

Overview of porous materials. Porous materials, such as compacted powders, clays, zeolites, silica gel, activated carbons, controlled pore glass, and Vycor glass, have been used in a variety of applications. They can be found in the chemical, oil and gas, food, and pharmaceutical industries as adsorbents, catalysts or catalyst supports for chemical reactions, humidity sensors, filters, and so forth. All the aforementioned materials are complex porous systems that have less than ideal geometries, with such defects as irregular pore shapes, wide pore size distributions, and interconnections between the pores.

More recently, there has been an increasing interest in porous materials with closer to ideal, well-controlled geometries, such as M41S silica materials, porous silicon, and porous alumina. The possible nanotechnological applications for these porous materials include DNA translocation, nanofluidic transistors, nanotemplates for nanostructured materials, optoelectronic devices, chemical and biological sensors, and even drug delivery systems. *See* ARTIFICIAL PHOTOSYNTHESIS; ELECTRONIC NANOPORES.

Moreover, porous materials are important for fundamental studies of confinement. The behavior of a few molecules confined within pores a few nanometers in diameter differs strongly from the bulk phase, because of the introduction of surface forces, dimensionality reduction, and other finite-size effects. In particular, new kinds of phase transitions appear (such as layering, wetting, and commensurate-incommensurate transitions), and well-known phase transitions (such as condensation; freezing; and liquid-liquid, solid-solid, and superfluid transitions) may become modified. The hysteresis and criticality of these phase transitions are also affected by confinement. In order to compare theoretical results for these phenomena with experiments, porous materials with well-known simple geometries are needed. Again, the same candidates as needed for nanotechnological applications appear: M41S silicas (especially MCM-41 and SBA-15), porous silicon, and porous alumina. A basic knowledge of confinement phenomena in porous materials is also useful in geophysics. Many rock and soil formations are porous, and their interactions with fluids such as water, pollutants, oil, and gas have relevance in different geological fields.

Porous alumina fabrication. Anodization is a process of electrolytic passivation of a metal (which acts as the anode of a cell) that is used to increase the thickness of the natural oxide layer on the surface of this metal. Anodization of aluminum, which gives rise to aluminum oxide (alumina) with a porous structure, has been studied for more than 50 years. However, because of previously poor control of both the pore size and pore regularity, it was not possible to utilize the porous structure.

In recent years, anodized porous alumina has become a popular template system for the synthesis of various functional nanostructures. This trend

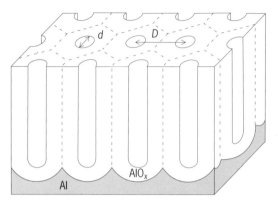

Fig. 1. Schematic representation of the ideally ordered porous structure in anodized alumina, a hexagonal close-packed array of cylindrical pores. The pore diameter *d* and pore distance *D* are the two main parameters that define this structure.

originated from the discovery of a highly ordered porous structure in anodized alumina, which was first achieved by Hideki Masuda and coworkers in 1995. Under appropriate voltages and electrolytes, the pores self-organize during their growth, leading to a hexagonal close-packed array of cylindrical pores that are perpendicular to the substrate (**Fig. 1**). The pores initiate from defects at the aluminum surface. A long anodization time allows the pores to self-organize into this ordered hexagonal array. In order to obtain this ordered structure from the surface and along the entire length of the pores, different techniques can be applied. One such technique is surface prepatterning using a molding process, which favors the growth of the pores from the imprinted texture. Another is the implementation of a two-step anodization: after a first anodization, the alumina is removed by selective acid etching, thus exposing an aluminum surface containing an ordered pattern of concave indentations (originated from the bottom of the growing pores), which in turn provide nucleation centers for the ordered formation of pores during a second anodization (**Fig. 2**).

Although the basic formation mechanism of self-ordering of pores is not completely understood, a simple explanation has been given by considering the interaction between the pores. Accompanying the formation of porous structures, the aluminum film is transformed into an aluminum oxide that has a reduced density, thus corresponding to a volume expansion. This volume expansion leads to repulsive forces between neighboring pores, promoting the formation of a hexagonal arrangement of pores (**Fig. 3**).

One important advantage of anodized alumina is the narrow distribution of the pore diameter d and the interpore distance D. Another is the tunability of both d and D, since it is well known that they are proportional to the applied anodization voltage. By adjusting anodization conditions, arrays of pores with diameters ranging from 10 to 200 nm and distances ranging from 20 to 500 nm have been obtained. Subsequent acid etching can be used to widen the pores without affecting the periodicity, thus decoupling the pore diameter d and the interpore distance D.

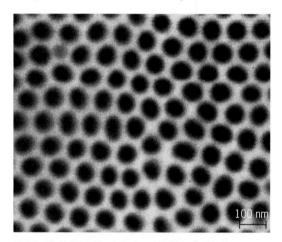

Fig. 3. Top-view scanning electron microscope picture of a porous alumina sample, anodized at 40 V in 0.3 *M* oxalic acid in two steps, followed by a pore widening in 0.5 *M* phosphoric acid. Pore diameter is 55 ± 6 nm, and pore distance is 100 ±10 nm. (*Courtesy of Chang-Peng Li*)

Applications. Anodized porous alumina has a broad range of applications. Membranes for micro- and ultrafiltration take advantage of the narrow pore size distribution of anodized alumina. The high surface area of the porous structure enhances gas adsorption and therefore is useful for gas sensing; for example, porous alumina sensors are used to sense humidity and ammonia with electrical detection and organic vapors with optical detection. There is also a huge potential application in biosensing or targeted drug delivery because the pore sizes can be tuned to match the characteristic sizes of the desired biomolecules. However, to achieve bioaffinity interactions in porous alumina, proper functionalization of the pore surfaces is required. Work in this area includes DNA detectors, which have been made by incorporating immobilized complimentary DNA into the pore walls; and urea sensors, which are based on immobilization of urease.

Porous alumina is also a good candidate for photonic crystals. Photonic crystals, which have

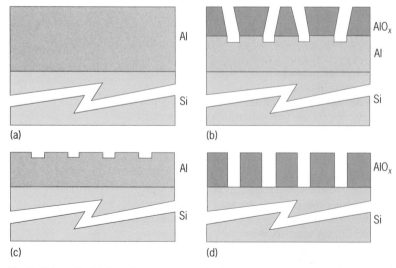

Fig. 2. Schematic drawing of a two-step anodization process to obtain a highly ordered porous structure. (*a*) Aluminum (Al) film on top of a silicon (Si) substrate. (*b*) First anodization. (*c*) Removal of the alumina (AlO$_x$) layer. (*d*) Second anodization.

generated increasing interest in recent years, are engineered periodic dielectric structures that are of the order of a wavelength. These structures modify the dispersion relation of incident light by producing an effect analogous to that of a crystalline structure on electrons, that is, generating photonic band gaps, energy regions where no photon modes are allowed. The periodicity of the porous structure in anodized alumina allows band gaps in the visible wavelength region. One of the most immediate applications of these artificial crystals is in optoelectronic devices whose performance is affected by unwanted spontaneous emission. They can also be used in the fabrication of tunable optical filters and microcavities, or lossless waveguides.

The most common use of anodized porous alumina is as a nanotemplate for the fabrication of functional nanostructures, taking advantage of the highly ordered porous structure. Metallic nanoporous membranes (composed of metals such as gold, platinum, or nickel), with the ordered structure transferred from the porous alumina using a variety of methods, can be fabricated in order to engineer properties suitable for electronic, optical, or micromechanical devices. Filling the high-aspect-ratio pores of the alumina with a metal (such as silver, nickel, bismuth, or ruthenium) leads to the fabrication of metal nanowires, which can be useful for photonic crystals, sensors, solar cells, or thermoelectric applications. If the pores have a low aspect ratio and are on a substrate, the alumina can be removed with selective acid etching after filling the pores with metal. The result is a regular array of nanodots over a macroscopic area on top of the substrate, which is potentially interesting for applications in magnetic storage if the filling metal is ferromagnetic. The maximal areal density for magnetic recording media is imposed by the superparamagnetic limit, the point at which the magnetic particles no longer have thermal stability. The limit in conventional longitudinal magnetic media ($\sim$70 Gbit/in.^{2} or 11 Gbit/cm^2) has been overcome by the use of antiferromagnetically coupled media and, more recently, by the use of perpendicular magnetic media. With the latter approach, a density of $\sim$250 Gbit/in.^{2} (40 Gbit/cm^2) has been reached in currently available hard drives. This density can be further improved with patterned perpendicular media, in which one bit of information corresponds to a nanoparticle with a single magnetic domain. It has been shown that ferromagnetic nanodots patterned with porous alumina templates can reach a density as high as $\sim$1 Tbit/in.^{2} (150 Gbit/cm^2). Moreover, this technique is much more economical than conventional lithography methods used for patterning of perpendicular media. Alumina nanopores can also be used as templates for growing carbon nanotubes or polymer nanotubes, which have attracted much attention in recent years for their unique properties.

Studies of new physics. Recently, anodized porous alumina has been proposed as a model system to understand the novel physical phenomena that result from confinement of molecules at the nanoscale, because of the nearly ideal geometry of the pores: cylindrical, with a narrow pore size distribution, and

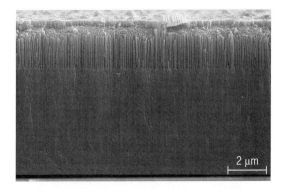

Fig. 4. Cross-sectional scanning electron microscope picture of a porous alumina sample with two different pore diameters along the pore length, 64 $\pm$ 10 nm at the top and 15 $\pm$ 3 nm at the bottom.

not connected to each other. Besides basic scientific interest, an understanding of these phenomena is necessary for engineering nanotechnological applications. In particular, porous alumina is being used to study capillary condensation (shifting of the vapor-liquid transition due to confinement), which depends on the pore diameter (the transition occurs at lower pressures for smaller pores). It is also being used to study the hysteresis associated with capillary condensation and evaporation, which depends on the homogeneity of the pore section and the possible interconnections among the pores. Furthermore, pore diameters can be tailored along the length of the pores to observe the effects of pore morphology on the hysteresis (**Fig. 4**). One such effect is the pore blocking effect, wherein the liquid at the narrowest pore section controls the evaporation of the liquid in the rest of the pore. These studies allow direct comparison of experiments with theoretical work, which is difficult because of the lack of ideal pore geometries in other porous systems. Another basic study concerns the wetting transition of polymer melts in alumina nanopores. Nevertheless, a large number of effects related to confinement remain to be studied in this promising model system.

For background information *see* CARBON NANOTUBES; ELECTROPLATING OF METALS; MAGNETIC RECORDING; METAL COATING; NANOSTRUCTURE; PHASE TRANSITIONS; SURFACE TENSION in the McGraw-Hill Encyclopedia of Science & Technology.

Fèlix Casanova

Bibliography. F. Casanova et al., Effect of surface interactions on the hysteresis of capillary condensation in nanopores, *Europhys. Lett.*, 81:26003, 2008; L. D. Gelb et al., Phase separation in confined systems, *Rep. Prog. Phys.*, 62:1573–1659, 1999; S. J. Gregg and K. S. W. Sing, *Adsorption, Surface Area and Porosity*, Academic Press, London, 1982; H. Masuda and K. Fukuda, Ordered metal nanohole arrays made by a two-step replication of honeycomb structures of anodic alumina, *Science*, 268:1466–1468, 1995; A. Moser et al., Magnetic recording: advancing into the future, *J. Phys. D Appl. Phys.*, 35:R157–R167, 2002; K. Nielsch et al., Self-ordering regimes of porous alumina: the 10% porosity rule, *Nano Lett.*, 2:677–680, 2002; E. Yablonovitch, Photonic bandgap structures, *J. Opt. Soc. Am. B*, 10:283–295, 1993.

Positive psychology: human happiness

Positive psychology as a "movement" began in 1998 when Martin Seligman proposed, during his term as president of the American Psychological Association, that psychology be just as concerned with what is right with people as it is with what is wrong. Approximately 20–30% of people in the United States suffer from a mental disorder at one time or another, and that slice of the population has received much attention in the psychology literature for many decades. Positive psychology sprang from a desire to rebalance psychology as a field so that, in addition to understanding the etiology and optimal treatment strategies for mental disorders, it would be possible to understand and better the lives of the other 70–80% of people who are not severely distressed.

Prior to 1998, several of the topics that are central to positive psychology—defining happiness, understanding what makes people happy, and learning how to make people happier—had been discussed under the banner of humanistic psychology. However, positive psychology is distinct from the humanistic movement in two key ways. First, humanistic psychology focuses largely on therapy, while positive psychology is a much broader field, integrating both theory and application, and encompassing many areas of psychology, including the social, educational, neuroscience, developmental, and clinical disciplines. Second, whereas humanistic psychology primarily makes use of qualitative research methods and single case studies, positive psychology is more firmly rooted in the experimental and longitudinal methods (analyses of data collected at different points of time). Thus, positive psychology is not a brand new field, but rather a reinvigoration of the spirit of humanistic psychology with a demand for rigorous empirical study of these topics. This "big tent" approach has had the effect of unifying researchers, scattered throughout the social sciences, who were already examining these topics.

Since its inception, the aim of positive psychology has been twofold: (1) to understand what it means to flourish, and what leads people, groups, and institutions to flourish; and (2) to design, test, and disseminate ways in which people, groups, and institutions can flourish more. Research findings relevant to each of these aims will be discussed below, along with some more general remarks about the future of the field.

Key theoretical findings in positive psychology. Four areas of growth in positive psychology can be analyzed: (1) defining and assessing well-being, (2) classifying and assessing character strengths, (3) understanding the nature of positive emotions, and (4) determining what makes a positive relationship.

Happiness. One of the first tasks set forth by positive psychology was to define and develop measures to assess happiness. In order to accomplish this goal, the scientifically unwieldy concept of happiness (useful only as a label for the field) was divided into three realms: (1) positive emotion, or the pleasant life, (2) flow, or the engaged life, and (3) meaning. Further research indicated that each of these paths uniquely predicts satisfaction with life and that, together, they have a synergistic effect that results in greater satisfaction (referred to as a "full life") than would be predicted by each separately. The Authentic Happiness Inventory (AHI), a 28-item self-report assessment, creates an overall well-being score by combining scores on pleasure, engagement, meaning, and accomplishment.

Character strengths. Paralleling efforts in psychiatry and clinical psychology to classify mental disorders, investigators in 2004 sought to classify human strengths. In this classification scheme, an initial list of hundreds of personality traits was narrowed down to 24 strengths that were ubiquitous and culturally valued. Concurrent with this research, the Values in Action Institute (VIA) Inventory of Strengths was developed, which is a 240-item assessment tool that provides users with a "strengths profile" ranking their endorsement of various strengths from highest to lowest. In studying the strengths profiles of individuals across 40 countries, it was found that individuals appear to be more similar than they are different: the average correlation between nations' strengths profiles was 0.80. In general, strengths such as kindness, gratitude, and open-mindedness are most commonly endorsed, with strengths such as prudence, modesty, and self-regulation being endorsed least often.

Positive emotion. Complementing the well-established area of negative emotion research, Barbara Fredrickson and colleagues conducted extensive experimental research on the nature of positive emotions such as joy, amusement, and contentment, and the impact that these emotions have on various aspects of human experience. Central to this line of research is the "broaden and build model" of positive emotions, which posits the following: whereas negative emotions narrow individuals' thought-action repertoires (that is, they restrict the range of thoughts and actions that an individual is likely to engage in), positive emotions broaden thought-action repertoires. The model further posits that a broadened thought-action repertoire promotes the building of abiding psychological capital. Positive emotions undo negative emotions, play an important role in resilience, and lead to enhanced creativity, but only in proper balance with negative emotions. (In this research, people who flourish, or display positivity, were found to have negative sentiments as well, but their negativity is outweighed by pleasant emotions and moods; an ideal ratio of positive to negative emotions was found to be no less than 2.9:1 and no more than 13:1.)

Close relationships. Much of the psychology of close relationships focuses on two topics: relationship dysfunction and how couples cope with negative events. Shelly Gable and colleagues conducted a series of studies examining the flip side of coping—how couples deal with positive events. Specifically, they identified one strategy for responding to good news from others that they call an "active-constructive" response: the person responds both positively and with genuine, visible enthusiasm. It was found that people who helped their mates savor life's victories, and whose mates responded in kind, were likely to be together several months later, and that this measure was more predictive of relationship success than was coping with negative events.

Applied positive psychology. Paralleling the progress in developing theory and assessment methods, applications of positive psychology have surfaced in a variety of domains. These include four important areas: (1) increasing happiness in nondistressed populations, (2) fostering optimal performance in the workplace, (3) augmenting education, and (4) treating mental disorders in clinical practice.

Increasing happiness. One of the most prominent and well-researched applications of positive psychology to date is the development of methods for increasing happiness in nondistressed populations (an effort spearheaded by researchers such as Sonja Lyubomirsky, Robert Emmons, and others). There is ample evidence for happiness set points, that is, baseline levels of happiness to which individuals are likely to return as they adapt to changes in the circumstances of their life, but there is also evidence that such set points can be offset by certain changes in habits. The majority of studies on happiness-increasing interventions involve the systematic testing of single exercises. To date, several of these exercises—for example, keeping a gratitude journal or a nightly record of good things that happened during the day, performing acts of kindness, and using one's strengths in new ways—have proven effective in randomized, controlled studies done both in person and over the Internet.

Work. Positive psychology has uncovered several factors that contribute to flourishing in the workplace. In several statistical research studies, employees report quantifiable increases in productivity, engagement, and life satisfaction when their work environment is restructured around their strengths. Meaning also seems to play a role: research suggests that employees are more motivated and productive when the positive impact that their work has on other people's lives is made more salient. Lastly, it has been observed that positive emotion is important to optimal functioning at work. Groups of employees who display more positive emotion have broader behavioral repertoires, greater resilience to adversity, and better productivity.

Education. Positive psychology has thrived for many years in higher education, at both the undergraduate and the graduate level, at many academic institutions across the United States. The practice of positive interventions (see "Increasing Happiness" above) has been an integral aspect of many positive psychology syllabi, with emphasis on merging education, research, and application of positive psychology principles in the classroom and in students' lives. There have also been efforts to implement positive psychology in secondary schools. Investigators are

currently in the process of evaluating a positive psychology curriculum, which focuses on character development and fostering positive emotion, at Strath Haven High School in Wallingford, Pennsylvania. The same research group is working with Geelong Grammar School in Victoria, Australia, to embed positive psychology across the entire curriculum.

Clinical practice. A related line of inquiry involves applications of positive psychology to clinical psychology, both as a way of modifying existing clinical practice and as the basis for developing new treatments. At least two positive therapeutic interventions—Michael Frisch's Quality of Life Therapy (QOLT) and Giovanni Fava and Chiara Ruini's Well-Being Therapy (WBT)—have been designed and implemented as adjuncts to existing therapy modalities, especially, cognitive therapy. Positive Psychotherapy (PPT) is a further stand-alone treatment for mood disorders that directly targets pleasure, engagement, and meaning. Preliminary research evaluating PPT has found it to be superior to control conditions in reducing depressive symptoms in both mild–moderately depressed and severely depressed populations, in both individual and group formats.

Outlook. In summary, there has been much progress in both the theory and the application of positive psychology, with achievements in defining and assessing happiness and character strengths, exploring the role of positive emotions, examining the factors that lead relationships to flourish, and applying positive psychology to individuals, groups, and institutions. The future of positive psychology will surely see further progress in many of these areas and should contribute to augmented human flourishing.

For background information *see* AFFECTIVE DISORDERS; DEVELOPMENTAL PSYCHOLOGY; EMOTION; PERSONALITY THEORY; PSYCHOANALYSIS; PSYCHOLOGY; PSYCHOTHERAPY; STRESS (PSYCHOLOGY) in the McGraw-Hill Encyclopedia of Science & Technology.

Acacia C. Parks; Martin E. P. Seligman

Bibliography. A. Duckworth, T. A. Steen, and M. E. P. Seligman, Positive psychology in clinical practice, *Annu. Rev. Clin. Psychol.*, 1:629–651, 2003; S. L. Gable and J. Haidt, What (and why) is positive psychology?, *Rev. Gen. Psychol.*, 9:103–110, 2005; S. Lyubomirsky, *The How of Happiness: A Scientific Approach to Getting the Life You Want*, Penguin Press, New York, 2008; M. E. P. Seligman, *Authentic Happiness: Using the New Positive Psychology to Realize Your Potential for Lasting Fulfillment*, Free Press, New York, 2002; M. E. P. Seligman and M. Csikszentmihalyi, Positive psychology: An introduction, *Am. Psychol.*, 55:5–14, 2000.

Predatory behavior in bacteria

Predation is a central ecological and evolutionary force. Predation shapes communities of organisms and ecosystems, impacts upon the behavior of both prey and predator, and influences evolutionary trajectories. Most of our knowledge about predatory interactions results from animal studies. If prokaryotes emerged before the eukaryotic lineages, around 3.8 billion years ago, then the first predators may have been prokaryotic organisms preying upon other prokaryotes. Here, for the sake of simplicity, all prokaryotes will be referred to as "bacteria."

Habitats and diversity. Predatory bacteria are bacteria that feed upon other bacteria and, in some cases, upon other microorganisms as well. Bacterial predators have been isolated or detected in soils, in the rhizosphere (the volume of soil around living plant roots that is influenced by root activity), in fresh, brackish, and marine waters, in extreme environments, and in the feces of various animals. Predatory bacteria live in the planktonic phase, preying upon suspended prey cells (in the planktonic phase, microorganisms are suspended in the water), as well as in biofilms formed by other microorganisms, preying upon these sessile bacteria (biofilms are microbial sessile communities composed of cells that are attached to a substratum; they are often involved in diseases and in equipment and infrastructure degradation). Their apparent population levels are usually low, but are probably underestimated; all measurements have been obtained by counting lytic plaques: prey cells inoculated at high concentration form an opaque cell lawn on petri dishes. When mixed with aliquots containing predatory bacteria, lytic, clear, and easily observable zones develop where the predators degrade the prey cells. The importance of predation between bacteria for bacterial mortality in nature is still unknown because of the large uncertainty about the levels of bacterial predatory activity in natural environments.

Although predatory bacteria belong to various phylogenetic (evolutionary) groups (see **table**), this diversity comprises but a small fraction of the huge microbial diversity discovered in the last several years. Until recently, knowledge of bacterial diversity was restricted by the ability to culture microorganisms in order to identify and characterize them. The development of molecular techniques has enabled the bypassing of this limitation and has led to the discovery of an enormous and as yet not exhausted prokaryotic diversity. Nevertheless, relatively few predatory prokaryotes have been described. This is due to the fact that the identification of a predatory interaction requires the observation of the phenomenon and does not solely rely on molecular features, the approach by which microbial diversity is mostly explored. Nevertheless, the development of culture-independent tools such as 16S rRNA gene-based PCR (polymerase chain reaction) primers and oligonucleotide probes for fluorescent in situ hybridization that are targeted to phylogenetic groups of predatory bacteria will certainly bring about new and more precise data on the diversity and sizes of these predatory populations in nature.

Types of predatory bacteria. Bacterial predators can be divided into two broad categories: facultative predators and obligate predators. Facultative predators, as the name implies, are organisms that can alternate between predatory and nonpredatory modes of feeding; that is, when adequate resources are present, they do not necessarily rely on predation to survive. In these organisms, predation and cell replication are not coupled. To date, all known facultative predators are saprophytes, meaning they can also feed on dead organic matter. Known facultative predatory bacteria belong to varied phylogenetic groups, suggesting that predation among bacteria may have evolved a number of times (see table). Among the facultative predators, the myxobacteria (gliding bacteria embedded in a layer of slime) are the best known, mostly because of their social behavior. Under adequate conditions, myxobacteria

Predatory and parasitic bacteria of other bacteria that have been isolated and at least partially characterized*	
Predation type	Predatory strategy
Facultative predators	
Agromyces ramosus[†] (Actinobacteria)	N.D.
Aristabacter necator (β proteobacteria)	N.D.
Cytophaga[†] spp. (Cytophagaceae)	Epibiotic
Cupriavidus necator (β proteobacteria)	N.D.
Ensifer adhaerens (α proteobacteria)	Epibiotic/Wolf pack?
Herpetosiphon[†] (Chloroflexi)	Wolf pack
Lysobacter (γ proteobacteria)	Wolf pack
Myxococcales (δ proteobacteria)	Wolf pack
Stenotrophomonas maltophilia[†] (γ proteobacteria)	N.D.
Streptoverticillium[†] spp. (Actinobacteria)	N.D.
Obligate predators and parasites	
Order *Bdellovibrionales* (δ proteobacteria)	Periplasmic
Family Bdellovibrionaceae:	
Bdellovibrio spp.	
Family Bacteriovoracaceae:	
Bacteriovorax spp.	
Peridibacter spp.	
Micavibrio spp. (α proteobacteria)	Epibiotic
Midichloria mitochondrii[‡] (α proteobacteria)	Periplasmic
Nanoarchaeum equitans (Nanoarchaeota)	Epibiotic parasite

*In parentheses: phylogenetic affiliation. N.D.: not defined.
[†]Not all strains in the phylum are predatory.
[‡]A parasite of mitochondria in *Ixodes* ticks.

congregate to form tightly regulated swarms that lead to the formation of fruiting bodies and to sporulation. Furthermore, predation efficiency and swarming appear to be linked. Facultative predators can attack a wide array of gram-positive (monoderms, that is, surrounded by a cytoplasmic membrane and peptidoglycan) and gram-negative bacterial cells (diderms, that is, surrounded by a cytoplasmic membrane and peptidoglycan and an outer membrane), yeast, and other fungi.

In obligate predatory bacteria, predation is essential for survival. Obligate predators must acquire their nutrients from their prey. They cannot replicate in the absence of prey and therefore, in these organisms, predation and replication are coupled. This implies that prey cells must be larger than the predator in order to sustain predatory cell multiplication. All known obligate predatory bacteria belong to the *Bdellovibrio* and like organisms (BALOs) [see table]. The BALOs form two main families, the Bdellovibrionaceae and the Bacteriovoracaceae. These groups exhibit large internal diversity, comprising organisms separated by large phylogenetic distances but having common morphologies and predatory behaviors. *Bdellovibrio bacteriovorus* belongs to the former group and is the most researched of the BALOs. Marine BALOs are exclusively found in the Bacteriovoracaceae, while freshwater and terrestrial BALOs are found in both groups. A third, phylogenetically unrelated BALO, the *Micavibrio*, was recently described. Strikingly, the BALOs solely attack and utilize gram-negative prey cells. Their phylogenies are not related to prey range, which may vary between strains within the same BALO species.

Predatory strategies. At least under laboratory conditions, bacterial predators reduce prey populations to very low levels in relatively short times by consuming them. Facultative and obligate predators exhibit distinct predatory strategies. Facultative predators may engage in "wolf pack," or group, predation. This strategy requires a high density of predators to ensure that the concentrations of secreted hydrolytic enzymes remain high enough to degrade the prey cell wall, followed by the release of prey cell contents (see **Fig. 1**). While direct contact with the prey may not seem necessary in this strategy, cell-to-cell contact between predatory swarms and prey cells may trigger predation. It also appears that single cells of at least some of these facultative predators attach to and excrete enough lytic enzymes to lyse prey cells. This latter behavior is called epibiotic, with membrane-bound hydrolytic enzymes possibly being involved in cell wall degradation. Epibiosis is also found in obligate predators, like in *Micavibrio* and in some *Bdellovibrio* strains (see table). Diacytosis or cytoplasmic invasion of the prey cell has only been described by electron microscopy in *Daptobacter*, an ill-defined bacterium. By far, the most common and better understood predatory strategy of obligate predation is periplasmic predation, exhibited by most BALOs. However, the various strategies described here may form more of a continuum of phenotypes rather than being strictly and formally separated.

Classical BALO cell cycle. In periplasmic predation, a small, free-swimming predatory attack cell bearing a long flagellum attaches to the prey's outer membrane (the outermost layer of the cell wall of gram-negative bacteria) and penetrates the periplasm, the space between the cytoplasmic and the outer membranes. Prey search involves chemotaxis, and pili (filamentous appendages other than flagella) and hydrolytic enzymes are needed for attachment to and then penetration into the prey cell. The prey cell wall is sequentially modified and degraded. At first, the invading BALO cell alters the prey cell wall: specific outer membrane proteins are subjected to proteolytic attack and the structure of the peptidoglycan is modified. These chemical alterations confer osmotic stability to the infected prey cell and render it immune to attack by further BALO cells as these cannot penetrate the modified cell wall. Infected prey cells are called bdelloplasts. The prey's respiratory functions are rapidly altered, with the predator killing its prey and not relying upon the latter' s metabolism. The bdelloplast provides a protective environment within which the predator initiates growth, with the prey's cytoplasm serving as the nutrient basis for cellular elongation and for the replication of its genetic material. Prey contents, such as nucleic acids and proteins, are sequentially degraded, solubilized, and utilized, rendering BALOs extremely efficient in producing biomass per unit of energy consumed. Exhaustion of prey contents leads to differentiation through the fragmentation of the filamentous cell into attack phase cells and flagellum synthesis. The remains of the prey are then lysed through the action of hydrolytic enzymes and the progeny is released (**Fig. 2**). According to the annotated genome, *B. bacteriovorus* possesses one of the largest complements of hydrolytic enzymes,

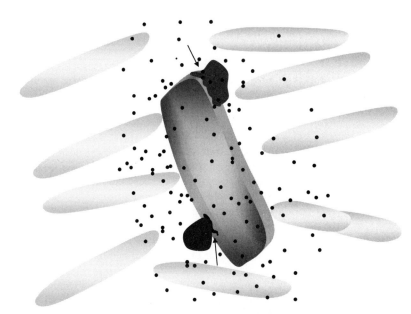

Fig. 1. Wolf pack attack. Predator cells (light gray) are in sufficiently large numbers to sustain high concentrations of lytic enzymes (black dots) and lyse the prey's cell wall. Arrows point to cracks in the prey cell wall, and the spilling of the cell's internal content.

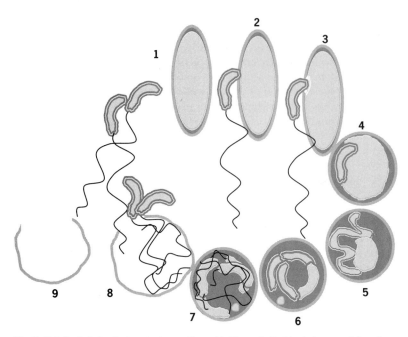

Fig. 2. *Bdellovibrio bacteriovorus* has a dimorphic life cycle that includes a periplasmic stage. 1. Attack phase cells. 2. Attachment to the prey's outer membrane. 3. Penetration of the prey's periplasm. 4. A bdelloplast is formed with an internalized predatory cell. 5. Growth and DNA replication of the predatory cell. 6. Septum formation. 7. Differentiation into progeny and flagellum synthesis. 8. Progeny. 9. Degradation of the prey envelope. Small cells: *B. bacteriovorus*. The long oval prey cell becomes rounded upon infection. Dark gray: periplasm; Light gray: cytoplasm.

with many remaining to be characterized. It is also the case that the larger the prey cell, the larger the number of released progeny cells. A full cycle may last between 2.5 and 5 h, depending upon the predator, the prey, and the environmental conditions.

It is noteworthy that host-independent (HI) BALO mutants can be rather easily isolated in the laboratory. These mutants grow axenically in rich medium, that is, in the absence of prey, and conserve the typical BALO dimorphic cycle as well as predatory capabilities. This suggests that the transition between facultative and obligate predation may involve a rather limited adaptive sequence.

Bacteria are effective models to study basic biological questions, especially due to their short generation times, large populations, the ease with which they and their growth conditions can be manipulated, and the tremendous amount of knowledge available. More specifically, bacterial predators and in particular BALOs are great models for studying predator-prey interactions, their dynamics and evolution, bacterial parasitism, and short developmental pathways. Their abilities to strongly reduce prey bacterial populations to very low levels, as well as their complement of enzymatic activities, hold the promise of attractive applications: it has been proposed that predatory bacteria could be implemented as biocontrol instruments in agriculture, as tools for the reduction of detrimental biofilms, as sources of enzymes, or as anti-infectious agents.

For background information *see* BACTERIA; BACTERIAL PHYSIOLOGY AND METABOLISM; BACTERIOLOGY; ECOSYSTEM; FOOD WEB; MICROBIAL ECOLOGY; MICROBIOLOGY; PARASITOLOGY; PREDATOR-PREY INTERACTIONS; PROKARYOTAE in the McGraw-Hill Encyclopedia of Science & Technology.   Edouard Jurkevitch

Bibliography. Y. Davidov and E. Jurkevitch, Phylogenetic diversity and evolution of predatory prokaryotes, in E. Jurkevitch (ed.), *Predatory Prokaryotes: Biology, Ecology, and Evolution*, Springer-Verlag, Heidelberg, 2007; D. Kadouri and G. A. O'Toole, Susceptibility of biofilms to *Bdellovibrio bacteriovorus* attack, *Appl. Environ. Microbiol.*, 71:4044–4051, 2005; C. Lambert et al., *Bdellovibrio*: Growth and development during the predatory cycle, *Curr. Opin. Microbiol.*, 9:1–6, 2006.

Progressive collapse of building structures

The collapse of the World Trade Center (WTC) in New York provided television viewers worldwide with a firsthand view of the structural phenomenon known as progressive collapse. Although this phenomenon is not new (for example, the Ronan Point apartment building in London collapsed some 40 years previously), the dramatic impact raised questions within the structural engineering community about exactly how the collapse happened, what was the governing mechanism, and how greater robustness might be provided to potentially vulnerable structures. The result has been heightened interest at the fundamental level, considerable research, and the prospect of more realistic representations of the key features of the phenomenon in design procedures.

Basic mechanics of progressive collapse. Progressive collapse is usually taken to mean the failure of a structure resulting from an initial localized event, such as an impact or explosion. Rather than just causing local damage in the immediate vicinity of the original event, the effects spread through a significant fraction of the structure, possibly even, as was the case with the WTC towers, resulting in a complete and spectacular collapse. **Figure 1** shows key stages, played out over a period of 2 h, for one of the WTC towers. **Figure 2** shows the partial collapse of Ronan Point, which spread from a small gas explosion on the 18th floor and which brought down one corner of the building. In contrast to these catastrophic incidents, a less-known aircraft impact on the Empire State Building in New York in 1945 resulted in very little structural damage, which was subsequently repaired. **Figure 3** shows severely damaged but still standing buildings in London after a WWII air raid. Why should the consequences of the initial event be so severe in some cases, yet relatively benign in others? How might lessons learned from the better performance of the latter be used in designing more robustness in the future?

Various explanations for the WTC collapses have been proposed, but the structural engineering community recognizes that the basic mechanics involved an initial incident that reduced the structure's ability to support its own weight, leading to a vertical collapse. Similar principles are used by demolition companies to "bring down" a building within its own footprint by a series of controlled explosions at key locations. Progressive collapse occurs when the vertical load-resisting system—for example, the

Fig. 1. Sequence in the progressive collapse of the World Trade Center.

framework of beams and columns, plus any assistance provided by cores or lift shafts—is damaged to the point that it cannot support the weight of the structure. Of course, the situation is likely to be complicated by the dynamic nature of the initial incident, other weakening effects such as fire, and the exact form of the overall structure and thus the paths available for the transfer of loads.

Design considerations. A direct consequence of Ronan Point was the introduction into the U.K. Building Regulations of the requirement to consider progressive collapse and to design certain beneficial features for structures above a certain height (typically taken as four stories). Essentially for steel-framed buildings, these regulations require the provision of ductile ties that would permit the structure to redistribute load away from the damaged area. Ductility, the ability to deform (and thus to absorb energy) without shedding load, plus redundancy, the provision of alternative load paths, should limit the effects of any local incident. However, these measures are essentially of the form "the structure should perform better if this is done than if it isn't." They do not allow the designer to make quantitative comparisons of different arrangements, as is usual when considering conventional gravity and wind loading. The provisions are essentially rules that should be followed, with little opportunity for the designer to either understand why or use his or her skills in devising the most appropriate solution for the particular structure under consideration.

Treating the problem rigorously remains, at least for the foreseeable future, impractical at the design level. Even in the assessment of previous failures, such as that following the WTC collapse, for which

Fig. 2. Progressive collapse at Ronan Point.

very substantial resources in terms of money and expertise were made available, it has not proved possible to allow for all important features. One convenient, yet practical, device is to assess in a threat-independent fashion, thereby removing consideration of the question: what type(s) of incident should be considered as the possible trigger for progressive collapse? This has led to the so-called column removal test, in which the ability of the structure to survive the removal of a single column, such as along one face of the building or at a corner, is assessed. **Figure 4** illustrates the concept. This approach may be implemented at a variety of levels of

Fig. 3. WWII air raid damage.

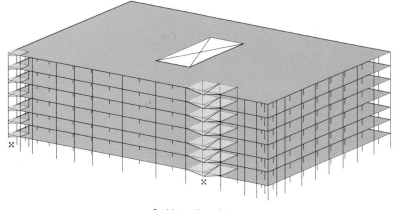

× Sudden column loss

Fig. 4. Column removal test for sustainability to progressive collapse.

sophistication. Ideally, any approach should recognize that events take place over a very short time scale and that the actual failure is therefore dynamic in nature. It should involve gross deformations, generating large strains and leading to inelastic material behavior as well as change of geometry effects. Failure essentially corresponds to an inability of the structure in its damaged state to adopt a position of equilibrium without separation at key locations. Moreover, the approach should also contain some form of analysis that permits quantitative comparisons between different arrangements to be made and be based on criteria of failure that reflect the actual physical nature of the problem.

Ideally, the approach should also be capable of implementation at a level of complexity compatible with the importance of the structure; that is, in a simple form for the preliminary work and/or for problems for which great precision is not required, and at a more sophisticated level when greater rigor is justified or more competitive solutions are sought.

Within the past few years, research teams in several centers around the world have been actively attempting to develop such approaches. Compared with the position just over a decade ago, when the topic was barely being considered, this represents a substantial change. However, as pointed out by W. Gene Corley, who led an investigation of the WTC collapse, a full treatment of all important effects has yet to be produced. In the meantime, greater appreciation of the potential hazards by both the professionals and the public should lead to more widespread adoption of construction practices that provide better levels of implicit robustness, that is, that incorporate features that increase resistance to progressive collapse. Where necessary, more scientific approaches are also available that permit quantitative comparisons between different trial arrangements to be made.

Outlook. Recent events worldwide have heightened public interest in the possibility of the progressive collapse of tall buildings, largely as a result of the threat of terrorist action. The response of the structural engineering research community has been to look carefully at the fundamental me-

chanics involved, with the aim of developing design treatments that properly reflect the key physical processes involved. While sufficient progress has been made that quantitative approaches based on sound structural principles are now emerging, the overall complexity of the problem is such that comprehensive solutions have yet to be devised.

For background information *see* ARCHITECTURAL ENGINEERING; BUILDINGS; LOADS, DYNAMIC; STRUCTURAL DESIGN; STRUCTURAL MECHANICS; STRUCTURAL STEEL; STRUCTURE (ENGINEERING) in the McGraw-Hill Encyclopedia of Science & Technology.

D. A. Nethercot

Bibliography. H. Griffiths, A. Pugsley, and O. Saunders, Collapse of Flats at Ronan Point, Canning Town, Her Majesty's Stationery Office, London, 1968; Robustness in Structural Engineering, *Struct. Eng. Int.*, 16(2), 2006; *Safety in Tall Buildings and Other Buildings with Large Occupancy*, Institution of Structural Engineers, London, 2002; *Structures and Extreme Events*, IABSE Symposium Lisbon 2005, 2005.

Regulation of fruit ripening

By botanical definition, fruits are seed-bearing organs of plants that function to protect and disseminate their contents. Most fruits can be classified as either fleshy or dry. The former are generally dispersed by organisms that consume the fruit and deposit the digestion-resistant seeds at another location. Seeds of dry or dehiscent fruit are typically dispersed by wind, gravity, or physical attachment to an animal.

Fruit ripening is a process unique to plants in which floral seed-bearing organs mature into fleshy structures that become attractive and nutritious to seed-dispersing organisms. Although the specific characteristics of ripening fruit vary among species, a number of general themes are exhibited in many fleshy ripening fruits. These attributes include accumulation of visually attractive pigments, production of aromatic volatiles, cell wall and cell turgor modifications leading to softening and juiciness, accumulation of sugars, and increased susceptibility to opportunistic pathogens responsible for the propensity of ripe fruit to rot.

Ripe fruit represent a significant and important component of dietary and nutritional uptake in humans and animals. Biochemical pathways leading to accumulation of nutrient compounds are regulated in many ripening fruits, including those affecting pro-vitamin A carotenoids (yellow, orange, red, or purple pigments), flavonoids (a series of widely distributed plant constituents related to the aromatic heterocyclic skeleton of flavan), and vitamins C and E.

Overripening and senescence lead to tissue decay and significant loss of production potential. Such losses are exacerbated in developing countries where access to modern storage systems is limited. Our understanding of the molecular and genetic basis of fruit ripening has expanded significantly in recent years and represents an opportunity to shed light on this novel area of plant biology while facilitating genetic strategies for increasing fruit availability and food security.

Tomato model. Molecular ripening studies have been performed on the fruits of numerous species, but the cultivated tomato (*Solanum lycopersicum*) has emerged as the primary model system for analysis of fleshy fruit ripening. Attributes driving adoption of tomato as the model ripening system include the facts that tomato has a short life cycle (approximately 3 months from seed to seed), possesses simple diploid genetics (having two complete chromosome pairs in a nucleus), is self-fertile, and is easily outcrossed (crossed with individuals of different strains, but usually the same species). In addition, the fruit undergo a rapid and dramatic ripening process, and exceptional germplasm (genetic material) resources have been characterized and are readily available. In this latter regard, a number of single-locus ripening mutations have been described, including the ripening-inhibitor (*rin*) mutation, which completely blocks ripening in the homozygous state and is also widely used as a heterozygous trait in commercial cultivars for extended storage life.

Single-gene mutations in the carotenoid biosynthetic pathway result in a range of color mutations from yellow through orange to red and are being exploited to both identify the underlying genes and to understand the regulation of carotenoid accumulation. Development of wild-species introgression (continued hybridization) lines facilitates quantitative trait loci (QTL) analysis and mapping, including efforts toward positional cloning of loci underlying traits of interest. Tomato is readily amenable to *Agrobacterium*-mediated stable or transient transformation and regeneration, making it a useful system for testing and analysis of function for genes of interest. Finally, the recent proliferation of tomato genomics data, in addition to an ongoing genome sequencing effort, has further positioned tomato as the optimal fruit species in terms of available molecular resources and opportunities.

Ethylene and ripening physiology. Plant hormones are important in fruit ripening, and many have been shown to play endogenous roles in ripening and the capacity to affect ripening when applied exogenously. Physiologically, ripening fruit are characterized as either climacteric or nonclimacteric, with the former undergoing elevated respiration, usually associated with ethylene synthesis, at the onset of ripening and the latter generally displaying neither of these characteristics. Tomato is a typical climacteric fruit in that respiration and ethylene increase dramatically with the onset of ripening. As is characteristic of climacteric fruit, inhibition of ethylene synthesis or perception in maturing tomato fruit inhibits ripening.

The critical role of ethylene in climacteric ripening was demonstrated in transgenic tomato plants via targeted independent repression of either of the final biosynthetic steps in ethylene synthesis—

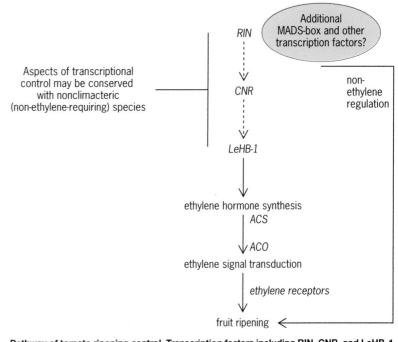

Pathway of tomato ripening control. Transcription factors including RIN, CNR, and LeHB-1 have been shown to be necessary for normal fruit maturation. Expression data suggest the possible hierarchy shown, but the dashed arrows indicate that this relative order remains to be proven. All three transcription factors are necessary for normal hormone synthesis, and at least *RIN* and *CNR* influence aspects of ripening control outside the realm of ethylene-mediated responses. As regulators of non-ethylene-mediated ripening activities, these transcription factors may represent conserved functions in nonclimacteric species.

specifically, the conversion of S-adenosyl-methionine (SAM) to 1-aminocyclopropane carboxylic acid (ACC) by ACC synthase (ACS) or the catalysis of ACC to ethylene by ACC oxidase (ACO). Similar repression of ethylene synthesis genes has been reported for a variety of climacteric species, including melon and apple, and with similar ripening inhibition effects (see **illustration**).

More recent efforts toward understanding the role of ethylene during ripening have focused on characterization of ethylene receptors and downstream signaling components originally defined in *Arabidopsis*. Although *Arabidopsis* does not produce fleshy ripening fruit, ethylene response can be readily assayed via the seedling "triple response" to ethylene (reduced/inhibited hypocotyl and root elongation; hypocotyl swelling; exaggerated curvature of the apical hook) and has thus facilitated the characterization of mutants and genes encoding ethylene signal transduction components. Comparison of tomato to *Arabidopsis* ethylene components indicates conservation of general mechanisms but different gene family sizes and expression patterns, including some predominantly fruit and ripening expressed genes in tomato. For example, members of the five-gene ethylene receptor family are fairly ubiquitous in expression among various *Arabidopsis* tissues, whereas two of the six tomato genes are strongly induced during ripening.

Nonclimacteric fruits, which by definition do not display elevated respiration and ethylene evolution, appear to fall into two general classes: those such as strawberry, in which ethylene appears to be unnecessary for normal ripening; and those such as members of the melon and pepper families, in which both nonclimacteric and climacteric members can be found. A useful example of the latter class is melon: Netted-skin cantaloupe types are typically climacteric, and smooth-skin types such as honeydew melons are nonclimacteric. In the case of melons, the nonclimacteric types typically ripen more slowly and produce fewer volatiles. These attributes, combined with their genetic similarity to climacteric types, suggest possibly simple genetic differences in ethylene synthesis or signaling genes. On the other hand, nonclimacterics such as strawberry may represent a different mode of ripening control compared to climacterics and, to date, no common molecular regulators of ripening have been characterized among them.

Recent insights into transcriptional control. The tomato *rin* ripening mutant mentioned earlier displays intriguing physiology. Specifically, although it fails to produce ripening-related ethylene and does not respond to exogenous ethylene by ripening, ethylene-regulated genes remain ethylene-responsive. This fact suggests that the *RIN* gene is responsible for regulating both ethylene synthesis and necessary but non-ethylene-mediated aspects of ripening and thus may represent a ripening regulator upstream of ethylene (see illus.). Cloning of the gene underlying the *rin* mutation revealed a MADS-box transcription factor whose expression is induced at the onset of ripening. In plants, MADS-box genes are typically associated with regulation of floral development and are thus consistent with a role in ripening. MADS-box genes remain to be functionally verified in nonclimacteric fruit such as strawberry, but it is noteworthy that a strawberry *RIN* homolog has been shown to be induced during strawberry ripening, indicating possible conservation of a common regulatory factor in climacteric and nonclimacteric species. MADS-box proteins are known to function as dimers or higher-order multimers, suggesting the possibility that additional MADS-box genes may function in concert with RIN to regulate ripening. Indeed, characterization of MADS-box gene phylogeny and expression in tomato indicates that additional members of the tomato MADS-box family are expressed in ripening fruit.

A number of single-locus ripening defective mutations with ethylene response characteristics similar to *rin* have been characterized in tomato and represent additional targets for elucidating mechanisms of ripening regulation. The tomato *Colorless non-ripening* (*Cnr*) locus was recently cloned and shown to harbor an epigenetic mutation in the promoter of a *SQUAMOSA PROMOTER BINDING* (*SPB*) protein gene that is also normally expressed during ripening. Gene expression analysis indicates that the *CNR* gene lies downstream of *RIN* while it is physiologically upstream of ethylene synthesis. It is worth noting that the phenotypes of *rin* and *Cnr* are common in terms of ripening inhibition but also display more subtle differences (including the colorless epidermis of *Cnr* fruit), indicating that they may

share overlapping but not necessarily identical regulatory targets. Like *RIN*, *CNR* sequences are found in expressed sequence tag (EST) collections derived from ripening fruit of additional plants, suggesting a function conserved in the fruit of diverse species.

The generation of tomato EST collections, an emerging genome sequence, and accumulating transcriptome (the set of RNA transcripts produced by the genome) data suggest a number of additional regulatory genes, including transcription factors that are expressed preferentially in ripening fruit and may contribute to the coordinated control of this process. One such gene that has been functionally verified in maturing tomato fruit is a homeodomain protein termed LeHB-1. Homeodomain proteins recognize short consensus repeat sequences that display dyad symmetry, and such sequences have in fact been observed in the promoter of the tomato fruit ethylene synthesis gene ACO1. Furthermore, recombinant protein derived from a cloned *LeHB-1* complementary DNA (cDNA) was shown to bind the ACO1 promoter in vitro, and ACO1 expression was repressed in transgenic tomato lines repressed in *LeHB-1* expression through RNA interference (RNAi; gene silencing). This result points specifically to a tomato transcription factor that mediates direct regulation of ethylene synthesis pathway genes, thus providing an additional component in the circuitry of ripening regulation. Direct regulation of *LeHB-1* by *RIN* or *CNR* remains to be verified, although its ability to directly bind the promoter of a gene involved in ethylene synthesis suggests that it may be downstream of these regulators (see illus.). Availability of additional tomato ripening mutants, ripening QTLs, and transcriptome data for maturing fruit provides numerous additional targets for further elucidation of steps in the ripening cascade prior to ethylene. As a clearer model of ripening control is populated with additional gene functions, opportunities for enhanced genetic control of ripening will expand and, with them, opportunities for a more nutritious and secure food supply.

For background information *see* CAROTENOID; ETHYLENE; FRUIT; FRUIT, TREE; GENE; GENETIC ENGINEERING; MUTATION; PLANT ANATOMY; PLANT HORMONES; SEED; SIGNAL TRANSDUCTION; TOMATO; TRANSCRIPTION in the McGraw-Hill Encyclopedia of Science & Technology.   Jim Giovannoni

Bibliography. R. Alba et al., Transcriptome and selected fruit metabolite analysis reveal multiple points of ethylene regulatory control during tomato fruit development, *Plant Cell.*, 17:2954–2965, 2005; C. Barry and J. Giovannoni, Ethylene and fruit ripening, *J. Plant Growth Reg.*, 26:143–159, 2007; J. Giovannoni, Fruit ripening mutants yield insights into ripening control, *Curr. Opin. Plant Biol.*, 10:283–289, 2007; Z. Lin et al., A tomato HD-Zip homeobox protein, LeHB-1, plays an important role in floral organogenesis and ripening, *Plant J.*, in press, 2008; K. Manning et al., A naturally occurring epigenetic mutation in a gene encoding an SPB-box transcription factor inhibits tomato fruit ripening, *Nature Genet.*, 38:949–952, 2006; J. Vrebalov et al., A MADS-box gene necessary for fruit ripening at the tomato *ripening-inhibitor* (*rin*) locus, *Science*, 296:343–346, 2002.

RFID in transportation and logistics

Radio-frequency identification (RFID) systems operate via radio waves in many different frequency ranges, with some of the most common being low-frequency (LF), high-frequency (HF), ultrahigh-frequency (UHF), and microwave. The frequency helps determine such things as range and method of communication (for example, between transmitter and receiver). In its simplest form, as shown in **Fig. 1**, an RFID system consists of a tag (attached to the product to be identified), an interrogator (reader), and one or more antennas attached to the reader. The reader must be connected to a computer either directly or through a network to provide control for the reader and to capture the data.

RFID systems can be classified by the frequency used and whether the system is active or passive. Passive tags not only transmit data via radio waves, but also are powered by the radio waves (an electromagnetic field) transmitted by the reader's antennas. A passive tag will remain powered only while it is within the read field. Once it leaves the read field, it stops transmitting and is essentially dormant. While it is in the read field, the powered tag will respond to the reader by reporting the data contained on the tag's chip. The reader then converts the returned signals into data that the computer can understand and store. Read ranges for passive tags are from a few inches (cm) for LF systems to 30 ft (9 m) or more for UHF and microwave systems. Unlike passive tags, active tags have an internal battery that powers the tag. Because active tags have their own power source, they do not need a reader to energize them (via radio waves). Instead, they can initiate the data transmission process. In addition, active tags have a longer read range, can store more data, and can perform some processing functions. But because they are battery-powered, they have limited lifetimes and are larger in size and more expensive than passive tags.

There are also significant differences between passive and active tags when it comes to communicating with the reader, because of the physics behind the two tag types. Passive tags require high signal strength from the reader, and they emit a low-power signal back. Conversely, active tags require only a low-power signal from the reader, and they can send back data with high power. The communication speed between the reader and a passive tag averages 20 ms, whereas an active tag communicates much faster, since the tag uses its own power to send data. However, active tags may have several integrated sensors onboard, and the large amount of data associated with these sensors can slow down the communication process significantly. The amount of information involved in the communication process between a passive tag and a reader is typically 128 bits. An active tag used for ID purposes also has typically 128 bits of data involved in communication, but it can carry much more data, if needed.

Contrary to popular belief, RFID is not a new technology. It has existed in various forms since the 1940s. In fact, an early version of RFID was used by the British during World War II to help identify friendly aircraft. Since then, RFID has been used in a number of applications, most of which have not garnered much public attention. For instance, many people use RFID-enabled automobile key fobs, employee identification badges, and toll-road passes without ever considering the technology at work. RFID's low-key status began to change in 2003 when Wal-Mart announced plans to have its suppliers start using the technology to streamline its supply chain (using passive UHF systems). Other companies soon followed, issuing similar mandates to their suppliers. The U.S. government even got into the act when the Department of Defense announced that its suppliers would start using RFID. Most of these organizations have used RFID in "backroom" applications, in which products are tracked at the pallet and/or case level as they move along the supply chain. The goal is better visibility of product movement and inventory levels, with an associated reduction in operational costs.

Recent RFID applications include retail and military supply chains and the movement of goods within factories, warehouses, and stores. Interestingly, the transportation/logistics industry has been largely overlooked by the RFID adoption efforts. However, this is changing because (1) transportation/logistics is needed to provide a complete supply-chain picture (that is, movement between facilities) and (2) RFID is providing benefits for transportation/logistics companies.

Transportation/logistics applications. There are many real and potential uses of RFID by transportation/logistics companies. Although RFID is used to improve the movement of people via public and private transportation systems, such as buses, toll roads, and commuter trains, the focus here will be on the movement of "things," involving maintenance, shipping/receiving accuracy and efficiency, and food quality.

Maintenance. Proper maintenance is key to the safe and cost-efficient use of equipment, such as trucks, trailers, and forklifts. Tire maintenance for trucks and trailers is one example of using RFID within the

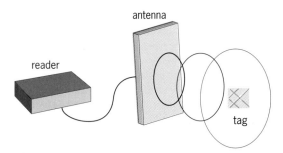

Fig. 1. Basic RFID components.

Fig. 2. RFID-enabled dock door.

Fig. 3. RFID-equipped forklift.

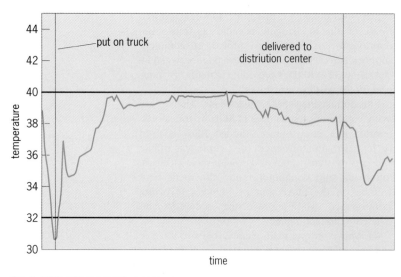

Fig. 4. RFID with temperature sensors.

transportation industry. Tires represent a large expense for transportation companies and are a major safety issue if they are not maintained properly. Without RFID, tires must be inspected and the data recorded manually to track things such as the number of miles traveled and the number of times they have been retreaded. Manually inspecting and recording the information for 18 tires per truck/trailer can be a laborious and hazardous task. With RFID, tags can be embedded in the tires to allow automatic reading of tire information as a truck/trailer enters a facility such as a warehouse yard. As the truck/trailer passes by a security checkpoint, for example, an RFID reader could read the information from each of its tires. Then, via database lookup, the system could determine the last time each tire was changed and the number of times each had been retreaded. Based on this information, the driver might be instructed to proceed to the warehouse for unloading or be redirected to the maintenance facility for tire changes. This scenario has already been deployed by several transportation companies, and it is improving the effectiveness and efficiency of tire maintenance.

Shipping and receiving accuracy and efficiency. RFID can reduce the errors in receiving via the process know as electronic proof of delivery. With barcode scanners, mistakes are made when the quantity and/or type of product are misidentified. For example, when a pallet of 48 cases of shampoo, consisting of 24 regular and 24 scented, is received at a distribution center, the receiving clerk mistakenly identifies the pallet as 48 cases of scented shampoo. This error causes an inventory inaccuracy for the receiving company and a perceived overage/underage situation for the transportation provider or supplier. With RFID, the product could be read automatically as it is unloaded through an RFID-enabled dock door (**Fig. 2**) or with a forklift equipped with an RFID reader (**Fig. 3**).

Distribution centers, manufacturers, and logistics operations often make mistakes by loading product on the wrong truck. With RFID, the system could send an alert (for example, visible or audible) to the person loading the truck that a mistake had been made. The alert could save money by preventing shipping errors and enhance inventory control.

In addition to improving shipping/receiving accuracy, early evidence suggests that RFID can reduce the time to receive products. Instead of scanning each case of product individually with a bar-code scanner, RFID-tagged products would be read automatically at a dock door. The process of receiving would not change drastically; that is, forklifts would unload the product as before. The only change would be in eliminating the need to scan the product manually. Thus, the process would become more efficient.

Food quality. In the previous examples, RFID was used to indicate "what it is" and "where it is." For food, RFID is also used to provide an indication of "how it is." As a product moves through the supply chain, the environment can change its condition, and thus its quality and safety level. Using RFID with special on-board sensors, such as temperature sensors, provides insight into the changing environmental conditions experienced by the product and the data necessary to determine to what extent those changes affected the product's quality or safety. Without sensors, one can get various single-point estimates of the environmental conditions (such as temperature at the time of loading and temperature at the time of delivery), but these lack visibility between point estimates. **Figure 4** is an example of a load of produce that was tracked with RFID from a manufacturing facility to a retailer's distribution center. Although the data could be examined many different ways, Fig. 4 shows the average temperature over time; everything seemed to be in order except for the first several hours in the manufacturing facility, when the product experienced temperatures below the acceptable limits [in this case, 32–40°F (0–4°C)]. After the product was loaded on the trailer, the product warmed to acceptable limits and stayed within the proper limits during transit. This information is valuable for transportation/logistics providers because if the product quality suffers, the transportation/logistics provider often is blamed and is held accountable financially. With the information provided by RFID, the source of the problem could be pinpointed.

Outlook. RFID holds much promise as a tool to improve the transportation and logistics industry. Although this article examined only a few applications, there are many others, such as asset utilization, yard management, and cross-docking optimization, that provide real and potential paybacks. As more retailers and suppliers adopt RFID, the transportation/logistics role of providing key visibility to product movement—what, where, how—will expand and become even more critical.

For background information *see* FOOD MANUFACTURING; MATERIALS HANDLING; SUPPLY CHAIN MANAGEMENT; TRANSPORTATION ENGINEERING in the McGraw-Hill Encyclopedia of Science & Technology.
Bill Hardgrave

Bibliography. D. Delen, B. C. Hardgrave, and R. Sharda, RFID for better supply-chain management through enhanced information visibility, *Prod. Oper. Manag.*, 16(5):613–624, 2007; K. Finkenzeller, *RFID Handbook: Fundamentals and Applications in Contactless Smart Cards and Identification*, John Wiley & Sons, New York, 2003; B. Glover and H. Bhatt, *RFID Essentials*, O'Reilly Media, Cambridge, 2006; B. Hardgrave, M. Waller, and R. Miller, *Does RFID Reduce Out of Stocks? A Preliminary Analysis*, White Paper, Information Technology Research Institute, Sam M. Walton College of Business, University of Arkansas, 2005.

Role of mechanics in biological materials research

Understanding biological materials and systems, which are intrinsically complex, requires a multidisciplinary approach. Engineering disciplines, such as mechanics, potentially can make significant contributions in this pursuit. Mechanics tools may be used to advance the fundamental knowledge of biological materials and systems in a number of ways, including understanding the fundamental mechanisms of biological systems, discovering new responses of biological materials to different stimuli, developing new tools for studying biological materials and systems, and creating hybrid engineered systems with unique features.

For many years, biomechanics researchers have studied biological materials at the tissue level. They treated biological materials as they treated engineered materials, that is, by characterizing the mechanical properties and establishing the constitutive relations (such as stress-strain and strain rate laws) of biological tissues. In recent years, researchers also have aimed at answering other fundamental questions, such as how mechanical stimuli influence the biochemical signaling pathways and how biological responses of cells (such as cell differentiation and migration) are affected by the mechanical environment.

Important issues. To study the complex behavior of biomaterials, especially the responses of the living cell to various stimuli and in different environment, researchers need to consider a number of important issues. (1) Multidisciplinary and multiphysics approaches must be used. A comprehensive study of biomaterials should involve coupled phenomena, such as biological-electrical coupling, biochemical-mechanical coupling, and thermal-biological coupling. Close collaborations between biologists, chemists, physicists, and engineers are critical. (2) Multiscale approaches must be taken, including multiple length and time scales. (3) The continuously evolving shape and structure of living materials should be taken into account in the study of biomaterial responses. To date, most physics and engineering methods are inadequate to handle evolving structures. New physics and engineering tools need to be developed. (4) The structural complexity of biomaterials needs to be taken into consideration. Not only do biomaterials exhibit structures with different characteristics, such as being one-dimensional (1D; for example, biomolecules), two-dimensional (2D; for example, membranes), three-dimensional (3D, for example, cells/tissues/organs), and four-dimensional (4D; that is, three spatial and one temporal dimension) in nature, but very often their structures are hierarchical in nature. New methods are needed to address structural complexity. (5) Another important issue is the efficiency versus robustness of biomaterials. Unlike engineered materials and systems, where the main concern is often the efficiency of the system, biomaterials are governed by their ability to survive, which requires robustness. And biological organs, such as the brain, are evolved from similar organs of their previous generations, which gives rise to inherent redundancy and inefficiency.

Understanding mechanisms of cellular behavior. A cell's sensing of, and responses to, forces and geometry is often transduced into biochemical

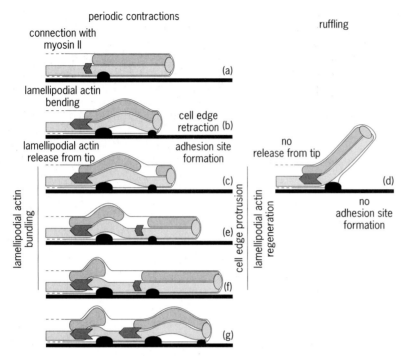

Fig. 1. Proposed mechanism of cell spreading and migration through periodic edge retraction and extension. The entire process includes (a) growth of lamellipodial actin (light gray) backward to an MII cluster (dark gray); (b) the MII pulls the lamellipodial actin to cause bending and cell edge retraction when an adhesion site (black "disks") is formed; (c) bending and adhesion at cell edge triggers release of lamellipodial actin from the tip; (d) the cell edge ruffles upon MII pulling when an adhesion site is not formed; (e–f) a new lamellipodial actin network grows to generate protrusion; and (g) another cycle begins by repeating step a. (*Adapted from G. Giannone et al., Cell, 128:561–575, 2007*)

signals that regulate vital cellular functions, such as cell growth, migration, differentiation, and apoptosis (programmed cell death). Although much remains to be learned about the general principles of these cell responses, some recent studies by biologists, mechanicians, and physicists have shed light on specific mechanisms.

Cell motility. Migration and spreading is one of the most basic functions of living cells. How and with what mechanisms the cells move about on a 2D surface or in a 3D environment is an intriguing subject to biologists and physical scientists. Cell motion involves cell shape change as a result of the evolution of the cytoskeleton, which, in addition to the static shape fluctuation of the cytoskeleton filament network, undergoes dynamic change by constantly lengthening or shortening the filaments through polymerization or depolymerization, respectively, by the addition or removal of proteins at their ends, a process regulated by biomolecules such as the tyrosine kinase Src, vinculin, and talin1. Through what mechanism(s) a cell extends its edges as the first step in migration or spreading was not known. It was first observed by G. Giannone and coworkers that such an edge extension process is not monotonic but rather has a periodic extension and retraction motion. Recently, they proposed a mechanistic model to explain this phenomenon. As shown in **Fig. 1**, the cell edge extension involves cyclic processes of lamellipodial actin contraction and lamellum actin resisting compression to produce lamellipodial actin, bending when the cell edge is pinned down by adhe-

sion sites. (Lamellipodia are thin, broad projections at the edge of a motile cell. The lamellum is the portion between the lamellipodium and the cell body.) Such bending, aided by the adhesion at the edge, triggers actin polymerization and release at the cell edge, which moves the edge forward. When the front of the extended cell edge (lamellipodium) reaches a certain characteristic length, lamellipodial contraction starts again, and the cell spreads a characteristic distance. This process repeats itself periodically and gives rise to cell motion or spreading. Two important conditions, both mechanical in nature, must be met in order to generate such cell motion. First, there must be pulling action (contraction) of the lamellipodial actin on the top and pushing (compression resistance) of the lamellum actin at the bottom to produce the local bending (Fig. 1b), which is necessary for further actin polymerization. Second, adhesion at the cell edge must be sufficiently strong to hold the cell edge down (Fig. 1c); without such adhesion, the cell edge will simply ruffle, and no cell motion would result (Fig. 1d).

Cell adhesion. Cell adhesion plays a central role in a cell's interaction with its surroundings. The biochemical understanding of cell adhesion is provided by the molecular interaction of the ligand-receptor pair. How many ligand-receptor pairs are needed to form a focal adhesion complex and how the shape of the cell is influenced by such focal adhesion site distributions are pertinent questions. L. B. Freund and coworkers were among the first mechanics researchers to try to understand cell adhesion. They modeled the cell membrane as a thin plate in contact with a flat substrate within a focal adhesion zone, and considered mobile binders (ligands in the cell membrane) that can move within the plate (**Fig. 2**). By considering the total free energy of the system, which includes the bending energy of the plate, the binding energy of the binders, and the entropic free energy of the binder distribution, they were able to determine the rate of

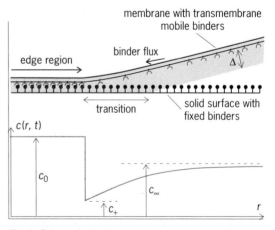

Fig. 2. Schematic drawing of the shape of the membrane near the adhesion zone in the upper diagram, and the distribution of binders with the density $C(r,t)$ depicted in the lower diagram, where c_0 is the density within adhesion zone, c_+ the edge density, and c_∞ the initial uniform density. (*Adapted from V. B. Shenoy and L. B. Freund, P. Nat. Acad. Sci., 102(9):3213–3218, 2005*)

spreading of a focal adhesion zone and the binder density distribution near the edge of the adhesion zone.

Other questions regarding cell adhesion include: What controls the size and shape of a focal adhesion complex? Is there a minimum or a maximum size of a focal adhesion cluster? Is it stable? V. B. Shenoy and L. B. Freund studied the shape stability of a cell membrane at an adhesion site by considering the diffusion of binders (ligands) in the membrane, and found that the circular shape of the adhesion site becomes unstable under perturbations for a zone radius significantly larger than the nucleation size. H. Gao and coworkers considered the stochastic process of molecular bonding within a focal adhesion cluster. They modeled the cell membrane and the substrate as two elastic half planes with different elastic properties (**Fig. 3a**), and used a coupled stochastic-elasticity model to evaluate the strength and lifetime of a cluster of molecular bonds as a function of cluster size. Although the results depend on material properties and other parameters, the general trend of their findings provides interesting insights. When the focal adhesion cluster size is too small, the stochastic nature of individual molecular bonds produces a low-strength adhesion zone with a short lifetime. When the size is too large, the focal adhesion complex leads to a cracklike singularity at the edge upon loading, therefore limiting the growth of the adhesion zone. These results show that for both bond strength (Fig. 3b) and lifetime (Fig. 3c), there is a window of focal adhesion zone size ($a_{min} < a < a_{max}$) within which the adhesion is both strong and stable, and beyond which the adhesion is not sustainable.

With the help of the focal adhesion complex, cells spread and migrate. But spreading has additional complications. Many cells—for example, fibroblasts—become elongated in shape as they spread. For a particular cell, the orientation of the elongation may change over time. However, the elongated cells may align themselves to a particular direction, and the orientation of the alignment is dependent on the straining of the substrate that the cell is attached to. Moreover, if the straining of the substrate is cyclic, the final alignment direction of the cells may depend on the frequency of the cyclic straining. Under static straining, fibroblasts were observed to gradually reorient themselves to align the long axis of elongation to the direction of stretching (**Fig. 4a**) Under cyclic straining of 1 Hz, fibroblasts and endothelial cells will gradually reorient to align the long axis of elongation perpendicular to the direction of stretching (Fig. 4b).

A. Safran and coworkers developed a model to study the possible mechanisms. They modeled the adherent, elongated cells as "elastic force dipoles that can change their contractile activity and orientation by reorganizing the focal adhesions and stress fibers in response to external forces" (**Fig. 5**). The force dipoles P can be visualized as the integration of focal adhesion forces near the edge of the cell. For elongated cells, there is experimental evidence

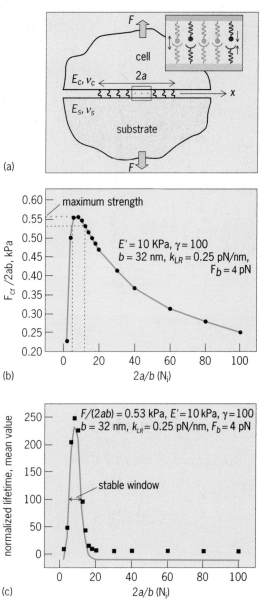

(a)

(b)

(c)

Fig. 3. Schematic diagram of (a) the idealized model of a focal adhesion complex between a cell and the ECM, showing cracklike behavior as the zone becomes large; (b) plot of the normalized strength as a function of normalized adhesion zone size; (c) plot of the normalized lifetime of adhesion as a function of adhesion zone size. Both b and c show a window of adhesion zone size in which stable adhesion is achieved. (*Adapted from J. Qian et al., Langmuir, 24(4):1262–1270, 2008*)

that the adhesion forces are oriented more along the stretching axis. The reorientation of the force dipoles is directly related to the ability of the cell's cytoskeleton to reform and reorganize, which occurs at different rates for different cells. Safran and coworkers further hypothesized that the cell will reorient itself by minimizing its total energy, characterized by the quantity $(P - P^*)^2$, where P^* is the optimal value of the cell force. Applying this principle, they were able to predict that under low-frequency loading, a cell prefers to align itself with the loading direction, while under high-frequency cyclic loading, the cell tends to align itself at an angle nearly perpendicular to the loading direction (Fig. 5b). This is a good

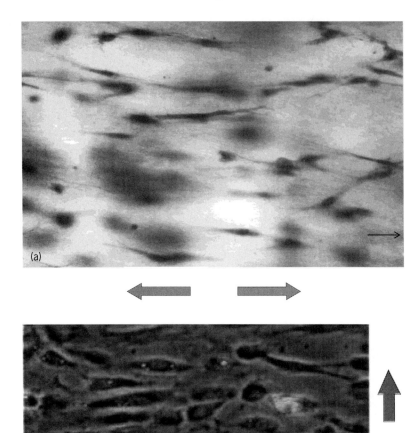

Fig. 4. Experimental results of the reorientation of cells under loading. Arrows indicate the direction of the stretching. (*a*) Fibroblasts under static loading realign along the stretching direction (*adapted from M. Eastwood et al., Cell Motil. Cytoskeleton, 40:13–21, 1998*). (*b*) Endothelial cells under cyclic loading of 1 Hz realign perpendicular to the stretching direction. (*Adapted from J. H.-C. Wang et al., J Biomech., 34:1563–1572, 2001*)

ing, as in the control case (a surface with a uniform coating of RGD peptide), a cell will become elongated in shape when it is placed on a surface with particles of 108-nm spacing (**Fig. 6**). This indicates that the formation of the focal adhesion complex by clustering of integrins in the cell membrane is somehow governed by a characteristic length scale. Below this characteristic length, the local spreading of the cell edge can always reach the next available adhesion site and establish focal adhesion complexes. Above this characteristic length, the cell edge needs to be overstretched to sense and reach the next available site, and it requires a coordinated geometry change (elongation) of the whole cell to establish such focal adhesion complexes. Their results also showed that the rate of spreading of the fibroblast on patterned surfaces is slower than that on nonpatterned surfaces, suggesting that additional biochemical "effort" is required for the cell to spread on patterned surfaces. Although it has been known that cell spreading and migration requires the establishment of focal adhesion complexes near the cell edge, that such establishment is governed by a characteristic length scale is a new discovery. This is the first time that the characteristic length has been measured quantitatively.

Nanotechnology has also provided opportunities to fabricate devices and tools that can be used to study biomaterials at small length scales and with high spatial and force resolutions. In particular, micro- or nano-electromechanical systems (MEMS or NEMS) are becoming more popular as enabling tools among biomaterial researchers. A. Chiba and T. Saif

example of rather simple physical principles (in this case, minimization of energy) being used to identify and understand the underlying mechanisms controlling the behavior of complex living systems.

Discovering new phenomena. In recent years, nanotechnology has provided numerous new tools for researchers. J. P. Spatz and coworkers fabricated on silicon wafers a nanoscale patterned surface with gold nanoparticles coated with RGD peptides (proteins that serve as receptors for integrins) to form selected adhesion sites. (Integrins are transmembrane receptor proteins of animal cells that functionally connect the inside with the outside by binding to the cytoskeleton and extracellular matrix.) The only parameter they considered was the spacing between the nanoparticles, which varied from 28–110 nanometers. While a rat fibroblast cell will spread uniformly (nearly circular in shape) when it is placed on a surface with nanoparticles of 58-nm spac-

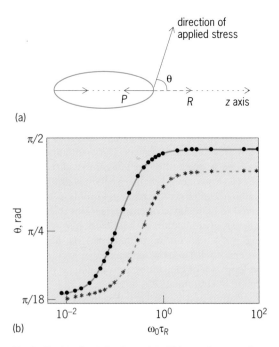

Fig. 5. The idealized dipole model of (*a*) an adherent cell under applied stress in a direction with a relative angle θ from the cell's elongation direction, where P is the force dipole and R is the reaction stress, and (*b*) the stabilized angle under cyclic loading as a function of the normalized loading frequency in the lower plot. (*Adapted from R. De et al., Nat. Phys., 3:655–659, 2007*)

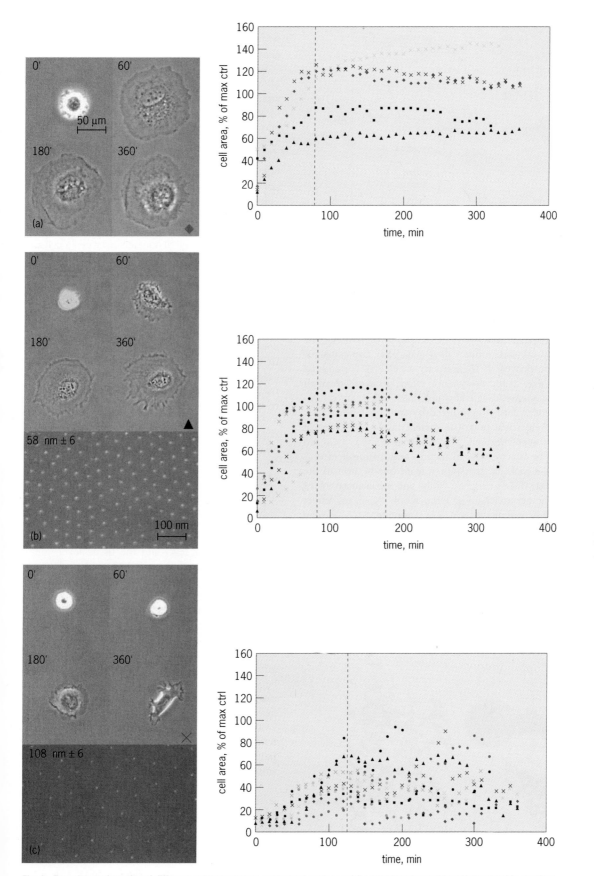

Fig. 6. Experimental results of different cell behavior on patterned surfaces: (*a*) control surface with uniform peptide coating, (*b*) surface with nanoparticles of 58-nm spacing, and (*c*) surface with nanoparticles of 108-nm spacing. The plots on the right show the measurements of cell shape over time. (*Adapted from E. A. Cavalcanti-Adam et al., Biophys. J., 92:2964–2974, 2007*)

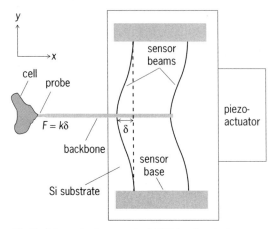

Fig. 7. Schematic diagram of a MEMS loading device applying tensile load to a cell. A similar device can be used to apply load to an axon in a neuron. (*Adapted from http://www.ccm.uiuc.edu/Workshops.html*)

used a MEMS device to study neurotransmission in neurons. Neuroscientists know that neuronal signals are transmitted from one neuron to the next at special contact points called synapses. The transmission process is triggered by an electrical potential change in the presynaptic axon, proceeds through the release of signaling molecules called neurotransmitters, and is completed by causing a postsynaptic electrical potential change in the dendrites of the receiving cell. The whole process is believed to be biochemical in nature, and during it electrical signals are transmitted from one neuron to the next. Out of curiosity, Chiba and Saif wanted to see whether mechanical force would have any effect on neurotransmission. They applied a tensile load to axons through a micro-loading MEMS device and measured the neurotransmission process (**Fig. 7**). They found that mechanical forces in axons must be present for initial assembly and subsequent plasticity of molecular components at synaptic terminals. They also measured the magnitude of the force naturally present in axons to be about 1 nN (10^{-9} newton).

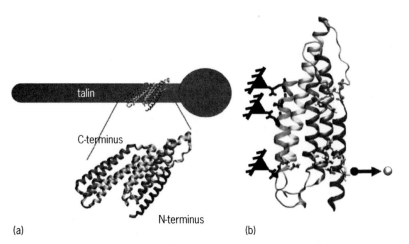

(a)   (b)

Fig. 8. The crystal structure of (*a*) a stable N-terminal five-helix bundle in ribbon representation superimposed on a hypothetical talin model. (*b*) Detailed view of the N-terminal five-helix bundle used in the simulations; the H1 polar side chains are pulled toward a dummy atom shown as the white sphere, and the H5 polar side chains are harmonica constrained as indicated by the black anchors. (*Adapted from S. E. Lee et al., J. Biomech., 40(9):2096–2106, 2007*)

This study shows how complex living systems are, and the coupled nature of the responses of living cells.

Tool development. Simulation tools that are capable of addressing the issues of coupled mechanical and biochemical processes have the true potential to affect research on biological materials. S. E. Lee and coworkers have provided an example of such a modeling and simulation tool. The goal of their work is to understand the mechanisms of how force is transduced into biochemical signals in a cell. In particular, they are interested in the process of vinculin (a membrane-cytoskeletal protein in focal adhesion plaques that is involved in linkage of integrin adhesion molecules to the actin cytoskeleton) recruitment during initial contacts between a cell and the extracellular matrix (ECM) due to tensile force. They used molecular dynamics (MD) simulations to study force-induced conformational changes in the vinculin binding site (VBS) in talin, an essential linking protein during initial contact. **Figure 8** shows their computational domain, with the constraint and applied load indicated in Fig. 8*b*. They demonstrated that the conformational change would expose the cryptic VBS to solvent upon applied force, causing the VBS to rotate and to release its vinculin-binding residues from the tight hydrophobic core.

A similar approach, using MD simulations to study conformational changes, was also used by W. Hwang and coworkers to study the force generation mechanisms of kinesin, the motor protein that is responsible for transporting "cargo" along microtubules in the cell. This type of modeling and simulation tool is becoming increasingly important in the study of cellular and molecular mechanisms of cell behavior.

To truly understand the behavior of biomaterial at the molecular level and the tissue level, multiscale simulation tools must be developed. These tools must be capable of dealing with multiple length scales and multiple time scales.

Engineering hybrid systems. It has been said that the cell is the ultimate chemical factory, since it manufactures all kinds of products with minimal energy consumption. How to use such capabilities to benefit society is one future research area. It has also been said that the cell is the ultimate microscopic machine. How to engineer a system that uses these machines is another emerging research area. M. Sitti and coworkers have envisioned a hybrid robotic system (**Fig. 9**) in which a swimming robot can be propelled by an array of bacteria attached to the back side of the robot. They demonstrated the feasibility of the concept by functionalizing polystyrene beads so that bacteria would attach to the end of the beads. They showed that these beads can be propelled by bacteria and can move around in a petri dish. Such hybrid systems could be used to accomplish a wide variety of different tasks, such as drug delivery, disease diagnoses, and data collection in hazardous environment.

Outlook. The study of biological materials at the cellular and molecular levels using physics, engineering principles, and the tools of mechanics

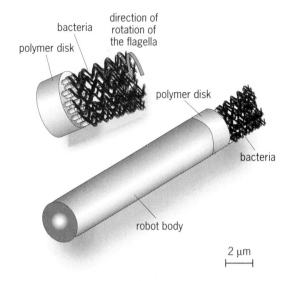

bacteria

polymer disk

direction of rotation of the flagella

polymer disk

bacteria

robot body

2 μm

Fig. 9. The concept of a bio-nonbio hybrid swimming robot, with bacteria as its propulsion system. (*Adapted from B. Behkam and M. Sitti, Proceedings of IEEE 2006 International Conference of Engineering in Medicine and Biology, pp. 2421–2424, 2006*)

is still at an early stage. To pursue these opportunities effectively and to enhance the impact of the research requires coordinating research efforts and focusing the available resources to tackle a few important and well-defined problems. Education is another way to have long-lasting impact. We urgently need to train the next generation of researchers so that they will be equally comfortable communicating with and working in both biology labs and engineering labs.

The societal impact of the successful research pursuit in this field is expected to lead to faster advances in scientific research and new engineering developments. It will have immediate applications in health care, the environment, and energy, addressing many urgent societal needs.

[Acknowledgement: The work is supported by NSF Grants No. DMR 05-04751 and CMMI 08-03692.]

For background information *see* BIOMECHANICS; BIOMEDICAL ENGINEERING; BIOPHYSICS; CELL (BIOLOGY); MECHANICAL ENGINEERING; MECHANICS; MOLECULAR BIOLOGY in the McGraw-Hill Encyclopedia of Science & Technology.   K. Jimmy Hsia

Bibliography. B. Alberts et al., *Molecular Biology of the Cell*, 4th ed., Garland Science, New York, 2002; M. Arnold et al., Activation of integrin function by nanopatterned adhesive interfaces, *Chem. Phys. Chem.*, 5:383–388, 2004; B. Behkam and M. Sitti, Toward Hybrid Swimming Microrobots: Propulsion by an Array of Bacteria, *Proceedings of IEEE 2006 International Conference of Engineering in Medicine and Biology*, 2006, pp. 2421–2424; D. Boal, *Mechanics of the Cell*, Cambridge University Press, 2002; E. A. Cavalcanti-Adam et al., Cell spreading and focal adhesion dynamics are regulated by spacing of integrin ligands, *Biophys. J.*, 92:2964–2974, 2007; Center for Cellular Mechanics Summer Course on Cell Mechano-Sensitivity; D. S. Coffey, R. H. Getzenberg, and T. L. DeWeese, Hyperthermic biology and cancer therapies: a hypothesis for the *Lance Armstrong Effect*, *JAMA*, 296:445–448, 2006; M. Dao, C. T. Lim, and S. Suresh, Mechanics of the human red blood cell deformed by optical tweezers, *J. Mech. Phys. Solids*, 51:2259–2280, 2003; R. De, A. Zemel, and S. A. Safran, Do cells sense stress or strain? Measurement of cellular orientation can provide a clue, *Biophys. J.: Biophys. Lett.*, doi:10.1529/biophysj.107.126060, 2008; R. De, A. Zemel, and S. A. Safran, Dynamics of cell orientation, *Nat. Phys.*, 3:655–659, 2007; M. Eastwood et al., Effect of precise mechanical loading on fibroblast populated collagen lattices: morphological changes, *Cell Motil. Cytoskel.*, 40:13–21, 1998; L. B. Freund and Y. Lin, The role of binding mobility in spontaneous adhesive contact and implications for cell adhesion, *J. Mech. Phys. Solids*, 52:2455–2472, 2004; H. Gao, Application of fracture mechanics concepts to hierarchical biomechanics of bone and bone-like materials, *Int. J. Fracture*, 138:101–137, 2006; G. Giannone et al., Lamellipodial actin mechanically links myosin activity with adhesion-site formation, *Cell*, 128:561–575, 2007; G. Giannone et al., Periodic lamellipodial contractions correlate with rearward actin waves, *Cell*, 116:431–443, 2004; G. Giannone et al., Talin1 is critical for force-dependent reinforcement of initial integrin-cytoskeleton bonds but not tyrosine kinase activation, *J. Cell Biol.*, 163:409–419, 2003; J. D. Humphrey, *Cardiovascular Solid Mechanics: Cells, Tissues, Organs*, Springer-Verlag, 2002; W. Hwang, M. J. Lang, and M. Karplus, Force generation in kinesin hinges on cover-neck bundle formation, *Structure*, 62–71, 2008; S. Jungbauer et al., Two characteristic regimes in frequency dependent dynamic reorientation of fibroblasts on cyclically stretched substrates, *Biophys. J.*, BioFAST, May 30, 2008, doi:10.1529/biophysj.107.128611; S. E. Lee et al., Force-induced activation of talin and its possible role in focal adhesion mechanotransduction, *J. Biomech.*, 40(9):2096–2106, 2007; J. S. Palmer and M. C. Boyce, Constitutive modeling of the stress-strain behavior of F-actin filament networks, *Acta Biomater.*, 4(3):597–612, 2008; J. Qian, J. Wang, and H. Gao, Lifetime and strength of adhesive molecular bond clusters between elastic media, *Langmuir*, 24(4):1262–1270, 2008; D. Sadava et al., *LIFE: The Science of Biology*, 8th ed., Sinaver Associates, Inc., Sunderland, MA, 2008; S. Siechen et al., Mechanical force drives assembly and plasticity of synaptic terminal, *Proc. Nat. Acad. Sci.*, manuscript under review; V. B. Shenoy and L. B. Freund, Growth and shape stability of a biological membrane adhesion complex in the diffusion-mediated regime, *Proc. Nat. Acad. Sci.*, 102(9):3213–3218, 2005; S. Suresh et al., Connections between single-cell biomechanics and human disease states: gastrointestinal cancer and malaria, *Acta Biomater.*, 1(1):15–30, 2005; J. Wang and H. Gao, Clustering instability in adhesive contact between elastic solids via diffusive molecular bonds, *J. Mech. Phys. Solids*, 56(1):251–266, 2008; J. H.-C. Wang et al., Specificity of endothelial cell reorientation in response to cyclic mechanical stretching *J. Biomech.*, 34:1563–1572, 2001; A. Zemel and S. A. Safran, Active self-polarization of contractile cells in asymmetrically shaped domains, *Phys. Rev. E*, 76:021905, 2007.

Role of the frontal lobe in violence

Numerous studies suggest that frontal-lobe dysfunction may serve as a significant risk for violent or aggressive behaviors. The frontal lobe, which constitutes approximately one-third of the cerebrum, extends from the anterior (forward) tip of the brain to the central sulcus. Its surface consists of several subregions, which have distinctive patterns of connections with other areas of the cerebral cortex, along with structures deep below the surface of the brain. Because of its unique composition, placement, and connectivity, the frontal lobe is thought to play a key role in various complex functions, including those related to aggression, violence, and antisocial behavior. Extensive research in multiple areas of the neurosciences—brain imaging, neurology, neuropsychology, and psychophysiology—continues to produce evidence that more and more clearly defines this role.

Brain imaging. Recent brain imaging studies have begun to suggest that frontal-lobe deficits play a crucial role in the neurobiological mechanisms underlying violent behavior. For example, structural magnetic resonance imaging (MRI) studies in violent individuals have revealed volume reductions of gray matter (nerve cell bodies), particularly in two regions of the frontal lobe—the orbitofrontal cortex (OFC; bottom, toward the middle) and the dorsolateral prefrontal cortex (DLPFC; top, toward the sides) [**Figs. 1** and **2**]. In addition, abnormalities of white matter (myelinated nerve fibers) in the frontal lobe have also been associated with violent behavior. Finally, functional brain imaging studies, using methods such as functional MRI (fMRI), single photon emission computed tomography (SPECT), and positron emission tomography (PET), have suggested a link between violence and frontal-lobe dysfunction. Specifically, studies have found abnormal neural activity in the OFC, medial frontal cortex, and anterior cingulate cortex (ACC) [Fig. 2] during cognitive tasks in vio-

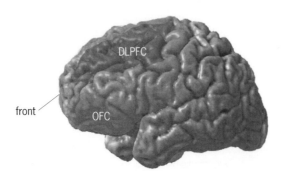

Fig. 1.   Lateral view of the frontal lobe with orbitofrontal (OFC) and dorsolateral (DLPFC) subregions.

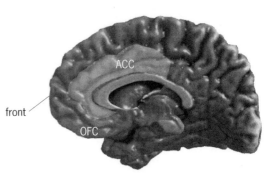

Fig. 2.   Medial view of the frontal lobe with orbitofrontal (OFC) and anterior cingulate cortex (ACC) subregions.

lent subjects compared to controls. Several PET studies have also reported that violent subjects showed significantly reduced glucose metabolism in the frontal cortex, particularly the anterior medial frontal region.

These findings echo the accumulated knowledge regarding the function of the frontal lobe. As a structure rich in inter- and intracortical neuron connections with other cortical and subcortical regions, the frontal lobe is closely linked to the emotional and behavioral components of violence. For example, the OFC is involved in processing the reward-punishment value of stimuli to guide behavior, thus being crucial for the control of violent acts. The DLPFC is critical in several executive functions, such as inhibitory control, which is necessary for the development of moral conduct and moral cognition. Finally, the ACC plays an important role in regulating the intensity of response to emotional stimuli, which when impaired gives rise to aggression and hostility. Structural and/or functional deficits in these key frontal areas may create a predisposition to violent behavior, and imaging studies suggest a critical involvement of the frontal lobe in the neural circuitry of violence.

Neurology. Although brain imaging studies provide evidence that antisocial and violent behavior is associated with structural and functional impairments in the frontal lobe, they cannot establish that frontal-lobe impairment causes violence. Some of the most striking evidence for a causal role of the frontal lobe in violence comes from descriptions of patients with damage to the frontal lobe who later develop antisocial behavior. One of the earliest reported cases is that of Phineas Gage, a railway foreman who had a tamping iron blown through his frontal lobe in an accident involving explosives in 1848. Gage survived the injury, recovering his physical and intellectual abilities, but his personality changed dramatically and he became markedly antisocial.

Similar personality changes have been observed in other case studies of frontal-lobe damage. Common features following injury include lack of empathy, difficulties with emotion regulation, impulsivity, disinhibited behavior, poor planning, and blunted emotions. Moreover, and consistent with the imaging literature, antisocial characteristics seem to develop particularly when damage occurs to the orbitofrontal region of the frontal lobe. For example, it has been found that aggressive and violent attitudes were heightened in Vietnam War veterans who had suffered lesions to the orbitofrontal region of the frontal lobe when compared to controls and individuals with lesions in other brain regions. Additionally, patients who incur damage very early in life (that is, before the age of 16 months) develop antisocial tendencies very similar to those observed in individuals who incur damage as adults, but the tendencies are often more severe and persist throughout development. Together, the evidence from these brain damage studies suggests that impairments in frontal-lobe functioning are involved in the development of antisocial or violent behavior.

Neuropsychology. Frontal-lobe activity is thought to be best represented by *executive functioning*, an umbrella term referring to the cognitive processes necessary for socially and contextually appropriate behavior and effective self-serving conduct. Neuropsychological tests have been undertaken to measure the behavioral expression of brain dysfunction. Test performance errors involving poor strategy formation, cognitive inflexibility, or impulsiveness all indicate deficits in executive functioning.

Neuropsychological examinations of executive functioning in antisocial behavior have traditionally focused on two areas: (1) categorical clinical syndromes (that is, antisocial personality disorder, conduct disorder, and psychopathy—a condition characterized by persistent antisocial, impulsive, and irresponsible behavior and a profound lack of empathy, loyalty, and guilt; frequently associated with violence); and (2) legal concepts/judicial status (criminality and delinquency). Quantitative review of these types of examinations indicates that antisocial individuals perform poorly on executive functioning measures relative to comparison groups. Additionally, specific executive functioning deficits have been reported in both incarcerated and non-incarcerated samples of violent persons, and have been found in relationships with various forms of aggression.

The literature related to executive functioning deficits in psychopaths (examined separately) has produced mixed results, and more recent neuropsychological investigations of psychopathy have focused upon differentiating DLPFC from OFC functioning. Orbitofrontal damage has been directly linked to poor decision making on a neuropsychological gambling task—findings that form the basis of the somatic marker hypothesis. (The somatic marker hypothesis suggests that behavior is guided by distinct bodily or emotional states connected, by learning, with certain stimuli. According to this hypothesis, when a negative outcome connected with a given response option comes to mind, an unpleasant "gut" feeling will be experienced. This alarm signal allows for the immediate rejection of a negative course of action and the selection of an alternative option.) Orbitofrontal pathologies appear more similar to psychopathic features, and recent neuropsychological evidence indicates that

psychopathy may be characterized more by OFC dysfunction than by DLPFC dysfunction. Additionally, specific types of psychopaths (that is, successful, uncaught psychopaths) have even demonstrated better performance on tests of DLPFC functioning than unsuccessful psychopathic and nonpsychopathic (normal) counterparts. Furthermore, in violent individuals, executive functioning deficits have been associated with difficulties in accurately interpreting facial expressions of emotions; indeed, specific frontal-lobe regions are thought to play a role in emotion processing. In sum, neuropsychological findings appear very much in line with evidence from brain imaging and neurological studies that implicates the frontal lobe in aggressive, violent, and antisocial behavior.

Psychophysiology. Psychophysiological studies, which examine brain-behavior relationships in the framework of physiological bodily responses, provide additional evidence of the role of the frontal lobe in violence and antisociality. Skin conductance (SC) activity, one important area of psychophysiological research, reflects very small changes in the electrical properties of the skin—with increased sweating leading to increased SC activity. SC is thought to be controlled via a neural network involving the prefrontal cortex and is considered a measure of the bodily state of arousal and responsivity.

The strongest research findings in this area are poorer SC classical conditioning and quasiconditioning in psychopaths, criminals, delinquents, and antisocials as compared to control groups in child and adult populations. Psychiatrically intact individuals are thought to develop a conscience that deters antisocial responding through successful classical fear conditioning, and failure to do so may predispose individuals to aggressive and antisocial behavior. More specifically, there is also evidence for SC underarousal (for example, reduced SC response frequency and/or lower SC level) in antisocial individuals. It has been theorized that individuals with chronically low levels of arousal may seek out stimulating events (including risky/antisocial acts) in order to increase their arousal to more optimal levels. Additionally, patients with orbitofrontal injuries do not show any anticipatory SC responses, which is consistent with the somatic marker hypothesis and further underscores the link between neuropsychological performance, electrical activity of the skin, and antisocial behavioral tendencies (that is, poor decision making).

The electroencephalograph (EEG) reflects the regional electrical activity of the brain recorded from electrodes placed at different locations on the scalp. Enhanced EEG slow-wave activity (reflecting low arousal) and abnormal frontal EEG asymmetry appear to be related to aggressive and violent behavior. The event-related potential (ERP), another psychophysiological measure, refers to averaged changes of electrical activity of the brain in response to specific stimuli. Reduced P300 ERP amplitudes (that is, ERP amplitudes peaking at approximately

300 ms after stimulus onset) have been consistently found in antisocial and aggressive individuals. For example, it has been reported that the number of violent offenses, but not nonviolent offenses, is associated with reduced P300 amplitude in inmates. It is worth noting, however, that P300 amplitude reduction may reflect a disposition toward externalizing problems in general rather than to specific behavior problems.

Conclusions. In aggregate, recent research from various branches of the neurosciences (brain imaging, neurology, neuropsychology, and psychophysiology) provides extensive and overlapping evidence for a link between dysfunction of the frontal lobe and aggressive, violent, and antisocial behavior. However, it must be emphasized that frontal-lobe impairment—represented by structural or functional deficits, localized brain damage, or behavioral or physiological correlates—should be considered as merely a risk factor for antisocial or violent behavior. Future research into this and other brain regions is clearly needed to understand the brain-violence relationship more definitively.

For background information *see* AGGRESSION; BRAIN; DEVELOPMENTAL PSYCHOLOGY; ELECTROENCEPHALOGRAPHY; EMOTION; MEDICAL IMAGING; NEUROBIOLOGY; PSYCHOLOGY; SKIN in the McGraw-Hill Encyclopedia of Science & Technology.

Robert A. Schug; Yu Gao; Andrea L. Glenn; Yaling Yang; Adrian Raine

Bibliography. E. M. Bernatet et al., Violent offending predicts P300 amplitude, *Int. J. Psychophysiol.*, 66:161–167, 2007; A. Damasio, *Descartes' Error: Emotion, Reason, and the Human Brain*, G. P. Putnam's Sons, New York, 1994; H. Flor, Cognitive correlates, in A. H. Felthous and H. Sass (eds.), *International Handbook on Psychopathic Disorders and the Law*, John Wiley & Sons, West Sussex, UK, 2007; J. Grafman et al., Frontal lobe injuries, violence, and aggression: A report of the Vietnam Head Injury Study, *Neurology*, 46:1231–1238, 1996; Y. Yang and A. Raine, Functional and structural brain imaging research on psychopathy, in A. H. Felthous and H. Sass (eds.), *International Handbook on Psychopathic Disorders and the Law*, John Wiley & Sons, West Sussex, UK, 2007.

Rosaceae (rose family)

Nonbotanists are often surprised to learn that the species that yield many of the most familiar and important temperate fruit crop plants, including apples, peaches, raspberries, and strawberries, are all classified by botanists in the same plant family, and that it is also the family to which some of the most beloved species of ornamental plants, namely, roses, belong and give their name. Many other plants of economic and ecological significance are also classified in this family (see **table**). Despite the apparent diversity represented by this array of species, the rose family has been recognized as a natural group since at least the late eighteenth century, when the botanical name Rosaceae was first applied to it. Members of Rosaceae share a distinctive floral feature—the presence of a hypanthium, a floral disk or cup formed from the fused bases of the sepals, petals, and stamens, which is sometimes fused to the ovary. The hypanthium is not a unique feature of Rosaceae, but its presence in combination with other characteristics, such as the presence of numerous (15 or more) stamens in the flowers of most species and the presence of cyanogenic glycosides (defensive chemicals that release hydrogen cyanide when parts of the plant are crushed) in many, has facilitated the long-standing recognition of this family. Nonetheless, as is inevitable with such a large and diverse group, modifications to both the membership of the family and its subdivision into smaller taxonomic groups (subfamilies, tribes, and subtribes) have occurred over the last 200 years as new evidence has come to light. In its current circumscription, the family includes about 90 genera, which together comprise 2000–3000 species.

Traditional classification. The current classification of flowering plants places Rosaceae in the order Rosales, which also includes the hemp family (Cannabaceae), the mulberry or fig family (Moraceae), the buckthorn family (Rhamnaceae), the elm family (Ulmaceae), the nettle family (Urticaceae), and several others. Fossil evidence suggests that Rosaceae may be as old as 90 million years, with the major evolutionary radiation of the family occurring 40–55 million years ago.

For much of the twentieth century, the most widely adopted classification of Rosaceae divided the family into four subfamilies, defined primarily by fruit type. In this classification, members of subfamily Rosoideae produce achenes (one-seeded, dry, and indehiscent fruits; for example, the "seeds" on the surface of a strawberry) or occasionally drupelets (one-seeded, fleshy fruits; for example, the sections of a raspberry); members of subfamily Amygdaloideae produce drupes (fleshy fruits with a single seed enclosed in a central stone; for example, a peach); members of Spiraeoideae generally have follicles (dry fruits that dehisce along one side); and members of Maloideae generally produce pomes (fleshy fruits with a papery core containing the seeds; for example, apples and pears). Biochemical and cytological characters have been shown to be more or less correlated with these traditional subfamilial circumscriptions. Thus, the sugar alcohol sorbitol is a constituent of phloem sap in members of the traditional Amygdaloideae, Maloideae, and Spiraeoideae, but not in most members of the traditional Rosoideae, and the presence of cyanogenic glycosides in the plant body shows a similar distribution. The base haploid chromosome number is 7 in most traditional Rosoideae, with a few instances of 8 or 9; in most traditional Spiraeoideae, it is 9, but with a few instances of 15 or 17; in Amygdaloideae, it is 8; and in Maloideae, it is 17.

Phylogenetic analyses. Beginning in the mid-1990s, several phylogenetic analyses of deoxyribonucleic acid (DNA) sequence data from both the plastid

---

Infrafamilial classification of Rosaceae†

Subfamily Rosoideae: worldwide, especially north temperate regions
　　Filipendula (meadowsweet)
Supertribe Rosodae
　　Rosa (roses), *Rubus* (raspberry, blackberry)
　Tribe Sanguisorbeae
　　Subtribe Agrimoniinae
　　　Agrimonia, Aremonia, Hagenia, Leucosidea, Spenceria
　　Subtribe Sanguisorbinae
　　　Acaena, Cliffortia, Margyricarpus, Polylepis, Poteridium,
　　　Poterium, Sanguisorba
　Tribe Potentilleae
　　Potentilla (cinquefoil)
　　Subtribe Fragariinae
　　　Alchemilla (lady's mantle), *Chamaerhodos, Comarum, Dasiphora,*
　　　Drymocallis, Fragaria (strawberry), *Potaninia, Sibbaldianthe, Sibbaldia, Sibbaldiopsis*
　Tribe Colurieae
　　Fallugia, Geum (avens)
Subfamily Dryadoideae*: primarily western North America
　　Cercocarpus (mountain mahogany), *Chamaebatia, Dryas, Purshia*
Subfamily Spiraeoideae
　　Lyonothamnus: Channel Islands off southern California
　Tribe Amygdaleae: Asia; Europe; Africa; North, Central, and South America
　　Prunus (almonds, apricots, cherries, peaches, plums)
　Tribe Neillieae: eastern Asia, western and eastern North America
　　Physocarpus (ninebark), *Neillia*
　Tribe Sorbarieae: eastern and central Asia, western North America
　　*Adenostoma** (chamise), *Chamaebatiaria, Sorbaria, Spiraeanthus*
　Tribe Spiraeeae: Asia, Europe, North and Central America
　　Aruncus, Holodiscus, Kelseya, Luetkea, Petrophyton, Sibiraea, Spiraea (bridal wreath), *Xerospiraea*
　Supertribe Kerriodae
　Tribe Kerrieae*: eastern Asia, western and eastern North America
　　Coleogyne, Kerria, Neviusia, Rhodotypos (jetbead)
　Tribe Osmaronieae: eastern Asia, western North America
　　Exochorda, Oemleria, Prinsepia
　Supertribe Pyrodae
　　Gillenia (Bowman's root): eastern North America
　Tribe Pyreae: North, Central, and South America; Europe; Asia; Africa
　　Kageneckia, Lindleya, Vauquelinia
　　Subtribe Pyrinae [= the traditional Maloideae]
　　　Amelanchier (shadbush), *Aria, Aronia, Chaenomeles* (flowering
　　　quince), *Chamaemeles, Chamaemespilus, Cormus, Cotoneaster,*
　　　Crataegus (hawthorn), *Cydonia* (quince), *Dichotomanthes, Docynia,*
　　　Docyniopsis, Eriobotrya (loquat), *Eriolobus, Hesperomeles, Heteromeles,*
　　　Malacomeles, Malus (apple), *Mespilus, Osteomeles, Peraphyllum, Photinia,*
　　　Pseudocydonia, Pyracantha (firethorn), *Pyrus* (pear), *Rhaphiolepis, Sorbus*
　　　(mountain ash, whitebeam), *Stranvaesia, Torminalis*

---

†Based on phylogenetic analysis of nuclear and plastid gene sequences (see illus.). Common names are given for genera that include species of notable economic or ecological importance, and geographic distributions of major lineages are indicated. Groups marked with asterisks (*) were formerly classified in subfamily Rosoideae. (*From D. Potter et al., Phylogeny and classification of Rosaceae, Plant Syst. Evol., 266:5–43, 2007*)

---

and nuclear genomes suggested that fruit type is a less accurate predictor of evolutionary relationship in Rosaceae than chromosome number (see **illustration**). Genera with haploid chromosome numbers of 9 that were traditionally assigned to Rosoideae because their members produce achenes were found to be more closely related to genera traditionally classified in Spiraeoideae (see table). Similarly, several genera with haploid chromosome numbers of 15 and 17 that were traditionally classified in Spiraeoideae because they produce follicles were found to be more closely related to pome-bearing members of Maloideae than to other spiraeoids. Biochemical and ecological characters also supported these relationships (see illustration).

Recently, several research groups from the United States, Canada, and Europe joined forces and combined their data to generate a hypothesis of phylogenetic relationships among most members of the family based on multiple gene sequences. The resulting phylogenetic analysis (see illustration) was used to revise the classification of Rosaceae at the subfamilial and tribal levels, with the goal of giving formal taxonomic names only to groups of genera that were strongly supported by all the data as being monophyletic, that is, including all descendants of a unique ancestor.

New infrafamilial classification. The new infrafamilial classification of Rosaceae, published in 2007, recognizes three subfamilies, Rosoideae, Dryadoideae, and Spiraeoideae. Rosoideae and Spiraeoideae are further divided into supertribes, tribes, and subtribes (see table). All genera formerly classified in subfamilies Amygdaloideae and Maloideae are now included in Spiraeoideae. Many of these groups, especially at the tribal level, correspond closely, if not exactly, to previously recognized infrafamilial taxa, but this is the first classification that explicitly recognizes only groups that are strongly supported as monophyletic. Because the analysis was based on evidence from multiple plastid and nuclear genes and because only groups that were strongly supported were given taxonomic recognition, it is expected that this new classification will be stable and robust.

Fruit type
- equivocal
- achene(s)
- drupelets
- follicles
- fleshy drupe(s)
- dry drupe
- capsule
- polyprenous drupe
- pome

ellagic acid

cyanogenic glycosides, sorbitol

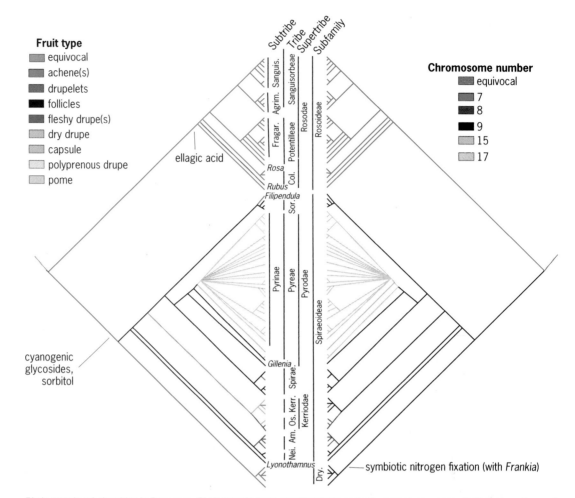

Chromosome number
- equivocal
- 7
- 8
- 9
- 15
- 17

symbiotic nitrogen fixation (with *Frankia*)

Phylogenetic relationships in Rosaceae. Phylogenetic trees resulting from analysis of gene sequences from four nuclear and four plastid genes for most genera of Rosaceae. These data have served as the basis for the revised infrafamilial classification of the family, as summarized in the table. (*From D. Potter et al., Phylogeny and classification of Rosaceae, Plant Syst. Evol., 266:5–43, 2007*)

The new classification follows the traditional taxonomic hierarchy, but differs structurally from many other infrafamilial classifications of plants in two ways. First, the rank of supertribe, not generally employed in plant classification, was used in three cases, allowing the incorporation of greater phylogenetic resolution than would be possible without the addition of this rank. Second, some genera were not included in higher-level groups if the result would be a tribe or subtribe consisting of just one genus. For example, *Filipendula* is included in subfamily Rosoideae but not in any tribe, *Rosa* and *Rubus* are both included in supertribe Rosodae but not in any tribe, and *Potentilla* is included in tribe Potentilleae but not in any subtribe, although the remaining genera are placed in Fragariinae. *Lyonothamnus*, which includes just one species found only on the Channel Islands off southern California and has opposite fern-like leaves that are unusual for the family, is included in subfamily Spiraeoideae but not in any tribe.

Phylogenetic analyses confirmed that chromosome number is a better predictor of evolutionary relationships in Rosaceae than is fruit type; however, a haploid chromosome number of 8 has been derived independently in several lineages (see illustration). Polyploidy is a common phenomenon throughout Rosaceae, with multiple ploidies occurring within many genera, and one entire group—tribe Pyreae in the new classification—is of ancient polyploid origin. The haploid chromosome numbers of the traditionally circumscribed subfamilies Amygdaloideae (8), Spiraeoideae (9), and Maloideae (17) have long intrigued botanists, since they show that members of Maloideae are clearly of polyploid origin and suggest the possibility that they might have been derived from ancient hybridization between an ancestral spiraeoid and an ancestral amygdaloid. Phylogenetic analyses of gene sequence data, however, have failed to support a close relationship between the maloid group and any extant genus with a haploid chromosome number of 8 and have consistently shown that the living group most closely related to Pyreae is the genus *Gillenia*, which consists of two species of perennial herbs from eastern North America. This relationship is reflected in the new classification by including *Gillenia* with Pyreae in the supertribe Pyrodae. These results refute the "wide-hybridization" hypothesis for the origin of Pyreae and suggest instead that the higher chromosome number of that group arose via hybridization and polyploidization among closely related species of an ancestral lineage with haploid chromosome numbers of 9 (the lineage that also gave rise to the living species of *Gillenia*), followed by a phenomenon known as dysploid reduction, in which the chromosome number was reduced from 18 to 17 (and further to 15 in the genus *Vauquelinia*). Within this group, species formerly classified in subfamily Maloideae because they produce pomes (or polyprenous drupes) are now assigned to subtribe Pyrinae.

Ecology and biogeography. The phylogenetic analyses reveal intriguing patterns of ecology and biogeography in Rosaceae. Symbiotic nitrogen fixation, via associations with actinomycete bacteria of the genus *Frankia*, has been observed only in Dryadoideae, in which members of all four genera have been reported to form root nodules that house the bacteria (see illustration). Association with rust fungi of the genus *Gymnosporangium* appears to be restricted to Pyrodae and provides a piece of nonmolecular evidence linking *Gillenia* with the genera of polyploid origin.

Phylogenetic analyses indicate a North American origin for the entire family, each of the three subfamilies, and supertribe Pyrodae. Many tribes provide examples of striking biogeographic patterns, including an eastern Asia–eastern and western North America pattern exemplified by tribes Neillieae and Kerrieae, a central and eastern Asia–western North America pattern in Sorbarieae and Osmaronieae, and potentially complex patterns involving multiple continents in groups such as Spiraeeae, Pyrodae, Amygdaleae, and each of the tribes of Rosoideae (see table). A series of detailed phylogenetic studies with thorough sampling of species within each of these groups will be required to tease apart the complexities of geographic patterns within the family.

For background information *see* BIOGEOGRAPHY; FLOWER; PHYLOGENY; PLANT EVOLUTION; PLANT GEOGRAPHY; PLANT KINGDOM; PLANT PHYLOGENY; PLANT TAXONOMY; ROSALES; SYSTEMATICS in the McGraw-Hill Encyclopedia of Science & Technology.

Daniel Potter

Bibliography. W. S. Judd et al., *Plant Systematics: A Phylogenetic Approach*, 3d ed., Sinauer Associates, Sunderland, MA, 2008; D. Potter et al., Phylogeny and classification of Rosaceae, *Plant Syst. Evol.*, 266:5–43, 2007; P. F. Stevens, Angiosperm Phylogeny Website, version 8, June 2007.

Satellite-based geodesy

The classical definition of geodesy is the "science of the measurement and mapping of the Earth's surface" by direct measurements, such as terrestrial triangulation, leveling, and gravimetric observations, and, in the past 50 years, also with space techniques, based primarily on the tracking of a wide range of artificial Earth satellites. The primary mission of geodesy can be defined as:

1. Establishment of a geodetic reference frame and determination of precise global, regional, and local 3D positions.

2. Determination of the Earth's gravity field and its related models, such as the geoid. (The geoid is, essentially, the figure of the Earth abstracted from its topographical features; the particular equipotential surface that coincides with mean sea level in the absence of currents, air pressure variations, and so forth, and that may be imagined to extend through the continents. This surface is everywhere perpendicular to the force of gravity.)

3. Measurement and modeling of geodynamical phenomena, such as crustal deformation, polar motion, and Earth rotation and tides.

Modern geodesy provides the foundation for all observations related to global change within the "Earth system." The importance of Earth observations, provided with increasing spatial and temporal resolutions, better accuracy, and decreasing turnaround time, should be seen not only in the context of scientific understanding of the Earth system, but also in support of fundamental societal activities, such as managing natural resources, environmental protection, human health monitoring, disaster prevention, and emergency response.

How space-based methods support the primary mission of geodesy. Since the launch of the first artificial satellite, SPUTNIK, in 1957, geodesy has evolved into a combination of a number of geosciences and engineering sciences. Several important research discoveries gave rise to the development of satellite geodesy. For example, by analyzing SPUTNIK's radio signals it was discovered that the observed Doppler shift could be converted to useful navigation information if the satellite's location in space were known.

An increasingly important contribution of satellite geodesy is satellite-based navigation, facilitated through radio-navigation signals transmitted by Earth-orbiting navigation satellites such as the Global Positioning System (GPS). GPS is now not only an indispensable tool for space geodesy, it has also revolutionized surveying and navigation, and is increasingly employed for personal positioning. Nowadays the phrase Global Navigation Satellite System (GNSS) is used as an umbrella term for all current and future global radio-navigation systems. Although GPS is currently the only fully operational GNSS, the Russian Federation's GLONASS is being replenished and will be fully operational by 2010, the European Union's GALILEO will be deployed and operational by 2013, and China's COMPASS is likely to also join the "GNSS Club" by the middle of the next decade. *See* SATELLITE NAVIGATION ADVANCES.

For satellites to serve as space-based reference points, their trajectories (or orbits) must be known. The precise computation of orbits is accomplished by geodetic techniques combined with orbital mechanics. The International Association of Geodesy

(IAG) has established services for all the major satellite geodesy techniques: International GNSS Service (IGS), International Laser Ranging Service (ILRS), and the International DORIS Service (IDS). These services generate products for users, including precise orbits, ground station coordinates, Earth rotation values, and atmospheric parameters. All have networks of ground tracking stations that also are part of the physical realization of the International Terrestrial Reference Frame (ITRF).

Precise Orbit Determination (POD) also requires the use of accurate geodetic models for gravity, tides, and geodetic reference frames. At the same time, POD's sensitivity to various types of geodetic parameters makes it a powerful tool in the refinement of geodetic models. Consequently, a satellite can be considered to act as a probe or sensor, moving in the Earth's gravity field, along an orbit disturbed by the gravitational attraction. Thus, measurements of directions, ranges, and range rates between terrestrial tracking stations (and sometimes also from other orbiting satellites) and the satellite targets provide, in addition to navigation parameters, a wealth of information about the size and the shape of the Earth and its gravity field. The first example of gravitational mapping by the use of satellites was the 1958 determination of Earth flattening from measurements to the EXPLORER-1 and SPUTNIK-2 satellites, followed by the discovery of the Earth's pear shape in 1959. The latest generation of gravity-mapping missions, CHAMP, GRACE, and GOCE, now are determining the features of the gravity field down to 100 km (60 mi) or so in size, on a monthly basis, and also monitoring changes in the gravity field with time due to mass transports such as resulting from the water cycle. This is an exciting time for gravimetric geodesy with the recent establishment of the International Gravity Field Service (IGFS) and the release of the most detailed and precise model, EGM2008, of the Earth's gravity field ever.

Aside from their navigation and precise positioning function, geodetic satellites serve as a remote-sensing tool. An example is satellite radar altimetry, a remote-sensing technology that can measure the ocean surface topography (TOPEX/Poseidon, Jason-1, and so on) or ice topography (for example, ICESat and CRYOSAT). Variations in ocean surface topography indicate gravitational variations due to undersea features, such as seamounts or trenches, as well as ocean circulation features from small eddies to basinwide gyres. The altimeters can also be used to measure parameters such as wave height, wave direction, and wave spectra. Other geodetic remote-sensing tools include differential interferometric synthetic aperture radar (DInSAR) satellites such as TerraSAR-X, ALOS, Radarsat-2, and others. These can detect changes in the shape of the topography as small as a centimeter with spatial scales of just a few meters.

Terrestrial and celestial reference systems. Modern geodesy is now equipped with an array of space technologies for mapping (and monitoring changes to) in the geometry of the surface of the solid Earth and the oceans, as well as its gravity field. However, the fundamental role of geodesy continues to include the definition of the terrestrial and celestial reference systems. These reference systems are the bedrock for all operational geodetic applications for national mapping, navigation, and spatial data acquisition and management, as well as for scientific activities associated with geodynamics and solid Earth physics, mass transport in the atmosphere and oceans, and global change studies.

The locations of points in three-dimensional space are most conveniently described by Cartesian coordinates x, y, and z. Since the start of the Space Age such coordinate systems are typically geocentric, with the z axis aligned with the Earth's conventional or instantaneous rotation axis. Because the Earth's center of mass is located at one focus of a satellite's orbital ellipse, this point is the natural origin of a coordinate system defined by the satellite-based geodetic methods.

The International Celestial Reference System (ICRS) forms the basis for describing celestial coordinates, and the International Terrestrial Reference System (ITRS) is the foundation for the definition of terrestrial coordinates. The definitions of these systems include the orientation and origin of their axes, and the scale, physical constants, and models used in their realization, such as, for example, the size, shape, and orientation of the reference ellipsoid that approximates the Earth's surface and the Earth's gravitational model. The coordinate transformation between these two systems is described by a sequence of rotations that account for precession, nutation, Greenwich apparent sidereal time (GAST), and polar motion, which collectively account for variations in the orientation of the Earth's rotation axis and its rotational speed. GAST and polar motion, also referred to as Earth rotation parameters, are monitored by geodetic techniques, while precession and nutation are described by respective models.

While a reference system is a mathematical abstraction, its practical realization through geodetic observations is known as a reference frame (or datum). The conventional realization of the ITRS is the International Terrestrial Reference Frame (ITRF), which is a set of coordinates and linear velocities (due mainly to crustal deformation and tectonic plate motion) of well-defined fundamental stations (for example, networks of stations of the IGS, ILRS, and IDS), derived from space-geodetic observations collected at these points. The ICRS is realized through the International Celestial Reference Frame (ICRF), which is a set of estimated position coordinates of extragalactic reference radio sources. At present, the ICRF is determined by Very Long Baseline Interferometry (VLBI), while the ITRF is accomplished by a combination of several independent space-based geodetic techniques, including VLBI, Satellite Laser Ranging (SLR), GNSS, and Doppler Orbitography by Radiopositioning Integrated on Satellites

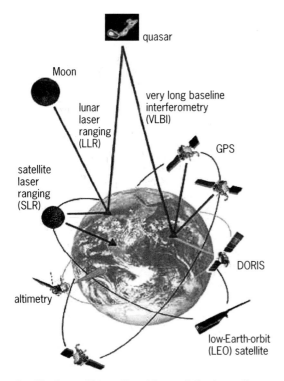

Combination and integration of the geodetic observation techniques. (*After H-P. Plag et al., The Global Geodetic Observing System, in K. Satake, ed., Advances in Geodetic Sciences 2008, World Scientific Publishing Company, Singapore, in press, 2008*)

(DORIS). The ITRF and ICRF are defined by the International Earth Rotation and Reference System Service (IERS). Combination and integration of the geodetic observation techniques are shown in the **illustration**.

Geodesy is facing an increasing demand from science, engineering, the Earth observation community, and society at large for improved accuracy, reliability, and access to geodetic services, observations, and products. Thus, a challenge that geodesy is now facing is to maintain the ITRF at the level that allows, for example, the determination of global sea-level change at the submillimeter per year level, as well as the determination of the glacio-isostatic adjustments due to deglaciation since the Last Glacial Maximum and to modern mass change of the ice sheets, at millimeter-level accuracy. Other applications include measurement of preseismic, coseismic, and postseismic displacement fields associated with large earthquakes at the subcentimeter accuracy level; early warnings for tsunamis, landslides, earthquakes, and volcanic eruptions, and millimeter- to centimeter-level deformation and structural monitoring. In response, the IAG is currently in the process of establishing the Global Geodetic Observing System (GGOS), which will unify all the geometric and gravity services of the IAG, in order to address the demands for improved geodetic products by society and the scientific community.

National mapping datums. Before satellite-based positioning systems were first used in the 1960s, na-

tional datums were nongeocentric. Typically a reference system that best fitted the shape of the Earth within the region covered by that datum was established; examples are the North American Datum 1927 (NAD27) and the Australian Geodetic Datum (AGD) 1966. A unified World Geodetic System became essential for several reasons, such as (1) the beginning of international space science and astronautics, (2) the lack of intercontinental geodetic ties, (3) the inability of the existing geodetic systems to provide a worldwide geo-data basis, and (4) the need for global maps for navigation and aviation, and military operations. Particularly since the advent of GPS, national datums around the world have been progressively redefined as geocentric reference systems, providing much higher reference frame accuracy, and ensuring that national geodatabases are compatible with GPS-derived coordinates. ITRF now provides the foundation for all national and regional reference frames, and facilitates a link of the local frames to each other.

World Geodetic System. World Geodetic System (WGS84) is a particular realization of a terrestrial reference system, created and maintained by the U.S. Department of Defense; it is a geocentric reference system used by GPS. Current geodetic realizations of WGS84 and the ITRF are consistent with each other at the few centimeter level globally.

Future developments and trends. The advent of space-based geodetic techniques put geodesy in transition by extending the scope of geodetic observing systems from providing the reference frame, as well as the methodology and tools for positioning and navigation, to a global dynamic system for Earth monitoring. It also modernized the primary mission of geodesy of establishing the reference frame by facilitating the development of geocentric global datums. This transition is therefore from a utility serving other geosciences to a "system of systems" that consists of the sensors (satellites), global reference frames, dynamic models, and observing systems that together create consistent data sets that are relevant to nearly all societal activities and numerous scientific and engineering applications. With the expansion of Earth-observing satellite-based systems, geodesy will continue to transform from a local service, aiding mappers and surveyors, to a global one supporting geophysical, oceanographic, atmospheric, and environmental science communities. Among the important applications that are possible only due to the availability of space-based geodetic techniques the following must be mentioned, as they will continue growing: sea-level monitoring and early tsunami warning, and atmospheric/climate remote sensing.

The increasing accuracy of navigation and positioning using satellite-based geodetic methods will continue to create new applications in transport, commerce (location-based services or infomobility, also known as telegeoinformatics), precision agriculture, construction, asset and natural resources monitoring, and so forth. Since the GNSS observations are inherently calibrated with respect to atomic time scales, GNSS will continue supporting time transfer and providing a time-base for synchronization of computer networks, telecommunication switches, electric grids, and so forth.

For background information *see* CELESTIAL REFERENCE SYSTEM; GEODESY; NUTATION (ASTRONOMY AND MECHANICS); PRECESSION OF EQUINOXES; SATELLITE NAVIGATION SYSTEMS in the McGraw-Hill Encyclopedia of Science & Technology.

Dorota A. Grejner-Brzezinska; Chris Rizos

Bibliography. H. Drewes (ed.), The Global Geodetic Observing System, *J. Geodyn.*, 40(4-5):355–502, November-December 2005; D. D. McCarthy and G. Petit (eds.), IERS Conventions, IERS Tech. note no. 32, 2003; G. Seeber, *Satellite Geodesy*, 2d ed., Walter de Gruyter, Berlin/New York, 2003; P. Tregoning and C. Rizos (eds.), *Dynamic Planet, Monitoring and Understanding a Dynamic Planet with Geodetic and Oceanographic Tools*, IAG Symposium, Cairns, Australia, August 22–26, 2005, Springer (Part VI, GGOS: Global Geodetic Observing System, pp. 701-766), 2007.

Satellite navigation advances

The Global Positioning System (GPS) has revolutionized navigation and precise timing applications for countless military and civilian users worldwide. Businesses, and indeed entire industries worldwide, have been created using the position and timing information provided by GPS, which was developed by the U.S. government. Users have come to expect highly reliable and accurate service from GPS, and as a result GPS has become the first truly global utility. The benefits of satellite-based navigation, which are demonstrated by GPS, have motivated other governments to develop their own independent Global Navigation Satellite Systems (GNSS). On the other hand, the limitations of standalone GPS are motivating innovative research and development to make position, velocity, and timing (PVT) technology even more accurate, reliable, and pervasive. Near-term advances in satellite navigation will be defined by the proliferation of new and modernized GNSS systems and the services they offer, along with the new and improved applications of GNSS technology that will provide more robust PVT services to users worldwide.

Global Navigation Satellite Systems. Besides GPS, a GNSS is operated by Russia, and such systems are being developed by the European Union, China, Japan, and India.

GLONASS. The Russian GNSS, named GLONASS, has been operating since 1983, almost as long as GPS, and is currently undergoing a revitalization and modernization program. GLONASS, originally developed by the Soviet Union, had deployed a constellation of 26 satellites in 1995. The dissolution of the Soviet Union caused serious funding shortfalls for GLONASS. As a result, the operational GLONASS constellation fell to a low of 7 satellites in 2001. However, through renewed political support and longer satellite lifetimes, the GLONASS constellation is once again being replenished, with plans for 24 satellites by 2010.

One key difference between GPS and GLONASS is the design of the two systems' signals (see **table**). GLONASS chose a frequency-division multiple access (FDMA) approach for its signals, while GPS uses a scheme called code-division multiple access (CDMA). In an FDMA system, each satellite transmits the same ranging code on a different frequency. CDMA systems transmit different ranging codes on the same frequency. To improve interoperability with other GNSS systems, Russia has announced that CDMA signals will be added to GLONASS satellites.

Galileo. The Galileo program officially began in 2003, with the goal of developing a civilian satellite-based navigation system controlled by the European Union. Galileo plans to offer several signals and services over a wide range of frequencies. In 2004, the United States and the European Union signed a major agreement to foster cooperation between GPS and Galileo, including the transmission of a common civil signal. This signal, known as E1 OS on Galileo and L1C on GPS, will be transmitted on the same center frequency of 1575.42 MHz and was designed to facilitate the use of signals from both systems simultaneously. GPS will transmit the L1C signal from its Block III satellites, currently in development. Once sufficient numbers of Galileo and GPS III satellites are transmitting this common signal, users in challenging environments, such as urban areas where visibility to many satellite signals is blocked by buildings, will benefit from greater availability of satellites from the combined GPS and Galileo constellations. This will provide a significant performance improvement over either individual system.

Compass. The People's Republic of China is also deploying an evolving global and regional satellite navigation system called Compass. Initial plans call for a Compass constellation consisting of 5 geostationary satellites (24-hour orbits, fixed over a specific location on the Equator), 3 inclined geosynchronous satellites (24-hour orbits, inclined to the Equator), and 27 medium Earth orbit (MEO) satellites (12-hour inclined orbits similar to GPS, Galileo, and GLONASS). When fully deployed, Compass will be a global system with both open services and restricted services for authorized users. China launched its first MEO Compass satellite in April 2007, transmitting signals on multiple frequencies in or near GPS and Galileo frequency bands. Compass has submitted multiple frequency filings with the International Telecommunication Union Radiocommunication Sector (ITU-R) to operate in these bands. Complete details of the Compass system and its signals have yet to be formally published by the Chinese government. However, because Compass intends to operate a GNSS using the same frequency bands as GPS, GLONASS, and Galileo, it can be expected to play an ongoing role in future GNSS developments and discussions.

GPS, Galileo, and GLONASS signals

System	Frequency	Signal	Signal description
GPS	L1 (1575.42 MHz)	C/A	Civil signal, open access
		L1C	Civil signal, open access, same spectrum as Galileo E1 OS
		P(Y)	Military signal, authorized access
		M	Military signal, authorized access
	L2 (1227.60 MHz)	L2C	Civil signal, open access
		P(Y)	Military signal, authorized access
		M	Military signal, authorized access
	L5 (1176.45 MHz)	L5	Civil signal, open access, same spectrum as Galileo E5a
Galileo	E1 (1575.42 MHz)	OS	Open service signal, with integrity messages, same spectrum as GPS L1C
		PRS	Publicly regulated service signal, authorized access
	E6 (1278.75 MHz)	CS*	Commercial service signal, authorized access
		PRS	Publicly regulated service signal, authorized access
	E5b (1207.14 MHz)	OS	Open service signal, with integrity messages
	E5a (1176.45 MHz)	OS	Open service signal, same spectrum as GPS L5
GLONASS	L1 (1602.6–1615.5 MHz)	SP	Standard-precision FDMA signal
		HP	High-precision FDMA signal
		L1CR	L1 CDMA signal near 1575.42 MHz, in development
	L2 (1240–1260 MHz)	SP	Standard-precision FDMA signal
		HP	High-precision FDMA signal
	L5	L5R	L5 CDMA signal near 1176.45 MHz, in development

*Commercial service (CS) message data may available on Galileo E1 OS, E5a, and E5b signals.

Quasi-Zenith Satellites System (QZSS). Japan is one of the world's most technologically advanced countries, where wireless and GNSS technology have been adopted on a massive scale. However, GNSS users in Japan, particularly those in dense urban environments such as Tokyo, must contend with very limited visibility to GPS satellites because of the tall buildings. Although these so-called urban canyon environments exist in any large city, the unique, compact geography of Japan has motivated authorities there to devise an innovative regional system known as the Quasi-Zenith Satellites System (QZSS) to address this problem for Japanese GNSS users. The term "zenith" refers to the point directly above one's head, and that is where at least one of the three planned QZSS satellites will be located at almost all times. At or near zenith, a satellite's signals will not be blocked by buildings, even in the densest urban areas. The first of three planned QZSS satellites is planned for launch in 2009 and will be placed in a 24-hour, slightly elliptical orbit, inclined 45° with respect to the Equator. This results in orbital geometry that maximizes QZSS signal availability at high elevation angles over the island of Japan. **Figure 1** depicts the ground tracks of the three QZSS satellites as well as the subpoints, the points on the Earth's surface directly below each satellite at a particular moment in time. In addition to its unique orbital configuration, QZSS will transmit a range of signals and services on frequencies common with GPS and Galileo.

India Regional Navigation Satellite System (IRNSS). The government of India is also developing its own independent system, known as the India Regional Navigation Satellite System (IRNSS), to meet specific regional needs. The Indian Space Research Organisation (ISRO) has developed a family of indigenous launch vehicles and has successfully launched

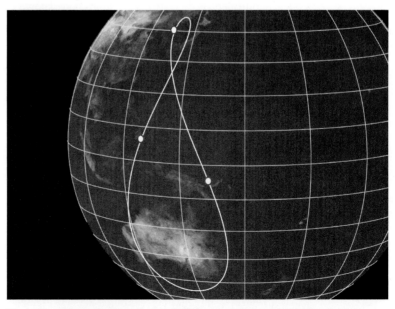

Fig. 1. **Quasi-Zenith Satellites System (QZSS) ground tracks and satellite subpoints.**

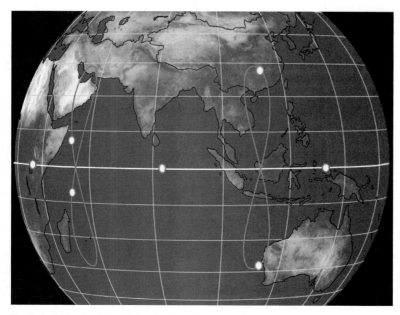

Fig. 2. India Regional Navigation Satellite System (IRNSS) ground tracks and subpoints.

several satellites in recent years. The IRNSS constellation will consist of 7 geostationary and inclined geosynchronous satellites with orbital geometry that assures at least 4 IRNSS satellites will always be in view over the Indian subcontinent and surrounding areas (**Fig. 2**). This will allow IRNSS users to navigate with IRNSS signals only, independent of other GNSS systems, though ISRO authorities are coordinating their activities with other GNSS systems to ensure signal compatibility.

Interoperability and compatibility. The government and industry sponsors of each of these GNSS systems originally envisioned them to operate independently of the others. However, the necessities of spectrum management—the fact that all these signals share the same relatively narrow radio frequency bands—have fostered considerable cooperation among GNSS providers. In fact, a United Nations body known as the International Committee on GNSS (ICG) has been established to promote this cooperation, which will ultimately benefit all GNSS users. However, ensuring that these systems are compatible (they will not unreasonably interfere with each other) and interoperable (appropriately designed receivers can use signals from multiple GNSS simultaneously) presents a unique set of engineering challenges.

Spacecraft technology advances. GNSS satellite technologies have made significant advances over the last 30 years. The space-borne atomic clock, perhaps the most critical enabling technology for GNSS, has improved significantly since the first generation flown on GPS and GLONASS satellites. Years of on-orbit experience have been applied to improve the frequency stability and reliability of atomic clocks, resulting in more accurate ranging signals and therefore better navigation and timing accuracy for all users. The newest GPS satellites include onboard integrity systems to detect problems such as atomic clock "frequency jumps," which can cause spurious

navigation signals, and to take corrective action in real time. GPS and other GNSS providers are also introducing new signals as part of modernization efforts. This presents new challenges for GNSS satellite engineers, who must design payloads to synthesize and efficiently transmit more signals than ever before.

Ground segment advances. The original GPS ground segment featured 6 monitor stations to "listen" to the GPS satellites and provide measurement data for computation of the satellite orbit and clock parameters, as well as 4 ground antennas to "talk to" the satellites, uploading commands and new navigation data messages. Unfortunately, the locations of the original 6 GPS monitor stations provided no visibility in the Southeast Pacific region. This introduced the possibility that a satellite could transmit bad measurements to users without being observed by the GPS Master Control Station (MCS). A new GPS ground segment was deployed in 2007 that incorporates 10 additional monitor stations operated by the National Geospatial-Intelligence Agency (NGA), as well as 8 Air Force Satellite Control Network (AFSCN) ground antennas. The extended network of monitor stations and ground antennas has eliminated coverage gaps in the original GPS network, and every GPS satellite is now monitored by no less than two monitor stations at all times (**Fig. 3**). The additional AFSCN ground antennas also mean that GPS satellites can now be commanded at any time. The inclusion of the NGA monitor stations has had the additional benefit of producing more accurate GPS orbit and clock estimates, which benefit all users.

User technologies. One of the most significant areas of GNSS technology advancement has been the integration of GNSS systems with other communication and navigation technologies to address some of the limitations of standalone satellite navigation. The success of GPS has led to the desire of many users to have PVT information available everywhere. This has led to the use of GPS and other GNSS systems in increasingly challenging environments. *See* ASSISTED GPS AND LOCATION-BASED SERVICES.

Military aviation GPS receivers have been coupled with inertial measurement technology for decades, and now inertial navigation systems (INS) are being applied in many other GNSS applications. An INS works by measuring linear and angular accelerations along each of three principal axes, and relating them to position and orientation by applying Newton's laws of motion (which must be applied in an inertial reference frame). After the INS initial position and orientation have been defined, the system integrates measured linear and angular accelerations to estimate the user's position and orientation in real time, though accuracy will degrade over time because of inertial "drift." A GNSS receiver can be coupled with the INS to establish initial position and to limit INS drift, in the process accurately calibrating the inertial system components. In a combined GNSS-INS device, the INS can also be used to aid the GNSS

Key: ◯ **GPS monitor stations**   ◉ **NGA monitor stations**

Fig. 3. Global Positioning System (GPS) and National Geospatial-Intelligence Agency (NGA) monitor station locations.

receiver to allow it to track very weak signals, or to enable the device to navigate for a limited time in environments where no GNSS signals are available. The advent of micro-electro-mechanical systems (MEMS), which include inertial sensors fabricated on silicon wafers at the nanometer level, offers the potential to reduce the size, weight, power consumption, and cost of MEMS-based GNSS-INS devices, opening the door to even more applications of GNSS-INS technology. *See* MEMS SENSORS.

Aviation applications. The tracking and control of an increasing number of aircraft in an increasingly complex airspace presents a difficult challenge for civil aviation authorities worldwide. GNSS technology is helping meet that challenge today and offers a number of promising future improvements. For decades, commercial and civilian aircraft in the United States have navigated the United States National Airspace (NAS) using a combination of ground-based radars and fixed navigation beacons known as VHF (very high frequency) omnidirectional range (VOR) and distance-measuring equipment (DME). Radar performance suffers from the fact that a line of sight between the radar facility and the aircraft is required, meaning that obstructions can block radar service to large areas. Radar positions are typically updated no faster than once every 5 s, and the accuracy of radar positioning decreases with distance. Many air-traffic control (ATC) delays due to weather or other problems are exacerbated by reliance on the ground-based radar/VOR/DME architecture, which requires large separation distances between aircraft to ensure safety.

A system known as Automatic Dependent Surveillance-Broadcast (ADS-B), currently undergoing field trials, combines GNSS and satellite communications technology and promises to improve this situation fundamentally. Aircraft equipped with ADS-B avionics compute their own positions using GPS or another GNSS and then transmit that information to ground-based receiver stations and to other aircraft. Data transmitted by the ADS-B-equipped aircraft include position, velocity, and unique aircraft identification information. ADS-B position is more accurate than radar and independent of distance from the ground station. Sharing this data among all aircraft in the sky will increase "situational awareness," allow closer safe spacing, and promote more efficient use of air routes. ADS-B data can be transmitted over greater ranges, allowing better control with fewer ground stations, and situational data will be updated at a much faster rate, assuring greater safety in dynamic situations where each second counts. This will translate into fewer delays and greater safety for the flying public and lower fuel consumption and lower operating costs for airlines.

For background information *see* AIR NAVIGATION; AIR-TRAFFIC CONTROL; ATOMIC CLOCK; INERTIAL NAVIGATION SYSTEM; MICRO-ELECTRO-MECHANICAL SYSTEMS (MEMS); MULTIPLEXING AND MULTIPLE ACCESS; SATELLITE NAVIGATION SYSTEMS in the McGraw-Hill Encyclopedia of Science & Technology.

Joe M. Straus; Thomas D. Powell

Bibliography. A. Daskalakis and P. Martone, A technical assessment of ADS-B and multilateration technology in the Gulf of Mexico, pp. 370–378, *Radar Conference, 2003*, Proceedings of the 2003 IEEE, May 5–8, 2003; B. Hofmann-Wellenhof, H. Lichtenegger, and E. Wasle, *GNSS, Global Navigation Satellite Systems: GPS, GLONASS, Galileo, and More*, Springer-Wien, New York, 2007; B. W. Parkinson and J. J. Spilker (eds.), *Global Positioning System: Theory and Applications*, AIAA Progress in Astronautics and Aeronautics, 2 vols., 1996.

Scientific workflows

From molecular biology and chemistry to astronomy, earth sciences, and particle physics, modern experimental science relies increasingly on the acquisition, manipulation, and processing of large amounts of data, and the systematic orchestration of computationally intensive simulations and analyses. Much of the analysis is expensive in terms of the data processing and storage resources required, and it sometimes runs for weeks and needs careful monitoring. It is also laboriously repetitive as scientists explore different settings for their algorithms, data sets are continually updated from instruments, and new information prompts the whole "in silico" (computer-simulated) experiment to be rerun and cross-checked.

Over the past six years, workflow technology has been increasingly adopted by scientists. A scientific workflow is the description of a process, often completely automated, that specifies the coordinated execution of multiple tasks. The tasks are software programs, run locally or remotely and increasingly published as Web services. Thus, workflows are a particular form of scripted distributed computing over service-oriented architectures. *See* SERVICE ORIENTED COMPUTING.

As an example, **Fig. 1** shows a workflow that is designed to be run by the Taverna workflow management system from a workflow library held at myExperiment.org. This workflow is used by bioinformaticians. It obtains genes from a public database (Ensembl), annotates them with Entrez (gene) and UniProt (protein) identifiers, and uses another public database (KEGG) to find the biological pathways for each gene. In this case, the tasks, also known as steps, nodes, activities, processors, or components (depicted as rectangles), represent either the invocation of a remote Web service (the databases), or the execution of a local program. Data flows along data links from the outputs of a task to the inputs of another, according to a predefined graph topology. The workflow defines how the output produced by one task is to be consumed by a subsequent task, referred to as "orchestration" of a flow of data.

A workflow is an accurate description of the scientific plan. The benefits of workflows to scientists are the rapid repeatability of complex operations through automation, an explicit and precisely accurate record of a scientific procedure that leads to more transparent and reproducible science, a way of reusing others' applications, resources, and workflows, and a means to avoid the details of executing a third-party application.

Workflows are executed by a scientific workflow management system (SWfMS). There is no SWfMS adopted by all, and likewise there is no one workflow language for describing workflows. More than 50 different systems are routinely used, either open source software such as Taverna, Triana, Kepler, and Knime, or commercial products such as Pipeline Pilot and InforSense KDE.

Workflow models. The workflow model describes the exact behavior of the workflow when it is executed. The features of different models are dictated by the types of application domains they are designed to support, as well as by the target user community. The models vary in several aspects.

Computation. Dataflow pipelines dictate that each processor be executed as soon as its data inputs are available, and processors that have no data dependencies among each other can be executed concurrently. They are used for integrating data from different sources, data capture, preparation and analysis pipelines, and populating scientific models or data warehouses. Control flows directly dictate the flow of process execution, using loops, decision points, and so on. These are used for controlling multiple sweeps of different parameter settings for simulations. Kepler (software) supports several different models of computation. Others layer control flow on top of data flow or are mono-model.

Resource type. The tasks can be high-level application programs familiar to the scientist, as in the Taverna system, or low-level scheduling and monitoring of jobs on a grid or compute cluster, such as the Pegasus system. Given a workflow description and a set of resource requirements (such as memory, processing power, operating system, and system libraries), Pegasus generates an appropriate, efficient, and executable workflow. The logic design of the workflow is decoupled from its execution environment, so as the pool of available resources changes, reassignment of the resources is hidden from the user.

Compatibility. Tasks may have been designed to be compatible or designed independently and be incompatible without transformation substeps (called shims).

Interactivity. Workflow execution may be wholly automatic or interactively steered by the scientist.

Adaptivity. The workflow design or instantiation may be dynamically adapted "in flight" by the scientist or by the system reacting automatically to changed environmental circumstances.

Abstraction. Most workflow models allow for some form of modular composition of workflows, for example, through hierarchical nesting.

Two scenarios illustrate the range of operational contexts for scientific workflows. The Pegasus system was used to manage computationally greedy earthquake simulations. A few developers in this planned collaborative project produced the necessary workflow designs then used by many scientists. The tasks were from a constrained set of codes incorporated with rigorous preparation procedures, mapped to run over a grid. The workflows, once run, were not repeated. Another project used the Taverna system to investigate parasite resistance in cattle by linking metabolic pathway and public gene expression datasets. The workflow designs were developed directly by a single scientist in an exploratory, iterative way. The components were unprepared, third-party publicly accessible databases executed at their host site. Workflows were constantly rerun as the underlying datasets were updated.

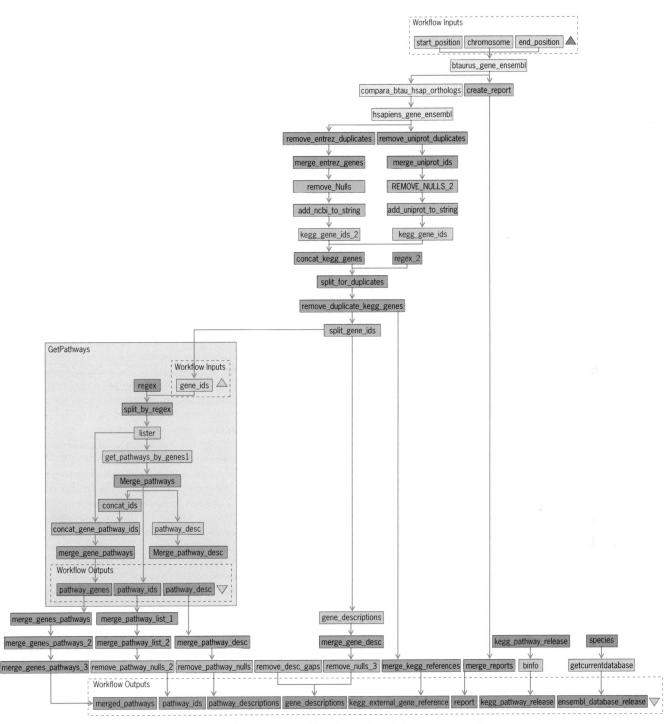

Fig. 1. A Taverna workflow that chains together several searches over different publicly available databases in the life sciences. (*From P. Fisher et al., 2007*)

All SWfMS aim to make their model intuitive for user scientists to understand, and to provide them with a language and facilities for workflow design without requiring a specific competence in computer programming. Striking a balance between simplicity and expressivity remains a challenge. Part of the unpopularity in the sciences of BPEL (Business Process Execution Language), despite its status as an industry standard for orchestrating Web services, is its complexity and the lack of available good, free software for designing its workflows.

SWfMS components supporting the workflow life cycle. SWfMS are software environments that provide users with a number of functionalities to manage the complete life cycle of a workflow (**Fig. 2**). Legacy code or new applications may need to be adapted in preparation for use as components in a workflow.

Workflow design is typically supported through a graphical user interface that allows users to compose workflows, from scratch by wiring together individual components, by using workflow

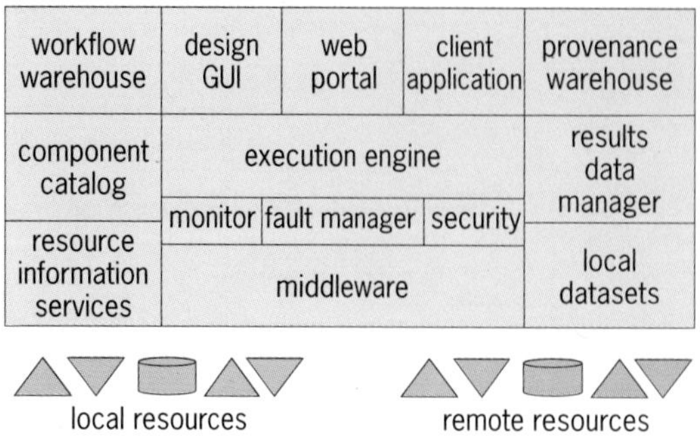

Fig. 2. The high-level components of a scientific workflow management system.

templates, or by reusing and composing existing workflows.

Workflow planning includes the validation of workflow correctness, that is, by static-type checking, allocation of resources to tasks (for instance, processing nodes on a grid cluster), task scheduling, various types of optimizations, and staging of input data prior to its processing on the nodes, as well as delivery of output to the user.

A workflow is executed using an enactment engine, which is a runtime environment similar to those that are normally associated with traditional programming languages. The engine controls the order of invocation of each of the tasks, monitors their execution, reacts in case of failures, manages the data and control flow, and logs operational metrics such as processor execution time and data transfer time.

Postmortem analysis of the workflow execution is designed to help users understand and interpret the results produced by the workflow as well as the details of its execution. Minimally, the SWfMS offers a debugging facility and gathers performance metrics. At a higher level, a provenance subsystem keeps an accurate historical record of the execution of the workflow. Provenance metadata traces the lineage between input data, intermediate data, and final results and keeps an auditable record of the processes used, their configurations, and their order.

Storage, collaboration, and sharing of components, workflows, patterns, and templates is done through libraries, warehouses, or registries. An example of a cross-system, public workflow repository is myExperiment.org, which was inspired by social networking sites. Discovering components and prior workflow designs needs semantically rich metadata.

A SWfMS is typically run over middleware that provides infrastructure for accessing the applications or resources consumed by the workflow and facilities, such as security and access control.

Challenges. A number of aspects of a workflow's life cycle pose challenging problems.

Simplifying design. Designing (reusable) workflows is difficult and takes skill and time. Languages need the salient features of the intended scientific analysis while removing as much of the complexity as possible. One approach introduces multiple process description languages specialized according to different application domains, expressed at a level of abstraction with which users are comfortable, and supported by dedicated visual editing environments. These specifications are then translated into actual executable workflows.

Dynamic adaptation. Dynamic changes to the workflow are necessary to deal with intermittently available resources, failures (especially for workflows that rely on external services for their execution), and user-based decisions that are based on observed intermediate results. Users should be able to re-execute portions of the workflow while modifying the data sources, the execution systems, and the workflow structure itself and exploit opportunities to reuse the results of past task executions rather than running the computation again.

Sharing and publishing. As workflows are rich scientific protocols that represent expert know-how, they need to be pooled, published, and curated just like scientific data, and to accompany the articles that arise from their use for truly reproducible science. Specific problems are managing workflow versions, and making it possible for users to repeat the execution of a workflow and compare results across large time spans (for example, years). Social issues include managing intellectual property and credit and peer review of workflows for accuracy and credibility.

Outlook. The coexistence of multiple workflow management systems is expected to be the norm. Rather than devising a single language or model, the community is working toward operational and provenance interoperability. This raises numerous challenges.

For background information *see* CLIENT-SERVER SYSTEM; DATABASE MANAGEMENT SYSTEM; DISTRIBUTED SYSTEMS (COMPUTERS); METADATA; PROGRAMMING LANGUAGES; SIMULATION; SOFTWARE; WORLD WIDE WEB in the McGraw-Hill Encyclopedia of Science & Technology.

Carole Goble; Paolo Missier; David De Roure

Bibliography. E. Deelman and Y. Gil, Managing large-scale scientific workflows in distributed environments: Experiences and challenges, *e-Science 2006, Second IEEE International Conference on e-Science and Grid Computing*, Amsterdam, December 4–6, 2006 (doi: 10.1109/E-SCIENCE. 2006.261077); D. De Roure, C. Goble, and R. Stevens, Designing the myExperiment virtual research environment for the social sharing of workflows, *e-Science 2007—Third IEEE International Conference on e-Science and Grid Computing*, Bangalore, India, pp. 603–610, 2007 (doi: 10.1109/E-SCIENCE.2007.29); P. Fisher et al., A systematic strategy for large-scale analysis of genotype-phenotype correlations: Identification of candidate genes involved in African trypanosomiasis, *Nucleic Acids Res.*, 35(16):5625–5633, 2007, doi:10.1093/nar/gkm623; Y. Gil et al., Examining the challenges of scientific workflows, *Computer*, 40(12):24–32, 2007, doi:10.1109/MC.2007.421; MyExperiment Virtual Research Environment.

Service-oriented computing

Service-oriented computing (SOC) is the computing paradigm that uses software services (or simply services) as fundamental elements for developing and deploying distributed software applications. Services are self-describing, platform-agnostic computational elements that support rapid, low-cost composition of distributed applications. Services perform functions, which can be anything from simple requests to complicated business processes. They allow organizations to expose their core competencies programmatically via a self-describing interface based on open standards over the Internet (or intranet) using standard, for example, XML-based (Extensible Markup Language), languages and protocols. Because services provide uniform and ubiquitous information distribution for a wide range of computing devices (such as handheld computers, personal digital assistants (PDAs), cellular telephones, or appliances) and software platforms (such as UNIX® or Windows®), they constitute the next major step in distributed computing. *See* SCIENTIFIC WORKFLOWS.

Service clients and providers. Services are offered by service providers and are consumed by service clients. Service providers are organizations that procure the service implementations, supply their service descriptions, and provide related technical and business support. Clients of services can be other solutions or applications within an enterprise or clients outside the enterprise, whether these are external applications, processes, or customers/users.

Service-oriented architecture. Service technologies help create complex customizable composite applications, leveraging any number of back-end and older (legacy) technology systems found in applications either local or remote to an organization. Key to this concept is the service-oriented architecture (SOA). SOA is a logical way of designing a software system to provide services to either end-user applications or to other services distributed in a network, via published and discoverable interfaces. To achieve this, SOA reorganizes a portfolio of previously siloed software applications and support infrastructure in an organization into an interconnected collection of services, each of which is discoverable and accessible through standard interfaces and messaging protocols. Once all the elements of an SOA are in place, existing and future applications can access the SOA-based services, as necessary. This architectural approach is particularly applicable when multiple applications running on varied technologies and platforms need to communicate with each other.

A logical view of the service-oriented architecture is given in the **illustration**, showing the relationship between the SOA roles and operations. First, the service provider publishes its service(s) with the discovery agency. Next, the service client searches for desired services using the registry of the discovery agency. Finally, the service requestor, using the information obtained from the discovery agency, invokes (binds to) the services provided by the services provider.

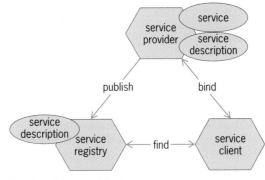

Service-oriented architecture.

In the illustration, the main building blocks of an SOA are the service provider, the service registry, and the service client. The services provider is the organization that owns the service and implements the business logic that underlies the service. From an architectural perspective, this is the platform that hosts and controls access to the service. The services provider is responsible for publishing the services it provides in a service registry hosted by a service discovery agency. This involves describing the business, service, and technical information of the service, and registering that information with the service registry in the format prescribed by the discovery agency.

The service client is the organization that requires certain functions to be satisfied. From an architectural perspective, this is the application that is looking for, and subsequently invoking, the service. The services client searches the service registry for the desired services. This effectively means, for example, discovering a Web service description in a registry provided by a discovery agency and using the information in the description to bind to the service.

The services discovery agency is responsible for providing the infrastructure required to enable the three operations in the service architecture as described in the previous section: publishing the Web services by the service provider, searching for services by service clients, and invoking the services.

Types of services. Services are stateless or stateful. If services can be invoked repeatedly without having to maintain context or state, they are called stateless; services that may require their context to be preserved from one invocation to the next are called stateful. The services access protocol is always connectionless. A connectionless protocol means that the protocol has no concept of a job or a session and does not make any assumptions about eventual delivery. Services come in two types: simple and composite services (commonly known as business processes). Business organizations can use a simple (discrete) service to accomplish a specific business task, such as billing or inventory control.

Simple services are usually stateless. Composite services involve assembling existing services that access and combine information and functions from possibly multiple service providers. Composite services are usually stateful. As an example of a composite service, consider a service that assembles collection of simple services to accomplish a specific business task, such as order tracking, order billing, and customer relationships management, to create a distributed e-business application that provides customized ordering, customer support, and billing for a specialized product line (such as telecommunication equipment and medical insurance).

Web services. Different forms of services are possible because they may be implemented on diverse access devices (such as servers, mobile devices, palmtops, handheld devices, and sensors) and accessed via different types of communications networks, including the Internet, cable UMTS (Universal Mobile Telecommunications System), Bluetooth, and networks.

A particularly interesting case is when the services use the Internet (as the communication medium) and open Internet-based standards. This results in the concept of Web services, which share the characteristics of more general services, but require special consideration as a result of using a public, insecure, low-fidelity mechanism, such the Internet, for distributed service interactions.

A Web service is a specific kind of service that is identified by a URI (Universal Resource Identifier) and exposes its features programmatically over the Internet using standard Internet languages and protocols, and can be implemented via a self-describing interface based on open Internet standards (for example, XML interfaces, which are published in a network-based repositories).

Web services standards are still evolving and are converging on the Simple Object Access Protocol (SOAP) for service communication, Web Services Description Language (WSDL) for service description, Universal Description, Discovery, and Integration infrastructure (UDDI) for registering and discovering services, WS-MetaDataExchange for allowing service endpoints to provide metadata to clients, and the Business Process Execution Language (BPEL) for service composition.

Application reuse and integration. Services provide a distributed computing infrastructure for both intra- and cross-enterprise application integration and collaboration. They help integrate applications that were not written with the intent to be easily integrated with other distributed applications and define architectures and techniques to build new functionality while reusing and subsequently integrating existing application functionality (possibly based on corporate information resources or legacy systems). They appropriately combine functionality internal to an organization with external application fragments possibly residing in remote networks. The unit of reuse with services is functionality that is in place and readily available and deployable as services that are capable of being managed to achieve the required level of service quality. This represents a fundamental change to the socioeconomic fabric of the software developer community that improves the effectiveness and productivity in software development activities and enables enterprises to bring new products and services to the market more rapidly.

To satisfy the above requirements, services should be technology-neutral, loosely coupled, and support location transparency. Services are invocable through standardized lowest-common-denominator technologies that are available to almost all information technology (IT) environments. This implies that the invocation mechanisms (protocols, descriptions, and discovery mechanisms) should comply with widely accepted standards. They must not require knowledge or any internal structures or conventions (context) at either the client or the service side. Services should be accessible by a variety of clients that can locate and invoke the services regardless of their location.

Service interface and implementation. One important aspect of services is that they place a sharp distinction between an interface and implementation part. The service interface part defines service functionality visible to the external world and provides the means to access this functionality. The service describes its own interface characteristics in terms of, the operations available, the parameters, data typing, and the access protocols and in a way that other software modules can determine what it does, how to invoke its functionality, and what result to expect in return. In this regard, services are contractible software modules in that they provide publicly available descriptions of the interface characteristics used to access the service so that potential clients can bind to it. The service client uses the service's interface description to bind to the service provider and invoke its functionality.

The service implementation part realizes a specific service interface whose implementation details are hidden from the users of the service. Different service providers using any programming language of their choice may implement the same interface. One service implementation might provide the functionality itself directly, while another service implementation might use a combination of other services to provide the same functionality.

It is important to distinguish between service interfaces and service implementations, because in many cases the organizations that provide service interfaces are not the same as the organizations that implement the services.

Functional and nonfunctional characteristics. Services are described in terms of a description language. A service description has two major interrelated components: its functional and nonfunctional characteristics. The functional description details the operational characteristics that define the overall behavior of the service, that is, defines the details of how the service is invoked, the location where it is invoked, and so on. This description focuses on details regarding the syntax of messages and how to configure the network protocols to deliver messages. The nonfunctional description concentrates on service quality attributes such as service metering and cost, performance metrics such as the response time or accuracy, security attributes, authorization, authentication, (transactional) integrity, reliability, scalability, and availability. Nonfunctional descriptions force the client to specify nonfunctional requirements that may influence which service provider it may choose.

For background information *see* CLIENT-SERVER SYSTEM; DISTRIBUTED SYSTEMS (COMPUTERS); INTERNET; METADATA; PROGRAMMING LANGUAGES; SOFTWARE; WORLD WIDE WEB in the McGraw-Hill Encyclopedia of Science & Technology.

Michael P. Papazoglou

Bibliography. M. P. Papazoglou, *Web Services: Principles and Technology*, Prentice-Hall, 2007; M. P. Papazoglou and D. Georgakapoulos, Special issue: Service-oriented computing, *Commun. ACM*, 46(10), 2003; Schmelzer et al., *XML and Web Services*, SAMs Publishing, 2002; S. Weerawarana et al., *Web Services Platform Architecture*, Prentice Hall, 2005.

Ship capsizing

Travel by sea is one of the oldest means of transportation, yet each year, vessel instabilities and capsizing claim cargo, vessels, and lives. Capsizing has been a concern since the days of the log canoe, and the study of capsizing is growing ever more interesting as high-speed craft, exceedingly large commercial ships, and innovative hull forms, all with new unique stability characteristics, are developed.

Static stability. Looking at capsizing in a purely static sense, that is, calm seas with no hydrodynamic effects, the forces balancing a ship at a heel angle, that is, the angle where the ship is balanced at some nonzero roll inclination, are buoyancy and weight.

Weight acts from the ship's center of gravity, *G*, towards the center of the Earth, and buoyancy acts upward through the ship's center of buoyancy, *B*, at the center of the displaced volume. By Archimedes' principle, the buoyant force equals the ship's weight for a ship that is neither sinking nor rising. As the ship rolls, the location of the center of buoyancy shifts. **Figure 1** depicts hydrostatic equilibrium for an upright position with no roll angle and a perturbed position with a small roll angle. Figure 1*a* is in static equilibrium as there is no net roll moment, whereas Figure 1*b* has a net restoring moment equal to the buoyant force (weight) times the righting arm, *GZ*, which is the shortest distance from the center of gravity to the line of action of buoyancy. For small roll angles, the line of action of buoyancy acts through a point called the transverse metacenter, *M*. The center of buoyancy, *B*, traces a circular arc centered at *M*, and *BM*, the distance from the center of buoyancy to the metacenter, is called the metacentric radius. The distance from the center of gravity to the metacenter, *GM*, is called the metacentric height. At small angles, a positive *GM* indicates positive initial stability, whereas a zero or negative value for *GM* (that is, center of gravity, *G*, above the metacenter, *M*) would yield neutral or negative initial stability.

At large roll angles the buoyant force no longer acts through the metacenter. As stated previously, the metacenter is defined by the line of action of buoyancy for small roll motions. At large roll, the center of buoyancy no longer traces a circular arc; therefore the line of action of buoyancy will not pass through the same fixed point at large roll angles. Instead, at large angles, we must go beyond the measure of initial stability given by *GM* and consider the righting arm, *GZ*. The plot of the variation of *GZ*

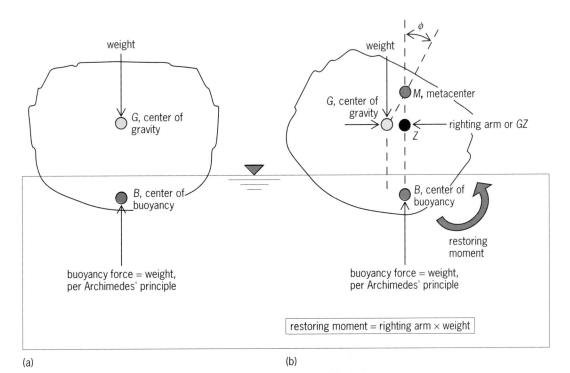

(a)   (b)

Fig. 1. Hydrostatic equilibrium for (*a*) an upright vessel, and (*b*) a vessel heeled to a small angle.

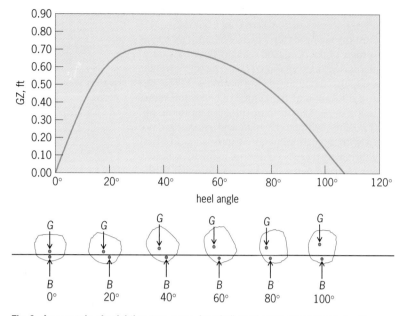

Fig. 2. An example of a righting-arm curve, for a ballasted sailng vessel loaded with passengers. Diagrams below the plot depict the center of gravity (*G*) and the center of buoyancy (*B*) at selected heel positions. 1 ft = 0.3 m.

over a range of roll angles is called the righting-arm curve, which characterizes the static stability of a vessel. **Figure 2** shows a righting-arm curve of a sailing vessel, along with diagrams depicting the center of gravity and center of buoyancy at selected heel positions.

If *GZ* is greater than zero, there is a positive righting moment returning the ship to its upright equilibrium. The slope of the *GZ* curve near the origin will be approximately equal to *GM*. This *GZ* curve shows a linear increase at small roll angles and an angle of maximum *GZ* at approximately 34°. As the roll angle increases beyond 34°, the *GZ* distance begins to decrease until reaching the angle of vanishing stability, which is approximately 107° for this vessel. Beyond the angle of vanishing stability, *GZ* becomes negative and a negative righting moment will force the vessel to an upside-down equilibrium, called capsize.

Dynamic stability. In reality, the effects of waves, hydrodynamic added inertia (A_{44}), and damping ($B_{44,i}$) impact the dynamics of the system. For example, a ship's righting arm when sitting on a wave is different than its righting arm in calm seas. Hydrodynamic added inertia, which can be thought of as the effective inertia of fluid that is accelerated with the body, and damping greatly influence the roll response of a ship. Additionally, coupling between the vessel motions in different degrees of freedom (surge, sway, heave, roll, pitch, and yaw) can result in interesting and unexpected motions. For example, to increase the modeling complexity from the static stability description above, yet still neglect coupling, one might write a single-degree-of-freedom roll and capsize model for a vessel operating in beam seas, that is, waves coming from approximately 90° off the bow of the ship, at the port or starboard side,

in the form given in the equation below. Here, ϕ is

$$(I_{44} + A_{44})\ddot{\phi} + B_{44,1}\dot{\phi} + B_{44,2}\dot{\phi}|\dot{\phi}| + C_{44,1}\phi$$
$$+ C_{44,3}\phi^3 + C_{44,5}\phi^5 = M_4(t)$$

the roll angular displacement, $\dot{\phi}$ and $\ddot{\phi}$ are roll angular velocity and acceleration, and I_{44} is the roll moment of inertia. In this equation, hydrodynamic added inertia and damping come about from the integration of pressure on the hull. Even in the absence of oncoming waves, oscillations of a surface ship will cause waves to radiate away from the body. The evaluation of these pressure loads results in terms proportional to acceleration and velocity, which yield what are physically interpreted as added mass and damping. Besides the hydrodynamic added inertia, A_{44}, and the damping coefficients in roll of order i, $B_{44,i}$ (with $i = 1, 2$), the equation also has stiffness coefficients in roll of order i, $C_{44,i}$ (with $i = 1, 3, 5$), which correspond to the righting-arm curve, *GZ*. The right-hand side of the equation, $M_4(t)$, which represents rolling moment due to the wave action, is often assumed to be sinusoidal, or comprised of a sum of sinusoids with differing amplitudes, frequencies, and phases dictated by an incident wave spectrum. Even the seemingly simple single-degree-of-freedom model for capsize given in this equation can yield chaotic responses. Progressively more complicated models will account for six-degrees-of-freedom motions, nonlinear coupling, waves coming from a range of directions, and so forth.

Further modeling complexity is added through consideration of other extreme weather conditions, and impaired or damaged stability. Vessels operating in freezing conditions may accumulate ice on exposed surfaces including railings, machinery, decks, and fishing equipment (**Fig. 3**). This topsides icing raises the center of gravity of the ship, and can cause an off-center load and moment, both of which increase the risk of capsize.

Other stability impairments include "downflooding," that is, taking on water through either nonwatertight openings or improperly secured hatches. This addition of water will increase draft, reducing freeboard and reserve buoyancy, likely resulting in asymmetric loading and destabilizing free-surface effects. The study of damaged vessel stability is a rich research field with investigators seeking to better understand and model the ingress of water in damaged conditions that again may result in off-center dynamic loading and in reduced reserve buoyancy, waterplane area, and stability.

Mechanisms of capsizing. There are a number of physical mechanisms through which an intact vessel may capsize, ranging from a pure loss of stability to multi-degree-of-freedom dynamic events. Three particularly timely mechanisms of capsize are resonant roll excitation, broaching, and parametric roll.

Resonant roll excitation. Much like a mass on a spring oscillated at its natural frequency, a vessel excited by waves close in frequency to the roll natural frequency may experience large roll response and capsize. Resonant roll motion is most commonly

Fig. 3. Icing aboard the *USCGC Alex Haley* operating in the Bering Sea. (*Photo by Lt. Tyson Scofield*)

seen in beam or stern-quartering seas, where "stern-quartering seas" refers to waves that approach the ship from somewhere between the stern and the port or starboard sides.

Broaching. Broaching is a yaw instability that may result in a rapid change of heading and potentially capsizing. The most common scenario for vessel broaching occurs when a vessel's speed matches the wave speed causing the vessel to "surf." If while surf-riding an unintentional rapid change in yaw and heading occur, the vessel can capsize. This phenomenon occurs as a result of coupled dynamics between surge, sway, yaw, and roll. Any kayakers who have attempted to surf-ride into shore have likely experienced a broach into a capsize at some point in their kayaking careers.

Parametric roll. Parametric roll gained attention after the quite costly parametric roll experience of the post-Panamax container ship the *APL China*. When the wave length is approximately equal to the ship length and the wave encounter period is around half the roll natural period, a Mathieu-type instability may be seen in vessels with large bow flare, large stern overhangs, and relatively wall-sided midsection. That is, the *GZ* curve itself will change, even in the absence of rolling motion. For example, as a ship moves through waves the waterplane area changes when a wave crest is at midships versus wave crests at bow and stern. The parametric resonance is the result of the restoring moment varying in time as a wave crest moves along the length of the ship. Therefore,

while the ship may be in head or following seas with exceedingly small initial roll motions, extreme roll responses can occur.

For background information *see* ARCHIMEDES' PRINCIPLE; BUOYANCY; CHAOS; FORCED OSCILLATION; MECHANICAL VIBRATION; SHIP DESIGN in the McGraw-Hill Encyclopedia of Science & Technology.
Leigh McCue; Charles R. Weil

Bibliography. V. L. Belenky and N. B. Sevastianov, *Stability and Safety of Ships: Risk of Capsizing*, Society of Naval Architects and Marine Engineers, New Jersey, 2007; V. L. Belenky, J. O. de Kat, and N. Umeda, Toward performance-based criteria for intact stability, *Mar. Technol. SNAME News*, 45(2):101, April 2008; M. S. Soliman and J. M. T. Thompson, Transient and steady state analysis of capsize phenomena, *Appl. Ocean Res.*, 13(2):82–92, 1991; Specialist Committee on Stability, Final report and recommendations to the 22nd ITTC, *22nd ITTC Proceedings*, 1999; N. Umeda and M. Hamamoto, Capsize of ship models in following/quartering waves: Pysical experiments and nonlinear dynamics, *Phil. Trans. Math. Phys. Eng. Sci.* 358(1771):1883–1904, 2000.

Simulation verification and validation

With the advent of digital computers in the middle of the twentieth century, simulation joined theory and experiment as an approach for predicting the behavior of engineering and natural systems. Advances in processor speed and in the ability to compute on a large number of processors at once (parallel processing) have resulted in the world's fastest computers doubling their computing power roughly every 14 months. These remarkable advances in computing power have led to an increased reliance on simulation in the decision-making process for researchers, project managers, and policy makers. The weaknesses of modern-day simulations are only now beginning to be understood. This fact should not be surprising given the relatively short 50-year history of simulation versus that of theory (about 300 years) and experimentation (more than 3000 years). Over the last two decades, verification and validation have emerged as a framework for rigorously assessing the accuracy and reliability of computer simulations.

While the term simulation could be broadly interpreted to include a wide variety of activities, this article will focus exclusively on those computational simulations that involve numerical solutions to partial differential equations, which are ubiquitous in the physical sciences and engineering. We further distinguish between a model and a simulation. The term model is used to represent the partial differential equations, initial conditions, boundary conditions, and any auxiliary relations or submodels that are used. Since these models rarely have closed-form mathematical solutions, it is generally necessary to use numerical methods to obtain approximate solutions using computers, that is, simulations.

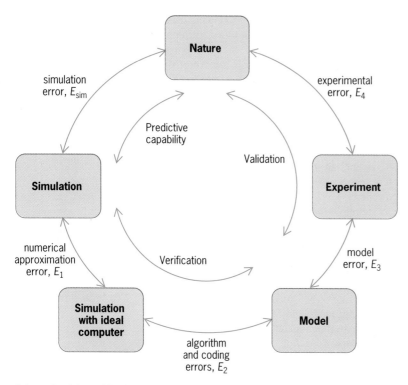

Schematic of the verification and validation process for computational simulation showing the different error sources.

These error sources represent numerical approximation errors (E_1), algorithm and computer programming errors (E_2), modeling errors (E_3), and experimental measurement errors (E_4). In order to estimate the accuracy and reliability of a simulation, each of these different error sources must be estimated (see **illustration**).

Numerical approximation errors (E_1). Solution verification deals with the estimation of the numerical approximation errors in computational simulation. The numerical errors that occur in every single simulation are round-off and discretization errors. Round-off errors are due to the fact that computers are able to retain only a finite number of significant figures, usually either 7 (single precision) or 14 (double precision). Discretization error comes from the fact that the domain of interest (usually a region of space or a period in time) must be broken into discrete subdivisions. For example, a climate simulation might decompose the earth's atmosphere into computational cells that are 15 km wide in the horizontal directions and 1 km high in the vertical direction. The numerical accuracy of the simulations will depend on the resolution (for example, the number of cells) used to decompose the computational domain. After the discretization step, the original partial differential equations have now been converted into a set of algebraic equations that can be solved on the computer. For nonlinear models (and large linear models), these algebraic equations cannot be solved exactly in an efficient manner, and are instead solved using approximate iterative methods. The difference between the current solution during this iterative procedure and the exact solution to the algebraic equations is the iterative error, another source of numerical error.

Discretization error is usually the largest numerical error and the most difficult to estimate. While there are many approaches for estimating discretization error, the simplest and most broadly applicable is Richardson extrapolation, which makes use of the fact that the discretization errors will be reduced as more and more computational cells are used (that is, as the mesh is refined) and as smaller time steps are used. For a representative cell length of h, the discretization errors generally reduce at a rate proportional to h^p, where p is the order of accuracy of the numerical approximation method. If this rate is either known or estimated from solutions on successively refined meshes, then the discretization error can be estimated by extrapolating to the limit as $h \rightarrow 0$. However, when using Richardson extrapolation, it is important to ensure that round-off and iterative errors are much smaller than the discretization errors to ensure that the extrapolation process is reliable.

Algorithm and coding errors (E_2). Ensuring that no algorithm or coding errors are present is called code verification. Algorithm errors can occur when a numerical algorithm is chosen that does not reproduce the exact solution to the model as the mesh is refined (that is, as $h \rightarrow 0$). Algorithm errors are related to algorithm consistency and stability, and can be

Sources of simulation error. In general, one would like to know the error in the simulation relative to the true value in nature. The simulation error can thus be written as Eq. (1), where f_{sim} is the simula-

$$E_{sim} = f_{sim} - f_{nature} \qquad (1)$$

tion result and f_{nature} is the true value found in nature. Knowledge (or estimates) of the simulation error can provide insight into the predictive capability of the simulation. However, this equation does not explicitly show the many possible sources of error, which can be broadly classified into mathematical errors (verification), physical modeling errors (validation), and experimental measurement errors.

In order to identify the different sources of error, it is helpful to expand the simulation error equation as Eq. (2), where $f_{sim}^{C \rightarrow \infty}$ is the result that one could

$$E_{sim} = \left(f_{sim} - f_{sim}^{C \rightarrow \infty} \right) + \left(f_{sim}^{C \rightarrow \infty} - f_{model} \right)$$
$$+ \left(f_{model} - f_{exp} \right) + \left(f_{exp} - f_{nature} \right) \qquad (2)$$

hypothetically obtain on an ideal computer with infinite speed and memory (this allows us to assume that all sources of numerical error related to computer round-off, domain discretization, and so forth are zero), f_{model} is the exact solution to the model, and f_{exp} is the result of an experimental measurement of this system. This expanded form of the simulation error equation highlights the possibility of error cancellation when comparing the results of a simulation to experimental data. A more compact form of the simulation error equation is given by Eq. (3):

$$E_{sim} = E_1 + E_2 + E_3 + E_4 \qquad (3)$$

particularly troublesome, since they can sometimes exhibit the proper convergence rates with mesh refinement while converging to the wrong solution. Coding errors are also called programming mistakes or simply bugs. For more general software applications, system-level testing is used to check whether or not the software produces the correct answer based on a given set of inputs. A key difficulty in applying such general practices to computational simulation software is that the output of the code is only an approximate numerical solution to the model, and thus there is no straightforward system-level software test that can be applied.

Both algorithm and coding errors can be addressed by instead solving a different problem for which the exact solution to the model is known. One approach is to solve a simplified problem for which exact analytic solutions to the partial differential equations are known. A more general approach is the method of manufactured solutions, wherein the differential equations themselves are altered in order to accommodate a chosen exact solution. In either case, the code is verified by examining whether the approximate numerical solutions produced by the code converge to the exact solution to the model at the proper rate (that is, at the formal order of accuracy p of the numerical algorithm).

Model error (E_3). The model error is the difference between the exact solution to the model (that is, the partial differential equation along with the initial and boundary conditions as measured in the experiment) and the experimentally measured value. The comparison between the model and the experimental result is conceptually and mathematically more difficult than estimating E_1 and E_2 for two reasons. First, the experimental result will contain uncertainty, both random (stochastic) measurement uncertainty and systematic (bias) measurement uncertainty. The random measurement uncertainty is usually characterized as a random variable, which can be represented as an empirical distribution function for the measurements. Second, critical experimental information, such as initial conditions, boundary conditions, and system characteristics, must be measured and provided as input data to the model so that a simulation can be conducted. This information is usually characterized by the experimentalist as either a random variable or an interval-valued quantity. These uncertain input data must then be propagated through the model, usually with Monte Carlo sampling, in order to compute the uncertain response quantities from the model for comparison with the experimental measurements. As a result, a quantitative comparison of the model results and the experimental results must deal with a difference measure either between cumulative distribution functions or between probability boxes, one from the experiment and one from the model. This new field of research is now referred to as the construction of validation metric operators. These operators serve a very different purpose from parameter estimation methods used in model calibration or model updating.

Experimental error (E_4). The difference between the experimentally measured value and the true value found in nature is the experimental error, which is the focus of the field of metrology. The value from nature can be considered either as a fixed but unknown quantity or as a random variable that is the result of a stochastic process due to initial and boundary conditions. The experimentally measured value f_{exp} always contains both random and systematic errors due to the measurement process. While the random errors can be better characterized by increasing the number of experimental measurements, it is the systematic errors that are most challenging to characterize. Systematic errors in the experimental measurements are also the most damaging, since they directly affect the model evaluation process or, worse, are transferred to the model during model calibration or model updating procedures. The design of experiments approach has proven very effective in quantifying both random and systematic errors. Using randomization and blocking techniques, one can convert systematic errors into random errors, thereby providing control over experimental error by increasing the number of measurements.

Predictive capability. The errors that arise in each phase of the computational simulation process cannot be computed directly because the ideal computer does not exist and the exact value of the model and the true value in nature are not known. These errors must therefore be estimated, and this estimation process converts these errors into uncertainties. The various sources of uncertainty affect the model's capability to predict the behavior of a system. In addition, the preceding discussion is appropriate for cases in which experimentally measured data are present. However, when computational simulations are used to predict the behavior of a system for cases in which experimental data are not available, then there is an additional source of uncertainty associated with the interpolation between or extrapolation from the model validation database. In addition, the model must also address other possible sources of uncertainty in the prediction, for example, uncertainties in the actual system of interest, its surroundings, possible environments (such as system failure conditions), and possible event scenarios of interest.

For background information *see* ALGORITHM; COMPUTATIONAL FLUID DYNAMICS; COMPUTER PROGRAMMING; DIFFERENCE EQUATION; DIFFERENTIAL EQUATION; EXTRAPOLATION; FINITE ELEMENT METHOD; INTERPOLATION; MODEL THEORY; MONTE CARLO METHOD; NUMERICAL ANALYSIS; PROBABILITY (PHYSICS); SIMULATION; STATISTICS; STOCHASTIC PROCESS; SUPERCOMPUTER in the McGraw-Hill Encyclopedia of Science & Technology.

Christopher J. Roy; William L. Oberkampf

Bibliography. American Society of Mechanical Engineers, *Guide for Verification and Validation in Computational Solid Mechanics*, ASME V&V, 10–2006, 2006; W. L. Oberkampf, T. G. Trucano, and C. Hirsh, Verification, validation, and predictive capability in computational engineering and physics,

Appl. Mech. Rev., 57(5):345–384, 2004; P. J. Roache, *Verification and Validation in Computational Science and Engineering*, Hermosa Publishers, Albuquerque, NM, 1998; C. J. Roy, Review of code and solution verification procedures in computational simulation, *J. Comput. Phys.*, 205(1):131–156, 2005.

Space-beamed solar power

Space-beamed solar power (SSP) is a concept based on a solar power satellite, that is, a solar panel orbiting high above the Earth that would be in almost continuous sunlight. A similar panel on the surface of the Earth would have greatly reduced power output as a result of cloud cover and night time. In geosynchronous Earth orbit (24-h period), a solar power satellite would appear stationary over a point on the Earth's equator, and would beam down power using either microwaves or lasers. On the ground the beam would be converted to usable electricity by a rectifying antenna or rectenna (**Fig. 1**). This system would emit no pollution or waste, and could provide 50–200 times as much power to the grid, for a given land area, than conventional terrestrial solar panels (assuming a cell efficiency of 10%, which is the current efficiency of most mass-produced cells).

The concept of solar power satellites was first published in 1968 by Peter Glaser, who was granted a U.S. patent for the idea in 1973.

Cost of construction and launch. Only approximate analysis has been performed on the costs of the technology, and estimates vary. A rough estimate of spacecraft manufacturing cost might be $1 billion per gigawatt. Launch costs would be additional and, although uncertain, would affect the economics. They depend on payload mass, with estimates ranging from 500 to 5000 metric tons per GW. However, today's launch costs would not apply as the world's launch capacity would increase orders of magnitude, and economies of scale drive down launch costs substantially. The most likely launch profile would involve the satellites injected into elliptical low earth orbit (LEO). Electric propulsion then would slowly and efficiently drive them to the final geostationary (GEO) location.

According to estimates derived from the Federal Aviation Administration Quarterly Launch First Quarter 2008 Report, the best current cost of launch to LEO would be on a Delta-4H class vehicle at approximately $2200/kg. It is entirely possible that this cost could be reduced by a factor of 10 or more using large-scale heavy launch vehicles (**Table 1**).

By comparison, the Three Gorges hydropower facility in China cost $26.5 billion, according to official sources, and has a capacity of 22.5 GW. Some observers claim the cost was actually higher and environmental costs were not accounted for.

In the early phases of deploying SSP systems higher costs would dominate. Once many units had been deployed, economies of scale would reduce both manufacturing and launch costs. In the early phases, SSP would serve remote areas where electricity costs are currently much higher than in urban areas. As SSP costs declined, it would begin to compete with conventional power generation systems.

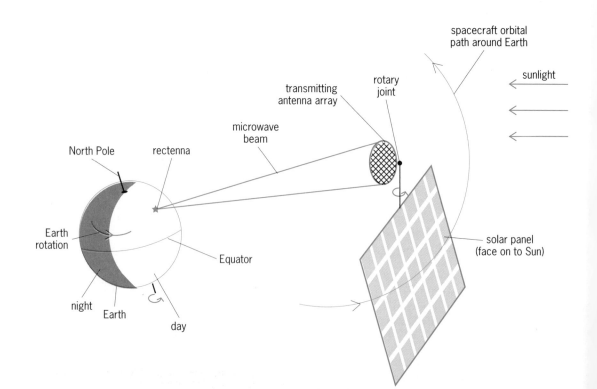

Fig. 1. Solar power satellite system. The microwave beam diverges but is shown here converging because of distance perspective.

TABLE 1. Solar power satellite cost estimates

Transmitted power, GW	Manufacturing cost, US$ billion	Low cost–low mass scenario			High cost–high mass scenario		
		Mass, metric tons	Launch cost, US$ billon	Total cost, US$ billion	Mass, metric tons	Launch cost, US$ billion	Total cost, US$ billion
1	1	500	0.11	1.11	5000	11	12
5.2	5.2	2600	0.572	5.772	26000	57.2	62.4
10	10	5000	1.1	11.1	50000	110	120
31	31	15500	3.41	34.41	155000	341	372
	Cost per kW, US$			1110	Cost per kW, US$		12000

Assumptions
Low launch cost   At 220 $/kg = 220,000 $/ton
High launch cost   At 2200 $/kg = 2,200,000 $/ton

System components. The basic technology for each of the key elements of a space power beaming system has been prototyped and tested in laboratories or field tests. No scientific breakthroughs are required. However, significant research and development would be needed to scale up the hardware, as it would be far larger than any spacecraft built to date. The system would be comprised of a number of components

Solar panel. This would most likely be a large flat array of photovoltaic cells, although thermal heat engines have been considered. A NASA/U.S. Department of Energy (DOE) study in 1980 considered two reference designs, one with silicon-based cells and the other with gallium arsenide cells. If we assume 10% efficient cells [photons to direct current (DC)], then the area of a 10-GW panel would be 75 million square meters or 8.65 km square (90 million square yards or 5.38 miles square). In order to achieve the very low mass required, so-called gossamer spacecraft technologies would be applied, including super thin solar cells. Currently solar cells less than 10 micrometers thick are available. The mass of the solar panel can be further reduced by using even thinner reflective concentrators which could take up several times the area of the heavier photovoltaic panel portions (**Figs. 2** and **3**).

Transmitter. This converts the dc power from the cells to a microwave beam, or alternatively is a laser. Microwave power amplifiers and tubes offer much higher efficiency and less heat dissipation than lasers. However, some lasers are more compact and lighter. The microwave frequency bands of greatest interest currently are two ISM (industrial, scientific, and medical) spectrum allocations at the 2.5- and 5.8-GHz bands. At those frequencies the efficiency of solid-state devices is good: 90% for diodes, 70% for power amplifiers, and 80% for klystron tubes. For a 10-GW solar panel, about 8 GW of microwave energy could be transmitted.

Radio-frequency (RF) antenna or laser mirror. This focuses the beam to make sure that all the radiation of the main spot beam falls within the rectenna collecting area. The wave nature of radiation causes an effect known as diffraction, which creates several secondary cones of radiation of increasing diameter concentric with the main beam known as side lobes.

The power density of the side lobes is much less than the main beam, and usually no effort is made to collect their radiation. If a laser is used, a mirror would be employed rather than an antenna, but the optical equations are fundamentally the same. A laser mirror would be much smaller than a microwave

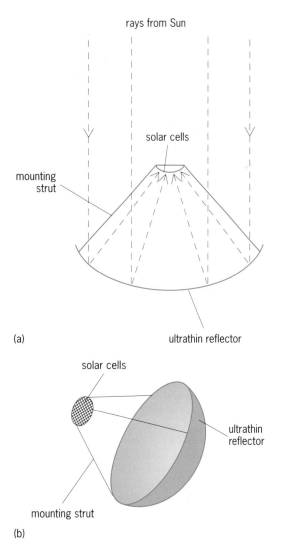

(a)

(b)

Fig. 2. Parabolic solar reflector (Cassegrain configuration). (*a*) **Cross section.** (*b*) **Perspective view. Solar cells are mounted on curved secondary surface to maximize efficiency.**

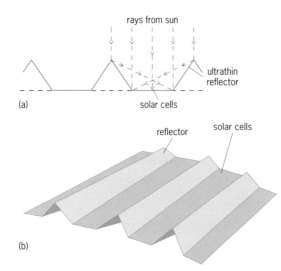

rays from sun

ultrathin reflector

(a)

solar cells

reflector   solar cells

(b)

Fig. 3. Grazing-incidence sunlight concentrator. (*a*) Cross section. (*b*) Perspective view.

antenna because the wavelength of the laser light is much shorter than microwaves. The size of the antenna or mirror is dictated by the Rayleigh diffraction criterion, which directly relates the main beam spreading angle to wavelength and aperture. In most microwave designs the antenna and transmitter portions would be built as a combined phased array of thousands of synchronized transmitter elements.

Steering system. In a manner similar to communications satellites, the solar panel would be maintained face-on to the Sun for maximum power output, while the microwave antenna or laser mirror would be rotated at a rate of one revolution every 24 h to point constantly at a fixed point on the Earth.

Ground element. The ground element would be either a rectenna (rectifying antenna) for a microwave system or photovoltaic (PV) cells for a laser system. The microwave rectenna receives the beam and converts the radiation to dc electric power, which can then be converted to alternating current (ac) and fed into a power grid. The rectenna comprises an array of antenna elements, typically half-wave dipoles, which feed into a set of rectifiers. The rectifiers are passive electronic circuits using high-frequency Schottky barrier diodes. The diodes can be silicon based for low-density microwaves or gallium arsenide for high density. If a laser beam is used instead of microwaves, the optical energy is converted to dc using either a band-gap matched photovoltaic array or a heat engine.

Conversion efficiency for a well-optimized microwave rectenna can reach 90%. Conversion of laser light would be much less efficient, around 20% using specially built PV cells, or less using thermal conversion. Lasers over wide frequency bands cause damage to the human eye, including blindness. "Eye safe" lasers are restricted to certain frequencies and intensity levels.

System assembly and construction. The system would be prefabricated in modules on Earth and assembled in orbit. These building blocks launched

into low Earth orbit would be combined into subassemblies, which then would be boosted from LEO to GEO by low-thrust electric propulsion. Deployment of the delicate ultrathin subpanels of solar cells would be done in orbit to avoid damage during launch. Then the subpanels would be connected together to form the final array. Similarly, the phased-array transmitter would be launched as modules, then assembled and calibrated in GEO. Robotic technology has advanced to the point where few if any astronauts would be involved in the space assembly process, and no orbital space station is needed.

System size. The size of the system depends on two principal factors: the frequency of the radiation for the power beam and the maximum allowable intensity of the beam at the rectenna, which is primarily a human-exposure safety consideration. The Rayleigh diffraction criterion determines the minimum size of the transmitter antenna on the spacecraft; the shorter the wavelength, the smaller the transmit antenna. This is expressed for small angles by the equation below where *a* is the beam spread-

$$a/2 = 1.22\,\lambda/D$$

ing angle in radians, λ is the wavelength, and D is the aperture of the antenna.

The size of the rectenna is dictated by a rule of thumb that the rectenna diameter should be approximately ten times the diameter (100 times the area) of the spacecraft transmitter antenna. This is because the space hardware is generally two orders of magnitude more expensive than ground hardware. Combining the "10 times rule" with the diffraction criterion results in the sizes shown in **Table 2**.

System performance. The output power of the rectenna is dictated by two factors: the intensity of the beam and the efficiency of the antenna elements and the rectifier circuits.

The restrictions for the beam intensity depend on the safety rules that apply. Microwave exposure rules vary widely from country to country. The United States Federal Communications Commission (FCC) rules equate to 5 mW/cm^2 in the range of interest, from 2 to 6 GHz, for controlled exposure averaged over 6 min. However, FCC rules provide for unlimited power intensity if physical barriers are present that eliminate the possibility of people entering the beam. How such rules would be applied to space-beamed power is currently a subject of debate.

In the 1980s it was thought that ionospheric heating would limit the acceptable intensity to 230 W/m^2

TABLE 2. Antenna and rectenna sizes for GEO			
Frequency	Transmit antenna	Rectenna diameter	Rectenna area
2.5 GHz	1.02 km diameter	10.2 km	82 km^2
5.8 GHz	0.67 km diameter	6.7 km	35 km^2

TABLE 3. Rectenna power output

Rectenna size		Intensity: 10W/m²		Intensity: 100W/m²	
Diameter, km	Area, km²	Power, GW	90% power, GW	Power, GW	90% power, GW
10.2	82	0.82	0.74	8.2	7.4
6.7	35	0.35	0.32	3.5	3.2

at 3 GHz. However, more recent testing has shown that the intensity could easily be doubled, and at 5.8 GHz the ionospheric heating is further reduced by 80%.

Table 3 shows power output for two different beam intensities. Thus, a total of 2439 satellites at 2.5 GHz (or 5673 satellites at 5.8 GHz) could supply the global energy production of 18 terawatts. In practice a mix of space power and terrestrial power is more likely.

Advantages and disadvantages. Energy costs around the world continue to increase unabated. Oil production is projected to peak and decline in the next decade or two. After that, other fossil fuels will continue to be the cheapest forms of energy for the rest of this century and into the next. These include coal and natural gas, and possibly offshore methane hydrates. However, there is increasing concern about the unprecedented acceleration in carbon emission from fossil fuels, and there is strong interest in alternatives. Currently, nuclear power is the only alternative to fossil fuels that could be deployed on a global scale, but that has long-term security and waste-disposal concerns. Terrestrial renewable energy sources continue to be expensive, provide undependable fluctuating power output, and are limited by geography in the scale at which they can be deployed, so space-based alternatives are worth considering.

The cost to launch hardware into space from Earth is still relatively expensive, and has hitherto prevented the large-scale development of space solar power. Several factors are making the concept more attractive; principally, the increasing costs of energy, improved lightweight materials, advances in microwave technology, and declining costs of space launch. Furthermore, there is concern amongst governments about heavy reliance on foreign sources of energy, and some are considering unprecedented investments to assure more secure domestic energy sources.

Unlike nuclear and fossil-fuel systems, space-beamed power would not generate any pollutants or hazardous waste. Fossil fuels emit enormous quantities of carbon dioxide (CO_2), which is widely suspected of causing global climate change. Space-beamed solar power could replace fossil fuel plants and eliminate the CO_2 emissions.

Space-beamed power offers advantages over terrestrial solar panels. On Earth, solar panels are usually fixed and do not track the Sun, so lose 50% of the energy, but a space solar panel is weightless, and can easily track the Sun. Also, because there is no wind or weather outside the atmosphere, the space panel can be very light and flimsy. The wide variations of power from terrestrial solar panels make them unsuitable for baseload grid supply, unless expensive storage systems are employed. Such storage systems usually rely on limited-life batteries such as lead-acid or nickel-cadmium types, which represent a major toxic waste disposal problem. Power beamed from space would be steady and continuous, unaffected by weather conditions, and with no variation during the day or outages at night. Hence the high cost of terrestrial power storage would avoided.

For terrestrial solar panels to supply the needs of the world, large land areas would have to be covered, and the loss of sunlight beneath the panels would destroy arable land and plant habitat on a large scale. Relative to terrestrial solar panels, microwave rectennas for space-beamed power would occupy much less area, and would be mostly transparent to sunlight.

In 2008, continuous global energy production rose to about 18 TW. To provide this amount using solar panels on Earth would require an area of about 2 million square kilometers (820,000 square miles), or about 4% of the Earth's land area. This assumes static panels that are 10% efficient, and alternating cloudy and sunny days averaged over the Earth.

The equivalent area of microwave rectennas for space-beamed power depends on the intensity used. The relevant FCC safety limit for uncontrolled public exposure to microwaves (OET Bulletin 65, 1997) is 1 mW/cm² or 10 W/m² averaged over 30 minutes. At 10 W/m², a total rectenna area of 200,000 square kilometers would be required, which would be 10% of the area of comparable terrestrial solar panels.

If public access is prevented by physical barriers, then the FCC permits higher power levels. At 100 W/m², 20,000 square kilometers would be required, or about 1% of the area of comparable terrestrial solar panels. If a beam density of 1000 W/m² were permitted (still less than sunlight), 0.1% of the terrestrial solar panel area would suffice. By comparison, the area of the state of New York is 122,000 square kilometers (47,000 square miles).

Current outlook. Research on solar power satellite technology continues at a minimal funding level. Most of the work currently is being performed in Japan and by the European Space Agency. In the United States, since 1997 there has been no specific budget for the technology in any federal agency. The only major study performed in the United States was in 1978–1980 when NASA and the DOE conducted a high-level evaluation. Then in 1997 NASA performed a smaller study called the "Fresh Look." In 2007 the National Space Security Office, within the

U.S. Department of Defense, performed an unfunded review of the state of the art and published an interim report. All the studies performed to date have been promising and showed that the concept is technically feasible, and offers potential advantages and benefits over other sources of energy.

For background information *See* DIFFRACTION; ELECTRIC POWER GENERATION; LASER; MICROWAVE; PHOTOVOLTAIC CELL; SATELLITE (SPACECRAFT); SOLAR CELL; SOLAR ENERGY; in the McGraw-Hill Encyclopedia of Science & Technology.   Charles Frank Radley

Bibliography.   P. E. Glaser, Power from the Sun: Its future, *Science*, 162:857–861, 1968; P. E. Glaser, Method And Apparatus For Converting Solar Radiation To Electrical Power, U. S. patent 3,781,647 December 25, 1973; G. A. Landis, *Reinventing the Solar Power Satellite*, NASA TM-2004-212743, February 2004.

Space flight

Space flight in 2007 moved forward in three dominant themes: commercial utilization of low and geosynchronous orbits, science missions into the solar system, and expansion of human presence in space toward exploration further outward from Earth's boundaries. Based on the number of launches to orbit plus the number of launched satellite payloads, the utilization of space, which had reached its lowest level since 1961 in 2004, remained at that level in 2005 and showed signs of reversing this trend in 2006, and in 2007 increased markedly. After staying at 55 total space launch attempts worldwide for 2 consecutive years, then climbing to 66 in 2006 (including four failed), the number of launches in 2007 reached 68 (including 3 failed attempts). The number of large commercial satellites (31) launched into space, including geostationary orbit (GEO), exhibited a robust rise for the second year (2006: 21, 2005: 19, 2004: 12), after a stagnant period during 2002–2004.

The National Aeronautics and Space Administration (NASA) continued building the *International Space Station (ISS)*. The space shuttle flew three highly successful missions to continue the station's assembly, and construction began on projects designed to send astronauts to the Moon, where they will establish a permanent outpost and prepare for eventual voyages to Mars. Space science missions were launched to Mars and the asteroid belt. Closer to home, Earth science satellites made a number of key discoveries, such as how waterways beneath an Antarctic ice stream affect sea level and the world's largest ice sheet.

As the United States space budget continued to stay its course on a relatively stable level with some upward adjustment for inflation, a major focus was the return to routine operations for the the space shuttle in steady support of the *ISS*. Russian launch services continued to dominate human flights, as in the previous year. International space activities extended their trends of modest public spending and

cautiously increasing launch services. Of the three failed launches (down from four in 2006; 2005: 3; 2004: 2), one was Russian (a *Proton-M*), one U.S. military and commercial (*Falcon-1*), and one a Sea Launch *Zenit 3SL* (60% foreign owned, non-U.S. carrier).

NASA went into the third year implementing the *Vision for Space Exploration*, mandated by President George W. Bush on January 14, 2004, America's long-term plan for returning astronauts to the Moon to prepare for voyages to Mars and other destinations in the solar system. The year included three space shuttle flights, further progress in preliminary design and procurement for America's next-generation spacecraft, and a number of scientific milestones. The fifth, sixth, and seventh shuttle missions to the *ISS* since the *Columbia* loss in 2003 continued a vigorous program of assembling the space station with delivery of large components and supporting spacewalks by crewmembers. Launched in August 2007, NASA's *Phoenix* spacecraft set out on a 8-months voyage to Mars, for a first-time landing in the north-polar region of the Red Planet in May 2008. In September, NASA's *Dawn* spacecraft, the first purely scientific U.S. mission powered by solar electric ion propulsion, set out on a deep-space voyage to rendezvous and orbit asteroids 4 Vesta and 1 Ceres located in the asteroid belt between Mars and Jupiter. On the Mars surface, the twin rovers *Spirit* and *Opportunity*, having long exceeded their expected design life, continued their exploration, amazing scientists with their findings. The *Cassini* spacecraft made more history in the planetary system of Saturn, the *Spitzer Space Telescope* continued its deep probes of the far reaches of the cosmos, and the *Hubble Space Telescope* added to its wealth of unique discoveries as another shuttle mission to service and upgrade its onboard systems, the fourth, was being readied by NASA.

The commercial space launch market dipped below the 2006 level, which had stayed below 2005 after some initial improvement following a 2004 decline, begun in 2003. Of the 65 (2006:62; 2005: 52) successful launches worldwide, 20 (31%) were commercial launches (2006: 21/34%, 2005: 23/44%, 2004: 19/36%, 2003: 20/33%) carrying 26 commercial payloads, compared to 33 in 2006 and 36 in 2005. In the civil science satellite area, worldwide launches totaled 10 (plus 16 microsats launched by Russia), two less than in the preceding year. NASA expanded cooperation with privately owned companies entering into the commercial space transportation enterprise, signing unfunded agreements with five companies and entered into funded agreements with two companies.

Russia's space program showed continued dependable participation in the buildup of the *ISS*. Europe's space activities in 2007 increased the number of flights of the Ariane 5 heavy-lift launch vehicle from five the previous year to six, bringing the number of successes of this vehicle to 36. China launched ten Long March rockets, four more than in 2006, Japan had two launches, four less than in 2006, India

TABLE 1. Some significant Space Events in 2007

Designation	Date	Country	Event
Progress M-59/24P	January 18	Russia	Crewless logistics cargo and resupply mission to the *International Space Station* (*ISS*) on a Soyuz-U rocket.
THEMIS 1-5	February 17	United States	Five identical "Time History of Events and Macro-scale Interactions during Sun-storms" probes to uncover the physics that powers the Earth's polar auroras.
Soyuz TMA-10 IISS-14S	April 7	Russia	Ninth *ISS* crew rotation flight on a Soyuz-FG, bringing the Expedition 15 crew of Fyodor Yurchikhin and Oleg Kotov, plus 5th SFP (Space Flight Participant) Charles Simonyi.
AIM	April 25	United States	"Aeronomy of Ice in the Mesosphere" satellite launched on a Pegasus-XL into a 600-km (373-mi) orbit to study noctilucent or "night-shining" clouds (NLCs).
Progress M-60/25P	May 12	Russia	Crewless logistics cargo/resupply mission to the *ISS*, on a Soyuz-U rocket.
STS-117 (Atlantis)	June 8	United States	*ISS* Mission 13A with the S3/S4 truss elements for assembly, *ISS* crewmember Clay Anderson to replace Sunita Williams on Expedition 15. Duration: 13d 20h 12 min.
Genesis-2	June 28	Russia	Demonstration flight of Bigelow Aerospace inflatable module for a future space station, on a Dnepr rocket.
Progress M-61/26P	August 2	Russia	Crewless logistics cargo and resupply mission to the *ISS*, on a Soyuz-U rocket.
Phoenix	August 4	United States	Launch of new Mars lander using Surveyor 2001 parts on a Delta-2, for a first-time landing in the north-polar region of the Red Planet in May 2008.
STS-118 (Endeavour)	August 8	United States	*ISS* Mission 13A.1 with a new truss element (P5) to resume station assembly. Crew included teacher-turned-astronaut Barbara Morgan. Mission duration: 12d 17h 57min.
Kayuga	September 14	Japan	Launch of SELENE (SELenological and ENgineering Explorer) Moon probe on an H-2A carrier.
Dawn	September 27	United States	Launch of NASA's ion-propelled probe on a Delta-2 to explore the asteroid belt, on a 4.8-billion-km (3.0-billion-mi) flight targeted for Vesta (2011) and Ceres (2015).
Soyuz TMA-11 IISS-15S	October 10	Russia	Tenth *ISS* crew rotation flight on a Soyuz-FG, bringing the Expedition 16 crew of Peggy Whitson and Yuri Malenchenko, plus the 6th SFP Muszaphar Shukor (Malaysia).
STS-120 (Discovery)	October 23	United States	ISS Mission 10A with the Node-2 "Harmony" module, *ISS* crewmember Dan Tani to replace Clay Anderson on Expedition 16; four EVAs. Total mission duration: 15d 2h 24min.
Chang'e-1	October 24	China	Launch of China's first probe to explore the Moon from lunar orbit, on a CZ-3 rocket.
Progress M-62/27P	December 23	Russia	Crewless logistics cargo/resupply mission to the *ISS*, on a Soyuz-U rocket.

accomplished three launches, and Israel launched one rocket.

The total number of people flown into space since 1958 (counting repeaters) in 2007 rose to 1039, including 113 women, or 463 individuals (46 women), in a total of 257 missions. Some significant space events in 2007 are listed in **Table 1**, and the launches and attempts are enumerated by country in **Table 2**.

TABLE 2. Successful launches in 2007 (Earth-orbit and beyond)

Country of Launch	Number of launches (and attempts)
Russia	25 (26)
United States (*NASA/DOD/commercial*)	18 (19)
People's Republic of China	10 (10)
Europe (*ESA/Arianespace*)	6 (6)
Japan	2 (2)
India	3 (3)
Israel	1 (1)
Zenit-3SL Sea Launch	— (1)
Total	65 (68)

International Space Station

During 2007, the *International Space Station* (*ISS*) marked the seventh anniversary of continuous crewed operations, during which NASA and *ISS* partner scientists have gathered vital information on the station that will help with future long-duration missions of the new exploration program, as the station's unique microgravity environment cannot be duplicated on Earth.

ISS goals are to establish a permanent habitable residence and laboratory for science and research, and to maintain and support a human crew at this facility. To those purposes, the *ISS* expands our experience in living and working in space, encourages and enables commercial development of space, and provides the capability for humans to perform unique long-duration space-based research in cell and developmental biology, plant biology, human physiology, fluid physics, combustion science, materials science, and fundamental physics. The *ISS* is already providing a unique platform for making observations of the Earth's surface and atmosphere, the Sun, and other astronomical objects.

The station is the largest and most complex international scientific project in history. The international partnership, drawing upon the scientific and technological resources of 16 nations [the United States, Canada, Japan, Russia, 11 nations of the European Space Agency (ESA), and Brazil], has together assembled a research facility, designed and produced in locations around the world, that now resides some 400 km (250 mi) above the Earth. The completed *ISS* by late-2010 will have a mass of 420 metric tons (925,627 lb), measuring 110 m (361 ft) end-to-end and 88 m (290 ft) long, with ~4000 m^2 (almost an acre) of solar panels to provide up to 110 kW power to six state-of-the-art laboratories.

Early in 2003, progress in *ISS* assembly was brought to a halt by the standdown of the space shuttles after the loss of the space shuttle *Columbia*. As an immediate consequence of the unavoidable reduction in resupply mission to the station, which now could be supported only by Russian crewless automated Progress cargo ships, station crew size was reduced from a three- to a two-person "caretaker" crew per expedition (also known as Increment), except for brief 10-day stays by visiting cosmonaut/researchers or commercial "tourists" arriving and departing in the third seat of Soyuz spacecraft. During 2006, as *ISS* entered its seventh year of operations as a staffed facility, three-person operation of the station was resumed with mission STS-121/*Discovery*, the first shuttle flight test of new inspection and protection techniques and systems, which also delivered the first European long-duration crewmember, to join the Expedition 13 crew, who had arrived in March 2006.

During 2007, there were nine flights to the *ISS*, including two crewed Soyuz missions and four Progress cargo flights. NASA launched three successful space shuttle missions in June, August, and October to deliver pieces of the *ISS*, allowing it to grow in size, volume, and power production. The electricity generated by the station and used aboard the outpost more than doubled. The station's six solar panels extended to ~2000 m^2 (more than half an acre) of surface area. NASA astronauts and Russian cosmonauts safely conducted 23 spacewalks devoted to building and maintaining the station.

Progress M-59. Designated *ISS-24P*, the first of four crewless cargo ships to the *ISS* in 2007 lifted off on a Soyuz-U rocket at the Baikonur Cosmodrome in Kazakhstan on January 17 and docked at the station on January 19. The drone carried ~2400 kg (5300 lb) of cargo for the *ISS* crews, consisting of propellant for the Russian thrusters, oxygen, spare parts, experiment hardware, and life-support components including food portions.

Soyuz TMA-10. *Soyuz TMA-10*/ISS-14S (April 7–October 21) lifted off at Baikonur on a Soyuz-FG rocket carrying Russian Expedition 15 crewmembers. The ninth *ISS* crew rotation flight by a Soyuz, it arrived at the station on April 9, docking smoothly at the Functional Cargo Block (FGB) nadir port. Twelve days later (April 21), the previous crew return vehicle (CRV), *Soyuz TMA-9*/13S, undocked from the Service Module (SM) aft port for a safe landing in Kazakhstan.

Progress M-60. Designated ISS-25P, the second of four crewless cargo ships to the *ISS* in 2007 lifted off on a Soyuz-U rocket at Baikonur on May 11, docking at the station on May 15. The resupply ship delivered 2325 kg (5125 lb) of cargo for the station crews, including propellant for the Russian thrusters, freshwater, oxygen, spare parts, repair gear, life support, and science experiment hardware as well as food.

Progress M-61. ISS-26P was the next crewless cargo ship, launched in Baikonur on a Soyuz-U rocket on August 2 and arriving at the station on August 5. The 26P drone delivered about 2.5 tons of cargo, including freshwater, oxygen, and food.

Soyuz TMA-11. *Soyuz TMA-11*/ISS-15S (October 10, 2007–April 19, 2008) was launched in Kazakhstan. The tenth crew rotation flight by a Soyuz, it carried Expedition 16 crewmembers from the United States and Russia plus Malaysian Sheik Muszaphar Shukor Al Masrie, the 13th Visiting Crewmember (VC) to visit the space station, flying under a Russia/Malaysia government-to-government offset agreement. *TMA-11* docked to the *ISS* on October 12, replacing the previous CRV, *Soyuz TMA-10*/14S, which undocked on October 21 at 3:14 am with Expedition 15 crewmembers plus the Malaysian Visiting Crewmember. Touchdown occurred ~340 km (210 mi) short (west) of the intended site near Arkalykh. The trajectory undershoot of the returning Soyuz Descent Module was due to a switch of the onboard computer to the (secondary) Ballistic Descent Mode (BS).

Progress M-62. ISS-27P, the fourth automated logistics transport in 2007, lifted off on its Soyuz-U rocket on December 23, docking at the *ISS* on December 26 with about 2.5 tons of cargo for the *ISS* crews.

United States Space Activities

Launch activities in the United States in 2007 showed a decline from the previous year. There were 18 NASA, Department of Defense, and commercial launches out of 19 attempts [2006: 22 of 23 attempts, 2005: 16 of 16, 2004: 19 of 19; 2003: 26 of 27 (loss of *Columbia*)].

Space shuttle. After the loss of Orbiter *Columbia* on the first (and only) shuttle mission in 2003 and the return of the space shuttle to the skies in Summer 2005 with the liftoff of STS-114/*Discovery*, more work was required on redesign of the foam-based insulation on the space shuttle's external tank, new sensors for detailed damage inspection, and a boom to allow astronauts to inspect the vehicle externally during flight in 2006. After another launch delay early in 2007 due to hail storm damage to the shuttle *Atlantis* on the launch pad, flights resumed in June to continue *ISS* construction assembly at a brisk rate. 2007 saw three shuttle flights, bringing the total number of shuttle missions since inception to 120.

STS-117. *Atlantis*, on its 28th mission, lifted off on June 8 on Mission ISS-13A, carrying the crew of Commander (CDR) Rick Sturckow, Pilot Lee Archambault, Mission Specialists Jim Reilly, Patrick Forrester, Steven Swanson, John "Danny" Olivas, and Flight Engineer (FE) Clay Anderson who replaced *ISS*

Expedition 14/15 FE Sunita Williams. Docking at the station was on June 10, and the station crew increased to ten.

Atlantis delivered two major truss segments, S3/S4, to the station. Their installation plus other maintenance, servicing, and inspection tasks were accomplished by the crew in four spacewalks (EVAs or extravehicular activities). A difficulty arose during the mission with the Russian computers that provide backup attitude control and orbital altitude adjustments. Russian specialists worked with U.S. teams, troubleshooting and restoring computer capabilities.

After undocking on June 19 (10:42 am), *Atlantis* spent some time flying solo, and then landed on June 22 at Edwards Air Force Base (EAFB) in California, concluding a 13-day, 20-h, 12-min flight covering 9.3×10^6 km (5.8×10^6 mi). The landing was diverted to California due to marginal weather at Kennedy. Williams became a new record holder for women in space, with 194 days, 18 h, 58 min.

STS-118. Shuttle *Endeavour* lifted off on August 8 on *ISS* Mission 13A.1, carrying a crew of seven. The launch returned *Endeavour* to active service after a 3-year hiatus for major modifications. The work, done at Kennedy Space Center, included addition of a "glass cockpit," a Global Positioning System (GPS) for landing, and the Station-to-Shuttle Power Transfer System (SSPTS), which allows the Orbiter to draw power from the space station, enabling an extended stay for the mission, 14 days in this case.

The payload comprised the S5 truss, a SPACEHAB module in the cargo bay, and the external stowage platform no. 3 (ESP-3) with a replacement control moment gyroscope (CMG). Four EVAs were conducted to install the S5 truss segment and other outboard equipment. The spacewalkers also replaced the CMG for a faulty one (CMG-3) on the *ISS*.

After undocking on August 19 (7:57 am EDT), *Endeavour* landed at Kennedy Space Center on the first opportunity after deorbit. It was the 119th shuttle mission, the 20th of *Endeavour*, and the 65th landing at Kennedy Space Center.

STS-120. Shuttle *Discovery* launched on October 23 on ISS Mission 10A, the 23rd assembly flight. *Discovery* docked on October 25, smoothly flown by Commander Pamela Melroy and Pilot George Zamka. Melroy and ISS Expedition 16 CDR Peggy Whitson made history in becoming the first female spacecraft commander to lead shuttle and station missions simultaneously (Whitson also holds the distinction of being the first woman to command a station mission). At the *ISS*, the STS-120 crew continued the construction of the station with the installation of the Node-2 module, named "Harmony," and the relocation of the P6 truss segment. There were four EVAs. During the second spacewalk, a $360°$ visual inspection of the station's starboard solar alpha rotary joint (SARJ) was added, which had shown increased friction for some time. An extra day was added to the mission between the fourth and fifth spacewalks to provide the crew off-duty time and equipment preparation for the fifth EVA. However,

the objective of the fourth spacewalk changed in order to repair a solar array, inadvertently torn during deployment, and the fifth spacewalk was transferred to the station crew to perform after the shuttle's departure.

Before the undocking on November 5 crew members transferred equipment and scientific samples to the shuttle, including the metal shavings from the SARJ for engineers to study and try to determine the cause of resistance in the starboard rotary joint.

Discovery landed on November 7. Mission elapsed time for 10A was 15 days, 2 h, 24 min, and 2 s, covering 10.1×10^6 km (6.25×10^6 mi). It was the 120th shuttle mission, the 34th flight of *Discovery*, and the 66th landing at Kennedy Space Center.

Advanced transportation systems activities. In 2007, NASA began laying the foundation for the future of space exploration. Construction projects across the agency supported the Constellation Program, which is developing next-generation spacecraft and systems to return astronauts to the Moon by 2020. All major contracts for the Ares I rocket were awarded in 2007. Construction got under way at the U.S. Army's White Sands Missile Range (WSMR) in Las Cruces, New Mexico, where NASA will hold the Constellation Program's first flight tests in 2008. At Kennedy Space Center workers erected a new lightning protection system at the Constellation launch pad. A new test stand for rocket engines is being built at NASA's Stennis Space Center (SSC) in Mississippi. In July, NASA signed a $1.2 billion sole-source contract with Pratt & Whitney Rocketdyne for the development of the J-2X engines that will power the upper stage of the Ares I crew launch vehicle and Ares V, its heavy-lift follow-on. In July, Boeing Space Exploration of Houston won a $515 million contract to produce the upper stage of the Ares 1, and in December Boeing also landed the $800 million contract for building and outfitting the avionics ring that will control the Ares I in flight. The *Orion* Crew Exploration Vehicle, which sits atop Ares I, is being built by Lockheed Martin Space Systems of Denver, put under contract in August 2006.

Global Exploration Strategy (GES). In May, 14 space agencies interested in working together on space exploration met in Italy and signed the Global Exploration Strategy, to determine, on a voluntary basis for each member, how NASA's Vision for Space Exploration, mandated by President George W. Bush in January 2004, Europe's Aurora program, and similar initiatives in Russia, China, India, Japan, and elsewhere can be melded into a cohesive global effort. The resulting document, "The Global Exploration Strategy: The Framework for Coordination" reflects a shared vision of space exploration focused on solar system destinations where humans may someday live and work.

Commercial Orbital Transportation Services (COTS). Through five new Space Act agreements, NASA signed unfunded agreements with SpaceDev, SPACEHAB, Transformational Space, PlanetSpace, and Constellation International, which indicated interest in developing orbital cargo transportation capabilities. Two funded agreements were furthermore signed with SpaceX (Space Exploration Technologies) and Rocketplane Kistler (RPK) for demonstrations under the COTS competition. As part of the COTS demonstrations, SpaceX successfully completed the Critical Design Review (CDR) for its first Falcon 9/Dragon mission.

Space sciences and astronomy. In 2007, the United States launched eight civil science spacecraft (one less than in the previous year): *THEMIS 1*, *THEMIS 2*, *THEMIS 3*, *THEMIS 4*, *THEMIS 5*, *AIM*, *Phoenix*, and *Dawn*.

AIM. The *AIM* (Aeronomy of Ice in the Mesosphere) satellite mission was launched on April 25 from Vandenberg Air Force Base (VAFB), Calif., on a Pegasus-XL launch vehicle to its orbit 600 km (373 mi) above Earth. *AIM* is the first satellite mission dedicated to the study of noctilucent or "nightshining" clouds (NLCs), also called polar mesospheric clouds (PMCs). At the end of the year, it had provided the first global-scale view of the clouds over the entire 2007 Northern Hemisphere season with an unprecedented resolution of 5 km by 5 km (3 mi by 3 mi) and was nearing completion of observations in the Southern Hemisphere season. Despite a significant increase in PMC research in recent years, relatively little is known about the basic physics of these clouds at "the edge of space" and why they are changing. They have increased in brightness over time, are being seen more often, and appear to be occurring at lower latitudes than ever before. The overall goal of the baseline mission is to determine why PMCs form and why they change. *AIM* is expected to last 2 or more years, during which time the instruments will monitor noctilucent clouds to better understand their variability and possible connection to climate change.

Dawn. After lifting off on a Delta-2 launch vehicle at Cape Canaveral Air Force Station in Florida on September 27, NASA's *Dawn* spacecraft began a 4.8×10^9-km (3×10^9-mi) odyssey deep into the asteroid belt, including exploration of asteroid Vesta in 2011 and the dwarf planet Ceres in 2015. *Dawn*'s science instrument suite will measure elemental and mineral composition, shape, surface topography, and tectonic history, and will also seek water-bearing minerals. In addition, the *Dawn* spacecraft and how it orbits Vesta and Ceres will be used to measure the celestial bodies' masses and gravity fields. The spacecraft's engines use a unique, hyperefficient system called ion propulsion, which uses an electrical charge to accelerate ions from xenon fuel to a speed 10 times that of chemical engines using only about 3.25 mg of xenon per second at maximum thrust. The three 30-cm-wide (12-in.) ion thrusters provide less power than conventional engines but can maintain thrust for months at a time. Each unit is movable in two axes to allow for migration of the spacecraft's center of mass during the mission. This also allows the attitude control system to use the ion thrusters to help control spacecraft attitude. With three units, of which only one thruster is operating at any given time, the ion propulsion system has enough thruster

lifetime to complete the mission and still has adequate reserve. *Dawn* uses ion propulsion for years at a time, with interruptions of only a few hours each week to turn to point its antenna to Earth. Total thrust time through the mission will be about 2100 days, considerably in excess of *Deep Space 1*'s 678 days of ion propulsion operation.

New Horizons. Humankind's first mission to the distant dwarf planet Pluto was launched by NASA on January 19, 2006, aboard an Atlas V rocket. Its flyby of Jupiter was a special highlight of 2007. *New Horizons* passed Jupiter on Febrary 28, riding the planet's gravity to boost its speed and shave three years off its trip to Pluto. The 478-kg (1054-lb), piano-sized space probe was the eighth spacecraft to visit Jupiter. However, a combination of trajectory, timing, and technology allowed it to explore details no probe had seen before, such as lightning near the planet's poles, the life cycle of fresh ammonia clouds, boulder-size clumps speeding through the planet's faint rings, the structure inside volcanic eruptions on its moon Io, and the path of charged particles traversing the previously unexplored length of the planet's long magnetic tail. From January through June, *New Horizons*' seven science instruments made more than 700 separate observations of the Jovian system—twice the activity planned at Pluto—with most of them coming in the eight days around closest approach to Jupiter. In October, the *New Horizons* spacecraft found hints that Jupiter's tiniest moons have been obliterated. The Long Range Reconnaissance Imager (LORRI) camera on *New Horizons* should have been able to spot moons down to a diameter of about 1 km (0.6 mi). But it saw nothing smaller than Adrastea, a 16-km-wide (10-mi) resident of Jupiter's faint ring system.

STEREO-A and STEREO-B. STEREO (Solar TErrestrial RElations Observatory) is the third mission in NASA's Solar Terrestrial Probes (STP) program. The twin *STEREO* spacecraft were launched on October 25 on a Delta-2 7925-10L rocket from Cape Canaveral Air Force Station (CCAFS) in Florida. The 2-year mission of the two nearly identical space-based observatories, one ahead of Earth in its orbit around the Sun (*STEREO-A*, for "ahead"), the other trailing behind (*STEREO-B*, "behind") on April 23 provided the first-ever 3D images and stereoscopic measurements of the Sun, to study the nature of its coronal mass ejections (CMEs), violent eruptions of matter from the Sun that can disrupt satellites and power grids. By revealing the CMEs' 3D structure, the *STEREO* satellite pair is a key addition to the fleet of space weather satellites, providing more accurate alerts for the arrival time of Earth-directed solar ejections with its unique side-viewing perspective. The two solar-powered observatories with three-axis stabilization, each with a launch mass of 620 kg (1364 lb, including propellant), were developed by the Johns Hopkins University Applied Physics Laboratory (APL) and communicate with its Mission Operations Center via NASA's Deep Space Network.

Gravity Probe B. Gravity Probe B (GP-B) was a NASA mission using four spherical gyroscopes and a telescope housed in a satellite orbiting 642 km (400 mi) above the Earth to measure with unprecedented accuracy two extraordinary effects predicted by A. Einstein's theory of general relativity (the second having never before been directly measured): (1) the geodetic effect—the amount by which the Earth warps the local spacetime in which it resides, and (2) the frame-dragging effect—the amount by which the rotating Earth drags its local spacetime around with it. The 3100-kg (6800-lb) spacecraft was launched on April 20, 2004, on a Delta-2 rocket. In 2005, almost 90 years after Einstein first postulated general relativity and after GP-B orbited Earth for more than 17 months, scientists finished collecting data. Fifty weeks worth of more than a terabyte of science data were downloaded from the spacecraft and relayed to a comprehensive computer database in the Mission Operations Center at Stanford University, California, where scientists began the painstaking task of data analysis and validation of the measurements collected from the gyros, telescope, and SQUID (Superconducting Quantum Interference Device) magnetometer readouts, until the liquid helium in the Dewar was exhausted on September 25, 2005. By end-2007, data analysis was still under way, and an agreement was reached with NASA to extend the data analysis phase through September 2008 and probably beyond. *See* GRAVITY PROBE B MISSION.

MESSENGER. NASA's *MESSENGER* (MEercury Surface, Space ENvironment, GEochemistry, and Ranging), launched on August 3, 2004, aboard a Delta-2 rocket from CCAFS, Florida, is scheduled to become the first spacecraft to orbit the planet Mercury, beginning in 2011. The approximately 1100-kg (1.2-ton) spacecraft is in a solar orbit, a 7.9×10^9-km (4.9×10^9-mi) journey that includes 15 trips around the Sun. On August 2, 2005, *MESSENGER* returned to Earth's vicinity for a gravity boost. Next, it flew past Venus in October 2006 and did so again on June 5, 2007. The second close pass of Venus changed the spacecraft's direction around the Sun and decelerated it from 36.5 to 27.8 km/s (22.7 to 17.3 mi/s), placing *MESSENGER* on target for a flyby of Mercury in January 2008 for a critical gravity assist needed to keep the spacecraft on track for its March 2011 orbit insertion, to begin an unprecedented yearlong study of Mercury.

MESSENGER will be the first probe to visit the innermost planet in almost 33 years. During its first Venus flyby, no scientific observations were made since Venus was at superior conjunction, placing it on the opposite side of the Sun from Earth, which resulted in a 2-week radio contact blackout between the spacecraft and its operators. The second encounter offered opportunities for new observations of Venus's atmosphere, cloud structure, and space environment. The spacecraft trained most of its instruments on Venus during the upcoming encounter. A total of three Mercury flybys, each followed about 2 months later by a course correction maneuver, will put the spacecraft in position to enter Mercury orbit in March 2011. During the flybys—January 2008, October 2008, and September 2009—*MESSENGER,*

the second spacecraft sent to Mercury after *Mariner 10* flew past it three times in 1974–1975 and gathered detailed data on less than half the surface, will map nearly the entire planet in color, image most of the areas unseen by *Mariner 10* in 1974–1975, and measure the composition of the surface, atmosphere, and magnetosphere. *See* MESSENGER MISSION.

Swift. NASA's *Swift* satellite, launched on November 20, 2004, aboard a Delta-2 rocket from Cape Canaveral, was designed and built with international participation (England, Italy) to solve the 35-year-old mystery of the origin of gamma-ray bursts (GRBs). These flashes are brighter than 10^9 suns, yet last only a few milliseconds. They had been too fast for earlier instruments to catch. Scientists now believe the bursts, distant yet fleeting explosions, are related to the formation of black holes throughout the universe: the "birth cries" of black holes. *Swift* has become a key tool also in studying galaxies, quasars, supernova, novae, black holes, and neutron stars in our galaxy, along with active stars, and even comets.

To track the mysterious GRBs, *Swift* carries a suite of three main instruments: the Burst Alert Telescope (BAT), the X-Ray Telescope (XRT), and the UltraViolet/Optical Telescope (UVOT). Updated orbital lifetime predictions for *Swift* indicate that the observatory may remain in orbit up to 2022.

GALEX. The *Galaxy Evolution Explorer* (*GALEX*), launched by NASA on April 28, 2003, on a Pegasus XL rocket from a L-1011 aircraft into a nearly circular Earth orbit, is an orbiting space telescope for observing tens of millions of star-forming galaxies in ultraviolet (UV) light across 10^{10} years of cosmic history, 80% of the way back to the big bang. Its telescope has a basic design similar to the *Hubble Space Telescope* (*HST*), but while *HST* views the sky in exquisite detail in a narrow field of view, *GALEX* is tailored to view hundreds of galaxies in each observation. Thus, it requires a large field of view, rather than high resolution, in order to efficiently perform the mission's surveys.

Spitzer Space Telescope (SST). Formerly known as SIRTF (Space Infrared Telescope Facility) and launched on August 24, 2003, the *Spitzer Space Telescope* is the fourth and final element in NASA's family of Great Observatories and represents an important scientific and technical bridge to NASA's Astronomical Search for Origins program. The observatory carries an 85-cm (33-in.) cryogenic telescope and three cryogenically cooled science instruments capable of performing imaging and spectroscopy in the 3.6–160 μm range. Its supply of liquid helium for radiative-cryogenic cooling was estimated postlaunch to last for about 5.8 years, assuming optimized operation.

The most remarkable discoveries of SST during 2007 included final definitive evidence that the universe's first dust—the celestial stuff that seeded future generations of stars and planets—was forged in the explosions of massive stars. These are the most significant clues yet in the longstanding mystery of where the dust in the very young universe came from. Scientists had suspected that exploding stars, or supernovae, were the primary source, but nobody had been able to demonstrate that they can create copious amounts of dust, until now. *Spitzer*'s sensitive infrared (IR) detectors have found 10,000 Earth masses worth of dust in the blown-out remains of the well-known supernova remnant Cassiopeia A.

In another discovery, astronomers have unmasked hundreds of black holes hiding deep inside dusty galaxies billions of light-years away. The massive, growing black holes, discovered by NASA's *Spitzer* and *Chandra* space telescopes, represent a large fraction of a long-sought missing population. Their discovery implies there were hundreds of millions of additional black holes growing in our young universe, more than doubling the total amount known at that distance. The findings are also the first direct evidence that most, if not all, massive galaxies in the distant universe spent their youths building monstrous black holes at their cores.

A third stunning IR image from *Spitzer* in 2007 showed the Seven Sisters, also known as the Pleiades, located more than 400 light-years away in the Taurus constellation. The star cluster was born about 100 million years ago and is significantly younger than the Sun.

RHESSI. RHESSI (Reuven Ramaty High Energy Solar Spectroscopic Imager, in honor of the late NASA scientist who pioneered the fields of solar-flare physics, gamma-ray astronomy, and cosmic ray research), launched on February 5, 2002, in 2007 continued its operation in Earth orbit, providing advanced images and spectra to explore the basic physics of particle acceleration and explosive energy release in solar flares. Since its launch the spacecraft has been very successful observing solar flares, which are capable of releasing as much energy as 10^9 one-megaton nuclear bombs.

Hubble Space Telescope. Seventeen years after it was placed in orbit, the *Hubble Space Telescope* (*HST*) continued to probe far beyond the solar system, producing imagery and data useful across a range of astronomical disciplines.

In 2007, the telescope saw a powerful jet from a supermassive black hole blasting a nearby galaxy, a never-before witnessed galactic violence that may have a profound effect on planets in the jet's path and trigger a burst of star formation in its destructive wake. Known as 3C 321, the system contains two galaxies in orbit around each other. Data from NASA's *Chandra X-ray Observatory* show both galaxies contain supermassive black holes at their centers, but the larger galaxy has a jet emanating from the vicinity of its black hole. The smaller galaxy apparently has swung into the path of this jet. This death-star galaxy was discovered through the combined efforts of both space and ground-based telescopes, including *Hubble*.

Also in 2007, *HST* provided strong evidence that white dwarfs, the burned-out relics of stars, are given a "kick" when they form. The sharp vision of Hubble's Advanced Camera for Surveys uncovered the speedy white dwarfs in the ancient globular star cluster NGC 6397, a dense swarm of hundreds of thousands of stars. Before the stars burned out as white dwarfs, they were among the most massive stars in NGC 6397. Because massive stars are thought to gather at a globular cluster's core, astronomers assumed that most newly minted white dwarfs dwelled near the center. *HST*, however, found young white dwarfs residing at the edge of NGC 6397, which is about 11.5×10^9 years old.

Astronomers using the *HST* have discovered a ghostly ring of dark matter in the galaxy cluster Cl 0024+17 that formed long ago during a titanic collision between two massive galaxy clusters. The ring's discovery is among the strongest evidence yet that dark matter exists. The existence of an invisible substance as the source of additional gravity that holds together galaxy clusters has long been suspected. Such clusters would fly apart if they relied only on the gravity from their visible stars. Although astronomers do not know what dark matter is made of, they hypothesize that it is a type of elementary particle that pervades the universe.

Closer to home, *Hubble* discovered that massive Jupiter is undergoing dramatic atmospheric changes that have never been seen before: Jupiter's turbulent clouds are always changing as they encounter atmospheric disturbances while sweeping around the planet at hundreds of miles per hour. But the new *Hubble* images revealed a rapid transformation in the shape and color of Jupiter's clouds near the equator, marking the entire face of the globe. Also in 2007, the Space Telescope Science Institute (STSI) in Baltimore, the science operations center for *Hubble*, entered a partnership with Google to produce "Sky in Google Earth," a new feature of the newest version of Google Earth, available free of charge.

Chandra Observatory. Launched on shuttle mission STS-93 on July 23, 1999, the massive (5870-kg/ 12,930-lb) *Chandra X-ray Observatory* uses a high-resolution camera, high-resolution mirrors, and a charge-coupled detector (CCD) imaging spectrometer to observe x-rays of some of the most violent phenomena in the universe which cannot be seen by the *Hubble*'s visual-range telescope. NASA formally extended the operational mission of *Chandra* from 5 years to 10 years in September 2001.

In 2007 *Chandra* astronomers discovered one of the fastest moving stars ever seen, a "cosmic cannonball" that is challenging theories to explain its speed. Astronomers used *Chandra* to observe a neutron star, known as RX J0822-4300, over a period of about 5 years. During that span, three *Chandra* observations clearly show the neutron star moving away from the center of the Puppis A supernova remnant, a stellar debris field created during the same explosion in which the neutron star was created about 3700 years ago. By combining how far it has moved across the sky with its distance from Earth, astronomers determined the neutron star is moving at over 1300 km/s (3 million miles per hour). At this rate, RX J0822-4300 is destined to escape from the Milky Way after millions of years, even though it has traveled only about 20 light-years so far.

The *Chandra X-ray Observatory* in 2007 also located an exceptionally massive black hole in orbit around a huge companion star, with intriguing implications for the evolution and ultimate fate of massive stars. The black hole is part of a binary system in M33, a nearby galaxy about 3 million light-years from Earth. By combining data from *Chandra* and the Gemini telescope on Mauna Kea, Hawaii, the mass of the black hole, known as M33 X-7, was determined to be 15.7 times that of the Sun. This makes M33 X-7 the most massive stellar black hole known. A stellar black hole is formed from the collapse of the core of a massive star at the end of its life. M33 X-7 orbits a companion star that eclipses the black hole every $3^1/_2$ days. The companion star also has an unusually large mass, 70 times that of the Sun. This makes it the most massive companion star in a binary system containing a black hole.

Chandra may also have cracked a 45-year mystery surrounding two ghostly spiral arms in the galaxy M106 (also known as NGC 4258), a spiral galaxy 23.5 million light-years away in the constellation Canes Venatici. In visible-light images, two prominent arms emanating from the bright nucleus and spiraling outward are dominated by young, bright stars, which light up the gas within the arms. However, in radio and x-ray images, two additional spiral arms dominate the picture, appearing as ghostly apparitions between the main arms. Data from *Chandra* and other instruments have now confirmed that these so-called anomalous arms consist mostly of gas that is being violently heated by shock waves.

Cassini/Huygens. NASA's 5.4-metric-ton (6-ton) spacecraft *Cassini* continued its epic 6.7-year, 3.2×10^9-km (2×10^9-mi) journey inside the planetary system of Saturn. It departed Earth on October 15, 1997 from Cape Canaveral, Florida, and embarked on a 7-year long, circuitous journey of several billion miles across the solar system to the planet Saturn.

In 2007, the spacecraft continued to return stunning imagery. Among else, *Cassini* discovered a narrow belt harboring moonlets as large as football stadiums in Saturn's outermost ring, probably remains of a larger moon shattered by a wayward asteroid or comet eons ago. Images taken by a camera onboard the spacecraft revealed a series of eight propeller-shaped "wakes" in a thin belt of the outermost "A" ring, indicating the presence of corresponding moonlets. On September 10, *Cassini* flew by Saturn's two-toned moon Iapetus and returned pictures that revealed a white hemisphere and a black one. The images showed a surface that is heavily cratered, along with the mountain ridge that runs along the moon's equator. Many of the closeup observations focused on studying the strange 20-km high (12-mi) mountain ridge that gives the moon a walnut-shaped appearance. On the moon Enceladus, *Cassini* observed jets of fine, icy particles spraying from the moon that originate from the hottest spots on the moon's tiger-stripe fractures that straddle the moon's south polar region.

In yet another stunning imagery of its radar instrument, the spacecraft found evidence for seas, likely filled with liquid methane or ethane, in the high northern latitudes of Saturn's moon Titan. One such feature is larger than any of the Great Lakes of North America and is about the same size as several seas on Earth. The radar imaged several very dark features near Titan's north pole, much larger than similar features seen before on Titan. The largest dark feature measures at least 100,000 km^2 (39,000 mi^2), but since the radar has caught only a portion of each of these features, only their minimum size is known. While there is no definitive proof yet that these seas contain liquid, their shape, their dark appearance in radar that indicate smoothness, and their other properties point to the presence of liquids. The liquids are probably a combination of methane and ethane, given the conditions on Titan and the abundance of methane and ethane gases and clouds in Titan's atmosphere. Titan is the second largest moon in the solar system and is about 50% larger than Earth's moon.

The *Cassini-Huygens* mission is a cooperative project of NASA, the European Space Agency, and the Italian Space Agency. The spacecraft successfully entered orbit around Saturn on June 30, 2004. On December 25, 2004, ESA's *Huygens* probe detached from the *Cassini* orbiter to begin a 3-week journey to Saturn's moon Titan. After 20 days and a 4×10^6-km cruise, the probe safely landed on Titan on January 14, 2005, becoming the first human-made object to explore on-site the unique environment of this moon, whose chemistry is assumed to be very similar to that of early Earth before life formed.

WMAP. NASA's *Wilkinson Microwave Anisotropy Probe* (formerly called the *Microwave Anisotropy Probe*, or *MAP*), launched on June 30, 2001, on a Delta-2, is now located in an orbit around the second Lagrange libration point L2. Its differential radiometers measure, with unprecedented accuracy, the temperature fluctuations of the cosmic microwave background radiation (CMBR). The CMBR is the light left over from the big bang, bathing the whole universe in this afterglow. It is the oldest light in the universe, having traveled across the cosmos for 14×10^9 years, and the patterns in this light across the sky encode a wealth of details about the history, shape, content, and ultimate fate of the universe.

Since its launch, *WMAP* has refined our understanding of the universe and its development. It's observations are a treasure trove of information, including at least three major findings: (1) new evidence that a sea of cosmic neutrinos permeates the universe; (2) clear evidence that the first stars took more than a half-billion years to create a cosmic fog; and (3) tight new constraints on the burst of expansion in the universe's first trillionth of a second. The light it measures lost energy as the universe expanded over 13.7×10^9 years and is now seen by *WMAP* in microwave frequencies. By making accurate measurements of microwave patterns, *WMAP* has answered many longstanding questions about the age, composition, and development of the universe.

ACE. The *Advanced Composition Explorer* (*ACE*), launched on August 25, 1997, is positioned in a halo orbit around the first Langrangian libration point L1, where gravitational forces are in equilibrium. The spacecraft and instruments are still working very well, with the exception of the SEPICA (Solar

Energetic Particle Ionic Charge Analyzer) instrument. The *ACE* spacecraft will continue operations through 2022.

During 2007, *ACE* continued to observe, determine, and compare the isotopic and elemental composition of several distinct samples of matter, including the solar corona, the interplanetary medium, the local interstellar medium, and galactic matter. With a semimajor axis of approximately 200,000 km (125,000 mi), its elliptical orbit affords *ACE* a prime view of the Sun and the galactic regions beyond, from a vantage point approximately 1/100th of the distance from the Earth to the Sun.

During the coming years multipoint data from *ACE*, *STEREO*, and other spacecraft will provide a new perspective on anomalous cosmic ray (ACR) studies by measuring the longitudinal structure and distribution of interplanetary coronal mass ejections (ICMEs), interplanetary (IP) shocks, solar energy particle (SEP) events, and suprathermal ions, and by correlating these in situ observations with multipoint imaging of the corona and inner heliosphere.

Ulysses. The joint European/NASA solar polar mission *Ulysses*, carried into space on October 6, 1990, by the space shuttle *Discovery* (STS-41), has already traveled 7×10^9 km (4×10^9 mi). On November 17, 2006, the spacecraft reached another important milestone on its epic out-of-ecliptic journey: the start of the third passage over the Sun's south pole. *Ulysses* is engaged in the exploration of the heliosphere, the bubble in space blown out by the solar wind. The first polar passes in 1994 (south) and 1995 (north) took place near solar minimum, whereas the second set occurred at the height of solar activity in 2000 and 2001. *Ulysses* carries a comprehensive suite of sophisticated scientific instruments. In addition to the first survey of the solar wind in four dimensions (three spatial dimensions and time), *Ulysses* "firsts" include first direct measurements of interstellar dust and neutral helium gas, first measurements of rare cosmic-ray isotopes, first surveys of the space environment above and below the solar poles, first measurements of so-called pickup ions of both interstellar and near-Sun origin, first in situ observations of comet tails at large distances from the Sun, and first observations of particles from solar storms over the solar poles. At end-2007, after almost four times its expected mission lifetime, the venerable spacecraft was approaching the end of its operations. Since its Jupiter flyby in 1992, *Ulysses* has been in a 6-year orbit around the Sun.

Voyager. The Voyager missions, now in their 31st year, continue their quest to push the bounds of space exploration. The twin *Voyager 1* and *2* spacecraft opened new vistas in space by greatly expanding our knowledge of Jupiter and Saturn. *Voyager 2* then flew by Uranus and Neptune, becoming the only spacecraft ever to visit these worlds. In 2006, *Voyager 1*, already the most distant human-made object in the cosmos, passed a major milestone when it surpassed the 100 astronomical units (AU) distance from the Sun (August 15), that is, the spacecraft, which launched nearly 3 decades ago, was 100 times more distant from the Sun than is Earth. At that time

Voyager 1 was about 15×10^9 km (9.3×10^9 mi) from the Sun. The Voyagers owe their longevity to their nuclear power sources, called radioisotope thermoelectric generators (RTGs). Both Voyagers are still working, 24 h a day, 7 days a week. The spacecraft are now traveling at a distance where the Sun is but a bright point of light and solar energy is not an option for electrical power. *Voyager 1* is at the outer edge of our solar system, in an area called the heliosheath, the zone where the Sun's influence wanes. This region is the outer layer of the "bubble" surrounding the Sun, and no one knows how big this bubble actually is. *Voyager 1* is literally venturing into the great unknown and is approaching interstellar space. Traveling at a speed of about 1.6×10^6 km (1×10^6 mi) per day, *Voyager 1* could cross into interstellar space within the next 10 years.

The *Voyager 2* spacecraft has followed its twin into the solar system's final frontier, a vast region at the edge of the solar system where the solar wind runs up against the thin gas between the stars, but has taken took a different path. In 2007, it entered the heliosheath on August 30 at a distance of 84 astronomical units. By comparison, Pluto is now about 32 AU from the Sun. *Voyager 2* actually crossed the boundary five times and was directly observed (unlike *Voyager 1*) making the passage three of those times. That is because the location of the termination shock is constantly changing in response to the Sun's activity. Coronal mass ejections temporarily push the boundary outward, for example, so that it washes back and forth over the spacecraft like a wave on the beach. An instrument on the spacecraft measured the abrupt slowdown of the solar wind that defines the termination shock. But the shock did not look the way it was expected to: Instead of seeing a very abrupt drop, the spacecraft observed a gradual slowing of the solar wind ahead of each crossing, followed by a relatively small drop at the termination shock itself. Mission scientists are not sure how to explain the gradual slowdown preceding the shock. Neutral atoms from beyond the termination shock may be interacting with the solar wind to produce speedy charged particles called cosmic rays, thereby sapping some of the wind's energy.

Because *Voyager 2* crossed the heliosheath boundary, called the solar wind termination shock, about 16×10^9 km (10×10^9 mi) away from *Voyager 1* and almost 1.6×10^9 km (1×10^9 mi) closer to the Sun, it confirmed that our solar system is "squashed" or "dented"—that the bubble carved into interstellar space by the solar wind is not perfectly round. Where *Voyager 2* made its crossing, the bubble is pushed in closer to the Sun by the local interstellar magnetic field.

One or the other of the spacecraft will become the first probe to reach interstellar space after a travel period scientists estimate to be about 7–10 years long. It is not clear which spacecraft will be first, even though *Voyager 1* is about 20 AU farther from the Sun than its sister spacecraft. *Voyager 1* is escaping the solar system at a speed of about 3.6 AU per year and *Voyager 2* is covering about 3.3 AU per year. Both spacecraft are expected to continue to operate and send back valuable data until at least the year 2020. The adventurers' current mission, the Voyager Interstellar Mission (VIM), will thus explore the outermost edge of the Sun's domain, and beyond.

Mars exploration. The main event in 2007 for NASA's Mars program was the launch of yet another crewless exploration probe, *Phoenix*, to the Red Planet, joining five other spacecraft currently studying Mars: *Mars Reconnaissance Orbiter, Mars Express, Mars Odyssey*, and two Mars Exploration Rovers. This is largest number of active spacecraft to study another planet in the history of space exploration.

Phoenix. Launched on August 4 on a Delta-2 from Florida, the *Phoenix* Mars mission is the first in NASA's Scout Program. During 2007, *Phoenix* coasted in its cruise phase, which lasts for approximately 10 months. During the cruise, the spacecraft has continuously verified the health of its scientific instruments and performed trajectory correction maneuvers (TCMs). The robotic lander arrived at the Red Planet on May 25, 2008, to begin a close examination of Mars' northern polar region. *Phoenix* will be the first mission to touch the planet's water-ice. Its robotic arm will dig into an icy layer believed to lie just beneath the Martian surface. The robot explorer will study the history of the water in the ice, monitor weather in the polar region, and investigate whether the subsurface environment in the far-northern plains of Mars has ever been favorable for sustaining microbial life. The *Phoenix* mission uses the Mars *Surveyor 2001 Lander*, built in 2000, but later administratively mothballed. The 2001 lander has undergone modifications to improve the spacecraft's robustness and safety during entry, descent, and landing.

Mars Reconnaissance Orbiter (MRO). MRO is a multipurpose spacecraft designed to conduct reconnaissance and exploration of Mars from orbit. It was launched on August 12, 2005, on an Atlas V launch vehicle and arrived at Mars on March 10, 2006, after a 4.8×10^8-km (3×10^8-mi) trip, taking more than 35 h to circle the planet in its initial very elongated (elliptic) orbit for subsequent aerobraking maneuvers to achieve a lower circular orbit. *MRO* began its its primary science phase in November 2006. In February 2007, it had already surpassed the record for the most science data returned by any Mars spacecraft, including more than 15,000 images from three cameras, more than 3,000 targeted observations by a mineral-mapping spectrometer, and more than 2200 observations with ground-penetrating radar.

Among its findings are some of the weirdest landscapes on Mars as well as more familiar-looking parts of the Red Planet: One type of landscape near its south pole is called cryptic terrain because it once defied explanation, but new observations bolster and refine recent interpretations of how springtime outbursts of carbon-dioxide gas there sculpt intricate patterns and paint seasonal splotches. In addition to radially branching patterns called spiders, which had been detected by an earlier Mars orbiter, other intriguing ground textures in the area appear in the new images.

Spirit (MER-A) and Opportunity (MER-B). By end-2007, the twin Mars rovers were 46 months into missions originally planned to last 3 months. The six-wheeled rover vehicle *Spirit*, launched on June 10, 2003, on a Delta-2/Heavy rocket, landed on January 3, 2004 (ET) almost exactly at its intended landing site in Gusev Crater in excellent condition. *Opportunity*, NASA's second Mars explorer and twin to *Spirit*, launched on July 7, 2003 (ET), also on a Delta-2/Heavy after a "cliffhanger" countdown, touched down on January 25, 2004, right on target on Meridiani Planum, halfway around the planet from the Gusev Crater site of its twin, also in excellent condition.

At end-2007, the twin rovers faced their greatest challenges yet: For nearly a month, mostly during July, a series of severe Martian summer dust storms affected *Opportunity* and, to a lesser extent, *Spirit*. Dust in the atmosphere over *Opportunity* blocked 99% of direct sunlight to the rover, leaving only limited and diffuse sky light to power it. Fortunately, the storm abated, and on August 21, *Opportunity* resumed driving toward Victoria Crater in Mars' Meridiani Planum region and began descending into the crater in September. At approximately 800 m (0.5 mi) wide and 70 m (230 ft) deep, it is the largest crater the rover has visited. On August 23, *Spirit* maneuvered into position for driving up to the top of a rock platform called "Home Plate," its long-term destination in a range of hills that were on the distant horizon from its landing site, and in September climbed onto a plateau of layered volcanic bedrock bearing clues to an explosive mixture of lava and water.

Late in 2007, NASA extended, for a fifth time, the activities of the two rovers, keeping the trailblazing mobile robotic pioneers active on opposite sides of Mars, possibly through 2009. By end-2007, *Spirit* had driven about 7.5 km (4.7 mi) and returned more than 102,000 images. *Opportunity* had driven 11.7 km (7.3 mi) and returned more than 94,000 images.

Among the rovers' many other accomplishments: *Opportunity* has analyzed a series of exposed rock layers recording how environmental conditions changed during the times when the layers were deposited and later modified: wind-blown dunes coming and going, water table fluctuating. *Spirit* has recorded dust devils forming and moving. The images were made into movie clips, providing new insight into the interaction of Mars' atmosphere and surface. Both rovers have found metallic meteorites on Mars. *Opportunity* discovered one rock with a composition similar to a meteorite that reached Earth from Mars.

NASA's next rover, the *Mars Science Laboratory*, is in development for launch in 2009.

Mars Odyssey. NASA's *Mars Odyssey* probe, launched April 7, 2001, reached Mars on October 24, 2001, after a 6-month and 4.60×10^8-km (2.86×10^8-mi) journey. Entering a highly elliptical orbit around the poles of the Red Planet, it began to change orbit parameters by aerobraking, reducing its ellipticity to a circular orbit at 400 km (250 mi) by end of January 2002. The orbiter is circling Mars, with the objectives of conducting detailed mineralogical analyses of the planet's surface from space and measuring the radiation environment.

Earth Science. In 2007, NASA launched five Earth science satellites, *THEMIS* 1–5, two more than in 2006, all of them dedicated to one mission: the study of Earth's auroras. (The *AIM* satellite, discussed above, can also be considered an Earth science satellite.) There were also two dual-use (civilian and military) Earth observation satellites launched on U.S. Delta-2 rockets: *COSMO-Skymed-1* (Constellation of Small satellites for Mediterranean basin observation) and *COSMO-Skymed-2* for the Italian Space Agency (ASI).

THEMIS 1-5. After launch of NASA's *THEMIS* (Time History of Events and Macroscale Interactions during Substorms) mission on February 17 on a Delta-2 rocket, the five *THEMIS* spacecraft were deployed into a "string-of-pearls" configuration on near-identical highly elliptical orbits with 31-h periods. *THEMIS* investigates what causes auroras in the Earth's atmosphere to dramatically change from slowly shimmering waves of light to wildly shifting streaks of color. Discovering the cause of the change will provide scientists with important details on how the planet's magnetosphere works and the important Sun-Earth connection. The five-craft *THEMIS* fleet may help scientists to determine why auroras are more common in the spring than at other times. The first science observations from the spacecraft were obtained on March 23, during a disturbance in the Earth's magnetic field known as a substorm. Their data confirmed the existence of giant magnetic ropes, and witnessed small explosions in the outskirts of Earth's magnetic field.

During 2007, the spacecraft were in a "coast phase," collecting information about the interaction of the solar wind and the Earth's magnetic field. At maximum distance from Earth (apogee), the spacecraft lie between the Sun and the Earth. On their way out to and back from apogee each orbit, the spacecraft routinely cross the outer boundary of the Earth's magnetic field and a shock wave that stands upstream from this boundary. Scientists are taking this opportunity to study the outer boundary, known as the magnetopause, and determine how the Sun's plasma and magnetic energy couples into the Earth's environment.

CloudSat. CloudSat is an experimental mission conducted jointly by NASA and the Canadian Space Agency (CSA) to study the effects of clouds on climate and weather with capabilities 1000 times more sensitive than typical weather radar, using millimeter-wavelength radar to measure the altitude and properties of clouds. This information is providing the first global measurements of cloud properties that will help scientists compile a database of cloud measurements, aiding in global climate and weather prediction models. The 999-kg (2202-lb) *CloudSat* was launched with the *CALIPSO* satellite by a Delta-2 rocket from Vandenberg Air Force Base on April 28, 2006, into a polar orbit at an altitude of 705 km (438 mi). Both spacecraft are part of a constellation of spacecraft called the "A-Train," including *Aqua*, *Aura*, and *PARASOL*, dedicated to studying the Earth's weather and environment, with *CloudSat* orbiting approximately 1 min behind *Aqua*. *CloudSat* completed its 22-month prime mission in 2008 (February 27) and is now in extended mission phase.

From its initial transition to operational mode on June 2, 2006, through the end of 2007, *CloudSat* has collected close to 9000 granules (orbits) of data, including 300 million radar profiles and about 45 billion individual radar bins (vertical measurements). The CloudSat Data Processing Center has distributed over 1.5 million product files, totaling 100 terabytes of data to scientists in 47 different countries. The new observations collected from *CloudSat* combined with other A-Train observations are beginning to shed new understanding on important climate processes, in particular about (1) cloud changes in the polar regions, and the effects of these changes on the energy balance of the Arctic, their relation to weather changes, and their role in sea ice change; (2) how frequently clouds rain and how much rain falls over the global oceans—thus offering insight into processes critical to the cycling of freshwater; and (3) how properties of clouds and precipitation together change with increasing aerosol, thus offering new insights into how aerosol might indirectly affect climate.

CALIPSO. Highly complementary with *CloudSat*, the *CALIPSO* (Cloud-Aerosol Lidar and Infrared Pathfinder Satellite Observation) satellite continued in 2007 to provide new insight into the role that clouds and atmospheric aerosols (airborne particles) play in regulating Earth's weather, climate, and air quality. *CALIPSO* was launched into orbit around the Earth along with *CloudSat* as part of the A-train, a constellation of Earth-observing satellites. *CALIPSO* provides the next generation of climate observations, including an advanced study of clouds and aerosols, drastically improving our ability to predict climate change and to study the air we breathe. Its payload includes three co-aligned nadir-viewing instruments: (1) the Cloud-Aerosol Lidar with Orthogonal Polarization (CALIOP, pronounced the same as "calliope") to provide vertical profiles of aerosol and cloud backscatter and depolarization; (2) an Imaging Infrared Radiometer (IIR) with three channels in the infrared window region optimized for retrievals of cirrus particle size; and (3) the Wide Field Camera (WFC), a moderate spatial resolution imager with one visible channel that provides meteorological context and a means to accurately register *CALIPSO* observations to those from MODIS on the *Aqua* satellite. These instruments are designed to operate autonomously and continuously, although the WFC acquires data only under daylight conditions. Science data are downlinked using an X-band transmitter system that is part of the payload. *CALIPSO* is a joint U.S. (NASA) and French (Centre National d'Etudes Spatiales-CNES) satellite mission with an expected 3-year lifetime.

GOES-N. The National Oceanic and Atmospheric Administration (NOAA)/NASA joint mission *GOES-N*, which was launched aboard a Delta 4 rocket from CCAFS, Florida, on May 24, 2006, into geosynchronous orbit of approximately 35,900 km (22,300 mi), continues to be available in its current "on-orbit storage" mode from where it will be able to more rapidly replace a failure of any existing operational GOES (Geostationary Operational Environmental Satellites) such as *GOES-12* at GOES-EAST or *GOES-11* at GOES-WEST, circa 2010. Later, *GOES-N* is aimed at becoming the primary U.S. hurricane-monitoring spacecraft. *GOES-N*, to be renamed *GOES-13*, is the latest in a series of Earth-monitoring satellites which provide the kind of continuous monitoring necessary for intensive data analysis. Being in a geostationary orbit allows GOES satellites to hover continuously over one position on the Earth's surface, appearing stationary. As a result, GOES provide a constant vigil for the atmospheric "triggers" for severe weather conditions such as tornadoes, flash floods, hail storms, and hurricanes.

NOAA-18. After its launch on May 20, 2005, on a Boeing Delta-2 expendable rocket, the *NOAA-18* environmental satellite for NOAA in 2006 continued to operate in excellent condition, circling the Earth in a polar orbit of 870 km (544 mi) altitude and 98.73° inclination. With the objective to improve weather forecasting and monitor environmental events around the world, *NOAA-18* continued in 2007 to collect data about the Earth's surface and atmosphere. *NOAA-18* has instruments used in the 1982-established international Search and Rescue Satellite-Aided Tracking System, called COSPAS-SARSAT. NOAA polar-orbiting satellites detect emergency beacon distress signals and relay their location to ground stations, so rescue can be dispatched.

Aura. *Aura* (Latin for "breeze"), launched from Vandenberg AFB on July 15, 2004, on a Delta-2 rocket, is NASA's third major Earth Observing System (EOS) platform, joining its sister satellites *Terra* and *Aqua*, to provide global data on the state of the atmosphere, land, and oceans, as well as their interactions with solar radiation and each other. *Aura*'s design life is 5 years with an operational goal of 6 years. The satellite flies in formation about 15 min behind *Aqua*.

During 2007, scientists continued to use *Aura* and other satellites for tracking different chemicals present in Earth's atmosphere. These data are giving researchers a more complete picture of the causes and effects of atmospheric pollution. The scientists combined atmospheric models with actual measurements of ozone, carbon monoxide, and nitrogen dioxide in Earth's lower atmosphere. The Ozone Monitoring Instrument on the *Aura* satellite measures the total amount of ozone from the ground to the upper atmosphere over the entire Antarctic continent.

ICESat. *ICESat* (Ice, Cloud, and land Elevation Satellite), also an Earth Observing System (EOS) spacecraft, is the benchmark mission for measuring ice-sheet mass balance and cloud and aerosol heights, as well as land topography and vegetation characteristics. Launched on January 12, 2003, on a Delta-2 Expendable Launch Vehicle (ELV) into a near polar orbit at an altitude of 600 km (373 mi) with an inclination of 94°, the spacecraft in 2007 continued to provide data from its one instrument, the

Geoscience Laser Altimeter System (GLAS). Scientists trying to understand the dynamics of the Earth are using the lasers of *ICESat* to measure the height of ice sheets, glaciers, forests, rivers, clouds, and atmospheric pollutants from space with unprecedented accuracy, providing a new way of understanding our changing planet. GLAS sends short pulses of green and infrared light though the sky 40 times a second, all over the globe, and collects the reflected laser light with a 1-m (39-in.) telescope, yielding elevations. It also fires a fine laser beam of light that spreads out as it approaches the Earth surface to about 65 m (210 ft) in diameter. On its way to the surface, those photons bounce off clouds, aerosols, ice, leaves, ocean, land, and more, providing detailed information on the vertical structure of the Earth system.

Aqua. Launched in May 2002, the 1750-kg (3858-lb) NASA satellite *Aqua*, formerly named EOS PM (signifying its afternoon equatorial crossing time), carrying six instruments weighing 1082 kg (2385 lb) designed to collect information on water-related activities worldwide, has been circling Earth in a polar, Sun-synchronous orbit of 705 km (438 mi) altitude. During its 6-year mission, *Aqua* is observing changes in ocean circulation and studies how clouds and surface water processes affect our climate. NASA and NOAA scientists, working with experimental data from *Aqua*'s Atmospheric Infrared Sounder, a high-spectral-resolution infrared instrument that takes 3D pictures of atmospheric temperatures, water vapor, and trace gases, are conducting research on improving the accuracy of medium-range weather forecasts in the Northern Hemisphere. Incorporating the instrument's data into numerical weather prediction models improves the accuracy range of experimental 6-day Northern Hemisphere weather forecasts by up to 6 h, a 4% increase. These data have now been officially incorporated into the NOAA National Weather Service operational weather forecasts.

POES-M (NOAA-M). The operational weather satellite *POES-M* (Polar-orbiting Operational Environmental Satellites-M) was launched from Vandenberg Air Force Base on a commercial Titan 2 rocket on June 24, 2002. The satellite, later renamed *NOAA-M,* is part of the POES program, a cooperative effort between NASA and NOAA, the United Kingdom, and France. It joined the *GOES-M,* launched in July 2001. Both satellites, operated by NOAA, in 2007 continued to provide global coverage of numerous atmospheric and surface parameters for weather forecasting and meteorological research. *NOAA-M* broadcasts data directly to thousands of users around the world, using its environmental monitoring instruments for imaging and measuring of the Earth's atmosphere, its surface, and cloud cover. Observations include information about Earth radiation, sea and land surface temperature, atmospheric vertical temperature, water vapor, and ozone profiles in the troposphere and stratosphere.

GRACE. Launched on March 17, 2002, on a Russian Rockot carrier, the twin satellites *GRACE* (Gravity Recovery and Climate Experiment), named "Tom" and

"Jerry," in 2007 continued to map the Earth's gravity fields by taking accurate measurements of the distance between the two satellites, using GPS and a microwave ranging system. This allows making detailed measurements of Earth's gravity field, which will lead to discoveries about gravity and Earth's natural systems with possibly far-reaching benefits to society and the world's population. Among other things, *GRACE* in its 5 years of operation may have found a crater deep under the Antarctic ice that may mark an asteroid impact greater than the one that doomed the dinosaurs, measured the seafloor displacement that triggered the tsunami of 2004, and quantified changes in subsurface water in the Amazon and Congo river basins. *GRACE* provides scientists from all over the world with an efficient and cost-effective way to chart Earth's gravity fields with unprecedented accuracy, yielding crucial information about the distribution and flow of mass within the Earth and its surroundings. The science data from *GRACE* consist of the intersatellite range change measurements, and the accelerometer, GPS, and attitude measurements from each satellite.

Data from the *GRACE* satellites in the first-ever gravity survey of the entire Antarctic ice sheet showed that the ice loss in Antarctica increased by 75% in the last 10 years due to a speedup in the flow of its glaciers and is now nearly as great as that observed in Greenland. The project is a joint partnership between NASA and the German DLR (Deutsches Zentrum für Luft- und Raumfahrt). *See* SATELLITE-BASED GEODESY.

Department of Defense space activities. Military space organizations in the United States continued their efforts to make space a routine part of military operations across all service lines. Highlights of military space in 2007 included the launch of the first U.S. Air Force (USAF) Wideband Global System (WGS) broadband communications satellite that marked a major step in military capabilities, on an Atlas-5 (October 11). There were eight successful military space launches carrying 15 payloads: one heavy Delta-4M vehicle, launching a DSP-23 early-warning satellite, two Delta-2 launchers carrying two new GPS IIR navigation satellites (-17/M4 and -18/M5); four Atlas-5 rockets, one with eight individual payloads (mostly nanosatellites), two with classified NROL (National Reconnaissance Office Launch) surveillance satellites, plus one with the WGS; and one Orbital Sciences Corporation (OSC) Minotaur launcher with an experimental satellite (NFIRE). The ninth launch, a failure, was the second Falcon-1 rocket from SpaceX, with a USAF demonstration payload.

Commercial space activities. In 2007, commercial space activities in the United States reached their lowest level in several years, to some extent due to the export restrictions imposed to the U.S. industry on sensitive technologies. Of the 19 total launch attempts by the United States in 2007, only one was a commercial mission (NASA/civil: 9; military: 9): a Delta-2 rocket with the *Worldview-1* imaging satellite, the world's only commercial satellite with a 0.5-m (1.6-ft) image resolution. The non-U.S./non-

Russian partnership of Boeing, RSC-Energia (Russia, 25% share), NPO Yushnoye (Ukraine), and Kvaerner Group (Norway) launched one Russian Zenit 3SL (SeaLaunch) rocket carrying the *NSS-8* (New Skies 8), the eighth comsat in a series intended to provide global coverage at C-band, but rocket and payload were destroyed at launch on January 30. The newcomer SpaceX sustained the second failure of its new Falcon 1 rocket during ascent to orbit on March 21 for the USAF.

Russian Space Activities

Russia in 2007 showed increased activity in space operations from 2006, launching 7 different carrier rockets. Out of 68 launch attempts worldwide in 2007, 26 space launches were attempted by Russia, again placing it in the lead of spacefaring countries including the United States. Its total of 26 launch attempts, of which 25 were successful, was one more than its previous year's 25 attempts (23 successful, same as in 2005 and 2004): six Soyuz-U, five Soyuz-FG (two crewed), three Proton-K, four Proton-M (one failed on September 6, carrying the Japanese comsat *JCSat-11*), one Zenit-2, one Zenit-3SL (failed sea launch, counted above under U.S. Activities), one Molniya-M, three Kosmos-3M, and three Dnepr-1. The upgraded Soyuz-FG rocket's new fuel injection system provides a 5% increase in thrust over the Soyuz-U, enhancing its lift capability by 200 kg (440 lb) and enabling it to carry the new Soyuz-TMA spacecraft, which is heavier than the Soyuz-TM ship used in earlier years to ferry crews to the *ISS*. Soyuz-TMA was flown for the first time on October 30, 2002, as *ISS* mission 5S. It was followed in 2003 by Soyuz *TMA-2* (6S) and *TMA-3* (7S), in 2004 by *TMA-4* (8S) and *TMA-5* (9S), in 2005 by *TMA-6* (10S) and *TMA-7* (11S), in 2006 by *TMA-8* (12S) and *TMA-9* (13S), and in 2007 by *TMA-10* (14S) and *TMA-11* (15S).

Russia is currently using three launch sites: Baikonur in Kazakhstan (45.6°N, 63.4°E), Plesetsk (62.9°N, 40.8°E), and Yasny Cosmodrome in the Orenburg region (51.0°N, 58.0°E).

The Russian space program's major push to enter into the world's commercial arena by promoting its space products on the external market continued in 2007. First launched in July 1965, the Proton heavy lifter, originally intended as a ballistic missile (UR500), by end-2007 had flown 251 times since 1980, with 16 failures (reliability: 0.936). Its launch rate in recent years has been as high as 13 per year. Of the seven Protons launched in 2007 four were for commercial customers: Canada's *Anik-F*, the huge U.S. *DirecTV-10*, Japan's *JCSat-11* (failed), and the Swedish/Danish/Teracom *Sirius-4*, plus three for the state/military carrying six *GLONASS-M* navsats (the Russian equivalent of GPS) and the *Raduga-1 #8* comsat. During 1985–2007, 199 Proton and 433 Soyuz rockets were launched, with 12 failures of the Proton and ten of the Soyuz, giving a combined reliability index of 0.965. Until a launch failure on October 15, 2002, the Soyuz rocket had flown 74 consecutive successful missions, including 12 with human crews on board; meanwhile, another

45 successful flights were added, including ten carrying 29 humans.

Besides the four commercial Proton launch attempts, there were three Dnepr-1 missions, one on April 17 from Baikonur with 16 civil microsats, the second on June 15 with the German civil *TerraSAR-X* from Baikonur, and the third on June 28 from the ISC Kosmotras Yasny Cosmodrome in southern Russia with the commercial *Genesis-2* Pathfinder payload, a successful technology demonstration for the inflatable Nautilus space station structure promoted by the commercial firm Bigelow Aerospace. After *Genesis-2*, two *Guardian* spacecraft will follow. (*Genesis* is a one-third-scale model of the Nautilus module, *Guardian* a 45% scale model.)

European Space Activities

After the decline of Europe's commercial activities in space to a low in 2004 leveled off in 2005, the new EC (enhanced capability) version of the Ariane 5, designed to lift 10 tons to geostationary transfer orbit, enough for two big communications satellites at once, allowed European industry a quick comeback with the successful first launch of an Ariane 5 ECA in 2005 (after its failure in 2002) as one of five flights (out of five attempts) of the Ariane 5-G (generic, 3 flights) and -ECA (2 flights) rockets (2004: 3), to which another five flights were added in 2006 and six in 2007, bringing its program total to 36. The six heavy-lift vehicles of 2007 carried a total of 12 satellite payloads, all of them commercial comsats (*Skynet-5A* and *-5B*, *Insat-4B*, *Astra-1L*, *Galaxy-17*, *Spaceway-3*, *Bsat-3a*, *Intelsat-11*, *Optus-D2*, *StarOne-C1*, *Rascom QAF-1*, and *Horizons-2*). Altogether, 180 flights had been successfully completed by Ariane rockets by end-2007, including 36 by the Ariane 5.

The development of the *Galileo* (GNSS) navigation and global positioning system by 15 European countries engaged in space, that received top-level approval in 2005, has run into delays and disputes. Originally planned to enable Europe to be independent of the U.S. GPS system starting in 2008, the newly formed consortium has yet to overcome basic difficulties, although the industry has made some moves in the right direction. *Galileo* (not to be confused with NASA's Jupiter probe) will consist of a constellation of 30 small satellites (27 operational, 3 backup) weighing 700 kg (1500 lb) each, placed in medium orbit (24,000 km or 15,000 mi) above Earth, with orbit inclination 55°. It will be independent of, but compatible with, the GPS system; that is, if the United States and Europe agree on cooperation at some future date, interoperability would be possible. *See* SATELLITE NAVIGATION ADVANCES.

In the human space flight area, while the *ISS* remains ESA's largest single ongoing program and its only engagement in human space flight, the European *ISS* share (totaling 8.6%) remains unchanged due to a top-level agreement signed by previous governments of the participating nations. A major event for ESA in 2007 was the launch of Italian Astronaut Paolo Nespoli in October, accompanying the Italian-built *Node-2 Harmony* on STS-120/ *Discovery* to the *ISS*.

France has a relatively large and active national space program, including bilateral (outside of ESA) activities with the United States and Russia. In 2007, the France/CNES-led project ATV (Automated Transfer Vehicle) named Jules Verne was readied for its inaugural flight to the *ISS* in 2008. In Italy, the Italian Space Agency ASI, created in 1988, participates in the *ISS* program through ESA but also had entered a protocol with NASA for the delivery of three multipurpose logistics modules (MPLM) for the *ISS*. Two MPLMs have already flown in space, *Leonardo* and *Raffaello*; the third MPLM is *Donatello*. Italy also developed the second *ISS* Node (Node-2), which docked to the *ISS* in October 2007. In addition, Node-3, developed at Thales Alenia Space in Turin, Italy, is scheduled to fly to the *ISS* in 2010, containing the most advanced life-support systems ever flown in space, an atmosphere revitalization system and other systems required to support a station crew of six. In Germany, the low governmental interest (unlike Italy's and France's) in this field continued in 2007. Germany is the second major ESA contributor after France but has essentially no national space program of its own remaining. Two more satellite-based radar reconnaissance systems were launched for Germany on Russian Kosmos-3M rockets, *SAR-Lupe 2* (July 2) and *SAR-Lupe 3* (November 1). When completed, SAR-Lupe will consist of five identical small satellites in a constellation in three orbital planes. A third German radar satellite, *TerraSAR-X*, was launched on a Russian Dnepr-1 (a converted R-36M missile) on June 15 for high-resolution imaging of the entire Earth surface using synthetic-aperture radar (SAR) technology, the first satellite implemented in a public or private partnership in Germany (cost shared by EADS Astrium GmbH and the DLR German Aerospace Center). Also in Germany in 2007, the ESA *Columbus* orbital laboratory, destined for the *ISS* in 2008, entered final checkout stages, with its Control Center (COL-CC) in Oberpfaffenhofen near Munich conducting operational tests with the *ISS* to verify its readiness.

Venus Express. In 2007, the 1240-kg (2734-lb) *Venus Express* spacecraft, launched on November 9, 2005 from the Baikonur Cosmodrome in Kazakhstan aboard a Russian Soyuz-Fregat launch vehicle, continued making the most detailed study of the planet's thick and complex atmosphere to date. The latest findings by the probe, which arrived at the planet on April 11, 2006, after a 153-day cruise, highlighted the features that make Venus unique in the solar system and provide fresh clues as to how the planet is, despite everything, a more Earth-like planetary neighbor than one could have imagined. Permanently covered in clouds, Venus has been a mystery for centuries. Although it is the planet nearest to Earth, it has proved extraordinarily difficult to study because of its curtain of clouds that obscures our view of its surface. Its surface temperature is more than 400°C (800°F) and the surface pressure is a hundred times that on Earth. The key to understanding Venus lies in its atmosphere, which *Venus Express* is studying. It is much thicker than Earth's and intercepts most of the Sun's energy before it can reach the surface. A second set of results from the probe concerns both the atmosphere's composition and its chemistry. *Venus Express* has taken compositional profiles of the atmosphere around the planet, and unambiguously confirmed the presence of lightning that can have a strong effect on the composition of the atmosphere itself. A third set of results is about the processes by which the atmosphere of Venus is escaping into space. They are driven by the solar wind—a stream of electrically charged particles given out by the Sun. As the solar particles collide with electrically charged particles near Venus, they energize the gases, stripping them forever from the planet. *Venus Express* has provided giant leaps in the understanding of all these phenomena, and found how Venus loses water due to its interaction with the solar wind. New measurements of heavy water in the atmosphere are also providing new clues on the history of water on the planet and its overall climate evolution.

Rosetta. ESA's comet intercept mission *Rosetta*, initially intended for a rendezvous with Comet 46P/Wirtanen but postponed, was launched on an Ariane 5 on March 2, 2004, to Comet 67P/Churyumov-Gerasimenko instead. The probe will rendezvous with the comet in 2014 and release a landing craft named *Philae*. It is hoped that on its 10-year journey the spacecraft will also pass by two asteroids, Steins and Lutetia. Along its roundabout route, *Rosetta* will enter the asteroid belt twice and gain velocity from gravitational "kicks" provided by close flybys of two planets. On arrival at 67P in 2014, *Rosetta* will enter orbit around the comet and stay with it on its journey in toward the Sun, to study the origin of comets, the relationship between cometary and interstellar material, and its implications with regard to the origin of the solar system.

Envisat. In 2007, ESA's operational environmental satellite *Envisat*, the largest Earth Observation spacecraft ever built, continued its observations after its launch on March 1, 2002, on the 11th Ariane 5. The 8200-kg (18,100-lb) satellite circles Earth in a polar orbit at 800 km (500 mi) altitude, completing a revolution of Earth every 100 min. Because of its polar sun-synchronous orbit, it flies over and examines the same region of the Earth every 35 days under identical conditions of lighting. The 25-m-long (82-ft) and 10-m-wide (33-ft) satellite, about the size of a bus, is equipped with ten advanced instruments (seven from ESA, the others from France, Great Britain, Germany, and Netherlands), including an Advanced Synthetic Aperture Radar (ASAR), a Medium Resolution Imaging Spectrometer (MERIS), an Advanced Along Track Scanning Radiometer (AATSR), a Radio Altimeter (RA-2), a Global Ozone Monitoring by Occultation of Stars (GOMOS) instrument, a Michelson Interferometer for Passive Atmosphere Sounding (MIPAS), and a Scanning Imaging Absorption Spectrometer for Atmospheric Cartography (SCIAMACHY). It is scanning the Earth similar to the way vertical slices are peeled off an orange as it is turned in one's hand. This enables *Envisat* to continuously scrutinize the Earth's surface (land, oceans, ice caps) and atmosphere, gathering a huge volume of invaluable information for scientists and operational users for global monitoring and forecasting to protect the planet.

SPOT 5. Launched on May 4, 2002, by the 112th Ariane 4 from Kourou (French Guiana), the fifth imaging satellite of the commercial Spot Image Company (CNES, 38.5%; EADS, 35.66%; Alcatel, 5.12%; IGN, 7.81%) in 2007 continued operations in its polar sun-synchronous orbit of 813 km (505 mi) altitude. Unique features of the SPOT system are high resolution, stereo imaging, and revisit capability. The *SPOT* satellite Earth Observation System was designed by the French Space Agency CNES, and developed with the participation of Sweden and Belgium. Starting in 2010, the *SPOT* satellites will be replaced by the PLEIADES program, a constellation of smaller, more agile satellites offering an improved spatial resolution of up to 0.7 m (2 ft).

INTEGRAL. ESA's *INTEGRAL* (International Gamma-Ray Astrophysics Laboratory), a cooperative project with Russia and United States of America, continued successful operations in 2007. Launched on October 17, 2002, on a Russian Proton rocket into a 72-h orbit with $51.6°$ inclination, a perigee height of 9000 km (5600 mi), and an apogee height of 155,000 km (96,000 mi), the sensitive gamma-ray observatory provides new insights into the most violent and exotic objects of the universe, such as black holes, neutron stars, active galactic nuclei, and supernovae. Its instruments have produced the first all-sky map of the 511-keV-line emission produced when electrons and their antimatter equivalents, positrons, collide and annihilate. With its sensitive detectors, *INTEGRAL* has accurately measured the hard cosmic x-ray background (CXB), responsible for creating the diffuse background glow spread throughout the universe, and it has confirmed that systems containing white dwarves contribute significantly to the galactic hard x-ray diffuse emission. The gamma-ray observatory has also found a rare class of anomalous x-ray pulsars, with magnetic fields 10^9 times stronger than the strongest steady magnetic field achievable in a laboratory on Earth, and has found massive stars by looking for their radioactive traces—radioactive decay gamma-ray lines from iron have been observed with the most significant detection to date, forcing re-evaluation of existing theoretical models. *INTEGRAL* has also discovered a new class of x-ray binary stars called supergiant fast transients—x-ray binary systems containing supergiant stars. *INTEGRAL* and other high-energy satellites have discovered that these transient systems are not as rare in the galaxy as initially thought. Although not designed for the purpose, *INTEGRAL* has also proved to be a great gamma-ray-burst "watchdog," providing instantaneous, accurate positions of gamma-ray bursts to other facilities within minutes or even seconds. In collaboration with NASA's *Rossi* X-ray Timing Explorer, *INTEGRAL* has also detected what appears to be the fastest spinning neutron star yet.

In November 2007, in recognition of its superb scientific output, the mission operations of Europe's flagship gamma-ray observatory *INTEGRAL* were extended until December 31, 2012.

XMM-Newton. Europe's *XMM* (X-ray Multi Mirror)-*Newton* observatory, launched on December 10, 1999, on an Ariane 5, is the largest European science research satellite ever built. Operating in an orbit of 113,946 × 7000 km (71,216 × 4375 mi) inclined at 40° to the equator, the telescope has a length of nearly 11 m (36 ft), with a mass of almost 4 metric tons (8800 lb). Using its three scientific instruments—a photon-imaging camera, reflection grating spectrometer, and optical telescope—it has obtained the first reliable measurement ever of the mass-to-radius ratio of a neutron star (EXO 0748-676). Among its discoveries, it characterized for the first time x-ray spectra and light curves of some classes of protostars (stars being born), and provided an unprecedented insight into the x-ray variability of the coronas of stars similar to our Sun. With its capability to respond as quickly as 5 h to target-of-opportunity requests for observing elusive gamma-ray bursts, this space observatory detected for the first time an x-ray halo around the bursts, where the halo appeared as concentric ringlike structures centered on the burst location. *XMM-Newton* is shedding new light on supernovae remnants, as well as on neutron stars. On the latter, an exciting discovery was that of a bow shock aligned with the supersonic motion of a neutron star (called Geminga), and the detection of hot spots indicating that the configuration of the neutron star's magnetic field and surface temperatures are much more complex than previously thought.

XMM-Newton has made breakthrough observations of a wide variety of compact objects, such as the first detection of an intermediate-mass black hole in globular cluster NGC 4472. This has direct implications for the formation and evolution theories of globular clusters in general. Thanks to its sensitivity at high energies, *XMM-Newton* has made the first and only direct probing of the central regions near a black hole, by sampling the presence of iron and the variability of its spectral fingerprints. The satellite's observations have also been fundamental in helping understand the physics of heavy subatomic matter (baryonic) in clusters of galaxies and in studying the dark matter component in clusters. *XMM-Newton* has given the first strong indication that very faint active galactic nuclei (AGN) are similar to the "normal" AGN population, and measured for the first time the size of the emission region of an AGN. Other major results include the progress in understanding the link between x-ray emission and luminosity of stars, as well as the relation between the x-ray emission and processes such as star accretion or collisions. The satellite has discovered the remnants of a new class of supernovae within the so-called Ia type, which are used as standard reference, or "candles," to determine stellar luminosity. *XMM-Newton* has also revealed x-ray emission in the Martian exosphere—the first definite detection of x-ray emission induced by exchange of electrical charges from the exosphere of another planet. As with *INTEGRAL*, in November 2007 the mission operations of *XMM-Newton* were extended until December 31, 2012, in recognition of its superb scientific output.

Mars Express. In 2007, *Mars Express* continued its operations in orbit around the Red Planet. *Mars Express*, Europe's entry into the ongoing and slowly expanding robotic exploration of the Red Planet from Earth as precursors to later missions by human explorers, was launched on June 2, 2003, from the Baikonur launch site by a Russian Soyuz/Fregat rocket. After a 6-month journey, it arrived at Mars in December. Six days before arrival, *Mars Express* ejected the *Beagle 2* lander, which was to have made its own way to the correct landing site on the surface but disappeared and was declared lost. The *Mars Express* orbiter successfully entered Martian orbit on December 25. Highly successful operations and stunning closeup imagery of the Mars surface went on during 2004–2007. Its instruments have shown that many of the upper layers of Mars contain water ice. They detected claylike minerals that form during long-term exposure to water, but only in the oldest regions of Mars. That suggested water flowed during the first few hundred million years of the planet's history only. When these bodies of water were lost, water then occasionally burst from inside the planet but quickly evaporated, and this phenomenon still may be happening in places. Residual water ice, in the form of a frozen lake, was discovered in the open Vastitas Borealis crater. After the 2007 eclipse (shadowed period) season ended on August 31, preparations were made for the 2008 eclipse seasons, with long eclipses during aphelion, starting in October. On October 2, *Mars Express* successfully performed a close flyby of the moon Phobos to within ~140 km (87.5 mi). The data from the Sub-Surface Sounding Radar Altimeter (MARSIS) for this close flyby, which are unique, were being analyzed. Starting November 18 and ending December 19, five maneuvers were executed to bring the spacecraft from its previous 11:3 resonance orbit to its new 18:5 resonance orbit. This change was necessary to guarantee a proper day-time/night-time distribution for future *Mars Express* observations. *Mars Express* continues to remotely explore the Red Planet with a sophisticated instrument package comprising the High Resolution Stereo Camera (HRSC), Energetic Neutral Atoms Analyzer (ASPERA), Planetary Fourier Spectrometer (PFS), Visible and Infrared Mineralogical Mapping Spectrometer (OMEGA), MARSIS, Mars Radio Science Experiment (MaRS), and the Ultraviolet and Infrared Atmospheric Spectrometer (SPICAM). *See* MARS EXPRESS.

Asian Space Activities

China, India, and Japan have space programs capable of launch and satellite development and operations. Kazakhstan is in early stages of joining the spacefaring community.

China. Probably the biggest news in 2007, stunning the world, was China successfully testing a kinetic antisatellite (Asat) weapon on January 11. This use of a mobile missile to destroy an aging weather satellite, the *Feng Yun 1C* (FY-1C) satellite at 537 mi (859 km) altitude, created more orbital debris than any previous event. It showed that China has mastered key space sensor, tracking, and other technologies important for advanced military space operations.

China, in effect, has two major space agencies: the PLA (People's Liberation Army) for crewed and military programs, and the CNSA (China National Space Administration) for civil or scientific projects. With a total of ten launches in 2007 (2006: 6; 2005: 5; 2004: 8; 2003: 6; 2002: 4; 2001: 1), China in 2007 continued to claim the third place of spacefaring nations before Europe, after Russia and the United States, having made worldwide headlines in 2003 with its successful orbital launch, by the PLA, of the first Chinese "Taikonaut" in the 4760-kg (10,500-lb) spacecraft *Shenzhou 5* ("Divine Vessel 5") on a 21-h mission, and in 2005 with the launch of the 7700-kg (17,000-lb) two-seater *Shenzhou 6* carrying two Taikonauts on a 75-orbit and 115 h 32 min flight, designed to further China's human spaceflight experience as it works toward developing a crewed space station and to serve as a symbol of national pride, demonstrating China's technological prowess.

The launch vehicle of the *Shenzhou* spaceships was the new human-rated Long March 2F rocket. China's Long March (Chang Zheng, CZ) series of launch vehicles consists of 12 differing versions, which by the end of 2007 had made 104 flights, sending 117 payloads (satellites and spacecraft) into space, with 92% success rate. China has three modern (but land-locked, thus azimuth-restricted) launch facilities: at Jiuquan (Base 20, also known as Shuang Cheng-Tzu/East Wind) for low-Earth-orbit (LEO) missions, Taiyuan (Base 25) for sun-synchronous missions, and Xichang (Base 27) for geostationary missions. Development of a less restrictive launch site, on the tropical island Hainan in the South China Sea, is under consideration.

A particularly interesting event was the launch of China's first lunar orbiter project, *Chang'e 1*, on October 24 on a CZ-3A, the third milestone in China's space technology after satellite and crewed spacecraft projects. *Chang'e 1*, based on a Chinese telecommunications satellite, will provide 3D images of the Moon's surface, chart elements on the Moon, measure the thickness of the lunar soil, and monitor the space environment between the Earth and the Moon.

Its ten major launches in 2007 demonstrated China's strongly emerging space maturity. On February 2, a CZ-3A launched the Chinese *Beidou-2A* (Big Dipper 2A) navigation satellite, part of the Compass Navigation Satellite System (CNSS) or Compass, followed on April 11 by a CZ-2C with the *Haiyang-1B* maritime surveillance satellite and on April 13 by a CZ-3A with *Beidou-M1*. On May 13, a CZ-3B launched *NigComsat-1* for Nigeria, on May 25 a CZ-2D the *Yaogan-II* remote-sensing satellite, on May 31 a CZ-3A the comsat *SinoSat-3*, on July 5 a CZ-3B the *ChinaSat-6B* comsat, on September 19 a CZ-4B

the *CBERS-2B* (China-Brazil Earth Resources Satellite 2B), and on November 11 a CZ-4C the *Yaogan-3* SAR satellite, winding up a record of 63 consecutive launch successes for the Long March, which in its two-stage 2C version has a lift-off weight of 192,000 kg (422,400 lb), a total length of 41.9 m (137.5 ft), a diameter of the rocket and payload fairing of 3.35 m (11 ft), and a low-Earth-orbit launching capacity of 1000 kg (2200 lb). The 3B version has a liftoff weight of 425,800 kg (936,760 lb), a total length of 54.9 m (180 ft), and a payload capability to low geosynchronous orbit of 4500 kg (9900 lb).

Japan. In 2007 Japan's space program had only two launches, both by the H2-A rocket, the nation's workhorse, raising its total number of launch attempts to 11 after returning the H2-A launch vehicle to flight in 2005 following its major failure in 2003.

In 2007, a major JAXA launch on an H2-A was the 2546-kg (6820-lb) [at launch] Selene (SELenological and ENgineering Explorer) Moon probe named *Kayuga* on September 14. On October 4. it went into a lunar orbit, inclined 95°, with 15 instruments including imagers, a radar sounder, laser altimeter, x-ray fluorescence spectrometer, and gamma-ray spectrometer, to study the origin, evolution, and tectonics of the Moon from space. *Kayuga* consists of three separate units: the main orbiter, a small relay satellite (R-Star), and the small VRAD satellite (V-Star). Also launched with *Kayuga* was *Micro-Labsat-2*, the second of Japanese microsatellites [typically ~53 kg (116 lb) after separation], employed as a teaching project for young engineers if the launcher has excess capacity.

The second H2-A launch from Tanegashima, on February 24, brought the heavy (1200-kg or 2646-lb) military payloads *IGS-4A* and *4B* (Intelligence Gathering Satellites) into low-orbit space, one optical (*4A*), the other with radar imaging (*4B*).

Also in the headlines in 2007 was Japan's Space Engineering Spacecraft *Hayabusa* ("peregrine falcon," or MUSES-C). Launched on May 9, 2003, from the Kagoshima Space Center in southern Japan on an ISAS solid-propellant M-5 rocket, the probe made a successful touchdown on the Asteroid Itokawa on November 11, 2005, and a second on November 25, to collect samples to bring back to Earth. *Hayabusa* then encountered serious technical difficulties that cast doubt on its ability to return to Earth. However, contact was reestablished in December 2005, and in March 2006 JAXA announced that communication with *Hayabusa* had been recovered and its position established at about 13,000 km (8000 mi) ahead of Itokawa. In June 2006, two out of four ion engines were reportedly working normally, which will be sufficient for the return journey to Earth, where its reentry capsule with the sample should arrive in June 2010, to be recovered by parachute at Woomera, Australia. Since February 2007, JAXA has been carefully preparing to start the nominal return trip to Earth using the remaining ion engine and one attitude control reaction wheel (as two of the three wheels are unavailable due to anomalies).

In its longer-range view, Japan's space agency JAXA is studying versions of a "new generation" launch vehicle, essentially a heavier lift version of the H-2A with 10–20% greater lift capacity than its predecessor, which would put it into the Delta-4 class. One area of great promise for Japan continues to be the *ISS* Program, in which the country is participating with a sizeable 12.6% share. Its $3 billion contributions to the *ISS*, launched starting in 2008, are the 15-ton pressurized Japanese Experiment Module (JEM) called *Kibo* ("hope"), along with its ancillary remote manipulator arm and unpressurized porchlike exposed facility for external payloads, and the H-2 Transfer Vehicle (HTV), which will carry about 6 metric tons (6.6 tons) of provisions to the *ISS* once or twice a year, launched on an H-2A beginning in 2009 or 2010. On May 30, 2003, the Mitsubishi-built JEM arrived at NASA's Kennedy Space Center (KSC) in Florida. *Kibo* was launched to the *ISS* on the space shuttle (STS-123, in March 2008), accompanied by Japanese astronaut Dr. Takao Doi, who had flown on the shuttle before, becoming the first Japanese to perform a spacewalk.

India. India's emerging space program, with three launch attempts in 2007, all successful, has come back strongly after only one attempt in 2006, when the first GSLV-F02 (Geosynchronous Satellite Launch Vehicle F02), carrying the *INSAT-4C* communication satellite, failed after liftoff on July 10 from the Satish Dhawan Space Center in Sriharikota and fell into the Bay of Bengal. The three 2007 launches from Sriharikota were executed by two PSLVs (Polar Space Launch Vehicles), one on January 10 with *CartoSat-2*, *SRE-1*, and *LAPAN-TUBsat* (Indonesia) and *Pehuensat* (Argentina), the other on April 23 with the small astrophysics mission *Agile* (Italy), plus one GSLV on September 2 with *INSAT-4C*, replacing the satellite lost in 2006.

India's main satellite programs are the INSAT (Indian National Satellite) telecommunications system, the *IRS* (Indian Remote Sensing) satellites for earth resources, the METSAT weather satellites, and the new GSat series of large (up to 2.5-tons) experimental geostationary comsats. India's main launchers today are the PSLV and the Delta-2-class GSLV.

India is working on plans to explore the Moon, with the announced intent to send a crewless probe there in the near future. ISRO calls the Moon flight project *Chandrayaan Pratham*, which has been translated as "First Journey to the Moon" or "Moonshot One." The 525-kg (1157-lb) *Chandrayaan-1* would be launched on a PSLV rocket. After first circling Earth in a geosynchronous transfer orbit (GTO), the spacecraft would fly on out into a polar orbit of the Moon some 100 km (60 mi) above the surface, carrying x-ray and gamma-ray spectrometers, and sending data back to Earth for producing a high-resolution digital map of the lunar surface. The project's main objectives are high-resolution photography of the lunar surface using remote-sensing instruments sensitive to visible light, near-infrared light, and low-energy and high-energy

x-rays. Space aboard the satellite also will be available for instruments from scientists in other countries. *Chandrayaan-1* is expected to be the forerunner of more ambitious planetary missions in the years to come, including landing robots on the Moon and visits by Indian spacecraft to other planets in the solar system.

Other Countries' Space Activities

The former Soviet Republic of Kazakhstan, an independent state since its declaration of sovereignty in 1990, is aspiring to join the select group of spacefaring nations. Home of Baikonur, the world's largest rocket launching site, Kazakhstan has been in lengthy political negotiations with Russia, the real owner and operator of Baikonur who has leased the 6717 km^2 (2593 mi^2) of the Cosmodrome from Kazakhstan. Baikonur comprises nine launch complexes with 15 launch pads, 11 assembly and test complexes, 34 technical centers, tanking stations, two airports, a factory for technical gases, two power trains with diesel aggregates for generating electricity, a 600-MW heating facility, 470 km (292 mi) of railroad tracks, and 1281 km (796 mi) of roadways. There are also several settlements, the largest being the city of Baikonur with 70,000 inhabitants. Today, Kazakhstan is making plans to use these facilities for two satellite launcher projects of its own, the solid-propellant rocket Ischim, launched from a MiG-31I fighter plane, similar to the U.S. Pegasus, and the carrier rocket Baiterek, a modified version of the new Russian heavy lifter Angara-A5, which is to take the place of today's Proton (first flight probably not before 2011). Kazakhstan's first satellite, the 850-kg (1874-lb) comsat *KazSat 1*, was launched in 2006 on a Proton-K in the presence of Kazakhstan's President Nursultan Nasarbayev and Russia's President Vladimir Putin. There were no launches in 2007.

In Latin America, Brazil has the most advanced space program, with capabilities in launch vehicles, launch sites, and satellite manufacturing, although with little funding. An agreement in 2004 with Russia concerned the expansion of cooperative efforts in space, including the joint development and production of launch vehicles, the launch of geostationary satellites, and the joint development and utilization of Brazil's Alcântara Launching Center (Centro de Lançamento de Alcântara, CLA) in Maranhão. There were no launches in 2007.

Israel is conducting modest space operations, with one launch of its Shavit rocket in 2007, carrying the *Ofeq-7* advanced technology remote-sensing satellite on June 11 from its Shavne launch site at Palmachin Air Force Base (also known as Yavne).

In Canada, the Canadian Space Agency (CSA) continued supporting work on its contribution to the *ISS* partnership, the Mobile Service System (MSS), consisting of the 1800-kg (3960-lb) Space Station Remote Manipulator System (SSRMS) *Canadarm2*, the Mobile Base System (MBS), and the Special Purpose Dexterous Manipulator (SPDM) "Dextre." Canada also has an active Canadian Astronaut Program for flights to the *ISS*. One of them, Dave Williams, flew to the *ISS* on STS-118/Endeavour in August 2007 as a Mission Specialist.

In 2007, *RadarSat-2*, Canada's second "eye-in-the-sky," was successfully launched for CSA on December 14 by Starsem on a Soyuz-FG from the Baikonur Cosmodrome. It has a SAR (Synthetic Aperture Radar) with multiple polarization modes and a highest resolution of 3 m (10 ft) with 100-m (330-ft) positional accuracy. *RadarSat-2* is a follow-on to *RadarSat-1*. Having the same orbit [798 km (496 mi) altitude sun-synchronous with 6 p.m. ascending node and 6 a.m. descending node], it is separated by half an orbit period (~50 min) from *RadarSat-1* (in terms of ground track, that represents ~12 days ground track separation). It is intended to fill a wide variety of roles, including sea-ice mapping and ship routing, iceberg detection, agricultural crop monitoring, marine surveillance for ship and pollution detection, terrestrial defense surveillance and target identification, geological mapping, land-use mapping, wetlands mapping, and topographic mapping. *RadarSat-1* meanwhile celebrated its twelfth anniversary in orbit. Launched on November 4, 1995, the sophisticated radar platform was expected to operate only 5 years, and the quality of images it captured exceeded the standards of the time. It is still operating and surpassing the standards.

For background information *see* ASTEROID; AURORA; CERES; CHANDRA X-RAY OBSERVATORY; CLOUD; COMET; COMMUNICATIONS SATELLITE; COSMIC BACKGROUND RADIATION; DARK MATTER; EARTH, GRAVITY FIELD OF; GALAXY, EXTERNAL; GAMMA-RAY ASTRONOMY; GAMMA RAY BURSTS; GEMINGA; HUBBLE SPACE TELESCOPE; INFRARED ASTRONOMY; ION PROPULSION; MARS; MERCURY (PLANET); MESOSPHERE; METEOROLOGICAL SATELLITES; MILITARY SATELLITES; MOON; PLEIADES; PLUTO; RELATIVITY; REMOTE SENSING; SATELLITE NAVIGATION SYSTEMS; SATURN; SCIENTIFIC AND APPLICATION SATELLITES; SOLAR WIND; SPACE FLIGHT; SPACE PROBE; SPACE SHUTTLE; SPACE STATION; SPACE TECHNOLOGY; SPITZER SPACE TELESCOPE; SUN; ULTRAVIOLET ASTRONOMY; VENUS; WILKINSON MICROWAVE ANISOTROPHY PROBE; X-RAY ASTRONOMY in the McGraw-Hill Encyclopedia of Science & Technology.   Jesco von Puttkamer

Bibliography. *Aerospace Daily*, various 2007 issues; *AIAA Aerospace America*, December 2007 issue; *Aviation Week & Space Technology (AW&ST)*, various 2007 issues; ESA *Press Releases*, 2007; NASA Public Affairs Office *News Releases*, 2007; *Space News* (various 2007 issues); various Internet sites.

Species and global climate change

Observations started by Charles Keeling at Manoa Loa, Hawaii, in the 1950s provided the indisputable evidence that anthropogenic carbon dioxide in the atmosphere has been increasing at an alarming rate. The current level is more than 380 parts per million by volume, which is already much higher than the natural range of 180 to 300 ppm in the past 650,000 years as recorded by ice cores. Carbon dioxide is a greenhouse gas, which, according to earlier calculations by Svante Arrhenius in the 1890s and confirmed by current numerical models, can cause a warming of our planet. Other greenhouse gases such as methane and nitric oxide have also been increasing, further exacerbating the problem. Meanwhile, since the beginning of the twentieth century, the surface air temperature of the Earth as measured by meteorological stations has increased by about $0.7°C$ ($1.2°F$). The data also show that the rate of increase in the last 50 years has been twice as great as the rate in the last 100 years, with the 10 warmest years on record occurring since 1995. Among the key issues associated with global warming is the fate of millions of species that inhabit the Earth. Warming disrupts and alters the ecosystems and therefore the diversity of plant and animal species in the systems. In this regard, the polar regions have been the center of attention, because climate signals in these regions are expected to be amplified by as much as three to five times as a result of ice–albedo feedbacks. (Albedo is the fraction of solar energy reflected by a surface. In ice–albedo feedback loops, melting ice resulting from warmer temperatures causes a lowering of the albedo, which leads to even more ice melting.)

Global analyses. Analyses of global data indeed show enhanced warming at high latitudes and especially the Arctic region. Associated changes in the ecosystems of these regions are already apparent in areas covered by sea ice, snow, permafrost, glaciers, ice sheets, and in surrounding areas. The survival rates of species in the polar regions are also relatively lower than in the tropics because there is much less species diversity in the polar regions. In high-diversity regions, such as the tropics, the ecologies are more interconnected; thus, when a species is removed from a system, the other species that rely on the former for food have the option to switch to other species within the system. However, in low-diversity systems, such an option may not be possible.

Satellite data have been used to evaluate changes in surface temperature in the polar regions from 1981 to 2007. Although only subtle changes are apparent from the 13-year averages, the difference maps show remarkable variations spatially in both hemispheres (**Fig. 1**). It is apparent that warming is much more dominant in the Northern Hemisphere than in the Southern Hemisphere, although both regions show evidence of cooling in some areas (parts of Russia in the north and eastern Antarctica in the south). Overall, the trend of average temperatures at latitudes higher than 60° N is $0.7°C$ ($1.2°F$) per decade in the Northern Hemisphere, while that at higher than 60° S is about $0.2°C$ ($0.36°F$) per decade in the Southern Hemisphere. The large variability in the changes reflects the complexity of the atmospheric circulation and suggests that the effect of climate change on ecosystems could vary strongly from region to region in each hemisphere.

A natural ecosystem, often referred to as a biome, represents a system consisting of a diversity of plants and animals that live in a natural balance.

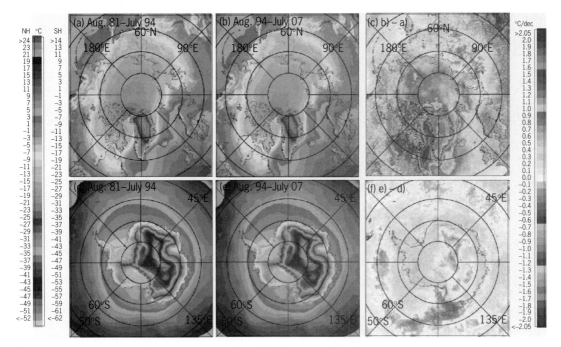

Fig. 1. Averages of surface temperatures from satellite AVHRR (Advanced Very High Resolution Radiometer) data from (*a*) August 1981 to July 1994 in the Northern Hemisphere, (*b*) August 1994 to July 2007 in the Northern Hemisphere, (*d*) August 1981 to July 1994 in the Southern Hemisphere, and (*e*) August 1994 to July 2007 in the Southern Hemisphere. Difference maps of the temperature averages in (*c*) Northern Hemisphere and (*f*) Southern Hemisphere.

Innumerable species of plants and animals that inhabit this planet have survived for a long time under such systems, which are considered a norm in the past. Recent warming effects, however, might be too abrupt for many species. The ecosystem also provides the food, water, nutrients, shelter, and other necessities of the various life forms. Hence, a sudden shift in the boundaries of the system (in space and time) may catch some species off-guard who are in search of food and cause such species to starve. The temperature images demonstrate subtle shifts in the temperature isotherms from one period to another, but even small shifts could cause considerable changes in the ecosystem boundaries when shifts occur at the threshold temperatures (e.g., near melt temperatures).

Aquatic ecosystems. Sea ice covers about 3–6% of the surface area of the Earth and is a key component of the Earth's climate system. Being a habitat and a source of food, it is an ecosystem in itself and could also influence the ecosystems of surrounding and other regions. In the Northern Hemisphere, the total ice area has been decreasing at a rate of about −4% per decade since 1979. More intriguing, however, has been the rapid decline in the Arctic perennial ice cover (the thick, multiyear ice floes that survive the summer melt) of around 11% per decade during the same period. In addition, a dramatic drop in the perennial ice cover occurred in 2007, with the areal coverage being 38% less than that of climatology estimates and 27% less than the previous lowest value in 2005 (**Fig. 2**). A continuation of this trend would mean a blue ocean in the Arctic in the summer and profound changes in the ecosystem of the region. In the Southern Hemisphere, the trend in the total ice

cover is positive, about +1% per decade, which is in part due to significant cooling in some areas such as the Ross Sea region, where the increase in the sea ice cover is greatest. In other areas, such as the

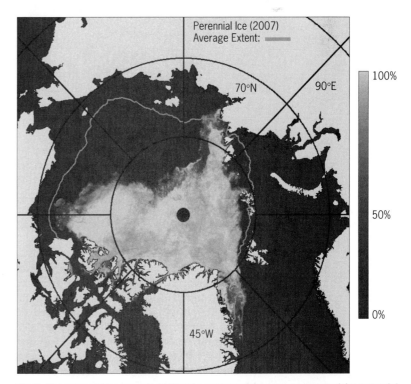

Fig. 2. The perennial ice cover in 2007 and a contour of the average extent of the perennial ice from 1979 to 2007. Trends in perennial ice area suggest that the perennial sea ice, which is a habitat of many microorganisms and some animals, will disappear within this century. Data represent percentage concentrations of sea ice as derived from the Advanced Microwave Scanning Radiometer (AMSR-E) onboard the EOS/Aqua satellite.

Fig. 3. Species in their climatic environments. (a) Penguins in the Antarctic. (b) Polar bear in the Arctic (*photo by Steven C. Amstrup*). (c) Caribou in western Greenland (*photo by Eric Post*). These are some of the many species that are threatened by climate change and the retreat of the sea ice cover.

Bellingshausen and Amundsen Seas, however, the sea ice cover is declining at a relatively fast rate. In the latter region, the populations of many species, including penguins, have been observed to be declining (**Fig. 3***a*).

The melt of sea ice also leads to the formation of a stratified upper layer in the ocean that provides an ideal environment for photosynthesis. The presence of sea ice is thus at least in part responsible for the occurrence of phytoplankton blooms and enhanced primary productivity at high latitudes. Phytoplanktons are at the bottom of the food web, and their abundance at high latitudes makes the region attractive for thousands of species of marine animals, including fish. The retreat of the sea ice cover would thus mean a shift in the boundary of such a highly productive region, leaving behind some species that cannot adapt (Fig. 2).

Sea ice is also a habitat of many microorganisms, including bacteria and algae. During ice freeze-up, these microorganisms accumulate within the water trapped between loose agglomerations of ice crystals (e.g., between platelets). The microorganisms are usually found at the ice bottom, where water channels are relatively larger and warmer and can be highly concentrated as the color of the ice turns brown. The algae at the bottom of sea ice serve as food for krill, a shrimplike creature, which is in turn the primary source of food for fish, penguins, seals, and whales. The decline in the ice cover in the Bellingshausen Sea has been used to explain the decline of the krill population in the region, which thus may have caused the observed decline of penguins and marine mammals in the region (Fig. 3*a*). Sea ice is also the platform that some animals use in searching for food. The best-known example is the polar bear, which feeds on seals which also use the platform as a hunting ground for fish through holes in the ice (Fig. 3*b*). Rapid retreat of sea ice in the Arctic spring (Fig. 2) has caused the ice platform to be inaccessible to many polar bears, depriving them of their ability to search for food.

Warming in the oceans also makes the coral reefs, which have some of the most diverse ecosystems on Earth, especially vulnerable. The coral reefs are famous for their magnificent colors, but these colors are actually provided by algae, called zooxanthellae, that have a symbiotic relationship with the reef. The reefs are built by the corals, which are sedentary animals that house the zooxanthellae that perform photosynthesis and provide nutrition to the corals. Coral bleaching occurs when the corals are stressed and they respond by expelling their symbiotic algae. This usually happens during heat spikes, which is when the ocean temperature warms up even just slightly higher than normal. Without the algae, which is the source of food for the reef, the latter eventually die.

Terrestrial ecosystems. Thawing of the permafrost in many parts of the Arctic region has been observed at the same time as the areal extent of snow has been declining. Permafrost thawing leads to a transformation of vegetation in the region from a predominance of tundras to taller and denser vegetation. Such shifts in vegetation may greatly reduce the breeding areas of many birds and the grazing areas of animals that depend on open landscapes that the tundras in the Arctic have been providing. Evidence of changing vegetation in the region is illustrated in normalized difference vegetation index (NDVI) maps (**Fig. 4**). The shift may cause sharp declines, if not extinction, of many polar species (Fig. 3*c*). Polar animals will also be threatened by the invasion of some animals that come from lower latitudes and cannot cohabit with the former. Because taller vegetations have lower reflectivities than tundras, the region will absorb more solar energy, causing even further warming in the region.

Early onset of snow melt in spring causes an early advent of vegetation in snow-covered regions, which in turn causes mismatch with the migratory patterns

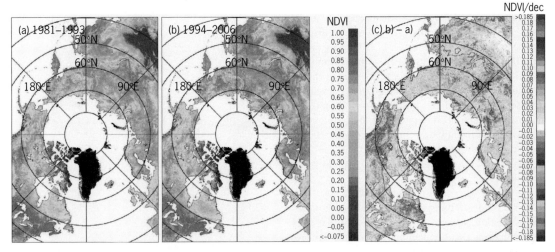

Fig. 4. Averages of the normalized difference vegetation index (NDVI) in summer (August) (*a*) from 1981 to 1993 and (*b*) from 1994 to 2006; (*c*) difference of the averages from (*b*) and (*a*). The line in the difference map represents the location of the edge of the discontinuous permafrost. The thawing of the permafrost has caused the NDVI in some regions of the Arctic to increase significantly. The NDVI is simply the ratio of the difference and the sum of the near-infrared and visible channels of the AVHRR sensor and has been shown to be strongly correlated to vegetation pigments.

of birds and animals. A warmer spring will also cause changes in timing of the flowering of plants, keeping them from being cross-pollinated by unsuspecting bees. The primary effect of the melting of glaciers and ice sheets is a rise in sea level, which can range from several centimeters to several meters. Many species that thrive in coastal regions and wetlands will likely vanish if they cannot migrate to higher ground. Glaciers and snow cover are also part of the hydrology of many regions and are the main sources of fresh water that most species (including humans) in these regions need to survive.

As global temperature continues to increase, the frequency of drought in some areas and flooding in other areas will increase. Drought can cause serious devastations to the ecology of a region as soil begins to dry up and become unproductive, and plant and animal life becomes impossible to maintain. In drought areas, the frequency and duration of fires are also enhanced, and the associated effects on vegetation and animal life can be very strong. Fires can completely change the structure of an ecosystem, and recovery in some places can take a long time. With warmer seas, hurricanes will also become stronger, causing coastal erosion, flooding, and the devastation of vegetation—all leading to further stress on species. There is currently a paucity of quantitative assessments, but in situ and modeling studies of the effects of warming on various species are among the many projects being carried out during the 2008 International Polar Year.

For background information *see* ALBEDO; ANTARCTICA; ARCTIC OCEAN; BIODIVERSITY; CLIMATE MODIFICATION; CLIMATOLOGY; ECOSYSTEM; ENDANGERED SPECIES; GLOBAL CLIMATE CHANGE; GREENHOUSE EFFECT; PERMAFROST; SEA ICE in the McGraw-Hill Encyclopedia of Science & Technology.

Josefino C. Comiso

Bibliography. ACIA, *Arctic Climate Impact Assessment*, Cambridge University Press, Cambridge, UK, 2005; J. C. Comiso, Arctic warming signals from satellite observations, *Weather*, 61(3):70–76, 2006; J. C. Comiso and C. L. Parkinson, Satellite observed changes in the Arctic, *Phys. Today*, 57(8):38–44, 2004; J. C. Comiso et al., Accelerated decline in the Arctic sea ice cover, *Geophys. Res. Lett.*, 35:L01703, 2008; I. A. Melnikov, *The Arctic Sea Ice Ecosystem*, Gordon & Breach, Amsterdam, 1997; J. Stroeve et al., Arctic sea ice decline: Faster than forecast, *Geophys. Res. Lett.*, 34:L09501, 2007; C. D. Thomas et al., Extinction risk from climate changes, *Nature*, 427:145–148, 2004; D. N. Thomas, *Frozen Oceans*, Natural History Museum, London, 2004; G. R. Walther et al., Ecological responses to recent climate change, *Nature*, 416:389–395, 2002.

Starlet sea anemone

There are millions of species living on planet Earth, yet experimental biologists focus on just a few dozen model systems. Ironically, these are often among the most mundane representatives of their respective evolutionary "families." The fruit fly, *Drosophila melanogaster*, has been studied in exquisite detail, but no comparable effort has been expended on the bombardier beetle, the mole cricket, or the luna moth, each of which is unquestionably more spectacular than *Drosophila*. Although model systems can seem pedestrian, they also offer a distinct advantage—when a small number of easily manageable experimental organisms are widely utilized, the scientific community can address fundamental biological questions with greater efficiency and economy.

The model systems approach assumes that biological processes are widely conserved among species; thus, lessons learned from the model are applicable to other organisms. This assumption has proven valid to a degree. For example, genetic studies on

Drosophila have advanced the understanding of human genetics. However, every organism represents a unique evolutionary history and a distinct ecological context; this limits the ability to generalize. Therefore, the model systems approach must be complemented with a biodiversity-based approach. Evolutionarily diverse organisms in ecologically diverse natural habitats must be studied, and the portfolio of model systems must be continually diversified, choosing new models not merely for their utility as laboratory tools, but also for the evolutionary and ecological insights that they can provide.

The starlet sea anemone, *Nematostella vectensis* (**Fig. 1**), is an emerging model system that complements existing model systems in terms of its ecology and evolution. As a basal animal, *Nematostella* provides a new perspective on early animal evolution. Furthermore, compared with the pantheon of major animal models (see **table**), *Nematostella* represents a number of important firsts, being the only estuarine species, the only species capable of complete bidirectional regeneration, and the only species capable of both sexual and asexual reproduction.

Nematostella. *Nematostella* is a small, inconspicuous sea anemone that lives partially burrowed in the soft sediments of estuarine habitats such as salt marshes and mudflats. It feeds on planktonic animals and worms. Although this anemone was first collected in England, genetic evidence suggests that the species is actually native to the Atlantic coast of North America. It was apparently transplanted to

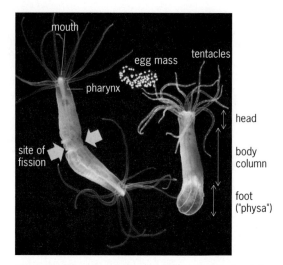

Fig. 1. Starlet anemone, *Nematostella vectensis*. An adult female (*right*) has just extruded an egg mass through her mouth. The major body regions along the oral–aboral axis include the head, the body column, and the foot. The two-headed individual (*left*) is undergoing a form of asexual reproduction known as polarity reversal. Fission will split the body column (*arrows*), producing two genetically identical individuals.

England and the Pacific coast of North America, presumably by human activity.

Utility as a model system. As with most established models, *Nematostella* is easy to raise in the laboratory. It can be cultured at high densities in small bowls of standing artificial seawater. It reproduces prolifically; well-fed females produce egg masses weekly, each containing tens to hundreds of eggs

Nematostella compared with major animal models			
Species	Higher taxon	Natural habitat	Reproduction and regeneration
Starlet anemone (*Nematostella vectensis*)	Phylum Cnidaria Class Anthozoa	Aquatic/coastal estuaries	Sexual: males & females; egg-laying Asexual: two forms of fission Regeneration: complete bidirectional
Insects Fruit fly (*Drosophila melanogaster*) Flour beetle (*Tribolium castaneum*) Honey bee (*Apis mellifera*)	Phylum Arthropoda Class Insecta	Terrestrial	Sexual: males & females; egg-laying Asexual: none Regeneration: none
Soil nematode (*Caenorhabditis elegans*)	Phylum Nematoda	Terrestrial/soil	Sexual: hermaphrodites & males egg-laying Asexual: none Regeneration: none
Purple sea urchin (*Strongylocentrotus purpuratus*)	Phylum Echinodermata	Aquatic/shallow coastal marine habitats	Sexual: males & females; egg-laying Asexual: none Regeneration: none
Ray finned fishes Zebrafish (*Danio rerio*) Medaka (*Oryzias latipes*)	Phylum Chordata Subphylum Vertebrata Class Actinopterygii	Aquatic/freshwater	Sexual: males & females; egg-laying Asexual: none Regeneration: none
Clawed frogs (*Xenopus laevis*) (*Xenopus tropicalis*)	Phylum Chordata Subphylum Vertebrata Class Amphibia	Aquatic/freshwater	Sexual: males & females; egg-laying Asexual: none Regeneration: none
Birds Chicken (*Gallus gallus*)	Phylum Chordata Subphylum Vertebrata Class Aves	Terrestrial	Sexual: males & females; egg-laying Asexual: none Regeneration: none
Rodents Mouse (*Mus musculus*) Rat (*Rattus norvegicus*)	Phylum Chordata Subphylum Vertebrata Class Mammalia	Terrestrial	Sexual: males & females; live birth Asexual: none Regeneration: none

(Fig. 1). The fertilized eggs develop into sexually mature adults in about 50 days.

Also like the established models, *Nematostella* is amenable to a myriad of molecular, cellular, biochemical, and genomic techniques. Unlike many marine invertebrates, it is easy to isolate deoxyribonucleic acid (DNA), ribonucleic acid (RNA), or protein. The technique of in situ hybridization has been optimized, so it is possible to visualize when and where genes are expressed in either embryos or adult anemones. It is possible to "knock down" the expression of individual genes using RNA interference, enabling investigations of gene function. In 2007, the starlet sea anemone became the first basal animal whose entire genome was sequenced.

Unique advantages as a model system. *Nematostella* exhibits a key experimental advantage that is not matched by the leading model systems (see table). When bisected transversely through the body column, the anemone rapidly undergoes complete bidirectional regeneration, producing two genetically identical individuals. This enables the rapid production of living genetic stocks. Aside from this practical advantage, regeneration itself is an issue of fundamental biological importance. If the molecular basis of this ability in *Nematostella* can be deduced, it may enable understanding of why vertebrates, insects, and nematodes cannot regenerate. Regeneration also can be used to evaluate the environmental tolerances of *Nematostella*—that is, the regenerative ability of different genetic strains can be compared under different conditions (varying temperature, salinity, etc.).

Relative to the major animal models in developmental biology, *Nematostella* exhibits a stunning degree of reproductive flexibility. The adult may arise via embryogenesis following sexual reproduction, or via two distinct forms of asexual fission termed physal pinching and polarity reversal. In physal pinching, the foot (or physa) pinches off, and it then regenerates a new body column and head. During polarity reversal (Fig. 1), a new head forms at the site of the existing foot, resulting in a "two-headed" creature. Fission then separates two fully developed offspring.

The reproductive and developmental flexibility of *Nematostella* raises a number of profound evolutionary questions that cannot be addressed using fruit flies, nematodes, or mice. How is this flexibility encoded in the genome, and why is it lacking in most other animals? How do the anemones decide whether to reproduce via sexual or asexual means? How does such reproductive flexibility affect the course of evolution?

The ecological context of *Nematostella* is also not represented among the major model systems. Estuaries are challenging habitats because of their natural variability in critical environmental parameters such as temperature and salinity. It follows that estuarine animals must have wide environmental tolerances—*Nematostella* is known to tolerate salinities ranging from 2 to 52 parts per thousand

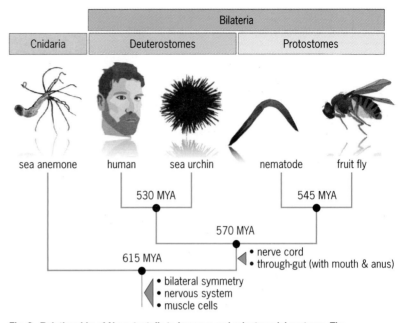

Fig. 2. Relationship of *Nematostella* to humans and select model systems. The approximate age of each common ancestor (*circles*) is shown in millions of years before the present (MYA). Triangles pinpoint the origin of select animal traits.

and temperatures ranging from −1 to 28°C (30 to 82°F). The genetic basis for such wide environmental tolerances is not yet understood, but the starlet sea anemone is an ideal model for investigating this question.

"Basal" animal. More than 99% of all living animals are members of the Bilateria (so named because most of its members exhibit bilateral symmetry). As a member of the phylum Cnidaria—a basal animal group that includes sea anemones, corals, sea pens, jellyfishes, box jellies, and hydras—*Nematostella* is among the small minority of animals that fall outside the Bilateria (**Fig. 2**). The Cnidaria and the Bilateria diverged some 600 million years ago and, following their divergence, they experienced vastly different fates. The ancestral Bilaterian gave rise to more than 1 million modern descendants encompassing an enormous range of shapes and sizes—from great apes to giant squids, from butterflies to blue whales. The ancestral Cnidarian gave rise to only 10,000 modern descendants, most sharing a similar "polyp" body plan (at least during part of their life cycle).

If the divergent fates of Cnidaria and Bilateria—one of the great mysteries of animal evolution—are to be understood, it is necessary to reconstruct their common ancestor. Recognizing shared features in such distantly related animals is difficult. Anemones and humans do not share any organs or body parts. However, research on *Nematostella* has revealed a number of surprising similarities with humans, including key aspects of body architecture and key features of the genome.

Deep conservation of body plan architecture. The overwhelming majority of bilaterians exhibit bilateral symmetry, that is, a single plane of mirror symmetry divides their bodies into symmetrical left and

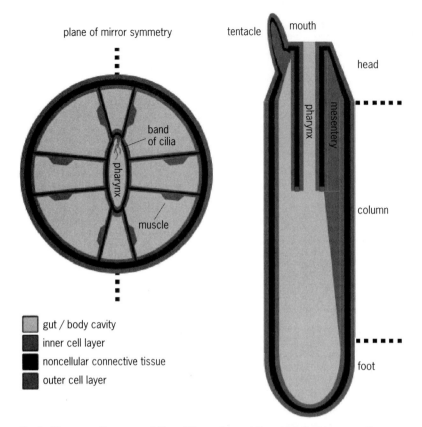

Fig. 3. Diagrammatic representation of the anatomy of *Nematostella*. A cross section through the pharyngeal region (*left*) reveals that there is a single plane of mirror symmetry bisecting the pharynx—that is, *Nematostella* is bilaterally symmetrical.

right halves. This type of symmetry requires the presence of two orthogonal body axes—the anterior-posterior (AP) axis, which passes from head to tail, and the dorsal-ventral axis, which passes from back to belly.

Bilateral symmetry is widely considered a key innovation of the Bilateria, partially responsible for their

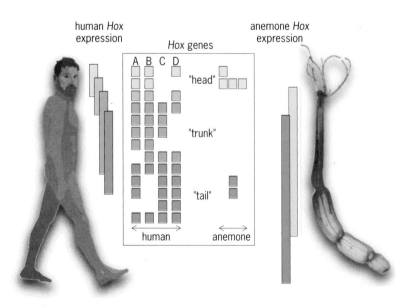

Fig. 4. A comparison of *Hox* gene expression along the main body axis of *Nematostella* and humans. Humans have 39 *Hox* genes; the anemone has only 6. In both species, corresponding *Hox* genes are expressed in comparable regions along the primary body axis (the anterior-posterior axis in humans or the oral-aboral axis in the anemone).

remarkable evolutionary success. Although most invertebrate zoology textbooks characterize Cnidarians as radially symmetrical (like cylinders), many corals and sea anemones (including *Nematostella*) exhibit some features of bilateral symmetry (**Fig. 3**).

The presence of bilateral symmetry in the Bilateria and the Cnidaria suggests that this feature could have existed in their common ancestor. If so, then both groups inherited the same genes required to establish two orthogonal body axes. In the Bilateria, *Hox* genes play a conserved role in establishing the anterior-posterior axis, and *Dpp* plays a conserved role in establishing the dorsal-ventral axis. *Nematostella* possesses these same genes. Based on where they are expressed in the larva during development, it appears that *Hox* genes may be involved in establishing its oral-aboral axis (**Fig. 4**), whereas *Dpp* may be involved in establishing a second orthogonal axis that passes through the animal's pharynx. This is compelling molecular evidence that bilateral symmetry may have predated the evolutionary origin of the Bilateria themselves.

Deep conservation of the genome. Despite its more ancient kinship with humans, *Nematostella* possesses a slightly greater number of genes that are implicated in human diseases than do the fruit fly or soil nematode. Furthermore, in some cases in which all three species possess a clear counterpart to a human disease gene, the anemone gene is most similar to the human gene. This counterintuitive finding is part of an emerging consensus that the genomes of fruit flies and nematodes have experienced extensive gene loss, while the genomes of humans and *Nematostella* have evolved in a more conservative fashion. As a consequence, *Nematostella* will prove valuable for reconstructing the early evolution of animal genomes, and it will provide insights into the functional evolution of proteins implicated in human disease.

For background information *see* ACTINIARIA; ANIMAL EVOLUTION; ANIMAL KINGDOM; ANIMAL REPRODUCTION; ANIMAL SYMMETRY; BILATERIA; CNIDARIA; GENE; GENETIC CODE; PHYLOGENY; REGENERATIVE BIOLOGY; SEA ANEMONE in the McGraw-Hill Encyclopedia of Science & Technology.   John R. Finnerty

Bibliography. J. A. Darling et al., Rising starlet: The starlet sea anemone, *Nematostella vectensis*, *Bioessays*, 27:211–221, 2005; J. R. Finnerty et al., Origins of bilateral symmetry: *Hox* and *Dpp* expression in a sea anemone, *Science*, 304:1335–1337, 2004; C. Hand and K. Uhlinger, The culture, sexual and asexual reproduction, and growth of the sea anemone *Nematostella vectensis*, *Biol. Bull.*, 182:169–176, 1992; N. H. Putnam et al., Sea anemone genome reveals ancestral eumetazoan gene repertoire and genomic organization, *Science*, 317:86–94, 2007; T. A. Stephenson, *The British Sea Anemones*, vol. II, The Ray Society, London, 1935; J. C. Sullivan and J. R. Finnerty, A surprising abundance of human disease genes in a simple "basal" animal, the starlet sea anemone (*Nematostella vectensis*), *Genome*, 50:689–692, 2007.

Syngas from biomass

The international price of oil per barrel has risen rapidly and has broken record levels. This results in high prices for gasoline and diesel fuel, and is a major reason for seeking alternative sources of transportation fuel, especially with potentially insecure supplies of imported petroleum. Other reasons include the environmental impacts of greenhouse gases, sulfur, heavy metals, and acid rain precursors emitted from burning fossil fuels. The production of alternative fuels from renewable lignocellulosic biomass from forest and agricultural growth and municipal solid waste is potentially a major means of alleviating the environmental impacts of fossil fuels, and there is hope for achieving this economically. One of the highest yielding methods for obtaining alternative fuels from biomass is through gasification and use of syngas (synthesis gas) to make diesel fuel, gasoline, jet fuel, or alcohols (although the capital costs are now high, this process should prove more sustainable and potentially more economical in the long term). As with transportation fuel, there is the same urgency for using more economical and environmentally benign fuels to generate power, and to provide heat for process energy in manufacturing and warmth for our comfort.

Syngas from biomass as a path to alternative fuels. Syngas is a generally accepted term for gas that is used to make fuels and chemicals synthetically. Synthesis gas (or syngas) and sometimes water gas are terms that are also commonly applied to gas as it comes from a gas producer or gasifier, before it has been cleaned sufficiently to use as a feedstock for further processing. However, in this article, the gas as it comes from the gasifier is called producer gas instead of syngas, and only the product that is used in further fuel refinement is called syngas.

Producer gas has been used since the early twentieth century as a fuel for internal combustion engines and is still used for that purpose today. Syngas from coal has been used steadily to make transportation fuel in South Africa for more than 50 years. In both syngas and producer gas, the main constituents are carbon monoxide (CO) and hydrogen (H₂).

Gasifier types. Producer gas is made in fixed-bed as well as in fluidized bed, entrained flow, and plasma arc gasifiers (*see* **illustration**). A fixed bed gasifier may be updraft, downdraft, or side draft. It may be air-blown or oxygen-blown. If the gasifier is air-blown, the producer gas will have high nitrogen content. For operating fixed-bed gasifiers, moisture content of the feedstock must be fairly low.

In an updraft or countercurrent gasifier, solid fuel enters a vessel from the top and comes to rest at the floor on the bottom of the vessel. Air or oxygen, and perhaps steam, enters at the bottom of the bed of fuel to react with the fuel and produce gas. This producer gas exits at the top of the gasifier, thus traversing counter to the direction of fuel feed. In a downdraft gasifier, the air or oxygen is blown in from the top and goes through the gasifier in the same direction as the fuel. In a side-blown gasifier, the air or oxygen is blown in from the side of the vessel. In a bubbling or stationary fluidized bed, there is a bed of inert material, usually sand. Combustion of fuel keeps the sand hot. Gasification fuel enters the bed from the top and producer gas leaves from the top. The inert bed is kept in motion by fluidizing recycled producer gas entering the bed from the bottom. A circulating fluidized bed has two beds: a combustion bed and a gasification bed. The inert material is circulated between the beds so that heated bed material from combustion is circulated to the gasification bed without nitrogen from combustion air. A cyclone separator may be used to remove the ash

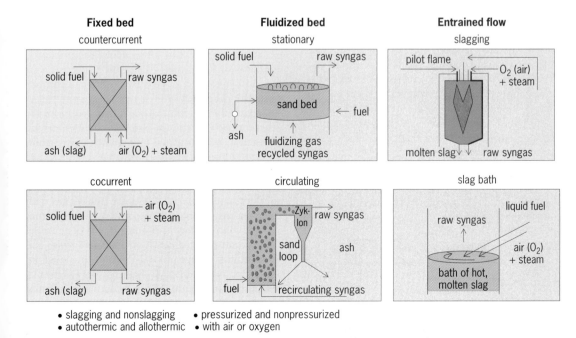

• slagging and nonslagging • pressurized and nonpressurized
• autothermic and allothermic • with air or oxygen

Common types of gasifiers.

TABLE 1. Comparable crude producer gas compositions for fixed bed gasifiers*

Component	Gas (vol. %) with wood and air-blown	Gas (vol. %) with MSW and oxygen-blown
Hydrogen	18.3	26.0
Carbon monoxide	22.8	40.0
Carbon dioxide	9.2	23.0
Methane	2.5	5.0
Hydrocarbon	0.9	5.0
Oxygen	0.5	0.5
Nitrogen	45.8	0.5

*Permission has been granted from Forschungszentrum Karlsruhe.

from combustion and gasification. In an entrained flow gasifier, there is no bed of solid material, but air or oxygen and steam enter the gasifier at high enough temperature and pressure to keep ash in a molten state; then, as producer gas is generated, it leaves from the molten slag. Plasma arc gasifiers subject the solid gasification fuel to high heat generated through electricity that is used to provide a carbon arc.

Some typical constituents of gas from a fixed-bed air-blown gasifier using wood chips as feedstock and a fixed-bed oxygen-blown gasifier using municipal solid waste (MSW) as feedstock are shown in **Table 1**. The air-blown gasifier produces a low calorific value gas (calorific value is the amount of heat generated by a given mass of fuel when it is completely burned), and the oxygen-blown gasifier produces a medium calorific value gas.

With fluidized-bed gasifiers, moisture contents of the feedstock may usually be higher and uniformity of feedstock particulate size is usually not as critical. Nitrogen contents of the producer gas are low, and a medium calorific value gas is produced. For several years, a circulating fluidized-bed gasifier operated at a large municipal power plant in Burlington, Vermont. Both the power plant and the gasifier operated with wood chips as the feedstock. After the developers of the gasifier determined that they had accumulated enough research data, they ceased operation at the power plant location. Typical producer gas constituents for the fluidized-bed gasifier are shown in **Table 2**. The high heating value (HHV) indicates that it is a medium calorific value gas.

TABLE 2. Crude producer gas compositions for a fluidized bed gasifier

Component	Gas (vol. %) with wood
Hydrogen	18.0
Carbon monoxide	47.0
Carbon dioxide	14.3
Methane	14.9
Hydrocarbon (ethylene and ethane)	5.8
HHV in MJ/m³	16.8 (455 Btu/ft³)

Entrained flow gasifiers can operate at higher pressures. This makes them well suited for higher capacity up to 1000 MW$_{th}$ (megawatt thermal).

There is a newer gasification process that is in early implementation stages for treating municipal solid waste. Known as plasma arc gasification technology, two plants have been operating in Japan. In this technology, electricity produces an arc at high temperatures. The hot plasma arc gasifies organic materials, and inorganic materials melt and subsequently solidify.

Formation of producer gas in gasifiers. Although plasma arc gasification is adapted mainly for turning municipal solid waste into heat energy for generating electricity, producer gas from gasifiers provides a promising feedstock for syngas to make alternative fuels, as well as containing heat that may be used for other purposes. To make syngas from producer gas, it is beneficial to use endothermic heat (heat absorbed from the environment) most effectively to make gas, and not have to unnecessarily discard exothermic heat (heat given off to the environment) from the gasification process. In some cases, much of this heat may be used in cleaning the producer gas and making liquid fuel from syngas.

The following reactions typically occur during gasification:

$$C + H_2O \rightarrow CO + H_2 + 118.4 \text{ kJ/mol}$$
$$C + CO_2 \rightarrow 2CO + 160.9 \text{ kJ/mol}$$
$$C + 2H_2O \rightarrow CO_2 + 2H_2 + 16.3 \text{ kJ/mol}$$
$$C + 2H_2 \rightarrow CH_4 - 87.4 \text{ kJ/mol}$$
$$CO + H_2O \rightarrow CO_2 + H_2 - 42.3 \text{ kJ/mol}$$
$$\text{(water gas shift reaction)}$$
$$CO + 3H_2 \rightarrow CH_4 + H_2O - 205.8 \text{ kJ/mol}$$
$$2CO + 2H_2 \rightarrow CH_4 + CO_2 - 248.4 \text{ kJ/mol}$$

A plus sign (+) indicates an endothermic reaction; a negative sign (−) indicates an exothermic one. From the standpoint of energy conservation, it is better to generate less methane (CH_4) and carbon dioxide (CO_2) in making producer gas. Both of these compounds are unusable in subsequent Fischer-Tropsch synthesis of liquid fuels. When the water gas shift reaction is applied, it produces a better hydrogen to carbon monoxide ratio. It produces energy, but it eliminates the heat value of the CO that is used for the reaction and it ties up carbon that could otherwise be used in making liquid fuel. Sometimes it might be more advantageous to add hydrogen that is made using renewable or nuclear energy instead of making it with steam. Generally, to use a gas for further product synthesis, it is also better to have gas with a higher energy value, depending on the optimum ratio of H_2 to CO for the product that is to be made.

From producer gas to syngas and liquid fuels. The method most often employed to make liquid fuels by using syngas is the Fischer-Tropsch process (a catalytic process to synthesize hydrocarbons and their oxygen derivatives by the controlled reaction of hydrogen and carbon monoxide). In applying the process, coal feedstock has usually been used. There is also a plant making liquid fuels from natural gas in

Malaysia. Fischer-Tropsch synthesis is used to provide a variety of end products depending on temperatures, pressures, and catalysts. These products may be mixtures of alcohols or straight-chain hydrocarbons. In the Fischer-Tropsch process, biomass may be substituted for fossil fuels. Besides the environmental advantage, wood has advantages over fossil fuels in containing more volatiles for gasification. Fossil fuels have a higher carbon density, and the best approaches for using biomass economically as a feedstock are being sought.

Besides having a higher energy density than wood and other biomass such as straw, paper, switchgrass, and corn stover, fossil fuels are concentrated at locations where they are found in the ground. To overcome the energy density problems and resultant higher transportation costs for feedstock, one approach is to transform the biomass into bio-oil (pyrolysis oil) and tar by heating it at a temperature around 500°C (932°F) in the absence of air. By thus increasing the energy density, the biomass may be carried more economically to a central location for further gasification.

The first task must be to transform the gas as it comes from the producer to syngas to use in the Fischer-Tropsch process. Crude gas is processed to remove water vapor, tars, organics, hydrocarbons, nitrogen, and carbon dioxide. Then, the clean gas, containing primarily H_2 and CO, is often processed in a shift reactor to react part of the CO to form additional H_2, so the final gas contains the proper ratio of H_2 and CO. In the shift reactor, additional CO_2 is formed and it again must be removed. Because the moisture condensed from the crude gas contains soluble organics, the stream must be cleaned for environmental reasons.

The Fischer-Tropsch process was first patented in 1923. It is characterized by catalytically converting syngas to alkanes and alcohols. The original Fischer-Tropsch reaction is defined by the following equation:

$$(2n+1)H_2 + nCO \rightarrow C_nH_{(2n+2)} + nH_2O$$

Since it was originally conceived, many refinements and adjustments have been made, and the term now applies to a wide variety of similar processes. The hydrocarbon products are further refined to various synthetic fuels including alcohols.

Other uses of producer gas and syngas. Producer gas has long been used for combustion in boilers and generation of electricity. If the gas is burned in boilers, it may generate steam to be used for process heat or space heating, or the steam may be used in steam turbines to generate electricity.

Another common use for producer gas after char and tar removal is for internal combustion engine fuel. Today such engines are used mainly for driving electrical generators, but during World War II they were used to drive cars, trucks, boats, and even tanks.

Another use of gas is for the operation of gas turbines. For this purpose, an improved clean gas such

as syngas, not producer gas, is needed. For the most efficient generation of electricity, a bottoming-cycle gas turbine is used. In this application, gas powers the gas turbine, and the waste heat from the gas turbine is used to make steam for a steam turbine to generate more electricity. This concept is called integrated gasification combined cycle (IGCC). If biomass is used to make the syngas, it is known as biomass integrated gasification combined cycle (BIGCC).

For background information *see* ALTERNATIVE FUELS FOR VEHICLES; BIOMASS; COAL GASIFICATION; COMBUSTION; ENERGY SOURCES; FISCHER-TROPSCH PROCESS; FLUIDIZED-BED COMBUSTION; FUEL GAS; GASOLINE; RENEWABLE RESOURCES; SYNTHETIC FUEL in the McGraw-Hill Encyclopedia of Science & Technology.   John I. Zerbe

Bibliography. M. A. Dietenberger and M. Anderson, Vision of the U.S. biofuel future: A case for hydrogen-enriched biomass, *Ind. Eng. Chem. Res.*, 46:8863–8874, 2007; T. B. Reed, Biomass gasification: Yesterday, today and tomorrow, pp. 50–53, in *Energy Generation and Cogeneration from Wood*, Forest Products Research Society Proceedings No. P-80-26, Atlanta, 1980; J. I. Zerbe, Thermal energy, electricity, and transportation fuels from wood, *Forest Prod. J.*, 56:4–12, 2006.

Synthesis of new nanocarbon materials

Nanocarbons, such as fullerenes and carbon nanotubes, are nanometer-scale substances made entirely of carbon atoms. These molecules have attracted the interest of scientists because of their unique structures in addition to their particular chemical and physical properties. They constitute a new form of carbon-based materials, like the better-known graphite and diamond. Ever since their discovery, nanocarbons have emerged as promising materials in a growing list of applications, including field-emission displays, radiation sources, solar cells, and hydrogen storage media. In addition to industrial applications, nanocarbons have also been shown to be promising candidates in the medical and pharmaceutical fields. For example, fullerene has been found to selectively recognize and attack cancer cells, making it an effective agent to fight against this disease. To meet the demands of these applications, chemists need to develop the tools and techniques to efficiently synthesize tailored nanocarbons.

Catalysis is a fundamental tool in organic synthesis. Some well-defined catalysts, such as those shown in **Fig. 1**, have enabled a variety of chemical transformations that were otherwise difficult, or even impossible, to achieve with traditional organic reactions. Well-defined catalysts can streamline and simplify synthesis strategies, achieving in a single step what previously would have required lengthy sequences of reactions. Catalysis thus contributes to improving reaction efficiency and reducing chemical waste. Although nanocarbons and catalysis originate from different fields of chemistry, both are

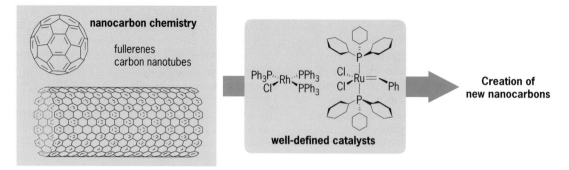

Fig. 1. Creation of new nanocarbons by well-defined catalysts.

in a rapid stage of development. The synergy between nanocarbon chemistry and well-defined catalysis opens new routes for creating new and interesting nanocarbon-based functional materials (Fig. 1). Recent advances have resulted in combined progress in these two areas of chemistry.

Chemical modification. Currently, the chemical modification of nanocarbons is the principal approach for preparing new carbon-based materials. Chemical modification can be use to tune the properties of nanocarbons by merging their features with those of organic molecules. For instance, nanocarbons are generally insoluble in common organic solvents because of strong π-π interactions. Increasing the solubility of nanocarbons is therefore of great importance for practical applications. Chemical modification provides a plausible solution to these problems. However, chemical modification methods

to date have been mostly limited to classical organic reactions. Well-defined catalysts can provide a new method for functionalizing nanocarbons.

As shown in **Fig. 2**, fullerene reacts with organoboron compounds in the presence of a rhodium catalyst to give unique fullerene derivatives that introduce various organic fragments on the fullerene surface. Previously, such fullerene derivatives could be synthesized only by using an excess of organometallic reagents, most of which are so reactive that they are easily decomposed by air and moisture. On the contrary, organoboron compounds are air- and moisture-stable, greatly simplifying the experimental techniques that use these reagents. Moreover, organoboron compounds are easily prepared, and large numbers of them are now commercially available. For these reasons, rhodium catalysis improves the reaction conditions and widens the

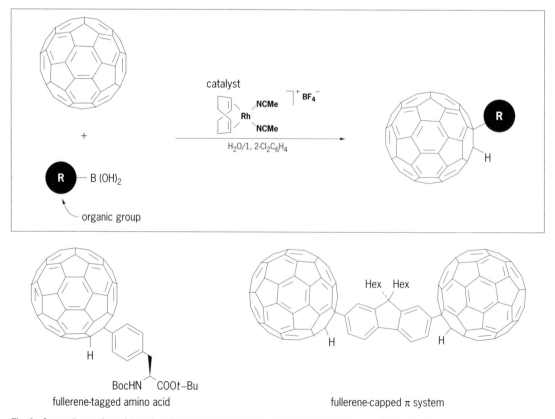

fullerene-tagged amino acid

fullerene-capped π system

Fig. 2. Organoboron-based functionalization of fullerene by rhodium catalyst. Boc = *t*-butyloxycarbonyl.

variety of fullerene derivatives that can be prepared. This method has enabled the preparation of various functionalized fullerenes. For example, when an organoboron compound prepared from phenylalanine is reacted, a fullerene-tagged amino acid is obtained (Fig. 2). Since amino acids are fundamental structural components of proteins, these new hybrid nanocarbons may prove to be useful in biological applications. When organic reagents bearing two boron centers were reacted with two equivalents of C_{60}, a two-directional reaction occurred, yielding an interesting fullerene-capped molecule (Fig. 2). This rhodium-catalyzed transformation not only broadens the scope of the functionalization chemistry of fullerenes by providing a new synthetic avenue to a wide variety of previously unexplored fullerene-based materials, it also opens new direction for catalysis in nanocarbon chemistry.

Nanocarbon synthesis. In addition to chemical modification of nanocarbons, several new nanocarbons have been synthesized directly by means of chemical and physical methods. In **Fig. 3**, sumanene, shown in structure (**1**), is a representative fullerene fragment that recently has been chemically synthesized. Sumanene is a bowl-shaped π-conjugated compound that is significant not only as a model of fullerene, but also for its own chemical and physical

properties. Presumably because of the high strain energy, all previous attempts to synthesize sumanene through various strategies had failed. However, in 2003, H. Sakurai and coworkers successfully prepared sumanene in a few steps from a simple organic compound under mild conditions with the assistance of a well-defined catalysis. The synthetic route is shown in Fig. 3, in which commercially available norbornadiene (structure **2**) was chosen as the starting material. First, the three molecules of norbornadiene were stitched to one another to give the benzene derivative (structure **3**). Then, an alkene-bridge exchange reaction of structure (**3**) constructed the basic bowl-shaped structure (**4**). Finally, removal of hydrogen from structure (**4**), using an oxidant, provided sumanene. Key to the success of this synthesis is the use of a ruthenium complex for the alkene-bridge exchange reaction and the effective construction of intermediate structure (**4**). The ruthenium complex was developed in the laboratory of Prof. Robert Grubbs and is one of the most popular catalysts in organometallic chemistry used for the alkene exchange reaction known as alkene metathesis, for which the Nobel Prize was awarded in 2005. In the entire synthetic process to sumanene, no harsh conditions were necessary. This elegant synthesis will encourage

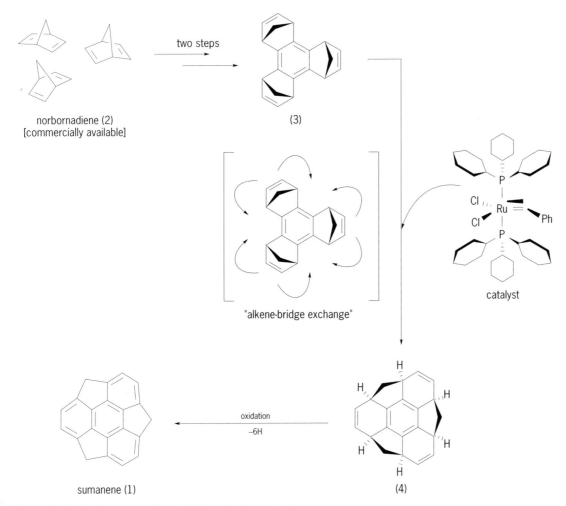

Fig. 3. Synthesis of sumanene (1) enabled by ruthenium catalyst.

practical applications of sumanene and its derivatives as functional nanomaterials. Other functionalized sumanene derivatives based on this synthetic procedure are also interesting intermediates in the chemical synthesis of other fullerenes and nanocarbons.

Outlook. Well-defined catalysts have enabled the chemical synthesis of several new nanocarbons. These results show that well-defined catalysts can be versatile and general methods for producing and functionalizing nanocarbons. As the demand for nanocarbons with unique structures and properties increases, considerable efforts to develop new methods for large-scale production will be required. Better understanding of the specific properties of unprecedented nanocarbons will lead to development of new nanocarbon materials and novel catalysts for their synthesis and functionalization. While the development of such catalytic transformation is in its infancy, well-defined catalysts can be expected to play an active role in the development of new nanocarbon-based materials.

For background information *see* ALKENE; ATOM ECONOMY; CARBON; CARBON NANOTUBES; CATALYSIS; FULLERENE; HOMOGENEOUS CATALYSIS; ORGANIC SYNTHESIS; RHODIUM in the McGraw-Hill Encyclopedia of Science & Technology.

Masakazu Nambo; Kenichiro Itami

Bibliography. R. H. Grubbs, Olefin-metathesis catalysts for the preparation of molecules and materials (Nobel lecture), *Angew. Chem. Int. Ed.*, 45:3760, 2006; N. Martin, New challenges in fullerene chemistry, *Chem. Commun.*, 2093, 2006; M. Nambo et al., Rh-catalyzed arylation and alkenylation of C_{60} using organoboron compounds, *J. Am. Chem. Soc.*, 129:8080, 2007; H. Sakurai et al., A synthesis of sumanene, a fullerene fragment, *Science*, 301:1878, 2003.

Table-top synchrotron systems

The need for advanced and higher-brightness light sources (that is, sources of electromagnetic radiation, particularly x-rays) is well documented by the recent creation of new large facilities such as SOLEIL, DIAMOND, MAX IV, and ALBA as well as the upgrade of the older facilities. The applications of light sources encompass all aspects of science, spanning the fields of physics, chemistry, biology, material science, electronics, catalysis, and medicine. An option to provide "more light" to this research community is to develop small laboratory sources beyond the standard and rotating anode sources. Recently, there has been a lot of activity worldwide to develop such compact "table-top" synchrotron systems. The term "synchrotron" in "table-top synchrotron" or "compact synchrotron" refers either to the high-brilliance features of synchrotron installation (where the brilliance is a measure of performance that is defined below) or to the manner in which the electrons are accelerated and kept on their path. The different ap-

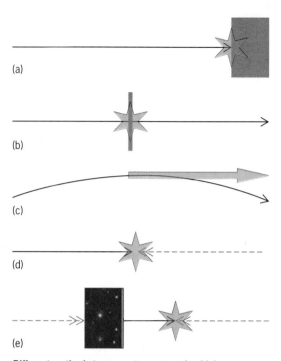

Different methods to generate x-rays using high-energy electrons (black arrow). (*a*) By bombarding a thick solid target (bremsstrahlung). (*b*) By bombarding and traversing a thin solid target (bremsstrahlung). (*c*) By bending the electron trajectory with a magnet field (synchrotron radiation). (*d*) By collisions with intense laser light (gray arrows, inverse Compton scattering). In the above cases, the electrons are generated using electron sources and accelerator techniques. In the case of laser wake-field acceleration (*e*), the electrons are created through the interaction of intense laser light with a gas phase (illustrated by the bright dots). The angular distribution of the emitted x-rays in all these methods (indicated by stars, and by the straight arrow in part *c*) strongly depend on the experimental implementations.

proaches that are currently followed are illustrated below.

Generation of x-rays. Since the days when x-rays were first discovered by W.C. Röntgen, the fundamental source of x-rays still has not changed much. X-rays are still most efficiently generated from high-energy electrons either through absorption and emission processes (like those that produce the characteristic Cu Kα lines) or by varying their velocity or direction. In the case of classical x-ray sources, a high current (300 mA) of 30–100 keV electrons is used to bombard a solid—and heavily cooled—target such as copper or molybdenum (**illus.** *a*). Part of the electron energy ($\sim$1%) is absorbed and converted into x-rays. Besides the x-ray lines characteristic for each target, there is also a broad background radiation present from a process called bremsstrahlung. When a rotating anode is used, the target is continuously rotated so that the heat dissipated at the point of the electron beam impact can be significantly reduced.

A variation on the above principle is to use much thinner targets, such that only a fraction of the electron energy is dissipated and the electrons can be transmitted through the target (illus. *b*). Specifically, when the electrons are already moving in a circular configuration, as is the case for a betatron or a synchrotron, then the same electron can come back several times and hit the target again. This principle has been used by H. Yamada and colleagues with 20-MeV electrons and by V. V. Kaplin and colleagues with 33-MeV electrons, respectively, in their designs of compact radiation sources (with an electron orbit radius of the order of 1 m or 3 ft).

Besides emission processes from high-energy electrons in solids, another means to generate x-rays is to deviate electrons from a straight-line trajectory, for instance by putting them into a magnetic field. The circular orbit of electrons in synchrotrons is a good example of this phenomenon, and this type of radiation is, strictly speaking, the only one that can be defined as synchrotron radiation, that is, electromagnetic radiation emitted by charged particles following a curved path in magnetic or electric fields (illus. c). To produce those high-energy x-ray photons (greater than 1 keV, corresponding to wavelengths less than 1.2 nm), the required starting electron energies are high (greater than 1 GeV), which, combined with the available magnetic fields ($\sim$0.2 T), leads to large-radius facilities (greater than 50 m or 164 ft). Nevertheless, it should be pointed out that even for the small compact sources (radius $\sim$1 m or 3 ft) mentioned above, a significant amount of low-energy (1–100 meV) synchrotron radiation is generated following the same mechanism, but radiating in the far-infrared range (wavelengths of 12 μm to 1.2 mm).

Another method to produce x-rays is through the scattering of electrons with photons in a process called inverse Compton scattering (illus. d). Due to the small electron-photon scattering cross section, a large flux of photons is required, which can be supplied only by high-power laser systems (>1 TW). Several groups are focusing their efforts in this direction. Compared to the large synchrotrons, these systems use electron beams in the energy range of 20–100 MeV, which makes compact designs ($\sim$1 m) with linear accelerators or recirculating electrons again possible. For instance, in order to produce 33-keV x-ray photons (38-picometer wavelength), it is possible to combine 1.16-eV photons (1.07-μm wavelength) with 43-MeV electrons.

Finally, the most recent developments are shown in illus. e, where first a laser beam is used to produce an intense plasma. Within such a plasma, very high electron energies can be generated ($\sim$1 GeV), and in a second step those electrons can again be scattered with photons, using the inverse Compton scattering mechanism. The combination of these two steps allows the building of an all-optical compact synchrotron system, as recently proposed.

The main advantage of table-top synchrotrons compared to rotating anodes is that they provide a good tunability in energy or wavelength. This capability is a key enabler for most spectroscopic methods, which is not available with rotating anodes. Even a low-flux or low-brilliance source with tunable wavelength provides unique capabilities. The main disadvantage of these systems is that they are quite new, they are not established outside their original laboratories, and specific optical elements may be needed to exploit their full potential.

Flux, brilliance, and x-ray spectrum. The five methods illustrated above each lead to a specific x-ray intensity and spectrum. The spectrum of the standard low-voltage (rotating) anodes consists of intense "characteristic" lines, such as Cu Kα, on top of a continuous bremsstrahlung background. In the best such systems (microfocus sources) available today, the brilliance can exceed 10^{11} photons/s/mrad^2/mm^2/0.1% bandwidth. The brilliance—as well as the flux—are quantities that can be used to compare the performance of the different sources. While the flux is defined as the total amount of x-ray photons emitted per second into an angular range of 1 mrad^2, the brilliance also takes into account the target area (mm^2) as well as the energy distribution of the photons in a bandwidth of 0.1% around the main energy of interest. However, not taken into account in these two definitions are the optical beam elements (mirrors, monochromators, and so forth) which will determine—together with the source parameters—the final amount of photons impinging on the sample.

The compact Mirrorcle$^©$ systems from Yamada and colleagues that use higher-energy electrons (20 MeV) have a targeted brilliance of 10^{14} photons/s/mrad^2/mm^2/0.1% bandwidth and a spectral range from about 0.1 kev to 20 MeV (6 femtometers to 12 nm). The brilliance of those systems is of the same order as that of second-generation synchrotron systems. Besides bremsstrahlung, other types of radiation mechanisms also can be used in such compact systems by a selection of the appropriate targets. Examples of such mechanisms are the transition radiation (0.1–2 keV, 0.6–12 pm) from amorphous targets and the quasimonochromatic parametric x-rays (8–30 keV, 40–150 nm) from crystalline targets. (Transition radiation appears when relativistic electrons cross the boundary between two media of different dielectric constant. Parametric radiation appears when relativistic electrons interact with the crystallographic planes of a single crystal.) And in the absence of targets, the circulating electrons provide low-energy synchrotron radiation.

Compared to the previous systems, the laser-based machines are intrinsically nearly monochromatic. The use of electrons with a specific energy and of photons with a specific wavelength leads to an x-ray beam that is essentially monochromatic. To change the x-ray energy, in order to collect a spectrum for instance, it is necessary to adjust either the electron energy or the laser photon energy in subsequent pulses. For both the Lyncean$^©$ and the MXISystem$^©$, the x-ray photon energy can be tuned over a broad range. In the former system, recirculating electrons are used, while in the latter system, a linear accelerator was chosen. In a single pulse (electron bunch interacting with a laser), a photon flux of the order of 10^8–10^{11} photons can be generated. The design goal of both systems is to provide more flux than second-generation synchrotrons.

As a comparison, the brilliance of the current state-of-the-art large synchrotron systems stands above 10^{20} photons/s/mrad^2/mm^2/0.1% bandwidth, and such systems can also provide a large spectral range, typically from the far infrared (1 mm) to hard x-rays (50 keV, 25 pm). An additional feature of those systems is that they can provide good temporal resolution. The high-energy electrons travel in well-defined bunches along the large-circumference ring with a repetition time in the range of 10^{-9} s, enabling experiments with such time resolution.

The whole field of tabletop synchrotrons is currently highly active and is undergoing rapid progression. Mature finished systems have not yet appeared outside of their development environment. In many cases, the optical elements (such as mirrors and monochromators) and the end-user stations need to be optimized or redesigned, and the all-optical systems are essentially still in the design phase.

For background information *see* BREMS-STRAHLUNG; COMPTON EFFECT; SYNCHROTRON RADIATION; X-RAY TUBE in the McGraw-Hill Encyclopedia of Science & Technology.

Jean-Pierre Locquet; Johan Vanacken

Bibliography. F. E. Carroll et al., Pulsed tunable monochromatic x-ray beams form a compact source, *Am. J. Roentg.*, 181:1197–1202, 2003; Z. Huang and R. D. Ruth, Laser-electron storage ring, *Phys. Rev. Lett.*, 80:976–979, 1998; V. V. Kaplin et al., Tunable, monochromatic x-rays using the internal beam of a betatron, *Appl. Phys. Lett.* 80:3427–3429, 2002; P. Rullhusen, X. Artru, and P. Dhez, *Novel Radiation Sources Using Relativistic Electrons: From Infrared to X-Rays*, Series on Synchrotron Radiation Techniques and Applications, vol. 4, World Scientific, 1998; H.-P. Schlenvoigt et al., A compact synchrotron radiation source driven by a laser-plasma wakefield accelerator, *Nat. Phys.*, 4:130–133, 2008; H. Yamada et al., Development of the hard X-ray source based on a tabletop electron storage ring, *Nucl. Instrum. Meth. A* 476:122–125, 2001.

Theory of temporal and spatial environmental design

Subjective preference is regarded as a primitive response entailing judgments that steer an organism toward maintaining life to enhance its prospects for survival. Cerebral hemisphere specialization in the human brain may play an important role in the independent effects of temporal and spatial factors on preference. The scale values of subjective preference of both the sound field and the visual field have been described in terms of time and space. Accordingly, a generalized theory of environmental planning incorporating these values for the left and right cerebral hemispheres (**Fig. 1**), respectively, has been proposed by blending the built environment and nature. In particular, examples of temporal design associated with the left hemisphere will be discussed.

Specialization of cerebral hemispheres for temporal and spatial factors of the sound field. The theory of subjective preference for the sound field has been described by using the temporal and spatial factors extracted from the respective correlation function mechanisms existing in the neural system, including activities for the monaural sound signal (autocorrelation function) and sound signals arriving at both ear entrances (interaural cross-correlation function). Recordings of brain activity over the left and right hemispheres, including the slow vertex response (SVR), electroencephalogram (EEG), and magnetoencephalogram (MEG), have revealed the evidence in **Table 1**, as described below.

Formulation of such a neurally grounded strategy for acoustic design has been initiated through a study of auditory-evoked electrical potentials (that is, the SVRs), which are generated by the left and right human cerebral hemispheres. The goal of these experiments was to identify the potential neuronal response correlations of subjective preference for the orthogonal acoustic parameters related to sound fields listed in **Table 2**. Particular ranges of the four factors preferred by most listeners were established by paired-comparison tests. SVRs were integrated for paired stimuli, obtaining the scale value of subjective preference. Typically, SVRs are greatest when stimulus patterns change abruptly, that is, when there is a contrast between the paired stimuli in which spatial and/or temporal factors change. The method of paired stimuli is therefore the most effective procedure because of this relativity of the brain response. The findings include:

1. The left and right amplitudes of the early SVR indicate that the left and right hemispheric dominance

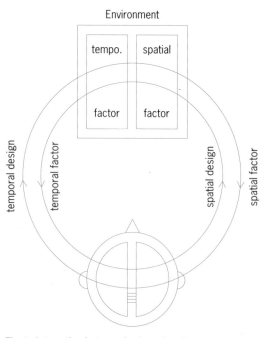

Fig. 1. Interaction between brain and environment created by incorporating temporal and spatial factors, which are associated with the left and right cerebral hemispheres, respectively. Such an environment created by each individual room may induce further creations, for example.

TABLE 1. Hemispheric specialization of temporal and spatial factors observed by analyses of the auditory evoked potentials (AEP), EEG, and MEG

Factors changed	AEP (SVR)	EEG, ratio of ACF τ_e values of alpha wave	AEP (MEG) N1m	MEG, ACF τ_e value of alpha wave
Temporal				
Initial time delay of first reflection, Δt_1	L > R (speech)*	L > R (music)		L > R (speech)
Reverberation time, T_{sub}	—	L > R (music)	—	
Spatial				
Listening level, LL	R > L (speech)	—		
Magnitude of interaural cross-correlation, IACC	R > L (vowel /a/) R > L (band noise)	R > L (music)†	R > L (band noise)	

*Sound source used in experiments is indicated in parentheses.
†The flow of EEG alpha wave from the right hemisphere to the left hemisphere for music stimulus in change of the IACC was determined by the cross-correlation factor $|\phi(\tau)|_{max}$ between alpha waves recorded at different electrodes.

results from the temporal factor Δt_1 (the initial time delay of the first reflection) and the spatial factors LL (listening level) and IACC (interaural cross-correlation), respectively. The sensation level (SL) or LL was classified as a temporal-monaural factor from a physical viewpoint. However, the results of the SVR indicate that the sound level is right-hemisphere dominant. Thus, SL or LL should be classified as a spatial factor. It is accurately measured by the geometric average value of the binaural sound energies arriving at both ears.

2. Results of the EEG for the cerebral hemispheric specialization of the temporal factors [that is, Δt_1 and T_{sub} (reverberation time)] confirmed left-hemisphere dominance, and that the IACC is a right-hemispheric factor. Thus, a high degree of independence between the left- and right-hemispheric factors may be achieved.

3. The scale value of subjective preference corresponds to the value of τ_e (the effective duration of the autocorrelation function, ACF) extracted from the ACF of the alpha wave (8–12 Hz) over the left hemisphere and the right hemisphere according to changes in the temporal and spatial factors of the sound fields, respectively.

4. Amplitudes of the MEG records confirm the left-hemisphere specialization for Δt_1.

5. The scale value of individual subjective preference relates directly to the value of τ_e extracted from

the ACF of the alpha wave of the MEG. Note that the amplitudes of the alpha wave in the EEG and the MEG do not correspond well to the scale value of subjective preference.

In addition to the above-mentioned temporal activities, on both the left and right hemispheres, spatial activities on the brain were analyzed by the cross-correlation function of the alpha waves of the EEG and MEG. The results show that a large area of the brain is activated when the preferred sound field is presented. This implies that the brain repeats a similar temporal rhythm in the alpha-wave range over a wide area of the scalp under preferred sound fields. Previously, it was known that the left hemisphere is mainly associated with time-sequential and function identification, and the right hemisphere is fundamentally concerned with spatial identification. The left-hemispheric specialization of speech signals has been reported by a number of authors using EEG and MEG records. However, when the IACC was changed using speech and music signals, right-hemisphere dominance was observed, as indicated in Table 1. Therefore, hemispheric dominance is relative, depending on which factor is changed in the comparison pair, and no absolute behavior could be seen.

Theory of subjective preference for the sound field.
It has been verified by a series of paired-comparison tests that four objective factors act independently

TABLE 2. Four orthogonal factors of the sound field and its weighting coefficients α_i in Eq. (2) as well as the most preferred conditions indicated by the symbol [ ]$_p$, which was obtained by the paired-comparison test with many subjects.

i	x_i	α_i	
		$x_i > 0$	$x_i < 0$
1	$x_1 = 20 \log P - 20 \log [p]_p$ (dB), $[p]_p$ is set at the center of the room	0.07	0.04
2	$x_2 = \log(\Delta t_1/[\Delta t_1]_p)$, where $[\Delta t_1]_p \approx (1 - \log_{10} A)(\tau_e)_{min}$*	1.42	1.11
3	$x_3 = \log(T_{sub}/[T_{sub}]_p)$, where $[T_{sub}]_p \approx 23 (\tau_e)_{min}$	$0.45 + 0.75 A$	$2.36 - 0.42A$
4	$x_4 = IACC$†	1.45	—

*$(\tau_e)_{min}$ is the minimum value of the effective duration of the running autocorrelation function (ACF) of the source signals. The effective duration of the ACF is defined by the duration τ_e at which the envelope of the normalized ACF becomes 0.1. A is the total amplitude of reflections.
†The most preferred value of IACC is obtained at a smaller value less than 0.4 (Y. Ando, 1985, 1998).

on the resulting scale values when two of the four factors are changed simultaneously. We may add scale values of subjective preference associated with the right and left cerebral hemispheric factors to obtain the total scale value S for the sound field, such as Eq. (1):

$$S = g_r(x) + g_l(x)$$

$$= [S_1 + S_4]_r + [S_2 + S_3]_l \qquad (1)$$

where $g_r(x) = [S_1 + S_4]_r$ and $g_l(x) = [S_2 + S_3]_l$ are the scale values associated with the right and left hemispheres, respectively. Experimental results, obtained with a number of subjects, yielded the following common formulas for the scale value in relation to each factor [Eq. (2)], where x_i are the normalized

$$S_i \approx -\alpha_i |x_i|^{3/2} \qquad i = 1, 2, 3, 4 \qquad (2)$$

factors and α_i are weighting coefficients, as listed in Table 2. Thus, scale values of preference have been formulated approximately in terms of the 3/2 powers of the normalized objective parameters, expressed in the logarithm for the parameters x_1, x_2, and x_3, with the real value $x_4 =$ IACC showing the maximum contribution on the scale value.

Applying the theory for concert hall design. To obtain an excellent sound field, a genetic algorithm was used to modify a shoebox-shaped space. The initial shape was 14 m wide, the stage was 9 m deep, the room was 27 m long, and the ceiling was 15 m above the stage floor. The sound source was 4.0 m from the front of the stage, 0.5 m to one side of the centerline, and 1.5 m above the stage floor. The front and rear walls were vertically bisected to obtain two faces, and each stretch of wall along the side of the seating area was divided into four faces. The walls were kept vertical to examine only the plan of the hall in terms of maximizing $\overline{S}_4$ due to the typical spatial factor, IACC, to determine the shape of the room. Each wall was moved independently of the other walls. Forty-nine seating positions distributed throughout the hall on a 2 × 4-m grid were selected. The moving range of each vertex was ±2 m in the direction of the line normal to the surface. In this calculation, the most preferred listening level was set for a point on the hall's long axis (central line), 10 m from the source position. The result of optimizing for $\overline{S}_4$ is shown in **Fig. 2**. The rear wall of the stage

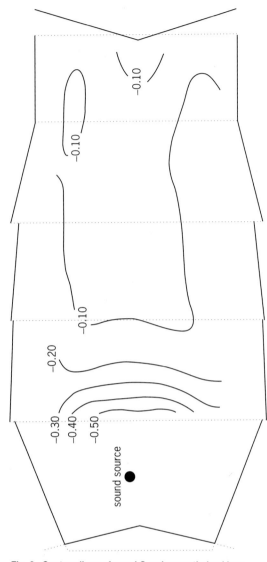

Fig. 2. Contour lines of equal S_4 values optimized by a genetic algorithm with the typical spatial factor, IACC.

and the rear wall of the audience area took on convex shapes. A practical application of this design theory was done in the Kirishima International Concert Hall (Miyama Conceru), in Japan, which was characterized by a "leaf" shape (**Fig. 3**). To enhance the satisfaction for individual listeners, a special room is available for testing each listener's subjective preference to allow the listener to select his or her best seating position in the concert hall.

TABLE 3. Left hemisphere specialization observed in EEG and MEG alpha waves in relation to the temporal factors of the visual field

Temporal factor changed	EEG, ACF τ_e value of α wave	EEG, CCF $\phi(\tau)_{max}$ of α wave	MEG, ACF τ_e value of α wave
Period of the flickering light, T	L > R* (SW)[†]	L & R[‡] (SW)	L > R (SW)
Period of horizontal movement of target, T	L > R (SW)	—	—

*The ratio of τ_e values of α wave in EEG, high to low preference, increased significantly on the left hemisphere.
[†]The $\phi(\tau)|_{max}$ value of α wave in MEG increased on the wide area of both hemispheres, when the scale value of subjective preference was high. The similar repetitive feature in the α wave on the wide area of brain relates to the preferred condition of vision.
[‡]Sinusoidal wave (SW) used to control the period T.

Application for the visual field. First, we will discuss the visual field in relation to the preferred temporal factors. The preferred sinusoidal period of a flickering light and moving a single target in both the vertical and the horizontal directions are almost the same, $[T]_p \approx 1.0$ s. The more preferred condition is obtained by introducing a fluctuation such as twinkling stars in addition to the preferred sinusoidal period of 1.0 s. The most preferred fluctuation of the flickering light is approximately expressed by $\phi_1 \approx 0.46$, where ϕ_1 is the amplitude of the first maximum of the ACF of the light. This signifies that the extreme conditions $\phi_1 = 0$ (random) and $\phi_1 = 1.0$ (periodic or sinusoidal) are not preferred, but instead a condition with a certain degree of fluctuation. Next, the most preferred spatial fluctuation of texture found is $[\phi_1]_p \approx 0.41$, similar to the preference given above. When we introduced normalized factors such that $x = T/[T]_p$ or $\phi_1/[\phi_1]_p$, then the scale value of subjective preference may be expressed commonly by Eq. (2). **Table 3** shows how left-hemisphere specialization is related to subjective preference by changing the temporal factor for the visual field.

At the preferred conditions of both auditory and visual fields, similar repetitive features are observed, represented by a prolonged value of τ_e that is extracted from the ACF of the alpha brain wave, and such repetition occupies a wider area of the brain.

Theory of planning environments incorporating spatial and temporal values. It is assumed that Eq. (1) for the sound field may be applied for all physical environments, such as in Eq. (3):

$$S = [g_1(x) + g_r(x)]_{\text{sound field}} + [g_1(x) + g_r(x)]_{\text{visual field}}$$
$$+ [g_1(x) + g_r(x)]_{\text{thermal field}} \quad (3)$$

In particular, this holds at least in the neighborhood of the optimal conditions of each physical factor, avoiding extreme physical environments such as listening to music at temperatures below $0°$C. The significant factors in designing the physical environments are listed in **Table 4**.

In architectural planning and design, we often forget the temporal factor. As shown in **Fig. 4**, the crucial factor in the temporal dimension of the

Fig. 3. Kirishima International Concert Hall with a ceiling of triangular panels.

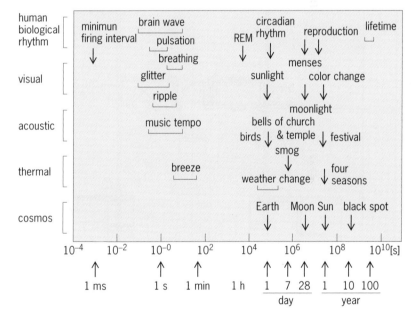

Fig. 4. Human biological rhythms and discrete periods of physical natural environments to be considered.

TABLE 4. Spatial and temporal factors of physical environments to be designed.

Physical environments	Spatial factors	Temporal factors
Acoustic (determined)	Listening level, LL	Initial time delay gap between the direct sound and the first reflection, Δt_1
	Interaural cross-correlation, IACC	Subsequent reverberation time, T_{sub}
Visual (proposed)	Lighting level, including color	Properties of movement function of reflective surface, T
	Properties of the reflecting surface	
	Spatial perception, including distance factor	
Thermal (assumed)	Sensors distributed on body	Air movement (breeze)
		Relative humidity
		Temperature

Note: In addition to these variables, characteristics of sources, location change of sources, and observer's activity should be taken into consideration.

TABLE 5. Typical examples of the temporal design for three stages of human life

Three stages of human life	Environment to be avoided	Environments to be designed
1 Body	Time-integrated effects of noise on unborn babies	
2 Mind	Time-integrated effects of noise on brain development	Concert hall acoustics (CHA) design based on auditory-brain model Seat selection enhancing individual satisfaction maximizes the individual preference
3 Creation based on personality		Creative workspace (CWS) for left- and right-hemispheric tasks

environment is the periodic cycle. For example, the shortest period (about 0.5 to 5 s) is related to brain waves, which are associated with perception of music and the twinkling stars as well as the glitter of leaves and ripples on water surfaces. Rapid eye movement (REM) of about 70–150 min, related to a basic rest-activity cycle, is associated with one session of a concert, lecture, or work. The circadian rhythm, deeply connected by sunlight with the Earth's rotation period, is associated with daily human activity. The week, created by social custom for work and leisure, is associated with the planning period of a concert, drama, or social activity without additional costs. Another period is concerned with the movement of the Moon. The revolution of the Earth around the Sun is associated with the color changes of leaves and annual festivals. The black spots on the Sun, which appear about every 11 years, may influence environments on Earth. The alternating generations of about 30 years and the life span of, say, about 90 years, may be considered in the planning of houses, in accordance with the individual schedule of life. In addition, there are longer cosmic periods.

The present theory suggests that these discrete periods should be explicitly recognized during the design process for any human environment. The passage of time in the designed environment should be as consciously considered as the three-dimensional organization of the space itself. There are specific dimensions of space, such as a room, house, building, region, urban, and so on.

Examples of temporal environmental design. All healthy creations that may contribute to human life for an order of a hundred or a thousand years have been based on the unique personality of the individual. It is proposed that the environment be designed for three stages of human time: (1) the time of body, (2) the time of mind, and (3) the time of creation based on a unique personality (**Table 5**). A well-designed environment would be a meeting place for art and science associated with both hemispheres, and in turn might help to discover further the individual personality as the minimum unit of society.

In order to demonstrate temporal design for three stages of life, a hillside house was built in Kirishima, Kyushu, Japan, about 700 m (2300 ft) above the sea, as shown in **Fig. 5**. For the first stage of time (body), the bedroom is designed with three small windows so that it suggests a cave with little natural light, making the body relax. For the second stage of time (mind), particularly for the periods of a day

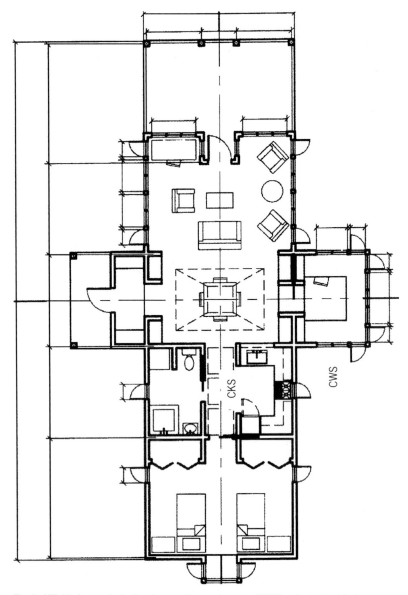

Fig. 5. Hillside house, including the creative work space (CWS) and creative kitchen space (CKS) designed by T. L. Bosworth and Y. Ando.

and the four seasons, windows in the living room, kitchen, and bathroom are carefully designed to provide light from outside and to look at trees and the large scale of a natural garden as well as the Sakurajima with its active volcano and the bay. Skylights fill the porch and table with natural light. The veranda is a joyful space for tea and food in the morning and afternoon.

A particular example of the third stage of life is the creative workspace (CWS), which has been designed to activate both cerebral hemispheres (**Fig. 6**). Eight office systems with this kind of space were introduced in 2002 to the Ando Laboratory, Kobe University. A CWS consists of three different panels specialized for the left- and right-hemispheric tasks and for integrating knowledge by an information-communication system. The left-hemispheric tasks include the temporal processes, such as writing, reading, speech, hearing, calculation, and logical considerations. The right-hemispheric tasks include the spatial processes, such as pattern recognition, space forming, drawing, painting, and making scale models.

Eight CWS users reported that the total qualities of this system were 2.5–15 times (mean value of 5 times) better than those of a one-dimensional working space, and the efficiency of work increased 2–15 times (mean value of 3 times). All the users reported that the efficiency of their scientific work improved more than 2 times ($p < 0.01$). And verbal and nonverbal materials created by a user, which were displayed on the walls around the three panels, could induce further creations (Fig. 1). A similar system for creative kitchen space (CKS) could be realized, for example, where the menus for a day and a week could be made by the left hemisphere.

There is evidence of new towns that are almost dead after 30 years because young people are not interested in moving to such places. It is worth noticing that the dimension of the head of newborn babies is relatively large because this part is initially developed in the body. If we consider the analogy of this, it is highly recommended that the facilities related to the human brain should be the first designed in house and urban planning, such as a museum, a concert hall, a library, a church, and an institution, which may act as an important role for the third life. The arms and legs, corresponding to the highways and communication systems, may be developed later on. Because of global climate change, a more dynamic and temporal environment design by blending human life should be considered.

For background information *see* ACOUSTICS; ARCHITECTURAL ACOUSTICS; BIOMAGNETISM; BRAIN; ELECTROENCEPHALOGRAPHY; GENETIC ALGORITHMS; HEARING (HUMAN); HEMISPHERIC LATERALITY; MUSICAL ACOUSTICS; PSYCHOACOUSTICS; VIRTUAL ACOUSTICS in the McGraw-Hill Encyclopedia of Science & Technology.
                                                     Yoichi Ando

Fig. 6. Example of CWS in the hillside house. The desk panel on the left-hand side is used for painting, drawing, and handicraft (the right-hemispheric tasks), and a front panel and a right panel are used for work with temporal processing, such as writing, calculating, and logical work (the left hemispheric tasks), and for the information system, respectively.

Bibliography. Y. Ando, *Architectural Acoustics: Blending Sound Sources, Sound Fields, and Listeners*, Springer-Verlag, 1998; Y. Ando, *Concert Hall Acoustics*, Springer-Verlag, Berlin, New York, 1985; Y. Ando, Concert hall acoustics based on subjective preference theory, in T. D. Rossing et al. (eds), *Springer Handbook of Acoustics*, Springer-Verlag, 2007; Y. Ando, Investigations on cerebral hemisphere activities related to subjective preference of the sound field, published for 1983–2003, *J. Temporal Des. Archit. Environ.*, 3:2–27, 2003; Y. Ando, On the temporal design of environments, *J. Temporal Des. Archit. Environ.*, 4:2–14, 2004; Y. Ando, Reviews on the temporal design for three stages of human life. most unlikely "time is money," but "time is life," *J. Temporal Des. Archit. Environ.*, 6:2–17, 2006; T. Bosworth and Y. Ando, Design of a hillside house in Kirishima with a small office, *J. Temporal Des. Archit. Environ.*, 6:18–22, 2006; J. Levy and C. Trevarthen, Metacontrol of hemispheric function in human split-brain patient, *J. Exp. Psychol. Human*, 2:299–312, 1976; R. W. Sperry, Chap. 1: Lateral specialization in the surgically separated hemispheres, in F. O. Schmitt and F. C. Worden (eds.), *The Neurosciences: Third Study Program*, MIT Press, Boston, 1974.

Titanosaur sauropod

Titanosauria was a diverse, globally distributed clade (a taxonomic group containing a common ancestor and its descendants) of herbivorous sauropod dinosaurs. [Sauropoda is a subgroup of the Sauropodomorpha subdivision of Saurichia (one of the two major monophyletic clades of Dinosauria, the other being Ornithischia).] During the Cretaceous Period (144–65 million years ago) when most other sauropod lineages became extinct, titanosaurs flourished and remained an integral part of terrestrial ecosystems. Body fossils from more than 40 valid titanosaur species are known from Late Jurassic to Late Cretaceous sediments and from every continent except Antarctica. The unique, wide-gauge trackways of titanosaurs extend their temporal range back as far as the Mid-Jurassic Period. In spite of their extensive temporal and geographic range, the titanosaur fossil record has been historically patchy, with many species known from only a few isolated bones. The past 10 years have witnessed a surge in professional and public interest in this enigmatic group of sauropods. Important new specimens have been identified that illustrate the cranial and postcranial morphology characteristic of the group, and permit recognition of a suite of locomotor features that depart in important ways from those of other sauropods. The titanosaur sample has increased to encompass the first embryonic titanosaur fossils and eggs, coprolites (fossil feces) that reveal titanosaur diets, and the first associated skulls and skeletons represented by an ontogenetic (developmental) sample. The origin of Titanosauria is well supported by these new data, although its species-level relationships remain poorly resolved. A broad phylogenetic perspective on titanosaurs is on the verge of being developed, which will enable more detailed aspects of their paleobiology, paleobiogeography, and evolutionary history to be investigated.

Historical background. The first titanosaur bones included two caudal vertebrae and a partial femur discovered in India in 1828 (14 years before the term "Dinosauria" was coined by Sir Richard Owen). It took almost 40 years for these bones to be described in the literature, and nearly 50 years for them to receive a name. In 1877, Richard Lydekker erected *Titanosaurus indicus*, a new dinosaur taxon, on the basis of these Indian bones. In his description, Lydekker recognized several diagnostic features of the vertebrae including procoely (ball-and-socket articulations in which the socket is at the front of the element) [**Fig. 1***a*]. Species of *Titanosaurus* were described in the late 1800s from Argentina, Madagascar, and Europe, and by the 1940s *Titanosaurus* and the family Titanosauridae soon swelled as new specimens were referred to the genus or named as closely related genera, most often based on the presence of procoelous caudal vertebrae. Essentially, *Titanosaurus* had become a "wastebasket taxon" that served as a catchall for any Cretaceous sauropod fossil with procoelous caudal vertebrae. In a recent revision of the 14 *Titanosaurus* species, including *T. indicus* (the type species), only five species were considered valid. The type species was based on obsolete characters and considered a *nomen dubium* ("doubtful name"), thereby forcing the abandonment of all coordinated rank-taxa (for example, Titanosaurinae, Titanosauridae, and Titanosauroidea). The unranked taxon Titanosauria

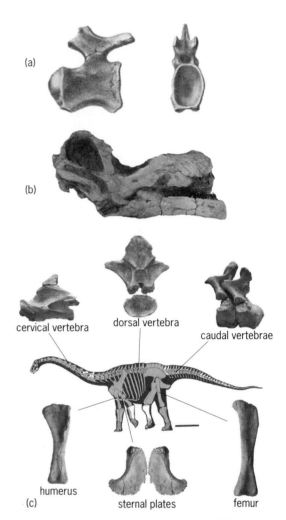

Fig. 1. Important anatomical milestones in titanosaur discovery. (*a*) Holotypic procoelous caudal vertebra of *Titanosaurus indicus* in side and front views, described in 1868. (*b*) The skull of *Nemegtosaurus mongoliensis*, the first discovered titanosaur skull, described in 1971. (*c*) The skeleton and representative postcranial elements of *Rapetosaurus krausei*, described in 2001.

remains valid, though, and a new system of standardized node and stem-based nomenclature for Titanosauria was proposed that allows standardization of names within a cladistic framework.

Titanosaur diversity and morphology. To date, titanosaur species represent ~34% of sauropod species and ~6% of dinosaur species (**Fig. 2**). More than 25 species are known from southern landmasses, and at least 15 are derived from northern landmasses. New titanosaur genera are being erected more prolifically than for any other group of sauropods, and recent reexamination of several historically debated species has placed them firmly within Titanosauria. In the past 10 years, several significant discoveries of new specimens have yielded key data for understanding titanosaur phylogeny and for detailed examination of titanosaur paleobiology (Fig. 1). These include (1) the first embryonic titanosaur remains and clearly identified eggs and nesting structures from Auca Mahuevo, Argentina; (2) *Rapetosaurus krausei*, the first titanosaur with associated cranial and postcranial remains; (3) the reevaluation of the skulls of the Mongolian taxa, *Nemegtosaurus* and *Quaesitosaurus*; (4) the discovery of phytoliths (fossilized plant silica bodies) and other plant fossils in the coprolites of titanosaurs from India; and (5) the discovery of nearly complete associated and articulated postcranial skeletons from South America (*Mendozasaurus, Epachthosaurus, Gondwanatitan, Bonatitan*), Asia (*Phuwiangosaurus, Tangvayosaurus*), India (*Isisaurus*), Europe (*Lirainosaurus, Ampelosaurus*), Africa (*Malawisaurus, Paralititan*), and North America (*Alamosaurus*).

The discovery of *Rapetosaurus krausei* clearly demonstrated the anatomy of a titanosaur skull and skeleton (Fig. 1c) and provided evidence for the inclusion of the isolated skulls of *Nemegtosaurus* (Fig. 1b) and *Quaesitosaurus* in Titanosauria. These three titanosaurs are characterized by a number of unique cranial characteristics including narrow tooth crowns, highly vascularized snouts, and novel connections between the quadrate bone and the base of the skull. The vertebral column of titanosaurs is also distinctive, although few complete series are known. Elongate cervical vertebrae with reduced neural arch lamination are characteristic for a number of titanosaurs including *Erketu* and *Rapetosaurus*. Dorsal vertebrae generally lose their hyposphene-hypantrum articulations, and a sixth sacral vertebra is added. As noted in the first specimens ascribed to titanosaurs, caudal vertebral centra are generally procoelous.

Changes in the titanosaur limb skeleton reflect the acquisition of a "wide-gauge" limb posture. Forelimb and hindlimb features related to this posture in titanosaurs include a humerus with a prominent deltopectoral crest and distal condyles (rounded bone prominences functioning in articulation) that are divided and exposed anteriorly, a prominent olecranon process on the ulna, medial deflection of the proximal third of the femur relative to the rest of the shaft, beveled distal femoral condyles, and a highly eccentric femoral midshaft cross section. This wide-gauge stance is also reflected in the ichnofossil record for titanosaurs. (Ichnofossils are trace fossils, that is, trails, tracks, or burrows made by an animal and found in ancient sediments such as sandstone, shale, or limestone.) Most sauropod trackways resemble those of other large animals with a parasagittal gait (in which the legs are oriented entirely under the body), with manus (forefoot) and pes (hindfoot) impressions approaching the trackway midline. Other sauropod trackways preserve manus and pes impressions that are more widely spaced. The variation in gauge width is inferred to be taxonomic: the widespread narrow-gauge stance is presumed to be primitive, and the wide-gauge stance is a derived (evolutionarily advanced) feature of titanosaurs.

Phylogenetic background. The first cladistic analyses of Sauropoda addressed higher-level relationships of the group, with little investigation into individual subgroups. Some of these early analyses supported the traditional dichotomy of sauropods into broad and narrow-tooth-crowned clades, and closely allied

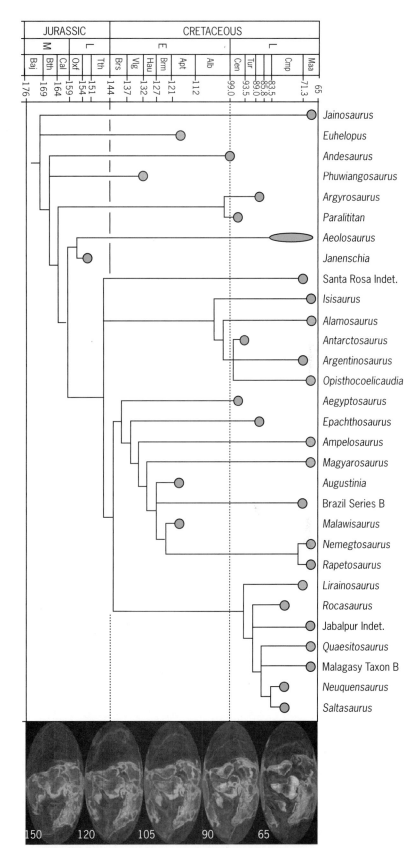

Fig. 2. Calibrated phylogeny of titanosaurs. Color circles indicate titanosaurs found in the southern hemisphere. Gray circles indicate titanosaurs found in the northern hemisphere. The global images illustrate the fragmentation of Pangaea during the evolutionary history of Titanosauria. (*Figure courtesy of Jeff Wilson, University of Michigan*)

titanosaurs with other taxa exhibiting narrow tooth crowns (for example, *Diplodocus*). Other analyses provided strong evidence that titanosaurs share closest ancestry with *Brachiosaurus*-like, broad-tooth-crowned taxa. This result has been corroborated in more recent cladistic analyses focused on lower-level relationships of Sauropoda. A few recent analyses focused specifically on titanosaurs have provided the foundation for more detailed in-group research, but have provided little resolution of titanosaur species-level relationships. Most of these analyses included 10 or fewer titanosaur species, and only 21 of the more than 40 different titanosaur species have been evaluated (Fig. 2). Of these, 11 appear in only one or two analyses, but 7 appear in more than half (>4) of the analyses. Clearly, it is difficult to compare the results of analyses including such disparate taxa, but some consistent signals pervade these cladograms. *Andesaurus*, a titanosaur genus from Argentina, is commonly resolved as the basalmost titanosaur. Most analyses place the African *Malawisaurus* as a basal titanosaur. The South American genus *Saltasaurus* is the only taxon included in all analyses to date and is always resolved as a derived titanosaur. These analyses suggest that there is general consensus on the basic framework for titanosaur phylogeny, although more than 30 valid titanosaur species still need to be accommodated by phylogenetic analysis.

Titanosaur paleobiology. Sauropods as a group are the largest terrestrial vertebrates ever to inhabit the planet, and titanosaurs include both the largest and smallest sauropods ever found. Popular misconceptions imagine that sauropods were evolutionary dead ends, that their giant body sizes were impossible to maintain, and that they were outcompeted and replaced by more specialized herbivores at the end of the Jurassic Period. These misconceptions stem largely from a lack of consideration for Titanosauria. The group diversified relatively late in dinosaur time and persisted until the end of the Cretaceous Period. They evolved in the context of dramatic global change as Pangaea (a postulated former supercontinent composed of all the continental crust of the Earth) fragmented into its constituent landmasses, and attained a global distribution within a few tens of millions of years (Fig. 2). Titanosauria experienced at least one evolutionary size reduction that may have led to uniquely small forms, as well as containing species at the theoretical maximum size for living life on land. Although many aspects of titanosaur paleobiology are only in the nascence of investigation, a few key observations are worth making at this time.

Titanosauria has the broadest range of ontogenetic and adult body sizes of any dinosaur. Newly hatched titanosaurs measured <0.5 m (1.6 ft) and weighed <10 kg (22 lb), and most species grew to adult sizes rivaling extant whales. Several titanosaur species may have attained adult lengths of over 30 m (98 ft) and weights of 30–100 tons, whereas other, purported dwarf species terminate growth at far smaller adult sizes [~5–7 m (16–23 ft), ~10 tons]. The growth strategies that permitted sauropods to attain such disparate adult body sizes remain poorly understood, but they can be elucidated through analysis of bone histology and inferred somatic growth rates. Preliminary data suggest that, like other sauropods, titanosaurs relied more upon accelerated growth rates than prolonged ontogenies (developmental stages) to attain their large adult body sizes.

Evidence for the reproductive biology of titanosaurs comes in the form of *in ovo* (in the egg) embryos preserved in megaloolithid (Megaloolithidae) eggs in nesting structures at Auca Mahuevo, Argentina. The taphonomy (fossil preservation) of this locality indicates that titanosaurs laid eggs in excavated depressions without burying them (**Fig. 3**). Clutches at Auca Mahuevo contain 20–40 eggs each, and evidence for site fidelity is indicated by the stratigraphic distribution of identical eggs and nest structures. Although nest attendance may be inferred by phylogenetic bracketing (phylogenetic inference, wherein unpreserved attributes of extinct organisms are inferred based on their phylogenetic relationships with living species), the close spacing of clutches at Auca Mahuevo may preclude significant parental care.

The diets that must have been an integral part of the rapid postnatal growth rates of titanosaurs and other sauropods have long been debated. Evidence on feeding heights for sauropods is usually based upon assumptions regarding neck flexibility and posture. Recently discovered coprolites attributed to titanosaurs from the Lameta Formation of India support conclusions based on biomechanical models, indicating that titanosaurs took advantage of a broad array of vegetative forage. Phytolith data derived from the titanosaur coprolites highlight diversified grass as an available and potentially important food for Cretaceous sauropods and indicate that titanosaurs probably took advantage of food at the ground level as well as that at shoulder height and above.

For background information *see* CRETACEOUS; DINOSAURIA; EXTINCTION (BIOLOGY); FOSSIL; GEOLOGIC TIME SCALE; PALEONTOLOGY; SAURISCHIA;

Fig. 3. Titanosaurs (based on *Rapetosaurus*) in the nesting ground at Auca Mahuevo, Argentina. (*Illustration courtesy of Mark Hallett*)

TAPHONOMY; TRACE FOSSILS in the McGraw-Hill Encyclopedia of Science & Technology.

Kristi Curry Rogers

Bibliography. L. M. Chiappe et al., Sauropod dinosaur embryos from the Late Cretaceous of Patagonia, *Nature*, 396:258–261, 1998; K. Curry Rogers, Titanosauria: A phylogenetic overview, in K. Curry Rogers and J. A. Wilson (eds.), *The Sauropods: Evolution and Paleobiology*, University of California Press, Berkeley, 2005; K. Curry Rogers and G. M. Erickson, Sauropod histology: A microscopic view on the lives of giants, in K. Curry Rogers and J. A. Wilson (eds.), *The Sauropods: Evolution and Paleobiology*, University of California Press, Berkeley, 2005; V. Prasad et al., Dinosaur coprolites and the early evolution of grasses and grazers, *Science*, 310:1177–1180, 2005; J. A. Wilson and M. T. Carrano, Titanosaurs and the origin of "wide-gauge" trackways: A biomechanical and systematic perspective on sauropod locomotion, *Paleobiology*, 25:252–267, 1999; J. A. Wilson and P. Upchurch, A revision of *Titanosaurus* Lydekker (Dinosauria-Sauropoda), the first dinosaur genus with a "Gondwanan" distribution, *J. System. Palaeontol.*, 1:125–160, 2003.

Transition to turbulent flow

The transition of laminar flow to turbulent flow has proven to be one of the most difficult problems in fluid dynamics. The importance of the topic stems from the fact that the transition has a major impact on all vehicles and devices that come in contact with fluid flow. For transportation systems, transition to turbulent flow results in increased drag and, in turn, increased fuel consumption and increased global warming. For aircraft, a reduction in 1% of the drag results in an increase of about 5–10% in the payload that can be transported. On the other hand, the transition is critical for the efficient performance of internal combustion engines because it enhances the mixing of fuel and oxidizer. Thus, there is a major economic incentive for understanding the mechanisms of transition from laminar to turbulent flows.

Mechanisms. In order to control the transition, it is important to understand its various mechanisms. This article will focus on transportation systems, with special emphasis on aircraft. The drag of an aircraft can be attributed to a number of factors. However, the dominant drag is the viscous drag, which results from flow sticking to the surface, or the no-slip condition. Because of the no-slip condition, there is a thin layer next to the surface, the boundary layer, where viscous effects dominate. The thickness of the layer is highly influenced by a parameter called the Reynolds number, which is a measure of the ratio of inertia forces to viscous forces. When this number is based on the chord length of a wing, it can range from 6 to 35 million depending on the size of the aircraft. The transition process involves free-stream oscillations, that is, fluctuations that are internalized, via a process called "receptivity," by the body boundary layer. These oscillations are amplified within the boundary layer at a rate that depends on their initial amplitude, the body surface condition, and the mean flow.

When the initial oscillations have a low amplitude, natural (that is, not forced) transition may occur. In this case, linear instability mechanisms dominate. (High-amplitude oscillations can lead to transition. However, as indicated below they belong to a different category of transition.) Typical instability modes are the (viscous) Tollmein-Schlichting (T-S) mode, which dominates when the flight Mach number, M (that is, the ratio of the velocity to that of the speed of sound), is less than 4; the Mack (compressibility) modes for $M \geq 4$; the cross-flow mode (wherein 3D mean flow over swept wings is excited by micrometer-sized surface roughness); and the Gortler mode (which occurs with concave streamwise curvature). The names associated with the various modes refer to individuals who discovered them. Moreover, the cause of the instability is indicated. Thus the T-S waves grow because of the viscous effects that dominate the flow next to a wall. The Mack waves are inviscid in nature, that is, waves where viscosity does not play a role in their growth. However, compressibility, which is a result of large variations in density, is responsible for the growth of these waves. In general, flows are three-dimensional; that is, their properties vary in all of the three dimensions. The cross-flow instability is dominant for swept wings, that is, where the leading and trailing wing edges are not normal to the fuselage. Finally, surfaces can be flat, convex, or concave. When the streamlines, which are the paths of flow particles, are concave, waves can grow and lead to Gortler instability.

Instabilities that are not a result of a linear instability mechanism are referred to as "bypass" mechanisms. These include a high-disturbance environment (that is, where the amplitude of the disturbance is not small); roughness; and attachment-line contamination, which is a result of waves emanating from the wing-root (that is, the place where the wing is joined with the fuselage) propagating along the leading edge of a swept wing that has a finite leading-edge radius.

Dependence on environmental conditions. An important reason for the complexity of the problem is that the transition is highly dependent on the environmental conditions. Thus, if we test the transition on a simple configuration like a flat plate in various wind tunnels, we will find that the onset and extent of the transition are facility dependent. This is because the geometry and air-supply characteristics are different for various facilities. This will result in different amplitudes and spectra for the various flow fluctuations that represent the difference between the actual flow properties and their mean values. Similarly, changing the plate surface conditions by changing the levels of surface roughness will result in a different onset and extent.

Prediction. There are a number of approaches that are used to predict transition onset and extent. These include the e^N method, which is based on linear stability theory (LST); parabolized stability equations (PSE); direct numerical simulation (DNS); and transition modeling. Linear stability theory is used to determine the conditions under which a wave of small amplitude can grow and lead to transition. The parabolized stability equations refer to the equations that are used to determine the onset of instability. These equations are more rigorous than those used in LST and thus do not require an empirical input to determine transition onset. Direct numerical simulation describes the approach that is based on the exact set of equations that govern fluid dynamics, and represents the most direct and accurate way to study transition. Transition modeling treats the transition by developing empirical models that attempt to capture the essence of the physics underlying transition.

The above methods, with the exception of transition modeling, are computationally intensive; that is, they require a great deal of computer time and storage. The e^N method, which is a linear method developed to model nonlinear phenomena, is highly popular in the aircraft industry. It is based on the assumption that, when the wave moves downstream of the location where it is neutrally stable (that is, where it does not grow), the transition will take place when it subsequently grows by a factor of e^N. The quantity N for airfoils, that is, sections of infinite wings, has to be provided by the user and can be deduced by wind tunnel tests. Typical values are 8 or 9. Transition on wings requires the specification of two parameters. In general, it takes four parameters to specify a wave propagating along a wing. However, solution of the stability equation yields two conditions. Thus, the two remaining parameters have to be specified in some manner. At present, there is no accepted procedure for determining these parameters.

Modeling approaches are extremely efficient and have proven to be rather accurate. Results that are typical of the success of these approaches are presented in **Fig. 1**, which shows the prediction of transition onset as a function of chord Reynolds number on an infinite swept wing where transition is a result of micrometer-size surface roughness. Another example pertains to transition prediction in the high-disturbance environment of a hypersonic wind tunnel where test data was obtained for an elliptic cone at Mach number $M = 8$. **Figure 2** shows the prediction of heat flux on selected locations on the cone surface.

Control. The desire to build more efficient airfoils has focused attention on laminar flow control (LFC), where the objective is to limit the growth of the linear disturbances, and on modifying the turbulence structure. Flow control can be passive or active. However, in order for any strategy to be successful, it must be part of the initial design process.

The various linear instability modes require different transition delay approaches. Thus, both the T-S and Mack modes can be damped by increasing the Mach number and maintaining a favorable pressure gradient, that is, one where the pressure decreases in the flow direction. Wall cooling is stabilizing for

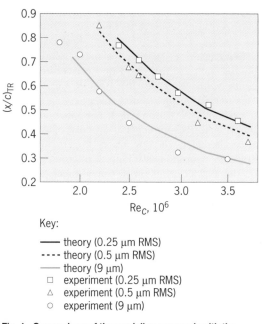

Key:

— theory (0.25 μm RMS)
- - - - theory (0.5 μm RMS)
— theory (9 μm)
□ experiment (0.25 μm RMS)
△ experiment (0.5 μm RMS)
○ experiment (9 μm)

Fig. 1. Comparison of the modeling approach with the experimental data of R. H. Radeztsky et al. The ordinate, $(x/c)_{TR}$, is the value of x/c at the location where the transition takes place, where x is the distance from the leading edge of the wing and c is the chord length of the wing. The abscissa, Re_c, is the Reynolds number based on the chord length. Quantities in parentheses in the key represent the size of the roughness elements. (*After E. B. Warren and H. A. Hassan, Transition closure model for predicting transition onset, J. Aircraft, 35:769–775, 1998; reproduced by permission of the American Institute of Aeronautics and Astronautics, Inc.*)

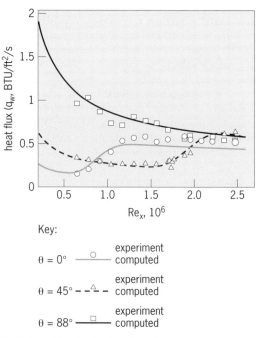

Key:

$\theta = 0°$ ○ experiment / — computed

$\theta = 45°$ - -△- - experiment / computed

$\theta = 88°$ □ experiment / — computed

Fig. 2. Heat-flux distribution along the rays of an elliptic cone. θ is the angle that a ray from the vertex makes with the line of symmetry. Re_x is the Reynolds number based on the distance from the vertex, where the Reynolds number per unit length L is $Re_x/L = 7.8 \times 10^5$/ft. (*After X. Xiao et al., Transitional flow over an elliptic cone at Mach 8, J. SpacecraftRockets, 38:941–945, 2001; reproduced by permission of the American Institute of Aeronautics and Astronautics, Inc.*)

T-S waves and destabilizing for Mack waves. On the other hand, cross-flow and Gortler instabilities are insensitive to Mach number and wall cooling. However, favorable pressure gradient is destabilizing for cross flow, while periodic discrete three-dimensional roughness elements are stabilizing. Finally, surface suction stabilizes all modes.

It appears that a system that employs suction near the leading edge and a configuration design that results in a favorable pressure gradient on much of the upper wing surface can result in greater cruise efficiency. The suction near the leading edge controls cross-flow instability on swept wings, while a favorable pressure gradient delays the growth of the T-S waves. A mild favorable pressure gradient can also control cross-flow instability. From a practical standpoint, suction raises questions about maintenance requirements and long-term structural integrity.

In a flight environment, stationary cross-flow instability (that is, cross-flow instability in which the wave does not travel) is the dominant instability. The most unstable wavelength can be determined from linear stability theory. As an alternative to suction, W. S. Saric and H. L. Reed, together with their associates, noted that cross-flow instability is sensitive to discrete three-dimensional roughness elements. Thus, unstable waves occur only at integer multiples of the primary disturbance wave number $k = 2\pi/\lambda$, where λ is the most unstable wavelength. Thus, spacing apart roughness elements with wave number k excites subharmonic disturbances with the result that unstable waves are prevented from growing. It was further found that the shape of the roughness elements is not important and holes can work equally well as roughness elements; the important thing is the spacing. However, the size of the roughness element should be such that it should not result in tripping the boundary layer to transition to turbulent.

In conclusion, a great deal of progress has been made in understanding the mechanisms of turbulence and, in turn, improving our abilities to predict onset and to design airplanes where laminar flow control is the goal. This, in conjunction with more efficient engines, is the only way to reduce fuel consumption. The present concern, however, is whether it can be economically maintained; it is not whether it is feasible.

For background information *see* ADAPTIVE WINGS; BOUNDARY-LAYER FLOW; FLUID MECHANICS; LAMINAR FLOW; REYNOLDS NUMBER; TURBULENT FLOW in the McGraw-Hill Encyclopedia of Science & Technology.
Hassan A. Hassan

Bibliography. D. Arnal, Numerical and experimental studies on laminar flow control, *Int. J. Numer. Meth. Fluids*, 30:193-204, 1999; D. M. Bushnell, Aircraft drag reduction—A review, *Proc. Inst. Mech. Eng.*, 217G:1-18, 2003; T. Cebeci, *Stability and Transition: Theory and Application*, Springer, 2004; Th. Herbert, Parabolized stability equations, *Annu. Rev. Fluid Mech.*, 29:243-283, 1997; R. L. Kimmel et al., Laminar-turbulent transition on a Mach 8 elliptic cone flow, *AIAA J.*, 37:1080-1087, 1999; A. Krumbein, Automatic transition prediction and application to three-dimensinal wing configurations, *J. Aircraft*, 44:119-132, 2007; F. R. Menter et al., Transition modeling for general purpose CFD codes, *Flow Turbul. Combust.*, 77:277-203, 2006; R. H. Radeztsky, Jr. et al., Effect of Micro-sized Roughness on Transition in Swept Wing Flows, AIAA Paper 93-0016, 1993; W. S. Saric and H. L. Reed, Toward Practical laminar Flow Control–Remaining Challenges, AIAA Paper 2004-2311, 2004; E. B. Warren and H. A. Hassan, Transition closure model for predicting transition onset, *J. Aircraft*, 35:769-775, 1998; X. Xiao et al., Transitional flow over an elliptic cone at Mach 8, *J. Spacecraft Rockets*, 38:941-945, 2001.

Tugboats

The modern tugboat is a very powerful and agile tool to assist ships in their regular maneuvers into and out of port, or when in distress. The development of very efficient steerable thruster systems in concert with unique and very stable hull forms and high-performance deck machinery enables today's tugs to tackle assignments that could not have been imagined a short generation ago.

Historical development. For more than a hundred years, tugboats have been used for moving oceangoing ships into and out of their loading berths. Since large ships generally lack maneuverability when traveling at slow speeds, and thus are very susceptible to wind and current forces, supplementary power is required to control them. That is still the tug's primary duty today, but its roles and responsibilities have expanded to include many other tasks.

The earliest tugs were steam-driven paddle wheelers, used to bring sailing ships into port. The screw propeller soon replaced the paddle wheel, and eventually the diesel engine replaced steam as the primary power source. From the early part of the twentieth century until well into the 1940s, the appearance of tugs changed little (**Fig. 1**). Wooden construction gave way to iron and eventually almost exclusively to steel in the 1950s.

As modern machinery developed, the domain of the tug became the world's oceans, where its role as a salvage vessel was much sought after. Large and

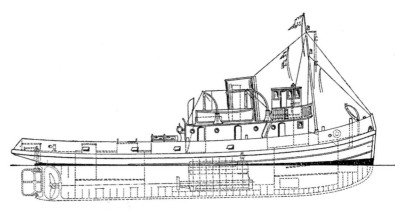

Fig. 1. Profile of early diesel-powered tug.

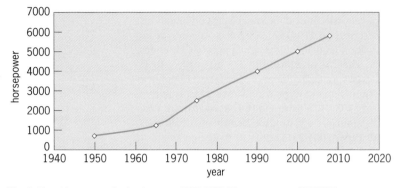

Fig. 2. Trend in average tugboat power, 1950–2008. I horsepower = 0.746 kW.

powerful tugs were posted close to areas of known marine hazards, and they were ready to offer service when needed, at a price. This trend continued well into the 1970s, when several ocean salvage tugs were built. Recently, the economics of deploying dedicated salvage tugs has become less attractive, and these big tugs are now used mostly for towing large barges and offshore rigs.

In the past decade, however, the tugboat world has witnessed dramatic changes. Oceangoing ships have become much larger than they were a generation ago, demanding tugs with greater power and stronger deck gear. Harbor towage accordingly has seen the development of several unique new design types. Increased concern over potential coastal pollution has led to the development of "escort" tugs.

The trend in average harbor-tug power over the past 60 years is illustrated in **Fig. 2**. There is a wide range in power depending on tug size and the specific locales, and these figures are representative of just the general market trend in harbor tugs during that period. It is clear, however, that tugs are steadily increasing in power to keep up with the demands of world shipping.

Modern tug types. Tugs today are generally categorized according to their primary tasks, as follows:

1. Harbor towing
2. Escort towing
3. Coastal towing
4. Offshore terminal support
5. Ocean towing and salvage
6. Construction service and support

Many tugs are designed and equipped to perform several of these functions, but such multipurpose vessels ultimately must compromise one or more of their primary functions in order to do multiple tasks reasonably well. There is also a completely different subset of tugs that handle barges, principally on the inland waterways of North and South America and in Europe. Generally referred to as push boats, these vessels fall outside the scope of this article. *See* BARGE TECHNOLOGY.

The harbor tug is generally a smaller vessel (32 m or 105 ft long, or less), equipped with heavy-duty towing equipment and fittings for handling ship's lines, and fendered for guiding large vessels (**Fig. 3***a*). In the past decade, many variations on this theme have emerged. The more successful new design con-

cepts include the "Rotor tug" with three independent Z-drive units, and the *Z-Tech*, which, through a unique hullform, provides almost identical thrust and speed ahead and astern, and combines the best features of the common "tractor" and "ASD" tug types described below (Fig. 3*b*).

Escort tugs, whose function is to control a tanker that has lost power or steering in a confined waterway, are specially designed to generate large forces working in what is known as the indirect mode (Fig. 3*c*). Acting obliquely to the line of travel of the attended tanker, an escort tug can, at speeds in excess of 6–7 knots (3–3.5 m/s), generate steering or braking forces far greater than those associated with its main propulsion power.

Coastal towing tugs are most common on the west coast of North America, hauling barges that are typically laden with various bulk products, often over long distances (Fig. 3*d*). Offshore terminal tugs perform many of the same sorts of tasks as harbor tugs, but do so in an open-water environment among oil fields (Fig. 3*e*). Duties include bringing tankers into oil loading buoys or FPSOs (floating production, storage, and offloading vessels), assisting with their mooring operations, and then pulling on a line astern to prevent the tanker from swinging on its mooring. Other duties include moving anchors, providing small parcel bulk and deck cargo movements, and personnel transfers.

Ocean towing tugs are large, seagoing tugs that are engaged typically in long-distance tows of barges with cargoes such as oil drilling platforms, jack-up rigs, or large modules for major construction projects. These tugs are also often the first called upon for major salvage projects.

The final category, construction support tugs, comprises typically smaller and simpler tugs that tow barges, handle dredges and pipelines, and assist with general foreshore construction projects (Fig. 3*f*).

Functional characteristics. The usual "measure of merit" of a tugboat is its bollard pull, defined as the maximum propeller thrust developed at zero speed of advance.

The major distinguishing characteristic of a tugboat, apart from its high power, is its duty cycle. For most harbor tugs, this is characterized by long periods of idle or low-power standby or maneuvering, followed by very short bursts of high power. Because of the need for this high power, tugs are also distinguished by very high power/displacement (weight) ratios. They are also exceptionally maneuverable, in order to move quickly from push mode to pull mode and from one position on a ship to another. Larger tugs, of course, have high power, but proportionately are not as extreme, nor generally are they as maneuverable. Long-distance tows also involve a duty cycle not unlike that of a major ship: many days at continuous load at near full power.

Major onboard technical systems. A tug is quite simply a platform for conveying power to move larger vessels. It does this usually through a straightforward power transmission system that generates thrust and then transmits that thrust to the attended vessel

(a)

(b)

(c)

(d)

(e)

(f)

Fig. 3. Modern tugboat types. *(a)* Modern harbor tug (United States). *(b) Z-Tech 6000*-class harbor tug (Port of Singapore). *(c) AVT*-class escort tug *Ajax* (Norway). *(d)* Coastal barge towing tug (Canada). *(e)* Offshore terminal support tug (France). *(f)* Construction support tug/workboat (United Kingdom).

either through a towline (when pulling) or through its fenders (when pushing).

Today, primary power in tugs is almost universally developed by diesel engines. In some larger tugs with many ancillary power demands, the power system might be diesel-electric, with multiple diesel-generator sets supplying power to large electric motors. At present, interest is developing in hybrid power systems (combined diesel/generator/battery systems) and the option of liquefied natural gas (LNG) and compressed natural gas (CNG) fuel systems, as the price of diesel fuel rises rapidly.

The propulsion system is the unique and dominant aspect of any modern tug. Whereas the previous generation of tugs was driven almost exclusively by conventional single or twin-screw propellers, today the majority of harbor and escort tugs feature fully azimuthing (360° steerable) drives. The two most common systems are known as Z-drives and Voith cycloidal propellers, respectively (**Fig. 4**). These drives are installed either in a tractor configuration, with the drives located about 30% of the length from the forward end, or in an azimuthing stern drive (ASD) configuration, with the drives aft in the normal propeller position (**Fig. 5**).

The Z-drive unit comprises a series of right-angle gears and shafts that enable the power to be transmitted from the engine to the propeller from entirely

(a)

(b)

Fig. 4. Propulsion systems. *(a)* Z-drive propulsion unit. *(b)* Voith cycloidal propeller.

within the hull, and a set of steering motors that then rotate the entire propeller assembly through 360° continuously.

The Voith cycloidal propeller is a more unusual device. Invented in the 1930s, it comprises a series of foil-shaped vertical blades that turn at relatively low speed in a rotating plate mounted in the bottom of the hull. As the plate rotates, the blades alter pitch in order to vector the thrust in the desired direction. The unique feature of this drive is that the direction of thrust can be changed almost instantaneously. It is, however, somewhat less efficient in terms of developing thrust than any screw propeller device.

The other major components on tugs are those used to transmit the tug forces, namely the winch and the fender system. Less common in modern tugs is a "tow-hook." Winches are either hydraulically or electrically driven, with brake capacity, line pull, and line speed varying according to the functional requirements. In most modern tugs, the towlines for ship

handling are all lightweight, very strong synthetic lines made from high-molecular-weight polyethylene (HMPE) fibers. These lines are spooled onto the winch and led up to the ship through very strong staples or tow bitts. Keeping the towline aboard the tug, rather than having a ship pass its line down, ensures that the tug crew knows that the line is of sufficient strength and in good condition.

Fender systems must transmit the full thrust of the tug into a ship without allowing any steel-to-steel contact, and must also spread the load to prevent damage to the ship's hull structure. A typical system in modern tugs has an upper "soft" cylindrical fender that provides the first point of contact. The load is then transferred to a lower "block" fender, which is stiffer, but helps to spread the load (**Fig. 6**).

Ancillary systems. Although the primary role of harbor tugs is ship handling, they also are often

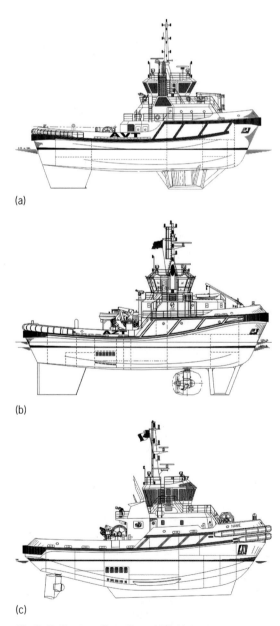

(a)

(b)

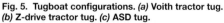

(c)

Fig. 5. Tugboat configurations. *(a)* Voith tractor tug. *(b)* Z-drive tractor tug. *(c)* ASD tug.

Fig. 6. Modern two-part fender system.

equipped to provide other services. Firefighting equipment is one of the systems most frequently found on tugs, but other common features are oil-spill response, oil recovery, and occasionally search and rescue equipment. Tugs serving the offshore industry are also frequently required to carry some deck cargo, typically pallets or small containers of stores for oil rigs.

For background information, *see* BOAT PROPULSION; MARINE MACHINERY; PROPELLER (MARINE CRAFT); SHIP SALVAGE; SHIP POWERING, MANEUVERING, AND SEAKEEPING in the McGraw-Hill Encyclopedia of Science & Technology.   Robert G. Allan

Bibliography. R. G. Allan, The evolution of escort tug technology ... fulfilling a promise, in *Society of Naval Architects and Marine Engineers, Proceedings of the 2000 Annual Meeting*, 2000; R. G. Allan, Tugs and towboats, in T. Lamb (ed.), *Ship Design and Construction*, 3d ed., vol. 2, Society of Naval Architects and Marine Engineers, 2004 pp. 49-1–49-30; M. J. Gaston, *The Tug Book*, Patrick Stephens Limited, 2002; H. Hensen, *Tug Use in Port, A Practical Guide,* 2d ed., Nautical Institute, 2003; B. Jürgens and W. Fork, *The Fascination of the Voith-Schneider Propeller: History and Engineering,* Koehlers Verlagsgesellschaft mbH, Hamburg, Germany, 2002.

Ubiquitination

All eukaryotic cells express a 78-amino-acid protein known as ubiquitin. This protein has the unique ability to be covalently attached onto other proteins in an enzymatic process known as ubiquitination (alternatively ubiquitylation). Ubiquitination is involved in the posttranslational regulation of a wide array of cellular processes, and different types of ubiquitin modifications exist that in turn can promote distinct protein fates. The most well understood role for ubiquitination is to target modified proteins for destruction by the large proteolytic enzyme complex known as the 26S proteasome. Ubiquitination is also involved in altering protein localization, promoting protein-protein interactions, regulating gene expression, and changing the activity of cellular enzymes.

Ubiquitin was originally named ubiquitous immunopoietic polypeptide for its expression in all the cells of the human body and also ATP-dependent proteolysis factor-1 (APF-1) for its function in protein degradation. Aaron Ciechanover, Avram Hershko, and Irwin Rose were awarded the Nobel Prize in Chemistry in 2004 for characterizing ubiquitin and describing ubiquitination enzymes. These scientists, together with Alexander Varshavsky, who described the first physiological functions of ubiquitin, are widely recognized for their fundamental discoveries on ubiquitination. In 2003, the U.S. Food and Drug Administration (FDA) approved the drug known as Velcade® (bortezomib) for certain types of cancer. As Velcade® blocks the degradation of all ubiquitinated proteins, current efforts aim to develop drugs that interfere with an individual protein's ubiquitination by targeting its specific ubiquitination enzymes. Such drugs may have many different applications—cancer therapy, anti-inflammatories, antivirals, and many more—reflecting the extensive involvement of ubiquitin and ubiquitination in physiology.

Process of ubiquitination. Ubiquitination involves the coordinated activity of a series of enzymes resulting in the ultimate modification of a protein with ubiquitin (**Fig. 1**). The activation of ubiquitin by the appropriately named ubiquitin-activating enzyme (also called E1) requires energy in the form of adenosine triphosphate (ATP). Once activated and transferred to the active-site cysteine residue of the E1 through a labile thiol-ester linkage, ubiquitin is then shuttled to the active-site cysteine of the ubiquitin-conjugating enzyme (also called E2) through a similar thiol-ester linkage. Ubiquitin ligases (also called E3s) have a critical function in protein ubiquitination—they recognize specific proteins to be modified and promote the covalent transfer of ubiquitin from E2 onto the E3-bound protein. The primary amine of lysine side chains is the most frequently modified site on proteins and occurs with the carboxy-terminus of ubiquitin. The linkage between ubiquitin and the protein is similar to a peptide bond between amino acids in a protein and has been called an isopeptide linkage to reflect this. As described in more detail below, ubiquitin can be synthesized into polymers—most frequently referred to as ubiquitin chains—in which ubiquitin molecules are attached between a lysine of one ubiquitin monomer to the carboxy-terminus of another. The process of ubiquitination is reversible as deubiquitinating enzymes (referred to as "dUbs") can disassemble ubiquitin chains and remove ubiquitin from proteins.

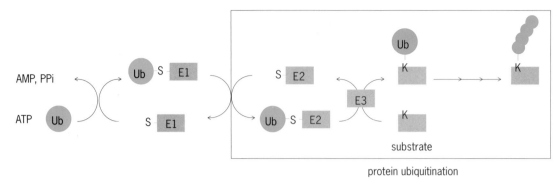

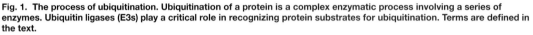

protein ubiquitination

Fig. 1. The process of ubiquitination. Ubiquitination of a protein is a complex enzymatic process involving a series of enzymes. Ubiquitin ligases (E3s) play a critical role in recognizing protein substrates for ubiquitination. Terms are defined in the text.

Ubiquitin ligases. The enzymes involved in ubiquitination function in many distinct pathways, reflected in the estimated 500 or more ubiquitin ligases encoded by the human genome. These enzymes, due to their dual role in recognizing specific protein substrates and promoting ubiquitin transfer, serve as a critical link between the ubiquitin system and cellular regulatory networks. The vast majority of ubiquitin ligases have not been studied in any detail yet, and the protein substrates that they recognize are unknown. Nevertheless, based on those that have been studied, several general features have emerged.

Two major classes of ubiquitin ligases have been identified based on the presence of conserved amino acid sequences that define how they participate in ubiquitination. Those containing a conserved cysteine- and histidine-containing fold that coordinates two zinc molecules—known as a "RING motif"—facilitate ubiquitin transfer directly from the ubiquitin-conjugating enzyme onto the lysine of the bound substrate. The other class of ligases contains a HECT domain, which forms a catalytic intermediate with ubiquitin on a conserved cysteine residue prior to its ultimate transfer onto substrate. Ubiquitin ligases either can be single proteins or require multiple proteins to function together as a complex. Posttranslational modifications such as phosphorylation are often a prerequisite for a protein to be recognized by its cognate ubiquitin ligase. Ubiquitin ligases themselves are often subjected to regulatory mechanisms to ensure appropriate ubiquitination, dependent upon various cellular signals and conditions. Examples of regulated expression, changes in subcellular localization, posttranslational modifications, and regulated assembly or disassembly of ubiquitin ligase complexes provide further mechanisms for spatial and temporal control of ubiquitination.

Ubiquitin-conjugating and ubiquitin-activating enzymes. The ubiquitin-conjugating enzyme family is considerably less complex than the ubiquitin ligases, with 33 members encoded by the human genome. These enzymes have a highly conserved catalytic core that contains the active-site cysteine residue that accepts ubiquitin from the ubiquitin-activating enzyme and is required for covalent ubiquitin transfer onto protein substrates. They are localized to distinct subcellular compartments such as the nucleus, cytoplasm, and endoplasmic reticulum outer member and can be expressed at different levels in the various tissues of the human body.

It was widely believed that the human genome encoded only a single ubiquitin-activating enzyme. However, a second ubiquitin-activating enzyme was discovered in 2007. These enzymes appear to have a general role in charging ubiquitin-conjugating enzymes, and their activity reflects the only energy-requiring step in ubiquitination.

Deubiquitinating enzymes. The process of ubiquitination is reversible through the activity of proteases known as deubiquitinating enzymes ("dUbs"). These enzymes—greater than 54 are encoded by the human genome—have diverse functions such as the processing of ubiquitin precursor proteins into individual ubiquitin molecules, proofreading of ubiquitin chains, and counteracting ubiquitination. Specific functions have not been attributed to many dUbs. A deubiquitinating activity is found associated with a subunit of the 26S proteasome, and it appears that deubiquitination of proteins occurs prior to their degradation. This process may promote efficient degradation of the protein and allows for ubiquitin molecules to be recycled.

Different types of ubiquitin modifications. There are different types of ubiquitination and currently much needs to be learned about how these are generated and the functions they serve in cellular signaling networks. Ubiquitin contains seven lysine residues (**Fig. 2**). The formation of an isopeptide linkage between a surface-exposed lysine residue on one ubiquitin molecule and the carboxy-terminus of another results in a ubiquitin chain of two molecules (diubiquitin). Given these different lysine residues and the potential for limitless numbers of ubiquitin molecules in a chain, it is easy to imagine an infinite variety of distinct ubiquitin chains. However, three major types of ubiquitin modification have been described to date.

The attachment of single ubiquitin molecules on proteins (known as monoubiquitination) is involved in regulating protein localization and gene expression. Cell surface receptors, proteins involved in the cellular response to deoxyribonucleic acid (DNA) damage, and histones are among proteins known to

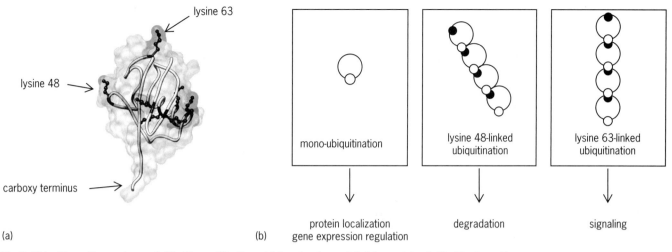

(a)

(b)

protein localization
gene expression regulation

degradation

signaling

Fig. 2. Ubiquitin and known types of ubiquitin modifications. (*a*) **A molecular model of the protein of ubiquitin shows the position of lysine residues (black) relative to the overall structure of the protein. Lysine residues with known functions (lysine 48 and lysine 63) and the carboxy-terminus (the part of ubiquitin important for its enzymatic transfer and attachment onto proteins) are indicated.** (*b*) **The three major types of ubiquitin modifications have distinct cellular functions.**

be monoubiquitinated. Ubiquitin chains generated through lysine 48 stimulate the degradation of modified proteins by the 26S proteasome, and other biochemical studies demonstrate that a tetraubiquitin chain is the minimum signal required. This type of ubiquitination appears to be the most prevalent in cells and has numerous functions where the rapid and irreversible inactivation of a particular protein through its destruction is needed. Ubiquitin chains generated through lysine 63-based linkages are involved in inflammation and the response to DNA damage. These modifications are thought to serve in a nonproteolytic signaling function by activating cellular protein kinases. Other types of ubiquitin chains may exist; however, their physiological significance remains to be elucidated.

Physiological functions of ubiquitination. The most well understood function of ubiquitin is its role in promoting the degradation of proteins by the 26S proteasome, but it also has nonproteolytic functions. In general, ubiquitination of proteins can be involved in attenuating and activating various cellular processes. Several established physiological functions of ubiquitination are summarized below.

Cell cycle regulatory proteins are ubiquitinated and subsequently degraded to control their abundance at appropriate times during the various phases underlying cell division. These proteins, such as the cyclins and cyclin-dependent kinase inhibitors, have distinct and defined roles during critical cell cycle events—that is, the ubiquitin-mediated degradation of these proteins plays an important role in the transition from one phase of the cell cycle to another.

Ubiquitination has roles in DNA replication and the cellular response to DNA damage. For example, the monoubiquitination of several proteins upon DNA damage promotes the assembly of a complex containing multiple proteins important for repairing altered DNA.

Cells utilize ubiquitination to target improperly folded or unfolded proteins for destruction by the

26S proteasome. These quality-control mechanisms may prevent problems that arise from defective proteins and help recycle amino acids for use in subsequent protein synthesis.

Ubiquitination is involved in the regulation of gene expression. Histones (proteins that associate with DNA to form chromatin) are subjected to reversible monoubiquitination, which regulates transcription. Proteins involved in the repression and activation of transcription are sometimes ubiquitinated to target them for degradation.

The various signaling pathways in the cell often involve ubiquitination. The activity of kinases (enzymes that modify proteins with phosphate) and phosphatases (enzymes that remove this modification) are sometimes regulated through ubiquitination. Once these enzymes are activated and their activities are no longer needed, they can be ubiquitinated and then degraded by the 26S proteasome to rapidly and irreversibly turn them off. Ubiquitination has also been shown to function in a nonproteolytic role by activating a protein kinase in response to extracellular stimuli. The monoubiquitination of some cell surface receptors is induced by binding of their ligands and this, in turn, promotes their internalization (by a process known as receptor-mediated endocytosis) and their subsequent targeting to lysosomes where they are degraded.

Aspects of the immune system are regulated through ubiquitination. Stimuli that ultimately trigger inflammation, a critical response to infection, require the ubiquitination of regulatory proteins. Lymphocytes, the specialized cells of the immune system that identify pathogens and other foreign materials, utilize ubiquitination to transmit intracellular signals upon activation.

Links between disease and ubiquitination. Alterations in normal protein ubiquitination are associated with different types of cancer. The tumor suppressor protein p53 is mutated in greater than 50% of all human cancers, and many of these mutations alter its ability to be ubiquitinated by the ubiquitin ligase known as MDM2. A protein known as BRCA1 (for breast cancer 1) structurally resembles a ubiquitin ligase. Mutations in the *BRCA1* gene are associated with increased susceptibility for breast cancer, and some of these mutations disrupt its ubiquitin ligase activity and therefore presumably promote the accumulation of its as yet unidentified target substrate, which is oncogenic. Colorectal cancers are sometimes associated with mutations in a gene known as *APC* (adenomatous polyposis coli). Many of these mutations disrupt the normal ability of the APC protein to regulate cellular proliferation by preventing the normal ubiquitination and destruction of signaling proteins. Changes in the abundance of a cell cycle regulatory protein known as p27 are associated with different types of cancer. Low levels of p27 and correspondingly high levels of its ubiquitin ligase (known as SCF^{SKP2}) are considered an indicator of poor prognosis for survival. The FDA-approved proteasome inhibitor Velcade® is used in certain cases of multiple myeloma and it blocks the degradation of all ubiquitinated proteins. Expanding the use of this drug to other types of cancer and searching for other proteasome inhibitors with different properties are currently under clinical investigation.

Disease-causing viruses often alter normal cellular ubiquitination to promote their own propagation or avoid detection by the immune system. For example, the human papilloma virus (HPV), a virus that causes genital warts and is associated with cervical cancer, encodes proteins that promote the ubiquitination and degradation of p53. The human immunodeficiency virus [HIV, the causative agent of acquired immunodeficiency syndrome (AIDS)] encodes several proteins that effectively hijack cellular ubiquitin ligases. These proteins target normally stable proteins for ubiquitination, leading to their degradation by the proteasome.

Neurodegenerative diseases such as Alzheimer's, Parkinson's, and Huntington's diseases are often associated with the formation of protein aggregates in the brains of afflicted individuals. Ubiquitin has been identified in these deposits. It is unclear if alterations in ubiquitination underlie the causes of these diseases or are a consequence of other physiological changes.

Cystic fibrosis, characterized by thick mucus production in the lungs and digestive system, is one of the most common genetic disorders and leads to premature death in afflicted individuals. Mutations in the cystic fibrosis transmembrance conductance regulator (CFTR) are associated with the disease and one class of these mutations promotes CFTR misfolding, leading to its ubiquitination and degradation and thereby eliminating CFTR's important functions.

In addition to these examples, ubiquitination has roles in inflammation, muscle-wasting disorders, diabetes and obesity, and cardiovascular diseases. It seems likely that further links between ubiquitin and human diseases will be uncovered as more ubiquitin ligases are studied in depth and as more proteins regulated through ubiquitination are identified.

Targeting aspects of ubiquitination for human therapies. Clear relationships have emerged between ubiquitination and a variety of physiological processes. Moreover, the changes in ubiquitination that are involved in several different human diseases are beginning to be understood. As a result, efforts are under way to focus on aspects of ubiquitination for the development of human therapeutics. The proteasome inhibitor Velcade® as a treatment for refractory and relapsed multiple myeloma currently represents the only FDA-approved drug related to ubiquitination. Other types of proteasome inhibitors are likely to be approved soon, and their use may likely expand to treat other types of cancer as more is understood about them. The specificity of ubiquitin ligases for particular protein substrates makes them attractive therapeutic targets and some of them have been subjected to extensive preclinical studies. The hope is that selective modulation of ubiquitination by targeting specific ubiquitination enzymes may allow for the development of novel drugs for the treatment of disease.

For background information *see* BIOCHEMISTRY; CANCER (MEDICINE); CELL BIOLOGY; ENZYME; HUMAN GENOME; MOLECULAR BIOLOGY; PROTEASOME; PROTEIN; PROTEIN DEGRADATION; PROTEIN ENGINEERING in the McGraw-Hill Encyclopedia of Science & Technology.   Matthew D. Petroski

Bibliography. R. J. Mayer, A. Ciechanover, and M. Rechsteiner (eds.), *Protein Degradation: Cell Biology of the Ubiquitin-Proteasome System*, Wiley-VCH, Weinheim, 2006; R. J. Mayer, A. Ciechanover, and M. Rechsteiner (eds.), *Protein Degradation: The Ubiquitin-Proteasome System*, Wiley-VCH, Weinheim, 2006; R. J. Mayer, A. Ciechanover, and M. Rechsteiner (eds.), *Protein Degradation: The Ubiquitin-Proteasome System and Disease*, Wiley-VCH, Weinheim, 2007; R. J. Mayer, A. Ciechanover, and M. Rechsteiner (eds.), *Protein Degradation: Ubiquitin and the Chemistry of Life*, Wiley-VCH, Weinheim, 2006.

Ubiquitous transportation network sensors

A ubiquitous network is one that provides services anywhere to anyone or anything on any device. Ubiquitous networks characteristically are mobile, wireless, and highly distributed. In transportation, a ubiquitous network would link drivers, vehicles, and infrastructure [such as traffic signals (lights), automatic toll collection, traffic advisory services, and media communication] by a network consisting of a very large number of wireless sensor nodes.

Ubiquitous transportation network sensors are a promising technology in intelligent transportation systems for monitoring the traffic stream on roadways as part of transportation management and control applications. Conventional traffic monitoring sensors use inductive loop detectors, image detectors, radar, and infrared detectors to collect traffic data, such as the number of vehicles passing the detectors, their average speeds, and their

Fig. 1. Example showing (*a–b*) intersection collision avoidance service based on (*c*) T-sensor nodes.

occupancy rates. Wireless communication technologies have been introduced in transportation information services to collect traffic data and provide real-time information to drivers. The communication media include the dedicated short-range communication system (DSRC), wireless access in vehicular environment (WAVE), communication air-interface for long and medium ranges (CALM), wireless broadband (WiMAX, WiBro, etc.), and wireless fidelity (Wi-Fi). Wireless sensor networks (WSN) are now being developed to replace the conventional detection systems on the roadway using ubiquitous transportation sensors.

Wireless sensor networks (WSN). Conventional traffic monitoring and control systems are expensive to install and maintain. WSN technologies promise to change that. Battery-operated sensor nodes for vehicle detection and wireless communication, measuring less than 10 cm in diameter, eliminate the need for wiring and reduce to a minimum the time and cost of roadwork required to install these systems. However, traffic monitoring and control applications pose a number of challenges related to WSN software and hardware design. Because the wireless sensor nodes used for vehicle detection are battery-powered, they must be designed to save energy. The energy constraint influences the choice of sensors, analog signal-conditioning circuitry, microcontroller, and wireless transceiver used in the design. An important point to consider is the environment in which the node is deployed. Even though a proper housing can shield the hardware from water and dirt, environmental factors, such as temperature, can have a significant effect on the sensor's analog signal-conditioning circuitry and battery performance. And rain or snow can limit the wireless communication range between the sink (data destination) and the sensor nodes. The sensor hardware is expected to operate reliably in temperatures ranging from −40 to 60°C, and the system should remain reliable over the entire temperature range. An application of telematics (information and communications technology) service for a developed ubiquitous sensor node for vehicle detection, the so-called T-sensor node, has been implemented in Korea at an outdoor test site in a real traffic environment. It includes collision avoidance in curved sections of the roadway and collision warning at intersections with and without traffic signals. An all-way stop control (AWSC) intersection system has been implemented, based on the T-sensor node, that warns drivers approaching the intersection of conflicting traffic by displaying a warning message on a variable message signboard, as shown in **Fig. 1**.

Communication media. To provide suitable ubiquitous sensor networks to support ITS and vehicle telematics services, including multimedia use, the current status of the related communication standards is reviewed in **Fig. 2** and **Table 1**.

Dedicated short-range communication (DSRC). The DSRC standard specifies the physical layer (PHY), data link layer, and application layer at 5.8 GHz for intelligent transportation systems (ITS) and telematics services; it is based on an open system interconnection (OSI) reference model. It is intended for use in ITS services, such as electronic toll collection systems (ETCS). The standard describes the communication mechanism and procedures between roadside equipment (RSE) and on-board equipment (OBE) that passes through the RSE communication zone, providing both point-to-point interactive communication and point-to-multipoint communication.

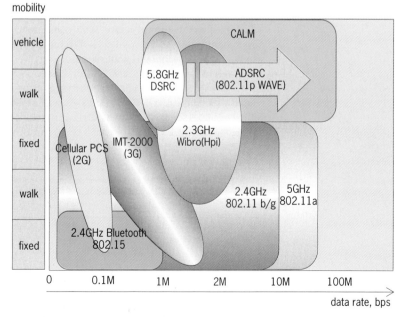

Fig. 2. Communication media standards.

TABLE 1. Operating parameters for various communications standards

Type	cellular	DMB	WiBro	Wi-Fi	DSRC	WAVE
Frequency (GHz)	0.8–1.7	0.2/2.6	2.3	2.4/5.8	5.8	5.9
Radio Range (Km)	20	100	3	0.1	0.1	1
Transmit speed (Mbps)	0.3	0.1	1	0.3–54	1	2–54
Mobility (Km/H)	5	60	60	—	160	200
Access time unit	s	—	s	s	ms	ms
Duplex		Full duplex	Half duplex	Full duplex	Full duplex	Full duplex	Full duplex

Wireless access in vehicular environment (WAVE). WAVE (IEEE 802.11p) introduces enhancements to DSRC that support (1) extremely short latency, measured in tens of milliseconds, (2) long range, up to 1 km (0.6 mi), while still supporting short ranges of a few meters, (3) very mobile devices, with operating speeds of up to 200 km/h (124 mi/h), and (4) extreme multipath, such as would be encountered by a car traveling on a highway alongside other cars and trucks and past large buildings and bridges. To provide complete interoperability at the application level requires standardization of the entire communications stack, not just of the lowest layers of the media access protocol (MAC) and PHY, as defined in the DSRC standard.

Communication air-interface for long and medium ranges (CALM). CALM is the International Organization for Standardization (ISO) approved framework for heterogeneous packet-switched communication in mobile environments (**Table 2**).

The scope of CALM is to provide a standardized set of air-interface protocols and parameters for medium and long ranges and high-speed ITS communication using one or more of several media, with multipoint and networking protocols within each media, and upper-layer protocols to enable transfer between media. This service includes the following communication modes.

1. Vehicle-infrastructure. These multipoint communication parameters are automatically negotiated, and subsequent communication may be initiated by either roadside or vehicle.

2. Infrastructure-infrastructure. The communication system may also be used to link fixed points where traditional cabling is undesirable.

3. Vehicle-vehicle. A low-latency, peer-peer network with the capability of carrying safety-related data, such as collision avoidance, and other vehicle-vehicle services, such as ad-hoc networks linking multiple vehicles.

WiMAX. WiMAX is a standards-based technology enabling the delivery of last-mile wireless broadband access as an alternative to wired broadband, like cable. WiMAX provides fixed, nomadic, portable, and mobile wireless broadband connectivity without the need for a direct line of sight with a base station. In a typical cell deployment radius of 3–10 km (2–6 mi), WiMAX Forum Certified™ systems can be expected to deliver capacity of up to 40 Mbps per channel for fixed and portable access applications

Wireless broadband (WiBro). WiBro is a wireless broadband Internet technology being developed by the South Korean telecommunication industry with the service name of IEEE 802.16e (mobile WiMAX) international standard. WiBro was devised to overcome the data rate limitation of mobile phones (for example, ADSL or wireless LAN). WiBro base stations will offer an aggregate data throughput of 30–50 Mbps and cover a radius of 1–5 km (0.6–3.1 mi), allowing for portable Internet use. In detail, it will be usable in moving devices at up to 120 km/h (74.5 mi/h). For comparison, LAN works at walking speed and mobile phones work at up to 250 km/h (155 mi/h).

Digital multimedia broadcasting (DMB). DMB is a method of multicasting multimedia content to mobile and portable devices, such as cell phones, by satellite or terrestrial services or a combination of the two. Some DMB-capable receiving devices can render content that is individualized to the location or the subscriber. Examples of multimedia broadcast content include mobile television, movie, video clip, music, and text message transmission. Most existing and proposed DMB services operate on a fee-based subscription model, although advertising has been suggested as a revenue source. A free, state-operated DMB service is available in South Korea.

Wi-Fi (IEEE 802.11). This is wireless network that uses radio waves, just as cell phones, television, and radio do. In fact, communication across a wireless network is a lot like two-way radio communication in that

TABLE 2. The scope of CALM standards being developed by ISO/TC204 WG16, Wide Area Communications

Medium	Microwave	Infrared	Millimeter wave
Scope	5–6 GHz microwave Support 3–27 Mbps (6–54 Mbps)	800–900 nm infrared Support 1–128 Mbps	Use 60–70 GHz millimeter wave Support latencies and communication delays in the order of milliseconds
	Range 300–1000 m	Range 300–1000 m	Support multisubcarrier operations
	Support latencies and communication delays in the order of milliseconds	Support latencies and communication delays in the order of milliseconds	

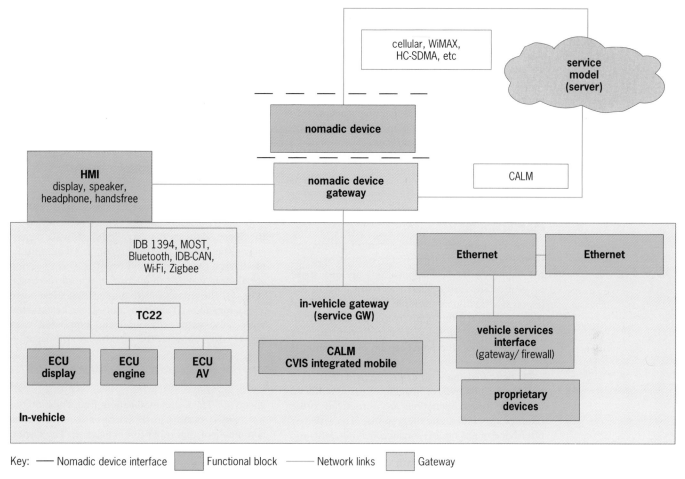

Key: —— Nomadic device interface   ▢ Functional block   —— Network links   ▢ Gateway

Fig. 3. Service scope of nomadic devices.

(1) a computer's wireless adapter translates data into a radio signal and transmits it using an antenna, and (2) a wireless router receives the signal and decodes it. It sends the information to the Internet using a physical, wired Ethernet connection.

3G Cellular. 3G cellular technologies enable network operators to offer users a wider range of advanced services, while achieving greater network capacity through improved spectral efficiency. Services include wide-area wireless voice telephony and broadband wireless data, all in a mobile environment. Typically, they provide service at 5–10 Mb per second.

Nomadic devices. ISO/TC204/WG17 is designed to facilitate the development, promotion, and standardization of nomadic and portable devices to support ITS services and multimedia use, such as passenger information, automotive information, driver advisory and warning systems, and entertainment system interfaces, for ITS service providers and ubiquitous transportation networks. This standard fosters the introduction of multimedia and telematics nomadic devices in public transportation and automobiles. Nomadic devices provide communications connectivity via equipment such as cellular telephones, mobile wireless broadband (for example, WiMAX), and Wi-Fi, and include short-range links, such as Bluetooth and Zigbee, to connect nomadic devices to the motor vehicle communications sys-

tem in ubiquitous transportation networks. These standards for the communications architecture and generic requirements enable connectivity between vehicle and infrastructure, or between vehicles, by using nomadic links within the vehicle (for example, Bluetooth) and devices introduced into the vehicle (such as music players, PDAs, etc.), including the provision of connectivity to the infrastructure, via mobile devices (2G/3G/mobile wireless broadband, etc.), as well as the support of application services within the vehicle and integration within the CALM architecture and in vehicle gateways.

Figure 3 illustrates the scope of the WG17 standard as it applies to (1) automotive information, (2) driver advisory and warning systems, and (3) entertainment system interfaces to motor vehicle communication networks.

For background information *see* DATA COMMUNICATIONS; INTERNET; LOCAL-AREA NETWORKS; TRAFFIC-CONTROL SYSTEMS; TRANSPORTATION ENGINEERING; WIDE-AREA NETWORKS; WIRELESS FIDELITY (WI-FI) in the McGraw-Hill Encyclopedia of Science & Technology.   Young-Jun Moon; Ray F. Benekohal

Bibliography. ISO/TC204 WG17, *The Use of Nomadic Devices to Support ITS Service and Multimedia Provision in Vehicles*, PWI 10992, 2008; A. N. Knaian, *A Wireless Network for Smart Roadbeds and Intelligent Transportation Systems*, 1999;

Y. J. Moon, J. Lee, and Y. K. Park, System Integration and Field Tests for Developing In-vehicle Dilemma Zone Warning System, Transportation Research Record 1826: 53–59 TRB, 2003; K. Sung et al., Collision warning system on a curved road using wireless sensor networks, *Proceedings of the IEEE 66th Vehicular Technology Conference*, VTC2007-Fall, Baltimore, October 2007; E. Zimmermann et al., An AMR sensor-based measurement system for magnetoelectrical resistivity tomography, *IEEE Sens. J.*, 5(2), 233–241, 2005.

Ultrasonic viscoelastic tissue measurements

Characterization of tissue mechanical properties, particularly the elasticity or tactile stiffness of tissue, has important medical applications because these properties are closely linked to tissue health. Palpation is a traditional example by which abnormal tissue is detected through sensing its stiffness with touch. However, palpation is subjective and may not be feasible when the target region of evaluation is deep within the body.

Elasticity imaging is an emerging field of medical imaging for noninvasive imaging of tissue stiffness. Generally, forced excitations are introduced to the tissue being studied, and the response is recorded by ultrasound or magnetic resonance imaging (MRI) and used to infer the tissue's mechanical properties. MRI-based techniques such as MR elastography (MRE), although very promising, are expensive and not widely available. Therefore, ultrasound-based elasticity imaging techniques may be more desirable.

An ideal ultrasound technique should meet three criteria. First, it should provide quantitative results. Many existing methods provide only a qualitative image (a relative mapping) of tissue stiffness. Such techniques are useful for detecting abnormal lesions but are insufficient for diffuse diseases such as liver fibrosis, in which there is no normal tissue to provide a background contrast. Such situations require quantitative techniques to provide measurements of stiffness in pascals. Second, an ideal technique should determine both tissue stiffness and viscosity because firstly tissue viscosity is another important index of tissue health state, and secondly if viscosity is neglected, stiffness is forced to account for its effects and bias can be introduced into stiffness estimation. Third, the technique should be compatible with current clinical ultrasound scanners to facilitate easy and large-scale application. Toward these aims, techniques have been developed that utilize ultrasound radiation forces to induce propagating waves in tissue. These waves are then measured using ultrasonic methods to quantify the stiffness and viscosity of tissue. A general name for these techniques is dispersion ultrasound vibrometry (DUV).

Methods. Various methods have been developed for studying large organs and arteries.

Shear-wave dispersion ultrasound vibrometry (SDUV). For large organs, shear waves are typically induced in

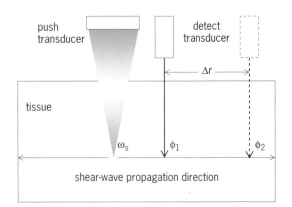

Fig. 1. Principle of dispersion ultrasound vibrometry (DUV).

the tissue, so the method is called shear-wave dispersion ultrasound vibrometry (SDUV). An amplitude-modulated (AM) ultrasound beam is used to generate a shear wave of frequency ω_s within the tissue (**Fig. 1**). The phases of the propagating shear wave at two locations Δr apart (ϕ_1 and ϕ_2) can be detected with pulse echo ultrasound and combined with Eq. (1) to calculate the shear-wave propagation

$$c_s(\omega_s) = \frac{\omega_s \Delta r}{\phi_2 - \phi_1} \tag{1}$$

speed c_s. Then the modulation frequency of the AM "push" beam can be changed to obtain measurements of c_s at other frequencies ω_s. The dispersion (variation of propagation speed with frequency) characteristics of c_s can be used to solve quantitatively for both the stiffness and the viscosity of the tissue.

For locally homogeneous tissue, the Voigt dispersion equation (2) can be used to solve for its

$$c_s(\omega_s) = \sqrt{\frac{2\left(\mu_1^2 + \omega_s^2 \mu_2^2\right)}{\rho\left(\mu_1 + \sqrt{\mu_1^2 + \omega_s^2 \mu_2^2}\right)}} \tag{2}$$

viscoelasticity, where ρ (assumed to be 1000 kg/m³ for all soft tissues), μ_1, and μ_2 are the density, shear elasticity, and shear viscosity of the tissue, respectively.

Arterial dispersion ultrasound vibrometry (ADUV). Equation (2) cannot be used to characterize an artery, because the artery's geometry can change the wave propagation speed and therefore must be properly accounted for. To do so requires solving a matrix differential equation. However, the wave speed can be calculated by measuring the phase change over a known distance for a selected set of frequencies, as described by Eq. (1). After the wave velocities have been obtained for different frequencies, the dispersion curve, which shows the variation of wave velocity with frequency, can be obtained. The decay rate of the wave amplitude is related to the viscous part of the elastic modulus. Because of the viscoelastic behavior of the vessel, the elastic modulus is a complex quantity that can be written as Eq. (3) where E^* is

$$E^* = E(1 + i\eta) \tag{3}$$

the complex modulus; E is the real modulus without viscoelastic effect; and η is the viscoelastic ratio, which is related to the decay rate μ by $\eta = 2\mu c_a/\omega_s$, where c_a is the wave speed in the vessel and ω_s is the frequency.

SDUV and ADUV are not imaging methods but rather point measurement techniques that provide quantitative results at selected points of measurement. Because the shear waves generated by the ultrasound push beam are relatively weak, their detection can be challenging because of the presence of cardiac and breathing motions. Therefore, echoes received by the detect transducer are demodulated with a special correlation method. A Kalman filter is then used to extract the shear wave speed at the frequency of the vibration with high accuracy and precision.

Applications. These methods have been used to evaluate liver tissue and arteries.

Liver evaluation. SDUV can be used for liver fibrosis staging. Liver cirrhosis can be caused by various chronic liver diseases and has very high prevalence (hundreds of millions of patients worldwide) and risk (50% five-year mortality). Earlier stages of fibrosis are treatable and reversible, and early diagnosis is therefore very important. Currently, the "gold standard" for liver fibrosis staging is liver biopsy, which is invasive and associated with limitations such as complications and sampling errors. Noninvasive alternatives such as serum marker, ultrasound, computerized tomography (CT), and MRI have limited sensitivity for earlier stages of fibrosis, when it may be advantageous to initiate therapy. Recently, it has been demonstrated that liver stiffness is an effective index for liver fibrosis staging. SDUV measurements in normal porcine (swine) liver in vivo will be presented to illustrate the feasibility of SDUV for this application.

The pulse sequence was modified in the swine experiment to improve the performance of SDUV. For the original AM push beam shown in Fig. 1, the sequence has to be repeated at different modulation frequencies to characterize dispersion. If, instead of a single frequency, tone bursts of push ultrasound are used, repeated for example at $\mathrm{PRF}_p = 100$ Hz (**Fig. 2**), where PRF is pulse repetition frequency, shear waves are generated that will contain not only the fundamental frequency at 100 Hz but also its higher harmonics at 200, 300, and 400 Hz, and so forth. Therefore, the push sequence needs to be applied only once, and dispersion information is in-

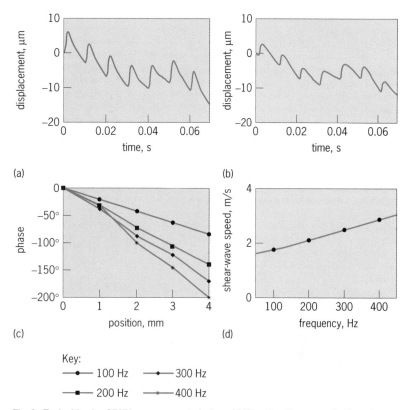

Fig. 3. Typical *in vivo* SDUV measurements in liver. (*a*) Vibration–time record at location 0 mm. (*b*) Vibration–time record at location 2 mm. (*c*) Vibration phase versus distance for shear waves with frequencies of 100, 200, 300, and 400 Hz, calculated from vibration–time records from five locations. (*d*) Shear wave speeds calculated from locations 0 and 4 mm in (*c*). 1 mm = 0.04 in., 1 m/s = 3.3 ft/s.

trinsically available. In addition, multiple detection beams can be applied between the push beams to avoid their interfering with each other.

A farm pig was anesthetized and mechanically ventilated. SDUV measurements were then performed on its liver through the pig's abdominal wall with one push transducer and one detect transducer, as shown in Fig. 1. Mechanical ventilation was temporarily suspended during SDUV measurements, and an electrocardiographic (ECG) signal was used to trigger SDUV measurements when cardiac motion was minimal during a heart cycle.

Figure 3*a* and 3*b* are examples of shear waves measured at two locations 2 mm (0.08 in.) apart along the shear wave propagation direction. Shear waves of the fundamental frequency (100 Hz) as well as its higher harmonics (at 200 Hz, 300 Hz, 400 Hz, and so forth) are clearly visible. The phases of shear waves at frequencies 100–400 Hz were estimated from the vibration–time records with the Kalman filter. Figure 3*c* demonstrates that the phases of the shear wave detected in this experiment are quite linear with distance for all frequencies. Shear wave speed, calculated using phase information at the locations 0 and 4 mm in Fig. 3*c*, is shown as data points in Fig. 3*d*. The curve is the least-mean-squares fit by Eq. (2), which gives $\mu_1 = 2.4$ kPa and $\mu_2 = 2.1$ Pa · s. If phases at two locations are used for speed estimation, the time required to make a single SDUV measurement is about 0.1 s.

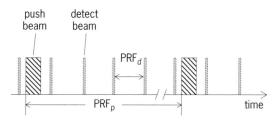

Fig. 2. Multifrequency pulse sequence for SDUV, characterized by pulse repetition frequencies for push beam (PRF_p) and detect beam (PRF_d).

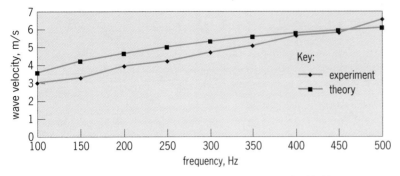

Fig. 4. Dispersion of wave propagation in an excised pig artery embedded in a tissue-mimicking gelatin phantom. 1 m/s = 3.3 ft/s.

Results of nine repeated SDUV measurements give a shear elasticity of $\mu_1 = 2.2 \pm 0.63$ kPa and a shear viscosity of $\mu_2 = 1.96 \pm 0.34$ Pa · s. These results are close to $\mu_1 = 2.06 \pm 0.26$ kPa and $\mu_2 = 1.72 \pm 0.15$ Pa · s reported for *in vivo* healthy human livers. The heating due to the push beam and its mechanical strength both fall within limits that render SDUV safe for *in vivo* human applications.

Artery evaluation. Cardiovascular disease (CVD) has been the leading cause of death in the United States. Several imaging techniques, including x-ray angiography, ultrasound, and MRI, can be used to image atherosclerotic vessels. However, many imaging techniques can measure only geometric properties such as luminal diameter, wall thickness, and plaque size; they cannot characterize the material properties of vessels. ADUV has been developed to measure local material properties of arterial vessels quantitatively. This capability may greatly improve the clinician's ability to identify and track subtle material property changes and thereby diagnose atherosclerosis at an earlier stage.

Figure 4 shows the dispersion curve for an excised pig artery embedded in a tissue-mimicking gelatin phantom. The measured wave speed is about 3.0 m/s (9.8 ft/s) at 100 Hz and 6.5 m/s (21.3 ft/s) at 500 Hz. Also shown in Fig. 4 is the dispersion curve calculated using the theory based on the previously mentioned matrix differential equation with

a Young's modulus of $E = 300$ kPa, a value that fits well with the experimental data.

As a result of the viscoelastic properties of the artery and its surrounding gelatin, the amplitude of the wave decays as it propagates. This decay reveals the viscoelastic characteristics of the artery and surrounding tissue. By measuring the decay rate over distance for each frequency, the viscoelastic ratio is calculated (**Fig. 5**). While the decay rate increases with the frequency, as expected, the imaginary part of modulus or the viscoelastic ratio decreases with the frequency and then remains almost constant for high frequencies.

Implementation. For clinical implementation, a single array transducer can be used for both push and detect functions, following the diagram in Fig. 2. The transducer can repeatedly transmit to one location to generate shear waves and intermittently switch to detection mode to monitor the propagation of shear waves. This pulse sequence makes ADUV and SDUV compatible with clinical scanners. The envisioned application of ADUV and SDUV, once implemented on clinical scanners, is as follows. First, a B-mode image of the liver or artery is obtained. Then a virtual biopsy is performed at locations selected interactively in the image. The wave speed can be measured with high temporal resolution in a few tens of milliseconds and with a spatial resolution of a few millimeters. Because one SDUV measurement takes only 0.1 s, multiple measurements can be made at different locations to obtain a more comprehensive evaluation of the organ. Then the appropriate mathematical model for wave propagation is used to determine the local viscoelastic parameters of the liver or artery noninvasively.

For background information *see* ARTERIOSCLEROSIS; BIOMEDICAL ULTRASONICS; CIRRHOSIS; ELASTICITY; ESTIMATION THEORY; MEDICAL IMAGING; MEDICAL ULTRASONIC TOMOGRAPHY; ULTRASONICS; VISCOSITY in the McGraw-Hill Encyclopedia of Science & Technology.

James F. Greenleaf; Matthew W. Urban; Xiaoming Zhang; Shigao Chen

Bibliography. S. Chen, M. Fatemi, and J. F. Greenleaf, Quantifying elasticity and viscosity from measurement of shear wave speed dispersion, *J. Acoust. Soc. Amer.*, 115(6):2781–2785, 2004; M. Fatemi and J. F. Greenleaf, Ultrasound-stimulated vibro-acoustic spectrography, *Science*, 280:82–85, 1998; L. Huwart et al., Liver fibrosis: non-invasive assessment with MR elastography, *NMR Biomed.*, 19:173–179, 2006; H. Kanai, Propagation of spontaneously actuated pulsive vibration in human heart wall and *in vivo* viscoelasticity estimation, *IEEE Trans. Ultrasonics, Ferroelectrics Frequency Control*, 52(11):1931–1942, 2005; X. Zhang et al., Noninvasive method for estimation of complex elastic modulus of arterial vessels, *IEEE Trans. Ultrasonics, Ferroelectrics Frequency Control*, 52:642–652, 2005; Y. Zheng et al., Detection of tissue harmonic motion induced by ultrasonic radiation force using pulse-echo ultrasound and Kalman filter, *IEEE Trans. Ultrasonics, Ferroelectrics Frequency Control*, 54:290–300, 2007.

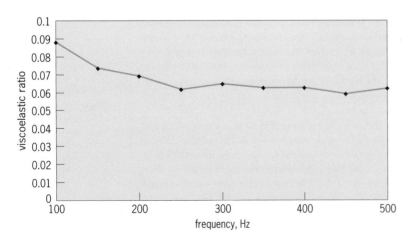

Fig. 5. Viscoelastic ratio η versus frequency for the excised artery in Fig. 4.

Underground processing of oil shale

Oil shale is a thinly layered, blocky to fissile (breaking into thin layers), sedimentary rock rich in preserved organic material that yields significant amounts of oil and gas when heated sufficiently. The global resource of oil potentially available from oil shale is substantially larger than the amount of oil produced historically. Potential reserves have been estimated at 2.8 trillion barrels of oil. However, recent revision of Chinese resource estimates may potentially add approximately 300 billion barrels to the global resource. All estimates of recoverable shale oil are uncertain at this time. Oil shale was commonly deposited in large, fresh to highly saline lakes or in shallow marine environments. The organic material consists of kerogen, a complex mix of fossilized remains consisting mainly of algae and cyanobacteria.

Technology for in situ (underground) production of shale oil. Previous investigations of the potential of oil shale included early experiments on underground or in situ methods. The U.S. Bureau of Land Management awarded six Research Development and Demonstration (RD&D) leases in 2006–2007 to companies developing oil-shale production processes, each tied to a specific production technology and a potential commercial lease area. This leasing program led to submittal of a diverse portfolio of production concepts, nearly all untested. Five leases in Colorado are focused on underground production of shale oil. In addition, other companies are working on private land in Colorado. All the approaches propose taking advantage of the coproduction of significant quantities of natural gas to provide the heat needed to generate oil from oil shale after an initial start-up phase. Most processes also envision recapturing heat from one block to preheat an adjacent block.

Shell In situ Conversion Process. Shell Exploration & Production Company is pilot-testing its In situ Conversion Process (ICP) on private acreage in northwestern Colorado. Shell has been developing this process for over 25 years, and is the only company to have successfully produced high-quality shale oil with high recovery efficiency. The ICP uses downhole heaters (presently electric resistance heaters) to heat the rock for several years to temperatures in the range from 300–400°C (**Fig. 1**). Slower heating than in surface retorts results in lower-temperature reactions that convert a large fraction of the kerogen to liquids and noncondensable gases without decomposing the dolomite in the rock matrix. The fluid-phase products are produced through traditional boreholes. The resultant liquids consist largely of light ($<C_{20}$) hydrocarbons, producing a high-quality mix of transportation fuels that require relatively modest upgrading in comparison to the products of traditional surface retorts. Shell also intends to contain the process behind a freeze wall, a technique borrowed from the mining industry that involves pumping a coolant into closely spaced wells on the perimeter of the heated block to freeze the groundwater, providing a barrier to the inward migration of groundwater as well as to the outward migration of hydrocarbons. Shell has agreements with China and Jordan to assess their oil-shale deposits for potential development using ICP. Shell was granted three RD&D leases by the Bureau of Land Management.

ExxonMobil Electrofrac™ process. ExxonMobil Upstream Research Company is currently experimenting with aspects of its Electrofrac process in the laboratory and owns private land in western Colorado. The process uses conventional technology to drill horizontal wells and fracture the formation to provide large continuous fractures (generally depicted as approximately vertical). Injecting electrically conductive proppant (particles) into the fractures will create large platelike heating elements that can heat the formation efficiently (**Fig. 2**). Reaction products will be produced through conventional production boreholes, and are expected to consist of high-quality light hydrocarbons. ExxonMobil expects that conversion of the kerogen will result in a substantial volume increase, which is likely to cause fracturing of the rock mass, thereby providing flow paths for the produced hydrocarbons in the otherwise relatively impermeable oil shale.

Chevron Recovery and Upgrading of Oil from Shale process. Chevron's Recovery and Upgrading of Oil from Shale (CRUSH) process also involves fracturing prior to the extraction of hydrocarbons. Chevron seeks to induce mainly horizontal fractures that would be contained within a limited productive horizon. Because rich oil-shale zones are generally impermeable, containment within the rich zones is expected to avoid issues of groundwater invasion and contamination. Chevron expects fracturing to be sufficiently complete to provide a zone of rubble into which a transport fluid will

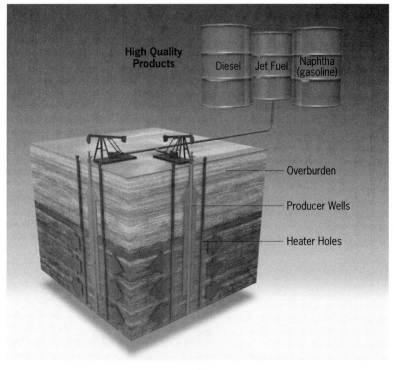

Fig. 1. Conceptual image of Shell ICP process. (*Courtesy of Shell EPW-Unconventionals*)

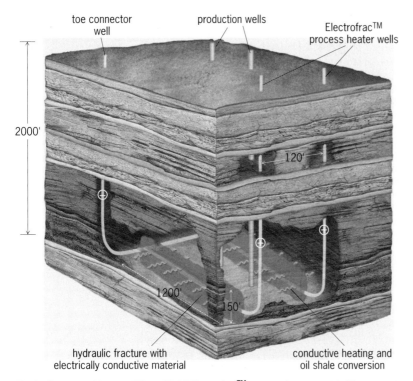

toe connector well

production wells

Electrofrac™ process heater wells

2000'

120'

1200'

150'

hydraulic fracture with electrically conductive material

conductive heating and oil shale conversion

Fig. 2. Conceptual image of ExxonMobil Electrofrac™ process (not to scale). (*Courtesy of ExxonMobil Upstream Research Company*)

closed-loop in situ retorting process with advantages of energy efficiency and manageable environmental impacts. The oil shale is heated with superheated steam or another heat-transfer medium through a series of pipes placed below the oil-shale bed to be retorted. Shale oil and gas are produced through wells drilled vertically from the surface and "spidered" to provide a connection between the heating wells and the production system (**Fig. 4**).

Other technological approaches to underground production. Schlumberger has recently purchased technology developed by the Raytheon Company and CF Technologies, Inc., that uses radio-frequency generators in combination with supercritical fluids injected into the formation to heat and extract the hydrocarbons from the oil shale. Other companies are also investigating radio-frequency and microwave heating, including Global Resources Corporation and Phoenix Wyoming, Inc. Independent Energy Partners, Inc., seeks to test a borehole fuel-cell heating system that would also generate power for surface operations and potential sale. Earth Search Sciences, Inc., proposes to inject superheated, pressurized air into the oil shale, extracting hydrocarbons as gases. Mountain West Energy LLC proposes a similar process using high-temperature gas (potentially natural gas) to heat the shale and sweep the products in the gaseous state to production wells.

Environmental issues for in situ conversion of oil shale. Two issues are considered likely to dominate in the development of oil shale by in situ processes: water and carbon dioxide. Air quality issues are likely to be less important for underground production than for surface processing, although large-scale

be injected, with carbon dioxide (CO_2) being the primary candidate, at elevated temperatures to extract liquid and gaseous hydrocarbons (**Fig. 3**).

American Shale Oil Process. The American Shale Oil, LLC (formerly EGL Oil Shale, LLC) approach is a

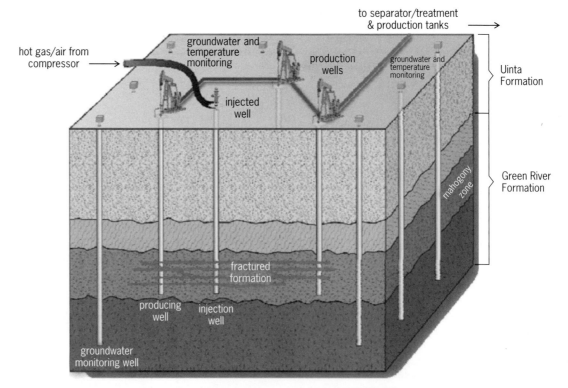

to separator/treatment & production tanks

hot gas/air from compressor

groundwater and temperature monitoring

production wells

groundwater and temperature monitoring

Uinta Formation

injected well

mahogany zone

Green River Formation

fractured formation

producing well

injection well

groundwater monitoring well

Fig. 3. Conceptual image of Chevron Recovery and Upgrading of Oil from Shale (CRUSH) process. (*Courtesy of Chevron Energy Technology Company*)

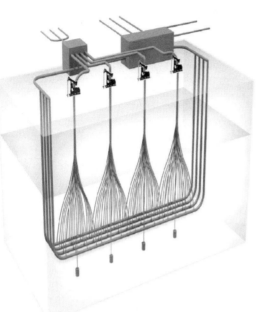

Fig. 4. Conceptual image of EGL underground production process. (*Courtesy of American Shale Oil, LLC*)

power generation could result in significant air emissions. One of the great attractions of oil-shale resources is their areal density. This is an important consideration that may make the impacts from surface disturbance more manageable than for the current widespread gas development activity in the same area. However, the socioeconomic impacts of an oil-shale industry are likely to be significant in this area, which is already strongly affected by oil and gas development.

At this point, no company has published an estimate of the water consumption needs of an in situ process. Other available estimates rely upon data from earlier efforts to commercialize oil shale. Shell has indicated that it intends to pump substantial quantities of water out of the rock before heating the underground. It also indicates an expectation of pumping water through expended ICP blocks to remove contaminants before the release of the freeze wall. Chevron indicated in its environmental assessment that it intends to confine production to a single thick and impermeable horizon, and to avoid fracture zones to ensure that the heated zone will not become a conduit for groundwater that might become contaminated by hydrocarbons or other contaminants released during underground production. EGL has also indicated plans to avoid aquifer zones for initial production.

Carbon dioxide production from underground production of shale oil comes from three sources: power-plant emissions, mineral breakdown, and oxidation of kerogen. Of these, emissions from power plants are likely to dominate unless power is generated in a carbon-neutral manner (such as wind, solar, or nuclear power). J. Boak calculated that a large-scale oil-shale industry (3 million barrels per day) using natural gas turbines to generate electricity for heating could release 100–300 million tons of

CO_2 to the atmosphere annually, depending strongly on the average grade of the oil shale. This amount is comparable to Saudi Arabian emission levels, but less than global CO_2 emissions from flaring of stranded natural gas. This production level may take decades to reach, so the issue must be considered as part of a wider discussion about global CO_2 emissions from hydrocarbon use.

Outlook. Oil shale is a resource that is at a critical point in its development. New approaches are being explored, and companies are defining processes that they will use to produce oil in an economical and environmentally sensitive manner. Many of these technical approaches will either fail or undergo radical changes, and actual production is still some way off. Research and development are proceeding with greater caution than in the past, but with considerable optimism about the place of this resource in the long-term global energy picture.

[Acknowledgements: The author gratefully acknowledges review by representatives from the major companies whose technology is described in this contribution.]

For background information *see* DOLOMITE; KEROGEN; OIL SHALE; PETROLEUM; PETROLEUM ENGINEERING; PETROLEUM GEOLOGY; SEDIMENTARY ROCKS in the McGraw-Hill Encyclopedia of Science & Technology.   Jeremy Boak

Bibliography. J. Boak, CO_2 release from in-situ production of shale oil from the Green River Formation in the Western United States, in J. Boak and H. Whitehead (eds.), *Proceedings of the 27th Oil Shale Symposium*, Colorado Energy Res. Instit. Doc. 2008-1, Colorado School of Mines, Golden (CD-ROM), 2008; J. R. Dyni, *Geology and Resources of Some World Oil-Shale Deposits*, USGS Sci. Investigation Rep., 2005–5294, U.S. Geological Survey, Reston, VA, 2006; R. H. Liu et al., The resources of China's oil shale and the prospect of its exploitation and utilization, in J. Boak and H. Whitehead (eds.), *Proceedings of the 27th Oil Shale Symposium*, Colorado Energy Research Institute Doc. 2008-1, Colorado School of Mines, Golden (CD-ROM), 2008.

Ventilation technology in deep mines

Mine ventilation is the application of principles of fluid dynamics to the flow of air in underground mine workings. Adequate dilution of diesel equipment emissions, fumes from explosives, methane emissions from coal strata, and dusts from cutting, mucking, hauling, and in-mine crushing are the essential considerations of a mine's ventilation system (**Fig. 1**). Quality of air is essential to all underground operations. Regardless of the material mined, ventilation always has to come first.

Three factors have made ventilation increasingly important. First, increasing mining depths due to continuing depletion of near-surface ore deposits worldwide. Second, an increase in the degree of mechanization in the mining industry. Third, more stringent safety and health regulations that demand

Fig. 1. Drawing showing surface facilities and extensive airways underground, forming a complex network of workings.

much higher environmental standards in all working environments. This requires improved control of both mine-wide and localized ventilation systems to maintain a safe underground mining environment.

The complexity of new mine ventilation design depends on a mine-specific layout, which, in turn, is determined by the specifics of an ore body, mining methods, equipment used, and mining regulations. Other important factors include high virgin rock temperatures, spontaneous combustion, dust control, and the requirement for ventilation automation. The ultimate goal is to provide a safe and productive underground environment, while maintaining low cost.

A safe underground working environment. The purpose of mine ventilation is to dilute, render harmless, and remove dangerous accumulations of pollutants from the underground working environment (flammable, explosive, and toxic gases; excessive moisture and temperatures; noxious and harmful gases, dust, smoke, and fumes). U.S. federal safety standards for underground mines mandate that the air where personnel work or travel shall contain at least 19.5% oxygen (O_2) and not more than 0.5% carbon dioxide (CO_2), and that the volume and velocity of the air current in these areas shall be sufficient to dilute, render harmless, and carry away pollutants. Other countries mandate similar standards.

The most commonly occurring pollutants underground are dust, methane (in both coal and non-coal mines), and diesel emissions. Hazardous concentrations of both dust and methane underground can be controlled by dilution (ventilation), capture before entering the host air stream (for example, methane drainage), or isolation (seals and walls).

Dust and methane explosions can be prevented or mitigated by eliminating ignition sources, by minimizing methane concentrations and coal dust accumulations, and by using passive and active barriers to suppress propagating explosions.

In coal mines, methane explosions can cause violent explosions of coal dust. To prevent such explosions, miners cover the floor, rib, and roof surfaces of mine openings with large quantities of inert rock dust, such as fine limestone dust. In the United States, rock dusting is mandated by and subject to federal safety standards.

The psychrometric factor caused by heat and moisture cannot be ignored. Heat can flow into work areas from wall rock, broken rock, hot fissure water, diesel and electric equipment, and hot surface conditions. The process of autocompression can also raise the underground work-area temperature. Deep-level operations must provide refrigeration to sufficiently offset the naturally occurring (that is, depth-gradient) heat so that workers and equipment can function. The cost of this function alone can rival that of supplying basic ventilation volumes for some deep operations.

How air moves underground. Air moves between two points because of the pressure difference between these points, which is usually provided by mechanical means (mine fans) and sometimes supplemented (or impeded) by natural ventilation. The latter is often unpredictable in both direction and amount, and cannot be relied upon as a pressure source.

The amount of air delivered within a ventilation system (Q, in ft^3/min or cfm) is dictated by the interaction between two variables: mine resistance (R) and mine pressure (H, in. water gage) according to Eq. (1).

$$Q = (H/R)^{1/2} \qquad (1)$$

For a given mine pressure (provided by the fan), lower (smaller) mine resistance will enable the fan to deliver more quantity. Conversely, for a given mine (thus with a given resistance), higher air quantity will require higher fan pressure in a square relationship: twice the air quantity will require four times the mine pressure.

Mine resistance is entirely dictated by the physical characteristics of the mine airways—it is directionally proportional to airway friction characteristics (K), average airway perimeter (O, in ft), and distance the air travels (L, in ft), and inversely proportional to the cube of average airway cross-sectional area (A, in ft^2).

$$R = KOL/5.2A^3 \qquad (2)$$

Mine resistance can be lowered by keeping the airways clean (thus lowering the K value), by providing larger airway cross-sectional areas where physically possible (thus increasing A), and providing additional airshafts (which will reduce air traveling distance, L, since the airshafts are closer to where fresh air is needed). All these are driven by

Fig. 2. TLT-Babcock axial flow GAF 31.5/16-1 controllable pitch fan in a Kentucky coal mine, Island Creek Coal Company. The fan shown, with a 2500-hp motor, has an optimum operating point of 15 in. water gage and 650,000 cfm. (Island Creek was acquired by CONSOL Energy in the late 1990s.)

economics. A new airshaft makes economic sense only when the savings from this airshaft (in terms of lower fan pressure and higher air quantity) exceeds the airshaft cost (capital and operating).

Mine fans. A mine fan is a linchpin of ventilation. It is in effect an air "pump," a machine that works by aerodynamic action to create a pressure differential to induce airflow. It converts mechanical energy to fluid energy to overcome resistance to moving air in underground airways.

Although all fans work by aerodynamic action, they are usually classified according to the path the air takes through the impeller and the resulting mechanism for pressure generation. The two most popular types of fans are axial-flow and centrifugal fans. The former is the most commonly used fan type in North America and continental Europe; the air flows along the axis of the fan shaft without directional change, with a screw or propeller producing airflow (**Fig. 2**).

A centrifugal fan is the squirrel-cage-type fan, with a large wheel or rotor in a curved casing (**Fig. 3**). Two separate and independent actions produce pressure: centrifugal force due to the rotation of the air, and kinetic energy imparted as the air leaves the tip of the impeller blades. The air is drawn into the center of a rotating impeller and is discharged radially into an expanding scroll casing, with the air entering the fan at right angles to the direction of discharge.

Most mine fans are on the surface, on top of either an intake airshaft (a blowing or push ventilation

system) or a return airshaft (an exhausting or pull system). A fan can also be installed underground between airshafts to act as a booster. Although most mines can be ventilated using a single fan, multiple fans (for example, fans on both intake and return airshafts or a push-pull system) are necessary when the ventilation system resistance becomes excessive and a single fan cannot provide the needed pressure. Depending on the size of the mine, main fans could range from 50 hp under 1–2 in. pressure (water gage) delivering 100,000 cfm of fresh air for a small operation in the midwestern United States to as large as 3500 hp operating under 20 in. (water gage), providing 750,000 cfm for a deep gold mine in South Africa.

Auxiliary fans ranging from 5 to 200 hp are commonly used to facilitate local or regional ventilation. They are flexible, are easy to move around, and can be located where fresh air is needed the most. In metal and nonmetal mines, auxiliary fans in combination with vent tubing are commonly used to provide the needed fresh air for long dead-end headings.

Over many decades, the trend has been away from large, surface-sited centrifugal units toward axial-vane units. Booster fans continue to have their place, but only as adjuncts within the large-scale, mine-wide system. Controlled recirculation with booster fans has been used to allow the mining of some otherwise unminable reserves.

Ventilation controls. While it is important for fans to provide adequate air at the fan, distributing the airflow to working faces effectively will always be a challenge. Airflow underground is distributed using control devices such as stoppings (line brattice, **Fig.** 4*a*; concrete block walls, Fig. 4*b*; or other channeling devices, for example, waste rock walls), overcasts (air bridges, much like highway overpasses, to keep intake air mixing with contaminated air), and airlocks.

Stoppings are permanent walls constructed of brick or other approved materials. They are built to separate and isolate different air courses in underground mines, such as the fresh air from the return air or from belt airways. Seals are substantially constructed according to federal standards. They are used to isolate worked-out areas of a mine that are no longer ventilated. Sealed areas cannot be entered by mine workers.

Overcasts and regulators are used for air splitting and properly adjusting the quantity of air flowing to various sections of the mine.

Atmospheric monitoring system. The atmospheric monitoring system (AMS) has been used in the U.S. mining industry since the early 1980s. When the Mine Safety and Health Administration (MSHA) conducted its first survey in 1984, it showed that only 38 mines were using this technology for monitoring mine fires. Mine operators continue to realize the benefits of improved fire detection capabilities, so the number of mines using the systems continue to grow. A 1992 survey showed that the number of mines using AMS had increased to 115. This number

Fig. 3. Two large centrifugal fans in parallel in a South African mine.

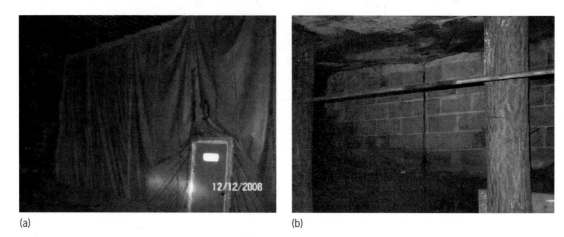

(a)   (b)

Fig. 4. In a large underground salt mine, (*a*) a large-size brattice and (*b*) concrete stopping used to course air underground. Both are commonly used in all underground operations.

had increased to 146, or 20.6% of active coal mines, in 2003 (**Fig. 5**).

The purpose of the AMS in nearly all instances is to detect fires in the belt haulage entry underground. Monitoring and control technology continues to change; it brings new features, higher speeds, greater safety, and lower cost. These developments enhance the utility and safety of mine ventilation monitoring systems. The addition of local-area network (LAN) workstations extends real-time access to a common database. New and automatic gas sensor calibration avoids the errors, cost, and risk of human intervention, while extending sensor life. Other improvements include higher-speed field I/O cards, automatic report printing, and a higher-voltage-rated power circuit barrier.

Today, sensors are able to discriminate between carbon monoxide (CO) produced by a fire and CO produced by diesel engines. This has allowed mine operators to reduce alarm levels (thus reduc-

ing the number of false alarms) and increase fire detecting capabilities. The recent development of a hydrogen-insensitive sensor provides the industry with a method of monitoring battery-charging stations without the problems associated with hydrogen interference.

Ventilation network simulation. A mine ventilation system consists of fans and miles of airways and control devices throughout the system, forming a complicated network. The proper design of such a system to provide the needed quantity and quality of air with an effective distribution system has always been a challenge. In recent years, this has been greatly facilitated by the use of network simulation programs.

Network simulation has been around since the late 1960s, but it was not until the 1990s, when computers became more powerful, reliable, and affordable, that simulation started to become readily available. Also, developments in simulation software have made it more powerful and user-friendly, facilitating the analysis and planning of ventilation systems, containment tracking, and refrigeration and cooling systems for full thermodynamic solutions in complicated networks. Computational fluid dynamics (CFD) analysis using high-speed computers can be used for airflow analysis in dead-end headings for optimized ventilation design.

Some more modern simulation packages have modules that determine airflow distribution, pollutant levels, and temperature distributions throughout the entire ventilation circuit and cooling network, all displayed in 3D full-color graphics, and with the capacity for real-time monitoring of underground conditions and remote control of ventilation on demand.

For background information *see* AIR COOLING; COAL MINING; COMPUTATIONAL FLUID DYNAMICS; FAN; HUMIDITY CONTROL; INDUSTRIAL HEALTH AND SAFETY; MINING; UNDERGROUND MINING; VENTILATION in the McGraw-Hill Encyclopedia of Science & Technology.   Jerry C. Tien

Bibliography. J. Burrows et al. (eds.), *Environmental Engineering in South African Mines*, The Mine

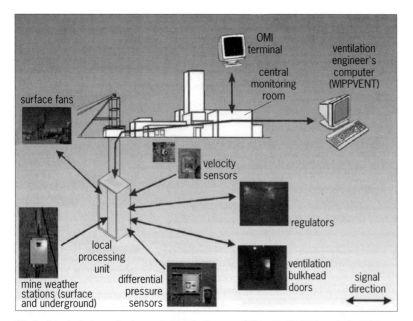

Fig. 5. A typical atmospheric monitoring system consisting of a central control center on the surface, a local processing unit underground, and various sensors.

Ventilation Society of South Africa, Johannesburg, South Africa, 1982; H. L. Hartman et al., *Mine Ventilation and Air Conditioning*, 3d ed., John Wiley & Sons, Inc., 1997; M. McPherson, *Subsurface Ventilation and Environmental Engineering*, Chapman & Hall, 1993; J. C. Tien, *Practical Mine Ventilation Engineering*, Intertec Publishing Corporation, Chicago, IL, 1999.

Virtual testing of structures

It is generally accepted that vehicles (cars, aircraft, trains) are tested before being released for public use, both by the manufacturers and on occasion by the appropriate regulatory authorities to "certify" the vehicle. These tests are often major undertakings; for example, the crash of a car into a barrier at motorway speeds or the test to failure of an aircraft wing. The tests are instrumented, with cameras to record the event, strain gauges to measure material deformation, and load cells to track the forces applied to the structure.

The phrase "virtual testing" describes the use of computer-based numerical analysis to predict how a structure might behave under the real-life conditions in such physical tests. The benefits of virtual testing arise from the fact that the structure, its component parts, the loads applied to it, and the results obtained from the virtual test exist only in the form of computer models. Virtual testing can therefore be highly cost-effective and efficient. The simulation can be applied to a range of structures, from individual material test pieces and individual components up to full mechanical assemblies such as the full-scale automobile crash tests and limit-load tests on aircraft.

How does it work? Software tools to analyze structural responses were first developed on a commercial basis in the aerospace industry in the 1960s. The capabilities have evolved since that time to the present day, where off-the-shelf software tools are routinely used in the design and development of a wide range of products, from beverage cans to automobiles, from mobile telephones to nuclear power stations.

Structural analysis is generally carried out using the finite-element (FE) method. The FE technique involves the discretization of the structural model into many finite elements, collectively called the mesh. The elements may be cubes, tetrahedrons, plates, or beams, depending on the topology of the structure, and for each element we can define functions that describe the stiffness, based on the physical properties and the material. The set of stiffness equations for the whole mesh can then be used to obtain a solution to a defined structural problem by establishing the equilibrium between the applied loads and the internal forces in the structure. Such analyses can run very effectively on modern personal computers, though large and more complex simulations will require significantly more computational power.

The results of the FE analysis can be extensive, comprising forces and moments, deflections, and

Fig. 1. An example of a finite-element model of a stiffened plate, showing contours of stress (indicated by variation of shading).

stresses and strains, and can extend to material damage and failure. It is this solution to a defined structural problem, with the availability of a wide range of detailed results, that provides the benefit to the engineer (**Fig. 1.**)

Who does it? Although originally pioneered by the aerospace and automotive industries, virtual testing was also adopted in the design of nuclear power plants and civil structures, and then extended into other domains. Today, almost any structure used in transport, consumer goods, and the built environment will have had some form of structural analysis carried out, to improve the understanding of the structure's response.

Benefits. The ability to carry out virtual testing provides significant advantages over an entirely physically based testing program. Perhaps the biggest benefits come in saving time and cost. While virtual tests can be complex to set up and require skilled analysts to execute, they are invariably cheaper to conduct than a series of physical tests. Automobiles and aircraft are expensive in themselves, even before the addition of the test rig and instrumentation, and so significant effort is spent on creating high-fidelity FE models and refining their predictive accuracy.

Once the computer model is created, the virtual test can be repeated again and again, with each analysis containing a variation of the test conditions or a design feature. Drop-testing of mobile phones, for example, requires the impact to be assessed with the phone in many different orientations, which is a simple modification to a computer model and can be readily automated. Similarly, the influence of the position of the car occupant on the effect of an airbag can be studied by extensive virtual testing, moving the occupant slightly for each new analysis. The only additional cost for repeated analyses is the computer time to process the simulation, which in general is small in comparison to the material and setup cost of a physical test.

The ability to rapidly repeat virtual tests with small design changes provides the opportunity to carry out design studies that explore a wider range of design variables than would be possible in a physical test. This can often lead to a better design, because there is less of a requirement to accept the first design that meets the target criteria.

Fig. 2. Finite-element model of an aircraft wing-box under a static load.

Once a design is established, virtual testing can also be used to carry out 'what-if' studies, subjecting the design to a wider range of loads and operating conditions than could be achieved with physical testing, given the normal constraints on time and cost. This helps to increase the understanding of the product's performance and behavior, again helping to develop an improved design. Aircraft structures are subject to a wide range of loading conditions, sometimes in varying combinations, and so virtual testing is used to study the load envelope much more extensively than could be achieved through physical tests (**Fig. 2**).

In general, the assessment of results from physical testing relies on diagnostic equipment and measurements that are determined in advance, and once the test is complete there is no further opportunity to extract more data, short of repeating the whole experiment. A virtual test, however, has a large amount of data created from the process of simulation itself, and so key parameters can be extracted at any time after the analysis, depending on the requirement.

A very important advantage of virtual testing is the ability to simulate events that cannot be tested physically. Cost, safety, and environmental impact can mean there are some scenarios that simply cannot be explored through anything other than simulation. Examples include the large-scale collapse of civil structures due to earthquake or explosive loading; the failure of nuclear power generation systems; and the impact of aircraft, ships, or trains in accidents.

Why is it difficult? Although the benefits of virtual testing are manifold, it is important to recognize that the technique is not without its difficulties. The key to successful virtual testing is to make the simulation as realistic as possible, which means building in to the computer model as many of the real-world effects as possible. This is by no means straightforward.

In the first instance it is likely that the computer model will be initially "perfect." That is, the dimensions will be accurate in relation to the design, the material will be entirely uniform and will behave precisely according to the constitutive law, the loads will be applied in a consistent and precise manner, and repeating the virtual test with the same input data will result in exactly the same outcome.

In reality, the structure and the test itself are often far from perfect, and any set of physical tests is likely to produce some scatter in the results. Even with the most stringent of quality assurance processes, it

is likely that the materials that make up the structure will have some degree of property variation caused by the manufacturing process, whether they are cast metal alloys, injection-molded plastics, or cured carbon-fiber composites. Incorporating these variations into a simulation is extremely difficult with anything other than a random distribution within a specified range. The dimensions of the real structure will vary within certain tolerances, due to the fundamental limitations in the way the components are manufactured and assembled. These tolerances are not always critical, but in some types of structure it is the small imperfections that can have a significant effect on the overall strength. Slight imperfections can initiate structural buckling and collapse at significantly lower loads than would be seen in a perfect structure. A computer simulation of a perfect structure ordinarily produces an overestimate of the critical buckling load, which is nonconservative and therefore undesirable from a design point of view.

For any product that is assembled from several components, the physical behavior will be influenced by the process of assembly. Processes such as bolting, welding, and adhesive bonding can generate stresses and distortions within an assembly that can subsequently affect the stiffness and strength of the overall structure.

Significant effort goes into improving the realism of the simulation model, taking into account all these imperfections. For an automotive crash, it is important that the underlying structures absorb the energy of impact in a controlled manner, minimizing peak accelerations on the occupant. This is normally achieved by forcing the structural components to undergo compressive crushing and failure rather than buckling and collapsing. The effects of residual stress near spot welds, variations in material thickness, and other dimensional imperfections are therefore extremely important.

Concerns. With the drive to minimize product design costs, and to reduce the time-to-market for new products, the question arises whether virtual testing will or should ever replace physical tests. In reality, good product design requires a detailed understanding of the product's behavior, and there is no doubt that, today, the highest level of confidence in that product knowledge comes from a combination of physical and virtual testing. Physical testing is essential to obtain the fundamental material properties of the product components. Whether the structure is composed of metals, concrete, or laminated composites, physical tests will be needed to generate data for input to the analysis model.

On the other hand, up-front simulation, in the early stages of design, can help to optimize future physical tests, assessing the best location for strain gauges and load cells, determining the likely loads on the jigs and fixtures, and helping to get the most from any physical tests that are ultimately carried out. The physical test on the final structure then provides real data that can be used to validate the computer simulation. It is important to obtain good correlation between the

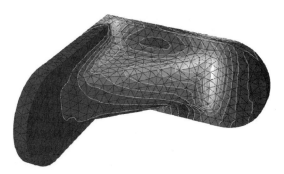

Fig. 3. Strain results from a finite model are used to correlate against strain gauge results from a test.

physical test and the analysis to build confidence in the simulation for future predictive analysis (**Fig. 3**).

Prospects. There is no doubt that the use of virtual testing will increase as the capabilities of the analysis software and computer hardware improve, and as engineers and regulatory authorities gain more confidence in the application of simulation. However, for the foreseeable future, it is likely that physical tests will continue to be used throughout the design process, from determining basic material properties through to the final certification test.

For background information *see* FINITE ELEMENT METHOD; SIMULATION; STRENGTH OF MATERIALS; STRUCTURAL ANALYSIS; STRUCTURAL DESIGN in the McGraw-Hill Encyclopedia of Science & Technology.

Alan Prior

Bibliography. M. A. Crisfield, *Non-linear Finite Element Analysis of Solids and Structures*, Wiley, vol. 1, 1991, vol. 2, 1997; T. A. Laursen (editor-in-chief), *Finite Elements in Analysis and Design* (an International journal for innovations in computational methodology and application), Elsevier, monthy; O. C. Zienkiewicz, R. L. Taylor, and J. Z. Zhu, *Finite Element Method*, vols. 1–3, 6th ed., Elsevier Butterworth-Heinemann, 2005.

Volumetric displays

Volumetric displays enable the depiction of three-dimensional (3D) images within a transparent volume (image space). Because such images can occupy three physical dimensions, a broad range of depth cues (by which we are able to judge the form and spatial arrangement of a three-dimensional scene), including oculomotor and parallax, are satisfied in a natural manner. When considered in terms of their inherent three-dimensionality, volumetric images possess many of the characteristics associated with traditional forms of sculpture and are formed from voxels (the 3-D equivalent of the pixel), the position of each voxel being defined as a point in a 3D space.

Subsystems. A volumetric display comprises three subsystems: image space formation, voxel generation, and voxel activation.

Image space formation. Image space formation represents the techniques used to implement the physical image space. In the case of the "swept volume" approach, an image space is formed through the rapid cyclic motion of a planar or curved surface. Either translational or rotational motion may be employed. Voxels are activated as the surface passes through the appropriate region of the image space. Alternatively, a "static volume" approach may be adopted, that places no reliance on mechanical motion, the image space being constructed from a material or arrangement of materials that is able to support the formation of visible voxels.

The presence of nontransparent components within an image space, an anisotropic refractive index, and boundary refraction can severely compromise freedom in viewing location and lead to preferential viewing directions, wherein an image exhibits greater clarity when positioned at a certain location within the image space and when viewed from a particular location. The form and dimensions of the image space can affect the display's suitability for use in particular applications, and in the static-volume technique the density of the image space may affect display portability.

Voxel generation. Voxel generation refers to the underlying physical technique or techniques used for the production of visible voxels. To date, practically all volumetric systems employ voxels that emit light generated within the display; relatively little work has been carried out on the production of voxels made visible through the scattering of ambient illumination. Isotropically emissive voxels give rise to translucent images and therefore the occlusion depth cue is absent. Although translucent images are desirable in some situations (for example, medical imaging, where internal structure is of interest), the formation of opaque images is an important goal.

Voxel activation. The voxel activation subsystem is responsible for the stimulation of the voxel generation process and thereby the production of the visible image. The voxel activation capacity N_a may be defined as the total number of voxels that may be activated during an image refresh period. If the time required to activate a voxel is T and the image refresh frequency is f, then the voxel activation capacity is given by Eq. (1), where P denotes the num-

$$N_a = \frac{P}{Tf} \qquad (1)$$

ber of voxels that can be activated simultaneously. Doubling the linear dimensions of an image space (without compromising intervoxel spacing) results in an eight-fold increase in the number of possible voxel locations (the voxel location capacity N_l). However, in order to discern the form and spatial separation of objects comprising an image scene, it is usual to ensure that the majority of the volume is void. The fill factor, ψ, denotes the percentage of available voxel sites that can be activated during an image refresh period, and is given by Eq. (2). A fill

$$\psi(\%) = \frac{N_a}{N_l} \cdot 100 \qquad (2)$$

factor considerably less than 1% is generally adequate; exhaustive addressing of all potential voxel

Fig. 1. Animated stick figure depicted on the cathode-ray sphere. The figure is able to run around the spherical image space and, by means of motion-tracking hardware, may mimic the movements of a physical person in real time, or for playback. Here a very small number of voxels is able to create a highly effective volumetric image. (*Reproduced by permission of J. Milo,* © *2000*)

sites is unnecessary, and a small fill factor reduces demands on the display subsystems (**Fig. 1**).

Swept-volume systems. John Logie Baird (who provided the earliest demonstration of practical television) pioneered swept-volume techniques in the late 1920s. Seminal research was undertaken by the Royal Signals and Radar Establishment in the United Kingdom during World War II, and since then a broad range of technologies have been explored.

A volumetric display system manufactured by Actuality Systems, Inc., supports a 100% fill factor. The Perspecta's® image space is created by the rotational motion of a planar screen (25 cm or 10 in. in diameter and rotating at 30 Hz; **Fig. 2**). Voxel activation is achieved by means of three spatial light modulators (SLMs) in the form of digital micromirror devices (DMDs), each comprising 1024 × 768 mirror elements. The DMDs generate spatially modulated red, green, and blue (RGB) beams that are projected onto the rotating surface. Image data is ordered into a set of 198 radial slices, and these are each mapped onto the screen via the DMDs. The use of DMDs supports high parallelism in voxel activation (in principle, $P \approx 786,000$), and where 198 radial slices are employed, the voxel activation capacity is approximately 156 million. (In fact, the Perspecta does not utilize all mirror elements, and $N_a \approx 10^8$.)

The formation of an image space by the rotational motion of a helical surface dates back to the pioneering work of R. Hartwig in the mid-1970s, and this basic approach continues to attract attention. In one system currently under development, a set of light-emitting elements that project upward is arranged in a series of concentric rings below the helical screen (**Fig. 3**). The helix rotates at a frequency in excess of the human critical flicker frequency, and the light-emitting elements rotate at a different angular velocity. As this angular velocity is increased, the number of elements within each ring may be reduced until

the limiting case is reached, and only a single element is responsible for voxel activation at each radial distance from the helix's axis of rotation.

Static-volume systems. The first proposal for a static-volume volumetric display dates back to a patent filed by E. Luzy and C. Dupuis in 1912, although it was not until the 1960s that this general approach began to receive significant attention. LightSpace Technologies, Inc., manufactures a static-volume display named the DepthCube®. The image

Fig. 2. Perspecta® display, manufactured by Actuality Systems, Inc. (*Reproduced by permission of Actuality Systems, Inc.,* © *2008*)

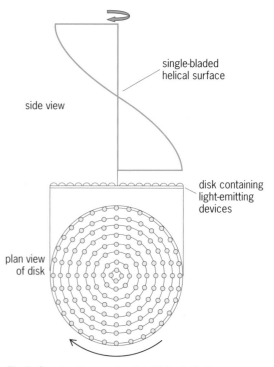

Fig. 3. Swept-volume system in which a helical screen sweeps out an image space and voxels are created by a set of light-emitting elements located below the screen. These are arranged in concentric rings. The number of light-emitting elements is significantly reduced by increasing the rate at which they rotate with respect to the helix.

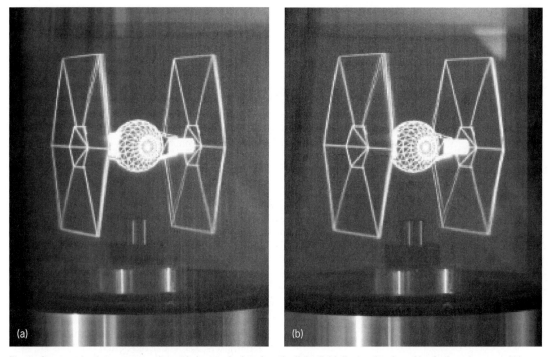

Fig. 4. Stereo pair of an opaque volumetric image depicted on the light-field display developed by Andrew Jones and his colleagues. The use of a stereo pair enables the binocular parallax depth cue associated with volumetric images to be preserved. The pair of images can be fused by slightly crossing the eyes. (*a*) Left image. (*b*) Right image. (*Reproduced by permission of T. Pereira, © 2008*)

space is formed from a stack of liquid-crystal panels that can be individually switched between transparent and scattering states. A high-speed projector is used to cast image slices into the image space. As each slice is projected, the appropriate liquid-crystal display (LCD) is switched into a scattering state and so the image slice emanates from a certain depth within the image space.

Two-step excitation of fluorescence, whereby visible voxels are created at the intersection of two directed nonvisible beams, provides an elegant approach to static-volume implementation. In the 1990s, I. I. Kim, E. J. Korevaar, and H. Hakakha verified that a gaseous medium would support the formation of voxels of appropriate intensity. Work has also been undertaken in employing this general approach in conjunction with an image space formed from rare earth lanthanides incorporated within a heavy-metal fluoride glass such as ZBLAN (composed of a mixture of zirconium, barium, lanthanum, aluminium, and sodium fluorides). However, this scheme has practical difficulties in terms of boundary refraction and image-space mass.

A further example of a static-volume approach employs a stack of passive light-scattering centers that are formed in a glass block (the scattering centers being created using a focused laser beam to induce small cracks). Image data is projected from below and the scattering centers are positioned so that each has a unique location in the horizontal plane. The variation of image visibility with viewing location is minimized by ensuring that the scattering centers are not positioned on a regular grid but are perturbed (in a known manner) from regular lattice positions.

Hybrid systems. Practically all volumetric systems give rise to translucent images and so fail to support image opacity. However, by way of example, a display recently developed by Andrew Jones and his colleagues brings together volumetric and multiview techniques, enabling opaque images to be formed (**Fig. 4**). The image space is formed through the rotational motion of a planar (mirrored) screen onto which image data is cast via a spatial light modulator. The screen is equipped with a holographic diffuser and the resulting projection geometry enables a particular view onto an image scene to be directed to a corresponding viewing location. Thus a set of views is computed, each corresponding to a different angular (equatorial) view onto the image space. The optical arrangement then ensures that each view is mapped to the intended vantage point. Although horizontal motion parallax is inherently associated with the displayed image, vertical parallax is absent. Consequently, when a viewer changes vantage point in a horizontal direction, a different view onto the 3D scene is observed. In contrast, view changes in the vertical direction do not result in a change of view. This issue can be overcome by head tracking, in which, for example, an ultrasound sensor system is used to determine the observer's location with respect to the image scene and change the geometry of the displayed image accordingly.

Interaction. Volumetric systems offer new opportunities for the visualization of complex 3D data sets. Additionally, they can support innovative interaction techniques. However, the presence of physical components comprising the image space precludes the insertion of physical objects into the

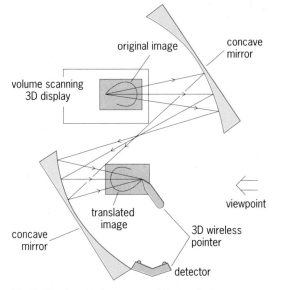

Fig. 5. Simple optical arrangement that projects a volumetric image into free space. This enables a haptic interaction probe to "touch" the image, thereby supporting more natural interaction. (*Diagram reproduced by permission of SPIE, © 1993*)

volume, which affects opportunities for haptic interaction. For a physical interaction tool or haptic glove to "touch" a volumetric image, it is necessary to project the image into a "free" image space. This allows the image and interaction spaces to coincide. Such a system was developed by K. Kameyama, K. Ohtomi, and Y. Fukui in the early 1990s, although at that time interest in haptic interaction opportunities was minimal (**Fig. 5**).

Opacity. The hybrid display outlined above supports image opacity but employs emissive (translucent) voxels. Opaque voxels (visible as a consequence of the scattering of external illumination) provide interesting opportunities for the formation of opaque images in which shadows and shading are achieved naturally via the interaction between the ambient source or sources of illumination and the voxels comprising the image. Advances in photochromic materials promise to support this display paradigm.

For background information *see* COMPUTER GRAPHICS; ELECTRONIC DISPLAY; FLUORESCENCE; LIQUID CRYSTALS; MICRO-OPTO-ELECTRO-MECHANICAL SYSTEMS (MOEMS); OPTICAL MATERIALS; VISION in the McGraw-Hill Encyclopedia of Science & Technology.

Barry G. Blundell

Bibliography. B. G. Blundell, *An Introduction to Computer Graphics and Creative 3-D Environments*, Springer-Verlag, London, 2008; B. G. Blundell, *Enhanced Visualization: Making Space for 3-D Images*, Wiley, New York, 2007; B. G. Blundell and A. J. Schwarz, *Volumetric Three-Dimensional Display Systems*, Wiley, New York, 2000; A. Jones et al., Rendering for an interactive 360° light field display, *ACM Trans. Graph.*, 26(3):Article 40 (10 pages), 2007; I. I. Kim, E. J. Korevaar, and H. Hakakha, Three-dimensional volumetric display in rubidium vapor, *Proc. SPIE*,

2650:274–284, 1996; S. K. Nayar and V. J. Anand, 3D display using passive optical scatterers, *IEEE Comput.*, 40(7):54–63, July 2007.

Whole genome association studies

Each copy of the human genome is made up of just over 3.2 billion pairs of nucleotides (the structural units of nucleic acids), comprising the individual "letters" that make up all deoxyribonucleic acid (DNA) sequences. The sequence is housed on 23 chromosomes, containing approximately 22,000 genes as well as the regulatory switches that turn those genes on and off. Every person carries two copies of the genome, one inherited from each parent, in each cell in the body (excluding the gametes). As the genome is passed from one generation to the next, changes occur in the DNA sequence at a very low rate; over time, some of these changes are lost, while others remain. As a result, the genomes carried by present-day humans contain a large number of genetic variants that arose as new mutations many generations ago. These old genetic variants are now shared across many genomes, making them common in the human population. These common variants may influence the risk of common human diseases and other traits, an idea that is referred to as the *common disease–common variant* (CDCV) hypothesis. *Whole genome* (or genome-wide) association studies (WGAS) are designed to test the CDCV hypothesis by examining the effect of common genetic variants on the risk of human disease, response to drug treatment, and other human traits.

Technological advances. The power of the WGAS approach was first enumerated by Neil Risch and Kathleen Merikangas in 1996. At that time, gene mapping typically involved conducting one of two types of genetic studies: linkage analysis, in which the inheritance patterns of genetic variants and traits were compared within families; and candidate gene studies, which required the identification of a promising candidate gene for study. Risch and Merikangas set out to examine the question of why these methods had been successful for simple Mendelian traits, that is, those that were caused by a single gene but had largely failed when applied to common diseases such as type 2 diabetes, heart disease, psychiatric disorders, and a number of cancers. They compared the statistical power of conducting a WGAS approach to that of a linkage analysis study, and showed that a WGAS analysis that tested as many as 500,000 genetic variants for association with disease would provide substantially greater statistical power than linkage analysis, even after statistical adjustment for conducting such a large number of tests. The difference in power was most pronounced for genetic variants that were common but had small to modest effects, specifically, those that increase the risk of disease by 10–50%. Although this research generated great excitement in the field of statistical genetics as a powerful approach to unraveling the genetic causes of human disease, the implementation of the WGAS approach required a number of significant technological advances that have now finally begun to bear fruit.

The first technological step toward WGAS was the completion of the Human Genome Project in 2000, which cataloged the entire human genome sequence. Prior to this point, only a small number of genes had been sequenced and even the total number of genes in the genome was unknown (Risch and Merikangas based their calculations on 100,000 genes). This first human genome sequence provided a scaffold that was necessary for WGAS.

Once the sequence had been determined, the next technological step needed for WGAS was to identify the common genetic variation present in the human genome. Single-nucleotide polymorphisms (SNPs), which are changes to individual nucleotides in the DNA sequence, are by far the most common type of variation in the human genome. The SNP Consortium, formed in 1999 to identify SNPs by comparing DNA sequences of 24 unrelated individuals, has cataloged more than 1.8 million of the more than 10 million common SNPs estimated to exist in the human genome.

By 2002 it had become clear that an organized effort was needed to produce a SNP map that could be used to conduct WGAS. This new effort, called the International Haplotype Map (HapMap) Project, endeavored to create a SNP-based haplotype map of the human genome. SNPs do not occur randomly along chromosomes, but instead often appear in groups known as haplotypes. The correlation between SNPs that occur on the same haplotype, called linkage disequilibrium, allows for researchers to genotype (measure) individual SNPs and predict the unmeasured SNPs using statistical methods. This statistical shortcut, termed indirect association, works best for common SNPs and can reduce the number of those that need to be genotyped from 10 million to around 0.75–1.5 million. The International HapMap Project validated more than 3 million SNPs in a total of 270 individuals from one of three continental populations: 30 parents–child trios of Yoruba from Africa; 30 parents–child trios of Europeans living in Utah; and 45 unrelated subjects of Chinese from Beijing combined with 45 unrelated subjects of Japanese from Tokyo. Multiple populations were chosen because linkage disequilibrium varies across populations, and some SNPs are population-specific.

The final, and perhaps most important, technological advance that has made WGAS possible was the development of large, parallel genotyping platforms or DNA chips. These platforms have drastically reduced the cost of genotyping, that is, reading the specific DNA sequence at the location of a known variant, and the time necessary to perform the genotyping. The latest DNA chips contain more than 1 million SNP probes and cost less than $500 per subject. At $0.0005 per genotype, the cost of genotyping has dropped nearly 1000-fold since 2002, making relatively comprehensive WGAS cost-effective. As such, the number of published reports that describe completed WGAS has also increased dramatically.

Conducting WGAS. In its simplest form, a WGAS approach involves identifying subjects with and without a disease (cases and controls), and genotyping their DNA on a DNA chip. Each point on the chip represents a specific SNP, and the frequency of each SNP is compared between the case group and the control group. If the frequencies between the two groups differ significantly, then that SNP is considered to be associated with the disease under study.

WGAS have been undertaken for a number of common human diseases across a range of human disease types, which include autoimmune diseases, various cancers, cardiovascular disease, eye diseases, and neurological diseases (see **table**). In addition to this

Phenotypes (diseases and traits) on which WGAS have been conducted	
Disease	Trait
Age-related macular degeneration (dry)	Blue vs. green eyes
Age-related macular degeneration (wet)	Body mass index
Amyotrophic lateral sclerosis	Bone mineral density
Asthma (childhood)	C-reactive protein
Atrial fibrillation/atrial flutter	Coronary artery calcification
Bipolar disorder	Diabetes-related insulin traits
Breast cancer	Echocardiographic traits
Celiac disease	Environmental confusion in the home
Colorectal cancer	Episodic memory
Coronary disease	Exercise treadmill test traits
Crohn's disease	F-cell distribution
Diabetic nephropathy	Fetal hemoglobin levels
End-stage renal disease	General cognitive ability
Exfoliation glaucoma	HDL-cholesterol
Gallstones	Heart-rate variability traits
Hypertension	Height
Inflammatory bowel disease/irritable bowel syndrome	Hemostatic factors
Late-onset Alzheimer's disease	Hip geometry
Lung cancer	HIV1 viral setpoint
Major cardiovascular disease (CVD)	Iris color
Multiple sclerosis	LDL-cholesterol
Myocardial infarction	Mean forced vital capacity (from two exams)
Obesity	Memory performance
Parkinson's disease	Methamphetamine dependence
Progressive supranuclear palsy	Morbidity-free survival traits
Prostate cancer	Neuroticism
Psoriasis	Nicotine dependence
Restless legs syndrome	QT interval prolongation
Rheumatoid arthritis	Recombination rate
Schizophrenia	Response to interferon-beta therapy
Stroke	Response to ximelagatran treatment
Systemic lupus erythematosus	Serum urate
Type 1 diabetes	Sleep duration
Type 2 diabetes	Thyroid-stimulating hormone
	Triglycerides
	Volumetric brain MRI
	Waist circumference traits
	YKL-40 levels

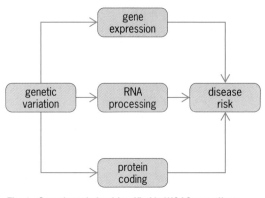

Fig. 1. Genetic variation identified in WGAS can affect disease risk by altering the regulation, processing, and gene product.

broad spectrum of disease types, WGAS have also examined disease-related traits such as lipid measures and measures of adiposity, non-disease-related traits including eye color and height, as well as response to drugs.

Genetic variants identified in these WGAS tend to have certain shared characteristics. First, by design, the variants identified are common. They have small effects on disease risk, usually encoding increases in risk of less than 50% per allele; they often have effects that are even more subtle. For example, of 10 SNPs identified in three large scans of type 2 diabetes, only one produced an increase in risk of greater than 25% per allele. These effects are quite small and would have been especially difficult to find using linkage analysis.

Second, the disease-associated variants tend to cluster in or near genes but often do not occur in the protein-coding regions of these genes. This suggests that gene regulation through the control of gene expression and ribonucleic acid (RNA) processing can play important roles in disease development (**Fig. 1**). Both mechanisms for interrupting gene function have been difficult to test using candidate gene approaches that focused on testing polymorphisms in the protein-coding regions of specific

genes. Notable exceptions to this generalization are the 8q24 locus, which affects prostate, breast, and colorectal cancer, and the 9p21 locus, which influences risk of coronary artery disease and myocardial infarction. These two loci are located far away from known genes, although they may act as distant regulatory elements that influence when certain genes are turned on and off.

Third, successful WGAS involve large samples, typically at least 2000 cases and a similar number of controls. With increasing sample sizes, acquired by combining the results from multiple WGAS, new genes with smaller risks can be identified. Height serves as the example for this type of analysis, for which 20 genes have now been identified by combining data from more than 90,000 individuals (height is collected for most WGAS). As the SNP density of DNA chips increases, additional risk variants that were not well covered on previous chips will likely be discovered.

WGAS and disease treatment. WGAS can improve disease treatment in two main ways: through a better understanding of the disease process and through studies of the genetic modification of disease treatment (**Fig. 2**). WGAS can identify new pathways involved in disease causation that provide new therapeutic targets. They can also help categorize subjects with the same disease into genetic subtypes. These subtypes may in turn have different paths of disease progression, warranting more aggressive treatment in one group compared to another. Genetic variation that is associated with disease risk or progression may also influence patients' response to treatment, allowing for a more tailored approach to disease therapy. Finally, treatment response can be influenced by pharmacokinetic and pharmacodynamic factors that affect drug processing within the body. Drug transport and metabolism can be influenced by genetic variation that has no direct effect on disease. For this reason, WGAS of treatment response will be an important part of translating the new findings of disease WGAS into clinical practice.

For background information *see* CANCER (MEDICINE); CHROMOSOME; DEOXYRIBONUCLEIC ACID (DNA); DISEASE; GENE; GENETIC MAPPING; GENOMICS; HUMAN GENOME; LINKAGE (GENETICS); NUCLEOTIDE; POLYMORPHISM (GENETICS) in the McGraw-Hill Encyclopedia of Science & Technology.

Eric Jorgenson

Bibliography. E. Jorgenson and J. S. Witte, Genome-wide association studies of cancer, *Future Oncol.*, 3(4):419–427, 2007; M. I. McCarthy et al., Genome-wide association studies for complex traits: Consensus, uncertainty and challenges, *Nat. Rev. Genet.*, 9(5):356–369, 2008; E. Pennisi, Breakthrough of the year: Human genetic variation, *Science*, 318(5858):1842–1843, 2007; N. Risch and K. Merikangas, The future of genetic studies of complex human diseases, *Science*, 273(5281):1516–1517, 1996; Wellcome Trust Case Control Consortium, Genome-wide association study of 14,000 cases of seven common diseases and 3,000 shared controls, *Nature*, 447(7145):661–678, 2007.

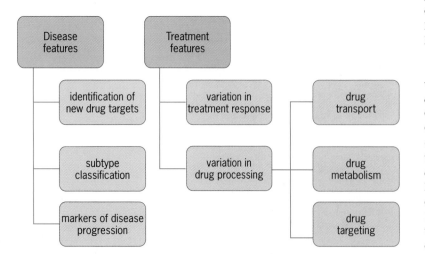

Fig. 2. Multiple pathways for WGAS to improve disease therapies.

WiMAX broadband wireless communications

Worldwide Interoperability for Microwave Access (WiMAX) is a growing family of standardized technologies for broadband wireless communications. WiMAX provides the ability to transfer high-speed data efficiently over long ranges and to support digital voice, data, and video services in challenging environments. There are two family trees: one for fixed broadband access, and one for mobile broadband access. These are separately defined by the Wireless Metropolitan Area Network (WMAN) working group of the Institute of Electrical and Electronics Engineers (IEEE) in publications 802.16d and 802.16e, respectively. While fixed WiMAX targets stationary terminals, mobile WiMAX targets mobile users, as well as portable and stationary terminals. WiMAX is intended to be the lowest-cost option to wide-area coverage offering high capacity.

The broadband wireless challenge that WiMAX is designed to address has many opposing forces. The process of worldwide standards adoption needs to accommodate different licensed spectrum allocations and bandwidths, single versus paired spectrum, extensions of legacy systems versus "greenfield" deployments (the installation of a new incumbent network), and fixed and mobile service offerings with various throughputs and quality-of-service ratings. This flexibility needs to be designed while maintaining the objectives of a standards-based system, worldwide adoption, and the entry of myriad vendors to drive competition and significant volume for silicon and devices, thus lowering the overall cost of service offerings. The WiMAX Forum was formed as a collaborative industry organization to promote and develop the WiMAX standards and, more importantly, certify equipment standards compliance and interoperability. While certification is an essential step, adoptions and deployments will primarily determine the volume and cost of service.

Services. Fixed and mobile WiMAX systems are both designed to support digital voice, data, and video services for non-line-of-sight applications, but are not compatible because of technological differences. Fixed WiMAX can be used by DSL (digital subscriber line) and cable operators to extend the reach of their wireline business with "last mile" wireless or rural deployments. It provides a cost-efficient option for rural or new greenfield deployments, with point-to-multipoint transmitters providing stationary data services. Because of its ability to accommodate long ranges of up to 50 km (30 mi) or high data rates of up to 70 megabits per second (Mbps), fixed WiMAX can be employed as backhaul by operators to consolidate data flow into a switched or Internet-Protocol-based (IP-based) network for routing (**Fig. 1**).

Mobile WiMAX is designed to support the much more challenging application of mobility, with its rapidly changing environment (that is, radio-frequency channel conditions), handoffs, and network roaming. To accommodate these channel conditions, more robust waveforms and coding are employed at the expense of range and throughput. Mobile WiMAX can support up to 10 Mbps over several miles, depending on the equipment and surrounding environment, with the higher throughputs at shorter ranges. Current mobile operators can use mobile WiMAX as an overlay to existing voice-plus-data networks to provide capacity relief for new data and video services. With currently increasing equipment availability, new service entrants can choose mobile WiMAX for their business case service offering.

Network architecture. IP is a common element in the current Internet infrastructure; therefore, there is a strong cost incentive to build WiMAX around this existing IP core. The WiMAX Forum's reference network model defines functional entities and the interfaces between them. Vendors can deploy different configurations of these functional entities as long as they conform to the interface requirements provided by the WiMAX Forum Network Working Group. A WiMAX Network is made up of terminal stations, an Access Service Network (ASN), and a Connectivity Service Network (CSN)[Fig. 1]. Terminal stations connect over the radio-frequency link to the Access Service Network through a base station. Radio resources, mobility, and quality of service are managed within the Access Service Network. An Access Service Network gateway connects to the Connectivity Service Network, which handles issues such as billing and the transition to an IP network like the Internet. This level of decomposition in a WiMAX network allows the involvement of multiple independent sources, thereby promoting competition and improving service.

Radio-frequency characteristics. WiMAX is designed to provide service for non-line-of-sight-link conditions between a base station and a terminal station. Because of this propagation channel, WiMAX transmits radio-frequency energy using an orthogonal frequency-division multiplexing (OFDM) waveform. OFDM is chosen because it is particularly robust against multipath, which occurs when transmitted energy arrives at the receiver antenna via multiple unique paths (that is, at different times

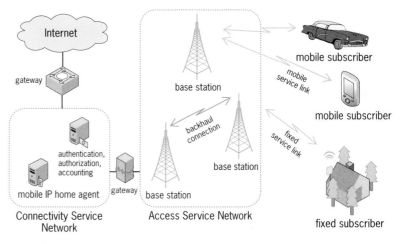

Fig. 1. **Fixed and mobile WiMAX services.**

WiMAX radio waveform characteristics		
	Fixed WiMAX	Mobile WiMAX
Radio waveform	OFDM256	S-OFDMA
Subcarriers	256	128, 512,* 1024,* 2048
Signal bandwidths, MHz	1.25, 1.75, 3.5,* 5, 7,* 8.75,10,* 14, 15	1.25, 1.75, 3.5,* 5,* 7,* 8.75,* 10,* 14, 15
Multiple access†	TDM	TDM/OFDMA
Duplexing	TDD/FDD	TDD
Data rate (minimum–maximum)	1–75 Mbps	1–75 Mbps
Frequency band, GHz	3.5, 5.8	2.3, 2.5, 3.5

*Denotes current requirements in existing WiMAX Forum certification profiles.
†TDM = time-division multiplex; OFDMA = orthogonal frequency-division multiple access.

and amplitudes). OFDM employs the simultaneous transmission of unique data on numerous frequency-spaced subcarriers through one channel to combat multipath. OFDM also provides high spectral efficiency (measured in data throughput per utilized bandwidth). Fixed WiMAX achieves efficiencies comparable to those of current new technology options, whereas mobile WiMAX can continue to improve current efficiencies with the advanced access techniques and antenna technologies described later in this article.

WiMAX also employs Adaptive Modulation and Coding, which dynamically varies both the modulation and the error-correction coding to optimize the error rate and throughput for a particular radio channel condition. Difficult channels use lower modulation and stronger error coding, which reduces user throughput, whereas cleaner channels use higher modulation and weaker error coding.

Fixed WiMAX and mobile WiMAX use different variants of OFDM. Fixed WiMAX uses OFDM256, which contains 256 subcarriers. Multiple users share a fixed WiMAX network through a predetermined allocation of time slots to each user. Mobile WiMAX uses scalable orthogonal frequency-division multiple access (S-OFDMA). S-OFDMA is scalable in the sense that as signal bandwidths change (for example, from 5 to 10 MHz while roaming), the number of subcarriers can also change (for example, from 512 to 1024). This wider bandwidth allocation allows more subcarriers to increase throughput. Mobile WiMAX can also partition the S-OFDMA subcarriers into groups

and allocate groups to specific users, adding another multiple access method. The network can separate download and upload traffic through time-division duplexing (TDD) or frequency-division duplexing (FDD). TDD, which is currently the only required mode in mobile WiMAX, does not require using a pair of frequency bands and therefore can operate on a single licensed frequency. A summary of the key physical characteristics of both radio waveforms in WiMAX is shown in the **table**.

Although the standards allow WiMAX to operate over a wide range of spectrum frequencies, it is most favorable for all systems to operate on the same licensed frequency spectrum. Though there is no globally licensed spectrum, 2.3, 2.5, and 3.5 GHz are used in the WiMAX Forum's certification profiles for the purposes of driving volume and facilitating cost reduction.

Antenna technology. WiMAX systems can use Advanced Antenna Systems (AAS) to improve network performance. Beamforming allows antenna behavior to be adjusted to match the channel optimally.

Figure 2 shows another AAS capability with multiple-input antenna techniques. One setup, called MISO (multiple input–single output), improves network robustness by transmitting the same data over distinct antennas, while another setup, MIMO (multiple input–multiple output), increases capacity by transmitting unique data over each antenna. AAS is not a certification requirement for fixed WiMAX, since high-gain antennas are commonly employed and radio-frequency link conditions are not as severe as they are on a mobile link. The WiMAX Forum has provided two phases of mobile WiMAX profiles called Wave 1, which requires S-OFDMA but not AAS functionality, and Wave 2, which includes AAS requirements.

Along with a robust waveform and advanced antenna techniques, the WiMAX Forum defines additional requirements that affect overall usage quality, including power management for battery conservation, resource allocation for efficiency and quality of service, and security. In resource allocation, an important distinction for WiMAX is that network resources are dedicated to a terminal after registration to avoid contention and to maintain required quality-of-service levels. This practice is necessary to help support voice and video services. For security, WiMAX can use the 128-bit Advanced Encryption System standard adopted by the United States

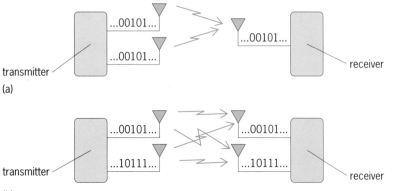

Fig. 2. WiMAX Advanced Antenna Systems (AAS) using multiple-input techniques.
(*a*) **Increased robustness using multiple input–single output (MISO).** (*b*) **Increased capacity using multiple input–multiple output (MIMO).**

federal government to provide confidentiality, and uses end-to-end authentication to prevent falsification of terminal or base station identity.

Current and proposed deployments. By mid-2008, there were more than 250 trials and deployments of WiMAX technologies. Certified WiMAX deployments had over 600,000 subscribers at the end of 2007, and this number was growing rapidly. In May 2008, a large-scale joint venture was formed by several major U.S. companies to deploy the first nationwide mobile WiMAX network, using the 2.5-GHz spectrum.

For background information *see* CRYPTOGRAPHY; FREQUENCY-MODULATION RADIO; INTERNET; MOBILE COMMUNICATIONS; MULTIPLEXING AND MULTIPLE ACCESS; WIDE-AREA NETWORKS; WIRELESS FIDELITY (WI-FI) in the McGraw-Hill Encyclopedia of Science & Technology.   Mark Gaudino; Andrew Wu

Bibliography.   J. Andrews, A. Ghosh, and R. Muhamed, *Fundamentals of WiMAX: Understanding Broadband Wireless Networking*, Prentice Hall, Upper Saddle River, NJ, 2007; D. Dobkin, *RF Engineering for Wireless Networks: Hardware, Antennas, and Propagation*, Newnes, Boston, MA, 2004; C. Eklund et al., *WirelessMAN: Inside the IEEE 802.16 Standard for Wireless Metropolitan Area Networks*, IEEE Press, New York, 2006; Y. Zhang and H. Chen, *Mobile WiMAX: Toward Broadband Wireless Metropolitan Area Networks*, Auerbach, Boca Raton, FL, December 2007.

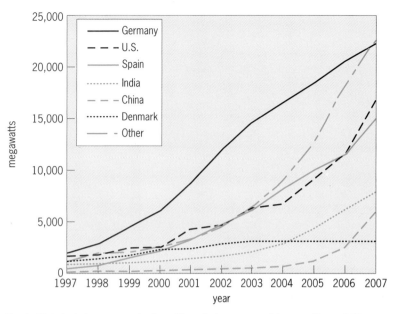

Fig. 1. Global wind-power production of top wind-power countries over the past 10 years. (*Data sources: Worldwatch, GWEC, AWEA, CREIA, and Earth Policy Institute*)

nents that can be installed quickly, which will have cost benefits from future manufacturing improvements and increased volume.

Here we will focus on the main engine of current wind electric power development (that is, the modern wind turbine) and describe these systems, their

Wind turbines

The use of wind energy by various devices to propel ships, pump water, grind grains, and perform other energy-intensive tasks has existed since the early Egyptians. The nineteenth and twentieth centuries brought fossil fuels and fossil-fuel engines, and wind had a minor role as an energy conversion source. Only since the oil embargo of 1973 has wind been rediscovered and started to emerge as an important energy source. Advances in large airfoil structure manufacturing, composite materials, computational aerodynamics, and machine design and control have produced larger, lighter, and more efficient and reliable wind energy power systems than ever before. These advances are reflected in wind-power production costs falling from 10 cents per kilowatt-hour (kWh) 10 years ago to 4.5 to 7.5 cents today, making these systems competitive with fossil energy power production. In **Fig. 1**, the rapid development of global wind-power is shown for the leading wind-power-producing countries. In the past few years, the growth rate of newly installed capacity has been approximately 30% per year, and the United States has led in these developments.

These developments, plus growing concerns about carbon dioxide (CO_2), global warming, and increasing fossil fuel prices, make wind energy very attractive. Its principal advantages include zero fuel cost, nondepleting supply, and minimal environmental impact. Also, wind systems are made of compo-

Fig. 2. Typical horizontal-axis high-power wind turbines. Such turbines can produce several megawatts of electric power depending on their size. (*Courtesy of the Department of Energy*)

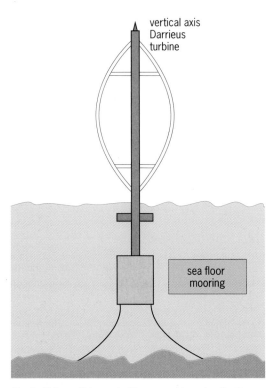

Fig. 3. Future offshore wind farms could employ floating vertical-axis wind turbines of the Darrieus type. Such systems would be more stable in ocean currents and waves and could be moved to avoid hurricanes.

characteristics, and the critical elements of their impressive performance.

Turbine configurations and components. Over time, many different wind energy conversion devices have been used, with most falling into two classes—horizontal or vertical—depending on the rotation axis of the blade rotor (**Figs. 2** and **3**). Among these,

the horizontal three-blade system of Fig. 2 is the configuration of choice for large wind-power producers. This is an upwind system with the rotor plane in front of the tower and a yaw system to keep the rotor into the wind. The rotor rotation speed is typically 20–50 rpm, and this must be geared up to 1000–3000 rpm to meet the requirements of the electric generator.

Immediately behind the rotor hub is an enclosure (the nacelle) that houses the drive train and control components (**Fig. 4**). The drive train consists of a low-speed shaft connecting the rotor to a two- or three-stage gearbox, followed by a high-speed shaft connected to the generator. In addition, there are extensive on-board controls that can change the orientation (or pitch) of the rotor blades, control the yaw, drive-train, and power components; and brake the rotor in possible runaway situations, such as high winds and power-grid outages.

Wind energy basics. Wind turbines are devices that convert wind kinetic energy to rotor shaft mechanical energy to produce electric energy through an on-board electric generator. The amount of wind power captured in this process is directly proportional to the product of the mass flow rate and the kinetic energy of the wind moving through the swept rotor area, as shown in the equation

$$P = \frac{1}{2} C_p \rho \, A V_w^3$$

where ρ is the wind density, A the swept area, V_w the approaching wind velocity, and C_p the power coefficient representing the fraction of the wind energy captured by the turbine rotor. Early studies by A. Betz of ideal stream flows and wind energy converters indicate a maximum possible C_p of 0.593. Although the Betz analysis is oversimplified, it gives a framework for analyzing conversion levels of real systems, where the predictions of C_p are more complex.

The equation indicates that wind power increases directly with the swept area of the turbine A, or the square of the rotor radius. For this reason, rotor blade lengths continue to increase over time. In 1999, maximum blade lengths were about 20 m; today they are approaching 70 m, with 100–120-m blades on the horizon. Also, the cube effect of the wind speed in the equation is critical in wind farm siting, since even a 10% advantage of one site over another in wind speed represents a 33% advantage in wind power.

Aerodynamic effects. A wind turbine has many functional elements, but none is more important than the rotor and its ability to convert the maximum amount of the available wind power, maximizing C_p. This is the domain of turbine blade aerodynamics.

In **Fig. 5**, the effects of aerodynamic forces on a section of an airfoil rotor blade are illustrated, where the rotor is moving in the vertical plane. The orientation of the blade to the vertical rotation plane is given by the pitch angle θ. As the blade rotates, it experiences not only the effects of the horizontal wind at speed V_w, but also a headwind of magnitude U from the rotation of the blade. As a result, relative to the blade, the wind is approaching at an angle of

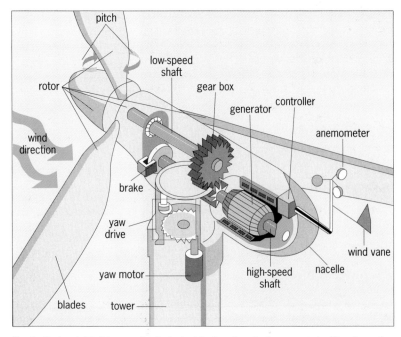

Fig. 4. Horizontal turbine mechanical, electrical, and control components. (*Courtesy of the U.S. Department of Energy*)

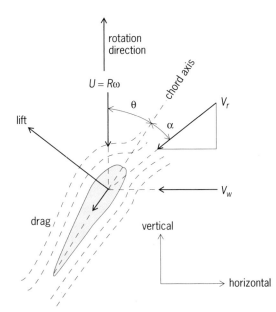

Fig. 5. Wind effect on turbine blade rotating in vertical plane at an angular speed of ω. Lift and drag forces at any radius R and pitch angle θ depend on the relative wind speed V_r and the angle of attack α, where $V_r = (V_w^2 + R^2\omega^2)^{1/2}$ and $\alpha = \tan^{-1}(V_w/R\omega) - \theta$.

attack α at a relative speed of V_r. Here, the specific values of α and V_r depend on U/V_w, with α increasing (decreasing) as V_w is increased (decreased) relative to U.

As the wind moves over the rotating blade, the blade experiences (1) a viscous drag force along the angle of attack and (2) a perpendicular lift force arising from flow-induced pressure differences across the upwind and downwind surfaces of the blade. As shown in Fig. 5, the lift force has a positive component in the direction of rotation and is sufficiently large to overcome the negative component of the drag force in this direction. The difference of these components is what drives the blade rotation and provides the torque that ultimately drives the electric generator.

The analysis of such aerodynamic effects on the turbine rotor blades is at the heart of turbine rotor design and ultimately leads to rotor performance characterization in terms of the power coefficient C_p. Current horizontal turbine rotors have C_p values approaching 0.5 compared to the Betz ideal value of 0.583. In general, high power coefficients are associated with high lift-to-drag ratio blade designs and designs that minimize the negative effects of the tip vortex wake.

Turbine performance and operation. Because of the cubic effect of wind velocity on power conversion (see conversion equation above), wind speeds commonly produce rotor power inputs in excess of the rated capacity of the turbine generator. To avoid this, modern wind turbines have controlled actuators to adjust the blade pitch and change the angle of attack, which adjusts the shaft power inputs (Fig. 5).

Figure 6 shows a typical power curve for a turbine operated between the cut-in and cut-out speeds. Below the cut-in speed, there is insufficient torque

to drive the generator. Above the cut-out speed, the wind loads could cause damage to the turbine components. Between the cut-in and rated speeds, the curve follows the conversion equation; above the rated speed, the pitch actuators are operable and keep the power output constant.

Turbine blade loads and fatigue. During rotation, the cyclic loads on turbine blades arise from several effects. First, because of ground shear effects, the wind speed increases with height. For turbine heights and rotor sizes today, this can mean a 20–30% difference in wind speed from the top of the rotor to its bottom. The corresponding drag and lift force differences would be even greater because of their nonlinear, high-power exponent dependence on V_r. These would be cyclic loads with frequencies of ω. Cyclic loads of similar frequency also arise from stationary and bow wakes of the tower, as well as from gravitational effects that cause cyclic bending moments on the blades.

Any fluctuations in the mean wind speed or turbulent bursts and vortices will expose the blades to a spectrum of loads and frequencies. When combined, all of these effects cause dynamic axial bending and torsional loads, simultaneously acting on the blades, and the frequencies can be much wider than simply ω. In a 20-year period with 4000 h of operation per year, the ω cyclic loads alone would expose the turbine blades to about 10^8 load cycles.

Exposure of material elements to such cyclic loads over long periods and load cycles can cause material fatigue and eventual failure, if the loads and load cycles are sufficiently large. The failure mechanism involves the formation of microcracks in stress concentration areas that grow into macrocracks, which ultimately lead to fracture surfaces and material failure.

Historically, turbine blade fatigue failure has been the primary failure mode of wind turbine systems. These failures can stem from flawed designs (excessive stress concentrations in areas that are not sufficiently reinforced) or from manufacturing and handling defects, including those created during long-distance transportation and installation. Even small material "bruises," which are not easily detected

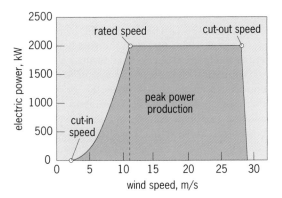

Fig. 6. Typical power curve for a wind turbine, indicating expected power production at different wind speeds. It also indicates the rated power and the rated wind speed as well as the cut-in and cut-out speeds.

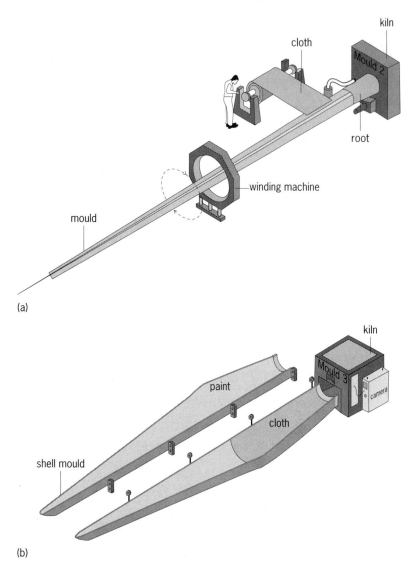

(a)

(b)

Fig. 7. Illustration of (*a*) spin-coating process for blade spar manufacturing and (*b*) molding process for blade shell manufacturing. (*Courtesy of Gamesa Corporation*)

visually, can be the origins of premature fatigue failure.

Turbine blade materials. Turbine blades are similar to aircraft wings, with an airfoil shell and internal support structures, which include longitudinal spars and other support elements (including transverse stiffeners and ribs) in critical stress areas. The spars are generally either special I-beam, channel, or box-shape web configurations that span the void between the upper and lower surfaces of the blade shell.

The airfoil shell materials are advanced glass fiber composites, with epoxy, polyester, or vinylester thermoset polymers as the continuous matrix. In addition, pre-prepared high strength-to-weight ratio composite sandwich sheets, with fiberglass and high-strength carbon fibers, are bonded to the shell in high-stress areas and often between the shell and the spar surfaces. Polyurethane foams are often used in these sandwich sections to lower the weight and, in some cases, to fill much of the airfoil core.

The turbine blade shells and spars are manufactured in large-scale spin-coating and molding pro-

cesses. Often the spars are spin-coated as in **Fig. 7a**. Here the fibers and fiber mats or fiber cloth sheets are impregnated with the liquid thermoset, spun onto the spar, and heated to cure the thermoset. Such processing is very efficient and produces very strong elements.

The blade shells are typically molded in upper and lower cavities, with the fiberglass mats or prepregs laid in first (Fig. 7*b*). Various methods are used to apply the thermoset, some involving repetitive hand layout and coating and others using vacuum or other methods, where the liquid thermoset is infused into the mold and the fiber mats. Thermal curing then follows, which produces the rigid shells. The spar and the sandwich sheets are then bonded in place to complete the blade. Since bonding surfaces are often the locations of fatigue failure, some manufacturers avoid this two-shell bonding by infusing the thermoset into a one-piece shell mold. The challenge of the latter is preventing incomplete thermoset infusion.

The layup and bonding steps are typically done by hand, and thus are always susceptible to quality-control issues and associated defects. Even in the vacuum infusion processes, voids can occur from incomplete infusion or nonuniform curing from the exothermic thermoset reactions. These defects are often not detected before the blades leave the plant and can compromise the blade over time under dynamic loading. Recently, one major manufacturer had to retrofit over 900 blades because of design or manufacturing problems. Such occurrences indicate the need for continuous improvement of the designs and the associated manufacturing processes.

Future turbine blades. Major improvements in next-generation turbine blades will largely come from material and manufacturing improvements. New material studies are looking at the replacement of thermosets with thermoplastics (nylon and other derivative polymers). If the fatigue characteristics can be improved, this would offer an attractive alternative to thermosets, particularly related to low-cost and high-quality manufacturing. Also, major chemical suppliers to the turbine industry are aggressively developing new and improved epoxies, other thermosets, and a new generation of bonding materials, as well as investigating the benefits of increased use of foams and nanomaterials.

The increased interest in wind should produce a new wave of technology development. Next-generation blades will be larger (100 m or more), stronger, lighter, more efficient, and cheaper. These developments will drive wind-power costs even lower, and wind energy will see even higher rates of development and application onshore and offshore.

For background information *see* AERODYNAMICS; AIRFOIL; COMPOSITE MATERIAL; ELECTRIC POWER GENERATION; ENERGY SOURCES; TURBINE; WIND; WIND POWER in the McGraw-Hill Encyclopedia of Science & Technology.   Raymond W. Flumerfelt;
                                                    Su Su Wang

Bibliography. O. L. Hanson, *Aerodynamics of Wind Turbines*, 2d ed., EarthScan, London, 2008; E. Hau, *Wind Energy—Fundamentals, Technologies, Application, Economics*, 2d ed., Springer-Verlag, Berlin, Heidelberg, 2006; G. Herbert et al., A review of wind energy technologies, *Renew. Sust. Energ.*, 11(6):1117–1145, 2007.

World Radiocommunication Conference (WRC) 2007

Under the aegis of the International Telecommunications Union (ITU) an international conference is held periodically to modify the treaty that governs the use of the radio spectrum resource on a global basis. These conferences are referred to as World Radiocommunication Conferences (WRCs). The most recent were held in 2000, 2003, and 2007. Another one is expected in 2011. In general this treaty consists of two basic parts: (1) The Allocation Table, and (2) The Radio Regulations.

The Allocation Table consists of over 400 designated bands of frequency spectrum. Each of these bands has been "allocated" to one or more of over 40 radiocommunication services. The Radio Regulations prescribe how these allocations should be used on a global basis. For the allocation of frequencies the world has been divided into three regions as shown in the **illustration.**

There are approximately 190 member countries of the ITU who are eligible to participate in World Radiocommunication Conferences. As an indication of the importance associated with these conferences, over 160 countries having nearly 3000 delegates participated in WRC-07. This conference was held in Geneva, Switzerland, the headquarters of the ITU, over a period of 4 weeks during October and November 2007.

Agenda. The agenda for a WRC is determined at the preceding WRC, it being usually the last item to be decided. Each such agenda is limited to those issues that are the most pressing from the standpoint of accommodating new technologies and services. WRC-07 had approximately 25 agenda items. Although it formally had 21, there were several items that had more than one issue embedded in them. The **table** provides an interpretive description of the issues on the agenda at this treaty conference.

There is a diversity of demands for adjustment in obtaining access to the global radio spectrum resource to accommodate new services and technologies. Some of these may be characterized as follows:

1. International Mobile Telecommunications (IMT) wanted new spectrum with more bandwidth for advanced mobile and Wi-Fi applications.

2. The Earth Exploration Satellite Service (EESS) was seeking 100 MHz of new bandwidth and interference protection from the transmissions of satellite services and terrestrial services in adjacent bands.

3. The Fixed Satellite Service (FSS) community wanted to revise the technical assumptions and procedures in the FSS Allotment Plan to bring it up to date. This Plan reserves FSS spectrum and orbit positions for each ITU country.

Other agenda items were not so clear cut, and in fact could be termed defensive in nature. This means that in respect to certain items the stated requirement was weak or nonexistent, but had been included for "political reasons." These conferences are not strictly technical in nature although there is a strong technical component that is developed over a period of 3 years of preparation.

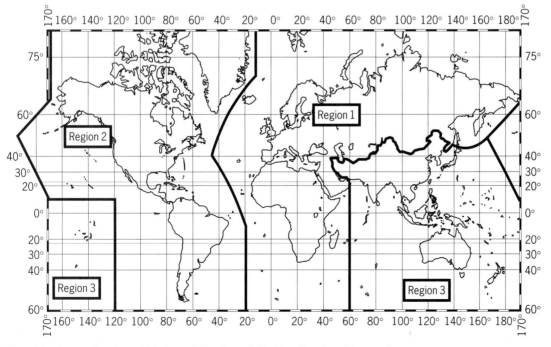

Map of the three regions into which the world has been divided for allocation of frequencies.

Agenda for WRC-07

Agenda item	Issue	Allocation/Regulation	Service(s)
1.2a	Sharing between services	10,600–10,680 MHz	Earth Exploration Satellite Service (EESS), Fixed (FS), Mobile (MS)
1.2b	Sharing between services	36,000–37,000 MHz	EESS/Fixed/Mobile
1.2c	Sharing between satellite services	18,000–18,400 MHz	Metsat Service, Fixed Satellite
1.3	Improved status for radars	9000–9200 MHz, 9300–9500 MHz	Radiolocation Service (RLS), EESS, Space Research Service (SRS)
1.4	Additional frequencies for international mobile telecommunications shared with other services	410–430 MHz 2700–2900 MHz 3400–4200 MHz 4400–4990 MHz	Fixed, Mobile Aeronautical Radionavigation Fixed Satellite Service (FSS) Fixed, Mobile
1.5	More spectrum for Aeronautical Mobile Telemetry (AMT)	4400–4940 MHz 5925–6700 MHz 5091–5150 MHz	Fixed, mobile FSS Aeronautical Radionavigation Service (ARS)
1.6a	Spectrum for new aerospace applications	108 MHz to 6000 MHz	Radionavigation Satellite Service (RNSS), ARS, Broadcasting (BS)
1.6b	Satellite services for civil aviation	4/6 GHz, 11/12/14 GHz, 20/30 GHz	Fixed Satellite Service (FSS)
1.7	Sharing MSS and Mobile Service (MS), Fixed Service (FS), Space Research (SR)	1668–1668.4 MHz, 1668.4–1675.00 MHz	Mobile Satellite Service (MSS), MS, FS, and SR
1.8	High-Altitude Platform Sharing (HAPS)	27.5–28.35 GHz, 31.0–31.3 GHz	Fixed and Fixed Satellite Services
1.9	Space Service/Terrestrial Sharing	2500–2690 MHz	FSS, Broadcasting Satellite (BSS), MS, FS
1.10	Update of FSS Allotment Plan	Appendix 30B of Radio Regulations	Fixed Satellite Service
1.11	Technical/regulatory for Broadcasting Satellite	620–790 MHz	Broadcasting Satellite Service
1.12	Coordination and notification/ procedures	Articles 9 and 11 of Radio Regulations	Fixed Satellite Service
1.13	More HF broadcasting spectrum	High-frequency allocations	Fixed Service, Mobile Service, Broadcasting Service (BS)
1.14	Operational Procedures for Maritime Distress and Safety (GMDSS)	Appendix 15 of Radio Regulations	Maritime Mobile Service (MSS)
1.15	Amateur service sharing	135.7–137.8 kHz	Fixed, Mobile, Maritime Mobile, Radiolocation
1.16	Operational, Maritime equipment	Article 19 of Radio Regulations	Maritime Mobile Service
1.17	Protection of existing service	1390–1392 MHz, 1430–1432 MHz	Terrestrial services
1.18	Fixed Service/Fixed Satellite Service	17.7–19.7 GHz	Fixed Service, FSS
1.19	Harmonized FSS for Internet	All FSS allocations	FSS
1.20	Protection of EESS (passive)	Five adjacent allocations (active)	FSS, MSS, Inter Satellite Service (ISS), BSS
1.21	Protection of radio astronomy	Adjacent satellite bands	RNSS, FSS, BSS
2.0	Incorporation of recommendations in the Radio Regulations	Article 5—the Allocation Table	Services to which recommendations apply

Under the ITU charter, each nation that has paid its dues has the right to vote at a conference such as this, and indeed voting has taken place at past conferences. However, given the procedural, time-consuming difficulty of conducting votes, there have been few at recent conferences. Instead, there is a strong motivation to reach consensus. Achieving consensus on difficult issues has increasingly been accomplished through the real-time meeting (during the conference) of regional groups. These include CITEL (Americas), Arab League, CEPT (Europe), APT (Asia Pacific Telecommunity), African Union, South African Group, and RCC (Eastern Europe and Russia). Throughout the WRC-07, more than at any previous conference, these regional organizations met frequently to forge compromise positions to obtain consensus.

Achievements. The WRC-07 was able to achieve through consensus agreements most of the objectives identified in its agenda. Exceptionally, no new spectrum was identified for high frequency (HF) broadcasting as it would have had to replace already existing critical fixed and mobile services. Otherwise, WRC-07 was quite successful in the following ways:

1. Globally allocations were identified for IMT in the following bands: 450–470 MHz, 698–862 MHz (Region 2 and some Region 3), 790–862 MHz (Regions 1 and 3), and 2.3–2.4 GHz. This resulted in 72 MHz of new global harmonized spectrum allocation for IMT/Wi-Fi applications as well as several hundred megahertz for such applications on a regional basis.

2. The successful incorporation into the maritime regulations procedures for use of new technology including that used for distress and safety within the framework of the Global Maritime Distress and Safety System (GMDSS).

3. The Fixed Satellite Service (FSS) Plan was modified to accommodate the latest technology and associated procedures to provide for more effectiveness in accessing the use of spectrum in the Plan, particularly by developing countries.

4. The spectrum for the Earth Exploration Satellite Service (EESS) was expanded and its protection in a number of bands was assured by the adoption of out of band emission standards. This service is critical to protection of the environment. The Meteorological Satellite Service was expanded by 100 MHz in all three ITU regions.

5. Civil aviation telecommunications was enhanced through improved spectrum status for radars at 9 GHz, and additional spectrum for aeronautical telecommand and the aeronautical mobile service.

6. Significant improvements were made to the procedures used to coordinate and register the frequencies used by satellites providing a variety of services.

7. The agreement on and agenda for the next World Radiocommunication Conference to be held in 2011, including consideration of spectrum for Electronic News Gathering, High Altitude Platforms, Unmanned Aircraft, Radio Navigation Satellite Service, and regulatory provisions for Software Defined Radio and Short Range Devices.

In summary the 2007 World Radiocommunication Conference provided the treaty vehicle for meeting the continued worldwide demand for access to radio spectrum resources to support continually changing requirements in the way of services to be supported. The details of the results of the conference may be found in Final Acts, WRC-07, Geneva, released by the International Telecommunication Union in April 2008. It is a testimony to the on-going viability of the 143-year-old International Telecommunications Union for providing the technical and regulatory mechanism for accommodating dynamic change.

For background information *see* RADIO SPECTRUM ALLOCATION in the McGraw-Hill Encyclopedia of Science & Technology.   Donald Jansky

Bibliography. International Telecommunications Union, *Final Acts, World Radiocommunication Conference (WRC-07)*, Geneva, 22 October–16 November 2007, 2008; National Telecommunications and Information Administration, *Manual of Regulations and Procedures for Federal Radio Frequency Management*, 2008 edition; *Radio Regulations*, vols. I–IV, International Telecommunications Union, 2004 edition; *Rules and Regulations of the Federal Communications Commission*, C RF 47, Part 2, Frequency allocations and radio treaty matters; general rules and regulations, 2007 edition.

Nobel Prizes for 2007

The Nobel Prizes for 2007 included the following awards for scientific disciplines.

Chemistry. The chemistry prize was awarded to Gerhard Ertl of Fritz-Haber-Institut der Max-Planck-Gesellschaft (Fritz Haber Institute of the Max Planck Society), Berlin, Germany, for his studies of chemical processes on solid surfaces. More specifically, he determined the catalytic mechanisms at gas-solid interfaces, including the Haber-Bosch process for making ammonia using nitrogen from air and an iron catalyst and the reaction of the platinum catalyst in the catalytic converter to convert carbon monoxide in automotive exhaust to carbon dioxide.

Ertl's experiments involved studying single layers of atoms under high vacuum (pressure) and contaminant-free conditions using a variety of techniques, such as auger electron spectroscopy, Fourier-transform infrared spectroscopy, high-resolution electron energy-loss spectroscopy, low-energy electron diffraction, photoemission electron microscopy, secondary ion mass spectroscopy, and ultraviolet photoelectron spectroscopy.

Reactions where the catalyst is present in a separate phase are known as heterogeneous catalysis. Usually, the catalyst is a solid, the reactants and product are in gaseous or liquid phases, and the catalytic reaction occurs on the surface of the solid.

The Haber-Bosch process is the major source of nitrogen used for fertilizer. In this process, nitrogen molecules react with hydrogen molecules to form ammonia under high pressure using a heterogeneous catalyst consisting of iron particles plus potassium hydroxide on an alumina and silica support.

In his studies of the Haber-Bosch process, for example, Ertl was able to show the correct structural mechanism and determine the energetics and kinetics of each step in the reaction, using a variety of the above-named instrumental techniques. He also discovered that potassium ions in the iron catalyst enhanced the absorption of nitrogen molecules at the surface by donating electrons to the iron atoms. In addition, he ran and studied the reaction in the reverse direction, starting with ammonia, to confirm his results.

In the course of his research, Ertl has not only developed the modern methods of surface chemistry, his thorough approach has set high standards for how such studies should be done.

For background information *see* ADSORPTION; CATALYSIS; ELECTRON SPECTROSCOPY; HETEROGENEOUS CATALYSIS; SURFACE AND INTERFACIAL CHEMISTRY; SURFACE PHYSICS; ELECTRON DIFFRACTION in the McGraw-Hill Encyclopedia of Science & Technology.

Physiology or medicine. The prize in Physiology or Medicine was awarded jointly to Mario Capecchi (University of Utah, Salt Lake City, Utah, United States, and Howard Hughes Medical Institute, Chevy Chase, Maryland, United States), Sir Martin J. Evans (Cardiff University, Cardiff, United Kingdom), and Oliver Smithies (University of North Carolina at Chapel Hill, Chapel Hill, North Carolina, United States) for their discoveries of the principles underlying the production of genetically altered mice by the introduction of specific gene modifications to embryonic stem cells.

Mice that develop from such modified stem cells possess precise, permanent, and heritable genetic changes. As the alterations often eliminate the function of a specific gene, such mice are often called "knockout" mice. The knockout (and "knock-in") mice produced with this gene-targeting technology have proved invaluable in gaining an understanding of the functions of a great many genes and in providing models for many varieties of human disease.

Prior to the work of these three researchers in the 1980s, work on gene function or targeted genetic manipulation on a cellular level was separate from the study of how the effects of missing or duplicated genes might manifest themselves in the whole animal. The work of Evans, Capecchi, and Smithies gave researchers the ability to make a single specific genetic change in a cell's genome and observe the results as the embryo developed into a mouse. Today more than 10,000 varieties of these genetically altered mice contribute to genetic and disease research in laboratories throughout the world.

Capecchi and Smithies began efforts to modify individual genes in the mammal genome by attempting to exploit a naturally occurring stage in cell division called homologous recombination. During homologous recombination, matching chromosome pairs (or "homologous chromosomes") line up next to each other prior to their separation and the sorting of the pairs into the two resulting daughter cells. In sexual reproduction, the pair is made up of a chromosome from each parent, each possessing alleles, or their own versions, of every gene. While the paired chromosomes are still adjacent to each other, homologous—or equivalent—stretches of their DNA can overlap and then can be exchanged, increasing the genetic variability within each chromosome.

A similar process had been discovered in bacteria more than 50 years ago. Methods for causing a bacterial cell to incorporate foreign genes into its genome were already in wide use in laboratories at the time Capecchi, Smithies, and Evans began their research. The same had never been done before with mammalian cells. Although an earlier technique existed for creating mammals with altered genes, these "transgenic" mice were produced by injecting genetic sequences directly into fertilized egg cells. The technique can result in mice with modified DNA in some of their cells; however, it is not possible to control where or to what extent the injected genetic sequences will be incorporated into the genome and thus it is impossible to target specific genes for change. In 1985, however, Capecchi and his research team demonstrated that mammalian tissue cells could incorporate new gene sequences into specific spots in their DNA when these segments were injected directly into the nucleus, which houses the chromosomes. He realized that the same technique could be used to make specific changes to the genome of a mammalian cell by causing it to take in stretches of DNA containing modified genes. Thus, he reasoned, altered genes could be incorporated into a genome, or by inserting stop sequences, existing genes could be switched off.

At about the same time, Smithies was experimenting with similar concepts in homologous recombination. He approached the work from a desire to treat genetic disease through the exchange of a functional gene for a faulty one. Interested in developing a method for treating blood diseases, Smithies discovered how to use homologous recombination in segments of the human genome coding for blood proteins. He was able to insert an altered version of the gene coding for the globin portion of the hemoglobin protein, which carries oxygen in the blood.

Thus, two of three essential principles were in place. Yet since both scientists were working with somatic (nonreproductive) cells, neither could produce genetic changes that could be passed down to offspring. For this they needed the advances being made by Martin Evans. Evans was also interested in eventually producing genetic changes in mammalian cells. Unsatisfied with cultures of embryonic tumor cells he was using, he began experimenting with the culturing of embryonic stem cells in mice. These are cells from an early stage of the embryo called the blastocyst, which have the ability to differentiate into any cell in the body. Unlike stem cells taken from developed mouse tissue that can only form elements of that particular tissue, a single embryonic stem cell can essentially produce all tissues within a mouse. Evans reasoned that a genetic change made in one embryonic stem cell could be multiplied and retained with each cell division and eventually distributed throughout the body of the developing mouse. To test this, he inserted viral DNA into a culture of mouse embryonic stem cells. He injected these modified cells into another blastocyst and implanted that embryo in a surrogate mouse mother. The resulting litter of pups contained viral DNA in cells throughout their bodies. Evans knew that the altered genome containing viral genes could also exist in sex cells, making the change heritable.

With the 1986 publication of Evans' seminal paper on his work, all the elements existed for the creation of animals with heritable genetic modifications. The three researchers began the exchange of techniques and ideas. Capecchi traveled to Evans' lab to learn his techniques for using embryonic stem cells and Smithies received a vial of stem cells from Evans. Both Capecchi and Smithies decided to experiment with mutations of a specific gene that codes for the hprt enzyme (hypoxanthine-guanine phosphoribosyltransferase). Only cells containing an intact version of the gene can survive in a particular culture medium. By 1987 Smithies showed he could successfully repair a faulty version of the *hprt* gene by inserting a new one when he produced cell cultures that survived on the special medium. At the same time Capecchi attempted a similar insertion of a gene that would allow altered cells to survive in the presence cultures infused with the antibiotic neomycin. This technique had little control over whether the gene in question was actually replaced, however.

Building on a technique of Evans' to produce neomycin survivors that could be distinguished from cells that did not take up a genetic stretch containing the new gene, Capecchi added an element to refine the selection of altered cells. He incorporated a gene causing cells to die in another type of medium if the gene had not aligned with and replaced the target gene through homologous recombination. The positive-negative selection technique allowed researchers to select only those cells with the desired gene replacement and inspired labs around the world to attempt the production of genetically modified mice. Several were successful by 1989.

Further advances have allowed researchers not only to turn genes on and off in mice, but to specify the tissue in which they do so or the time in the mouse's life in which the change takes effect. The combination of this suite of discoveries and research techniques spawned a broad field of study and made the knockout mouse a cornerstone of biomedical research.

For background information *see* CHROMOSOME; CROSSING OVER (GENETICS); DEOXYRIBONUCLEIC ACID (DNA); DEVELOPMENTAL GENETICS; GENETIC CODE; GENETICS; MEIOSIS; MUTATION; RECOMBINATION (GENETICS); SOMATIC CELL GENETICS; STEM CELLS in the McGraw-Hill Encyclopedia of Science & Technology.

Physics. The physics prize was awarded to Albert Fert of the Université Paris-Sud and Unité Mixte de Physique CNRS/THALES (CNRS/THALES Joint Physics Center), both in Orsay, France, and Peter Grünberg of the Forschungszentrum Jülich (Jülich Research Center) in Germany, for the discovery of giant magnetoresistance.

The property of giant magnetoresistance (GMR) has driven the rapid shrinkage of computer electronics technology since its independent discovery by Fert and Grünberg in 1988 and commercial application in 1997. As the devices themselves shrink and the data on their hard disks become ever more densely packed, the microscopic areas that carry the magnetic signal become smaller and the signal itself becomes weaker. GMR technology allowed the development of the more sensitive read heads necessary to detect and distinguish the changes between the stored magnetic bits of information and then efficiently convert them to an electric current.

The read heads that use GMR themselves have undergone a dramatic miniaturization. The materials used are composed of layers only a few nanometers thick. The increased sensitivity has allowed the data-storage areas of the resulting devices to become exceedingly small and data rich. The reduction in electrical resistance within these read heads is also significantly greater than that possible with those heads using standard magnetoresistance, resulting in significantly faster and more efficient computing. GMR is now in ubiquitous use for the reading of densely packed magnetic data stored on the compact hard disks of an abundance of electronics, from slim notebook computers to matchbook-sized music players.

In 1856 William Thomson (Lord Kelvin) discovered that he could alter the properties of metals in the presence of a magnetic field. Magnetoresistance refers to this change in the electrical resistance of a material when subjected to a magnetic field. In general, electric current flows through a material as the electrons within it move in one direction. Resistance is produced when the electrons encounter irregularities in the material and scatter. In a magnetic material this scattering depends strongly on the electron spin. A magnetic field aligns the spins of most of the electrons in the material, giving it a net magnetic moment called the magnetization. As a result, the degree to which moving electrons are scattered depends on whether their spins point in the same or the opposite direction as those of most of the electrons in the material. Thomson observed that the material then "acquires an increase of resistance to the conduction of electricity along, and a diminution of resistance to the conduction of electricity across, the lines of magnetization."

Thomson was able to induce only modest changes in his metals. However, in the 1970s the technology was developed for producing metal layers only a few nanometers thick, allowing the possibility of inducing much greater changes in electrical resistance in the presence of a magnetic field. For example, if a layer of nonmagnetic material is sandwiched between two layers of magnetic material, the total electrical resistance to electrons moving through the system will be relatively low if both magnetic layers are magnetized in the same direction, since electrons that can move easily through one magnetic layer can move easily through the other as well. However, if the magnetizations of the magnetic layers are opposed, an electron that can travel easily through one layer is likely to be scattered by the other, so that the total resistance will be higher. If the magnetization of one of the layers is held fixed while the other is influenced by the magnetic fields on a hard disk, the current passing through both layers will depend on those fields and the device will thus function as a sensitive read head—the smaller the scale of the device, the greater the sensitivity. These principles are the basis of the emerging field of spintronics.

Fert and Grünberg each used semiconductor stacks composed of alternating films of iron (magnetic material) and chromium (nonmagnetic material) only a few atoms thick. Grünberg and his team built a simple sandwich of a layer or two of chromium atoms enclosed by just two or three layers of iron atoms. They observed a 10% change in the resistance of their material. Fert's team alternately released low-pressure clouds of gaseous iron and chromium into a vacuum chamber, slowly building up structures with as many as 30 layers of a few atoms each. With these thicker stacks Fert succeeded in changing the resistance by 50% and thus named the effect giant magnetoresistance. With an improvement in the process used to create the layers, called sputtering, industrial production of such layered read heads became possible. The incredible sensitivity and commercial viability of the process has led GMR technology to become the industry standard.

For background information *see* MAGNETIC RECORDING; MAGNETIZATION; MAGNETORESISTANCE; SEMICONDUCTOR; SPUTTERING in the McGraw-Hill Encyclopedia of Science & Technology.

Nobel Prizes for 2008

The Nobel Prizes for 2008 included the following awards for scientific disciplines.

Chemistry. The chemistry prize was awarded jointly to Osamu Shimomura (Marine Biological Laboratory, Woods Hole and Boston University Medical School, both in Massachusetts, United States), Martin Chalfie (Columbia University, New York, United States), and Roger Y. Tsien (Howard Hughes Medical Institute, University of California, San Diego, United States) for the discovery and development of the green fluorescent protein, GFP.

Few research tools have so illuminated the understanding, literally and figuratively, of the internal processes and development of cells over time as much as the green fluorescent protein, often called GFP. Borrowed from a jellyfish and since then inserted into countless types of cells and organisms—from bacteria to plants, animals, and fungi—GFP and its variants, produced from modified versions of its gene, have allowed researchers to visualize and follow the largely invisible activities of chemical compounds, proteins, and organelles inside living cells, and even track several processes at once. Requiring only illumination by ultraviolet or blue light, the protein will fluoresce and indicate the location or movements of any compound, cell, or tissue to which it is attached without causing any apparent harm to the organism.

Unlike other glowing proteins or compounds, GFP can function entirely independently of a given cellular environment. It requires no particular proteinaceous partner, catalyst, or supply of energy-rich molecules to give off light. Previously known light-producing proteins or compounds required a partner, each modifying the structure of the other to produce light, or were built in several steps. In contrast, GFP actually self-catalyzes in the presence of ultraviolet (UV) light or blue light to produce its green flourophore (fluorescent molecule). Earlier fluorescent tags had to be physically attached to the protein of interest or to antibodies for the protein. Then the cell would be induced to take up the tags, a process which did not provide consistent results. However, a single gene codes for GFP. Thus, the GFP gene can be relatively easily joined with a piece of an organism's native gene sequence and inserted into the genome of a cell or embryo. The cell or tissues into which it develops will then consistently produce the protein whenever the gene sequence is expressed, fluorescing when exposed to blue or UV light.

The winners of the 2008 chemistry prize were responsible for three separate stages in the development of this tool.

Shimomura was just a young scientist tasked with identifying the light source of a jellyfish, *Aequorea victoria*, when he discovered a strange green fluorescing protein after isolating its blue bioluminescent protein companion from the same animal. Oddly, although the bioluminescent protein he had derived from tissues at the edges of the jellyfish bell glowed blue with the protein that he later named *aequorin*, the live jellyfish glowed green. Eventually he found that a second protein in the animal could take up the energy of that blue light, change its own shape, and emit the energy as green fluorescence. He called the new protein green fluorescent protein.

More than 25 years later, Chalfie, who studied development and genetic expression in a microscopic nematode worm, learned about the jellyfish protein and recognized its tremendous potential to reveal previously hidden processes if inserted in another organism. The roundworm, which has only 959 cells as an adult, shares a third of its genes with humans. Chalfie realized that attaching the GFP gene to given gene sequence, such as that of a promoter (a gene switch), would allow him to visually follow the destination of any proteins produced by the gene sequence and observe the tissues into which they were incorporated. After Douglas Prasher, of Woods Hole Oceanographic Institute, identified and cloned the GFP gene, Chalfie and his team managed to do just what he had envisioned. They inserted the GFP gene behind a promoter in the nematode genome, making its natural product into a fluorescent tag. The tag illuminated the neurons in which the gene was active through the walls of the worm's nearly transparent body. This new tool was ideal for the mapping of the entire nematode, from gene expression and protein production, to the development and destination of each cell.

Later Tsien pieced together just how the GFP protein worked, showing that oxygen catalyzed a known reaction of three amino acids to form the chromophore. But more importantly, Tsien took this basic fluorescing protein to new heights of utility with genetic changes to its amino acid structure that allowed the altered protein to light up in a artist's color wheel of different hues. He created variants that shone more brightly in blue light than the original, which had fluoresced more strongly in UV light, a wavelength that can damage living cells. Later he developed variants that produced cyan, blue, and yellow colors; paved the way for others to develop a red marker; and redesigned the red to such colors as MPlum, mStrawberry, and mCitrine. Given the new color palette of glowing markers, researchers could now follow multiple processes in an organism, tracking complex protein interactions and the development of all cells in a tissue.

Today the GFP protein and its variants have been used for diverse applications, such as detecting arsenic in contaminated wells or changes in cellular organelles in response to drug treatment, tracking tumor cell migration, observing the movements and morphological changes in the Golgi apparatus or mitochondria, quantifying exact quantities produced of a protein of interest, or mapping the development and destination of specific cells in the body.

For background information *see* BIOLUMINESCENCE; FLUORESCENCE MICROSCOPE; GREEN FLUORESCENT PROTEIN; HYDROIDA; PROTEIN ENGINEERING; VIDEO MICROSCOPY in the McGraw-Hill Encyclopedia of Science & Technology.

Physiology or medicine. Harald zur Hausen (German Cancer Research Centre, Heidelberg,

Germany) received half the prize for his discovery of the causal connection between human papilloma viruses and cervical cancer. Françoise Barré-Sinoussi (Institut Pasteur, Paris, France) and Luc Montagnier (World Federation for AIDS Research and Prevention, Paris, France) shared the other half for their discovery of the human immunodeficiency virus (HIV).

HPV. Cervical cancer is the second most frequent cancer in woman, with approximately 600,000 cases diagnosed worldwide each year and 270,000 deaths, primarily in places where screening programs are not widely available. Harald zur Hausen's hypothesis that human papilloma virus (HPV) plays a role in this cancer contradicted the views that prevailed when he proposed it. However, his research was eventually able to associate two strains of HPV with the genomes of cervical cancer tumor cells. It furthermore led to an understanding of the mechanisms of virally induced carcinogenesis, and by making the oncogenic HPV strains available to others, he facilitated development of vaccines that are highly effective against them.

zur Hausen's early work on Burkitt's lymphoma demonstrated that the DNA of Epstein-Barr virus (EBV) persisted in the genomes of human tumor cells and probably could effect tumor growth in these cells as well as in another type of tumor, an epithelial carcinoma. zur Hausen then hypothesized the connection between cervical cancer and HPV, which led to the search for DNA within the genomes of the tumor cells of cervical cancer. Over many years, his group analyzed numerous papilloma cells, benign and cancerous, discovering thereby numerous strains of HPV. They finally succeeded in cloning a new strain, HPV 16, from cervical cancer cells and determined that it was present in approximately 50% of such biopsies. They then discovered HPV 18, which is associated with nearly 20%. There are more than 140 identifiable strains, with 80 having been genetically sequenced. Forty HPV strains are associated with genital infection, with strains 16 and 18 conveying the highest risk for the development of cervical cancer. A vaccine is now in use that can provide greater than 95% protection against types 16 and 18.

HIV. In the early 1980s, a new immunodeficiency syndrome was observed, which subsequently became known as acquired immunodeficiency syndrome or AIDS. Epidemiological studies indicated that it was rapidly spreading and occurring worldwide. It was transmitted through sexual contact as well as across the placenta from infected mothers to the fetus, and through the exchange of blood and blood products. Then began the search for the pathogen responsible for the disease. Françoise Barré-Sinoussi and Luc Montagnier in France examined cells taken from the swollen lymph nodes of patients in the early stages of AIDS. They found evidence of a mammalian retrovirus in the cell cultures grown from the samples: the enzyme reverse transcriptase, particles budding from infected cells, and other indications. Subsequent investigations revealed many of the properties of the suspected

new virus: the virus was present in patients with both early and full-blown cases of AIDS; it was a 90–130 nanometer retrovirus in the lentivirus subfamily; and, it targeted CD4* T cells, showing cytopathic effects on T cells of the immune system.

Investigations in the United States, in particular in the laboratory of Robert Gallo, established the causal connection between a virus they had been studying and AIDS. The French and U.S. groups subsequently agreed that they were working on one and the same type of virus, which in 1985 was formally recognized as Human Immunodeficiency Virus Type 1 (HIV-1).

For background information, *see* ACQUIRED IMMUNE DEFICIENCY SYNDROME (AIDS); ONCOGENES; RETROVIRUS; TUMOR VIRUSES in the McGraw-Hill Encyclopedia of Science & Technology.

Physics. Yoichiro Nambu (Enrico Fermi Institute, University of Chicago, Illinois, United States) was awarded one half of the Nobel Prize in Physics for the discovery of the mechanism of spontaneous broken symmetry in subatomic physics. The other half was awarded jointly to Makoto Kobayashi [High Energy Accelerator Research Organization (KEK), Tsukuba, Japan] and Toshihide Maskawa, Kyoto Sangyo University; and Yukawa Institute for Theoretical Physics (YITP), Kyoto University, Japan for the discovery of the origin of the broken symmetry that predicts the existence of at least three families of quarks in nature.

In modern physical theory, the fundamental laws of physics are said to possess a high degree of symmetry. That is, transformation of the systems they describe might change such laws, but in fact do not. For example, a basic concept of the theory of special relativity is that transformations moving the universe at constant velocity do not change the laws of physics. The assumption of symmetry has played a key role in determining the equations of fundamental theories from general relativity to quantum chromodynamics. However, under diverse circumstances, certain physical systems exhibit less symmetry than their governing laws themselves, and the study of such systems can provide important insights into natural phenomena. This might occur, for example, when a system is not stable when the solutions to its governing equations are most symmetrical and the system assumes a less symmetrical but stable state. In such a case, physicists speak of spontaneous symmetry breaking. The less than full symmetry of a crystalline lattice or preferential alignment of the spins of electrons within ferromagnetic materials are examples of this as well as the observation that spontaneous symmetry breaking tends to be associated with the appearance of order.

Nambu was studying asymmetries in superconductivity in the 1960s when he was able to model the occurrence of spontaneous symmetry breaking at the subatomic level. His work contributed to the refinement of the standard model of particle physics, for example in the explanation of the effects of the strong nuclear force. His mathematical work was especially useful in subsequent theoretical studies on the acquisition of mass by subatomic particles. This theory assumes the collapse of symmetry as the universe cooled following the big bang, with particles acquiring mass as they encountered drag in the so-called Higgs field. Physicists hope to provide proof of this picture by demonstration of the postulated "Higgs boson" with the Large Hadron Collider. *See* LARGE HADRON COLLIDER.

Two types of symmetry that physicists have studied involve (1) invariance of physical laws following spatial transformation by inversion of all spatial directions, which is known as parity or *P*, and (2) invariance following replacement of every particle by its antiparticle, which is known as charge conjugation or *C*. The combined operation, known as *CP* symmetry, would, for example, turn a left-handedly spinning neutrino into a right-handedly spinning antineutrino. Since it is believed that equal amounts of matter and antimatter were created after the big bang, the question follows as to why they did not annihilate each other. Instead, we have a universe consisting of matter. This leads to the assumption that violation of *CP* symmetry must have existed. In the 1950s, the theoretical studies of Tsung-DaoLee and Chen Ning Yang led to experiments that demonstrated the violation of parity (*P*) in weak nuclear interactions. And, in 1964 James Cronin and Val Fitch observed that *CP* symmetry was violated in the decay of particles known as *K*-long mesons. The standard model as it existed at that time could not explain this *CP* violation. The theoretical studies of Kobayashi and Maskawa predicted that the violation could be explained if the standard model was extended to assume the existence of three new quarks, the so-called charm, bottom, and top quarks. These particles were subsequently demonstrated in experiments. Their models also predicted symmetry violation in another particle, the *B* meson, which was verified in experiments in 2001.

For background information *see* HIGGS BOSON; PARITY (QUANTUM MECHANICS); QUARKS; STANDARD MODEL; SYMMETRY BREAKING; SYMMETRY LAWS (PHYSICS); TIME REVERSAL INVARIANCE; WEAK NUCLEAR INTERACTIONS in the McGraw-Hill Encyclopedia of Science & Technology.

Contributors

Contributors

The affiliation of each Yearbook contributor is given, followed by the title of his or her article. An article title with the notation "coauthored" indicates that two or more authors jointly prepared an article or section.

A

Abbott, Dr. Dallas H. *Columbia University, Lamont Doherty Earth Observatory, Palisades, New York.* PLUMES AND BANDED IRON FORMATIONS—coauthored.

Aleman, Dr. André. *BCN Neuroimaging Centre, University Medical Centre Groningen, the Netherlands.* HALLUCINATIONS.

Allan, Mr. Robert G. *Robert Allan Ltd., Vancouver, British Columbia, Canada.* TUGBOATS.

Alvarez, Dr. Pedro. *Instituto de Astrofisica de Canarias, La Laguna, Tenerife, Spain.* GRAN TELESCOPIO CANARIAS.

Anderson, Dr. Jason S. *Faculty of Veterinary Medicine, University of Calgary, Alberta, Canada.* ORIGINS OF MODERN AMPHIBIANS.

Ando, Prof. Yoichi. *Professor Emeritus, Graduate School of Science and Technology, Kobe University, Japan.* THEORY OF TEMPORAL AND SPATIAL ENVIRONMENTAL DESIGN.

Andrews, Dr. Peter. *Department of Palaeontology, Natural History Museum, London, United Kingdom.* EVOLUTION OF AFRICAN GREAT APES.

Aston, Dr. Kenneth I. *Department of Animal, Dairy, and Veterinary Sciences, Center for Integrated Biosystems, Utah State University, Logan.* CLONING RESEARCH—coauthored.

Awan, Dr. Shakil A. *Department of Bioengineering, Imperial College London, South Kensington Campus, Bagrit Centre, United Kingdom.* MEMS SENSORS.

B

Baab, Dr. Karen. *Department of Anatomical Sciences, Stony Brook University, New York.* FOSSIL HOMININS FROM DMANISI.

Bar-Ad, Dr. Shimshon. *School of Physics and Astronomy, Tel Aviv University, Israel.* FEMTOSECOND PHENOMENA.

Barrow, Dr. Jeffery R. *Department of Physiology and Developmental Biology, Brigham Young University, Provo, Utah.* GENE TARGETING.

Basu, Dr. Ananda. *Division of Endocrinology, Metabolism, and Nutrition, Mayo Clinic, Rochester, Minnesota.* DIABETES AND ITS CAUSES—coauthored.

Beebee, Ms. Dorothy M. *International Mushroom Dye Institute, Forestville, California.* HISTORY AND ART OF MUSHROOMS FOR COLOR.

Benekohal, Prof. Rahim F. *Department of Civil and Environmental Engineering, University of Illinois, Urbana-Champaign.* UBIQUITOUS TRANSPORTATION NETWORK SENSORS—coauthored.

Bennett, Dr. Kevin B. *Department of Psychology, Wright State University, Dayton, Ohio.* HUMAN-COMPUTER INTERFACE DESIGN—coauthored.

Blackwell, Prof. Garston. *Department of Mining Engineering, Queen's University, Kingston, Ontario, Canada.* GEOGRAPHIC INFORMATION SYSTEMS FOR MINE DEVELOPMENT—coauthored.

Blundell, Dr. Barry G. *Faculty of Design and Creative Technologies, Auckland University of Technology, New Zealand.* VOLUMETRIC DISPLAYS.

Boak, Dr. Jeremy. *Colorado Energy Research Institute, Colorado School of Mines, Golden.* UNDERGROUND PROCESSING OF OIL SHALE.

Bossart, Dr. Gregory D. *Marine Mammal Research and Conservation, Harbor Branch Oceanographic Institute, Florida Atlantic University, Ft. Pierce, Florida.* EMERGING DISEASES IN MARINE ANIMALS.

Bouteiller, Dr. Laurent. *Laboratoire de Chimie des Polymères, Université Pierre et Marie Curie, France.* POLYMERS THROUGH NON-COVALENT BONDING.

Bowcutt, Dr. Kevin G. *The Boeing Company, Phantom Works, Huntington Beach, California.* HYPERSONIC VEHICLES.

Brescoll, Dr. Victoria L. *Department of Psychology, Yale University, New Haven, Connecticut.* EPIDEMIC OF OBESITY.

Bridgeman, Dr. Bruce. *Department of Psychology, University of California, Santa Cruz.* CONSCIOUSNESS AND CONSCIOUS AWARENESS.

Bringloe, Mr. Thomas. *The Glosten Associates Inc., Seattle, Washington.* BARGE TECHNOLOGY.

Browne, Dr. Brandon L. *Department of Geological Sciences, California State University, Fullerton.* MINERALS TEXTURES IN LAVAS.

Brzezinska, Dr. Dorota A. *Department of Civil and Environmental Engineering and Geodetic Science, Ohio State University, Columbus.* SATELLITE-BASED GEODESY—coauthored.

C

Campbell, Dr. Carla. *Division of General Pediatrics, Children's Hospital of Philadelphia, Pennsylvania.* CHILDHOOD LEAD EXPOSURE AND LEAD TOXICITY.

Carter, Mr. Austin R. *Department of Physics, Ohio State University, Columbus.* ORGANIC-INORGANIC INTERFACES—coauthored.

Casanova, Dr. Fèlix. *Physics, University of California, San Diego, La Jolla.* POROUS ALUMINA.

Chen, Dr. Shigao. *Mayo Clinic College of Medicine, Rochester, Minnesota.* ULTRASONIC VISCOELASTIC TISSUE MEASUREMENTS—coauthored.

Chicarro, Dr. Agustin. *Research and Space Science Department, European Space Research and Technology Centre (ESTEC), European Space Agency (ESA), Noordwijk, The Netherlands.* MARS EXPRESS.

Cohen, Prof. J. John. *Department of Immunology, University of Colorado Medical School, Denver.* HYGIENE HYPOTHESIS.

Comiso, Dr. Josefino C. *Cryospheric Sciences Branch, NASA Goddard Space Flight Center, Greenbelt, Maryland.* SPECIES AND GLOBAL CLIMATE CHANGE.

Conrad, Dr. Jack. *Department of Vertebrate Paleontology, American Museum of Natural History, New York, New York.* EVOLUTION AND INTERRELATIONSHIPS OF LIZARDS AND SNAKES.

Curry Rogers, Dr. Kristi. *Departments of Biology and Geology, Macalester College, St. Paul, Minnesota.* TITANOSAUR SAUROPOD.

D

Dalton, Prof. Larry R. *Department of Chemistry, University of Washington, Seattle.* ELECTRO-OPTIC POLYMERS.

Davies, Prof. Richard J. *Department of Earth Sciences, Durham University, United Kingdom.* LUSI MUD VOLCANO.

de Heer, Prof. Walter. *School of Physics, Georgia Institute of Technology, Atlanta.* GRAPHENE.

Delaney, Prof. Terrence P. *Department of Plant Biology, University of Vermont, Burlington.* PLANT DEFENSE AGAINST PATHOGENS.

Demirbas, Dr. Ayhan. *Sila Science, University Mah, Trabzon, Turkey.* BIOETHANOL PRODUCTION.

De Roure, Prof. David. *School of Electronics and Computer Science, University of Southampton, United Kingdom.* SCIENTIFIC WORKFLOWS—coauthored.

Disteche, Dr. Christine M. *Department of Pathology, University of Washington, Seattle.* DOSAGE COMPENSATION OF THE ACTIVE X CHROMOSOME—coauthored.

E

Epstein, Prof. Arthur. *Department of Physics, Ohio State University, Columbus.* ORGANIC-INORGANIC INTERFACES—coauthored.

Ernst, Prof. Mary Fran. *Director of Forensic Education, Saint Louis University School of Medicine, Missouri.* MEDICOLEGAL DEATH SCENE INVESTIGATION.

Esses, Dr. Steven J. *Department of Medicine, Mount Sinai School of Medicine, New York, New York.* IMMUNE RESPONSE IN INFLAMMATORY BOWEL DISEASE—coauthored.

Evans, Dr. Lyndon. *CERN, Genève, Switzerland.* LARGE HADRON COLLIDER (LHC)—in part.

Everitt, Prof. Francis. *Gravity Probe B Project, W.W. Hansen Experimental Physics Laboratory, Stanford University, California.* GRAVITY PROBE B MISSION.

F

Fazio, Mr. Albert. *Director, Memory Technology Development, Technology and Manufacturing Group, Intel Corporation, Santa Clara, California.* NONVOLATILE MEMORY DEVICES.

Finnerty, Dr. John R. *Department of Biology, Boston University, Massachusetts.* STARLET SEA ANEMONE.

Flumerfelt, Dr. Raymond W. *Department of Chemical and Biomolecular Engineering, University of Houston, Texas.* WIND TURBINES.

Freeland, Dr. Howard. *Institute of Ocean Sciences, Department of Fisheries and Oceans, North Saanich, British Columbia, Canada.* ARGO (OCEAN OBSERVING NETWORK).

Freeling, Dr. Michael. *Department of Plant and Microbial Biology, University of California, Berkeley.* DIFFERENTIAL DUPLICATE GENE RETENTION.

Frei, Dr. Heinz. *Physical Biosciences Division, Lawrence Berkeley National Laboratory, California.* ARTIFICIAL PHOTOSYNTHESIS—coauthored.

Fujita, Dr. Terunori. *Mitsui Chemicals, Inc., Chiba, Sodegaura, Japan.* NON-METALLOCENE SINGLE-SITE CATALYSTS FOR POLYOLEFINS—coauthored.

G

Gao, Dr. Yu. *Department of Psychology, University of Southern California, Los Angeles.* ROLE OF THE FRONTAL LOBE IN VIOLENCE—coauthored.

Garrett, Prof. Chris. *Department of Physics and Astronomy, University of Victoria, British Columbia, Canada.* IN-STREAM TIDAL POWER GENERATION.

Gaudino, Mr. Mark. *Executive, MaXentric Technologies LLC, Fort Lee, New Jersey.* WIMAX BROADBAND WIRELESS COMMUNICATIONS—coauthored.

Gibson, Dr. George. *Xerox Corporation, Webster, New York.* ADVANCES IN ELECTROPHOTOGRAPHY—coauthored.

Giovannoni, Dr. Jim. *Robert W. Holley Center, USDA-ARS, Cornell University, Ithaca, New York.* REGULATION OF FRUIT RIPENING.

Glenn, Dr. Andrea L. *Department of Psychology, University of Pennsylvania, Philadelphia.* ROLE OF THE FRONTAL LOBE IN VIOLENCE—coauthored.

Goble, Prof. Carole. *Department of Computer Science, University of Manchester, United Kingdom.* SCIENTIFIC WORKFLOWS—coauthored.

Goodwin, Dr. William. *Department of Forensic and Investigative Science, University of Central Lancashire, Preston, United Kingdom.* NEANDERTHAL DNA—coauthored.

Greenleaf, Dr. James F. *Mayo Clinic College of Medicine, Rochester, Minnesota.* ULTRASONIC VISCOELASTIC TISSUE MEASUREMENTS—coauthored.

Griffin, Dr. Dale Warren. *Florida Integrated Science Center, U.S. Geological Survey, Tallahassee, Florida.* DESERT DUST-STORM MICROBIOLOGY.

Grima, Prof. Joseph N. *Department of Chemistry, Faculty of Science, University of Malta, Malta.* AUXETIC MATERIALS.

Groff, Dr. Rachel. *Department of Medicine, University of Colorado School of Medicine, Aurora.* HUMAN PAPILLOMAVIRUS: IMPACT OF CERVICAL CANCER VACCINE.

Grzywacz, Robert K. *Department of Physics and Astronomy, University of Tennessee, Knoxville.* ALPHA AND PROTON RADIOACTIVITY ABOVE TIN-100—coauthored.

H

Hall, Eric S. *U.S. Environmental Protection Agency, National Exposure Research Laboratory, Research Triangle Park, North Carolina.* GROUND-BASED MEASUREMENT OF SOLAR ULTRAVIOLET RADIATION.

Hammond, Dr. Scott M. *Department of Cell and Developmental Biology, University of North Carolina, Chapel Hill.* MICRORNA.

Hardgrave, Dr. Bill. *Director of the Information Technology Research Institute, University of Arkansas, Fayetteville.* RFID IN TRANSPORTATION AND LOGISTICS.

Hartgens, Dr. Fred. *Department of Surgery, Sports Medicine Centre Maastricht, University Hospital Maastricht, the Netherlands.* ANDROGENIC-ANABOLIC STEROIDS AND ATHLETES.

Hassan, Prof. Hassan A. *Department of Mechanical and Aerospace Engineering, North Carolina State University, Raleigh.* TRANSITION TO TURBULENT FLOW.

Hilton-Taylor, Dr. Craig. *Red List Unit, International Union for Conservation of Nature, Cambridge, United Kingdom.* EXTINCTION OF SPECIES.

Hirsch-Kreinsen, Prof. Dr. Hartmut. *Chair of Economic and Industrial Sociology, University of Dortmund, Germany.* ECONOMICS OF LOW-TECH INNOVATION.

Holland, Prof. Patrick L. *Department of Chemistry, University of Rochester, New York.* NITROGEN FIXATION.

Hoover, Dr. Stephen. *Xerox Corporation, Webster, New York.* ADVANCES IN ELECTROPHOTOGRAPHY—coauthored.

Hsia, Dr. K. Jimmy. *Department of Mechanical Science and Engineering, University of Illinois at Urbana-Champaign.* ROLE OF MECHANICS IN BIOLOGICAL MATERIALS RESEARCH.

Hubbard, Dr. Dennis. *Department of Geology, Oberlin College, Ohio.* CORAL REEF COMPLEXITY.

Huntley, Dr. John Warren. *Department of Geosciences, Virginia Polytechnic Institute, Blacksburg.* PHANEROZOIC PREDATION INTENSITY AND DIVERSITY.

I

Isley, Dr. Ann E. *Department of Physical Sciences, Kutztown University of Pennsylvania.* PLUMES AND BANDED IRON FORMATIONS—coauthored.

Itami, Prof. Kenichiro. *Department of Chemistry, Nagoya University, Japan.* SYNTHESIS OF NEW NANOCARBON MATERIALS—coauthored.

J

Jansky, Dr. Donald. *Jansky-Barmat Telecommunications Inc., Washington, D.C.* WORLD RADIOCOMMUNICATION CONFERENCE (WRC) 2007.

Jiang, Dr. Yu. *Department of Pharmacology, University of Pittsburgh School of Medicine, Pennsylvania.* MTOR.

Jiménez-Guri, Dr. Eva. *Developmental Biology Group, DCEXS, Universitat Pompeu Fabra, Barcelona, Spain.* BUDDENBROCKIA—coauthored.

Johnson, Dr. Paul A. *Geophysics Group, Los Alamos National Laboratory, New Mexico.* IMAGING MECHANICAL DEFECTS.

Jorgenson, Dr. Eric. *Institute for Human Genetics, University of California, San Francisco.* WHOLE GENOME ASSOCIATION STUDIES.

Jurkevitch, Prof. Edouard. *Department of Plant Pathology and Microbiology, Faculty of Agricultural, Food, and Environmental Quality Sciences, Hebrew University of Jerusalem, Rehovot, Israel.* PREDATORY BEHAVIOR IN BACTERIA.

K

Kaneyoshi, Hiromu. *Mitsui Chemicals, Inc., Chiba, Sodegaura, Japan.* NON-METALLOCENE SINGLE-SITE CATALYSTS FOR POLYOLEFINS—coauthored.

Kessler, Dr. André. *Department of Ecology and Evolutionary Biology, Cornell University, Ithaca, New York.* PLANT DEFENSE AGAINST HERBIVOROUS INSECTS.

Kimbrough, Sue. *U.S. Environmental Protection Agency, National Exposure Research Laboratory, Research Triangle Park, North Carolina.* AIR POLLUTION MONITORING SITE SELECTION—coauthored.

Kota, Dr. Arun K. *University of Pennsylvania, Department of Materials Science and Engineering, Philadelphia.* POLYMER NANOCOMPOSITES WITH CARBON NANOTUBES—coauthored.

Kowalick, Prof. Thomas M. *President, Click, Inc., Southern Pines, North Carolina.* MOTOR VEHICLE EVENT DATA RECORDER.

Kurs, André. *Department of Physics, Massachusetts Institute of Technology, Cambridge.* MID-RANGE WIRELESS POWER TRANSFER.

L

Lande, Mr. Stig. *Kebony, AS, Porsgrunn, Norway.* FURFURYLATED WOOD—coauthored.

Liu, Dr. Rui Hai. *Department of Food Science, Cornell University, Ithaca, New York.* PHYTOCHEMICALS: HUMAN DISEASE PREVENTION AGENTS.

Llewellyn Smith, Prof. Chris. *JET, Culham Science Centre, Oxfordshire, United Kingdom.* ITER (INTERNATIONAL TOKAMAK EXPERIMENTAL REACTOR).

Locquet, Prof. Jean-Pierre. *Department of Physics and Astronomy, Katholieke Universiteit Leuven, Belgium.* TABLETOP SYNCHROTRON SYSTEMS—coauthored.

Lux, Dr. Rick. *Xerox Corporation, Webster, New York.* ADVANCES IN ELECTROPHOTOGRAPHY—coauthored.

M

Martin, Prof. Jonathan E. *Department of Atmospheric and Oceanic Sciences, University of Wisconsin-Madison.* EXTRATROPICAL CYCLONE OCCLUSION.

Martin, Dr. Lane W. *Materials Science Division, Lawrence Berkeley National Laboratory, California.* MULTIFERROICS—coauthored.

Mayer, Dr. Lloyd F. *Department of Medicine, Mount Sinai School of Medicine, New York, New York.* IMMUNE RESPONSE IN INFLAMMATORY BOWEL DISEASE—coauthored.

McCue, Dr. Leigh. *Department of Aerospace and Ocean Engineering, Virginia Polytechnic Institute and State University, Blacksburg.* SHIP CAPZING—coauthored.

McNutt, Jr., Dr. Ralph L. *Space Department, John Hopkins University Applied Physics Laboratory, Laurel, Maryland.* MESSENGER MISSION—coauthored.

Meyer-Lindenberg, Dr. Andreas. *Central Institute of Mental Health, Mannheim, Germany.* DARPP-32—coauthored.

Miller, Dr. Norman L. *Climate Science Department, Berkeley National Laboratory, Berkeley. California.* HYDROLOGICAL CONSEQUENCES OF GLOBAL WARMING.

Missier, Dr. Paolo. *School of Computer Science, University of Manchester, United Kingdom.* SCIENTIFIC WORKFLOWS—coauthored.

Moon, Dr. Young-Jun. *Research Fellow, Department of Land Transport and Advanced Technologies, The Korea Transport Institute (KOTI), Ilsan-Gu, Koyang City, Kyonggi-do, Korea.* UBIQUITOUS TRANSPORTATION NETWORK SENSORS—coauthored.

Moore, Dr. Michael J. *Department of Biology, Oberlin College, Ohio.* ANGIOSPERM PHYLOGENETICS.

N

Nambo, Masakazu. *Department of Chemistry, Nagoya University, Japan.* SYNTHESIS OF NEW NANOCARBON MATERIALS—coauthored.

Nandy, Dr. Debashis K. *Division of Endocrinology, Metabolism, and Nutrition, Mayo Clinic, Rochester, Minnesota.* DIABETES AND ITS CAUSES—coauthored.

Nethercot, Prof. David A. *Department of Civil and Environmental Engineering, Imperial College London, United Kingdom.* PROGRESSIVE COLLAPSE OF BUILDING STRUCTURES.

Nguyen, Dr. Di Kim. *Department of Medical Genetics, University of Washington, Seattle.* DOSAGE COMPENSATION OF THE ACTIVE X CHROMOSOME—coauthored.

O

Oberkampf, Dr. William L. *Consulting Engineer, Albuquerque, New Mexico.* SIMULATION VERIFICATION AND VALIDATION—coauthored.

O'Connor, Dr. Susan. *Research School of Pacific and Asian Studies (RSPAS), Australian National University, Canberra, Australia.* EARLIEST SEAFARING.

Ogilvie Robichaud, Ms. Christine. *Civil and Environmental Engineering, Duke University, Durham, North Carolina.* ASSESSING RISKS FROM NANOMATERIALS.

Ojemann, Dr. Jeffrey. *Department of Neurosurgery, University of Washington, Seattle.* EPILEPSY.

Okamura, Dr. Beth. *Department of Zoology, National History Museum, London, United Kingdom.* BUDDENBROCKIA—coauthored.

Ovchinnikov, Dr. Igor. *Department of Molecular and Cell Biology, University of Connecticut, Storrs.* NEANDERTHAL DNA—coauthored.

P

Papazoglou, Prof. Michael P. *Chair of Computer Science and Director of INFOLAB, Tilburg University, The Netherlands.* SERVICE-ORIENTED COMPUTING.

Parks, Dr. Acacia C. *Department of Psychology, University of Pennsylvania, Philadelphia.* POSITIVE PSYCHOLOGY: HUMAN HAPPINESS—coauthored.

Parmesan, Dr. Camille. *Section of Integrative Biology, University of Texas, Austin.* ASSISTED MIGRATION FOR SPECIES PRESERVATION.

Pau, Dr. Stanley. *College of Optical Sciences, University of Arizona, Tucson, Arizona.* MINIATURIZED ION TRAPS—coauthored.

Petroski, Dr. Matthew D. *Signal Transduction Program, Burnham Institute for Medical Research, La Jolla, California.* UBIQUITINATION.

Pfützner, Dr. Marek. *Institute of Experimental Physics, University of Warsaw, Poland.* MULTIPLE-PROTON RADIOACTIVITY.

Pierce, Dr. Marcia M. *Department of Biological Sciences, Eastern Kentucky University, Richmond.* DISEASE OUTBREAKS IN LIVESTOCK DUE TO GLOBAL WARMING; NAEGLERIA FOWLERI INFECTIONS DUE TO DROUGHT; PANDEMIC OF 1918.

Pikaard, Dr. Craig S. *Department of Biology, Washington University, St. Louis, Missouri.* NUCLEAR RNA POLYMERASES IV AND V.

Posey, Ms. Shannon M. *Department of Psychology, Wright State University, Dayton, Ohio.* HUMAN-COMPUTER INTERFACE DESIGN—coauthored.

Potter, Dr. Daniel. *Department of Plant Sciences, University of California, Davis.* ROSACEAE (ROSE FAMILY).

Powell, Dr. Thomas David. *Navigation Division, The Aerospace Corporation, Los Angeles, California.* SATELLITE NAVIGATION ADVANCES—coauthored.

Prior, Dr. Alan. *Dassault Systemes-SIMULIA, Cheshire, United Kingdom.* VIRTUAL TESTING OF STRUCTURES.

Pynn, Prof. Roger. *Department of Physics, Indiana University, Bloomington.* NEUTRON SCATTERING.

Q

Quigg, Dr. Chris. *Theoretical Physics Department, Fermi National Accelerator Laboratory, Batavia, Illinois.* LARGE HADRON COLLIDER (LHC)—in part.

R

Radley, Charles Frank. *Micro Aerospace Solutions, Inc., Portland, Oregon.* SPACE-BEAMED SOLAR POWER.

Raine, Dr. Adrian. *Department of Criminology and Psychiatry, University of Pennsylvania, Philadelphia.* ROLE OF THE FRONTAL LOBE IN VIOLENCE—coauthored.

Ramesh, Prof. Ramamoorthy. *Department of Materials Science and Engineering and Department of Physics, University of California, Berkeley.* MULTIFERROICS—coauthored.

Rawson, Dr. Robert B. *Department of Molecular Genetics, University of Texas Southwestern Medical Center, Dallas.* CHOLESTEROL AND THE SREBP PATHWAY.

Ritzwoller, Prof. Michael H. *Department of Physics, University of Colorado, Boulder.* AMBIENT NOISE SEISMIC IMAGING.

Rizos, Prof. Chris. *School of Surveying and Spatial Information Systems, University of New South Wales, Sidney, Australia.* SATELLITE-BASED GEODESY—coauthored.

Rose, Dr. John. *Smith & Nephew Orthopaedics, Inc., Memphis, Tennessee.* ABSORBABLE ORTHOPEDIC IMPLANTS—coauthored.

Rowell, Dr. Roger M. *Biological Systems Engineering, University of Wisconsin, Madison.* HEAT-TREATED WOOD.

Roy, Dr. Christopher J. *Department of Aerospace and Ocean Engineering, Virginia Polytechnic Institute and State University, Blacksburg.* SIMULATION VERIFICATION AND VALIDATION—coauthored.

Royse, Dr. Daniel J. *Department of Plant Pathology, University of Pennsylvania, University Park.* OYSTER MUSHROOMS.

Ryan, Dr. Matthew J. *Bioservices, CABI Europe UK, Egham, Surrey, United Kingdom.* FUNGAL SOURCES FOR NEW DRUG DISCOVERY—coauthored.

Rykaczewski, Dr. Krzysztof P. *Physics Division, Oak Ridge National Laboratory, Tennessee.* ALPHA AND PROTON RADIOACTIVITY ABOVE TIN-100—coauthored.

S

Salehi, Dr. Abraham. *Smith & Nephew Orthopaedics, Inc., Memphis, Tennessee.* ABSORBABLE ORTHOPEDIC IMPLANTS—coauthored.

Sanmartin, Dr. Isabel. *Department of Biodiversity and Conservation, Real Jardin Botanico, CSIC, Madrid, Spain.* DISPERSAL VERSUS VICARIANCE.

Sarkar, Dr. Nandita. *NOAA, Environmental Research Division, Pacific Grove, California.* MIXED LAYERS (OCEANOGRAPHY).

Sarpeshkar, Rahul. *Department of Electrical Engineering and Computer Science, Massachusetts Institute of Technology, Cambridge.* NEUROMORPHIC AND BIOMORPHIC ENGINEERING SYSTEMS.

Scanlon, Dr. Michael J. *Department of Plant Biology, Cornell University, Ithaca, New York.* APICAL DEVELOPMENT IN PLANTS—coauthored.

Schneider, Dr. Marc H. *Faculty of Forestry and Environmental Management, University of New Brunswick, Fredericton, Canada.* FURFURYLATED WOOD—coauthored.

Schug, Dr. Robert A. *Department of Psychology, University of Southern California, Los Angeles.* ROLE OF THE FRONTAL LOBE IN VIOLENCE—coauthored.

Schurr, Dr. Jürgen. *Electricity Division, Physikalisch-Technische Bundesanstalt (PTB), Braunschweig, Germany.* IMPEDANCE UNITS FROM THE QUANTUM HALL EFFECT

Seligman, Dr. Martin E. P. *Department of Psychology, University of Pennsylvania, Philadelphia.* POSITIVE PSYCHOLOGY: HUMAN HAPPINESS—coauthored.

Sessions, Dr. Benjamin R. *Department of Animal, Dairy, and Veterinary Sciences, Center for Integrated Biosystems, Utah State University, Logan.* CLONING RESEARCH—coauthored.

Sinha, Dr. Deepen. *ATC Labs, Chatham, New Jersey.* AUDIO COMPRESSION.

Sleight, Prof. Arthur W. *Department of Chemistry, Oregon State University, Corvallis.* NEGATIVE THERMAL EXPANSION MATERIALS.

Slusher, Dr. Richart E. *Georgia Tech Quantum Institute, Georgia Institute of Technology, Atlanta.* MINIATURIZED ION TRAPS—coauthored.

Smith, Dr. David. *Bioservices, CABI Europe UK, Egham, Surrey, United Kingdom.* FUNGAL SOURCES FOR NEW DRUG DISCOVERY—coauthored.

Solomon, Dr. Sean C. *Department of Terrestrial Magnetism, Carnegie Institution of Washington, Washington, D.C.* MESSENGER MISSION—coauthored.

Soutis, Prof. Constantinos. *Department of Aerospace Engineering, The University of Sheffield, South Yorkshire, United Kingdom.* COMPOSITE MATERIAL SYSTEMS AND STRUCTURES.

Srinivasan, Dr. Venkat. *Lawrence Berkeley National Laboratory, California.* ADVANCED BATTERIES.

Stanley, Dr. George D., Jr. *Department of Geosciences, University of Montana, Missoula.* CORALS AND OCEAN ACIDIFICATION.

Steinhardt, Prof. Paul J. *Department of Physics, Princeton University, New Jersey.* CYCLIC UNIVERSE THEORY.

Stephan, Dr. Douglas W. *Department of Chemistry and Biochemistry, University of Windsor, Ontario, Canada.* METAL-FREE HYDROGEN ACTIVATION AND CATALYSIS.

Straus, Dr. Joe M. *The Aerospace Corporation, Los Angeles, California.* SATELLITE NAVIGATION ADVANCES—coauthored.

T

Takagi, Dr. Shinobu. *Research and Development, Novozymes Japan Limited, Mihamaku, Chiba, Japan.* KOJI MOLD (ASPERGILLUS ORYZAE).

TenWolde, Mr. Anton. *Forest Products Laboratory, USDA Forest Service, Madison, Wisconsin.* MOISTURE IN HOUSES.

Thorley, Ursula. GEOGRAPHIC INFORMATION SYSTEMS FOR MINE DEVELOPMENT—coauthored.

Tien, Dr. Jerry C. *Department of Mining Engineering, Missouri University of Science and Technology, Rolla.* VENTILATION TECHNOLOGY IN DEEP MINES.

Timp, Prof. Gregory. *Beckman Institute, University of Illinois, Urbana.* ELECTRONIC NANOPORES.

U

Urban, Dr. Matthew W. *Department of Physiology and Biomedical Engineering, Mayo Clinic College of Medicine, Rochester, Minnesota.* ULTRASONIC VISCOELASTIC TISSUE MEASUREMENTS—coauthored.

V

Vallero, Dr. Daniel A. *U.S. Environmental Protection Agency, National Exposure Research Laboratory, Research Triangle Park, North Carolina.* AIR POLLUTION MONITORING SITE SELECTION—coauthored.

Vanacken, Prof. Johan. *Department of Physics and Astronomy, Katholieke Universiteit Leuven, Belgium.* TABLE-TOP SYNCHROTRON SYSTEMS—coauthored.

van Diggelen, Dr. Frank. *Broadcom Corporation, Santa Clara, California.* ASSISTED GPS AND LOCATION-BASED SERVICES.

von Puttkamer, Dr. Jesco. *NASA Headquarters, Office of Space Flight, Washington, D.C.* SPACE FLIGHT.

W

Wang, Prof. Pao K. *Department of Atmospheric and Oceanic Sciences, University of Wisconsin-Madison.* JUMPING CIRRUS ABOVE SEVERE STORMS.

Wang, Prof. Su Su. *Department of Mechanical Engineering, University of Houston, Texas.* WIND TURBINES.

Weare, Dr. Walter W. *Physical Biosciences Division, Lawrence Berkeley National Laboratory, California.* ARTIFICIAL PHOTOSYNTHESIS—coauthored.

Weil, Mr. Charles Russell. *Seawolf Engineering, Blacksburg, Virginia.* SHIP CAPZING—coauthored.

Weinberger, Dr. Daniel R. *Clinical Brain Disorders Branch, Genes, Cognition, and Psychosis Program, National Institute of Mental Health, Bethesda, Maryland.* DARPP-32—coauthored.

Westin, Dr. Mats. *EcoBuild, SP Sveriges Tekniska Forskningsinstitut, Boras, Sweden.* FURFURYLATED WOOD—coauthored.

White, Dr. Kenneth L. *Department of Animal, Dairy, and Veterinary Sciences, Center for Integrated Biosystems, Utah State University, Logan.* CLONING RESEARCH—coauthored.

Wikle, Dr. Thomas A. *Department of Geography, Oklahoma State, Stillwater.* DIGITAL ELEVATION MODELS.

Willmann, Dr. Matthew R. *Department of Biology, University of Pennsylvania, Philadelphia.* PLANT JUVENILITY.

Winey, Prof. Karen I. *Department of Materials Science and Engineering, University of Pennsylvania, Philadelphia.* POLYMER NANOCOMPOSITES WITH CARBON NANOTUBES—coauthored.

Woodward, Mr. John B. *Department of Plant Biology, Cornell University, Ithaca, New York.* APICAL DEVELOPMENT IN PLANTS—coauthored.

Wörheide, Dr. Gert. *Department of Geobiology, Geoscience Centre Göttingen, Germany.* PALEOGENOMICS AND BIOCALCIFICATION.

Wu, Mr. Andrew. *Engineering, MaXentric Technologies LLC. Fort Lee, New Jersey* WIMAX BROADBAND WIRELESS COMMUNICATIONS—coauthored.

X

Xu, Dr. Chunhui. *Department of Biology, Chinese University of Hong Kong.* ENGINEERED MINICHROMOSOMES IN PLANTS—coauthored.

Y

Yang, Dr. Yaling. *Department of Psychology, University of Southern California, Los Angeles.* ROLE OF THE FRONTAL LOBE IN VIOLENCE—coauthored.

Yu, Dr. Weichang. *Department of Biology, Chinese University of Hong Kong.* ENGINEERED MINICHROMOSOMES IN PLANTS—coauthored.

Z

Zak, Dr. Paul J. *Center for Neuroeconomics Studies, Claremont Graduate University, California.* NEUROECONOMICS.

Zerbe, Dr. John I. *Forest Products Laboratory, Forest Service, USDA, Madison, Wisconsin.* SYNGAS FROM BIOMASS.

Zhang, Dr. Xiaoming. *Mayo Clinic College of Medicine, Rochester, Minnesota.* ULTRASONIC VISCOELASTIC TISSUE MEASUREMENTS—coauthored.

Index

Index

Asterisks indicate page references to article titles.

O